THE
CELLULAR
RADIO
HANDBOOK

THE
CELLULAR
RADIO
HANDBOOK

A Reference for
Cellular System Operation

Fourth Edition

NEIL J. BOUCHER

A Wiley-Interscience Publication
JOHN WILEY & SONS, INC.
New York · Chichester · Weinheim · Brisbane · Singapore · Toronto

This book is printed on acid-free paper.©

For ordering and customer service, call 1-800-CALL WILEY.

Library of Congress Cataloging-in-Publication Data is available.

Boucher, Neil J.
    The cellular radio handbook : a reference for cellular system operation / Neil J. Boucher.—4th ed.
        p.   cm.
    "A Wiley-Interscience publication."
    ISBN 0-471-38725-8 (cloth : alk. paper)
    1. Cellular telephone systems—Handbooks, manuals, etc. I. Title.
    TK6570.M6 B68   2000
    621.3845′6—dc21   00-027086

Printed in the United States of America

10 9 8 7 6 5 4 3 2 1

CONTENTS

PREFACE TO THE FOURTH EDITION

The first edition of *The Cellular Radio Handbook* was released in mid-1980, at a time when all commercial systems were analog. It soon became an industry reference, and was well-accepted. The second edition, which was released in June 1992, had little mention of digital systems because implementation was running late and no substantial networks existed.

The third edition was released in January 1995 when a large number of GSM systems were installed, but most were still having teething problems. By 2000, the GSM systems have gone on to become dominant, with CDMA offering a strong, but belated challenge.

The new edition is written from the perspective of a digital world, and covers the new technologies that are shaping the future of the industry. The original format has proven to be successful and is maintained. The intention is to present in a concise manner, good engineering practices, and sufficient theory to enable an understanding at a professional level. Mathematics has been used sparingly, but some topics require more extensive mathematical treatment. Where possible, these more technical sections have been dealt with in separate chapters and can be by-passed without any real loss of continuity.

Many people have assisted me with the preparation of this work, but I would like to extend a special thanks to the following people:

Ian Nicholson, of Telstra

Stuart Jeffrey, of Synacom

Marc Rolfes, a consultant based in Chicago

Stewart Fist, communications journalist with *The Australian*

Cindy Wishart, marketing/executive assistant with Stealth

Greg Delforce, engineering manager at Filtronics.

Last, but not least, I would like to thank my wife, June, for her patience and help with the proofreading of this book.

NEIL J. BOUCHER

Maleny, Australia
December 2000

ABOUT THE AUTHOR

Neil Boucher is a communications engineer with more than 20 years experience in mobile and cellular networks. He has worked in various senior engineering positions on cellular networks, both state- and nation-wide in extent. His responsibilities have included system design, installation, maintenance and operations. It is from this broad background that the present book is derived.

He is fluent in a number of computer languages and has produced a Windows software suite for cellular on mobile engineering, which encompasses most of the traffic, RF and power engineering that a cellular engineer is likely to encounter. In all, there are more than 100 routines. This package is sold commercially and is available through the author.

Mr. Boucher is the author of a number of other technical books on mobile communications including *The Trunked Radio and Enhanced PMR Radio Handbook* and the *Paging Technology Handbook*, both published by John Wiley & Sons, Inc. He has written dozens of technical papers for various trade publications.

Currently a free-lance mobiles technology consultant, Mr. Boucher's onside interests include flying, sailing, classic cars (and classic car rallys) and astronomy.

He can be contacted by email at NBoucher@ozemail.com.au

THE
CELLULAR
RADIO
HANDBOOK

1

WHAT IS CELLULAR RADIO?

Cellular radio is the fastest growing area of telecommunications. In almost every installation throughout the world, the operators' initial forecasts of customer demand have needed constant upward revision. In recent times, the field has been extended to include Personal Communication Service/Personal Communication Network (PCS/PCN), and in developed countries the number of cellular lines rivals wireline connections. The mobile telephone, however, largely sells itself.

The introduction of digital cellular in late 1992 brought an attendant increase in hardware and system complexity. The dramatic growth in the industry has led to a worldwide shortage of skilled personnel, and with the industry doubling in size every two years, this problem has no solution in sight.

Today, most new telephone installations in countries with mature cellular systems are mobile. Forecasts for the next decade indicate that this trend will continue.

To date, the major manufacturers have borne most of the responsibility for establishing new installations and training staff. In these endeavors, they have generally done a commendable job. The strain on their resources, however, is beginning to show, and it is obvious that, in the future, operators will need to become more self-sufficient.

There are numerous mobile systems in existence today that provide telephone access. Cellular radio was originally conceived as a means of providing high-density mobile communications without consuming large amounts of spectrum. The earliest cellular-like proposals date back to the American Telephone and Telegraph (AT&T) proposals of the 1940s for high-density mobile systems. In 1968, AT&T submitted a proposal for a cellular system to the Federal Communications Commission (FCC).

The original concept involved containing a group of frequencies within a "cell," reusing the frequencies in the same vicinity, but separating them in space to allow reuse without serious interference. The hardware needed to implement such a system did not become available until the late 1970s and by then the "cellular" concept (that is, frequency reuse in cells) was accepted as a sound frequency planning tool.

Prior to the first cellular-telephone systems, a number of automatic mobile-telephone systems existed, usually having only one transmit-and-receive site. Such systems sometimes had around 2000 subscribers and were characterized by high sites, with relatively high transmit power and deviation. These systems usually operated in the very high frequency (VHF) or low ultra high frequency (UHF) bands, and from single cells, could achieve a radius of operation of 20 km or more. They were not cellular systems in the modern sense even though they were often fully automatic.

Cellular radio is different things to different people. To many investors it is seen as a potential gold mine. High growth rates, decreasing costs, and value to the community at large have caused the value of many companies that own cellular franchises to rise substantially in the last few years. But even with regulated tariffs, few cellular companies have been able to achieve a positive cash flow. Achieving profitability, however, is relatively easy.

To the cellular operator, cellular radio is a whole new world where the technologies of radio, switching, transmission, and computing merge into a single system. But the operator must be proficient in all of these areas or pay dearly for expertise when it is required. There is also a high cost for mistakes and bad decisions caused by a lack of experience. New operators often find the spectrum of skills required to be a successful player quite daunting.

Once the technical skills have been addressed, the new operator must then face the areas of marketing, finance and accounting, advertising, and public relations. The successful operator will be the one who also masters these basic business skills.

Cellular radio is no place for amateurs. A minimum of two years full-time experience is needed to produce an effective cellular engineer or marketing professional. It takes a lot longer to become an expert. Getting a wide spectrum of experience is difficult. Large operators can rotate their staff to widen employee exposure. Small operators often lack the necessary available positions for employees to move within the company; however in small systems there are often many opportunities for diverse experience.

EARLY CELLULAR SYSTEMS

Early cellular radio systems were designed for frequency reuse and had low capacities, which were thought adequate for future demand. The earliest cellular system, NAMTS, the Tokyo metropolitan system that started service in 1979, came with a basic capacity of 4000 subscribers and was expandable to 8000 subscribers. Initially it did not feature a handover capability; this feature first appeared in the Australian NAMTS System in 1981. In 1981, the NMT450 system was introduced with a basic capacity of 10,000 subscribers and was expandable to about 20,000 subscribers.

Each of these systems operated in the 400-MHz band and used about 180 frequencies. It was possible to have about 4000 subscribers without frequency reuse.

In 1967, the Japanese Telecommunications Laboratory (NTT) developed a 400-MHz cellular frequency plan for cellular radio, but it was never implemented commercially.

Later systems—Advanced Mobile Phone Service (AMPS) (Chicago, 1983), Total Access Communications System (TACS) (1985), and NMT900 (1987)—were designed to operate in the 800- and 900-MHz bands and use from 666 to 1000 frequencies. It was assumed that they would usually operate in a multiple frequency-reuse environment (that is, interference environment). Therefore, these systems came with an inherent protection against interference.

MOBILE AND TRUNKED RADIO

Because of the high initial costs of implementing a full cellular network, these early systems targeted the top of the market (car telephones). These networks were the beginning of modern cellular mobile-phone systems. They could reuse frequencies and could "hand off" a mobile call from one cell to a more appropriate one as the call moved out of range of the original cell. Until 1987, these two features were sufficient to define a cellular-telephone system.

However, in recent times, the emergence of powerful trunked radio systems has made this distinction inadequate. A trunked radio system is one that allows dynamic assignment of radio access channels to a group of mobiles.

A simple mobile radio system consists of a mobile and a repeater (see Figure 1.1). Car A transmitting on frequency F_1 is received by the repeater and rebroadcasts to all other mobiles on frequency F_2.

A trunked radio or cellular system works in much the

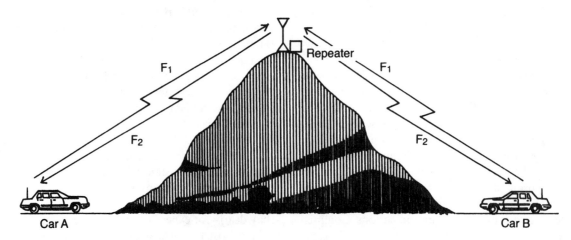

Figure 1.1 Duplex mobile repeater.

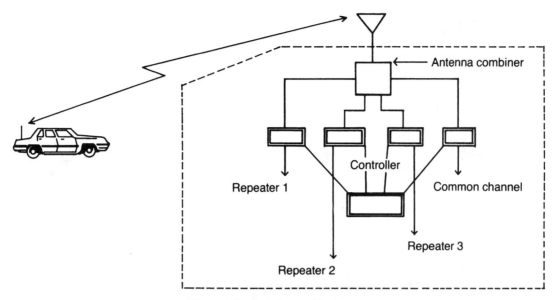

Figure 1.2   A typical trunked system.

same way, except that the mobiles, instead of having only two pairs of frequencies, have a group of frequencies and some logic control to ensure the right frequency is used. Logic in the base and in the mobile enables a free channel to be selected and switched to automatically (see Figure 1.2).

Consider a typical mobile-originated call:

1. The mobile calls (by a data transfer) on a common channel and requests the use of a free channel.
2. The controller either assigns a free channel, indicating that the channel is assigned to the mobile via the common channel, or the controller places the request in a queue (only in some trunked systems) until a channel becomes available.
3. The mobile switches to the assigned channel and conversation can take place, as with a simple mobile repeater.

As trunked mobile systems evolved, they used more sophisticated hardware and software so that frequency reuse and call handoff are now sometimes a feature of a trunked radio system. Furthermore, automatic telephone access and calls strictly to individual mobiles are also common.

The main distinction between a sophisticated trunked radio system and a cellular system is that the latter is usually 2-channel simplex (although some full duplex call capability may be a feature) while a cellular system is full duplex. Another distinction is that cellular systems were designed to operate in an interference environment where frequency reuse was seen as a limiting factor. Trunk mobile radio wasn't designed with this feature.

Trunked radio systems permit group calling; that is, the simultaneous calling of all mobiles in a particular group, which can be quite large. Cellular systems don't ordinarily permit group calling. This is not a technical limitation, but originally one imposed by the FCC and other regulatory commissions to limit competition between cellular and land mobile systems. iDEN is an exception to this. Therefore, cellular systems are designed to handle conventional telephone traffic and trunked systems are designed primarily to handle dispatched mobile traffic.

CELLULAR SYSTEMS

A cellular-radio system is structured differently than a land mobile system. A basic cellular system is illustrated in Figure 1.3.

The heart of a cellular system is the cellular switch. It is called a full-availability switch (it can connect any inlet to any outlet). The cellular switch connects base stations to the public switched telephone network (PSTN) and base stations to each other as required. The cellular switch makes these connections using trunk routes to the PSTN.

A feature of cellular architecture is the continuous monitoring of the call progress and the ability to reconfigure the system quickly so that switching occurs without disturbing the user.

Handoff times vary from about 0.5 to 1 second in older systems, to about 100 milliseconds (which is barely noticeable) in new systems. The mobile shown in Figure 1.3, traveling from base 1 to base 2, will eventually drive out of range of base 1. The system senses the decreased

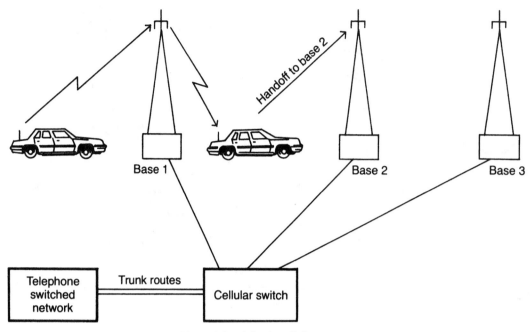

Figure 1.3 A basic cellular system.

field strength and instructs the surrounding bases to look for the mobile and report back on field strength (in some cases the reverse happen; the mobile scans the available base stations). If one base reports a higher field strength, the system can instruct the mobile to change channels and handoff to the best base (in this case, base 2).

Early systems used stored program control (SPC) switches, which were designed primarily as land-line switches. Generally, it was found that the processor load had been underestimated and that a processor upgrade was necessary to handle the extra demands of the cellular system. Modern cellular switches have processors that are well suited to the system demands and the number of ports supported.

The cellular switch can either be a purpose-designed switch or a modified telephone switch. The switch will be processor-controlled. The processor may handle many functions other than switching, including customer validation, call monitoring, system diagnostics, and interconnection with other cellular switches and base stations. Base stations have some local intelligence, but they are essentially controlled repeaters.

No strict definition of cellular radio exists and PCS/PCN can be regarded as a digital cellular system at a higher frequency, but all cellular systems have the following features:

- Frequency reuse
- Ability to handoff a mobile from cell to cell according to signal field strength and/or noise requirements

- Multicell and multibase configurations
- Access to a fixed telephone network with mobiles receiving calls on an individual basis only (group calling is only available using a switched network)
- Ability to work in a controlled interference environment

The familiar hexagon pattern that has popularly become associated with cellular radio is shown in Figure 1.4. To understand cellular radio, it is important to understand what it isn't. Therefore, it is worthwhile exploring the hexagon theme a little further.

It would be possible to create this hexagon system in the real world if:

- All sites were to have the same antenna systems (that is, all omni or all sectored)
- The terrain were perfectly flat (no forests or large buildings)
- All antenna heights were identical

This type of terrain is called a desert! A "desert" cellular system that satisfies these conditions is illustrated in the cell pattern in Figure 1.5. With an analog cellular system, this is as close as you can get to an hexagonal pattern. Digital systems permit rigid definitions of boundaries by triangulating the position of the mobile.

Because of the extreme symmetry of this type of terrain, it has been possible for mathematically inclined

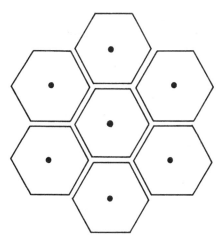

Figure 1.4   The hexagon frequency pattern of cellular radio.

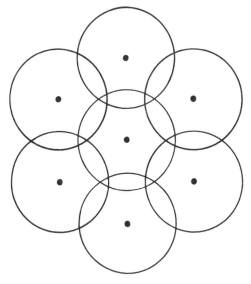

Figure 1.5   Cellular patterns on a flat earth.

engineers to write volumes on how to obtain high-density cellular systems by regular cell division and range contraction. Various cell plans have been devised that optimize frequency reuse (minimize the ratio of carrier to interference) for these theoretical systems. It is assumed that this approach can be translated directly into a real-world environment so that the most efficient concept on paper will also be the most efficient in practice. The hexagon approach can be used as a starting point, but the real environment should determine the actual system configuration.

Some planning engineers go to great lengths to place sites precisely within a hexagonal pattern. They fail to realize that the radio patterns are far from hexogonal and so the trouble taken to precisely locate sites is mostly futile.

Nominally hexagonal patterns sometimes appear in nature—as can be seen in Figure 1.6, which shows the rock formations at The Giants Causeway in Northern Ireland. The hexagons are from one-half to one meter across. The patterns were formed as a result of cooling

Figure 1.6   Nature's hexagons at The Giants Causeway in Northern Ireland.

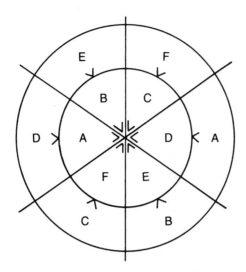

Figure 1.7   The Stockholm-Ring cell pattern.

fractures, and, as can be seen, are three dimensional. A closer inspection shows, however, that as well as hexagonal shapes, four-, five-, seven- and eight-sided faults are found. Despite these irregularities, these rock formations are in fact much closer to a regular hexagonal pattern than most cellular systems are.

Many other configurations are possible, one of the most successful models being the Stockholm Ring, illustrated in Figure 1.7. However, it is only applicable to low-deviation NMT systems.

This model allows all channels to be used at a common high-density center (usually a city center). It was used in the Stockholm NMT450 system. This model has since been used successfully elsewhere.

2

WORLD SYSTEM STANDARDS—A HISTORY

There are a number of different cellular systems used around the world today. The lack of a uniform standard has hampered the development of cellular systems and of roaming (that is, using the mobile phone in other than the home network) in particular. Global System for Mobile Communications (GSM), the pan-European digital system, is a partial solution to the uniformity problem. Roaming between Europe and the United States is not really practical because the two continents reserved different blocks of frequency for mobile-telephone services.

PRE-CELLULAR SYSTEMS

Early Attempts

Provision of subscribers local loop dates back a long way. As early as 1950, single-channel VHF systems were used in the United States to provide wireless local loop (WLL). The equipment used valve technology and proved too expensive to maintain. In most major western cities from about the mid-1950s onward, there was a "radio phone" service that provided links to a manual operator. These services were generally over-subscribed (with typically 50 users per channel, very expensive and had long waiting lists. Even earlier, HF radio links were being used in Australia and elsewhere to provide trunk and subscribers routes for remote communities (again with manual PSTN interconnection).

Over the ensuing period, many attempts were made at providing WLL and at least up until the turn of the century few ever operated profitably.

IMTS

This mobile telephone system was used extensively in the United States and has been used elsewhere. The phones are vehicle mounted only and have power outputs of up to 40 watts. The system does not feature frequency reuse and is based on the concept of a large single cell. There is no handoff between cells, but because the cells are so large the user does not see this as a serious problem.

The number of channels allocated for Improved Mobile Telephone Service (IMTS) across the United States was variable, so that the customer capacity of the systems also varied. As no frequency reuse was employed in a given geographical region, the spectrum efficiency of this system was poor.

In Chicago only 23 channels were available, and it was assumed that this number was sufficient to serve 1150 customers, or 50 customers per channel. It can be seen that the customer-to-channel density is much higher than that used in cellular systems where 20 customers per channel would be more usual. This means that a lower grade of service (or higher call fail rate) was to be expected on the IMTS.

New IMTS systems were being installed as late as 1991. IMTS could be set up more quickly and at less cost than a cellular system in the 1990s. IMTS mobile equipment is expensive relative to the cost of cellular phones, which is due to the low production volumes. Late versions of IMTS have integrated paging systems.

The structure of an IMTS system is shown in Figure 2.1. Notice that it has a number of incoming telephone lines, separate from the outgoing lines allocated one per channel. This simplifies the internal switching, which has to be done only for incoming calls, as the outgoing ones

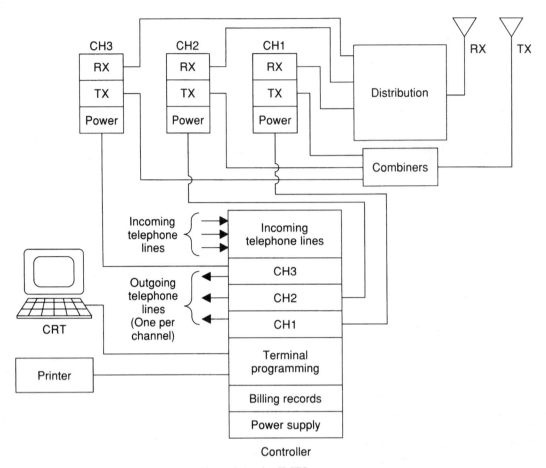

Figure 2.1 An IMTS system.

merely seize a line. The telephone lines may be either standard subscriber lines (which will require Dual Tone Multifrequency (DTMF) overdialing) or Direct Inward Dialing (DID).

NAMTS

The NAMTS system originally evolved in Japan and was operational in 1979. In its original form, it was not really a cellular system because it lacked handover capability. By 1981 handover had been added and the system was truly cellular.

The mobiles on this system were around $5000 in 1981, although this was not so much a problem because switch capacities were only 4000, and the system was really only targeted at the rich.

The network mostly operated at around 500 MHz and could be ordered in various bands. It usually had a rather narrow duplex spacing; in Australia it was

10 MHz. This meant that bulky duplexers were part of the baggage of the system.

The networks were capable of using 120 channels and had transceivers (see Figure 2.2a and 2.2b) that had a 25 watt output. It was generally intended that up to 16 transmitters be combined into a single antenna.

The algorithm for handoff was rather straightforward. If the signal got below 10 micro volts for a total of 8 seconds continuously, then a new base site would be sought.

In order to minimize eavesdropping in one version of the system, echo suppressors were used at the base station sites to take out the side tone; this resulted in only the base station TX path being sent to air. However it also cancelled out the sidetone in the receivers, which had to be modified to "artificially" induce the tone at the mobile end.

The system itself was not cheap and cost about $25,000 per channel (this figure being a network average for a system of about 3000 users).

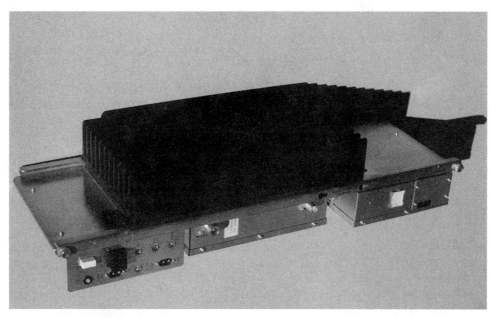

Figure 2.2a   An NAMTS 25W TX.

Radiocom 2000

This French system, Radiocom 2000, was in many ways similar to the IMTS system, and was, in fact, a hybrid trunk radio/cellular system. In particular, it used DTMF signaling and did not have handoff capability. It operated in the 200- and 450-MHz and 900-MHz bands. It was launched in 1985 and used 256 channels. The ultra high frequency (UHF) version was used countrywide, while the very high frequency (VHF) version was used in Paris, Lyon, and Marseilles.

In all, 280 payphones were installed on the TGV (*Train à Grande Vitesse*) the French railways high-speed trains.

The Radiocom system was extended to the 900-MHz band in 1989 to cater to the increasing demand, in addition to the standard RS2000.

France had a customized NMT system known as NMT-F, which in 1991 had 65,000 subscribers. The system ultimately had about 250,000 subscribers.

These systems have since closed.

NET$_z$ B & C WEST GERMANY

Germany initially used a non-cellular, 160-MHz car telephone system (B-Net$_z$) (without handoff) alongside its digital cellular C-Net$_z$, 450-MHz system. The C-Net$_z$ had 237 channels and a nominal capacity of about 150,000 customers. In 1989, there were 120,000 custom-ers, and considerable problems have been associated with that many subscribers.

C-Netz was used in a few other countries (notably South Africa where it was also problematic). It can be regarded, along with the NMT Systems, as the precursor of GSM.

CT2

In the early 1990s there was a big rush in the telecom-munications community to install a cellular look-alike called CT2. In principle it was just a lower powered version of the cellular systems already in place. It was digital (which in 1990 meant it must be good), and in principle it could support very large numbers of sub-scribers [due mainly to the low (ERP) used and the sub-sequent short range, which made all cells minicells]. The voice quality was rather good, and mostly significantly better than today's digital cellular offerings. But the customers never saw it the way the marketing people wanted them to.

CT2 as a network system has ceased to be. Singapore Telecom's Callzone, France Telecom's BiBop, and PTT Netherlands Greenpoint, the last bastions of CT2 net-work, have all closed.

As a business failure, the CT2 networks will be case studies for marketing schools for years to come. But CT2 was doomed from the outset. Few serious engineers ever expected it to succeed. With virtually no range (actual range was between 10 and 100 meters) it was

Figure 2.2b An NAMTS base station channel receiver.

Figure 2.3 The Voxon cordless phone.

economically impractical to offer ubiquitous coverage; even wide-area patchy coverage was beyond it. In many countries (such as the United Kingdom), poorly thought out legislation crippled the system; to reduce the competition to the cellular providers, only outgoing calls were permitted, while incoming calls were banned.

There was worse on the standards front. There was CT2, CT2+, CT3 (which later became synonymous with DECT), which meant that manufacturers had to produce too many standards and the efficiencies of scale were lost. Just to make things a little tougher for the technology, frequency allocations varied from country to country.

CT2 was seen as a "poor man's cellular," but because an enormous number of sites were needed for even fair coverage, the actual network costs were as high as they would have been for cellular. The handsets were a little cheaper than cellular phones, but not by much. Such an inferior system was bound to fail.

CT2 phones, however, still survive as cordless telephones and wireless private automatic branch exchange (PABX) handsets. In this role, mainly because of the good voice quality, small size, and long battery life, they may be around for a long time to come.

More recently versions of CT2 in the 900 mHz band, featuring frequency hopping, have appeared which give good voice quality and range. See the Voxson cordless phone shown in Figure 2.3.

JAPAN

Japan started an 800-MHz cellular system in 1979 in the Tokyo metropolitan area. The system had 600 channels and used 25-kHz channel spacing.

In order to cater to the ever-increasing demand for capacity, a second system using 6.25-kHz channel spacing with 2400 channels was later introduced. This 6.25-kHz system includes these additional features:

- Diversity reception in mobile and base stations
- Dynamic channel assignment
- Compatibility with the earlier wide-band system (through dual-mode mobile systems)
- Digital speech encryption

The system originally used a 17-cell omnicell plan, but this was later reconfigured to N = 14. The base stations have a 5-watt transmitter power and are installed 16 to a rack.

By 1988 there were 520 mobile bases and 4200 radio channels. Over half of all subscribers were in Tokyo. JTACS, a Japanese version of TACS, was introduced in 1988. This was later enhanced to a half-channel system called ETACS.

Until 1988 the only cellular provider in Japan was the NTT (the government-owned wireline company), but in that year new common carriers (NCC) were permitted to operate. There are two NCCs, Nihon Ido Tsushin and Daini Denden Inc. By 1990, of the national total of 700,000 subscribers, the NCCs had 13 percent of the market.

Some of the early Japanese mobile systems were sold outside Japan to places including Australia, Hong Kong, Singapore and Kuwait, but there appears to have been little attempt to market any systems, except the colorful PHS system (which is discussed later in the text) since about 1985.

NMT450

This system became available in 1981 and featured fully specified system interfaces. It was the forerunner of a number of world-wide systems seen in Fig. 2.1. It specified, to the level of the switch-to-base station and switch-to-switch, a standard signaling protocol (as well as the base-station-to-mobile interface). This standardization, missing from other systems, enabled equipment from different manufacturers to be interconnected (well sort of). Thus, a base station from one supplier could be directly connected to a switch from another supplier.

But soon after its introduction, NMT450 systems were installed in various countries on different frequencies and with different channel spacings. This made them incompatible from a subscriber's viewpoint. Different frequencies were chosen to suit local spectrum availability. The channel spacing, initially 25 kHz, was altered to 20 kHz in some systems to obtain more channels (an increase from 180 to 225) in the same spectral bandwidth. The earliest bases had separate active and passive RF components, which had the disadvantage that a minimum-sized installation (a few channels) still needed

at least two racks. Figure 2.4 shows an early NMT450 base station in Jakarta, Indonesia. The two outside racks contain the receivers and transmitters; the combiners are housed in the middle.

The transceivers are the links to the mobiles. They are sometimes fully frequency-agile (that is, they are software-programmed to a particular channel, but older versions may need mechanical tuning). Once installed they ordinarily act as a fixed-frequency transceiver. In some systems from some suppliers, a voice channel can be designated as a standby control channel and may change frequency automatically when a control channel failure is detected. In other systems, the transmitters and combiners can be retuned remotely so that "channel borrowing" or the reallocation of the channel frequency can occur during periods of blocking on the regular frequency. This is used in some high-density systems.

Because it is difficult to predict the traffic distribution, it is advantageous to have channel equipment that can be moved easily from site to site. One of the earliest attempts at this was in the NMT450 system in Kuala Lumpur, Malaysia, in which each of the four racks was a self-contained set of four channels. As equipment became smaller, eight channels per rack became standard.

The earliest systems released around 1980, operated on 450 MHz. These systems had a relatively small number of channels—usually 180 to 225—and were not designed for extensive frequency reuse. The maximum capacity of such systems is thus between 20,000 and 30,000 users in most city areas.

These 450-MHz systems were not designed to use a handheld, but at least one manufacturer produces one and transportables are readily available.

NMT900

The NMT900 system is essentially an NMT450 system moved in frequency to 900 MHz. Although many more channels are available (1000 or 2000 using interleaving), the paging channel baud rate is not increased so that these channels cannot really be used effectively.

Some improvements were made, including the addition of receiver antenna diversity and a voice compander. The low deviation of 4.7 kHz was retained, however, resulting in significantly poorer performance in rural and in-building environments, as well as reduced interference immunity when compared to the other 800- and 900-MHz analog systems.

Virtually all remaining NMT450 networks are struggling for survival, mainly because of the high cost of mobiles and network and limited competition between vendors. In October 1999 the NMT MoU group con-

Figure 2.4 This NMT450 system, circa 1988, in Jakarta, Indonesia, has two active RF racks and two separate multicoupler racks. (Photo courtesy of L.M. Ericsson.)

sidered a digital 450-MHz standard to allow migration from the current analog service.

The NMT450 service often will have better coverage than its GSM counterpart, and so it may be viewed as offering a different service. Since most NMT systems are in Europe, and many of these are owned by existing GSM operators, a possible outcome is to adopt a GSM450 MHz solution.

However, there are a few other offerings. Radio Design from Sweden is proposing a DNMT solution that is basically a solution similar to DAMPS and offers analog/digital dual mode. Alternatively, Qualcomm has proposed a code division multiple access (CDMA)/GSM structure that offers a GSM-MAP with a CDMA air interface. Since Qualcomm's sale of its infrastructure division to Ericsson in March of 1999, there have been serious doubts that Qualcomm can offer more than ideas, and support for this proposal is accordingly

reserved. Despite that, Qualcomm has demonstrated a working system based on GSM switching and IS-95 (CDMA) air interface to Vodaphone in the UK in 1998. Qualcomm has additionally proposed the alternative of ANSI-41, with CDMA.

New Switch

At a time when most NMT450 networks were struggling, Ericsson announced in 1999 that it had released a new, small, full-featured switch called the Micro MTX for NMT450.

ITALY

Italy started its wide-area mobile network in 1974 and introduced a cellular system known as RMTS in 1985.

TABLE 2.1   Technical Comparison of Four Analog Cellular Systems

Specification	AMPS	TACS/ETACS	NMT900	NMT450
TX Band	800 MHz	900 MHz	900 MHz	450–470 MHz
Channel separation	30 kHz	25 kHz	25/12.5 kHz	25/20 kHz
Duplex separation	45 MHz	45 MHz	45 MHz	10 MHz
Channels	832	920*	1000 (1999)	180/225
Modulation type	FM	FM	FM	FM
Peak deviation	±12 kHz	±9.5 kHz	±4.7 kHz	±4.7 kHz
Compander	2 : 1	2.1	2.1	No
	Syllabic	Syllabic	Syllabic	
Possible cell plans	4, 7, 12	4, 7, 12	7, 9, 12	7
Control channel modulation	FSK	FSK	FFSK	FFSK
Control channel deviation	±8 kHz	±6.4 kHz	±3.5 kHz	±3.5 kHz
Control channel code	Manchester	Manchester	NRZ	NRZ
Ctrl. chn. capacity (subs)	77,000	62,000	13,000	13,000
Transmission rate	10 kbit/s	8 kbit/s	1.2 kbit/s	1.2 kbit/s
Competitive operators allowed[†]	Yes	Yes	No	No
Interexchange handoff	Yes	Yes	Yes	Yes
Diversity	Yes	Yes	Yes	No
Subscribers in service 1990	6.3 M	1.4 M	0.6 M	0.600 M
Voice privacy available	Yes	Yes	No	No
Roaming between different service area	Yes	Yes	Yes	Limited due to different channel spacing and frequencies of operation

* Excludes GSM reserve channels.

[†] Due to control channels being exclusively available to each operator.

This network had 98,000 subscribers in 1990 and operated in the 450-MHz band. The RMTS had countrywide coverage provided by 350 base stations.

ADVANCED CELLULAR SYSTEM (ACS)

The Comvik advanced cellular system (ACS) system was placed in service in 1983 in Sweden and later introduced into Hong Kong. This system operated on the 400- and 800-MHz bands. It featured distributed intelligence and relied on the mobile unit to control more of the switching functions than is the case in other cellular systems. For this reason the Comvik ACS system had small and relatively cheap base stations.

The handoff is initiated by the mobile unit, which begins to scan for a free channel once the field strength drops below 3 microvolts. When indicated, the mobile will begin a scan of free channels at other sites, and if a channel is available, it will request a handoff. The system was limited to 16 channels per site.

When first turned on, a mobile scans for a free mobile terminating (MT) channel and stays on this channel until a handoff is indicated. Calls are initiated by changing to a mobile originating (MO) channel.

The ACS was designed in the 1970s and was meant to make a system comprising a number of switches, each with a capacity of 96 base-station channels and approximately 2000 subscribers. The subscriber memory had a capacity of 20,000 plus 5000 roamers.

Mobiles had a 2-MHz duplex bandwidth with 25-kHz channel spacing. This means that a maximum of 80 channels are available before frequency reuse.

The system was configured to interface with the PSTN at a 2-wire level, as shown in Figure 2.5. Notice that the switch acts as a controller and is not in series with the mobile circuit.

AMPS

Advanced Mobile Phone Service (AMPS) is by far the most successful of the analog systems, and until early 1999, was the dominant system in North America, although that position has been seriously challenged by the digital alternatives.

The AMPS systems, which originated as North American specification and were first operational in 1983, rapidly became a de facto world standard. Variants of AMPS, with the main variation being frequency, TACS and JTACS (the Japanese version of TACS),

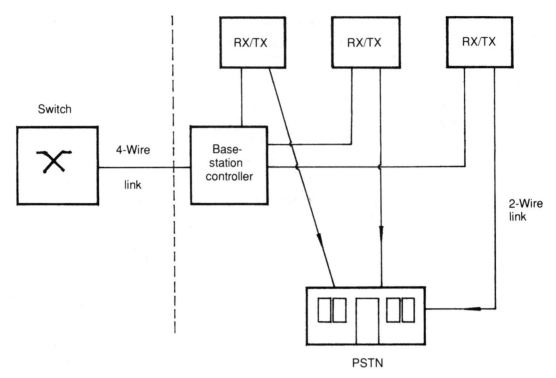

Figure 2.5 The ACS system interfaces at the 2-wire level to the PSTN.

were also successful, but were mostly (but not exclusively) confined to areas where frequency restrictions dictated their use.

AMPS has proved to be a robust system, adequately serving large, high-density cities in the United States, at penetration rates far higher than was originally envisaged. This was achieved mainly through the use of microcells, which permit VHF reuse rates.

The AMPS specification is essentially an air-interface one, and this allowed manufacturers to produce equipment that was compatible only to the extent that any mobile would work on any system. It soon proved that this was a serious weakness of the standard, especially when roaming was attempted, and so the IS-41 standard was written to allow intersystem operability.

Table 2.2 shows a typical AMPS base-station transceiver rack specification. Figure 2.6 shows a Motorola AMPS sector base station with 40-channel capacity.

TABLE 2.2 Typical AMPS Base-Station Transceiver Rack Specification

Parameter	Specification
Frequency stability	±1 PPM
RF sensitivity	−116 dBm @ 12 dB SINAD
Expander	1:2 attack time 3 ms recovery 13.5 ms ± 20% (per EIA PN 1377)
Audio de-emphasis	−6 dB for 300 Hz to 3000 Hz (per EIA PN 1377)
Audio distortion	2.5% at telephone interface
TX spurious	To meet EIA PN 1377
Compressor	2:1 attack time 3 ms recovery 13.5 ms ± 20% (per EIA PN 1377)
DC power	24-volt nominal (neg. earth)
Power/channel	8 amps
Equipment floor load	800 kg/m²
Operational temperature range	−4°C to 45°C

Figure 2.6   A Motorola AMPS sector base station with 40-channel capacity, circa 1989.

AMPS was the first 800-MHz system produced: it was released in late 1983. TACS is essentially an "enhanced" AMPS system.

THE MOTOROLA SC9600

The debate over which digital option will be dominant in the United States is still unresolved in 2000, but the situation was even less clear in 1994 when Motorola brought out the ultimate answer—the SC9600, which can support all of the contending systems. The SC9600, depicted in Figure 2.7, shows a 3-rack system, which can handle TDMA, AMPS, NAMPS, and CDMA in one rack. The system is composed of one rack of transceivers and processors, one site interconnect rack to connect to antennas and test subsystems, and one wide-band power amplifier, which combines all of the formats. In some ways this concept has been derived from the JTACS/JDC system.

The PA is driven by two preamps, which provide some parallel redundancy. It is made up of 10-watt amplifier modules [the size chosen because of the cost-effectiveness of bipolar device radio frequency (RF) technology at this power level]. Additional amplifiers are added as needed to cater to the total number of channels in use. These amplifiers have a peak power capability of 10 times the average, which permits them to operate efficiently even when all the channel power vectors are in phase. The ampliers are coupled in such a way that only 0.25-dB loss occurs in the coupling.

Centralization includes a PA that can be remotely tuned and a central Operations and Maintenance Center-Radio (OMC-R).

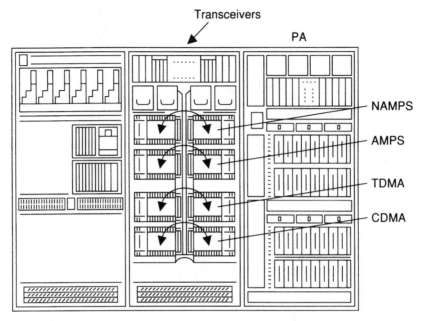

Figure 2.7 The Motorola SC9600.

In the basic configuration of three frames a capacity of 96 channels or 320 CDMA channels can be accommodated.

Trunking efficiency is improved by including a transcoder to digitally compress the voice channels. The connection back to the switch compresses as follows:

CDMA	4:1	
TDMA	4:1	
Analog	2:1	(ADPCM)

A mobility manager moves much of the call management to the base station, so that channel allocation, collection of subscriber traffic data, and billing all go to a central controller.

THE AT&T AUTOPLEX SYSTEM 1000

The AT&T Autoplex uses wide-band power amplifiers configured in a ring, as shown in Figure 2.8. This configuration simplifies retuning because it can be done at the low-powered preamplifier stage and it eliminates expensive and bulky combining equipment.

The tuning functions that are remotely controlled in the Autoplex are channel frequency assignment, SAT tones, antenna connections, and radio-to-trunk links (which are digital). This means that system reconfigurations can be done from a central site, with minimal interruption of service.

ANALOG CALL CHANNEL CAPACITY

The paging capacity of analog systems differs substantially. Table 2.3 shows the paging capacities where mobile call messages are assumed to be single words occupying 50 percent of the control channel availability.

The difference between the TACS and AMPS paging rate (of 8 and 10 kbps) is due to the difference in channel bandwidth (25 and 30 kHz, respectively).

NMT900 uses idle voice channels for paging; it does not use dedicated paging channels. This is advantageous in rural areas with small systems where it is not necessary to allocate a dedicated paging channel; in systems with high traffic, this results in reduced system capacity. The lack of dedicated control channels precludes competitive operators using the same system in any one area, because operator-specific control channels are not available.

ANALOG FREQUENCY BANDS

The major analog systems differ in two basic ways: range and spectrum efficiency. The first of the (800- to 900-MHz) systems designed was AMPS; it was released in late 1983. TACS followed in 1985 and NMT900 in early 1986. The "newer" systems (TACS and NMT900) emphasized conservation of frequency and the availability of many channels. Various techniques were used but, most significantly, the frequency deviation of the fre-

The linear amplifiers

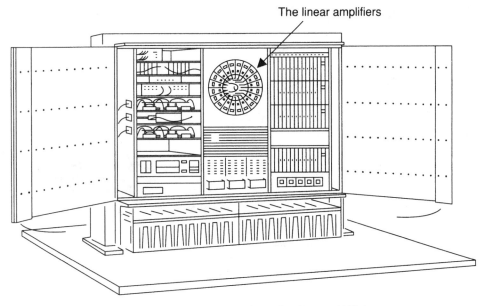

Figure 2.8 The AT&T Autoplex System 1000.4.

TABLE 2.3   Paging Capacities

System	Pages/Hour
CDMA	4,536,000
AMPS	77,760
TACS	62,000
NMT900/450	13,000

quency modulation (FM) was progressively reduced. For operation above the noise threshold, this decrease in deviation was accompanied by a decrease in range, as shown in Table 2.4.

Table 2.4 compares the range from co-sited bases of the four major systems. Although operation below the threshold negates the advantages of higher deviation, it is generally expected that cellular mobile systems will operate in high-quality signal-to-noise (S/N) conditions. Notice that in the later TACS and NMT900 systems, the range progressively decreases.

An additional but relatively minor link budget difference is caused by the smaller aperture of a 900-MHz antenna relative to an 800-MHz antenna of the same gain. This gives a net overall gain of about 0.6 dB to the 800-MHz system, which results in a small increase in range.

NMT450 does not have high deviation but, because it operates at about one half the frequency of the other systems, it has a usable path loss 6 dB greater than those

systems. Its range is consequently similar to AMPS and TACS.

Table 2.5 shows the spectrum used by the various 800- and 900-MHz cellular systems. Notice in particular the frequency overlays between the ETACS and AMPS/ AMPS extended band, which involves 10.4 MHz in the region 894.4 to 872 MHz.

Table 2.6 gives the specifications of the various system subscribers units.

WORLD CELLULAR CONNECTION RATES

The total number of subscribers on the cellular networks worldwide has grown dramatically, as the figures below and in Tables 2.7 show. Falling prices, smaller and better mobiles, as well as rapidly increased coverage has contributed to the spectacular growth.

The total number of subscribers is

1989	4.3 million
1990	8.6 million
1993	27.7 million
1998	200 million
1999	300 million
2000	400 million

AMPS FREQUENCIES

The AMPS A band is spread over three frequency groups, and the B band is spread over two groups.

TABLE 2.4   Effect of Deviation on Coverage*

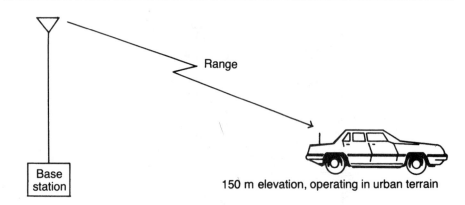

Range

Base station

150 m elevation, operating in urban terrain

System/Type	Range (km)	Range as % of AMPS	Coverage (km²)	Deviation (kHz)
AMPS	10	100	312	12
TACS	8	85	200	9.5
NMT900	6	60	113	4.7
NMT450	10	100	312	4.7

* Range determined by Okumura method for urban regions for high voice quality (30 dB S/N).

TABLE 2.5   Frequency Bands Used by Analog Cellular-Radio Systems

FREQ. (MHz)	BASE RX	BASE TX

(Frequency band chart showing BASE RX and BASE TX allocations)

820

824 / 826 — AMPS A EXPANSION

826 — AMPS A

836 — AMPS B

845

845 / 846.5 — AMPS A EXPANSION

846.5 / 849 — AMPS B EXPANSION

860 / 870 — JTACS (860, 870)

860
868
870 — AMPS A EXTENSION

880 — AMPS A

AMPS B

872 — ETACS A

880 — ETACS B

888
890 — TACS A

890 — AMPS A EXTENSION
891.5 — AMPS B EXTENSION
894.4

897.5 — TACS B

905

TACS RESERVE

915 — JTACS

925

917 — ETACS A

925 — ETACS B

933

935 — TACS A

942.5 — TACS B

950

ETACS RESERVE

960

TABLE 2.6 Subscriber Analog Unit Specifications Compared

Specification	AMPS	TACS	NMT450	NMT900
Channels	832	1320	180	1000/2000
TX frequency	824–849	872–905	453–457.5 (& other bands)	872–905
Channel separation	30 kHz	25 kHz	25/30 kHz	25 kHz/12.5
Frequency stability	± 2.5 ppm	± 2.5 ppm	± 5 ppm	± 5 ppm
TX power	3 watts	2.8 watts	15 watts	3 watts
Voice deviation	12 kHz	9.5 kHz	4.7 kHz	4.7 kHz
Data deviation	8 kHz	6.4 kHz	3.5 kHz	3.5 kHz
Receiver sensitivity*	12 dBc @ −116 dBm	20 dBp @ −113 dBm	20 dBp @ −113 dBm	20 dBp @ −113 dBm
Adjacent channel selectivity	60 dB	55 dB	70 dB	70 dB
Spurious rejection	60 dB	65/55 dB	70 dB	70 dB
Intermodulation	65 dB	65/55 dB	67 dB	67 dB
Signaling method	Manchester	Manchester	FFSK	FFSK
Speed	10 kbps	8 kbps	1.2 kbps	1.2 kbps

* *Note:* Because of differing measurement techniques these figures cannot be directly compared. Sensitivity is about equal for all systems.

Base-Station TX Frequency for Channel "N"

1. For N between ch1 and ch799. This includes the original A and B band—the A band has the channels 1 to 333 and the B band has 334 to 666. Also included is the extended A band (ch667 to ch716) and the extended B band (ch717 to ch799).

$$\text{Frequency} = 870 + 0.03 \times N$$

2. For N between ch991 and ch1023 (extended A band)

$$\text{Frequency} = 839.31 + 0.03 \times N$$

The control channels for AMPS are channels 313–333 for the AMPS A band and 334–354 for the AMPS B band.

TACS FREQUENCIES

TACS frequencies, like AMPS, are spread over a discontinuous range of spectrum. The frequencies are related to the channel numbers as indicated below:

Base-Station TX Frequency for Channel "N"

1. For N between ch1 and ch600

$$\text{Frequency} = 934.9875 + 0.025 \times N \text{ MHz}$$

2. For N between ch1329 and ch1968 (the ETACS band)

$$\text{Frequency} = 883.7879 + 0.025 \times N \text{ MHz}$$

TABLE 2.7 Ovum Forecasts (1998) from January 1999 to 2004, in Units of One Million

	1999	2000	2001	2002	2003	2004
United States and Canada	77.7	97.3	128.6	148.2	163.3	173.2
South and Central America	22.0	28.5	34.5	45.4	59.1	72.3
Europe	96.6	125.6	150.6	178.5	206.5	230.8
Africa	3.1	5.5	7.1	8.5	10.3	12.4
Middle East	7.5	9.8	12.4	15.3	19.0	26.7
Central Asia	27.3	49.1	58.1	70.9	77.6	84.8
Asia Pacific	73.2	87.6	105.2	116.0	128.0	139.5
Totals	307.7	404.4	495.5	582.8	661.8	741.3

The base-station receive frequencies are 45 MHz lower than the TX frequencies.

The control channels for TACS can be system-specific, but they need to be selected from a set of channels, the lowest channel number of which is;

23, 98, 173, 248, 323, 398, 473, 548, 1396, 1471, 1546, 1621, 1771, 1846, and 1921.

The extended version of TACS, which intrudes well into the AMPS band, is known as universal TACS, UTACS, and occupies the frequency band base RX 880–905 MHz and base TX 935–950 MHz.

NAMPS

Motorola introduced an interim system, NAMPS, which split the existing AMPS 30-kHz channels into 3×10 kHz for additional capacity. This system was trial-tested in a number of places, but generally got a lukewarm reception. Most Motorola phones today are NAMPS compatible. This system is covered more extensively later.

DIGITAL

Digital systems will be covered only briefly here, as each of the systems is covered in some detail in dedicated chapters.

GSM

GSM first emerged commercially in 1991 and after a troubled start, eventually in 1998 became the dominant system worldwide. This dominance was driven more by marketing excellence rather than any technical qualities.

By mandating that all European countries must use GSM, a critical mass was put in place that provided both the incentive and the market to drive the development of the technology. GSM is used in most countries in the world, but outside Europe and Africa, it mostly operates in competition with other technologies.

The switching platform used by GSM has become the model for all future systems as it has been developed with consideration given to pan-European working. This virtually ensures worldwide interworking and compatibility.

DAMPS

GSM is a most complex specification; much of this complexity could have been avoided, and a "cut-down" version of GSM, namely DAMPS, has been developed. DAMPS is mostly GSM stripped of all the messy bits. DAMPS may well have succeeded had it not been for the fact that is got caught between two powerful alternatives. As DAMPS evolved, GSM had already gained a strong foothold outside the United States. In the United States, CDMA development was offering a new technology that would change the way cellular radio was delivered on the air interface. Many could not decide and most waited to see what CDMA could do.

CDMA

Like all new technologies, CDMA was introduced with somewhat too much haste and far too much hype. When the early systems in Hong Kong and Korea proved to be "just another cellular system," many people lost interest. In the United States, the debate about technologies raged on, resolving little.

CDMA, like all the systems before it, gradually solved the teething problems but it soon became obvious that the claim of being equal to or better than analog had once again fallen short. CDMA originally claimed a range that was equal to or superior to AMPS. The tricky part about this claim is that it is sometimes true. In high multipath environments, CDMA can match AMPS, but in low multipath it generally cannot.

Voice quality is a highly subjective thing; but like all digital systems CDMA leaves something to be desired. It is however arguably the best of the digital offering.

3G

All that happens after the current generation of digital technologies is called 3G. The announcement that the 3G technologies would all use a CDMA modulation really came as no surprise. Whatever its shortcomings, CDMA is the best digital air interface protocol that has been demonstrated. Also it is virtually certain that most 3G platforms will have a GSM switching platform, so it can be said with some confidence that the 3G systems will have the best of the digital offerings.

The challenge that 3G has set is to provide high speed data over a noisy cellular environment. This will not be easy. High gross data rates, for what they are worth, are themselves hard to achieve over mobile networks. Getting low error rates (which is the same thing as high throughput) is much harder again. The next decade will be most interesting.

3

BASIC RADIO

Radio was first postulated in a roundabout way in 1873 by Maxwell (who found mathematically that a wave which was both electric and magnetic, and traveled at the speed of light should exist), demonstrated in 1888 by Hertz (who commented that it was an interesting phenomenon, but of no practical consequence), and used for practical communications in 1895 by Marconi. Radio is an electromagnetic phenomenon and radiates as photons. It belongs to the family of radiation that includes X-rays, light, and infrared (heat) waves. The different categories of radiation differ in frequency, as shown in Figure 3.1. They also differ in energy and ability to propagate through different media.

BASIC ELEMENTS

All practical radio systems can be reduced to the basic scheme shown in Figure 3.2. For the purposes of telecommunications the radio spectrum is divided into frequency bands, virtually all of which have some application for communications. The bands are defined in Table 3.1.

The names reflect the development of technology, with relatively low frequencies, such as shortwave, being called "High Frequency" (HF). In the days when this band was named, most practical transmitters where mechanical and these frequencies were obtained only with great difficulty. (Although some spark transmitters were capable of generating GHz frequencies, because the detectors were so crude, the range was limited to meters, and these frequencies were not regarded as being of practical use.) Consequently, as the upper limits were pushed away, new superlatives for "Very High" (VH) needed to be used.

Virtually all cellular applications are in the ultra HF (UHF) band, although the microwave links can be in the extremely HF (EHF) band.

At the lower end there is the extremely low band (ELF) which includes the power supply frequency (50 or 60 Hz).

Transmitter

A transmitter consists of two basic parts: a modulator and a carrier. Figure 3.2 shows an analog transmitter with an audio input that is converted into the form to be transmitted (in this case, via a microphone to the modulator). A radio frequency generator generates the radio energy that will carry the signal. This generally consists of an oscillator (which produces the initial signal) and a number of amplifier stages (which amplify the level to that required at the antenna). A modulator mixes the signal to be transmitted with the radio frequency signal (called the carrier) in such a way that the signal can be decoded at a distant receiver.

Receiver

The receiver in Figure 3.2 gets a signal from its antenna, which also receives a number of unwanted signals. The tuned circuit tunes out all but the wanted signal, which is then demodulated (decoded) by the demodulator.

The very simple receiver illustrated in Figure 3.2, consisting only of a tuned section and a demodulator, is known as a *tuned radio frequency* (TRF) receiver. Until about 1930 most receivers were of this type. Such a receiver is seen in Figure 3.3. This is a 5-valve set with 3 tuned RF stages followed by two transformer coupled

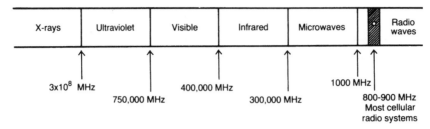

Figure 3.1 Electromagnetic spectrum.

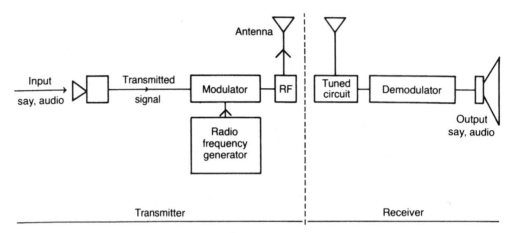

Figure 3.2 A simplified analog transmitter/receiver with an audio AM modulation.

TABLE 3.1   The RF Communications Bands

Frequency Range	Name	Abbreviation	Main Uses
Less than 300 Hz	Extremely low	ELF	Submarine
300 Hz–3 kHz	Infra low	ILF	Voice band
3 kHz–30 kHz	Very low	VLF	Audio band
30 kHz–300 kHz	Low	LF	Broadcast/long range
300 kHz–3 MHz	Medium	MF	Broadcast/long range/shortwave
3 MHz–30 MHz	High	HF	Shortwave band
30 MHz–300 MHz	Very high	VHF	Mobile radio/paging/VHF TV
300 MHz–3 GHz	Ultra high	UHF	Mobile/microwave/TV
3 GHz–30 GHz	Super high	SHF	Microwave/satellite
30 GHz–300 GHz	Extremely high	EHF	Microwave
300 GHz–3000 GHz	Tremendously high	THF	Experimental

audio stages. The 1927 U.S. designed Hazeltine TRF set is still used daily in my home. It works fine in the daytime but lacks selectivity for night time reception. The downside is the tuning, which uses three separate dials that make the task a bit like finding a safe combination.

The TRF was replaced as the main commercial receiver by the *superheterodyne*, commonly known today as the *superhet*, shown as Figure 3.4. The name was derived from the word *heterodyne*, which means the beating of two signals together to produce the sum and difference frequencies; (CW or *continuous wave* Morse code signals were decoded by beating the incoming signal with one of almost the same frequency so that an audio tone equal to the difference frequency was heard). The tuning and signal amplification in the early superheterodynes was done at supersonic frequencies of around 50 to 60 kHz (the heterodyne frequency), and hence the word "super." These frequencies were chosen because the early valves worked better at these lower frequencies than at RF (approximately 1 MHz). The

Figure 3.3   The Hazeltine TRF receiver, designed in 1927 in the United States. This receiver uses 3 independent dials for tuning.

heterodyne frequency is now known as the *intermediate frequency* or IF, and in modern receivers is well above the supersonic range—typically 455 kHz or 10.7 MHz.

Even today, considerable advantages are had by doing a lot of the amplification and tuning at relatively low frequencies where technology is well-developed and components are inexpensive.

The superhetrodyne uses a mixer to down convert the carrier. In the process of mixing, the mixer will present not one but two carrier frequencies to the IF stage. Let's assume that IF frequency is 500 kHz and the mixer os-

cillator frequency is 800 Mhz. As a result of mixing, the 800.5-MHz carrier and the one that is 799.5 MHz, both will be presented to the IF amplifier as the "wanted" 500-kHz signal (i.e., there are two frequencies 500 kHz away from the IF frequency). The second, unwanted frequency is known as the *image frequency*. To prevent this from being a problem, it will be necessary to pre tune the carrier to eliminate the unwanted signal. Let's assume the wanted signal was the 800.5 MHz and a rejection of 60 dB was required. This would require that that the preselector filter had a Q of around 400,000; a

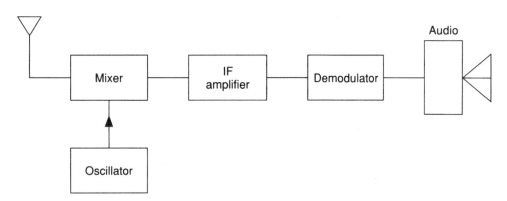

Figure 3.4   A basic superhetrodyne receiver.

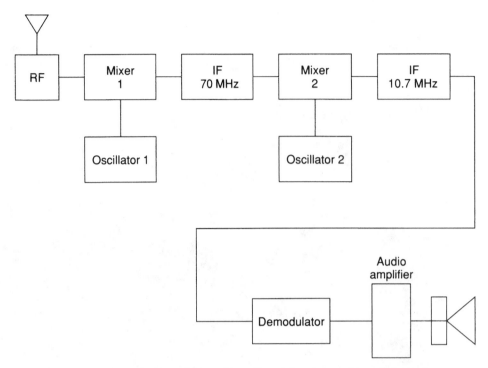

Figure 3.5 A typical superhet configuration of the type used in cellular radios.

figure that even a superconducting filter would not (readily) be able to achieve! To make it easier, we could instead of a 500-kHz IF, start with a 10.7-MHz IF (a common commercial IF frequency) so now the image frequency is 21.4 MHz away. This image can be rejected to the extent of 60 dB with a Q of only ~19,000; still asking a bit much. However if the IF frequency is further increased to 70 MHz, a more manageable Q of just more than 3000 is required. However for 800-MHz systems, the image frequency (which is now 140 MHz away) will be outside the passband of the RX preselector filter, so much of the rejection will occur before the RF stage. This approach is what is in fact done in cellular RX design.

Although a wideband RF stage like the 70 MHz just discussed will have good image rejection, it will not have good selectivity and so often a second stage of IF at a lower frequency again will be used to attend to this problem. In this case, you have a double conversion superhetrodyne; this technique is commonly used in communications receivers, some of which are even triple conversion.

A final improvement that is characteristic of modern receivers is the addition of an RF stage, which consists usually of a single stage, low-noise amplifier to improve the signal-to-noise (S/N) performance of the set. In fact, it will be seen in the chapter on noise performance, that in a normally operating superhet, it is the noise figure of that RF stage that will virtually determine the overall receiver performance.

A receiver incorporating all these elements will be found in the typical cellular receiver in both base stations and in mobile phones—shown in Figure 3.5.

Portable FM radio was first developed by Motorola for the US Army Signal Corps during World War II. The first unit was a backpack weighing around 16 kilograms and having a range of about 30 kilometers.

Modulator

The modulation system used in most analog cellular radio systems is known as frequency modulation (FM). Digital modulation is usually phase modulation. In FM modulation the frequency of the carrier is varied proportionally to the signal to be transmitted. A typical FM modulator is shown in Figure 3.6.

The audio input varies the bias on the varactor (a solid-state variable capacitor, illustrated in Figure 3.6), which in turn changes the frequency of the tuned circuit. The maximum amount that the frequency can deviate from its central carrier frequency is called the peak deviation.

The S/N performance of FM systems is very high, provided the noise level is reasonably low. FM systems with wide deviation have better S/N performance than those with narrow deviation.

Some analog systems use phase modulation, particularly for data transmission. Phase modulation is closely related to frequency modulation and can be derived

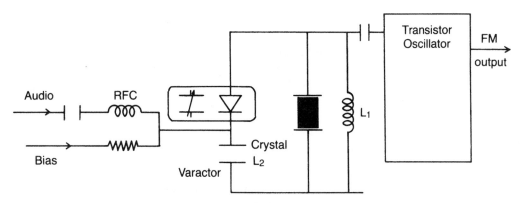

Figure 3.6 Typical FM modulator.

from it by passing the signal through a simple differential circuit before frequency modulation (see Figure 3.7).

Frequency shift keying (FSK), at relatively low speeds (for example, the Manchester code at 6.4 kbits is for TACS and 8 kbits for AMPS on the control channel), is often used for data, because it has better S/N performance than FM at low signal levels. This enhances signaling in areas of poor reception.

DYNAMIC CHANNEL ALLOCATION

Because cellular systems use many channels, it is necessary for the mobile to automatically switch to the correct channel. This is done by sending an instruction (data) that indicates the channel number required. The mobile system then switches to the channel indicated by using synthesized tuning, a system where the frequency of the oscillator is numerically compared to the required frequency and adjusted by a "phase-locked loop" until the two frequencies match.

NOISE AND SIGNAL-TO-NOISE PERFORMANCE

All radio systems are ultimately limited in range by noise. When the intrusion of noise is such that an acceptable signal can no longer be obtained, then the system is said to be noise-limited.

Medium-wave (broadcast band) and short-wave broadcasts operate in a very noisy environment; background noise limits the performance in the broadcast/shortwave bands. The very high frequency (VHF) and ultra high frequency (UHF) bands where cellular radio operates are relatively quieter and most of the noise is generated in the radio frequency preamplifier of the receiver itself. Regardless of how well designed the receiver is, there is a theoretical noise power level that, at a given temperature, cannot be improved upon. This is because of the thermal noise, generated by the movement of atomic particles (in the receiver and most particularly in the first radio frequency amplifier). This noise is proportional to the operating temperature. Hence, the antenna and RF amplifier stages will generate thermal noise continuously. For this reason high-quality receivers, such as radio telescopes, operate their input stage RF amplifiers at liquid-nitrogen temperatures or lower.

In order to perceive a relatively noise-free signal, the incoming signal must exceed the noise level by a respectable margin, known as the signal-to-noise ratio (SNR). For cellular radio systems, this level is usually regarded as about 12 dB for marginal reception and 30 dB for good-quality conversations.

Signal-to-noise ratio is usually expressed as:

$$S/N = \frac{\text{Signal level}}{\text{Noise level}} \text{ (usually expressed in dB)}$$

Because modern receivers closely approach the theoretical noise limits for their operating temperatures, it can easily be deduced what minimum received signal

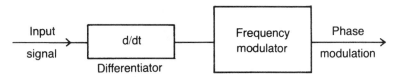

Figure 3.7 Phase modulation from frequency modulation.

level is required to achieve satisfactory signal-to-noise ratio.

Thus when we speak of a 39-dBμV/m boundary level for an Advanced Mobile Phone Service (AMPS) system, this is equivalent to specifying the point at which the signal-to-noise ratio is regarded as satisfactory.

A "satisfactory" level of signal-to-noise for cellular systems is usually regarded as one where the noise is just noticeable. For public mobile radio (PMR), a usable but noisy level is often regarded as satisfactory. In FM systems, the noise usually occurs as sharp clicks (sometimes known as "picket fence" noise because it is similar to the sound produced by dragging a stick along a picket fence).

In digital systems the RF signal level still determines the signal quality, but the quality itself is measured in bit error rates (BER). It is standard practice to take the standard as a raw error rate of 2 parts in 100 as the minimum acceptable level [this is the digital equivalent of a Signal to Noise and Distortion (SINAD) level of 12 dB]. However, for good-quality digital signals the BER is very low indeed, so much so as to be almost meaningless for speech. What is more important is the *margin* above the minimum BER. This is usually taken to be something of the order of 15–20 dB.

Digital systems degrade rapidly when the minimum BER is approached, and a much sharper boundary is established than is the case with analog.

dBs

Human senses perceive power approximately logarithmically. For example, doubling the energy level of a sound pulse produces only a 3-dB increase in the perceived level—and that increase is only just noticeable. The term dB was introduced to define relative power levels logarithmically.

The term dB is used often in radio systems and can be a major source of confusion to the uninitiated because of the large number of different units of dBs. Essentially, the dB level is the log of a power ratio: dBm, the most common form of dB, is the power of the system measured compared to 1 milliwatt. Mathematically, this can be expressed as:

$$\text{Power dBm} = 10 \log \left[\frac{\text{Power (in watts)}}{(0.001)} \right]$$

$$\text{Power dBm} = 10 \log \left[\frac{\text{Power (in milliwatts)}}{1} \right]$$

Thus

$$1 \text{ watt} = 10 \log \frac{1}{0.001} \text{ dBm} = 30 \text{ dBm}$$

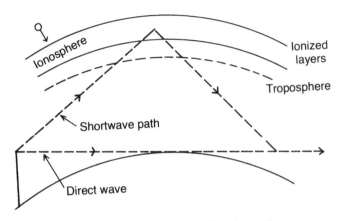

Figure 3.8   Waves reflecting off the ionosphere.

dBμV/m is a unit of field strength that compares the measured level with 1 μV/m (1 microvolt per meter).

Mathematically, this is:

$$\text{dBμV/m} = 20 \log \left[\frac{\text{field strength in microvolts per meter}}{1} \right]$$

Note: 20 is the multiplying factor here because the terms being used are voltage, not power. Voltage squared gives the power ratio.

PROPAGATION

Radio waves propagate at the speed of light (299,800 km/sec, or approximately 300,000 km/sec). Medium- and high-frequency waves can propagate very long distances by reflecting off the ionosphere, as shown in Figure 3.8. This is the way shortwave propagates around the world.

At higher frequencies (above about 50 MHz), the troposphere/ionosphere absorbs the waves instead of reflecting them, so that the predominant mode is the direct wave. This is the mode that generally applies to cellular radio systems. However, ionospheric-related propagation does sometimes occur on cellular systems, and interference (or sometimes useful propagation over distances of 200–300 km) has been noted worldwide; at least in analog systems—digital systems could not decode such a signal because of timing constraints.

The direct wave is not limited to line of sight; in fact, a good deal of refraction (bending of the path of propagation) occurs. This enables the transmissions to extend well beyond line of sight. Diffraction (bending around obstacles) also occurs, allowing the path of the wave to extend around obstacles. The ability to refract and diffract decreases with increasing frequency, but is still

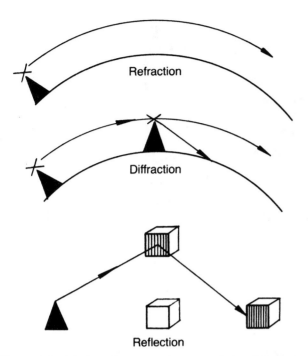

Figure 3.9 Modes of propagation at cellular frequencies. Super-refraction, or ducting, is a sporadic phenomenon responsible for propagation over very large distances under certain atmospheric conditions.

is significantly enhanced by the ability of the radio system to reflect into most areas that are inaccessible via a direct path.

These three modes of propagation are summarized in Figure 3.9.

Radio base stations are effectively radio concentrators that serve a large number of customers using a few channels. Typically, 20–30 customers can be handled on each channel (depending on the calling rate). The base station is always modular in construction. Figure 3.10 shows the basic components of a base station. These elements of the base station shown in Figure 3.10 are discussed in the following sections. Figure 3.11 shows a rural base station with 120-degree sector antennas.

RADIO CONTROLLER

The radio controller is the interface between the mobile switch and the base station, and operates similarly to a remote-subscribers switch. Under the control of the mobile switch, the radio controller selects (switches) the radio channels as required. It also supervises various system parameters, including alarm conditions and radio field strength.

ANTENNAS

Antennas used in cellular radio are usually gain antennas, meaning that they have gain, compared to the simplest form of antenna, the dipole. A dipole is shown in Figure 3.12. The simplest vehicle-mounted antenna with

most significant at the lower cellular radio frequencies (800–900 MHz).

A third property, reflection, is significant at all cellular radio frequencies. Coverage in high-density city areas

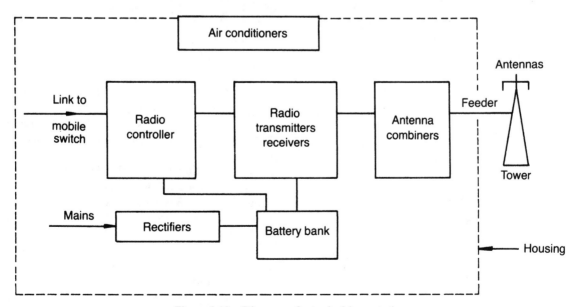

Figure 3.10   Diagram of a typical base station.

Figure 3.11 A base station with a purpose-built tower and transportable hut.

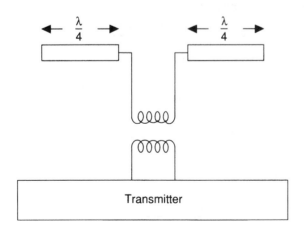

Figure 3.12 Dipole antenna.

the same gain as a dipole antenna is the quarter-wave antenna, illustrated in Figure 3.13. A mobile gain antenna is usually easily recognized by its loading coil, as shown in Figure 3.14. The loading coil is usually visible as a "bump" in the antenna.

Mobile (vehicular) cellular antennas are usually between 3 and 4.5 dB gain while handhelds have gains around 0 dB. Base-station omnidirectional antennas, which stack many radiating elements in series, are often 9 dB in gain; unidirectional base-station antennas can have gains as high as 17 dB. Very high gain antennas are only practical in fixed locations, because they are also physically large and must be exactly vertical to operate satisfactorily. Sometimes, however, they are deliberately tilted down to limit the base-station range.

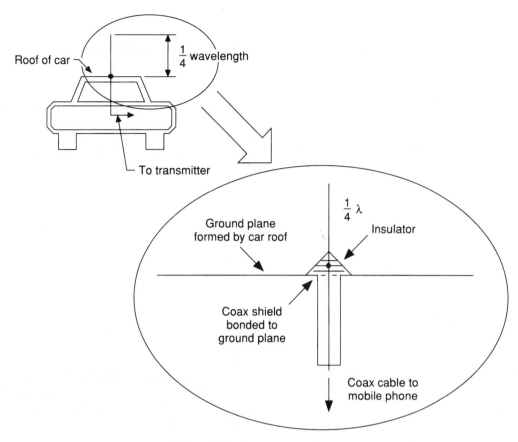

Figure 3.13   Quarter-wave antenna.

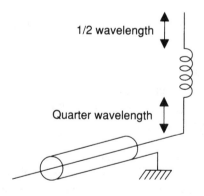

Figure 3.14   Three-dB gain antenna for roof mounting.

MOBILE TRANSMIT POWER AND HEALTH

The radiation from a radio system is non-ionizing (distinguishing it from radioactive decay products); the main effect on the human body is a rise in temperature.

In cellular systems, the mobile's power is usually limited to a few watts. It is generally believed that these powers and the radio frequency radiation levels do not pose any health hazard. However, it is not recommended to hold or touch the antenna of a mobile in use. Some authorities have expressed doubts about the wisdom of long-term use of handheld units.

Contrary to popular belief, it is not the high frequencies used by cellular radio (800–900 MHz) that are inherently most harmful to humans. At the lower frequencies—around 100 MHz—the body can become resonant and, therefore, very absorbent. Hence, these lower frequencies are potentially more hazardous. Early experiments using small laboratory animals pointed to relatively more harmful effects at higher frequencies. However, the small size of the animals, which gave the animals a high resonant frequency, probably accounted for these results.

4

PLANNING—AN ESSENTIAL NETWORK FUNCTION

Very few cellular networks are well planned. Most are installed with almost indecent haste, with a lot of corners cut and too many compromises made. The situation in cellular has been exacerbated by consistent underestimation of demand and the high capital investment required, which causes management to be very impatient with system engineers who would prefer to do the job properly rather than just quickly. As a result badly planned cellular networks are the rule not the exception. They are characterized by

- Almost exclusive use of pre-existing sites (for wireline operators)
- Lack of in-house survey facilities
- Poor performance in a frequency reuse environment, particularly with regard to dropped calls and interference (which has the characteristics of cross talk)
- Poor coverage
- Inefficient use of resources
- Dropped calls
- Excessive blocking

FLEXIBILITY

Flexibility is the key to cellular system planning. If you rely on marketing forecasts, the business plan or other sources of soothsaying, you are bound to have problems. The cellular industry is growing too fast and too uncertainly for anyone to be able to forecast the future with even reasonable accuracy. The system must be planned and designed to allow for any and all reasonable contingencies. It must—above all—be flexible.

TIME FRAME

The time frame for planning of a properly planned system is such that the planning should be at least a year ahead of the implementation. This time frame is dictated mainly by lead times on hardware and sites. Some particular lead times are

- Site acquisition and town planning approvals, 6–12 months
- Construction of a major switch building, 12–18 months
- Microwave and switching equipment deliveries, 6 months
- Antennas, rectifiers, and batteries, 4 months
- Base station equipment [radio frequency (RF) channels, etc.], 3 months

These are the lead times that are considered "normal." Under "exceptional" circumstances, which seem to occur so often that they seem to be the rule rather than the exception, longer delays can be experienced.

One of the problems engineers have with managers who refuse to plan, is that when the crunch is really on they can always get "something" up and running in much shorter time frames than those listed above. When this is done it will result in some built-in compromises. Some of these compromises include

- Less than optimal base-station location
- The site may not be suitable for future expansions
- Future frequency reuse may be limited
- The desired level of redundancy may not be built in

- Equipment purchased because of its availability may not be compatible with the rest of the network, and may require a stock of dedicated spares and test equipment. Also, staff training and future support problems may not have been considered
- The cost of the badly planned network will be higher
- The future cost to tidy up the compromises may be overlooked

With continued crisis management, the network becomes a patchwork that is expensive, inefficient, and difficult to maintain. However, management often doesn't bother too much about these "technical" details and the next expansion is handled the same way, and so the cycle continues. Management comes to believe that planning is not necessary and that short time-frames can be delivered by "good" management.

RADIO SURVEYS

Knowing the coverage and potential interference problems is the key to a good cellular design. Very few cellular managers/engineers understand this—consequently there are very few well-designed cellular systems to be found. Less than one-half of all cellular systems could be regarded as well designed! (If you think that statement is a little harsh, then obviously you haven't been using your cellular phone very much.) Some are very poor.

A good network can only be designed by good engineers who know its coverage and interference habits intimately. For that reason it is essential that the *designer* be a *user* of the system, since this is an excellent way to learn the performance.

Radio surveys should be done in-house and be at least one year (but preferably two) ahead of the implementation timetable. This will allow for the acquisition time, site preparation, and a safety factor so that alternatives can be found if necessary. If computerized radio coverage prediction techniques are used, it is possible that this time frame could be reduced by a few months.

SITE ACQUISITIONS

Site acquisitions are very time-consuming, considering that after obtaining either a contract to buy a site or a long-term rental contract, it will then be necessary to obtain one or more of the following:

- Tower construction permits
- Aviation department clearances
- Local government building approvals
- Frequency permits for microwave links
- Road access permits (if transportable huts are used)
- Connection of electric power

Although construction personnel can sometimes do wonders, at least six months should be allowed for completion. Most operators of large systems have found that it is cost-effective to have dedicated personnel for property acquisition. These people can negotiate with the property owners and follow through on approvals and permits.

Opposition from local authorities that the installation will be an eyesore can be expected. However, with some attention to design, a base station—even though still a little prominent—need not be altogether out-of-character with its surroundings, as can be seen in Figure 4.1. Here

Figure 4.1 Base stations need not clash with the surroundings. This base station seems to blend with the surroundings.

even the 30-meter monopole seems to harmonize with the surroundings. (See also disguise antennas in Chapter 12.)

Sometimes a quick fix can be a future liability on a grand scale. Inaccessible sites that require helicopter access or four-wheel drives can be ideal microwave repeater and relay sites, but the cost of maintaining them can be prohibitive. Don't be lured into thinking that a microwave system that has a mean time between failures (MTBF) of 10 years can be left unattended. Don't forget that the batteries, solar cells, buildings, towers, and even the site vegetation require regular attention. I know of such sites that cost over $1 million per year to maintain.

FREQUENCY PLANNING

Frequency reuse is the essence of cellular radio [except for code division multiple access (CDMA)]. Good frequency planning starts with well-chosen sites. Although some designers make a lot of fuss about site location relative to a hypothetical hexagonal grid, a much more critical consideration is coverage and reuse potential. Fortunately a system frequency retune can be done relatively easily, and so bad designs can often be corrected at a reasonable cost.

The frequency plan should not be done with the sole consideration being the "as-built configuration." Planned future cells should be allowed for and their future allocations reserved. This permits an orderly expansion and will minimize retunes.

TRUNK NETWORK PLANNING

The general rule for trunk capacity is that there is never enough. No matter how generous you thought the initial installation was in regard to future trunk provisioning, it is always too little. Fiber optics and relatively low-cost, high-capacity digital microwave have made this problem a little easier, but it still has not gone away.

While some companies still use 2- and 8-Mbit (or T1), more forward-thinking ones are doubting if 34-Mbit is adequate and have standardized on 140 Mbit for main trunk routes. The extra cost is minimal for fiber optic, but the shorter hops (and hence additional repeaters and towers) for microwave have to be justified.

It is necessary to always consider future alternative trunk routes as well as route diversity. The route diversity should include the provision for future nodes. All sites, no matter how remote, may one day be needed as drop and insert points. Don't design this capability out of the system.

Where applicable, alternate routing can be used to reduce the cost of the trunk network *and* improve its survivability.

ROOM TO GROW

Poorly planned systems rarely have built-in capacity for future growth. This can be a particularly expensive problem if building and site plans do not allow for expansion. How often do we encounter a site that is difficult to expand, just because the initial building was located so as to prohibit growth? How often do site managers lament that the site next door was not purchased when it became available? Or even that the tower was not designed to carry the number of dishes that are now needed? As seen in Figure 4.2 towers can easily become fully loaded, and the one illustrated really has little room for any expansion. A good rule of tower design is to estimate the future demand as accurately as possible, and then double the estimate.

Towers can present difficult problems for the installer if the site and site access have not been well planned.

Figure 4.2 This tower is fully loaded with an array of microwave dishes and Yagi antennas. Towers become full all too easily.

Not only does a tower require a good deal of space to build and erect but the foundations for the footings require extensive excavations, which can be potentially hazardous to the pre-existing structures.

All of these problems result from shortsighted thinking and lack of planning.

OTHER FACILITIES

Some cellular operators will intend to expand into related areas such as paging, trunk mobile, backbone provisioning, and voice messaging. If your operation does intend to expand, leave room for the equipment. This can be done either by dimensioning the rooms a little larger or by designing equipment rooms that can be easily expanded. Non-structural walls improve flexibility.

Often opportunities will arise to rent tower and equipment space to other users. This can be a good additional source of income, but if you intend to do this, you must plan the access for the "other users" so that they can install and maintain their equipment without disruption to your service, and preferably without access to sensitive areas like the switch. Some operators have a self-contained "other users" room that has no access to the main equipment rooms. Increasingly, shared sites are being mandated by governments.

5

CELL SITE SELECTION AND SYSTEM DESIGN

Cell site selection is the process of selecting good base-station sites. The process is half art, half science. To the customer, the most vital feature of a cellular system is its coverage. Therefore, it is important for the system to deliver what the customer wants—good coverage within a logically defined service area. The service provider is usually interested in both extensive frequency reuse (except in small towns) and good coverage. The selection of the best cell sites is essential. Often this part of cellular-system design is poorly considered and results in poor system performance.

DESIGN OBJECTIVES

The design objective should be to cover the service area without any serious discontinuities as economically as possible; and, where applicable, to allow for future frequency reuse. In most technical journals, coverage is defined as 90 percent of the area covered for 90 percent of the time. Were this "objective" to be achieved literally, the coverage would indeed be poor (that is, 81 percent of the service area would provide adequate field strength for successful calls; 19 percent of the target area would be substandard). In fact, the 90 percent/90 percent coverage standard means something quite different. The 90 percent/90 percent definition means 90 percent of the area should be covered at any time or, alternatively, that 90 percent of the regions achieve a defined standard at any point in space and time.

The FCC has specified a field strength of 39 dBμV/m average [for Advanced Mobile Phone Service (AMPS); code division multiple access (CDMA) coverage is similar] as the boundary of a cell. Although this figure is realistic, it is a compromise. In practice, the boundary field strength for acceptable service is a function of the terrain. The real objective is to obtain a signal-to-noise (S/N) ratio comparable, but not necessarily equal, to a land-line telephone, which is usually accepted as 30 dB (some European authorities use 20 dB, which is below the quality of a very poor telephone line). Land-line systems, however, generally achieve 40 dB or more.

As a guide, Table 5.1 lists more realistic boundaries. No firm rule for urban central business district (CBD) can be formulated, but it is generally accepted that a 20–30-dB margin over mobile levels is required for handheld use in multistory buildings. Good handheld coverage is defined as a signal level yielding a voice quality that is usable without discomfort in buildings from the ground floor up, excluding elevators and their immediate vicinity (2 meters).

ASSUMPTIONS AND LIMITATIONS

All new systems need good handheld coverage, a fact that provides a good starting point. Individual cellular operators will probably have their own preferences for buildings to house cellular bases, such as telephone switching centers, buildings designated as radio telephone sites, or maybe just a few particular buildings to which access can easily be arranged. The cellular-system designer should accommodate these preferences, but not at the expense of sound design.

A cell plan (that is, $N = 4$, 7, or 12) should be chosen on the basis of the long-term customer density, noting that high density favors the 4-cell system. Initially, however, the 4-cell system is both more costly and less toler-

TABLE 5.1 Boundary of Cell Sites for AMPS Systems

Area	Recommended Field Strength AMPS/NMT450/CDMA	NMT900	GSM*
Urban CBD	60 dBμV/m	68	72
Suburban	39 dBμV/m	47	51
Rural	34 dBμV/m	41	46

* Still not widely agreed upon.

ant of site selections that do not approximate the theoretical cell plan.

Defined Coverage

Before a design is implemented, the number of bases to be used should be defined based on the techniques outlined in Chapter 25, "Budgets."

Defining Boundaries

Radio waves propagate according to natural laws that have little to do with city boundaries and customer service areas. Therefore, there is little value in defining a precise area to be covered until the "natural" boundaries defined by the propagation characteristics are known. Of course, it is still practical to define certain areas as essential for coverage, leaving the "fine-tuning" of the actual boundaries as flexible as possible.

When boundaries must be decided, the decision should be made in consultation with engineering and marketing staff once alternative boundaries (those possible, given the resource constraints) are known. At coverage boundaries, high sites become increasingly attractive in gaining an economic population of customers in a given service area. These sites are usually in low-density suburban or rural areas.

Usually, the physical boundaries of a city are poorly defined (in terms of customer density). At the "edges," a number of options are available. For example, one base to the north of a city may provide a potential 1000 km² coverage along a sparsely populated highway, while the same base to the south may cover only 300 km² but includes two important towns. For this reason it is important that the system designer have good communication with the marketing staff.

While the designer should be free to select sites that provide continuous coverage in high-density regions, the engineering staff must recognize that there are no clearcut "best" designs and that many alternate solutions, with different coverage, could be equally viable. Indeed, detailed marketing studies can be undertaken to "resolve" the problem.

Note: It is important to provide your marketing staff with a clearly marked map showing the expected coverage of the alternatives. Without this information, they have no foundation on which to draw their conclusions.

SUITABLE SITES

All selected sites must have reasonable access for installation and maintenance, and suitable accommodation (usually a minimum of 18 m²) with a ceiling of 0.6 meter above the equipment rack. Equipment racks are usually 2.2, 2.7, or 2.9 meters high, depending on the manufacturer. It is important that equipment access is available and that provision is made to accommodate heavy crates of equipment. Access for maintenance must also be provided. If the site is located in a rural area, power and reasonable road access should be available. All of these improvements can be expensive if they have to be provided after initial installation.

Links to the base stations (either radio or physical) must be provided, and power is always essential. Consequently, the designer should undertake a site visit to ensure suitability before pursuing the design too far. If microwave links are used, the path back to the switch (or nearest hub) must be considered. Remember that base-site antennas can be large and 12 (or more) sector antennas with dimensions of 0.5 m × 3 m (approximately) may need to be fitted. Although it is usually impossible to guarantee the life of a site, some sites are more vulnerable than others and it is best to avoid those where new building that will obscure the radio path may occur.

The site needs access to commercial power (about 400 watts/carrier, including air-conditioning), and usually some provision for an emergency power plant must be made. Hospitals are often good sites because they are usually relatively low buildings with good clearance around them (parking lots and gardens). They also have emergency power. On some tall buildings where emergency power is not available, it may be feasible to run a cable down the building to an external socket that can be powered by a portable generator. This saves cost, space, and possible objections to a permanent installation.

Because a cell site is a substantial building and, usually has, a prominent antenna-support structure, the zoning of the site must be appropriate. Although it is often possible to get city planning approval to build base sites in residential areas, the process is uncertain and slow. Areas zoned for industrial or commercial use are usually easier to acquire for cellular purposes.

The availability of a site will be determined by local building codes, neighborhood environmental attitudes (particularly if a substantial antenna structure is involved), as well as any limitations imposed by the site owner.

If rental property is being considered, leases should be for at least five years, preferably with an option to extend the lease. If possible, it is advisable to house on-site equipment in transportable buildings, which will minimize costs in the event of lease termination. Also, an installation should be done with the expectation that at some future date you will need to move the equipment.

Where an antenna support structure must be built, especially in a residential area, the following questions will be posed by the local residents:

- Will it interfere with the TV reception in the area?
- Is the tower safe?
- Will the microwave make me sterile?
- How big is the tower?
- Will it cause cancer?

Of course, there are usually no problems, but the questions will be asked, and the design engineer should be ready with the answers.

JOINT USER SITES

For most cellular operators, the days of intense competition to get a little more coverage than the competitor are over. During start-up coverage was everything, but most established systems have adequate coverage, and new cell sites are needed mainly for capacity. There could be significant savings if the operators were able to get together and share the costs of sites, towers, and infrastructure. Separate lockable buildings can be provided, but all else could be shared.

Existing sites could either be excluded, cross-shared, or sold to a third company, of which both (or all) carriers are shareholders. Cost savings could easily amount to 30–40 percent of the total cost of a base station.

While sharing is desirable for cellular applications, it is probably imperative for PCS/PCN.

GETTING A STARTING POINT FOR THE DESIGN

Inasmuch as the choice of suitable sites is an interactive one, the sooner a desirable central site can be identified, the better. A starting point for the design must be established. Usually, that point will be the primary site selected for the central business district (CBD) coverage. The primary CBD site is a good starting point because it can be selected to cover the CBD regardless of other coverage requirements. Also, CBD sites are difficult to acquire and once one has been found, it is a good idea to make it fit into the final pattern.

In small towns, frequency reuse may not be necessary, so high "broadcast" sites can be chosen. Therefore, height can be used to select a good first site. This type of design is very different from high-density designs and is a good deal easier. As handheld use is contemplated, it is often necessary to find a prominent central urban site to provide handheld coverage in CBD buildings. This is particularly true if all other sites are more than 4 km from the CBD.

In big cities the opposite is true: a selected site(s) should offer good handheld penetration while at the same time containing the coverage. It may be possible to use surrounding buildings as radio-path shields. If frequency reuse is a consideration, the ability to provide adequate frequency reuse should be a major consideration. It should be noted that although placing the antenna in a situation where it is surrounded by tall buildings, as shown in Figure 5.1, is a good way of confining the radiation for that antenna, the surrounding buildings will cause reflections that will seriously degrade the back-to-front isolation of the sector antennas used. This can itself be a problem if there is to be frequency reuse within that confined area.

When tall buildings are used, as in Figure 5.2, virtually all mobiles in the area will interfere in this antenna. As the front-to-back ratio in real life (as distinct from free space values) is usually only about 10 dB, even interference from the back of the antenna will be a

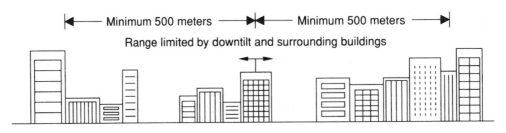

Figure 5.1 Preferred frequency reuse.

Range limited by downtilt only

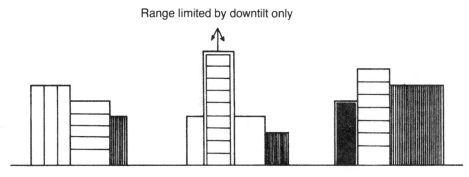

Figure 5.2   Poor frequency reuse.

problem. This configuration is good for coverage but bad for frequency reuse.

Frequency reuse would eliminate the tallest buildings (even considerable antenna downtilt will not help much) as CBD sites. A good choice might be to use a smaller building with a clear area of about 500 meters from the next obstructing building.

Site availability is a major limitation. Many ideal sites have uncooperative owners. The designer often has little choice, particularly with inner-city sites.

For some rooftop installations, weight may also be a problem.

SPECIAL CONSIDERATIONS

Significant reflections occur from large buildings. As a result, some of the front-to-back ratio immunity of sectored antennas will be lost. For this reason it is essential to survey sectored sites facing tall buildings, paying particular attention to the area outside the nominal cell area. The front-to-back ratio of an antenna that may nominally be 22 dB can be reduced to 6–15 dB because of reflections (see Figure 5.3). This reduction can result in interference problems from behind the cell.

Figure 5.3 shows one way of exploiting natural or man-made boundaries to improve frequency reuse. Assuming that the buildings form long rows following the coastline, as shown in Figure 5.3, then the buildings can be effectively serviced from a seashore site. At the same time, the buildings shield the site to permit effective frequency reuse behind them. Experience shows that ship-to-land cellular communications are usually made via the highest base station rather than the one nearest the sea. This fact should be considered when deciding if reflected signals are likely to be a source of interference. Use by sea vessels is generally limited, and reflections are not likely to cause problems with sea coverage. Where significant sea traffic is anticipated, a high base should be dimensioned accordingly. Line-of-sight propagation losses (rather than mobile environment losses) should be assumed. Because of reduced multipath, a lower field strength of 32 dBμV/m for AMPS/CDMA & 44 μV/m for GSM 900 will suffice. This lower figure results in significantly greater coverage. Although with digital sys-

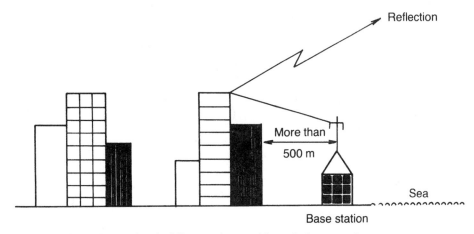

Figure 5.3   Using buildings and natural boundaries to confine coverage.

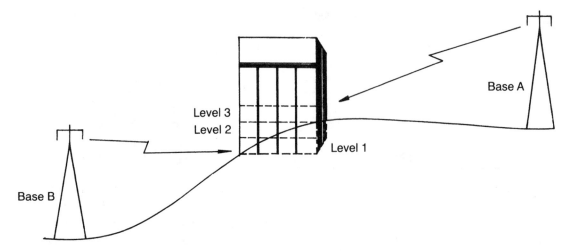

Figure 5.4   Land topography can cause dead spots. This building is partially underground from some directions. Note that the building may not be underground with respect to base B, but the first two levels of the building are underground with respect to base A.

tems over-water range is likely to be propagation-time limited rather than field-strength limited. For the path loss calculations, the boat antenna can be considered to be at 3 meters.

These same reflections can, however, be a serious source of interference in high-density areas when the reflected signal reappears behind the cell site. Downtilt minimizes this problem.

Special problems occur when buildings are not built on flat ground and excavation is necessary. Figure 5.4 shows such a building. The building in Figure 5.4 is, in effect, underground from some directions. For example, the building is clearly underground with respect to base A. The first two levels probably will not have coverage from that base, but they will achieve coverage from base

B. Many buildings are constructed this way, particularly in hilly cities, and they can make good handheld coverage extremely difficult.

Using these guidelines, a central site can be selected according to availability, cost, access, and the site's building potential. Having selected the pivotal site, the next step is to survey the site to determine its actual coverage. (For more information, see Chapter 6, "Radio Survey.")

In Figure 5.5, the poor city penetration in the northeasterly and northwesterly directions implies significant obstructions in these directions. Additional city sites may be necessary to remedy this situation. Sites should be selected from a combination of visual site inspections and map studies to provide continuous coverage.

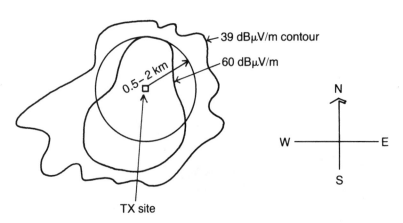

Figure 5.5   The surveyed field strength of the first selected site.

WHAT THE CUSTOMER WILL ACCEPT

There have been experiments conducted in the past on conventional wireline systems to determine just how good the customer's subjective evaluation of the quality of service is. If a group of subscribers are connected to a switch with access grade of service (GOS) of 0.01 (meaning that one call in one hundred will fail to complete because of lack of access to switching equipment— 0.01 is a typical value for a suburban subscriber) and that GOS is suddenly reduced to 0.1 (one in ten calls fail), there will be a flood of complaints. However, should the GOS be reduced gradually over a period of months, the subscriber becomes acclimatized, and the GOS may get as bad as 0.2 (one in five calls fail) or even worse, before a significant number of customers will complain.

Major system reconfigurations always bring in subscriber complaints. It doesn't really matter too much what the change is, it seems as if once the subscribers are aware that something has changed there is an increase in awareness of system performance and the complaints will come in.

In the experience of the author some small systems that perform dismally (at least one call in five failing to mature for system reasons) have acceptably low levels of customer complaint. These systems have something in common; they are either single monopoly systems or they are working alongside an equally dismal competitor, so the subscriber has nothing to compare the performance against. They think that "radio" systems are always this bad! Poor-performance systems are usually managed by "managers" who don't understand the engineering and assume everything is fine as long as the customers are not complaining.

There is just one other factor that needs to be considered when trying to understand the limits of customers tolerance, and that is that about 50 percent of all calls will fail to mature even in a good system. This percentage might seem very high, and it is, but consider the nature of the calls:

- Calls to mobiles frequently receive the message "that the mobile is unattended or turned off" and in fact it probably is!
- Calls to land-line parties are often unanswered.
- Calls made by dialing while driving often result in wrong numbers.
- Correctly dialed calls (particularly in big cities) often are routed to "This number has been changed. . . ."
- Overseas calls seem to fail as a matter of course, especially if they are not directed to the major industrial countries.

- Calls are often terminated by the dialer before completion.
- The called party is busy.

In this environment a badly designed network can often operate without excessive complaints for a long time. Subscribers' complaints are no judge of the performance of the network, and good engineering standards should be kept at all times. When a poorly designed network finally reaches the stage where the complaints begin to rise significantly, it will be very expensive to reconfigure. Undoubtedly many base stations will require relocation, new bases may be needed, and some may become redundant. The single most common problem in a big system will be sites that are too high and so radiate too far. In a small system the biggest problems may be patchy coverage due to inappropriately located sites and to overload at some of the sites. New sites are expensive and time-consuming to acquire. If the old sites cannot be reused, they may be difficult to dispose of or may have long leases that have to be honored. While all this expensive relocation is going on the GOS will drop even further, and the complaints will rise again.

A well-engineered system will be designed to quantifiable performance standards and these standards will be met. It is not unusual that very large systems that are well engineered perform better than much more modest ones. The reason is that the big systems *must* be well engineered to work at all. The operators usually know the importance of good engineering, and if they don't, they will soon be left behind by their competitors.

Doing It the Hard Way

Today there are many engineers who are "designing" cellular networks with planning tools who simply don't understand the underlying algorithms used by those tools. As a result, a lot of very poor designs are produced. Every design engineer should get some experience of doing it manually; then they can understand what the design tool is telling them.

MAP STUDIES

The first approximation of coverage is made from a map study or computer prediction. This process is only the first step in a design and should not be confused with the design itself. Cellular design is highly iterative.

The following approximation methods are recommended:

- CCIR. "Recommendations and Reports of the CCIR," 1982, Volume V, *Propagation in Non-Ionized Media*, Report 567-2

- Yoshihisa Okumura *et al.*, *Review of the Electrical Communication Laboratory*, Volume 16, Nos. 9–10 September–October 1968, NTT, Japan
- Computer techniques

The first two methods are essentially based on similar empirical techniques and yield almost equivalent (but not identical) results. These methods use a series of curves that were derived from field studies of ranges obtained from transmitters on various frequencies and from various elevations. They have become classic studies that are widely used. In fact, algorithms derived from these and other similar studies are often used in the computer models. Their main attractions are simplicity and proven utility. These methods will be used later in the chapter.

The Hata Model

The Hata model is a computerized implementation of the Okumura studies (as detailed above) in Japan.

COMPUTERIZED TECHNIQUES

Computerized methods using digitized maps can produce a higher degree of accuracy, but are very costly. The accuracy of good computer techniques is still only ±6 dB (compared to ±10 dB for manual methods).

Computer prediction methods normally assume a two-dimensional path between the transmitter and receiver. In real-life propagation, however, contributions to the far-field pattern are made by reflected, scattered, and refracted paths that are in other planes. Wave scattering produces a spatial spectrum that is a complex holographic image of the surface causing the scattering. It is generally not practical to take these effects into account. Therefore, there is a limit to the degree to which any calculations can reflect reality.

The detail needed to accurately determine the path profiles is also very large. If a city of 2000 km^2 is characterized by 100-m × 100-m areas, then there are 2000 × 10 × 10, or 200,000 such areas. A minimum representation would contain two pieces of information about each area, namely height above mean sea level and type of terrain (that is, urban, rural, open, or water), and thus would contain a total of 400,000 pieces of information. It is also desirable to store information about surface clutter (whether man-made, like buildings, or natural, like trees) and its height and distribution. This information can be the most difficult to obtain and to update. Moreover, for computer systems to be useful, they must not be limited to one city.

Thus, a large digital database and data storage in the high-megabyte range is necessary. The acquisition of these data is costly and is, ultimately, the limiting factor. Unless a database is available from other sources (for example, mapping authorities), the cost of producing one may be too high to be undertaken by a cellular operator. Most computer-prediction databases lack surface-clutter information and so are of limited use.

Despite the apparent limitations, computerized techniques have found favor among many designers, and most major operators and suppliers have an in-house system (which has generally been customized). These systems can generally do more than just forecast propagation. Most can graphically depict composite coverage, which has been derived from the files of a number of individual site predictions. As an extension of this, composite interference maps can be produced. These can be particularly useful for gaining an overview of a system's performance. A number of these forecasting systems are available for sale. Typically, the cost will be about $100,000 to $200,000. Most will run adequately on a high-end PC.

HANDHELDS

The measurement of field strength inside buildings reveals high standard deviations (meaning a large variability in the readings) and is not recommended for practical survey because it is difficult to do and does not yield useful results. However, on-site tests with handhelds, after the installation of bases, can be informative.

Street-level field strengths can be used to characterize levels inside buildings as an "average" building loss. Losses vary from 10 to 35 dB at ground level and are a function of floor level (decreasing by about 0.5 to 2 dB per floor). The loss measured from window or door increases as the building is entered, at a rate of about 0.6 dB per meter. The actual loss depends on the type of building, its size, and the amount of steel reinforcement used in its construction. Thus, in earthquake-, hurricane-, and tornado-prone areas, losses are likely to be greater. Tokyo has average building losses of 28 dB, while US cities generally average 20 dB.

Some cities were planned with wide streets, most of which run at right angles to each other, with many open areas such as parks and squares. Other cities just grew with roads following cart tracks. Planned cities offer less impediment to cellular propagation, and spaces between buildings act as large and fairly efficient waveguides. In these cities, handheld penetration at considerable distance (up to 4 km at 800 or 900 MHz) can be achieved, particularly if the terrain is relatively flat. Propagation along the direction of a major road can be 6 dB better (at the same distance) than at right angles to that direction. In unplanned cities, a significant reduction in range can be expected.

For these reasons, it is not possible to have a universally applicable rule. It is advisable to survey each new city or region in order to get a clear understanding of the possible range before drawing conclusions from results obtained in other, different regions. Handheld coverage from 0.4 to 6 km (at 800 or 900 MHz) has been reported for cities of similar size with different construction and features. A street-level field strength of 60 dBμV/m (70 dBμV for GSM 800) will generally be sufficient to ensure adequate handheld coverage inside normal office buildings at ground level.

With their low power outputs, handhelds represent the real challenge of cellular radio frequency (RF) design and also represent a new environment. Design rules need to account for the fact that handhelds are not usually used in high, multipath environments. Because handhelds are also the "weakest link" in the cellular path, a system designed for handhelds is generally adequate for vehicle-mounted units.

Base-Station Sensitivity and Improved Handheld Performance

In many instances it is worthwhile to actively improve the uplink budget by improving the base-station sensitivity. There are a number of ways to achieve this. First, diversity can be used. Diversity takes advantage of the fact that two different radio paths are likely to have a fading pattern that is mostly independent. So two antennas separated in space by sufficient distance (for cellular purposes this usually means a few meters) will receive signals that have traveled along different paths, and therefore the received signals will have a low correlation. By combining these signals appropriately (using combining diversity, a process which sums the two signals in phase), a gain of around 6 dB can be achieved relative to the gain of a single antenna. A simpler form of diversity (switched diversity) simply chooses the "best" antenna at any given time and switches to it, which will result in an approximately 3 dB gain.

Another form of diversity that is currently popular is polarization diversity. Because this can be done without spatial separation, and two diversity antennas can be built into one antenna radome, it is very attractive from a clutter point of view. It works on the principle that a signal that has been refracted or reflected will undergo a polarization change as a result. Hence signals that arrive in different polarizations have undergone different paths along the way, and so will be uncorrelated. The cross-polarized antennas are usually polarized at two angles

each 45 degrees off the vertical. Gains of about 4.5 dB are obtainable from this technique.

Low-noise amplifiers (LNA), mounted as close as possible to the antenna, will improve the link budget, typically by about 3–5 dB, but this may be offset somewhat by higher intermodulation levels. Nevertheless, with good filtering and good (low) noise figure amplifiers significant sensitivity gains can be achieved.

Good quality high-Q pre-selector filters can significantly enhance base-station sensitivity. By reducing adjacent channel interference (and hence reducing the desensitizing effect), quality filtering can enhance the receiver (RX) sensitivity by several dB (the amount depending greatly on the type and magnitude of the adjacent channel signals).

Smart antennas can significantly improve the received signal level, while at the same time reducing unwanted signals. The net result can be a very significant improvement in the overall link budget.

A PROGRAM FOR CALCULATING RANGE OR PATH LOSS

A useful BASIC program to enable the quick calculation of range and path loss can be found in Appendix A. Based on CCIR Report 567-2, the program was written to enable determination of coverage from most cellular sites. The range of validity of the algorithm is limited to

- Frequency: 450–2000 MHz
- Base-station effective height: 30–200 m
- Vehicle antenna height: 1–10 m
- Range: 1–20 km
- Buildings/land: 3–50 percent

These limits should not cause any problems in the majority of cellular applications. Note that use beyond these limits (particularly for range) is not advised, as the results will not be accurate.

GETTING TO KNOW THE TERRAIN

Most problems with cellular systems begin with poor design. Poor design begins with misuse of computer models. Even the best models need calibration for the *actual* terrain. Uncalibrated computer models will work best in rugged terrain with plenty of hills and prominent obstructions. They are at their worst when the terrain is flat with a lot of low-level clutter, such as trees and houses. Frequently problems arise because the designer applies a model from a previous design to a new area

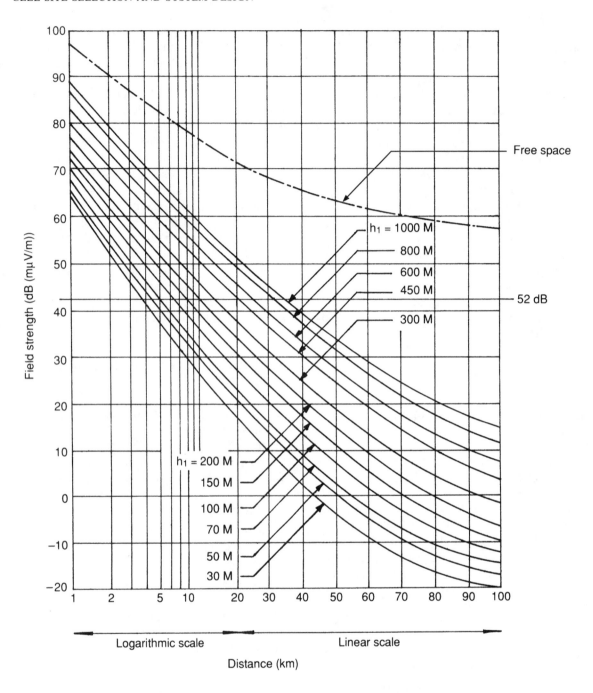

Frequency ≈ 900 MHz; Urban area 50% of the time; 50% of the locations; $h_2 = 1.5$ m

Figure 5.6 Field strength (dBμV/m for 100-W ERP) in an urban area. (Source: CCIR Report 567-2.)

without accounting for the different terrain characteristics. With many years of experience the designer can get a feel for the terrain, and should recognize right away if it is unusual or different from what has been encountered before.

MANUAL PROPAGATION PREDICTION

A number of empirical studies have produced algorithms that can be used to determine the far-field strength as a function of effective radiated power (ERP) and range.

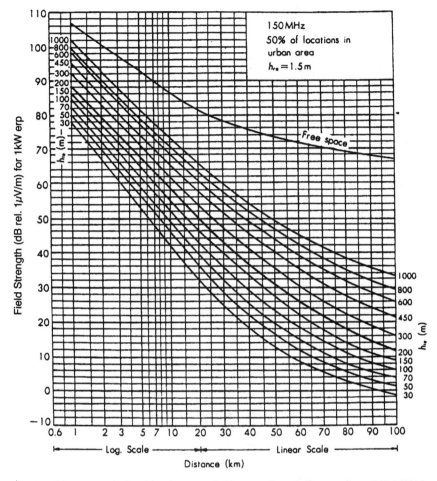

Figure 5.7 Okumura relationships between field strength and distance for a 150-MHz band.

Generally, these results were produced after an extensive series of propagation tests in one or, at most, a few countries.

Studies done in different countries (with different terrains) revealed substantial variations in path attenuation over similar terrain. One of the main sources of discrepancy is in the description of topography. A hill in Venice may be compared to a molehill in San Francisco!

Notice that Figure 5.6 is calibrated for *urban* areas. For typical suburban areas, the range will be improved by a factor of about 7 dB, and for rural areas at least 12 dB. To calibrate the curve take a few measurements in the area being studied and find a correction factor (in dB) to get the curve to correspond to the measured range. A correction of as much as 28 dB can be obtained in very flat rural areas. It should be noted, however, that once the factor has been applied the curves will yield good values for future studies at different antenna heights and for different ranges without further recalibration.

THE OKUMURA STUDIES

A landmark study by Yoshihisa Okumura with Ohmori, Kawano, and Fukuda, and printed in the *Review of the Electrical Communications Laboratory* (Volume 16, Numbers 9–10, September 1968), was one of the first attempts to quantify, in a statistical way, propagation in the mobile environment for frequencies from 150 MHz to 2000 MHz. The study was done in Japan and was conducted from a van, moving at 30 km/h, the results being measured on recording paper. The equipment had a high sensitivity of −125 dBm, and a dynamic range of 50 dB.

Referring to some early attempts to get something definitive from their studies, they report on page 831: "Some propagation curves hitherto published were not definitive, because they took in all sorts of terrain irregularities and environmental clutter haphazardly so were not very useful for estimating the field strength or service

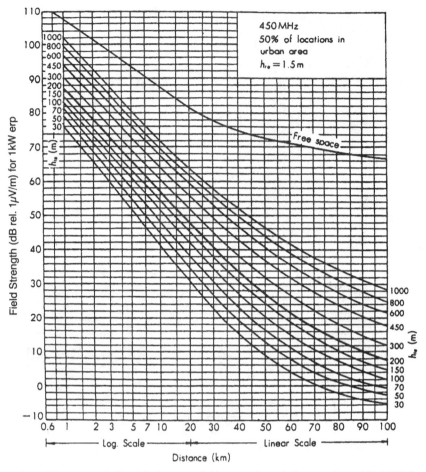

Figure 5.8 Okumura relationship between field strength and distance for a 450-MHz band.

area adapted to the real situation.'' What this is saying is that the real world is a mess, and too hard to generalize. To get around this they went on to define types of terrain and to compare like with like. While this is consistent with the reductionist scientific method, the engineers and computer modelers who came along afterwards to use these results generally ignored the context in which they were derived. It would be fairly rare that the service area of a cellular radio site would be entirely one terrain type; more likely it would be a mixture of types, like rolling hills, flat terrain, urban, and plains, with the population density and foliage varying considerably. Despite the shape of the real world, many computer algorithms accept only one terrain description for the service area of a site. The consequence of using the information out of context is wrong answers!

Used intelligently (more of this later), and not blindly in a computer model, the results of the Okumura study can give some good insights into RF propagation. Figures 5.7 to 5.10 to give the Okumura relationships between field strength (referenced to a transmitter with a 1-kW ERP) and distance in kilometers.

Correction Factors

The Okumura curves refer to the signal level in 50 percent of locations, and is therefore referring to the *median* signal level. The field strength at any range is near to being log normally distributed, and so is virtually symmetrically distributed about the median. In that case, the mean is equal to the median. It should be noted that the easiest thing to measure is the mean, but only because of the additional computations, and so time, needed to calculate the median or other decile value.

It is more usual today to design for 90 percent coverage in 90 percent of locations, and this upper decile value will give a figure that more realistically reflects a consistent standard of coverage.

The relationship between the mean and the upper decile value is given by:

$$\text{Mean} = \text{Level}(90\%/90\%) + 1.28\rho$$

where ρ = the standard deviation. A fairly simple relationship, were it not for the fact that the standard devi-

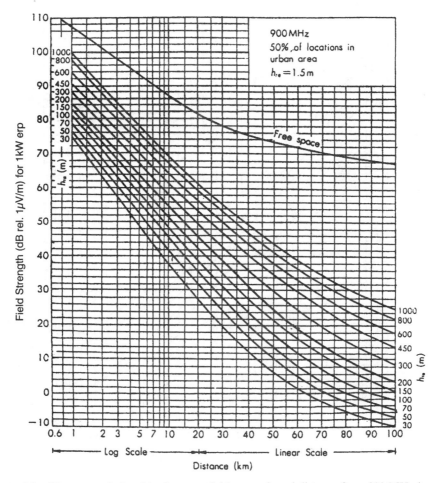

Figure 5.9 Okumura relationships between field strength and distance for a 900-MHz band.

ation of the level measurements is itself a function of the terrain. The values are given in Table 5.2.

To get a consistent upper decile value of field strength now requires that a range of median (mean) field-strength values be used in place of a single value. Again this is something that is allowed for in very few computer models.

There are a number of other correction factors that need to be applied to the Okumura model, and for-tunately most of these are fixed values.

Terrain Factor

This factor is a constant level in decibels that allows for the different attenuation of terrains encountered. It is best determined by measurement and ranges from around −20 to +20 dB.

ERP

The plots given are for a 1-kW transmitter, and the level type needs to be increased by $30 - 10 \times \log$ (ERP).

The results obtained by Okumura were summarized in a formula that gives the path loss in the Hata model, which states that

$$L = 69.55 + 26.16 \log f - 13.82 \log H_b - a(H_m)$$
$$+ \log R(44.9 - 6.55 \log H_b)$$

where

$L =$ path loss in dB

$f =$ frequency in MHz $(150 < f < 1500)$

$H_b =$ height of the base-station antenna

$H_m =$ height of the mobile antenna

$R =$ the range in kilometers $(1 < R < 20)$

$a(H_m)$ is a correction factor for the terrain that is height dependent

All logs to base 10

For a nominal antenna height of 1.5 meters, the values of $a(H_m)$ are

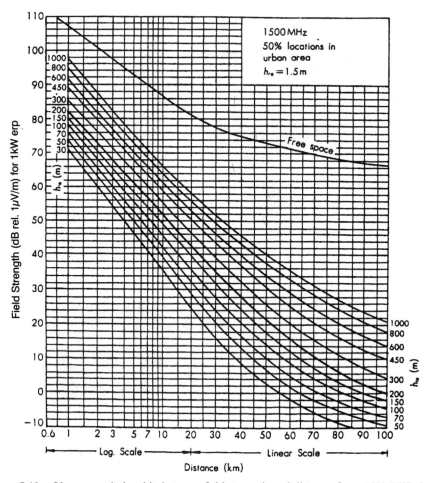

Figure 5.10 Okumura relationship between field strength and distance for a 1500-MHz band.

Urban: 0 dB
Suburban: −9.88 dB
Rural: −28.41 dB

THE COST 231 WALFISCH/IKEGAMI MODEL

The Cost 231 model is a combination of deterministic and empirical modeling and considers the loss to be made up of three components: the free-space loss, roof-to-street diffraction and scatter loss, and multiscreen loss. These three components are summed to arrive at the final loss.

This model is widely used in Europe and is designed for short-range services such as GSM and the personal communications network (PCN).

$$\text{The loss } L = L_f + L_r + L_m$$

where

L_f = free space loss

L_r = roof diffraction and scatter loss

L_m = multiscreen loss

and

$$L_f = 32.4 + 20 \log R + 20 \log f$$

TABLE 5.2 Field Strength Deviations for Various Terrains

Terrain	Standard Deviation (dB)
Urban	8–12
Suburban	6
Flat suburban	3–5
Rural	3
Water paths	1.5

which is simply the equation for free-space path loss

$$L_r = -16.9 - 10 \log W + 10 \log f + 20 \log \delta H_m + L_o$$

where W is the street width in metres

$$\delta H_m = H_r - H_m$$

and

$$H_r = \text{height of roof}$$

$$L_o = -10 + 0.354\theta \qquad 0 < \theta < 35$$
$$= 25 + 0.075(\theta - 35) \quad 35 < \theta < 55$$
$$= 4 - 0.114(\theta - 55) \quad 55 < \theta < 90$$

where θ is the angle of incidence relative to the street level. Finally

$$L_r = L_b + K_a + K_d \log R + K_p \log f - 9 \log w$$

where

$b = $ distance between buildings along radio path

$$L_b = -18 \log(1 + \delta H_b) \quad H_b > H_r$$
$$L_b = 0 \quad H_b < H_r$$
$$\delta H_b = H_r - H_b$$

$$H_b > H_r$$
$$K_a = 54 - 0.8\Delta H_b \qquad R > 500 \text{ m}, H_b < H_r$$
$$= 54 - 1.6\delta H_b \cdot R \qquad R, 500 \text{ m}, H_b < H_r$$

Both K_a and L_b increase the path loss for lower base-station heights:

$$H_b > H_r$$
$$K_d = 18 - 15 \cdot \delta H_b/\delta H_r \qquad H_b < H_r$$

$$K_p = 4 + 0.7(f/925 - 1) \qquad \text{for a midsized city}$$
$$= 4 + 1.5(f/925 - 1) \qquad \text{for the CBD}$$

The model is constrained to the following:

$$800 \text{ MHz} < f < 2000 \text{ MHz}$$
$$4 \text{ m} < H_b < 50 \text{ m}$$
$$1 \text{ m} < H_m < 3 \text{ m}$$
$$20 \text{ m} < R < 5000 \text{ m}$$

THE CAREY REPORT

The FCC originally based its planning for cellular service on report no. R-6406, "Technical Factors Affecting the Assignment of Facilities in the Public Mobile Service" by Roger Carey (June 1964). The report was based on a study done at 450–460 MHz. This report used data from a 1952 empirical study of TV field strengths at 50 percent of locations for 50 percent of the time. It was realized that mobiles would require a design for a field strength that was reliable for at least 90 percent of locations 90 percent of the time, and so a 14-dB "fudge factor" was added to the curves. The original data assumed that 25 dBµV/m would permit a mobile to operate in 50 percent of locations for 50 percent of the time, so adding the 14 dB yielded the familiar (in the United States, at least) 39 dBµV/m for AMPS mobiles.

Given its dubious application of rigor, you may wonder, Why not 40 dBµv/m instead of 39 dBµV/m? In fact, I prefer to use 40 dBµV/m and have found in practice it serves well. Those making application for AMPS in the United States, however, were required by the FCC to submit service areas based on 39 dBµV/m. A reasonable approximation of the Carey curve can be found from Figure 5.6 by reading off distance against the 40 dBµV/m line.

If a standard series of published curves (for example, as in Figure 5.6) are used to determine range, it is often wise first to "calibrate" the curves by measurement. In this process, a field-strength survey is undertaken to enable a detailed comparison between reality and the model. If a correction of more than 3 dB is necessary, additional measurements should be taken in various terrains to improve the accuracy.

Not all systems are created equal, and the time division multiplex access (TDMA) systems in particular suffer from poorer link budget than the analog or code division multiple access (CDMA) systems. The relative disadvantage of the major systems is shown in Table 5.3.

TABLE 5.3 Adjustment for Various Systems

Adjustment for the Same Performance from Different Systems	
	ADD (dB)
TACS	2
NAMPS	3
NMT900	6
GSM900	12
GSM1800/1900	18
DAMPS	8
CDMA	0
CDMA1900	6
AMPS	0

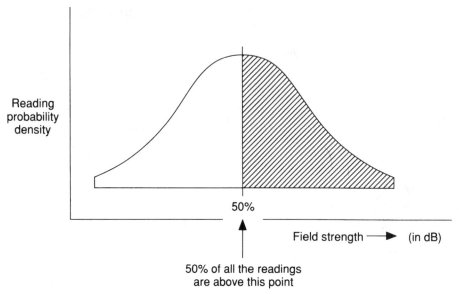

Figure 5.11 Field strength is log-normally distributed. The shaded area represents the probability that field strength will exceed X. The value thus obtained is really the median value of the log of the field strength and will differ from the actual mean field strength. In practice, this discrepancy is generally ignored.

ESTIMATING BASE-STATION RANGE

Let's assume you want to estimate the range for a typical cellular site; let's also assume a 20-watt TX transmitter power. The cellular system is designed for handheld coverage, and the system range is required in a suburban environment using an AMPS system. You can use Figure 5.9 to make the estimate. Proceed as follows:

1. Correct for actual ERP. The ERP of a 20-watt TX is about 50 watts (depending on feeder loss, antenna gain, etc). The graph is drawn for 1000-watt ERP, and so a suitable correction factor needs to be applied for the 50-watt ERP actually being used. Therefore,

$$\text{Factor} = +10 \log \frac{1000}{50}$$
$$= +13 \text{ dB}$$

Thus, the 39-dBμV/m boundary (for AMPS) will be located on this curve at $39 + 13 = 52$ dBμV/m.

2. Draw the graphs for the field strength attained at 50 percent of locations and times; that is, the field strength that is exceeded in 50 percent of all readings (average values). Field strength is a log-normally distributed variable and is illustrated in Figure 5.11.

3. Draw a line through 52 dBμV/m on the graph to obtain the 39-dBμV/m contour. You can now obtain the range in an urban environment by reading off the range against this line. A correction must then be made for the actual environment (unless it is urban).

4. Find the curve corresponding to the base-station height (h_1) with respect to the surrounding terrain.

5. Read off the range corresponding to the type of terrain (Table 5.1); for example, for urban terrain, base-station height $h_1 = 30$ m, range = 4 km.

6. Plot this coverage on the map, making adjustments for local terrain; for example, hills form natural boundaries.

7. Plot the 39-dBμV/m (50 percent, 50 percent) contour (which is 52 dBμV/m on the graph) or other, as applicable.

Select sites that look likely to provide good continuous coverage with respect to the central site and then survey them. Note that this process is not used to select sites. It only determines which sites should be surveyed. From the survey results, it will become evident which sites are useful and which are not.

Figure 5.6 is drawn for urban environments defined as 15 percent of the land occupied by buildings. In very dense CBDs, this percentage can be much higher, and in rural areas it can be zero. A correction factor,

$S = 30 - 25 \log \alpha$, can be used, where α equals percent of building to total land area. This percentage can be applied to terrains different from suburban terrains.

If a correction factor is applied, the new level in Figure 5.11 for 39 dBμV/m (50 percent, 50 percent) is (39 dBμV/m − S). Notice that at $\alpha = 15$ percent (that is, in an urban environment), S = 0, so that S may take positive or negative values.

Finer distinctions between terrain types are offered by the Okumura paper, but because this whole technique is, at best, approximate, additional corrections are normally not needed except for very unusual terrain. Such terrain includes very flat land (reclaimed swamps, subtract 27 dB or S = 27), water or part water paths (see Okumura), and large hills (treat as absolute boundaries).

TERRAIN DEPENDENCE AND STANDARD DEVIATION

A set of propagation curves such as Figure 5.6 can be drawn for only one type of terrain. Very flat land will permit much greater range, as will land that has little in the way of obstructions.

The terrain will change the shape of the distribution of field strengths in a way that can be measured by the standard deviation of the measurements. If the standard deviation is calculated from the actual measurements, Table 5.4 will be a good guide.

The fact is that an "average" reading of field strength is a rather crude measurement of signal quality, and a lot more information can be obtained by also getting the standard deviation. Extensive testing by the author on AMPS CDMA systems has led to the conclusion that the 39-dBμV/m figure used as a "standard" is okay only in suburban areas. As the actual aim is to design a system with 90 percent/90 percent coverage, it is useful to find the corresponding field strength. The mean, standard deviation, and 90 percent/90 percent level are linked for a normal distribution by the relationship:

$$\text{Mean} = L(90\%/90\%) + 1.28 \times \rho$$

TABLE 5.4 Standard Deviations for Various Terrains

Terrain	Standard Deviation (dB)
Urban	8–12
Suburban	6
Very flat suburban	3–5
Rural	3
Water paths	1.5

where

$$\rho = \text{the standard deviation}$$

From this it can be established that the actual 90 percent/90 percent figure in suburban regions ($\rho = 6$ dB) corresponding to 39 dBμV/m average is 31.3 dBμV/m (90%/90%). Given that it is easier to measure the mean field strength, this relationship can be used to calculate the mean that will give an *equivalent* 90 percent/90 percent performance for terrains other than suburban.

Consider a rural area where $\rho = 3$ dB. To achieve 31.3 dBμV/m at 90 percent/90 percent it will be necessary to have a mean of $31.3 + 1.28 \times \rho$, or 35.4 dB (average).

This figure will in fact be *conservative* because the areas of lower standard deviation also have less extreme multipath, and so the noise generated will be less, even relative to areas of higher standard deviation.

Variable Terrain

Where the terrain is variable within one cell coverage area, the plot of the coverage should be done in sectors. Consider the prediction of the coverage for the area in Figure 5.12. There are three discrete types of terrain seen by the cell site. The propagation over each type will be quite different. If the CBD is substantial and the cell site is more than 4 km away, it should be regarded as an absolute barrier.

Sea Paths

Sea paths offer the best radio paths. As boats move slowly (compared to wavelengths/second), they suffer very little from multipath noise. Boats should be equipped with antennas mounted higher than 1.5 meters (the level normally assumed for vehicles) unless, of course, handhelds are used. Even then, they would mostly be used at heights greater than 1.5 meters. These sea paths can provide any given grade of service (measured as S/N) at a lower field strength than land paths.

A field strength of 32 dBμV/m + C can be regarded as adequate over sea, although a usable service is available at 25 dBμV/m + C and sometimes even lower (depending on antenna height) note C is the correction factor in Table 5.3. When at sea, the most prominent (highest) land site will be the one with the highest field strength rather than the one closest to the boat. Ranges of 50 km (over unobstructed paths) are typical and up to 100 km are not uncommon under favorable circumstances (GSM systems and some CDMA systems are limited by timing constraints to smaller paths).

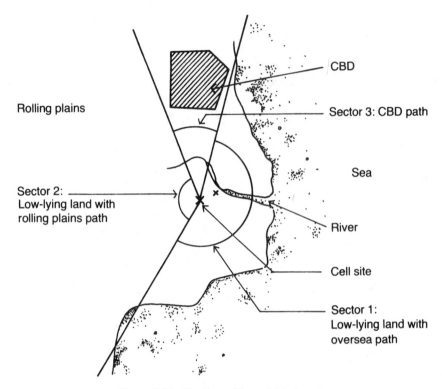

Figure 5.12   Dealing with variable terrain.

Microcells

Increasingly microcells and minicells are becoming more common in high-density cities, where they are needed for traffic purposes. These can be either very low-powered base stations or, more often, low-powered repeaters. The graph of Figure 5.6 does not allow for such small cells, but Figure 5.13, which has been derived from field measurements by the author, can be used for microcells. The graph is based on 1-watt ERP and so will be 30 dB

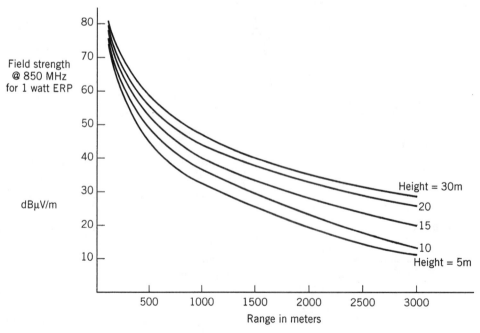

Figure 5.13   Field strength in dBμV/m for 1-watt ERP microcells, 800 MHz.

down on the graph in Figure 5.6. It is otherwise compatible and the same correction factors can be applied. For 1800/1900 MHz decrease the field strength (FS) by 3 dB.

ESTIMATING THE EFFECT OF BASE-STATION HEIGHT

Often it is not possible to get survey results at the desired height. This may be because the original survey was done at a height other than the one chosen for the final installation; or in rural cases it may be that no survey mast is available that equals the height as the proposed tower, or you may simply want to estimate the effect of reducing the height of an existing antenna.

A simple relationship that applies fairly well in practice is that the field-strength (FS) correction factor depends on height as

$$FS = 20 \times \log(\text{new height/survey height})$$

It can readily be seen that doubling or halving the antenna height will require a correction of 6 dB to *all* the measurements.

Remember that the height in question is the height above ground level (and not simply the tower height) and that if extreme height changes are made, local obstructions may come into play and invalidate the rule. Where there is doubt, resurvey.

MULTI-CELL SYSTEMS

Having chosen and surveyed the central site(s), the next task is to plot its actual coverage. Using the expected subscriber density, determine the required cell radius for the adjacent cells. This procedure will provide a good indication of base-station height (lower bases for smaller cells).

Next, select six sites roughly equidistant from the central site and, by map studies or computer studies, plot the expected coverage of the new sites. The map study should be regarded as the first stage in site selection. That is, the map should be used to identify sites that are worth surveying and to eliminate those that are not. Because a survey takes about two weeks per site, this preliminary screening can save time and money.

Figure 5.14 shows the results of such a map study (dotted coverage contours). Some sites have insufficient coverage to be worth consideration. The sites that hold promise, namely sites A, B, C, D, and E, are then surveyed and the results are plotted.

COMPUTER-AIDED DESIGN

Most systems today are designed by computer-aided design (CAD) tools that take a lot of the hard work out of RF planning. CAD RF planning tools are now widely available on UNIX or PC platforms. The UNIX based tools tend to have some advantages, most of which are historically based, because the earlier tools were all UNIX, and most of the development has been concentrated on this platform. Today, there is really no inherent advantage for UNIX.

Over the last few years the tools have become better, but often the designs that they produce have not. Some of the tools are difficult to calibrate and some users simply ignore the need for calibration. It is important to realize that the planning tools are based on algorithms that are in effect broad approximations, and that there are wide variations in the "nature" of different terrains.

Calibration

The most important feature of an RF planning tool is the ease with which it can be calibrated. Some tools require a set of "sample data" that is fed into the program for local calibration. Typically, this input requires that a local area of about 100 linear meters be "characterized" by a number of measurements. This type of "calibration" is highly suspect, particularly as there is no assurance that the next 100 meters will be necessarily similar, and additionally there is the problem that the reading itself is time-variant; come back a few hours later and the reading will be different. It is best to avoid this kind of tool. The best method of calibration is one that can take continuous measurement data directly from a test drive and use it directly.

There is little point in using a mobile as a calibration device, because it was never designed to be a measuring instrument and its accuracy is not high, either in an absolute sense or from the point of view of time consistency. Sometimes there is no choice but to use a "mobile-phone"-based measuring set. In this case it is essential that the mobile be calibrated before use and if at all possible more than one mobile be used (to compare measurements). Any inconsistencies in the mobiles should lead to the results being discarded; bad information is often worse than none at all, and measurements that are suspect can lead to expensive design errors. Be aware also that most "measuring" devices based on mobiles are only linear over a range of about 70 dB, and care must be taken that all measurements are in that range, unless of course the correction factors have been applied.

It is well worth the expense of buying a proper field-strength meter (or measuring receiver), and no serious

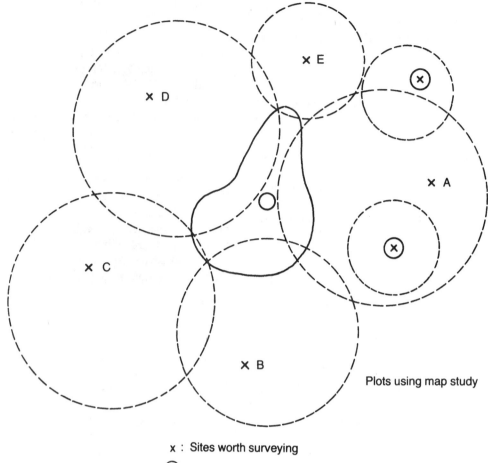

x : Sites worth surveying

ⓧ : Sites rejected on map study

Figure 5.14 Map studies of proposed sites.

designer should be without one. However, if it is to be used with a CAD package, then it is necessary to ensure compatibility with that package (a method of direct data input is most desirable).

SURVEY PLOTS

Survey results are only as good as the equipment and techniques used. Therefore, large discrepancies between predicted and surveyed coverage should ring alarm bells that perhaps something is wrong with your equipment. Assuming that satisfactory explanations for any discrepancies can be found, the selection process is continued by selecting more sites to provide the remaining coverage, conducting more map (computer) studies and re-surveying until adequate coverage has been found.

Figure 5.15 shows that sites B, D, and E look prom-

ising, but sites A and C are inadequate. However, before totally dismissing sites A and B, the large discrepancy between the predicted and actual coverage should be examined and the reason for the discrepancy identified.

Because continuous coverage is important, the design proceeds much like putting together a jigsaw puzzle—starting from a fixed point and working outward. Placing six equal-coverage cells around a central cell gives a good approximation of the regular hexagon pattern. This method more closely approaches the theoretical "equal cell size" pattern than does the approach used by some designers, who attempt to locate sites according to their physical location on the cellular grid.

Were sites to be precisely located (for example, on a flat surface) according to a cellular grid, they would only produce an approximate optimum pattern (equal hexagons). In the real world, however, obstructions such as hills, buildings, rivers, and foliage near sites that the located *exactly* according to their theoretical hexagonal

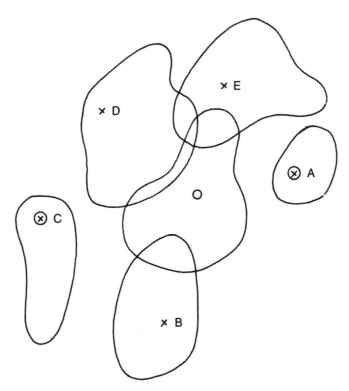

x : Suggested site with potential

Ⓧ : Site not suitable

Figure 5.15 Survey results plotted.

pattern positions may produce patterns that do not resemble a regular configuration.

Site selection involves compromise, particularly with site availability. Fortunately, the designer can tailor coverage by using height, power, and antenna patterns to achieve efficient spectrum reuse.

Once the seven cells (or most of the seven cells) have been selected, the cell pattern and its orientation is established. That is, the hexagon pattern (if used) is derived from the site selection, not vice versa. The hexagon pattern derived from Figure 5.15 is shown in Figure 5.16.

Many designers proceed by starting with the hexagon pattern and selecting sites from the pattern. Where uniform terrain exists, the result can be much the same as deriving the pattern from the terrain. However, the concept of visualizing each cell as surrounded by six equal-coverage adjacent cells is easier and leads to a more flexible approach to site selection. Once a network has matured, a more valid approach is to adopt a flexible plan to provide service to areas that need it and to dispense with hexagons completely.

FILING SURVEYS

The design can be seen to proceed iteratively. Therefore, it is unwise to discard any survey result as totally unsuitable. Unused survey results can be useful in filling gaps in the current or future design. Conversely, as the design evolves, sites that once appeared ideal may become redundant. Thus, all survey results have the potential to become part of the jigsaw puzzle.

When filing surveys, remember that all surveys should be conducted using the same units (dBμV/m is recommended. See Chapter 9, "Units and Concepts of Field Strengths," for more information). This will help you avoid confusing results. In addition, all survey maps should be filed with the following data clearly marked:

- Date of survey (very important) and name of surveyor
- Location of site, owner, name, and phone number of contact person
- Survey antenna height (above ground level and above sea level)

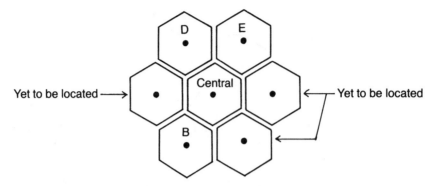

Figure 5.16 Hexagon pattern.

- Cable loss (preferably measured because connector losses are difficult to calculate)
- Transmitter power
- Antenna gain (usually 6 dB)
- Correction factor used to convert actual ERP to nominal ERP (that is, all readings are corrected to the ERP of an actual cellular base station)

These details may prove essential in the future if discrepancies are detected.

MAPS AND MAP TABLES

The maps used to plot survey results should be about 1:100,000, certainly not larger than 1:250,000 or smaller than 1:50,000. Topographic projections should be used for coverage prediction. However, street maps that show street names in detail are usually best for recording individual survey results, before they are plotted at a scale suitable for system studies. Actual coverage plotted on street maps is useful for recording detail coverage of individual bases. The final result should be an easy-to-read map designed for use by the subscriber. The map shown in Figure 5.17 is the coverage map for the Extelcom system, which was prepared by the author.

A map table of at least 2 m × 2 m (or larger) is needed to adequately handle coverage maps, given the scale needed to ensure sufficient detail for survey results. Once a large number of maps are stored, it is essential to have a suitable cataloging system. Maps also need adequate storage and a map file (a file that can store full-size maps without folding) is best.

Transparencies can be used to completely overlay the maps to depict coverage. Felt pens used to mark the transparencies should be of the "whiteboard" erasable type. A very handy accessory is a set of map weights. Without weights, maps can be awkward to handle.

DESIGNING FOR CUSTOMER DEMAND

Continuous coverage is an essential element of any design. However, expandability for the future must also be built in. The system operator has an obligation (as well as a financial incentive) to meet the demand for services. The ultimate capacity of the system and its ability to expand should be included in the design. A good designer will produce a design that does not place constraints on future expansion. And a good designer will also know that forecasts are purely hypothetical and will build in maximum flexibility for the system. This flexibility should include the ability to expand rapidly for unexpected demand and the possible unavailability of previously selected base-station sites. A cellular system has many parameters that can be utilized to extend or contract base-station coverage to temporarily cover unforeseen problems, provided the capacity is available. A minimum of six months should be allowed to obtain additional channel equipment.

If a site must be moved to allow for future expansion or improved coverage, then it is possible that the design was too limited in scope. Predicting future customer demand is difficult. All good designs should allow for substantial errors in forecast demand.

DETERMINING CHANNEL CAPACITY

The capacity of a fully equipped system can be determined approximately by using the first equation below. It should be noted that this equation is valid only for systems and not for individual bases, because sectored and small bases have different capacities per channel to a large omni base. However, given the usual mix of cell sizes and types in a typical system, this equation has proved valuable in practice. Where the calling rate is

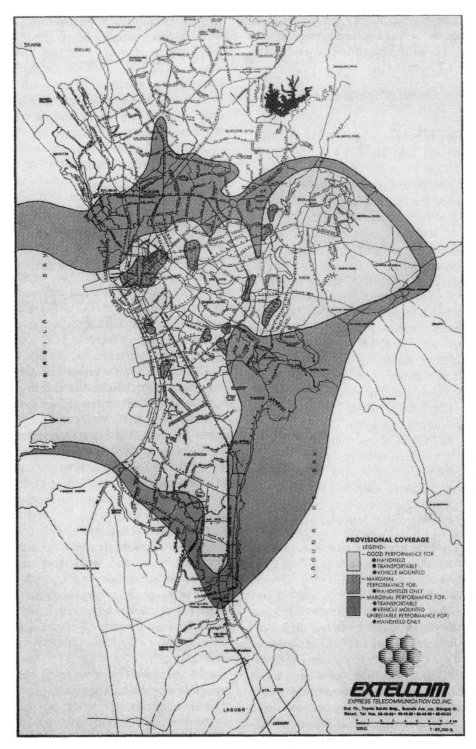

Figure 5.17   Extelcom system map.

unknown, 10 milli–Erlangs (mE) can be assumed with a system capacity of 30 subscribers per channel.

$$C_A = N_C \times 20 \times \frac{300}{CR}$$

where

N_C = number of speech channels

C_R = calling rate in mE

C_A = capacity in subscribers

The maximum value of $N_{C\,max}$ at any one site depends on the system type, the selected cell pattern, and the spectrum available. To be consistent with a particular cell plan ($N = 1, 4, 7, 12$), then:

$$N_{C\,max} = \frac{\text{Total number of speech channels}}{N}$$

For example if an operator was using GSM and had 5 MHz of spectrum and used an $N = 4$ pattern, then the total number of channels is $8 \times 5000/200 = 200$. In order to accommodate three carriers in the 900-MHz band it has been usual for the operators to initially be given only 5 to 7.5 Mhz. For an $N = 4$ plan and assuming three sectors, there will be $4 \times 2 \times 3$ or 24 of those channels reserved for control purposes, and so the number of speech channels at an average site is $176/4 = 44$ channels. And in each of the sectors this means an average of 14.666 channels/sector. Obviously we can't have a fractional number of channels, but also for GSM the channels come in modules of 8, so for speech channels there may be 6, 14, or 22, and so we conclude that the average number of channels/sector is 14. This relatively low number of channels will reduce the circuit efficiency; this is something that is not usually a problem in the analog systems [where 25+ circuits per sector ($N = 4$) is the norm] or CDMA (approximately 30 and $N = 1$). In the PCS/PCN spectrum allocation is more generous and this again is not a problem.

More elaborate computer models exist to determine capacity; the author's software suite Mobile Engineer contains a batch traffic processing and forecasting model which can take the existing traffic data and project the network forward for any system to any capacity. This approach is preferable if available.

BASE CAPACITY

Channels can be redeployed from low-usage bases to high-usage bases, but base stations are fixed and expensive to move. In a real system, the traffic is distributed so that some bases reach capacity before others, and therefore the actual capacity is about 70 percent of the theoretical base-station capacity. In a new system (one with no traffic history), it is safer to assume 50 percent of the maximum capacity (achievable if all bases were full). This is because a mature system has less uncertainties about its traffic distribution.

INCREASING CAPACITY

The Commonwealth Scientific and Industry Research Organisation (CSIRO), an Australian government body has developed a planning tool called Frequency Assignment by Stochastic Evolution (FASE). FASE optimizes frequency assignment to cells to minimize networkwide interferences, eliminate local trouble spots and increase call carrying capacity. The tool runs on a PC and looks at vast numbers of cell frequency assignments (on the order of hundreds of thousands) to find an optimum frequency assignment for each cell. The actual number of possible assignments in a reasonably sized network is too large to allow the consideration of all combinations, so a very efficient allocation technique has been developed that takes into account many network constraints. The system is claimed to have increased capacity of the Melbourne network by 30 percent, and some smaller networks by up to 100 percent. For details contact Stephen.Pahos@dbce.csiro.au

REAL-TIME FREQUENCY PLANNING

In most systems a frequency plan is developed in the office and put into place in the field. As new sites are added, some changes to pre-existing frequency allocation can be made, and with time even the best designed system becomes non-optimum. At this time a complete network re-tune takes place and the process starts again.

No fixed frequency plan can be truely optimum, because the "best" plan at any one time will depend on the actual usage at that time. A true optimum can, however, be approached if each site has the capability of selecting the best channel to use on a call-by-call basis; while in principle this concept is simple, there are a few complications.

First, to be able to get all the channels at any site, it will be necessary to use a broad-band linear amplifier for the transmitter and a wide-band receiver. Both of these options are achievable, at reasonable cost.

Next we need an algorithm and methodology that allows the base station to select the "best channel." To consider what is needed look at Figure 5.18. The interference at the base station from other mobiles is readily determined by a fast scanning receiver at each base-station site. It will be necessary, however, to scan *all* the channels that the base station can select from, and to do this rapidly (within a few seconds maximum).

What is not so easy is determining the interference to the mobile. The mobile will obviously have a scanning capability, and it will be necessary for the system to access the levels of all base sites scanned. However, the mobile will only be scanning the control channels. That means that its report will identify which sites may be a problem, but it will have no information on which channels are in use. Because of this it will be necessary to separately obtain in real time information on which channels are in use at every site. This means that the

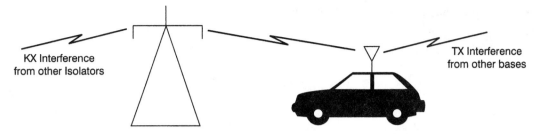

Figure 5.18   The interference modes that need to be considered for a frequency allocation.

decision making will involve networking of information between sites.

While it is relatively easy to find the optimum allocation for any one mobile, that allocation may itself be non-optimum for some other mobile already in use. In particular there is the potential for the new allocation to cause interference to other mobiles already on the network that are using either co-channel or adjacent channel frequencies. A true optimum therefore would require the recalculation of the C/I of all co- and adjacent channel mobiles with each new mobile assigned onto the network. Since this effect would be different for every frequency allocation, in order to find the true optimum it would be necessary to do the pairing calculation for every mobile on the network. Clearly, this is too much of a calculation overhead, and so some limiting assumptions must be made. One obvious simplification is to limit the world view to the immediate neighboring cells. Another simplification is to select a non-optimum assignment that meets pre-defined criteria. This kind of problem seems to be ideal for an artificial intelligence approach.

Nortel's frequency-hopping technique allows almost all frequencies to be used at all sites. By optimizing in real time, the capacity can be increased significantly. In the Irish Esat Digifone network, tests in 1998 claimed a fivefold increase in capacity.

Metapath Software International (MSI) of the UK has a software-based optimization tool called Planet Maxxer, which uses a relationship database and network data to identify the best and worst performing sites. This allows resources to be directed to where they are most efficient.

CUSTOMER DENSITY

Customer density expressed as customers/km^2 or Erlangs/km^2 can be determined only after the system has matured. Before commissioning the system, this information is very difficult to quantify, and even the most careful estimates can be in error by factors of about 2 to 5. Thus, the designer has to work in a very uncertain environment. This means that the design should be flexible enough to be reconfigured to account for actual (as opposed to forecast) traffic demand.

Fortunately, it is very easy to relocate channels in modern cellular systems by moving channels, provided there are sufficiently correctly located base stations. GSM systems can only be expanded in multiples of 8 voice channels, and CDMA in multiples of around 30, but analog systems are themselves usually expanded in modules of 8, 10, or 16 channels, because the racks needed for the expansions are modular. The extra equipment needed to expand beyond the module size includes additional antennas and should be considered at major sites. The uncertainty is particularly serious in the CBD. Here the base-station configuration, not channel allocation, should allow for traffic levels of about two times the forecast traffic for new systems.

If channels are to be moved from base site to base site, then extra racks must be purchased so that channels can actually be moved. It is a good idea to purchase and install an additional spare rack for every major base station to allow for a quick expansion.

DETERMINING BASE STATIONS IN THE CBD

To determine the number of base stations required in the CBD, perform the following calculation.

Use the estimated CBD proportion of traffic (50 percent if unknown) and define on a map the boundaries (approximate) of the CBD. Calculate the number of bases required in that area using the following equation. The number of channels/sector $N_S =$

Assuming a 200 channel GSM system

$$N_S = \frac{\text{Total channels}}{N \times \text{sectors}} = \frac{200}{4 \times 3}$$

$$= 16.66 \text{ channels/sector}$$

However voice channels will amount to only 14 sector, so

$$N_S = 14$$

or for a 3 sector system

$$\text{channels/base} = 3 \times 14 = 42,$$

for a 5-MHz GSM station with 200 channels and $N = 4$ (3 sector)

where

$$C_R = \text{calling rate in mE}$$
$$C_A = \text{total customers}$$
$$N_C = \text{total number of channels}$$
$$N_B = \text{number of bases}$$
$$N_S = \text{average channels/base}$$
$$N_B = \frac{C_A/F}{300/C_R \times N_S}$$

Note: $300/C_R = \text{customers/channel}$

where

$$F = 0.5 \text{ for new designs}$$
$$F = 0.7 \text{ for extensions to existing designs}$$

Note: Handheld coverage will require that base stations in high-density urban areas be 2 km (or less) apart in earthquake-, hurricane-, and tornado-prone areas, or 4 km elsewhere.

OMNI CELLS

Omni cells have a significantly higher traffic capacity than an equivalent sector cell, particularly when a small number of channels (less than 20) are used. Omni cells are also simpler and cheaper to construct than sectored cells for any given number of channels. An omni-configured system thus costs significantly less per customer than a sectored installation. When frequency reuse is not a significant factor, using omni cells can be a valuable way to (temporarily) increase the capacity of a new cellular system.

In general, it will be necessary ultimately to sector the omni cells except for small service areas. The reduced handheld talk-back ability of an omni cell may result in poorer handheld performance and so may not be a good idea in CBD areas.

ANTENNAS

It is normal that two diversity antennas will each be connected to each sector RX. Transmit antennas usually carry only 15 or 16 channels, so one antenna can generally carry all digital TXs. To provide maximum flexibility, base stations should not be installed without reserve antenna capacity for analog systems. Delivery times on antennas, cables (feeders), and couplers can be about three months.

SYSTEM BALANCE

In the original cellular-system specifications, an effort was made to achieve a balanced system and, as will be shown, this balance has been achieved reasonably well. A balanced mobile system is one where the speech path between the mobile and the base, and vice versa, are of the same quality. The quality is measured as a signal-to-noise ratio. In a cellular system where all data transfers involve a "handshake" (that is, all instructions require a reply) there is little to be gained by having a better signal path in only one direction. The communicable range is limited by the loss along the least-loss-tolerant path. In general, the most vulnerable path will be talk-out (that is, from the relatively low-powered handheld mobile to the base station).

The limiting factor for a handheld is talk-out where the radiated power is relatively small. In general, the talk-out path loss allowance of a base station equals or exceeds that of a handheld, so any increases in base-station power will not improve handheld coverage. Vehicle-mounted mobile installations have some small advantages with increased base-station power.

Increased base-station power causes increased interference in the co-channel and adjacent-channel mode, so the gains are not necessarily worthwhile, except where frequency reuse is not a consideration. When such interference occurs, data corruption is to be expected. Despite adequate available field strength, some calls cannot be made or received in these areas.

Figure 5.19 shows the parameters that contribute to overall path loss. In the direction of an omnidirectional base to the mobile, the maximum path loss is determined by base-station ERP. ERP is usually limited by local regulations. In this example, 50-watt ERP is assumed:

$$50 \text{ watts} = 10 \log 50{,}000 \text{ dBm} = +47 \text{ dBm}$$

Net allowable path loss from a base station (assume AMPS)

$$47 \text{ dB} + \text{minimum receive level for 18 dB,}$$
$$S/N = 47 + 101 = 148 \text{ dBm}$$

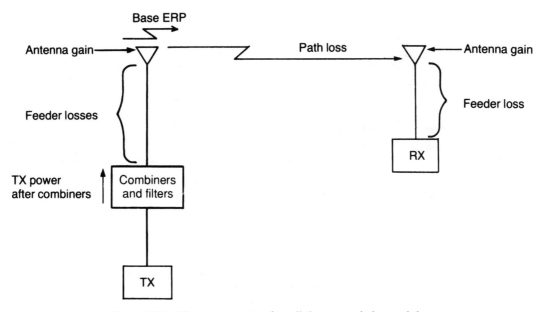

Figure 5.19   The components of a cellular transmission path loss.

Note: Received level for S/N (at antenna) = 18 dB is here assumed to be −101 dBm in 50 percent of locations. Various values are used for the sensitivity receive level, including −116 dBm (maximum sensitivity) and −95 dBm (= 39 dBμV/m at the mobile). This will not affect the balance calculations, although it may lead to different allowable path losses because differing criteria are set.

The paths from a vehicle-mounted mobile to omni-directional and sectored bases will yield different allowable path losses. In this instance, there are a few additional factors, as shown in Table 5.5 and Table 5.6.

As you can see, for systems where path loss becomes the limiting factor, at losses of 141 dBm the handheld will fail to provide a satisfactory path to the base. As the range is increased to a path loss of 148 dB, the vehicle-mounted unit will begin to experience an inadequate path from the base station. Increasing the base-station ERP by 3 dB for a sectored antenna (that is, to 100 watts) to improve the balance the vehicle-mounted ERP does not improve handheld performance. Sectored antennas, however, significantly increase handheld performance and will balance the path. Using 100-watt ERP will improve mobile performance by 3 dB when sector-receive antennas are used.

Exactly the same principles apply to all other cellular systems. All that changes is the path budget, so this calculation can be used for any system.

TABLE 5.5   Allowable Path Loss for a Vehicle-Mounted Mobile, Uplink; Downlink = 148 dB

	Omni Base	Sectored Base
TX Power = 3 watts = 10 log 3,000 dBm	35 dBm	35 dBm
Feeder loss	−3 dBm	−3 dBm
Antenna gain	+3 dBm	+3 dBm
Base receiver antenna gain	+9 dBm	+17 dBm
Base feeder loss	−3 dBm	−3 dBm
Base diversity gain	+6 dBm	+6 dBm
System gain	47 dBm	55 dBm
Allowable path loss for S/N = 18 dB	+101 dBm	+101 dBm
Net allowable path loss for handheld	148 dB	156 dB

TABLE 5.6   Allowable Path for a Handheld, Uplink; Downlink = 148 dB

	Omni Base	Sectored Base
TX Power = 0.6 watt = 10 log 600 dBm	28 dBm	28 dBm
Feeder loss	0 dBm	0 dBm
Antenna gain	0 dBm	0 dBm
Base receiver antenna gain	+9 dBm	+17 dBm
Base feeder loss	−3 dBm	−3 dBm
Base diversity gain	+6 dBm	+6 dBm
System gain (sum of above)	40 dBm	48 dBm
Allowable path loss for S/N = 18 dB	+101 dBm	+101 dBm
Net allowable path loss for handheld	141 dB	149 dB

SECTORED ANTENNAS AND SYSTEM BALANCE

Sectored base-station antennas have higher gains than omni-directional antennas and can improve the talk-back performance of a handheld to the point of balance with the base station. This only occurs, however, in the direction of the main lobe where the gain is maximum.

The gain of a sector antenna (17 dB) is 8 dB higher than an omni antenna (9 dB), so a balanced system to handhelds can be achieved.

System balance calculations need not be done in detail for each base-station design (unless required by regulation), because the results will all be within a few dB of each other. The purpose of these calculations is to ascertain the optimum base-station TX power, which is that power just assures (or nearly assures) a balance. Because of interference, too much TX power is counterproductive.

Consider the case where omni antennas are replaced by sectored antennas with a +8-dB net gain. If the TX power is not reduced, the ERP will also increase from 50 watts by 8 dB, which means an ERP of $50 \times 10^{0.8} = 315$ watts! This ERP probably exceeds the legal limits and will cause serious interference.

To balance the transmission paths, the TX power should be reduced by $10^{0.8}$ (or by a factor of 6.3). In practice, it is usual to reduce the base-station TX power to between 5 and 10 watts when using high-gain sector antennas. This range is used because the actual gain of the sectored antenna varies from 17 dB to 10 dB as a function of the angle with respect to the center of the antenna.

500-WATT ERP RURAL SYSTEMS

In Rural Service Areas (RSAs) in the United States, the FCC has authorized ERPs as high as 500 watts for AMPS systems. These higher powers are designed to balance the system for vehicle-mounted units only. The 500-watt limit can be used in rural areas that are at least 38 km (24 miles) from a metropolitan statistical area (MSA). It is envisaged that this higher power will be used along highways or other areas where elongated coverage is useful.

The system gain can be balanced by using higher-gain antennas in the receive path than in the transmit path.

A typical highway coverage base station may have a 1.8-meter (6-foot) diameter parabolic dish transmit antenna with a gain of 18.9 dBd, and 3.7-meter (12-foot) receiver grids with a gain of 25.4 dBd. The gain difference of 5.5 dB almost balances the system.

DESIGNING PCS NETWORKS

The design of PCS networks requires a slightly different approach to cellular radio. Base stations in PCS have a coverage ranging from 100 to 300 meters in the CBD. The base stations themselves are very small, consisting of a box measuring less than 30 mm × 30 mm × 30 mm. Installation consists of screwing the hardware to the wall and running a power lead, telephone cables, and an antenna cable. The antenna installation usually involves mounting a mobile antenna (of the type used on an automobile) at a convenient point near the base.

MICROCELLS

Microcells are typically placed in areas of high traffic density such as shopping malls and downtown streets. Mostly they will be deployed below the roof level and have a targeted coverage of 200–300-m radius. The base stations have low power outputs and they are designed to cover local hot spots.

Microcells need to be small, durable, easily deployed and cheap. They can be supplied by the network's main infrastructure supplier, although there are an increasing number of after-market suppliers. Com Dev Wireless Systems have a range of microcells for AMPS, TACS/ETACS, NMT900, and GSM, which come complete with their own 18- and 23-GHz microwave links. These cells are designed to use existing cells as distribution hubs for their deployment.

The power output of the microcell is small and decreases as the number of channels increases. While a 2-channel analog site can be rated at 3 watts/channel, a 16-channel site is a mere 0.05 watt/channel.

The microcell is meant to be powered from a regular main supply and consumes only 380 watts.

Microrepeaters

It will often be the case that coverage is required inside buildings (such as airports, train stations, campus buildings, convention centers, and industrial facilities) and tunnels for which a base station is not an economical alternative. A solution to this problem, which is becoming widely adopted, is to use a fiber-optic repeater.

The repeater is set up to directly repeat the carrier of a base station. A hub unit (see Figure 5.20) is set up at the base station, and is connected directly to the base station RF. The hub then sends the RF signal along a fiber-optic cable to one or more distant repeater sites, which amplifies and repeats the original signal.

An implementation of this system by CI Wireless that works for CDMA, GSM/PCS 1900, IS-136, AMPS,

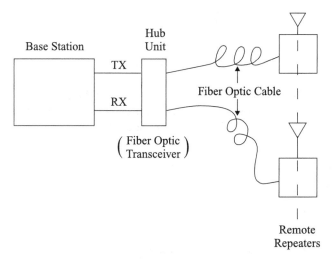

Figure 5.20   Fiber-optic repeaters.

specialized mobile radio (SMR), and enhanced SMR (ESMR) is depicted in Figure 5.21. Known as the Eko-Cel, the hub supports up to four remote repeaters and has been designed to allow a one-person installation. EkoCel is also dual band and can operate simultaneously in the 1.8- to 2.0-GHz region, as well as at

800 to 960 MHz, and has an output power of up to 8 watts.

Handover Algorithms

The successful operation of a microcell depends on the algorithm used for the handover. Generally these will be supplier specific, but a good idea of what is required can be seen in the Motorola GSM approach. The algorithms ensure the following:

- Users remain in either macro or micro cell environment if possible
- Emergency handover is available in the case of signal degradation
- Street-corner handovers should be deferred until the user has turned the corner
- Fast-moving vehicles should not use microcells
- Avoid handovers in cases of rapidly varying signal strengths in the vicinity of microcells

No Limit

With careful use of microcells, there is no real limit to the capacity that can be carried, provided the handover problems can be managed.

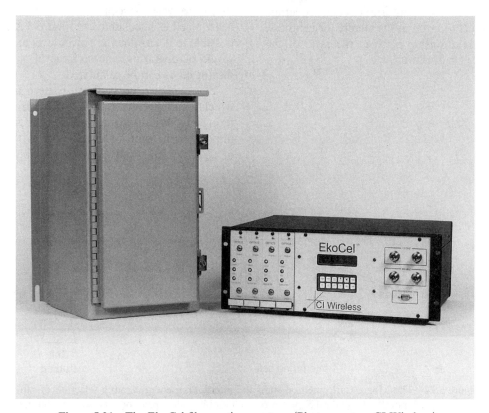

Figure 5.21   The EkoCel fiber-optic repeaters. (Photo courtesy CI Wireless.)

Picocells

Picocells are simply an evolution of microcells and are those with a design range of less than 100 meters. These mostly will be found providing in-building coverage, and may be used as much for signal strength as for capacity reasons.

As picocells are probably destined to be mounted in doors, and as their power requirements are small, the objective of the system provider is to make them increasingly small and unobtrusive. Picocells are available in a choice of colors to permit them to blend with the background.

MODELING CELLULAR SYSTEMS

The real-world cellular networks are very complex, and from one system to another there can be so many variables that impinge on propagation that it can be difficult to draw generalized conclusions. The mathematicians who model cellular networks dispense with the real-world messiness and work on models based on a flat earth where the propagation is neatly confined in hexagons. Although any relationship between these models and the real world will be largely coincidental, valid conclusions can be drawn about the relative performance of different cell patterns and configurations.

The simplification from uniform circular coverage areas of equal radius to hexagons is merely for neatness, and it leads to a relationship between the ratio of the reuse distance to cell radius of:

Equation 5.1

$$D/R = \sqrt{(3 \times N)}$$

where

D = reuse distance

R = cell radius

N = reuse number of cells per cluster

The constraint that the clusters must be symmetric (that is, all cells will be equidistant from their reuse cells) is met if N can only take values such that;

$$N = (k + l)^2 - kl$$

where

k and l are positive integers

This is equivalent to saying that N can only take the values such that $N = 3, 4, 7, 9, 12, \ldots$ if the pattern is to be symmetric. Other values of N will produce systems in which the distance between various cells of the same type will be asymmetric and so the interference will be different for different cells. This does not mean that such patterns are necessarily precluded, and at least one $n = 2$ reuse pattern has been proposed.

One of the most useful things that can be studied with these models is how the various cell patterns compare for frequency reuse density. The most usual way to study this is to simulate the system on a computer in Monte Carlo trials. This means that mobiles are randomly assigned to various locations and the path loss to the nearest cell is calculated; see Figure 5.22. From this it is possible to determine the power level at which the phone would be operating, and so its level of interference into distant bases can be calculated.

This is shown in the 7-cell pattern in Figure 5.23, in which a mobile operating in the central D cell is interfering in the more distant reuse cell. Knowing that the important measure of interference is the relative value of C/I, it is then only a matter of applying a propaga-

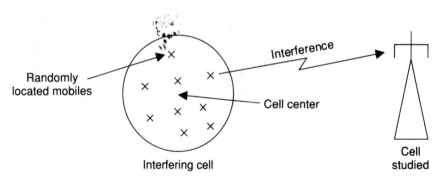

Figure 5.22   The Monte Carlo method sums the interference energy from a selection of randomly spaced mobiles. Note the TX power of each mobile will depend on its distance from the cell center.

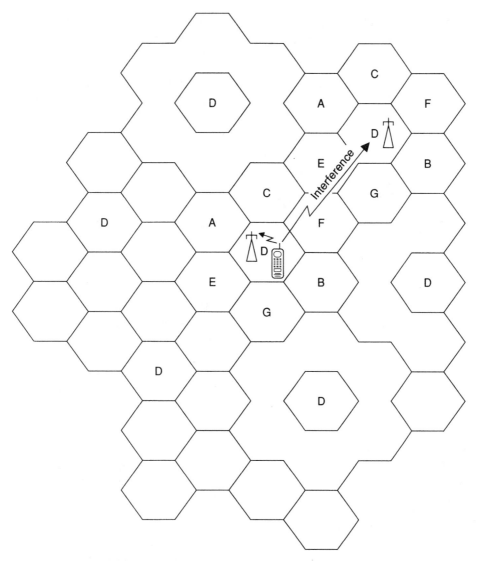

Figure 5.23 The interference mode that determines the usable C/I. Notice that each D cell is surrounded by six co-channel D cells.

tion relationship to find the level of interference. For example, if the relationship

Equation 5.2

$$F = k/d^4 \text{ (a widely used approximation)}$$

where

$$F = \text{field strength}$$
$$k = \text{constant}$$
$$d = \text{distance}$$

is used, then it will be easy to get the interference signal

level at the distant base. By repeating this process over a large number of different locations for the mobile, the result will be a spread of values for interference. The 90 percentile value of this interference is often taken as the reference. The same simulation will give the 90 percentile value of the wanted signal, and so a value for C/I can be calculated. Table 5.7 lists some C/I values for some common configurations.

If the values of R and D are substituted for d in Equation 5.2, then

$$I = k/D^{3.5}$$
$$C = k/R^{3.5}$$

Then

TABLE 5.7 *C/I* **Values for Various Cell Patterns**

Cell Pattern	90% C/I	D/R
12-sector omni	20.8	6
7-cell omni	14.2	4.58
7-cell, 3-sector	20.8	4.58
7-cell, 6-sector	24.9	4.58
4-cell omni	8.3	3.46
4-cell, 3-sector	15.0	3.46
4-cell, 6-sector	19.4	3.46
3-cell, 3-sector	12.5	3.0
3-cell, 6-sector	16.7	3.0
2-cell, 6-sector	12.4	3.0

Source: Land-Mobile Radio System Engineering by Garry C. Hess.
Note: Based on cells of 6-km radius, base stations 30 meters high, with log-normal shadowing.

Equation 5.3

$$C/I = (D/R)^{3.5}$$

In a large cellular system there will be multiple interferences, so that the net interference will be the sum of all the co-channel cell signals. It can be shown from geometry that for any cell in a regular hexagonal pattern there will be six co-channel cells symmetrically located around it (see Figure 5.23). Farther away there will be more co-channel cells, but as mobile propagation follows approximately an inverse fourth-power law, the contribution of the more distant cells will be negligible. So the conclusion is that the net interference will be six times the value derived from considering one co-channel cell.

As an example consider a 7-cell omni pattern where D/R is 4.58. If it is assumed that the propagation follows the inverse 3.5 law, then the C/I will be $1/6 \times (4.58/1)^{3.5}$, or $48.5 = 10 \times \log(48.5) = 16.8$ dB. If the calculation is repeated for a 4-cell omni, the result will be 11.1 dB. Although the actual values vary by a few dB from those in Table 5.7, the relative values (a difference of about 5.8 dB) are the same. This is because the basic propagation conditions as assumed in Table 5.7 are for a particular cell.

The values for sectored antennas are found by considering the level of interference from each co-channel cell *separately*, because the actual interference from each cell will vary according to the antenna gain in the direction of the interferer.

A value of $C/I = 18$ is considered suitable for AMPS systems, and are from 8–15 dB the values for the TDMA systems.

When designing a system it is practical only to use cell patterns that have a theoretical C/I greater than the system C/I. An AMPS system has a C/I of 18 dB, so the smallest cell pattern that is applicable is the 4-cell, 6-sector pattern. In the early days of GSM there were a lot of proposals to use $N = 3$ cell patterns because a C/I of 12 or better was anticipated. However today most are $N = 4$ or $N = 7$.

The choice of a cell pattern is very much a compromise. In theory, the lower the N number the greater the traffic efficiency and the greater the frequency reuse. However, some consideration must be given to the difference between the real world and the way the C/I figure is obtained. There is no guarantee that a real system using a 3-cell, 6-sector pattern could achieve an intrinsic C/I of 16.7. What we probably can say from the theoretical study is that a 7-cell, 3-sector system would be more robust in the face of interference than the 3-cell, 6-sector system, and because of this it will experience less blocking and cross talk. Blocking and cross talk, in turn, will reduce capacity.

In particular, Table 5.7 was calculated for the situation in which all co-channel cells are equidistant from the central cell. In practice, this virtually never happens, as some will be closer and some farther away. The closer ones, it can be shown, will contribute relatively more to the C/I than the more distant ones so that the net C/I will be reduced by asymmetry.

It is also worth noting that these results are for a specified cell site height and cell radius. In most studies a propagation model based on the Okumura results, the CCIR studies or similar will be used, and these will give results that depend on cell radius, terrain type, and base-station height.

One thing that does emerge from these studies is that the simple, predetermined fixed-cell plan is not the most efficient way of achieving a maximum level of frequency reuse. Systems that can allocate channels in real time to achieve the lowest level of interference are more efficient. This requires base stations that can measure the interference present and dynamically allocate any frequency to any channel (perhaps with some limits based on combiner intermodulation products). Even better spectral efficiency can be obtained if the base stations have intelligent algorithms, which can learn from previous allocations to select channels that minimize handoffs and interference during calls. This kind of algorithm could even learn to take advantage of the few dB difference that is found from channel to channel and even on the same channel at different frequencies in its home base. Once this is implemented, there is effectively no frequency or cell plan because the configuration varies in real time.

SITE ACQUISITIONS AND SAFETY

While there have been many thousands of studies of the health effects of RF radiation, and none have conclusively proven that a problem exists, the handling of the subject by the media has encouraged a general fear among the general public. It is becoming all too common that a new site is greeted with howls of protest by local-resident groups.

The most common reaction from the operators is to ignore the protests and rely instead on the legislation to enforce the right to install. This largely exacerbates the future problems, as there are few signs that the protesters are going away, and if anything they are becoming more vocal.

Design engineers can do a lot to minumize these problems by selecting sites that are least likely to cause protests.

In general, it is best to stay away from schools and residences, not because of any health risk, but because of the potential for an emotional response. Locating on existing buildings and structures, will draw less attention than a monopole or tower. Co-siting with other services on an existing tower often goes unnoticed.

Where it is imperative to locate on or near an existing residence, consideration should be given to using low-profile or even camouflage antennas. Avoid where at all possible locating an antenna near a patio, window, or directly above a balcony.

It pays to be prepared for public opposition and it is a good idea to have a prepared information package on the subject of EMR. Information packages are available from the Cellular Industry Association, and often from the regulatory authority in each country. These are good reference documents on the subject.

While some people advocate that the cellular operator provide speakers in public forums, in practice this can be difficult. The organizers of the public protests are mostly not interested in any "facts" other than the fact that a base station is being located near them and that they want it removed.

SATELLITE MAPPING

In remote regions and some developing countries it is not unusual that up-to-date maps, or indeed in some cases even any maps at all, are available. In this case satellite mapping may be the only information available.

In recent times satellite-derived mapping data have become quite affordable, and because the data are collected on request, it is assured that they will be current. This is particularly important when considering terrain clutter, which can change dramatically over a few years.

Techniques have been developed to enable data from the satellite photographs to be digitized using an automated parallax method, to produced digitized terrain data accurate to around 10 meters. Typically the three-dimensional data coordinates may contain 30 million sets of positional and height data for a 60-km area.

When deciding whether to purchase raw mapping information or the alternative of digitized data ready for the computer, account should be taken of the considerable investment needed to process the mapping data. Except for large-scale users, it generally will be preferable to have the data delivered already processed.

Most computerized propagation tools have many terrain models built into them, and it is possible to include information about the terrain type with the positional data. For wireless applications six terrain types are ordinarily enough to classify the area: water, forest, bare land, farming land, urban, and suburban. Within these categories there are many variants and mostly it will be necessary to classify the type attenuation by acquiring field data over typical stretches of land.

6

RADIO SURVEY

A radio survey is the process of measuring the propagated radio field strength over an area of interest. It is an essential part of the cellular radio site-selection process. Many radio-survey techniques exist, but not all yield consistent and satisfactory results. A radio survey is necessary as a design aid and as a maintenance tool. As a design aid, it helps determine potential coverage of a proposed base-station site. As a maintenance tool, a radio survey confirms continued satisfactory coverage.

A radio survey usually uses a field-strength measuring receiver located in a vehicle to measure the field strength. Sometimes the reciprocal path (that is, the path from the mobile to the base station) is measured instead. Both measurements are mathematically equivalent. (However, this does not apply to satellite mobile links where the send and receive paths are totally uncorrelated.)

When measuring field strength, it is important to note that what is being measured is a statistical variable and that the measurement technique must allow for this.

Three factors operate together to produce the measured field strength: path loss (free space), log normal fading, and Rayleigh fading. Figure 6.1 illustrates the free (that is, unobstructed) path loss. This loss is the most significant in microwave links, but it is only one of the losses in the mobile environment.

The free space path loss P_L is given by

$$P_L = 20 \log(42 \cdot d_{km} \cdot f_{MHz}) \text{ dB}$$
$$= 32.5 + 20 \log f_{MHz} + 20 \log d_{km} \text{ dB}$$

where

$$P_L = \text{path loss in dB}$$
$$d = \text{distance}$$
$$d_{km} = \text{distance in kilometers}$$
$$f_{MHz} = \text{frequency in megahertz}$$

Figure 6.2 shows log normal fading. This process is called log normal fading because the field-strength distribution follows a curve that is a normally distributed curve, provided the field strength is measured logarithmically.

Multipath, or Rayleigh, fading is a salient feature of mobile communications and, to some significant extent,

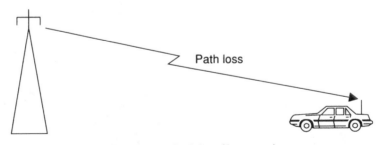

Figure 6.1 Path loss (free space).

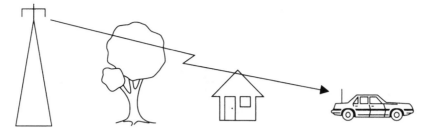

Figure 6.2 Log normal fading that is due to obstruction is known as "shadowing" or "diffraction losses."

limits the coverage of mobile systems when the mobile is moving in a multipath environment. It is not such a dominant factor in handheld mobile usage but, in low-field-strength areas, it can be detected by variations in noise levels as the receiver is moved. Figure 6.3 illustrates multipath fading.

An empirical formula for the cumulative effect of these three types of fading is given in *Recommendations and Reports of the CCIR*, 1982, Volume V, Report 567-2 as

$$P_L = 69.55 + 26.16 \log f_{\mathrm{MHz}} - 13.82 \log h_1 - a(h_2)$$
$$+ (44.9 - 6.55 \log h_1) \log d_{\mathrm{km}} \ \mathrm{dB}$$

where

$$P_L = \text{loss in dB}$$

$$f_{\mathrm{MHz}} = \text{frequency in megahertz}$$

$$h_1 = \text{base-station antenna height in meters}$$

$$h_2 = \text{receiver antenna height in meters}$$

$$a(h_2) = (1.1 \log f - 0.7)h_2 - (1.56 \log f - 0.8)$$

$$d_{\mathrm{km}} = \text{distance in kilometers}$$

where 15 percent of the area is covered by buildings (that is, an urban area).

This formula is based on field experience. Experience dictates that in different terrains some or all of the coefficients must be recalibrated. For general use, the formula should be regarded as accurate to ± 10 dB.

It is often said that in the mobile environment, the loss is inversely proportional to distance to the fourth power. You can see that this is consistent with the formula by looking at the last term $(44.9 - 6.55 \log h_1) \cdot \log d_{\mathrm{km}}$. This is the only term that is a function of d_{km}. If, for example, $h_1 = 30$ meters (the base-station antenna height), then this term becomes $(44.9 - 6.55 \log 30) \times \log d_{\mathrm{km}} = 35.2 \log d_{\mathrm{km}}$.

Because this is an expression for loss in dB, it can be rewritten in the form

$$\mathrm{Loss} \propto \frac{1}{d_{\mathrm{km}}^{3.52}}$$

This reduces the relatively complex formula to approximately d_{km}^4; the fourth-power relationship holds exactly at an antenna elevation of 5.6 meters.

In the far field, where all these loss modes are operating, the field strength varies with distance and time. At any one instance, the field strength can be shown as illustrated in Figure 6.4.

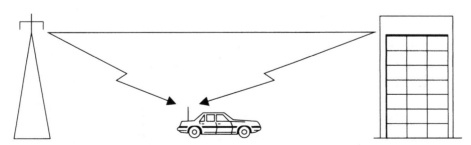

Figure 6.3 Multipath, or Rayleigh, fading is produced as a result of interference patterns between the various signal paths in a multiplan environment.

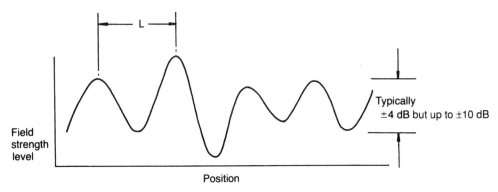

Figure 6.4 Standing-wave pattern.

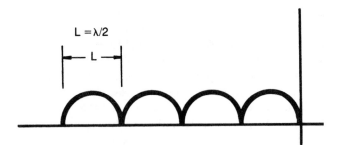

Figure 6.5 Standing-wave pattern caused by reflections off a plane surface.

STANDING-WAVE PATTERNS

The limiting case of standing-wave patterns is one produced by a reflecting plane at right angles to the line of propagation. A standing wave produced by a wave incident on a plane reflecting surface (such as a wall) produces the familiar $\lambda/2$ standing-wave pattern shown in Figure 6.5.

Other forms of interference generally produce interference patterns with a wavelength greater than $\lambda/2$. The distance L between the waves is such that $\lambda/2 < L$, but L can take any larger value. In practice, however, $L \approx \lambda/2$ can be taken as the worst case. Notice that $\lambda/2 = 0.16$ m (16 cm) at 900 MHz, or 8 cm at 1800 MHz.

MEASURING FIELD STRENGTH

As a mobile radio must necessarily operate over its entire service area, the field strength at a point becomes meaningless in terms of the overall performance of a mobile receiver. Consequently, individual spot readings are also meaningless.

Some operators have tried to solve this problem by using a field-strength meter in a moving vehicle and "guessing" the average level. This also yields meaningless results, as you can see from the structure of a typical field-strength meter, as shown in Figure 6.6.

The field-strength meter consists of a receiver, which has some way to access the intermediate frequency (IF) limiter or automatic gain control (AGC) drive stage to measure the output of that stage via a meter. A log-law amplifier gives a usable output in dB. The meter may be a conventional moving-coil meter or a digital one.

Because these field-strength meters are designed to operate in a point-to-point environment, the smoothing capacitor C (or its mechanical equivalent) is usually incorporated to even out small fluctuations due to log normal fading, and it likely has a time constant of 0.5–2 seconds.

When such a meter is confronted with a rapidly varying signal, it tends to follow the peaks, as indicated in Figure 6.7.

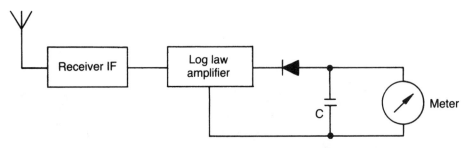

Figure 6.6 Basic field-strength meter.

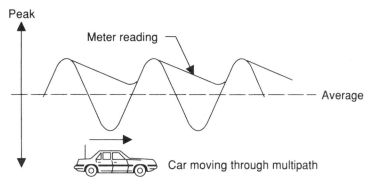

Figure 6.7 Detector response.

Because the relationship between the peak value and the average value from such a device is dependent on the depth of the fade and the multipath frequency, no firm relationship between these values can be said to exist. Thus, a sampling method that samples at a sufficiently high rate to measure the actual standing-wave is necessary.

If this standing-wave pattern were totally uncorrelated, the necessary sampling speed would be the Nyquist rate (two times the pattern frequency), but, because of the existence of a correlation, in practice it has been demonstrated that about one quarter of that rate yields results that are accurate to ± 1 dB.

SAMPLING SPEED

To calculate the required sampling speed in a mobile environment, consider a vehicle moving at 100 km per hour through a standing-wave pattern of a 900-MHz transmission.

The wavelength of that pattern is

$$C/F = \frac{300,000,000}{900,000,000} = 0.333 \text{ meters}$$

where

$$C = \text{speed of light}$$

$$F = \text{frequency}$$

In the worst instance, the standing-wave pattern is repeated every $\lambda/2$ m (as in the case of a reflection of 180 degrees from a wall).

$$\text{So } \lambda/2 = 0.333/2 = 0.1665 \text{ meters}$$

$$100 \text{ km per hour} = 27.7 \text{ meters/sec}$$

The Nyquist sampling rate requires two samples per pattern interval, or one sample every 0.1665/2 meters. Thus, the Nyquist sampling rate is $27.7/(0.1665/2) = 332$ samples/sec. As already mentioned, about one quarter of that rate (or 80 samples/sec) would suffice, but for 1800 MHz 160 samples/sec would be required.

Quite a number of commercially available field-strength meters sample much slower than 80 samples/sec and will consequently return inaccurate readings if used in a fast moving vehicle. Notice that the inaccuracy will increase as the sample speed decreases and that, by its nature, the error will be randomly distributed.

MODERN SURVEY TECHNIQUES

Modern survey equipment is based on the principles of the system illustrated in Figure 6.8. The radio receiver can be a specialized communications receiver with a wide dynamic range of both level and frequency, or it can be a special-purpose receiver such as a mobile telephone. If the receiver is not designed as a measuring receiver, some limitations will occur due to the limiter time constants.

Mobile phones are widely used as survey devices but they are not really suited to the purpose.

The two main criteria for the receiver are that the radio frequency (RF) level is a monotonic (single-valued) function of the RF level and that the limiter output has a bandpass characteristic of about 200 Hz.

Most modern receivers meet the requirement that the RF level is monotonic. Some older wide-band communications receivers with a measuring capability use switched attenuators to increase their dynamic range. Often the switching-in of these attenuators causes discontinuities in the output level, which can render them unsuitable for survey.

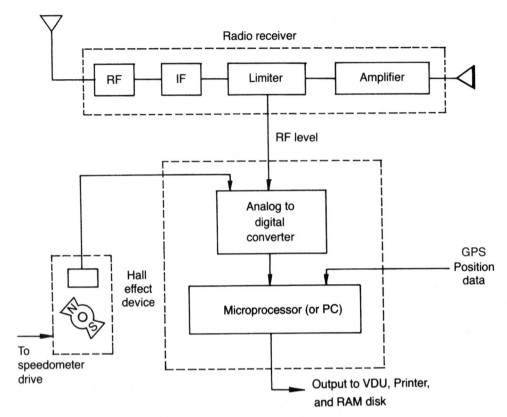

Figure 6.8   Basic survey block diagram.

The response time of the limiter of a communications receiver is deliberately damped so that the device operates only to compensate for level variations that ordinarily result from fading, but the response frequency is limited so that RTTY, MORSE, and other digital signals can be passed normally. This damping can be as elementary as a simple resistance/capacitance (RC) bandpass filter, or it may be more sophisticated.

Figure 6.9 shows the limiter/automatic gain control (AGC) response curve. If it was originally planned that

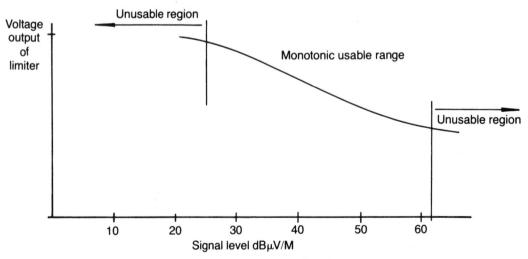

Figure 6.9   Usable portion of limiter/AGC response curve.

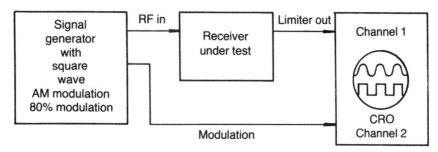

Figure 6.10 Testing the limiter response time of a survey receiver.

the receiver was to operate in a fixed location, this damping may limit the response time of the limiter to the extent that it is unsuitable for survey.

The easiest way to determine suitability is to input an amplitude modulation (AM) square-wave modulated carrier and compare the input with the limiter drive as the frequency is increased. Provided the output tracks fairly well up to 100 Hz, the receiver will be effective. Figure 6.10 shows the limiter response time of a survey receiver.

Most measuring receivers track only up to 50–100 Hz, although a few go a little higher than this. Notice that the inability to sample to at least 50 Hz results in damping errors of the same nature as those of a simple field-strength meter previously discussed.

In a digital field-strength meter, the output of the limiter is read by an analog-to-digital (A/D) converter, which samples the limiter levels. As previously explained, the sample rate depends on the RF frequency and vehicle velocity. Figure 6.11 shows a typical A/D converter.

A single high-speed A/D converter chip measures the limiter output voltage and outputs the level to a data

bus. Usually, a multiplexer chip is used to allow multiple inlets to be sampled in turn.

Receivers not specifically designed as measuring receivers may well have a band-pass characteristic similar to the one shown in Figure 6.12. You can see that the limiter drive does not respond to low-frequency signal variations. Such a receiver will probably "see" Rayleigh fading but will not respond to log normal fades. Such a receiver may also have a square-wave response, as illustrated in Figure 6.13. Notice that the behavior of the receiver in Figure 6.13 seems to be limited mainly by a simple *RC* network.

These receivers also have a response to large-level changes that saturate very readily at excursion of about ± 4–6 dB. Thus, they generally underestimate the standard deviation and the decile value.

The speedometer pulses or Differential Globalized Positioning System (DGPS) can be used to ensure that samples are taken regularly over fixed-distance intervals. This is done to avoid the sample biases that can occur when, for example, the survey vehicle is held up at one spot (a traffic light, for example), and so records many

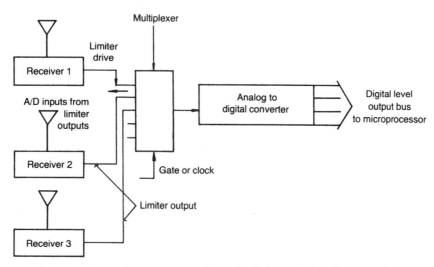

Figure 6.11 An A/D converter with multiple inputs from a bank receiver.

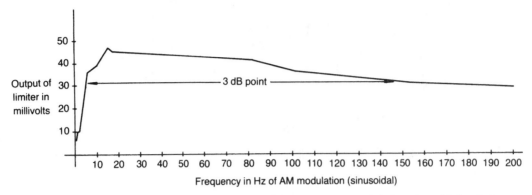

Figure 6.12   Frequency response (expressed in Hz of AM modulation-sinusoidal) of the limiter voltage with respect to a constant carrier level change. (An ideal measuring receiver has a flat response.)

readings at that one point, which then give a distorted average value for the area.

The speedometers in many modern vehicles are driven by a Hall-effect device (see Figure 6.8) that has a sufficiently high output level to be read directly by an A/D card.

Position data can also be read and recorded along with the readings.

SAMPLING INTERVAL

Within localized regions, the *average* field strength as measured from a distant transmitter can be shown to be approximately constant over intervals of 50 m to 500 m. Within these localized regions, the Rayleigh and log normal fades occur; the log normal mode generally dominates, although significant Rayleigh fading may occur, particularly in built-up areas.

Studies by Okumura *et al.* have shown that the local field strength in an area bisected by a 50–500 meter path can reasonably be accurately described by a single mean (average value) and a standard deviation. Figure 6.14 shows an area represented by sample measurement.

Hence, an adequate record of regional field strength can be obtained by sampling at the Nyquist rate and obtaining the mean and standard deviation over a 50–500-meter path. This is the basic approach used by most computerized field-strength measuring devices, where the sample intervals can vary from 10 meters to 1000 meters. In practice, it can easily be shown that sample averages, taken in groups over intervals of 50–500 meters, yield consistent results to a few dB.

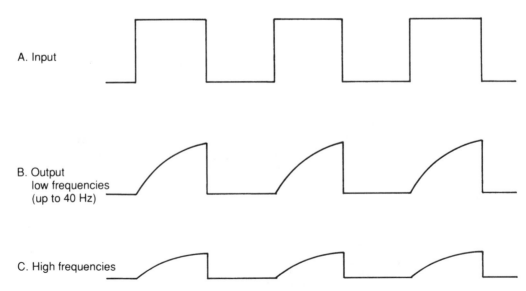

Figure 6.13   The limiter drive output response of a commercial wide-band receiver that was not designed to measure a square wave.

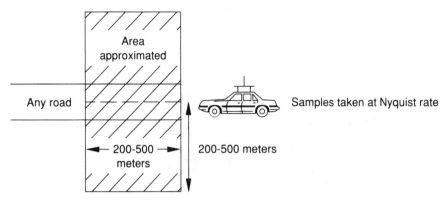

Figure 6.14 Area represented by sample measurement.

Figure 6.15 illustrates the decile method, a commercial measuring system. The illustration shows the structure of the Radio Survey Master system from Telstra, which was developed by the author (Figure 6.16). The system has an external trigger (based on the speedometer pulses) that ensures that samples are taken at fixed intervals. The system records mean, standard deviation, decile value, the number of samples taken, and local time. The decile value is often used and favored by some because it gives a more conservative estimation of mo-

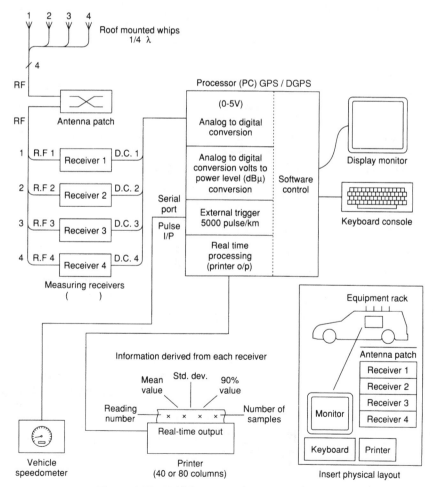

Figure 6.15 Vehicle-mounted survey equipment.

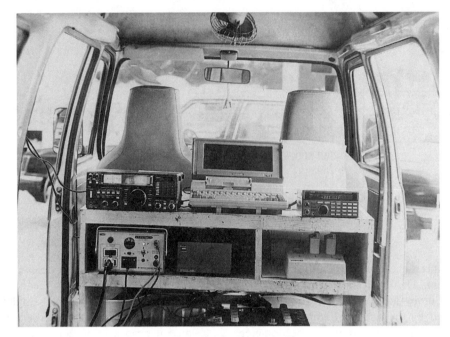

Figure 6.16 A survey vehicle with equipment as described in Figure 6.15.

bile performance than the average value—especially in areas of deep multipath.

In Figure 6.17, the upper 10-percent decile corresponds to T; this means that 90 percent of the readings are above the level T dBμV/m.

The 90-percent level is calculated using an iteration method or bubble sort. If the average is also needed from this sample, you should note that, although decile levels can be determined with equal accuracy, as an absolute level or the log of the level (dBs), the values must be converted to absolute values (μV/m) before a true average can be calculated.

In high-multipath areas with deep fades (high standard deviation), a measured mean value does not tell much about the extremities of the readings and particularly about the minimum (and hence, noisy) locations.

Taken with the standard deviation, more can be extracted, but the decile method eliminates the need to know about the standard deviation and allows the use of a single standard-design field strength.

Figure 6.18 shows an example of log normal distribution of field strength, which is what Figure 6.17 would look like if a very large number of samples at small intervals were taken.

For example, if a field strength of 39 dBμV/m *average* with a standard deviation of 6 dB (in a suburban area) is considered the objective, this can be equated to 39 dBμV/m − 1.28 × 6 = 31.32 dBμV/m for 90 percent of readings to define the boundary corresponding to 90 percent of locations having 39 dBμV/m average 90 percent of the time.

Because in lower multipath regions the 90-percent reading moves closer to the mean (and conversely, further away in high multipath regions), this method adjusts to the multipath environment in a way that an average reading cannot.

For example, if the field-strength measuring equipment was set as above (31 dB for 90 percent of readings as in the sample above) and it were to move from a suburban area ($\sigma = 6$ dB) to a rural area where $\sigma = 2$ dB, then the measured field strength would correspond to 31 + 1.28 × 4 = 36.1 dBμV/m average. That is, at this lower multipath level, an adjustment is automatically made for the lower noise level.

Clearly this method gives a more effective measure of signal quality than an average reading. It is not often done, however, because it requires somewhat more

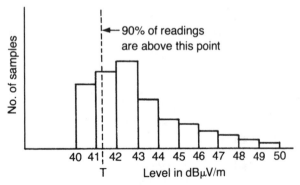

Figure 6.17 Histogram of sampled data.

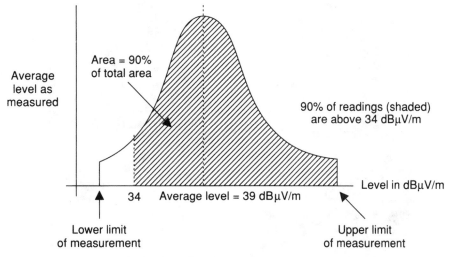

Figure 6.18   Lognormal distribution of field strength in 50 percent of locations and time.

"number-crunching" power than is needed to obtain an average reading.

For a more detailed discussion on these methods, see Chapter 9, "Units and Concepts of Field Strength."

REVERSE PATH SAMPLING

Some surveyors prefer to use the transmitter in the mobile and locate the receiving equipment in the base station. This has the advantage of less clutter in the vehicle, but has the disadvantage of requiring a very good position-location system that can relate readings to position.

A further disadvantage is that real-time outputs cannot normally be obtained. Thus, the mobile survey team does not have continuous direct contact with the receiving site and has no feedback on the survey's progress. This can lead to many problems, including time wasted surveying areas clearly out of range and time wasted surveying when equipment may be faulty.

When the measuring equipment is in the vehicle, the operators have a real-time measurement from which to solve these problems. Either way of measuring, however, if done effectively, gives equivalent results. Do not confuse the fact that the actual cellular transmit-and-receive hardware does not necessarily give balanced path losses with the concept that the radio paths themselves are reciprocal. Figure 6.19 shows that the paths *are* reciprocal and that it doesn't matter which way they are measured.

USING WIDE-BAND MEASURING RECEIVERS

Wide-band measuring receivers usually have poor sensitivity (typically 1–2 µV for 12 dB SINAD). Because surveys are often done on test transmitters with rather low effective radiated power (ERPs) (compared to the cellular base station), measurements are often limited by receiver performance.

It is often useful to acquire a low-noise ($NF < 1$ dB) amplifier for the band being surveyed. Because of inter-

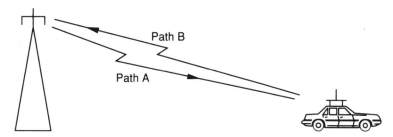

Figure 6.19   Because the signal traversing a path to the mobile uses the same path (almost) as the signal on the path from the mobile, any losses are equal. Measuring either path yields equivalent results.

Figure 6.20   A survey vehicle can be fitted with a metal ground plane attached to ski bars. A number of antennas can be mounted on this good-quality ground plane.

modulation susceptibility, it is best to only use such an amplifier at low signal levels and to switch it out when it is not needed. If the survey vehicle is equipped with a cellular phone (or other transmitting device), observe the effect on the performance of the measuring system; the transmitting device may desensitize the wide-band amplifier. This must be done in a low signal area when the mobile phone TX is high and the receiver level is low.

MULTIPLE RECEIVER ANTENNAS

Because of cross-coupling, if multiple-survey-receiver antennas are used, they should be spaced at least two wavelengths apart. Because most cellular sites are sectored, multiple receivers will effectively increase survey

efficiency. When surveying omni sites, use a second channel as a check on the integrity of the first channel.

The best way to mount multiple antennas is on a ground plane made by suspending a sheet of aluminum between the bars of a roof-rack, as shown in Figure 6.20. This not only gives a very good ground plane, but it also avoids the necessity of drilling holes in the roof of the vehicle. The whole assembly is easily detachable when the time comes to replace the vehicle.

SURVEY TRANSMITTERS

It is preferable to have the survey transmitter on the same frequency and with the same ERP as the final cellular base. However, this is frequently not practical,

Figure 6.21 A Plexsys 45-watt survey transmitter. (Photo courtesy of Plexsys.)

particularly because transmitters of high-power outputs on 800–900 MHz (an even more so at 1800/1900 MHz) are rather inefficient and are also somewhat difficult to come by. However, at least one company, Plexsys, makes a dedicated survey transmitter of 45 watts output at these frequencies, as illustrated in Figure 6.21.

For temporary installations (such as surveys), it is usually much more convenient to use lower-power transmitters (especially if they are being run from batteries) and small antenna feeders that are easier to handle.

The stationary transmit antennas used for surveying are preferably 6 dB. This avoids the need to get the antennas accurately vertical (as would be necessary for higher-gain units) and means that they are physically reasonably small and manageable. Lower-gain antennas reduce the ERP to a point where it might be difficult to measure far-field regions.

Combining all these factors, you will probably find that the survey ERP is about a magnitude lower than a cellular base station. This can be easily corrected mathematically, but places some constraints on the receiver signal-to-noise (S/N) performance because it means that the receiver must be functional at very low received-signal levels.

Figure 6.22 shows a typical configuration of survey equipment. The ERP of the transmitter is (in dB)

$$10 \log T(\text{watts}) - L + A_t$$

where

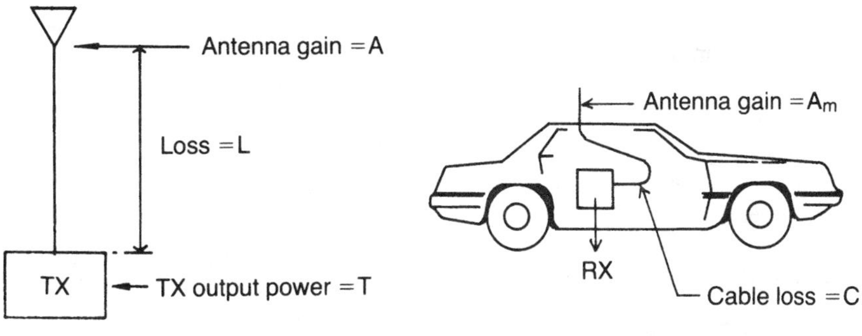

Antenna gain $= A$

Loss $= L$

TX

TX output power $= T$

Antenna gain $= A_m$

RX

Cable loss $= C$

Figure 6.22 A typical survey equipment configuration.

$$T = \text{transmitter power in watts}$$

$$L = \text{transmitter cable loss in dB}$$

$$A_t = \text{antenna gain in dB}$$

To convert the field strength measured by this arrangement to an equivalent cellular system level, it is necessary to correct for ERP and receiver gain. Thus, if the cell site has an ERP of E watts, the correction for the transmission system is

$$10 \log E - 10 \log T + L - A_t$$

Including the passive receive gain, a correction factor of

$$10 \log E - 10 \log T + L - A_t - A_r + C$$

where

$$A_r = \text{receiver antenna gain}$$

$$C = \text{receiver cable loss}$$

must be applied.

Generally, the availability of transmitters on frequencies below 500 MHz is much greater than those above that frequency. It is also common for 500 MHz transmitters to have power outputs from 10 watts to 25 watts, compared to 800-MHz units, which are usually limited (mostly by regulations) to about 10 watts. In most respects, frequencies from 400 MHz to 500 MHz accurately enough model propagation at 800–900 MHz if a 2-dBμV/m allowance is made for slightly lower path losses at the lower frequency.

Be cautious if this is done; the 2-dB correction factor applies only to field strengths measured in dBμV/m. In other units (dB, dBμV, μV), due allowance must be made for the aperture of the antenna. (See Chapter 9, Equations 9.1–9.4, under "Relationship Between Units of Field Strength at the Antenna Terminals.")

It is not good practice, however, to extrapolate these results to 1800/1900 or beyond. If the terrain is very hilly, it will be necessary to use the actual base-station frequency, as the propagation will be very frequency dependent for this terrain.

For survey purposes, the transmitters must be rated for continuous operation; you should note that most two-way radios with press to talk (PTT) switches are not so rated. Generally, it is necessary to de-rate the output power of such transmitters to 20 percent if continuous operation is expected. Even then, mounting a computer fan over the heat sink to provide adequate ventilation is a good idea because the difference between continuous and non-continuous rated transmitters relates to the ability of the RF power amplifiers to dissipate heat.

Notice also that continuously rated mobiles are normally duplex and have a duplex coupler to the antenna. This normally has a 3-dB loss (meaning 50 percent of the output power is lost in the duplexer), so the power efficiency is low. This can particularly be a problem when the survey station runs on batteries (as is often the case). The duplexer can be bypassed to increase ERP or save battery power as required.

When batteries are used to power the survey transmitter, it is a good idea to purchase a timer to switch on the transmitter just before work begins in the morning and switch it off in the evening. It is also worth noting that nothing shortens the life of lead–acid batteries faster than a complete discharge. Unattended battery-powered transmitters should have an automatic cutoff that activates when the battery voltage reaches 11 volts (or 1.8 volts/cell). This technique usually allows three days of operation from a pair of truck batteries and saves many visits to the site. Commercially available domestic power timers with clockwork timers usually serve this purpose adequately and have the advantage of being very cheap.

MOUNTING SURVEY ANTENNAS

The mounting of survey antennas can be an exercise to tax the ingenuity of the surveyor. The antenna needs to be mounted rigidly, but also should be able to be erected cheaply and quickly. Commercially available hydraulic masts of up to 50 meters can be an option. The smaller masts (30 meters or less) can easily be carried on the roof of a medium-size sedan, and can be erected in less than an hour by two operators. Problems can occur during high winds, when it can be difficult to get the mast up and to keep it vertical. The bigger masts are usually (but not always) vehicle mounted and sometimes even have electric pumps. These can be erected by one person in a few minutes.

Cranes and cherry pickers can be used to raise the antenna. They are convenient but not cheap, costing about $80 per hour. This can double the cost of a survey. Cranes present an additional problem related to how to mount the antenna. As the crane gets higher (and so more nearly vertical), the crane boom will come nearer to the antenna and interfere with the pattern. Ideally the antenna should be mounted above the crane boom, but this is not always possible. Because of this the crane should be significantly taller than the maximum survey height so that a boom angle that keeps the boom well away from the antenna can be used. Unfortunately bigger cranes cost more. Keeping the antenna vertical is easier with a cherry picker than with a crane.

Balloons have been used by some surveyors, and they can be an effective way of getting height. However, because balloons tend to sway, only low-gain antennas can be used. It is preferable to use dipole, unit-gain antennas on balloons, as these have broad radiation patterns that can be a long way off-vertical before the far field is sufficiently distorted to be a problem. Experience shows that balloons perform best if tethered to only one rope. Multiple ropes can help stabilize the balloon in low winds, but will tend to push it downward as the wind speed picks up.

Helicopters have been used, but at $300 or more per hour they are extravagant and probably justified only when studying undeveloped mountaintop sites.

Portable Survey Equipment

Until recently serious RF testing of the network required a van full of equipment to accurately measure and document the network performance. While this is still the preferred mode for operators, it is possible today to get most of the data-collecting power of the van into a small case with a miniature receiver, Global Positioning System (GPS), and notebook computer. This not only allows greater flexibility in the means of transport, but it even allows indoor testing with the same accuracy of the van.

AUTOMATIC POSITION-LOCATING SYSTEMS

The GPS system consists of 24 satellites plus two spares in six orbital planes at 20,200 kilometers. Each satellite transmits on two L-band frequencies, L1 (1,575.42 MHz) and L2 (1227.6 MHz). All satellites transmit on exactly the same frequencies.

Each satellite passes over a control station twice a day where it receives updates on its location and its clocks are synchronized.

The L1 transmission carries the precise P code information and a coarse and acquisition code (C/A code). The L2 transmission carries only the P code, which is a complex code and can be changed from time to time. The P code is generally encrypted (when it is known as the Y code), and civilian access is limited to the C/A code.

The satellites transmit using spread spectrum. GPS antennas need to have omni-directional coverage with the main lobe pointing skyward. Cross-polarized patch antennas are commonly used for this.

GPS signals effectively need line-of-sight, and signals can be lost under trees and in buildings. The ground receive level is only -132 dBm, which means that the receivers must be extremely sensitive. The receivers today are typically 12 channel, and they select the best signal. Reflections from buildings can make operations

in the central business district (CBD) difficult. Interference from other RF systems can be a problem, especially when the GPS is often used in close proximity to mobile communications equipment. In fact, one of the reasons that GPS has trouble indoors is that the radiation from fluorescent lights can exceed in level the wanted signals.

Because the GPS system was designed basically a military one, the civilian accuracy was originally as a result of deliberate downgrading low through a process called selective access (SA). This meant that a stationary position typically was seen to move at a speed of 2 kmph in random directions. The overall object of the SA algorithm was to make the GPS receiver read a location that is with 100 meters of the true location for 95% of the time. The basic accuracy of civilian GPS equipment *without* SA is 10 to 15 meters. Early in 2000, SA was turned off.

To improve the reliability and robustness of the civilian GPS system, newer satellites will include two new frequencies at 1227.6 MHz and 1176.45 MHz. Operations at these new frequencies are expected to begin in 2003. This measure will give significant protection from interference.

GPS for Clocking

Every GPS satellite contains two rubidium and two cesium clocks. These clocks are cross-referenced to ground-based atomic clocks and they are calibrated against the universal coordinated time (UTC). The satellites themselves are accurate to better than one nanosecond.

When the U.S. military restricted full access to the GPS network by a system known as SA, the accuracy that a civilian clock can obtain was around 100 nanoseconds. Without SA, the accuracy would be about 10 nanoseconds. There are other factors such as atmospherics that further restrict the civilian accuracy to about an additional 10 nanosecond error.

One way to improve the accuracy is to average the timing from a number of satellites, and the improvement so obtained can be seen in Figure 6.23. With eight satellites available the accuracy can improve by a factor of 3.

Code division multiple access (CDMA) requires a very accurate clock, but it also requires that that clock be stable. The GPS signal jumps and so with the SA encoding could not itself provide a reference for the CDMA networks. Rubidium clocks, costing about $3000 each, can provide the sort of stability needed for CDMA, but apart from the high initial cost, they require regular maintenance and replacement. Rubidium clocks provide a reference accuracy of around one part in 10^{10}.

A high-quality quartz-crystal oscillator can provide the short-term stability needed for CDMA, but the cen-

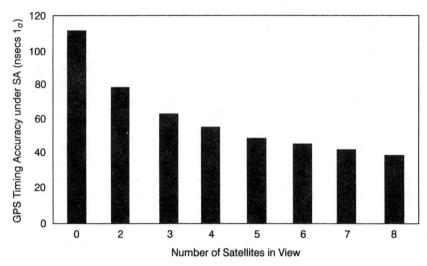

Figure 6.23 The improvement in GPS clock accuracy with multiple satellites.

ter frequency will drift with time. Using GPS, and averaging the signal over long periods of time, the accuracy can be improved by a factor inversely proportional to the time interval of the average. Over very long periods of the order of 1000 seconds, the GPS clock can provide a time accuracy of about one part in 10^{10}.

A purpose-built GPS clock with a double-crystal oven, will look something like the one in Figure 6.24. The 10-MHz output provides a reference frequency that can be used by most frequency-sensitive test instruments such as signal generators, test sets, and frequency counters.

Differential GPS

SA activation originally meant that most reasonably precise systems needed DGPS. However, even with SA

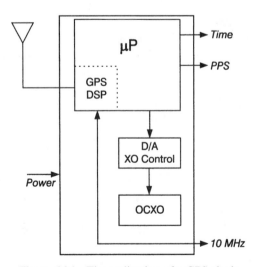

Figure 6.24 The realization of a GPS clock.

turned off, the basic accuracy of GPS is only around 10–15 meters. DGPS can provide millimeter accuracy. Differential GPS uses a known fixed location in the vicinity of the survey area that determines its own local error in location and transmits this information to receivers in the area. The term *differential* comes from the fact that what was originally transmitted was a current deliberate timing error, which is turn is a differential of the time.

The correctional information can be sent over cellular or trunked radio systems and is sometimes superimposed on an FM radio carrier. There have been a number of inventive attempts to commercialize access to DGPS information. Higher accuracy is sought by many uses, including radio surveys, taxi companies, delivery and transport services, mining, and farming.

In the United States the entire coast and many river systems are covered with DGPS.

For some applications, like farming, DGPS can be used to lay out fields and to ensure that crops are sown in straight lines. Local stand-alone transmitters with an accuracy of 1 to 2 cm are a viable option.

Using differential GPS submillimeter accuracy is possible, but typically services offer accuracies of around 1–5 meters.

How It Works Each satellite has on board four highly accurate atomic clocks. It also has a data base (which is regularly updated) of the other satellites in the constellation that can be accessed by the GPS receiver to reduce the scanning time.

The GPS works on the principle that it is possible to calculate an exact position in three dimensions if the distance to three reference points is known. The basic resolution is about ±10 meters. The system uses the delay of the transmission from each satellite to determine the distance, and it reads the location of the satellite from the satellite broadcasts.

The range R_i as calculated in Equation 6.1 can be found for each satellite within range. Provided at least three such satellites are within range, the system, to fully determine its location, merely needs to solve the three simultaneous equations.

Equation 6.1

$$R_i^2 = (X - X_{si})^2 + (Y - Y_{si})^2 + (Z - Z_{si})^2$$

where

R_i = Range to satellite i
(determined by propagation delay)

X, Y, Z = Three-dimensional coordinates
of the receiver

X_{si}, Y_{si}, Z_{si} = Three-dimensional coordinates of the
satellite (as broadcast)

It is a little more complex when you realize that the range, as determined by the propagation delay, will have an error proportional to the time error of the reference clock in the mobile. Thus, if the time base of the mobile receiver is ΔT seconds in error, then the error in the range calculation will be $R_E = \Delta T \times C$ (where C = the speed of light).

To avoid the need (and expense) of all receivers carrying atomic clocks, a fourth equation containing the time error can be solved to allow for the inaccuracy of the receiver's clock.

So, the coordinate Equation 6.1 is replaced with the following equation and solved as four simultaneous equations that are now independent of the accuracy of the mobile clock.

Equation 6.2

$$(R_i - R_E)^2 = (X - X_{si})^2 + (Y - Y_{si})^2 + (Z - Z_{si})^2$$

An accurate mobile time base is no longer required.

Since Equation 6.2 has four unknown variables, it requires a fourth equation to solve it. Thus, if reasonably priced timepieces are to be used at the receiving end, it is necessary to simultaneously obtain data from at least four satellites.

The GPS uses 24 satellites with 18 active units in six different orbits plus one spare in each orbit. The satellites are in an inclined orbit that takes them over any point in their path once every 12 hours because a narrow baseband means less total noise power.

The satellites use spread-spectrum techniques with a signal bandwidth of 2 MHz, but an inherent baseband of 100 Hz. This enables the system to work down to −163 dBm (a very low power density).

Semi-Automatic Position Location

Semi-automatic position location involves using the survey map with a digitizer board. When a reading is required, the surveyor touches a pen to the map location, which automatically takes a reading and records (accurately) the location coordinates. This method is rarely used today, and has largely been superceded by GPS and DGPS methods.

Manual Position Location

Unless a very good position-location system is used, it is necessary to use two people in a survey environment. When two people are used, the manual location does not overtax the non-driver; this is both the cheapest and most accurate method.

This method is still useful in remote areas where DGPS is not available and for small systems where the investment in sophisticated survey equipment is not justified.

In normal suburban environments, it is recommended that position fixes be street intersections and that reading numbers be marked on the map in real time. Figure 6.25 illustrates the path of a typical survey run. Samples are taken at the points marked 1, 2, and so on, and the run numbers are marked on the map as the samples are taken. A conventional street map in black and white is preferred, so that the run numbers are easily seen when marked off with red pen. At some later time, the field-strength readings are transferred to the map, using another readily visible color.

PREPARATION OF RESULTS

If the results are manually collected, they should be transferred to a street map. The map should clearly show the following information:

· Date of survey and site surveyed
· ERP of transmitter
· Surveyed frequency
· Antenna gain
· Any correction factors

The run number and field strength should be marked on the map in different colors and it should be clearly in-

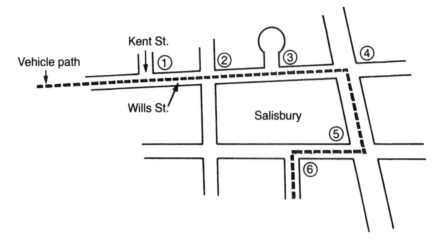

Figure 6.25 Map identification of samples.

dicated which is which. In order to make the results easier to visualize, it is a good idea to draw the service-area contour [say, 39 dBµV/m for Advanced Mobile Phone Service (AMPS)] and the CBD handheld contour (60 dBµV/m for AMPS).

Because frequency reuse is usually an important cel-

lular consideration, it is advisable to survey down to the 20 dBµV/m contour (the level at which interference becomes service affecting) (see Figure 6.26). This defines the region where frequency reuse is not practical without sectoring. Notice that if sectoring is contemplated, an additional interference immunity of approximately

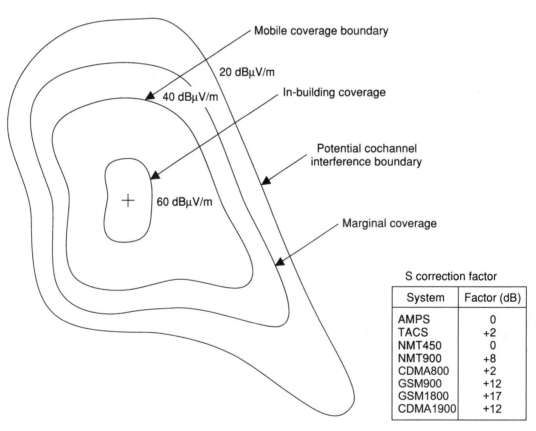

S correction factor	
System	Factor (dB)
AMPS	0
TACS	+2
NMT450	0
NMT900	+8
CDMA800	+2
GSM900	+12
GSM1800	+17
CDMA1900	+12

Figure 6.26 Interpreting field-strength contours.

10–15 dB will be obtained, and then the interference boundary of the cell approximately equals the coverage boundary.

In order to see how various cell sites will work together, it is a good idea to transfer the coverage contour to a sheet of transparent plastic (if manual) or to a computer graphic, so that you can clearly see the overlapping of different cells.

Some map or computer studies will have been done before selecting sites to be surveyed. The survey results should always be compared with the predictions and discrepancies explained. A large discrepancy means either a problem in the forecasting technique or a problem with the survey procedure. The comparison can be done quickly and will frequently highlight problems.

It is advisable to have a second person check all correction factors. Use the table in Appendix E to assist with this conversion. At cellular frequencies, the cable loss from the mobile antenna to the receiver port is about 3 dB and cannot be neglected.

SPECTRUM CHECK

While the survey antenna is on site, it is advisable to scan the whole band to be used from the survey site. Other services using the band will then be detected. For example, microwave links are occasionally found in the cellular band.

Notice that microwave links directed away from the site may be difficult to detect and the presence of any foreign carrier in the band is cause for concern. Spectrum analyzers are not usually sensitive enough for this check and cannot always distinguish between legitimate cellular traffic and another mobile type. The analyzer will, however, give a clear indication of the nature and extent of high-level systems, such as microwave link systems, and systems with sporadic occupancy. Simply listening to a wide-band receiver will sometimes give a good indication of the nature of the traffic if it is analog.

CONFIRMING COVERAGE

For turnkey projects, the operator often specifies the coverage as being at a certain quality for, say, 90 percent of the area for 90 percent of the time. Suppliers often undertake to guarantee the stated coverage.

How can an operator confirm that the supplier's design meets the specification? To do this for a real city is almost impossible, but to see how it might be done, consider the case of an imaginary town, called "*Square Town*," which is shown in Figure 6.27.

If the operator had specified that the coverage should be such that calls could be made from 90 percent of locations for 90 percent of the time (most unwisely, as this means 10 percent of the area *is totally* unserviceable), then it would be possible to proceed by attempting to make calls at various locations around the city and determining that the success rate is 90 percent or better. The results would not be conclusive, however, unless every street was sampled at the Nyquist rate (every 0.083 meters). If the town was 2 km × 2 km, then the

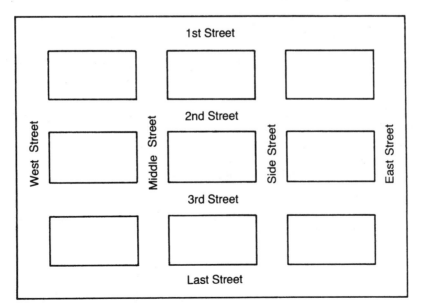

Figure 6.27   Square Town city map.

number of samples would be 200,000, or 8 streets × 2 km/0.083 × 10³ call attempts. Statistical sampling methods could also be used to reduce the sample size (that is explained in the next example).

If the specification called for a minimum field strength—say, *F*, in 90 percent of locations for 90 percent of the time—then there are two practical ways of measuring this value. First, you could calculate the corresponding mean from the *estimated* standard deviation for a field strength of *F* at 90 percent of the time/ locations and measure the mean field strength (sampling at or about the Nyquist rate). This should be done on a street-by-street basis and, provided 90 percent of all readings are above the calculated average, then the criteria are satisfied. Or, second, you could use a measuring set that can return the field strength above which 90 percent of all readings occur.

The second method is equivalent but more accurate than the first method. The whole city can then be measured as a single entity. If the measurements on this basis yield field strength of *F* or higher, then the criteria are satisfied.

The difficulty of performing the above measurements in a real city can be enormous. A real town, for example, might consist of 50 percent developed land and 50 percent rugged hills. An operator might find inadequate coverage in the city and complain that 20 percent of the built-up area (measured as above) is below standard coverage. The supplier could counterclaim that all of the undeveloped area is covered at or above the standard level and therefore 90 percent of the city is covered at or above the specified level.

Alternatively, the city to be tested can be divided into a grid, and a sufficient number of coordinates can be selected to satisfy that samples taken in these regions will yield the average field strength to a given degree of confidence. This method has, in fact, been used by a number of system operators, but it has some serious limitations.

First, the sample length should be defined to be, for example, 100 meters, sampling twice every wavelength. The random selection will probably yield points that are not directly measurable (that is, a path diagonally across a city block may be chosen). In order to avoid bias (either positive or negative), it is then necessary to generate a set of unambiguous rules to select the nearest practical path—both its start and end. Next, again to avoid bias, as the interpretation of the selected path would still be quite open (to the survey team), it would be necessary to have some rules that would prevent choosing a biased sample space during measurement. One method might be to increase the sample run to the whole city block. With careful control, this method may be the most practical way of ascertaining coverage standards.

Thus, coverage guarantees for most real cities are not worth much and probably could not be effectively challenged in court. I know of no case where the guarantee has resulted in the guarantor paying any damages or costs associated with unsatisfactory coverage, and yet less-than-expected coverage is a frequent complaint.

SURVEYING AS A MAINTENANCE TOOL

Very few operators fully appreciate the value of surveying as a maintenance tool. Provided the original design and acceptance-survey results are well documented, the survey equipment can be very effectively used for future maintenance.

No matter how careful the original design/surveys were, in any city of significant size it is not practical to survey every street. Soon after start-up, the operator will probably discover some areas covered that would not have been expected to be, and also some areas not covered for which good coverage was anticipated. Indirectly, therefore, some interpolation/extrapolation of measured result is needed.

When complaints come in (as they will) about coverage, the first reference is the original survey map. When, however, the area from which the complaint comes was not originally surveyed in detail, it may be necessary to look further. A survey vehicle should be sent to the area and a detailed survey of the suspect area taken. Always check nearby areas that were originally surveyed to confirm that the coverage has not deteriorated since the original acceptance survey. If there are discrepancies between the original survey and the check, then a base-station fault can be expected.

Sometimes you will find that adequate field strength is present but that calls still cannot be successfully made. This can probably be traced to co-channel or adjacent-channel interference. As a matter of routine maintenance, it is a good idea to spot-check the field strength from each base station on a yearly basis (or more often if complaints warrant it).

Remember that field-strength measurements only check the transmitter path; if complaints are coming in and the field strength is adequate and interference has been eliminated (this can often be detected by the sound of the control channel—a channel suffering interference will often sound very fuzzy compared to one that is working normally on analog systems; in digital systems it can be very noisy), then the problem likely is with the RX path.

Where diversity is being used, it is practical to temporarily disconnect one antenna and use it to transmit a carrier that can be measured to confirm the integrity of the RX path. Diversity reception can disguise quite acute receive-antenna problems.

SOME NECESSARY PRECAUTIONS FOR RADIO SURVEY

The following considerations are important precautions for a radio survey: air-conditioning, possible errors, and equipment stability.

Air-Conditioning

The survey vehicle should be air-conditioned to reduce errors due to temperature-sensitive drifts in the receivers and A/D card. Air-conditioning also helps the operators stay alert so they are more likely to pick up potential sources of errors.

Errors can occur for a large number of reasons and, where real-time output of results is available, those results should be checked for consistency as the survey progresses.

Possible Errors

Possible errors and their sources are considered in the following list:

- Receiver out of calibration. This should rarely occur with a good receiver, but can be quite a problem when cheaper (or older) mobile radios or mobile phones are used as the measuring device. Frequent checks (weekly on suspect receivers and monthly on quality-measuring receivers) of calibration against a good signal generator are necessary. This will frequently be a problem if the survey receivers are modified mobile phones. It doesn't make sense to spend tens of thousands of dollars on survey equipment and then use a $100 mobile phone for the receiver.
- Faulty antenna. The antenna should be free of visible defects and have a low voltage standing-wave ratio (VSWR). It should occasionally be checked against a few other similar antennas to confirm gain, using different feeders to the receiver.
- Faulty or damaged feeders. Use the same checks as for antennas.
- Mobile receiver de-tuned or simply not tuned to correct frequency. Double-check that the receiver is tuned to the correct frequency. The mobile can also be desensitized by other transmitting apparatus in the vehicle, particularly in regions of low field strength. Beware of transposing the TX and RX channels when setting the frequency.
- Wrong calibration table used. This often becomes a problem when the units of the signal generator output are not the same as the units used for field strength (for example, the signal generator is cali-

brated in dBμ and the field-strength equipment is calibrated in dBμV/m). Converting units in the field often results in error. Always have a correction table handy (for example, use the table in Appendix E).
- Test-receiver output is voltage/temperature dependent. Always check the test receiver for temperature and voltage sensitivity. This is more likely to be a problem with old receivers and mobile phones, but all sets should be checked before being placed into service.

    The voltage regulation on a car battery may not be good, particularly if the receiver gets its power from a source distant from the battery where significant voltage drops may occur. Be cautious of voltage drops from indicators and brake lights.
- Insufficient settling time. Even good-quality measuring sets should be powered up half an hour before beginning measuring. Most of the drift occurs in the first 10 minutes after the set is switched on.
- Inaccurate records of transmitter base. Keep good records of the survey conditions, particularly for a temporary test transmitter. In particular, record the following:

    1. Power output at start (measured)
    2. Feeder loss (cable should be calibrated)
    3. Antenna gain, antenna height
    4. Frequency of test transmitter
    5. Power at end of test (remeasure to ensure no drift has occurred)
    6. Date, and TX site name

Failure to record any of these details could render the data useless in the future.

Equipment Stability

It is most important that all items in a moving vehicle be securely fixed. It is usually necessary to mount the receiver and other survey hardware in a rack. This rack should be padded and constructed in such a way that it does not interfere with the vision of the driver (this usually limits the rack height).

It is best if the equipment can be mounted beside the operator, where, in the event of an accident, it is unlikely to come in contact with the operator. The most dangerous mounting position is in front of the operator, where a collision will throw the operator into it. This is particularly true of a visual display unit (VDU).

An internal master switch for battery power also should be provided, as should a fire extinguisher (CO_2 type). For security, it is best if the rack can be completely covered so as to hide the hardware it contains.

When it is necessary to leave the vehicle parked in the street for some time, the sight of a few expensive measuring receivers and some computer hardware is likely to attract unwanted attention.

Use notebook computers. Because they are much smaller, notebooks can be placed safely on the operator's lap, requiring only an up-and-down movement of the head. An equipment rack mounted at the side of the operator places some strain on the operator's neck muscles when he or she is viewing the keyboard or screen. The operator must a have clear field of vision forward to allow for proper street identification and to minimize the effect of car sickness. The motion sickness that results from operating survey equipment is similar to that caused by reading in a moving vehicle. It is such a problem for some people that they are unable to do this task effectively.

Quality of Service Surveys

While it has always been possible to use design survey equipment to do network quality surveys, there are some limitations. First, the network design survey equipment is meant to be used by engineers and the results from it need to be interpreted for marketing. Often these surveys were not being done, and frequently communications with marketing has been poor.

Since the early days, marketing people have been doing field testing, which often amount to no more than making a number of calls from various locations and reporting on the quality. These tests are often adequate for small systems where changes occur slowly, but they tend to become meaningless in systems with hundreds of base stations where new sites and reconfigurations are occurring daily. This is exacerbated by the need to do comparative studies with competitor networks, which may be just as big and which may use completely different technology.

Survey instruments designed to evaluate quality of service, will generally evaluate the audio quality of both the up- and downlink. The International Telecommunications Union (ITU) has a subjective mean opinion score (MOS) model that can be used for the quality comparison. Comarco Wireless Technologies offers a product that can simultaneously monitor eight separate channels, any one of which can be AMPS, N-AMPS, TACS, IS-136 (800 and 1900), CDMA (800 and 1900), GSM (900, 1800, 1900), and iDEN.

7

CELLULAR RADIO INTERFERENCE

In cellular radio systems, which have frequency reuse (and that includes most systems), some interference is inevitable. Equipment designers have allowed for interference and have incorporated many elaborate countermeasures into this environment. The symptoms of interference range from dropped and blocked calls to cross talk. To the user, the effect can mean blocking of system access, noise, or even intelligible voice (in analog systems), which is much like a crossed line on a wireline system. Actual crossed lines in the public switched telephone network (PSTN) link or the cellular operators own system can cause a similar effect (analog cross-talk may be intelligible). The system designer also must be aware of the nature, causes, and control of interference to achieve effective frequency reuse.

Most major cities have an interference problem, which is the major cause of customer dissatisfaction and a contributing factor to system cost, since blocked channels cannot be used for service. During off-peak periods interference will not be a problem because the interfering carriers come from the operation of the system itself. It is during the busy period, when the capacity is most needed, that interference limits the system capacity.

About one half of the interference seen is unavoidable, but the other half is due to bad design. This bad design is evident today in almost all major cities. Cellular antennas placed far too high on towers, poles, hill-tops, and buildings are all too obvious. Desperate attempts to counteract poor locations are evidenced by the use of excessive downtilt that makes the antenna assembly look deformed. This technique, while it may reduce interference, leads to the generation of a lot of local dead spots since downtilt reduces coverage. In the case

of excessive downtilt $(15°+)$, dead spots can be generated in the direction of the antenna beam main lobe. The most effective way to reduce interference is to lower the antenna height. The Racal-Vodaphone London site shown in Figure 7.1 is an example of how it should be done! Where reuse is a priority, first get the antennas as low as is consistent with near-end terrain clearance. This is what you will see around London, and there should be a lot more of it in high-density areas.

Most bad design can be traced to the use of inappropriate sites, which were selected because

1. The operators already owned them, and they could readily be established
2. They were more easily obtained than better sites
3. The designer selected sites with a "broadcast" mentality
4. Excessive downtilt

Often the problem is a failure to evolve the network rather than just expand it. When a cellular system is first installed it is usually designed to cover the city with a minimum number of sites and at the minimum cost. Because of this the initial design, which is often done by the suppliers, will include some relatively prominent sites. After the initial installation it is very frequently the case that the design is taken over by less-experienced people who virtually replicate the original site selection.

Poor planning, which fails to allow an adequate time to find, survey, and acquire suitable sites, can lead to a situation where the selection is based almost entirely on *availability* rather than *suitability*.

Figure 7.1 In high-density areas like this London site, it is essential to keep the antennas low.

FREQUENCY REUSE INTERFERENCE

Interference can occur in many ways, but the most significant interference in cellular radio is from a mobile unit to a distant cell on the same frequency. Figure 7.2 shows this form of interference.

Interference is not usually noticed directly by the users, as the safeguards built into cellular systems generally allow the interference to be detected and the affected channel to be temporarily taken out of service. This will happen when co-channel or adjacent channel interference is detected.

Interference is normally assumed to become objectionable at carrier-to-interferer ratios of 18 dB or less on analog systems 12 dB in Global System for Mobile Communications (GSM) and 3 dB in code division multiple access (CDMA). In low multipath environments (such as stationary handhelds) it becomes a problem only at much lower levels. This lower level allows successful operation in a more severe interference environment. In systems with multiple frequency reuse, the interference level at which the channel is blocked can be reduced to increase the traffic capacity, but only at the expense of more noticeable interference.

Cross talk and call dropouts can result from co-channel (same-channel) interference. These types of interference are usually associated with high sites and handheld use in high-rise buildings or mobiles operating from hilltops. For the same reason, operation of handhelds in aircraft is not encouraged. Actually the fiction that mobile phones cause interference to aircraft navigation equipment is an urban myth deliberately spread by cellular operators.

Increasing frequency reuse initially increases system capacity, but as the problems of interference increase, further frequency reuse decreases the system capacity. The objective is to maximize frequency reuse with minimum interference.

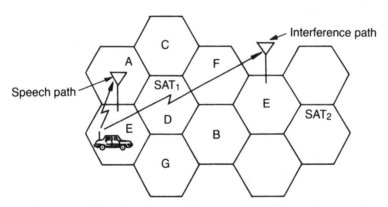

Figure 7.2 The most significant interference is from a mobile unit to a distant cell on the same frequency.

CO-CHANNEL INTERFERENCE

Co-channel interference is the most frequently encountered form of cellular interference, and fortunately it is the easiest to monitor and control. Provided the original survey was done in sufficient detail as described in Chapter 6, there will be sufficient information available to determine the co-channel interference potential. To a lesser extent this problem can be looked at using propagation tools.

Some of the better planning tools will permit detailed studies of co-channel interference using the field survey data. This is superior to using the planning tools to predict co-channel interference, which being far field, will be predicted with a lower accuracy than will be the case for the service area. In most places today the lower valued C/I ratio applicable to handhelds can be used for this study.

ADJACENT-CHANNEL INTERFERENCE

Other forms of interference can also occur, notably adjacent-channel (a channel either one channel space above or below the reference channel) interference. Adjacent-channel interference on the control channel corrupts data and causes call failure.

Adjacent-channel interference can be managed by the frequency planning techniques discussed in Chapter 8, "Cell Plans," on cell planning. This problem often requires reassignment of frequency of one or more cells. The problem occurs from interference between base stations on adjacent channels in certain areas, and it is recognized by high incoming-call (to the mobile) failure rates in certain areas (where the field strengths are sufficiently close to cause data corruption). Except where the problem is most severe, once a call is established, it usually proceeds normally, although some blocking may be noticed.

The two most effective ways to limit adjacent-channel interference are antenna downtilt and antenna height reduction. These methods can restrict the range without drastically reducing near-end power (and hence building penetration).

INTERFERENCE FROM OTHER SYSTEMS

In most countries there will be at least two competitive systems, and while they have been designed to operate in the same region, interference between systems can occur. In the AMPS system it is generally accepted that channels up to four channel numbers apart can cause adjacent channel interference. In these systems there will be two regions where this may potentially be a problem. They are at the A/B boundaries at channels 716 and 717 and the control channels 333 and 334. It would be advisable for operators who have this problem (and that's the majority) to coordinate with the competitor on the use of these channels.

Cellular systems are designed to operate in an environment of interference. One way the system reduces interference is by instructing the mobiles to reduce power when they are sufficiently close to the base station to perform adequately at lower power levels. A similar facility is available on the NMT systems. Table 7.1 shows the power reduction levels in an AMPS system.

GSM requires a significant guard band between adjacent systems and problems often occur when this is not implemented.

It is not unusual in new systems that the cellular band reserved for the system has previously established users in it (often microwave systems). Where these can be identified, the cellular operator must make the necessary arrangements to move the offending systems to other frequencies. This is ordinarily done at the cellular operator's expense (except where the occupiers are illegal).

Some interesting effects occur when unidentified "other users" remain on the band. Most systems in this frequency band are point-to-point microwave links, and therefore operate continuously and are highly directional, thus limiting their area of interference.

Figure 7.3 illustrates an interesting problem that occurred with an AMPS system located near a microwave link (also owned by the cellular operator) that used mobile transmit frequencies corresponding to those used at a neighboring cell. In this example, the problem occurred when a vehicle at cell B requested a handoff attempt. If the vehicle was traveling away from cell B when a handoff attempt occurred, the signal-measuring receiver at cell A would be asked to measure the field strength of the vehicle. It would report a strong signal (mostly due to the microwave link) [because only the radio frequency (RF) carrier level is measured], when the signal was in fact from the microwave system. The switch would then request a handoff attempt to cell A. The mobile would attempt to handshake with cell A on the specified channel and would find no carrier. The call would then fail. If the mobile was roaming toward cell A, an effective handoff could perhaps occur. The solution was to change the microwave link to a new frequency.

INTERMITTENT AND MOBILE INTERFERENCE

This type of interference is the most troublesome because it is very difficult to locate the offending source. There is

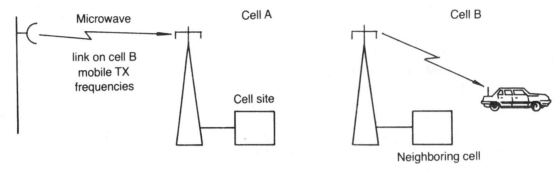

Figure 7.3 A false handoff report can occur if interference on the same frequency as the mobile is detected during a handoff attempt. In this example, cell A measures a signal from a microwave system on the same frequency as the scanned mobile, wrongly reporting ideal handoff conditions.

really no effective way of tracking an offending mobile in the urban environment, except with a great deal of patience and elaborate tracking equipment.

The intermittent nature of these transmissions means that you must spend a good deal of time waiting. Once transmission commences, the most effective method is to use two vehicles, each equipped with direction-finding and two-way communication equipment, to locate the target, preferably by homing in from different directions and using the intersection of the directional vectors to pinpoint the target. Remember, though, that the propagation mode is such that accurate fixes are nearly impossible and that it is necessary that the target transmit for some considerable time (usually hours).

Doppler-effect devices are available that use four receivers and give the bearing of an interferer directly via a digital readout.

Less difficult to locate, but still not easy, are intermittently operated links or repeaters. The first step is to measure the field strength from a number of different cellular bases and then triangulate the position. The next step is to obtain a portable field-strength measuring receiver and, equipped with a Yagi antenna, set out in search of the target. Unless the sources are very close and hence easy to locate, you should allow at least a week to locate an interferer.

INTERFERENCE FROM NON-CELLULAR SYSTEMS

The bands immediately adjacent to the lower cellular bands are variously occupied by ultrahigh frequency (UHF) TV, microwave, and other mobile services. Generally the frequency separation of these systems will ensure immunity (within a few hundred meters), but they can still sometimes interfere due to intermodulation.

Most likely it will be a local high-power transmitter that is likely to cause intermodulation. It is most likely to occur in the receiver multiplexer, the RF preamp of which may become overloaded by a local, strong out-of-band signal, which will cause an interference in the $2f_2 - f_1$ mode (that is a mixing of the second harmonic of the interferer with the incoming wanted signal, which in turn produces an in-band signal). Fortunately this is fairly rare.

Sometimes out-of-band interference can be from reasonably high-powered HF systems, which interfere directly with the cellular base at the multiplexer (MUX) level. This will often be due to earth loops, but may sometimes be due to direct propagation. The author has had experience of one base station that had to be lined with metal inside and out, have all cables screened, and all grounding replaced before system outages due to a nearby high-frequency (HF) transmitter ceased. Even though these measures cured the outages, high levels of RF could still be measured at many different places within the base station.

There are essentially three kinds of external interference. They are

- Transmitter noise
- Receiver desensitization
- Intermodulation

Transmitter noise is caused by a nearby transmitter, which may be adjacent to the channel experiencing interference, and the side-lobes of the transmitter spill over with sufficient energy to cause problems in the receiver. This noise is essentially in-band and so can be reduced only by placing filters at the transmitter to cut down the unwanted emissions or by increasing the distance between the offending transmitter and the affected receiver.

Receiver desensitization is caused by strong out-of-band signals, which appear at the input of the receiver. Because these do not represent any problem at the source, they can be cured only at the receiver. Filtering the offending frequency will often be the easiest solution, but if this fails it will be necessary to increase the distance between the receiver and the desensitizing transmitter. This can often be a difficult thing to identify, particularly when the interferer is intermittent. It should be noted that commissioning tests to measure the receiver sensitivity will probably not identify this problem because it will most likely receive the interfering signal via the antenna, which is not even connected during the sensitivity testing.

INTERMODULATION

Intermodulation will occur whenever two different carriers appear together at a non-linear junction. Typical non-linear junctions or devices that can serve as mixers are

- Corroded connectors
- Bolted joints between tower sections
- Antenna clamps
- Tower guy wires
- Metal fences
- Chains
- Rusted metal
- Drink cans
- Rusted clamps, bolts, debris

The general form of the intermodulation product is

$$I_M = N \times A \pm M \times B$$

where

$$I_M = \text{intermodulation frequency}$$

$$A = \text{the frequency of carrier } A$$

$$B = \text{the frequency of carrier } B$$

N and M are whole numbers

The "order" of the intermodulation product is the sum of $M + N$.

So, for instance, if transmitter A is at 860 MHz and another is at 890 MHz, the third-order intermodulation product $2 \times A - B$ is $2 \times 860 - 890 = 830$ MHz and falls in the AMPS/CDMA band. A higher-order product might be $4 \times A - 5 \times B$, or a ninth-order product.

Experience has shown that the third-order product is usually the most troublesome. This is because a third-order product derived from transmitters on like frequencies is liable to fall into the same band (as it did in the third-order example above). It can easily be shown that second-order products will fall a long way out-of-band unless they are formed by a combination of frequencies far removed from the band affected by the intermodulation (and for this reason it may sometimes be hard to pick).

Products above third order most likely to fall in-band are similarly the odd ones, 5, 7, 9, etc., but as the energy of the product decreases rapidly with the order, it is increasingly less likely to cause a problem.

The intermodulation in a cellular system will almost always be intermittent. This is because channels are mostly only activated as needed, so the intermodulation will only be present when the offending pair of channels are active. A further complication is that the *magnitude* of the intermodulation product causing the problem will vary significantly with changes in level of the carriers causing it (and of course the level of these carriers vary in real time). It may be that the offending carriers are active but the levels are such that the intermodulation product can be missed.

Before an intermodulation problem can be cured it must be isolated. The source can usually be located by using a Yagi antenna and simple field-strength meter. Because it is almost certain to be fairly close, the search area will be limited. Once it is located determine the frequencies involved (with a spectrum analyzer) and exactly where the mixing is occurring. It may be occurring

- At the receiver
- At a transmitter port
- At some other place

A word of caution here! The field-strength meter is operating in an environment where there are multiple transmitters and the signal levels are high. There is a serious risk that the intermodulation that the field-strength meter sees is developed in its own RF stage. If this is the case, the signal may appear to be omni-present. In this case, it is worth trying an attenuator at the field-strength meter input port.

If the mixing occurs at the receiver, it can only be cured by reducing the level of the intermodulating signals reaching the receiver. This will ordinarily be done by filtering, using cavity filters to block the incident frequencies, or by increasing the distance between the transmitter and receiver. This type of interference is most likely to be third-order products from adjacent frequencies that are efficiently delivered to the receiver by the antenna. Receiver mixing intermodulation can easily be identified by placing a 3-dB pad before the re-

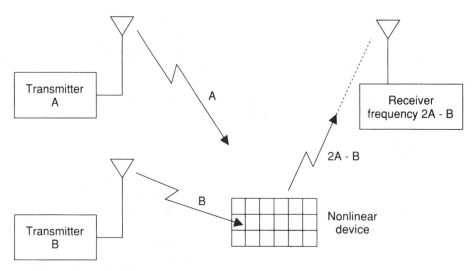

Figure 7.4 Intermodulation produced by a non-linear device.

ceiver [or low noise amplifier (LNA), if it is separate], which will result in a 6-dB drop in the level of inter-modulation if the "mixing" is occurring in the receiver.

Transmitter intermodulation occurs at the transmitter output, or in the combiners. It must be eliminated at the transmitter itself and the most likely way to do this is with filtering.

If the intermodulation occurs elsewhere, the problem is that it may be occurring almost anywhere, or may be at a number of places at once, such as the joints on a tower. If it is localized it will be much easier to track down. This kind of intermodulation will produce an RF signal in-band for the affected receiver (see Figure 7.4), and so no amount of filtering will assist. The source must be tracked down and eliminated.

The search can be a painful process of elimination. The first move should be to clear the site of all debris. A physical inspection of the equipment should be undertaken to check for any signs of deterioration in the antennas, connectors, or cables. Don't overlook metal corrosion of the tower, clamps, and support structures. Almost any part of the systems can be suspect. Connectors, lightning protectors, grounding, cables, and even the filters that have been used to reduce intermodulation can themselves be the culprits.

Some cures that have been known to work include the replacement of rusty mesh fences, tack welding joints on steel plates or sometimes tower sections, improving grounding and eliminating ground loops, and removal of rusty chains and guywires (see also Chapter 10, "Filters").

It is a good idea to do an intermodulation study on all sites; this is especially true if other RF systems are used at the site (whether they are cellular or not). A number of vendors supply intermodulation study software and it is a part of the Mobile Engineer Suite.

INTERFERENCE BETWEEN AMPS AND GSM BAND SYSTEMS

AMPS, CDMA and GSM band systems (which includes NAMPS/TDMA/E-TDMA/GSM) do not co-exist well, mainly because they have the base transmitter (TX) and receiver (RX) back to front with respect to each other. AMPS/CDMA systems have base station TX frequencies ranging from 870 MHz to 890 MHz. The lower GSM base-station receive frequencies intrude into the AMPS TX band. Obviously, where the two are operating together some agreement will have to be reached on spectrum sharing so that no intrusion occurs. It is also advisable to have a guard band of at least 2 MHz to 4 MHz.

For co-located sites it will be necessary for the GSM system to have sharp filters to keep out the AMPS TX. It has also been found that for co-located GSM/AMPS systems it may be necessary to put filters in the AMPS receivers (as well as the GSM receivers), and this may occasionally be necessary when AMPS/GSM combinations are received.

It is worth noting that the reverse problem may occur when mobiles are used in close proximity. A TACS/GSM mobile operating in the vicinity of an AMPS mobile *may* be transmitting on frequencies very close to the AMPS receiver, and so may cause desensitization. In this case, there is little that can be done about the problem. For more detail on this problem see Chapter 10.

INTERFERENCE INTO NON-CELLULAR SYSTEMS

One of the most common fears of a potential base-station site landlord is that the cellular system will interfere with his TV reception. Although this fear is common, the situation is very rare. Harmonic filters and precautions to minimize intermodulation have made actual cases of interference to third-party equipment rare. Of course, TDMA mobile phones tend to interfere significantly with nearby equipment, but this is because of their close proximity and it should not be inferred from this that base-station interference also will result.

IMPROVING FREQUENCY REUSE

In small cities and rural areas, frequency reuse will probably rarely be a problem, so prominent high sites can be used to maximize coverage. In urban areas, however, every effort must be made to maximize frequency reuse.

Reuse is enhanced by using lower N numbers, and for example reducing the cell pattern from $N = 7$ to $N = 4$ will improve the frequency reuse, but will place tighter constraints on frequency planning. Of course, the ultimate in this game is CDMA where an $N = 1$ plan is the norm.

Lowering base station heights is probably the most effective way to ensure good frequency reuse. Down-tilting is widely used, but few designers seem to understand it and it can cause more problems than it solves, particularly by generating unexpected "dead" zones, and at the same time doing little to reduce interference.

The easy cop-out is to use a dual-mode design, using personal communication service/personal communica-

tions network (PCS/PCN) in the high-density areas and 800/900 MHz in the suburban and rural areas.

BLOCKING

The easiest way to determine the magnitude of frequency reuse problems is to look at channel blocking information (channels out of service due to interference). This information is normally available from the system housekeeping data; it should be kept below 5 percent.

You can choose a trade-off between blocking (which may mean lost calls) and increased interference (which occurs as cross-talk and spurious noise). This trade-off is controlled by adjusting the C/I limit for blocking, which is a user-definable system parameter.

USE OF TERRAIN AND CLUTTER

The primary aim of a cell site is to provide coverage for its service area. With careful planning, it is possible to achieve this goal while minimizing cell coverage outside the intended service area. Using local terrain and clutter can be an effective way of containing a cell. Effectively using local terrain and building obstructions can greatly enhance frequency reuse, if base sites are placed so that far-field propagation is reduced. Placing a base station behind a hill greatly increases the chances of successful frequency reuse. Similarly, placing the base station so that buildings form a natural shield can also be very effective (see Figure 7.5).

In very high-density areas, effective frequency reuse is more critical than coverage.

USE OF SECTOR ANTENNAS

Sector antennas are important tools in minimizing interference. Figure 7.6 illustrates the basic principle of sector antennas. In this example, consider a mobile roaming

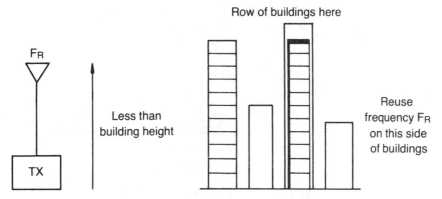

Figure 7.5   Use of obstructions to enhance frequency reuse.

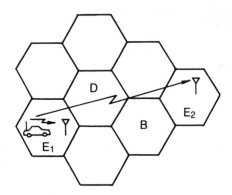

Figure 7.6   Omni-directional interference mode.

from cell E_1 to E_2, both of which use the same frequencies. If the cell sites are omnidirectional and E_1 is relatively high, the mobile might roam well into cell D before a handoff occurs. All the time it is getting closer and closer to E_2 and the probability of interference is increasing.

As shown in Figure 7.7, sectoring E_1 and E_2 can improve things considerably. When the vehicle begins its journey (at the maximum distance from E_2), the situation is the same as for the unsectored site. However, when the mobile passes the base station on the way to E_2, a handoff is initiated to cell E_B. The mobile is now using cell E_B at site E_1, but is behind the antenna for cell E_B at site E_2. This gives the added protection of the effective front-to-back ratio of the antenna (usually 10–30 dB) against interference. Note that the effective front-

to-back ratio is almost always less than the value quoted by the antenna supplier (which is true in free space only). When the mobile finally hands off to cell E_2, it uses cell E_A and again interferes with cell E_1 only via the back of the antenna.

Interference protection can thus be increased by use of sectored antennas and the degree and type of sectors. Antenna sectors as narrow as 60 degrees can be used effectively.

7-CELL PATTERNS

The 7-cell sector is a commonly used pattern for sectorized cell sites. Such cell sites are often recognized by the characteristic triangular mounting platform seen in Figure 7.8. This sector site has sectorized transmit and receive antennas. There are a few variations on this pattern. In its simplest form both transmit and receive antennas are 120 degree, and there is one receive antenna. More commonly two receive antennas are used for diversity.

4-CELL PATTERNS

It can be shown theoretically that significantly higher cellular densities can be obtained using a 4-cell pattern rather than a 7- or 12-cell pattern. In practice, the gains are not as large as they are in theory, but 4-cell patterns are measurably more efficient than 7-cell patterns.

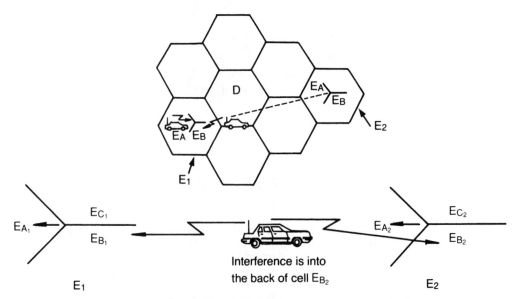

Figure 7.7   Sectoring to enhance frequency reuse. Antennas are orientated in the same direction for maximum interference protection.

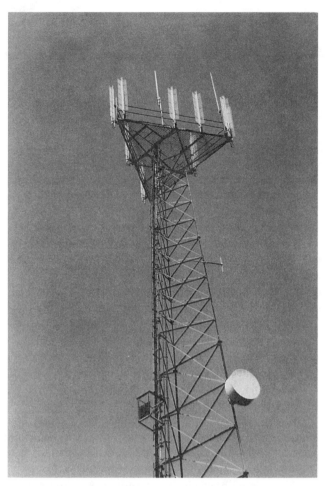

Figure 7.8 A 7-cell sectorized site, which has the familiar triangular mounting platform.

Figure 7.9 A British Telecom (Cellnet) base station in County Down, Northern Ireland.

The 4-cell pattern requires more critical placement of cells sites; its efficiency is reduced if this cannot be attained. Despite these limitations, it is generally more efficient overall in Erlangs/km to use 4-cell patterns and in customers/circuit.

The 4-cell sector pattern in the Motorola version, uses six 60-degree sector antennas mounted so that a full 360 degrees is covered. In this proprietary system, the transmit antenna is an omni, and a pooling of the receivers over the sectors allows for a traffic efficiency approaching that of an omni base. This system is particularly neat for the installer and can be quite aesthetic as seen in the example in Figure 7.9.

CHANNEL BORROWING

The channel-borrowing technique allows dynamic channel reassignment of a channel in a cell that has some of its "normal" channels blocked due to interference.

It assumes that channels can be retuned remotely on command. Channel borrowing can increase traffic densities by 10 to 20 percent, but it is costly to implement and uses a good deal of processing power. The technique is sometimes called "Dynamic Channel Assignment."

POWER REDUCTION

Reducing the emitted power of a base station will decrease the transmitter range, but it must be appreciated that the most severe cause of interference is from the mobile to the base receiver. Because of this it may sometimes be necessary to reduce the receiver sensitivity if channel blocking is not to occur. While it is the C/I ratio and not the absolute field strength that determines interference on a channel, it is still possible for a distant carrier to block out a free channel at relatively low absolute RF levels. The system parameters can first be used to increase the level at which this blocking occurs, but in very high levels of interference it may be necessary to introduce some attenuation in the RF receive leads.

ANTENNA HEIGHT

Simply mounting the antenna at a lower level is probably the most effective way to decrease the effective cell radius and increase the chances of frequency reuse. When cells are located on buildings, it is generally very difficult to reduce the antenna height; you should consider this when selecting building sites. However, modern panel antennas blend in with the building well and can be mounted flush with the building walls at any height.

Tower-mounted antennas, on the other hand, have a wide scope for varying the level of the antenna and even of locating different sectors at different levels. The conventional triangular mounting structure for cellular antennas presuppose all sectors mounted at the same level. This is often appropriate, but in some cases (particularly where the local terrain is relatively flat) it can restrict the degrees of freedom available to the designer.

As discussed earlier, downtilt is effective against co-channel and adjacent-channel interference, but it is not particularly effective against mobile-to-base station problems. For more information, see "Effective Use of Downtilt," later in this chapter.

COPING WITH INTERFERENCE

In all systems with significant frequency reuse there will be some interference. It will become evident to the system operator in the form of "foreign carrier detected" alarms. This will usually identify both the affected base station and channels. Once the interference becomes severe [ranging between 2 percent in areas of moderate frequency reuse to 5 percent in major central business district (CBDs)] remedial action is needed.

From the alarm reports, it should be possible to identify the channels that are experiencing interference most frequently. The next step is to go to the base station with a spectrum analyzer to determine the magnitude of the problem. The channels should be examined one at a time (unless equipment is available to look individually at several channels) and during the peak hour. The channel(s) under study should be removed from service for the testing period so that any signals measured will be interference originated.

Both the signal strength of the interference and its source need to be determined. It will be possible to get an idea of the level of the interference by using a spectrum analyzer. However, as the most likely source of RF is mobile originated, it can be expected to be erratic and transient. This means that the utility of the spectrum analyzer is limited.

A more effective way of determining the level of interference is to monitor the signal level over a few days with a digital monitor that registers the average level of interference and its duration. Naturally the averaging would not include the times of no interference.

Once the level of the interference has been established, you will know how much it needs to be reduced to bring it to a tolerable level (which is usually taken to be a level that produces less than 2 percent blocking).

The options that can be tried to reduce the interference are

- Reduce the antenna height
- Reduce TX power *and* RX gain
- Use downtilt

If it is possible the best option is to reduce the receive antenna height. Reducing the height by 50 percent will reduce the average received level by around 6 dB. So if the antenna was mounted at 30 meters, then lowering the receive antenna to 10 meters will reduce the interference by 12 dB.

It is preferable to also reduce the TX antenna height, but this will not result in the level of improvement that can be achieved by lowering the receive antenna. It is, however, important to maintain the system balance.

The ease with which the antenna can be lowered will depend a lot on the structure on which it is mounted. Towers are ideal, but some problems could be experienced with monopoles and buildings, both of which are best suited to top-mounting.

Monopoles can easily be designed to allow for future lowered mounting levels, but it may be difficult to convince the average building owner that six or so sector antennas would not look too out of place half-way down the building. In this case perhaps TX/RX power reduction is the answer.

The TX power can be reduced by around 10 dB using the in-built level controls, although the actual range of power reduction will be manufacturer-specific. Most manufacturers have a range of TX power amplifiers typically spanning 5, 25, 50, and 100 watts. It may be necessary to fit a lower power unit. Notice that the alternative of placing attenuators in the line is not cost-effective in the long run because of the effect of the dissipated power, as seen by the increased power bill.

RX received levels can easily be reduced by 3 dB to 6 dB by simply removing the diversity (with the actual reduction being dependent on the type of diversity used). However this should only be done as a temporary expedient. Additional reductions can easily be achieved by placing attenuators in the line between the first RF distribution amplifier and the antenna. However, fitting a simple attenuator is the recommended way.

It is not recommended that lower gain antennas be used to reduce the RX/TX levels, as low-gain antennas have wide lobes and so can receive interference from more sources. It is best to stay with high-gain antennas if possible.

While on the subject of antennas, one possible source of poor performance in terms of dropped calls—which may not in fact be due to interference, is the case of high-gain antennas mounted less than rigidly on the supporting structure. As the antenna moves both vertically and horizontally it will move the coverage pattern with it. Mobiles operating on the fringe of the coverage area may move intermittently in and out of coverage, resulting in dropped calls. Antennas with gains of 13 dB plus are routinely mounted by securing one end only. This probably will not cause problems when downtilt is not used, but as will be seen later in this chapter even small amounts of downtilt can cause big reductions in service area. Where antennas have significant downtilt and their mounting allows them to move in the wind, then it is possible that mobiles operating on the edge of the coverage zone can be flicking in and out of service with every gust.

When downtilt is used make sure the assembly is rigid!

The third of these options, downtilt, is the least effective way of interference reduction. Downtilt is effective in reducing the coverage area and so increasing the cellular density. Interference is, however, predominantly into the receive antenna because the angle of incidence of interference is erratic (the more distant the source the more likely it has been reflected or refracted). It will not be significantly reduced by antenna tilt.

ANTENNA TYPES USED IN CELLULAR RADIO

The most commonly used antenna patterns are illustrated in Figure 7.10.

HANDHELD BENEFITS

Using sectored antennas can improve handheld coverage by overcoming, to some extent, the limited talk-out power of a handheld. This is due to the additional antenna gains achieved with sectored antennas.

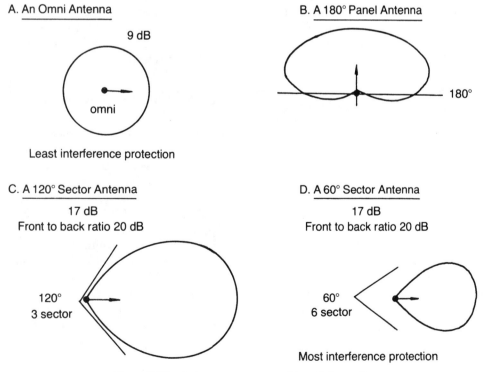

Figure 7.10   Antenna patterns used in cellular radio.

LEAKY CABLES

Using leaky cables to feed buildings, tunnels, and large complexes from inside rather than from an outside antenna provides better-quality service and improved frequency reuse, because the building itself is an effective shield against interference. This technique is more attractive as cells get smaller.

SYSTEM PARAMETERS

Cellular systems have a number of system parameters that can be modified to alter the performance of a cell (although it is often attempted). The following parameters can improve frequency reuse:

- Signal level for handoff. Raising this level reduces the effective size of the cell.
- Signal-to-noise level before handoff is attempted. Increasing this level causes handoff to occur earlier.
- Signal-to-noise (S/N) level before releasing a call. This can be used to extend or contract the effective range.
- Level at which mobile is instructed to increase/ decrease power. This can be effective in small cells. Causing earlier power reductions reduces interference.
- Signal-blocking level is the level at which interference blocks channels. Increasing this level can reduce congestion due to channel blocking, but if it is adjusted too low, it can lead to data corruption and hence lost calls.

These parameters are loaded into the base stations and can be adjusted by a simple data entry. Each base should be studied separately to determine the usefulness of changes in these levels. Those most likely to benefit are those in congested areas, those experiencing interference, those requiring extended range, and those from which interference originates.

EFFECTIVE USE OF DOWNTILT

Downtilt is one of the most frequently used techniques for producing small cells and achieving good frequency reuse. It works well if used only on small cells (about 500 meters radius) and with carefully calculated downtilts. It is not effective for large cells (although it is often attempted).

Downtilt effectively reduces

- Cell coverage to provide for higher traffic densities
- Co-channel and adjacent-channel interference to other cell sites
- Channel blocking from co-channel mobiles

However, downtilt does not improve handheld coverage and may in fact reduce the effectiveness of handhelds in the normal service area of the cell.

Small and new systems generally should not use downtilt until it is proven necessary because of the attendant range reduction of the cell and reduced handheld performance even at moderate ranges.

Large angles (greater than 5 degrees) of downtilt are rarely justified and will usually not be effective except in very small cells. Before downtilting an antenna, determine its original coverage (by survey) and after applying the calculated downtilt, resurvey to be sure that patchy coverage has not resulted from the change. Downtilt will decrease the far-field strength dramatically; the designer must be aware of the consequences when employing this technique.

Inadvertent downtilting of high-gain omni-antennas (greater than 6 dBd) of more than a few degrees can cause significant shortfalls in performance. This can occur when the installer fails to properly align the antenna, or when the installation is poor and the antenna tilts mechanically with time. Storms and other external influences can also cause such problems. Those familiar with lower-gain antennas (6 dBd and less) are aware that this problem is not very pronounced with such antennas. Caution is needed with any antenna that has a gain greater than 6 dBd and is tilted.

DOWNTILT AND HOW IT WORKS

For the purposes of illustration, consider an idealized corner reflector with a pattern as shown in Figure 7.11. Figure 7.12 shows the radiation pattern that occurs when this antenna is downtilted by B. From Figure 7.11 you can see that the far-field ERP (that is, the horizontal ERP) is now H. H is normally expressed as a fraction of the field strength measured as the E or H components with respect to the value along the main axis (that is, $0 \le H \le 1$). Thus, the far-field attenuation by downtilt is $20 \log H$ dB.

The near-field (and the far-field) field strength in practice is highly dependent on local clutter, particularly in high-density environments. For the purposes of this idealized study, we must, for the time being, assume zero clutter.

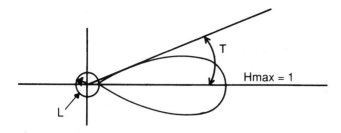

T is the angle of the tangent to the main lobe.

L is the gain outside the main lobe.

Figure 7.11   Vertical radiation pattern of a gain antenna.

The downtilted antenna will always have a relative gain over a vertically mounted antenna in the middle field. This can be seen in Figure 7.13.

Now, consider the gain of the downtilted antenna compared to the vertical one measured in the horizontal plane. As a function of the Angle M (the angle from the antenna to the receiver)

The gain at $M = 0°$

Vertical antenna $= L$

Tilted antenna $= H$

Relative gain of a tilted antenna in dB $=$
$20 \log H - 20 \log L = 20 \log H$

Notice that $H \leq 1$, so the relative gain of the downtilted antenna is negative or zero at $M =$ measured at any other angle M, the difference in gain equals

$$20 \log G_V - 20 \log G_T$$

where

$G_V =$ gain of a vertical antenna measured at $M°$
to the horizontal. (see Fig 7.13)

$G_T =$ gain of a tilted antenna measured at $M°$
to the horizontal.

$B =$ the angle of downtilt.

You can see in Figures 7.14 and 7.15 that this difference (the relative gain of the downtilted antenna) increases as M increases and equals zero at the value of M where the two patterns intersect symmetrically (point I). This angle is such that $M = B/2$. The downtilted antenna has a relative positive gain that peaks and finally equals zero at the angle $B + T$, where $T =$ the angle to the lobe tangent.

The important points to note are

1. Far-field loss $= 20 \log H$ (and hence interference protection factor $= 20 \log H$).
2. Handheld gains are only realized in the angular range $B/2$ to $B + T$; that is, at an angular range of $h/\tan(B/2)$ (where $h =$ base station height). Outside that range the field strength decreases. The problem of downtilt is now apparent.

If reasonable handheld range is sought, $B/2$ ($B =$ downtilt angle) must be small. However, a small $B/2$ gives a low far-field loss (and hence poor interference immunity) unless the beam width is very narrow.

Thus, the interrelated parameters are height and downtilt. Table 7.1 shows the effect of height and downtilt on small cell operation. Because of these conflicting factors, where frequency reuse is important, it is better to use lower antenna heights and less downtilt than higher antennas and greater downtilt. Remember also that the interference received from the back of the antenna increases with height gain and is virtually independent of downtilt. This is a big problem in high-density areas.

To estimate the trade-off, note that

$$\text{height gain} \propto 20 \times \log (\text{height})$$

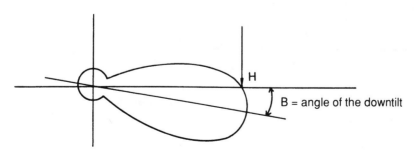

Figure 7.12   Pattern of a downtilted antenna.

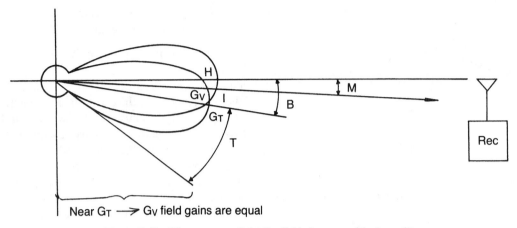

Near $G_T \longrightarrow G_V$ field gains are equal

Figure 7.13 The pattern of the far-field changes with downtilt.

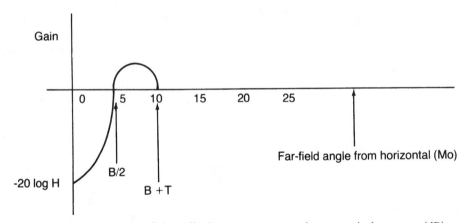

Figure 7.14 Relative gain of downtilted antenna compared to a vertical antenna (dB) as a function of angle of elevation to receiver.

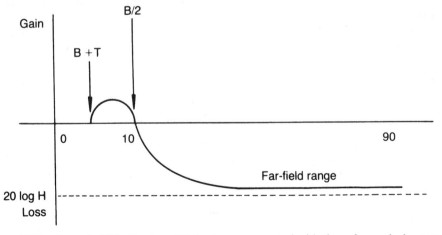

Figure 7.15 The gain (dB) of a downtilted antenna compared with that of a vertical antenna (dB) as a function of distance from the antenna.

EXAMPLE   **101**

TABLE 7.1   Effect of Height and Downtilt on Small Cell Operation

| Parameter | Height and Downtilt Effect on Small Cell Operation | |
	Pros	Cons
Height	Increased height will increase far-field losses by use of greater downtilt (for same range).	Increased height will increase far-field interference by height gain. Increased height will also reduce interference immunity of base-station receivers.
Angle	Increased downtilt will increase far-field losses.	Increased downtilt will reduce handheld range within the service area.

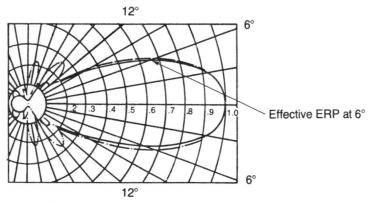

Figure 7.16   Radiation pattern of typical 17-dB unidirectional antenna.

and far-field loss can be calculated by first determining the handheld range required, then using $(B/2) = \arctan$ (height/range) (where B = downtilt). From B, use the radiation pattern to find H = far-field loss.

EXAMPLE

Consider the antenna pattern in Figure 7.16. The objective is to achieve good handheld coverage at 1 km from a 30-meter site. As shown in Figure 7.17, you can see that

$$B/2 = \arctan(30/1000) = 1.7°$$

$$B = 3.4°$$

The far-field strength reduction available from this antenna would be

$$20 \times \log H = 1.1 \text{ dB}$$

which is hardly worth having.

Also note that at about 10 degrees downtilt there is a null in the direction of the main lobe, meaning that any

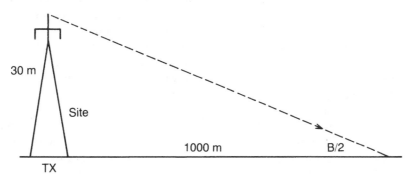

Figure 7.17   Range calculation for a downtilted antenna.

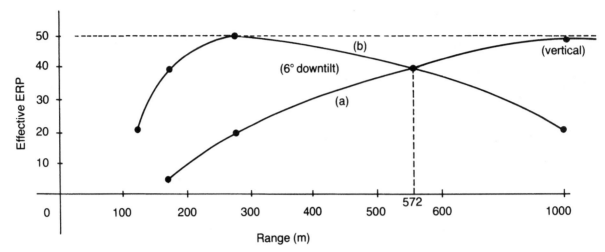

Figure 7.18 Effective ERP versus distance from 30-meter transmitter for curve (a) vertical antenna and (b) antenna at 6° downtilt.

downtilt beyond 10 degrees means that the radiation in the forward direction is from the side lobes only.

Same Antenna at 60 Meters

The same antenna at 60 meters can be shown to require a downtilt angle of 6.8 degrees (to achieve the same 1-km range) and a far-field loss of 8.6 dB. The height gain is 6 dB, and the net far-field loss is 2.6 dB.

Notice, however, that the 6-dB height gain is also added to the gain of the antenna in the backward direction. Thus, the antenna is significantly more vulnerable to receiver interference and hence to channel blocking in high-density areas.

Although larger cells from high sites do not really benefit much from downtilt, smaller cells with moderate elevations do. Consider the field-strength plot of the original 30-meter site with 6-degree downtilt, as shown in Figure 7.18. You can see that the two systems have an equal field strength at a distance of only 572 meters ($B/2 = 3°$). At smaller distances, the downtilted antenna has a relative gain, but except for very marginal places (like elevator shafts and basements), most areas within 570 meters are likely to be well covered without downtilt. Any improvement in coverage close to the antenna and attributable to downtilt is marginal.

The most useful effect of downtilt is the relative reduction in the far-field strength (in this case, to 21 watts ERP or 3.8 dB, which is still not substantial). Notice, however, that one more degree downtilt (to 7 degrees) increases the far-field loss to 9.1 dB, while decreasing the equal field-strength range to 490 meters.

TABLE 7.2   The Range to the Center of a Beam for an Antenna at Various Downtilts and Heights

Downtilt (degrees)	Antenna Height		
	30 m (m)	50 m (m)	70 (m)
0	Infinity	Infinity	Infinity
2	860	1400	2005
4	430	710	1000
6	280	475	660
8	210	350	500
10	170	280	400

DOWNTILT IN PRACTICE

Downtilt is most effective for very small cells (of about 500–1000 meters radius) when used in combination with downtilt angles that approach half the beam width of the vertical propagation pattern. Downtilt provides greater opportunity for frequency reuse but not improved handheld coverage, as is often mistakenly believed.

Downtilt should be the method of last resort to improve frequency reuse. It will severely limit range, as can be seen from Table 7.2, which shows the range of the downtilted beam center point as a function of height and angle of tilt.

Interference reduction is achieved by careful cell-site selection, antenna height and handoff control, frequency planning, and more recently some exotic techniques like dynamic channel allocation. Downtilt should rarely exceed 10°. Unless considerable care is taken, it is likely to lead to local dead spots in coverage. *Use it with care!*

8

CELL PLANS

The objective of a cell plan is to cover the service area as economically as possible, while allowing for maximum flexibility for future frequency reuse. As an aid to visualizing frequency patterns, the hexagonal cell plan has been devised. In the real world, however, radio waves do not propagate to hexagonal boundaries, and cells often overlap. Indeed, cell overlap is sometimes an intentional part of the pattern design. This real-world untidiness means that some conceptual mental gymnastics may be necessary to relate the theory to practice.

BASIC CONSIDERATIONS

The most common cell plans involve splitting the available spectrum into 3, 4, 7, or 12 frequency cells, each using a hexagonal grid to define the relative positions of the base sites. Such cell plans define a "tidy" and systematic frequency-reuse pattern, although most cellular systems do not *require* the use of such patterns. It is possible to devise a workable system that is virtually "non-cellular"; that is, a system that uses large numbers of channels at one or two sites or that is cellular but has very irregular channel configurations. Some designers plan a non-cellular system to meet objectives such as lower cost. As discussed later, in "Non-Reuse Plans," this technique should be reserved for small remote service areas.

Because most cellular systems are prone to adjacent-channel interference, the first requirement of a system is that adjacent channels, as far as possible, should not be located in neighboring cells (for example, those to which a handoff can be attempted).

Cell site selection should not be unduly influenced by

something as hypothetical as the hexagonal cell plan (see Figure 8.1). Use the cell plan to help determine the most suitable frequency to be used at a site selected by taking into account real-world constraints.

Theoretically, cell plans with the smallest number of cells can achieve the highest subscriber densities but are more sensitive to interference and hence to cell site location and design. For example, a 4-cell plan cannot avoid adjacent-channel interference without using sector antennas (if frequency reuse is contemplated). For this reason, it is not unusual, even where a 4-cell plan is the ultimate goal, to start with a 7-cell plan, which is more flexible and can even incorporate some omni bases.

A 7-cell plan has some cells that are both physically adjacent and have adjacent channels. The adjacency problem can be confined to three cells only, and sectoring these three may be necessary. In smaller systems and in areas of irregular terrain it is often possible to avoid sectoring altogether. The 12-cell plan is designed to use omni-directional cells and does not have the immediate adjacency problem.

Because each of these cell plans is based on a hexagonal grid, it is possible to change from one cell plan to another by simply rearranging frequency assignments. Doing so, of course, may not be so simple on a working system.

7-CELL PATTERN

A common cell plan uses 7 cells, which can be further subdivided to form a 21-cell (sectored) plan. Figure 8.2 shows a commonly adopted 7-cell pattern. As you can see from Figure 8.2, despite an effort to separate adjacent channels, the pairs D, C and D, E are still adjacent.

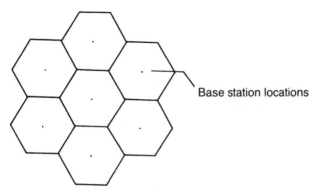

Figure 8.1 The hexagonal grid where each cell (hexagon) contains a group of frequencies.

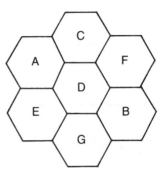

Figure 8.2 Cell distribution. Cells that are alphabetically adjacent are also adjacent in frequency (note that *A* is also adjacent to *F*).

Careful frequency planning can substantially reduce the cost of a new installation. For example, the adjacent channel pairs D, C and D, E will probably ultimately require that the sites be sectored for separation. For relatively small systems, however, sectoring in the early stages can be avoided. These sites can be installed as omni-directional sites provided that lower frequencies are assigned to the D cell and higher frequencies are assigned to the C and E cells (or vice versa).

4-CELL PATTERN

The 4-cell pattern is really a 4/12 or 4/24 pattern. That is, the cells are grouped into four frequency plans with either 12 or 24 sectors. The 4-cell sector pattern uses either six 60-degree antennas or three 120-degree antennas at a common site. This arrangement produces the 24-sector (6 sectors by 4 cells) and 12-sector (3 sectors by 4 cells) patterns, respectively. Figure 8.3 shows the 24-cell pattern.

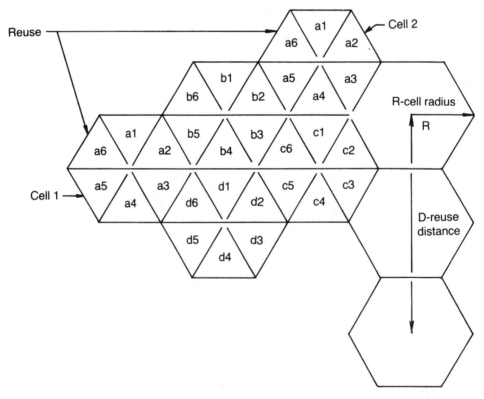

Figure 8.3 4-Cell sector pattern (24 cells).

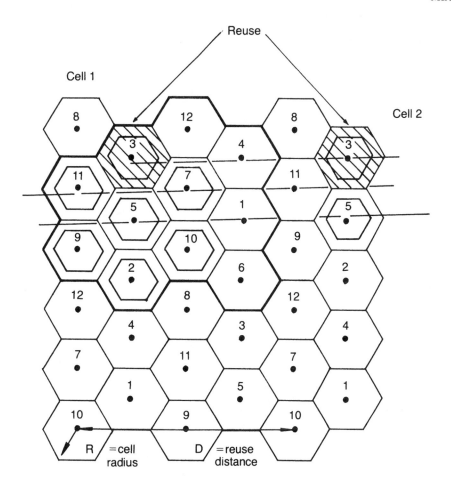

Cell-site location

Figure 8.4   12-Cell pattern.

12-CELL PATTERN

Less commonly used is the 12-cell pattern, which has good carrier-to-interference ratios even when the sites are omni-directional. The individual cells in this configuration are very small, however, and the traffic efficiency is correspondingly low. Figure 8.4 shows the 12-cell pattern.

THE STOCKHOLM RING MODEL

The Stockholm Ring model has been used on NMT450 systems, originally in Stockholm, and later in Jakarta, Malaysia, and Bangkok. This technique was designed to respond to high traffic densities in central Stockholm. Using quite different principles from the conventional hexagonal pattern, the Stockholm Ring attempts to increase the CBD capacity by using all available frequen-

cies at a central site. NMT450 has 180 frequencies and the plan divides those channels into 6 by 60-degree co-sited sectors, as shown in Figure 8.5. Co-siting all frequencies implies that good adjacent-channel separation is possible. This requirement, of course, limits this technique to the low-deviation systems, NMT450 and NMT900 (without channel interleaving only).

Figure 8.6 shows the next expansion outward, using another concentric ring with all the frequencies rotated 120 degrees. As cell expansion occurs away from the central business district (CBD), wider antenna patterns accommodate the lower densities efficiently.

MIXED PLANS

When a 4-cell plan is used to obtain high density, it can be economical to design the network to gradually taper to a 7-cell plan at the city boundaries, and maybe evolve

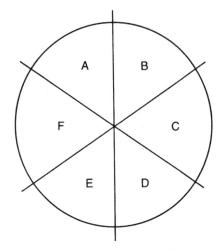

Figure 8.5 Inner cell shown with 6 by 60-degree sector cells using all 180-degree frequencies.

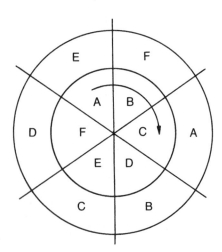

Figure 8.6 First expansion pattern.

into a 12-cell plan farther away. There are no standard techniques for carrying out this type of graduated plan, but careful frequency allocation and a few fortuitous hills can make this technique a little less daunting.

It is not necessarily optimum to follow any of the conventional frequency plans. In fact, those plans can be regarded as rules for simplifying the frequency management. An optimum allocation need not be constrained by such rules. There is a limitation on how close frequencies can be at any one site, but this is a function of the cavities and coupling devices, which are designed to isolate the channels. Although combining devices exist that can couple channels together no matter how close they are in frequency, a penalty is paid in terms of loss. The closer the frequencies to be combined are to each other the greater the loss. This means that plans with lower N numbers will have more radio frequency (RF) power losses.

A truly optimum plan will be limited only by the number of channels that can be placed at a site, as determined by the combining equipment specifications. In a large system it is possible to allocate the channels on the basis of traffic and then to allocate frequencies in a manner that minimizes the interference.

One method of doing this has been pioneered in Hong Kong by Motorola. The planning is done by a semi-intelligent computer program that optimizes the frequency reuse on the basis of the least interference. The computer has the measured interference matrix. It then looks at all the possible frequency uses at each site and determines the net interference. This is done using some algorithms that speed up the process (it is not practical to consider the set of all possible frequency allocations),

but also introduce a degree of randomness into the process so that successive runs on the same data will not necessarily produce the same result. It should be noted that this model has no need of concepts like frequency plans or N numbers as it is attempting to find the ultimate cell plan for the given base station configuration. Here we have the ultimate mixed plan. It is said to produce a network with about 10 percent more capacity than can be achieved with conventional planning. However, the model is still being refined.

Increasingly the need for greater density is driving the use of non-formal planning. Because the real-world radio propagation does not occur in hexagonal patterns, any system that supposes that they do must be non-optimum.

Planning tools often have the facility to find and/or use sites that bear no relationship to any formal plan. Complete frequency reuse at all sites is currently available on some systems (supplier dependent), where by the frequencies are dynamically allocated in real time [this is in fact a specification in Digital European Cordless Telecommunication (DECT)].

AMPS

There are some rather complex considerations in cell planning for Advanced Mobile Phone Service (AMPS) systems and these are considered here in detail. If each cell had seven-channels, they could be allocated as shown in Table 8.1. Such frequency planning works only for bases that are at less than 50 percent of maximum

TABLE 8.1   Channel Allocation

Cell	D1	C1	E1
Channels	4	192	194
	25	213	215
	46	234	236
	67	255	257
	88	276	278
	109	297	299
	130		
Control channel	316	315	317

Channel allocation of low channel numbers for the D cell and high channel numbers for the C and E cells will avoid adjacency problems for small systems.

channel capacity, and it can delay the need for expensive sectoring for a few years.

Some base stations can be programmed to preferentially choose odd or even channel numbers so that adjacency problems are unlikely to occur whenever less than half the channels are in use. These two techniques —cell subdivision and use of odd or even channels—can be combined.

Sectoring further subdivides channel groups and prevents the worst effects of adjacent-channel interference. Tables 8.2 and 8.3 show the channel groups subdivision.

As shown in Figure 8.7, once the site is sectored, the adjacency problem is lessened by the antenna patterns.

Sectoring can be visualized to occur either at the cell edge (as shown in Figure 8.7) or at the cell center (as shown in Figure 8.8). Figure 8.9 shows a cell split based on site D by maintaining the central cell as a D cell while rotating the rest of the pattern counterclockwise by 120 degrees. The new cells have sides one-half the size of the original cells.

SAT CODES

An AMPS system has Supervisory Audio Tones (SAT) codes (6-kHz tones of frequencies 5910, 6000, and 6030) that are transmitted on the speech channels. These tones are used to identify the local cell traffic from adjacent interfering traffic. The SAT is generated by the base station and looped back via the mobile circuitry.

Allocating SAT codes is very simple, following the rule that adjacent cell clusters have different SATs. There are only three SATs: SAT 0, 1, and 2. As Figure 8.10 shows, each cluster intersects only two other cells.

At the cell junctions (marked with triple lines) shown in Figure 8.10, the condition that all adjacent cells must have a different SAT is sufficient to determine all SAT assignments once any two have been designated.

TABLE 8.2   AMPS A-Band

	A_1	B_1	C_1	D_1	E_1	F_1	G_1		A_2	B_2	C_2	D_2	E_2	F_2	G_2		A_3	B_3	C_3	D_3	E_3	F_3	G_3
Control Ch	313	314	315	316	317	318	319		320	321	322	323	324	325	326		327	328	329	330	331	332	333
Voice Ch	1	2	3	4	5	6	7		8	9	10	11	12	13	14		15	16	17	18	19	20	21
	22	23	24	25	26	27	28		29	30	31	32	33	34	35		36	37	38	39	40	41	42
	43	44	45	46	47	48	49		50	51	52	53	54	55	56		57	58	59	60	61	62	63
	64	65	66	67	68	69	70		71	72	73	74	75	76	77		78	79	80	81	82	83	84
	85	86	87	88	89	90	91		92	93	94	95	96	97	98		99	100	101	102	103	104	105
	106	107	108	109	110	111	112		113	114	115	116	117	118	119		120	121	122	123	124	125	126
	127	128	129	130	131	132	133		134	135	136	137	138	139	140		141	142	143	144	145	146	147
	148	149	150	151	152	153	154		155	156	157	158	159	160	161		162	163	164	165	166	167	168
	169	170	171	172	173	174	175		176	177	178	179	180	181	182		183	184	185	186	187	188	189
	190	191	192	193	194	195	196		197	198	199	200	201	202	203		204	205	206	207	208	209	210
	211	212	213	214	215	216	217		218	219	220	221	222	223	224		225	226	227	228	229	230	231
	232	233	234	235	236	237	238		239	240	241	242	243	244	245		246	247	248	249	250	251	252
	253	254	255	256	257	258	259		260	261	262	263	264	265	266		267	268	269	270	271	272	273
	274	275	276	277	278	279	280		281	282	283	284	285	286	287		288	289	290	291	292	293	294
	670	671	672	673	674	675	676		677	678	679	680	681	682	683		684	685	686	687	688	689	690
	691	692	693	694	695	696	697		698	699	700	701	702	703	704		705	706	707	708	709	710	711
	712	713	714	715	716	717																	
							991		992	993	994	995	996	997	998		999	1000	1001	1002	1003	1004	1005
		1006	1007	1008	1009	1010	1011	1012		1013	1014	1015	1016	1017	1018	1019		1020	1021	1022	1023		

TABLE 8.3   AMPS B-Band

	A₁	B₁	C₁	D₁	E₁	F₁	G₁	A₂	B₂	C₂	D₂	E₂	F₂	G₂	A₃	B₃	C₃	D₃	E₃	F₃	G₃
Group	1	2	3	4	5	6	7	8	9	10	11	12	13	14	15	16	17	18	19	20	21
Control Ch	334	335	336	337	338	339	340	341	342	343	344	345	346	347	348	349	350	351	352	353	354
Voice Ch	355	356	357	358	359	360	361	362	363	364	365	366	367	368	369	370	371	372	373	374	375
	376	377	378	379	380	381	382	383	384	385	386	387	388	389	390	391	392	393	394	395	396
	397	398	399	400	401	402	403	404	405	406	407	408	409	410	411	412	413	414	415	416	417
	418	419	420	421	422	423	424	425	426	427	428	429	430	431	432	433	434	435	436	437	438
	439	440	441	442	443	444	445	446	447	448	449	450	451	452	453	454	455	456	457	458	459
	460	461	462	463	464	465	466	467	468	469	470	471	472	473	474	475	476	477	478	479	480
	481	482	483	484	485	486	487	488	489	490	491	492	493	494	495	496	497	498	499	500	501
	502	503	504	505	506	507	508	509	510	511	512	513	514	515	516	517	518	519	520	521	522
	523	524	525	526	527	528	529	530	531	532	533	534	535	536	537	538	539	540	541	542	543
	544	545	546	547	548	549	550	551	552	553	554	555	556	557	558	559	560	561	562	563	564
	565	566	567	568	569	570	571	572	573	574	575	576	577	578	579	580	581	582	583	584	585
	586	587	588	589	590	591	592	593	594	595	596	597	598	599	600	601	602	603	604	605	606
	607	608	609	610	611	612	613	614	615	616	617	618	619	620	621	622	623	624	625	626	627
	628	629	630	631	632	633	634	635	636	637	638	639	640	641	642	643	644	645	646	647	648
	649	650	651	652	653	654	655	656	657	658	659	660	661	662	663	664	665	666			
						717	718	719	720	721	722	723	724	725	726	727	728	729	730	731	732
	733	734	735	736	737	738	739	740	741	742	743	744	745	746	747	748	749	750	751	752	753
	754	755	756	757	758	759	760	761	762	763	764	765	766	767	768	769	770	771	772	773	774
	775	776	777	778	779	780	781	782	783	784	785	786	787	788	789	790	791	792	793	794	795
	796	797	798	799																	

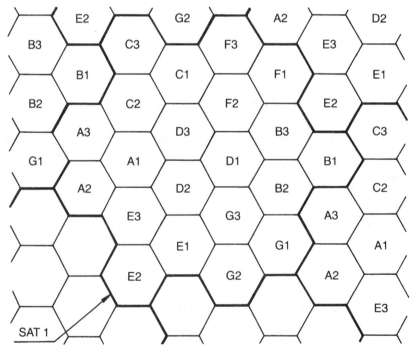

Figure 8.7 Sectored sites cell plan (sectored at the cell edge). Adjacent channels are now isolated by the antenna patterns. The orientation and separation provide the interference immunity required.

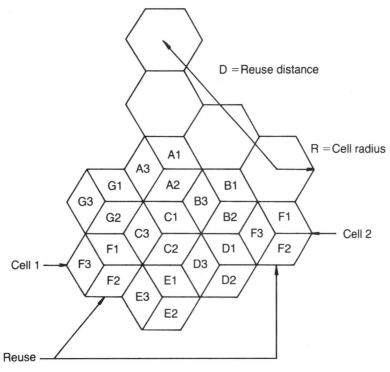

Figure 8.8 7-Cell pattern (sectored at the cell center).

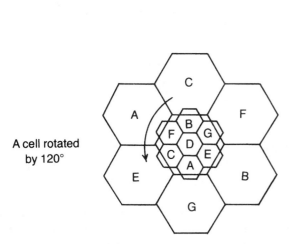

Figure 8.9 Cell orientation during a cell split can be seen as a 120-degree counterclockwise rotation. Note how the small cell C has moved with respect to the larger cell C.

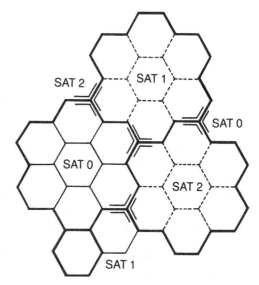

Figure 8.10 SAT are allocated so that any three adjacent cell groups are different.

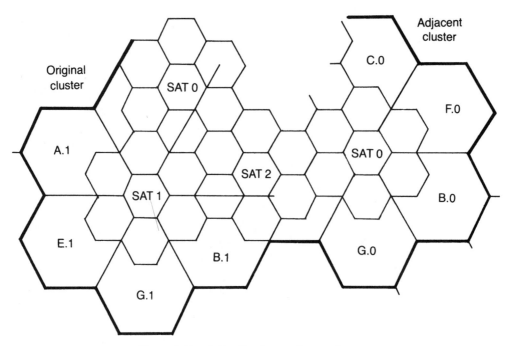

Figure 8.11 Cell split of two adjacent clusters.

SAT CODES AFTER CELL SPLITTING

Allocating SAT codes with cell splitting is also quite simple. As Figure 8.11 shows, cell splitting occurs around the central cell. The new cell takes on the SAT of the old cell, but what happens to the SAT of the old cell? To answer this question, it is necessary to consider splitting the center cell of the adjacent cluster. If the new clusters at the center of the two original clusters retain their old SAT codes, then the new cluster between them must be the third SAT code.

Figure 8.12 shows the new SAT codes that result from replacing the SAT codes of the old cells with the SAT code at the center of the overlay split cell.

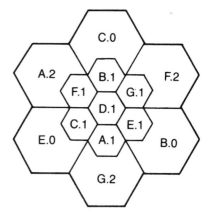

Figure 8.12 New SAT codes for the old cells.

DIGITAL COLOR CODES

The Digital Color Code (DCC) is a similar identification tag to SAT, except that DCC applies to control channels. There are four DCC codes: 0, 1, 2, and 3. As shown in Figure 8.13, randomly allocating three of the codes to adjacent cell clusters determines the pattern for the rest.

Each cell can be visualized as being at the center of a 7-cell cluster (that is, surrounded by six other cells). If the D cell is first considered to be at the center of a cell cluster (cluster 1) and DCCs are randomly assigned, then $D = 3$, $C = 0$, $F = 1$, and $B = 2$. By repeating the pattern around the D cell, it follows that $A = 2$, $E = 1$, and $G = 0$.

This method can now be extended to all other cells. If the B cell is now regarded as being at the center of a new 7-cell cluster (cluster N), by applying the same pattern, the assignments for cells X, Y, and Z can be found. This is continued until all DCCs are allocated ($X = 1$, $Y = 3$, and $Z = 0$).

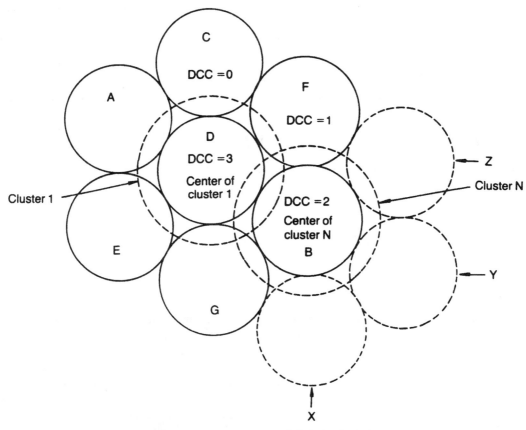

Figure 8.13 Allocation of digital colorcode.

Cell expansion proceeds in much the same way as described earlier, except that instead of the center cells taking on the DCC of the "old" cell, all center cells take on the same DCC. This means that the reorganization of the first cell can be random but that future cells are locked into the first one. Again, the old cells take on the DCC of the overlay control cell.

This section is not relevant to non-AMPS systems.

9

UNITS AND CONCEPTS OF FIELD STRENGTH

There are a diversity of units of field strength in use in radio frequency (RF) engineering today. Mobile engineers tend to prefer the unit dBμV/m, but some still use units like dBm, which have their origins in the landline network. This chapter seeks to clarify the usage of the different units and the concepts behind the measurement of field strength.

In free space, the energy of an electromagnetic wave propagates through space, and, according to an inverse-square law, with distance. The energy is dispersed over the surface of an ever-increasing sphere, the area of which is R (where R is the sphere radius). The total energy is constant because there is no loss in free space. The energy measured in watts/square meter (or any other units) will be constant in total (over the wavefront) and so the energy per unit area will vary as $1/R^2$. Notice that this equation holds for all frequencies. This concept is illustrated in Figure 9.1.

The total energy within a solid angle is a constant at any radial distance from the origin.

In a mobile-radio environment, the signal is attenuated much more rapidly than in free space and follows approximately an inverse fourth-power law with distance. There is a common misconception that this attenuation increases rapidly with frequency. It will be shown later in this chapter that, although the attenuation is an increasing function of distance, it is not a very strongly frequency-dependent function.

However, because the capture area, or aperture, of an antenna decreases directly with frequency, the energy captured by the antenna is directly a function of frequency. Thus, a quarter-wave antenna at 450 MHz is twice as long as a 900-MHz antenna and so can capture more energy from a field with the same intensity. This difference in capture area or aperture is what mainly accounts for the better long-range performance of lower-frequency systems.

Of course, this discussion must be limited to a frequency band where the propagation mode is similar. If the range 150–2000 MHz is considered, then the assumptions will generally hold.

Mathematically, the effective aperture or capture area A, of an antenna is

$$A_{\text{eff}} = \frac{\lambda^2 G}{4\pi}$$

where

$$\lambda = \text{wavelength}$$
$$G = \text{antenna gain}$$
$$A_{\text{eff}} = \text{effective aperture}$$

Therefore, if the wavelength is increased by a factor of 2, the effective capture area will increase by a factor of 4 for antennas of constant gain. The energy received will also increase by a factor of 4. Thus, the power collected by a 900-MHz dipole antenna compared to a 1800-MHz dipole antenna is $10 \log 4 = 6$ dB higher in a field of the same intensity. The same result can be derived by visualizing the antenna as immersed in an electric field of V volts per meter. A longer antenna is swept by more lines of field strength and so induces a higher voltage.

If a 900-MHz antenna is compared to a 450-MHz antenna of the same type, the 450-MHz antenna is twice as long and will intercept twice the electric field poten-

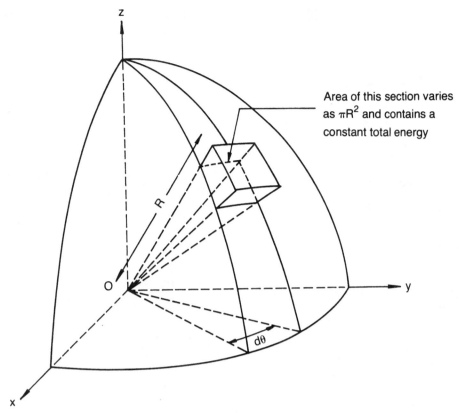

Figure 9.1 The originating energy from the origin, "O," is dispersed uniformly over a spherical surface as it propagates in space.

tial. The resulting increase in received field strength is 20 log 2 or 6 dB.

The effective aperture of an antenna of the same gain (for example, 3 dB) will also slightly affect the relative performances of 800-MHz and 900-MHz systems.

Code division multiple access (CDMA) systems have a center or mid-frequency of approximately 867.5 MHz, and the Global System for Mobile Communications (GSM) systems have a central frequency of 927.5 MHz (although both of these frequencies are outside the respective system frequencies). By comparing the aperture we can obtain the relative gain, as shown here:

$$\text{The relative aperture is then } \left[\frac{927.5}{867.5}\right]^2$$

$$\text{or a relative gain of } 10\log\left[\frac{927.5}{867.5}\right]^2$$

$$= 0.6 \text{ dB relative gain for the CDMA system}$$

Field strength is a measure of power density at any given point. The main units of field strength are dBμV/m, dBμV, μV, and dBμ. The dBμV/m unit measures power density of radio waves. The other units measure received

energy levels. Most mobile-radio engineers prefer the dBμV/m unit, while microwave engineers prefer dBm, and bench technicians use μV. This preference is partly historical and partly the practicality of the units in the different fields.

To visualize how these units relate, consider Figure 9.2, which shows a test dipole in the plane at right angles to the direction of propagation. It measures the electric field strength in the direction of the antenna.

The units dBμV, dBm, and μV measure the power or voltage received by a dipole antenna in the field. Voltages in mobile equipment are usually measured at 50 ohms, but any impedance can be used. These units are defined as

$$\mu V = \text{voltage at transformer terminal into } 50 \text{ ohms load in microvolts}$$

$$\text{dB}\mu V = 20\log\frac{\mu V}{1 \text{ microvolt}}$$

$$P = \text{Power measured at the transformer } (\text{impedance independent})$$

$$\text{dBm} = 10\log\frac{P}{1 \text{ milliwatt}}$$

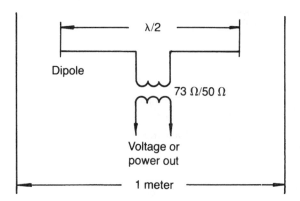

Figure 9.2 The concept of measurement of field strength in μV/m using a dipole.

The unit dBμV/m is the voltage potential difference over 1 meter of space, measured in the plane at right angles to the direction of propagation and in the direction of the test antenna. It is defined as follows:

$$dB\mu V/m = 20 \log \frac{\text{voltage potential per meter}}{1 \text{ microvolt}}$$

To see how all these units relate to each other in a real environment, a case study based on the work of Okumura *et al.* (a classic work on mobile propagation) will help.

Consider a typical base station that has a reference height of 200 meters and a transmitter Effective Radiated Power (ERP) of 100 watts. Table 9.1 lists the received field strength for various transmitter frequencies in the far field; for example, 10 km.

From the table, it should be clear that the actual receiver signal varies enormously with frequency, even though the transmitter power, site, and antenna height remain fixed. Therefore, if the units μV, dBm, or dBμV are selected, the results of a survey of one frequency

cannot easily be translated to frequencies that are significantly different. However, if dBμV/m is selected, only a few dB separate the readings. Furthermore, if additional sites are studied (that is, different heights for the transmitter, and different distances), this relationship is retained. Thus, any field strength measured in dBμV/m measures energy density at a given point and is dependent on frequency only to the extent that atmospheric and clutter attenuation is dependent on frequency. Therefore, it is possible to use results from a survey done at one frequency to draw conclusions about another if an allowance of 2 dB per octave is used. Some caution should be exercised when using these broad generalizations. However, they can be most useful approximations of coverage.

RELATIONSHIP BETWEEN UNITS OF FIELD STRENGTH AT ANTENNA TERMINALS

Assuming a 50-ohm termination, a dipole receiving antenna (unity gain) and a zero-loss feeder, the relationship between units of field strength at the antenna terminals is as follows:

$$E(\mu V/m) = \frac{\mu V}{39.3924} \times FREQ \text{ (MHz)}$$

Equation 9.1   Starting with dBm all into 50 Ω

$$\mu V = 2.236 \times 10^5 \times 10^{dBm/20}$$

dBμV/m

$$= 20 \log (5.676 \times 10^3 \times FREQ \text{ (MHz)} \times 10^{dBm/20})$$

$$dB\mu V = 20 \log (2.236 \times 10^5 \times 10^{dBm/20})$$

Equation 9.2   Starting with μV all into 50 Ω

$$dBm = 20 \log \frac{\mu V}{2.236 \times 10^5}$$

$$dB\mu V = 20 \log \mu V$$

$$dB\mu V/m = 20 \log (\mu V \times FREQ \text{ (MHz)}/39.3924)$$

Equation 9.3   Starting with dBμV/m all into 50 Ω

$$dB\mu V = dB\mu V/m - 20 \log \frac{FREQ \text{ (MHz)}}{39.3924}$$

$$dBm = 20 \log \left(\frac{10^{dB\mu V/m/20} \times 1.76168 \times 10^{-4}}{FREQ \text{ (MHz)}} \right)$$

$$\mu V = \frac{39.3924 \times 10^{dB\mu V/m/20}}{FREQ \text{ (MHz)}}$$

TABLE 9.1   Field Strength at 10 km for a 100-Watt TX at 20 Meters in A Suburban Environment

Freq (MHz)	dBμV/m	μV	dBm	dBμV
150	49	71	−70	+37
450	47	17	−82	+25
900	45	8	−89	+18
1800	43	94	−97	9·8

Note: This table was derived from a paper by Okumura *et al.* entitled "Field Strength and Its Variability in VHF and UHF Land-Mobile Radio Service," *Review of the Electrical Communication Laboratory,* Vol. 16, Nos. 9, 10, Sept.–Oct. 1968, pp. 835–873.

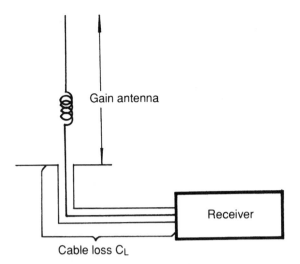

Figure 9.3 Actual field-strength measurement.

Equation 9.4   Starting with dBμV all into 50 Ω

$$\mu V = 10^{dB\mu V/20}$$

$$dBm = 20 \log \frac{10^{dB\mu V/20}}{2.236 \times 10^5}$$

$$dB\mu V/m = dB\mu V + 20 \log \frac{FREQ \ (MHz)}{39.3924}$$

CONVERSION TABLES

Because it is very easy to make a mistake when applying the formulas to translate between units, the conversion table in Appendix E can be very helpful. This table is useful for calculating the relationship between the variables illustrated in Figure 9.2. If a different antenna impedance is considered (a 300-ohm folded dipole antenna, for example), then the results cannot be used directly. Similarly, in a real-life environment, the signal levels will probably be measured at the receiver input, as shown in Figure 9.3, and the necessary corrections must be applied.

The relationship between the variables must be adjusted by the antenna gain minus the cable loss. At 900 MHz, using a 3-dB antenna, the cable loss is about 3 dB and thus the correction factor approaches zero. At other frequencies, this approximate relationship will not hold.

STATISTICAL MEASUREMENTS OF FIELD STRENGTH

In point-to-point radio, field strength is a one-dimensional variable of time. Most of the time variance is due to log-normal fading, and the nature of the signal variability is well documented. Because it is a simple function of time, field strength in point-to-point radio can be easily understood and measured. The situation is somewhat more complex in the mobile RF environment. Thus the field strength in the point-to-point environment is a simple function of time as shown here:

$$F = f(t)$$

The situation is somewhat more complex in the mobile environment where the field strength also varies with location (space), and so the measured value is a four-dimensional statistical variable

$$F = f(t, x, y, z)$$

or, if $(x, y, z) = L =$ location, then

$$F = f(t, L)$$

The real-life measurement of the field strength of an area is the collective result of a number of measurements made in different points in space and time, as illustrated in Figure 9.4. Thus, if a measure of field strength is required to typify an area, then its statistical nature dictates the measurement is made. Let's assume a measurement of the average field strength along a 500-meter section of a road is required. If all samples are taken at the Nyquist rate (in space at locations $L_0, L_1, \ldots, L_m$) and at one instant (T_0), the result would be an average value at a particular time, T_0.

Mathematically this can be expressed as

$$F_0 = \frac{\sum\limits_{i=0}^{m} f(T_0, L_i)}{m + 1}$$

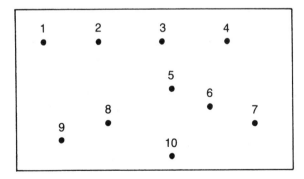

Figure 9.4 The measurement of field strength over an area is made at unique points in space and time. Actual readings vary with time (due to log normal fading) and space (due to log normal and Rayleigh fading). Any statement about the field strength in this area is a statement about a variable of space and time.

($m + 1 =$ the number of readings taken at positions $L_0, L_1, \ldots, L_m$).

In practice making simultaneous measurements is difficult and the individual space measurements would be made at different times ($T_0 \cdots T_n$). The average value in time and space of the field strength in that region is then given by F as shown here:

$$\text{Average field strength} = \frac{\sum_{K=0}^{n} F_K}{n + 1}$$

where $F_K =$ value of the Kth measurement.

Virtually all real measurements are an average (or some other statistical measure) of samples in different locations in space and time.

Other statistical measures are widely used. For example, all readings can be recorded and separated into two equal groups depending on level. Thus, it is possible to obtain the value above (or below) which 50 percent of the recorded samples occur. This is known as the median value (not the same as the average value). If all readings were taken at a single instant, the median value F_{50} can be found and the result would be at time T_0

$$F_{50} = \text{median } \{f(T_0, L_i)\} \quad \text{for } i = 0 \text{ to } m$$

where

$F_{50} =$ the level below which 50 percent
of the samples 0 to m lie

Most published data on mobile RF propagation plots the field strength as the median 50 percent/50 percent level. This is probably because what was measured was the mean, which for a normally distributed sample will be the same as the median. The mean is the easiest thing to measure and so the 50 percent/50 percent plots predominate.

For a number of distributions the mean and median are equal, although this is not necessarily so for all samples. As an example of how the average and median values can differ, consider the following readings:

$$25, 29, 33, 33, 35, 55$$

The average is

$$\frac{25 + 29 + 33 + 33 + 35 + 55}{6} = 35$$

But the median is 33 (because half of the sample is above 33).

In practical terms, as survey readings are taken at different points in space and time for a given region, the average of a set of readings in a region is the true time/space averaged measurement. Because time fluctuations of field strength are usually fairly fast relative to measurement periods, this assumption holds fairly well. Log normal fades produce field-strength variations with a periodicity of a few seconds, while a measurement is usually taken over 1- to 5-minute intervals. Therefore, measurements of field strength taken from a moving vehicle and used to calculate F_{50} will result in the field strength for 50 percent of locations and 50 percent of the time.

As the number of samples in space equals the number of samples in time, it is also reasonable to use the same technique to obtain the 70 percent/70 percent or 90 percent/90 percent values. However, because F_{50} is a function of space and a different one of time, a simple survey technique cannot be used to directly obtain the value for 90 percent of locations for 50 percent of the time. To obtain this value it would be necessary to obtain the standard deviation of the time-dependent variation independently (approximately the standard deviation of the log normal fade).

Interpreting a cellular requirement for the field strength to be, for example, 32 dBμV/m for 90 percent of locations and 90 percent of the time is a difficult task, because the concept has never been adequately defined. In particular, the concept originally applied to small regions of space and its application to service areas is vague. It might mean that within the service area only 10 percent of all readings will be below 32 dBμV/m and this is how most people think of it. However, the goal of a system designer is not to have 10 percent of the service area substandard. Probably what is really meant by this is that 10 percent substandard coverage is a worst-case scenario and is an upper limit rather than a design parameter. One would not expect, for example, a batch of resistors with 10-percent tolerance to be *designed* to be 10 percent out of tolerance. The 10-percent limit merely means the manufacturer's tolerance or worst case is 10 percent.

For the Advanced Mobile Phone System (AMPS) system, a 39-dBμV/m field strength is the accepted design level at the boundary. When it is the boundary that is defined to be 39 dBμV/m, it is clear that within the area defined by the boundary the average level will be higher than this. So because a 39-dBμV/m average in practice is a field strength that will ensure a "reasonable" service for 90 percent of locations *and* 90 percent of the time at the cell boundaries, it follows that the quality of area covered should be significantly better than 90 percent of locations *or* time.

Some European administrations using Nordic Mobile Telephone (NMT) systems define the boundary to be the 20-dBμV/m level. This level has some interesting implications. At this level conversations are only just intelligible and calls are likely to be dropped. It is below the effective service level.

At 20 dBμV/m the noise is similar to that of the familiar 12-dB SINAD used to measure mobile sensitivity. At this level the processing gains of systems of different deviation are equalized, and a low-deviation system such as NMT900 will perform as well as a high-deviation system such as Total Access Communications System (TACS).

Using the 20-dBμV/m yardstick, it is possible to conclude erroneously that the NMT900 system will perform as well as an TACS/AMPS system in the rural environment. Measuring the subjective performance at 20 dBμV/m will also lead one to conclude that at this level the two systems are more or less equal in performance. However, unless they are also equal in deviation, then this is the *only* level at which they are equal subjectively. As the measured field strength rises, so the subjective performance of the higher-deviation systems will rise more quickly. At 39 dBμV/m the higher-deviation systems will be performing significantly better than the lower ones at the same level. For any given field strength a 450-MHz system, being of lower frequency, will outdistance a 900-MHz system.

Except in very hilly terrain, the performance of AMPS/TACS is roughly the same as NMT450 in rural applications. For digital systems, processing gain is important and in this case, CDMA has about a 10-dB advantage over GSM.

For digital systems, it is far better to design for a field-strength boundary than it is for a BER. The reason is that the abrupt fall-off in BER as the threshold is approached makes field measurements of threshold both difficult and non-repeatable.

CONCLUSION

It is much simpler and less ambiguous to define the field strength of the *service area boundaries*, using AMPS as an example, as 39-dBμV/m average (approximately = 39 dBμV/m median) or 32 dBμV/m in 90 percent of locations and times. Any specification wherein the time and space variables are not equal is nearly impossible to realize in practice (that is, 90 percent of places, 95 percent of the time is physically meaningless unless what is meant by each parameter and how it can be measured are clearly defined).

10

FILTERS AND COMBINERS

TRANSMITTER COMBINERS

At most rented sites today there is a "per antenna" rental charge, which might well be as high as $10,000 per year. This can be a very significant cost to the operations of a large system, and can be the driving force to keep the number of antennas down. Even if the operation is at a site owned by the operator, competition for the best antenna locations, wind loading, and ascetics may all be factors mitigating against too many antennas.

It is not possible to simply directly connect several transmitters into a single antenna as shown in Figure 10.1. One reason that this cannot be done is that both the antenna and the transmitters are designed to work into a particular load impedance (usually 50 ohms), and connecting the transmitters in parallel would play havoc with the impedances. Further, in this configuration each transmitter would effectively be delivering power to every other transmitter, resulting in wasted power and incredible intermodulation problems.

What is needed is a way of delivering the power directly to the antenna so that only a minimal amount of power is fed back to the other transmitters, and that the impedance at the point of combining remains at 50 ohms.

If the transmitters were sufficiently far apart in frequency, a simple series-tuned circuit might do the job of preventing the transmitters from interacting, and the circuits could be coupled by an impedance transformer. Such an impedance transformer can be a quarter-wave transformer, which is made up of a small matching quarter wavelength of coaxial cable, as seen in Figure 10.2. The matching section has a characteristic impedance equal to the square root of the product of the input

impedance (50 ohms) and the output impedance (25 ohms), or 35.5 ohms. When connected together this transmitter combiner would be as shown in Figure 10.3. It is easy to see that this method could be extended to any number of transmitters: simply by adding another quarter-wave transformer to the antenna lead, as shown in Figure 10.4, to connect four transmitters. This process can then be repeated as necessary for additional transmitters.

Although the configuration in Figure 10.3 would rarely be used in practice to combine two transmitters, in principle it could be, and since in essence (as we shall see later) it is the *same* as the more common cavity filters, it is worth exploring the basics of this circuit just a little more.

As stated before, the inductance–capacitance (LC) combiner requires some considerable frequency separation, but how much is enough? To begin we have to consider the sharpness of the LC circuit involved. A measure of this is its Q (or quality factor). The Q is a measure of the efficiency of the tuned circuit and is defined as

$$Q = 2 \times \pi \times (\text{maximum energy stored/energy dissipated per cycle})$$

For the frequencies we are concerned with here, nearly all the losses are due to the coil resistance, and the Q can be calculated as

$$Q = 2 \times \pi \times L/R$$

where R = the coil resistance and

$$L = \text{the coil inductance.}$$

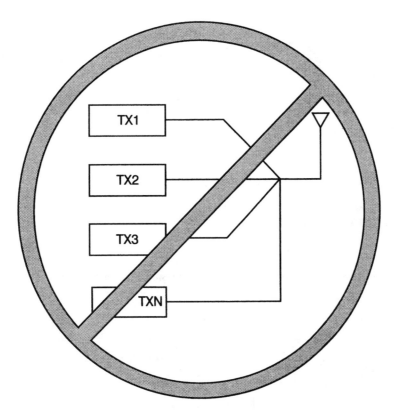

Figure 10.1 You can't do this.

Typically for such a coil the Q will be about 200. We can now use another parameter, the bandwidth of the tuned circuit to see just how broad the tuning of the circuit is. For a tuned circuit we can define the bandwidth to be the point where the power through the circuit is 50% (3 dB) down (see Figure 10.5). The frequencies fL and fH indicate the lower and higher frequencies for which the power throughput will be within 3 dB of the

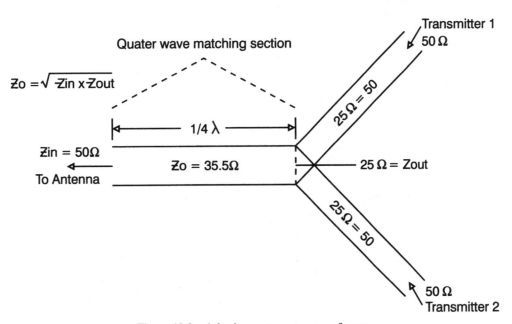

Figure 10.2 A basic quarter-wave transformer.

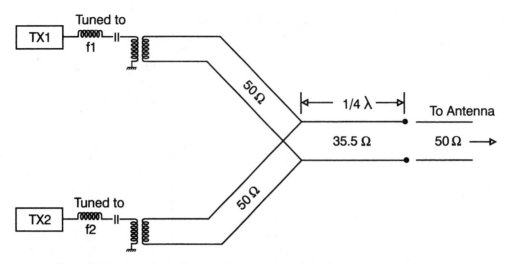

Figure 10.3   A method of connecting two transmitters into a common antenna.

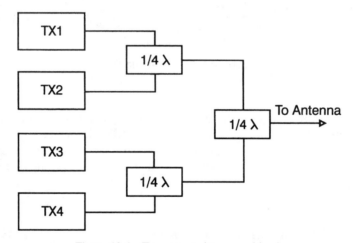

Figure 10.4   Four transmitters combined.

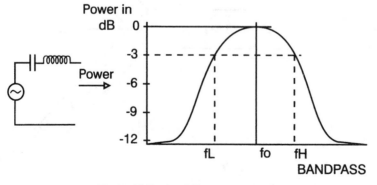

Figure 10.5   An *LC* response curve.

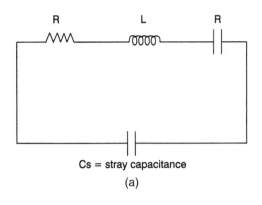

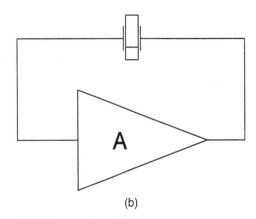

Figure 10.6   (*a*) The equivalent circuit of a crystal. (*b*) A crystal oscillator.

power at the resonant frequency. A convenient relationship is that

$$fH - fL = fo/Q$$

In other words, the bandwidth is inversely proportional to the Q. The concept of Q can be extended to other more complex tuned circuits and is a good indicator of the efficiency of the device.

So, if we assume a transmitter that operates at 800 MHz, then the bandwidth for a Q of 200 is 400/200, or 4 MHz. This tells us that the second channel must be at least 4 MHz away. In practice it would be a lot further because 3-dB isolation is hardly enough. So is it practical to use such a filter? If we consider a transmitter at 2 MHz, then we similarly find that the bandwidth is only 2/200 MHz or 10 kHz wide. Thus the problem comes down to the fact that at higher frequencies, higher Qs are required to achieve a reasonably sharp filter.

CRYSTAL FILTERS

Crystals are essentially slabs of quartz that are cut to a predetermined resonant frequency. The equivalent circuit of a crystal is shown in Figure 10.6*a*, and it is a series-tuned circuit, with a parallel capacitance, which is made up of the stray capacitance of the crystal holder and leads.

Because it is possible to make the resistance, R, small, and the ratio of L to C large, the crystal can have a very high Q, being typically of the order of 100,000. The most likely application for a crystal in a cellular system is in the clocking system where the high stability of L and C ensure good frequency stability in an oscillator, and hence in the clock. To derive an oscillator from a crystal, all that is needed is to use the crystal to apply positive feedback to an amplifier, as seen in Figure 10.6*b*.

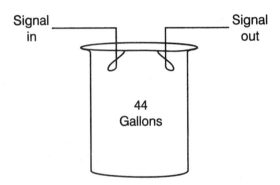

Figure 10.7   A 44-gallon drum as a resonant cavity.

RESONANT CAVITIES

Resonant cavities are routinely made with Qs of the order of 5,000–10,000. With this sort of Q it is possible to do some serious filtering even at high ultrahigh frequency (UHF) frequencies. A simple resonant cavity is, for practical purposes, a series-tuned circuit and works in the same way as the circuit in Figure 10.3. It consists of a metal container, usually of simple geometric construction (drums or box shaped are the most common). The resonance occurs because the internal dimensions are *electrically* some odd multiple of a quarter wavelength at the frequency concerned. All that is needed additionally to the cavity is a means of getting power into the cavity, and a means of extracting the signal. At UHF frequencies a single half-turn loop is generally sufficient to do this. Figure 10.7 shows the construction of a simple cavity.

In principle, any cavity, including a 44-gallon (200-liter) drum could be used as a filter. There are a number of reasons why a 44-gallon drum would not make an ideal filter. The most immediately obvious one is that it is too bulky. Since currents are flowing in the skin of the drum, causing energy losses, it might have been better if it were

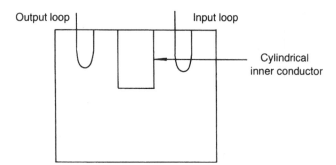

Figure 10.8 The interior of a simple cavity.

made of copper or some better conductor than its steel construction. And speaking of steel and the magnetic fields that will be produced by those currents, steel is a high-loss material for magnetic fields at radio frequency (RF) frequencies. Temperature variations will cause changes in the dimensions of the drum, and we have no compensation mechanism.

Figure 10.8 shows a simple resonant cavity used in UHF applications. This simple cavity has a resonant frequency that can be adjusted by the physical dimensions of the inner conductor. Such cavities are not often used because they are very temperature-dependent and cannot be adjusted for different frequencies.

More commonly, a few simple enhancements are made to improve the frequency stability and to give some degree of control over the resonant frequency.

A practical cavity should be constructed so that ther-

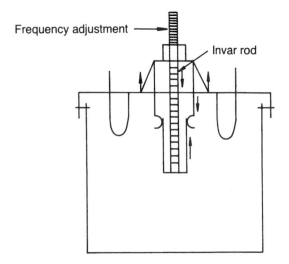

Figure 10.9 A practical cavity with temperature compensation and some flexibility in resonant frequency. The arrows indicate the direction of expansion of the components when heated.

mal expansion effects have automatic compensation. Figure 10.9 shows a cross section of a practical cavity. In this figure, you can see that by suitably dimensioning the components, it is possible to neutralize the effects of thermal expansion. You can adjust the dimension of the internal conductor (and hence the resonant frequency) by turning the adjustment screw at the top.

Cavities are usually made of low electrical-loss, high thermal-conductivity metals such as copper, brass, or aluminum. The thermal dissipation of the cavity depends on its loss, but a 3-dB loss cavity connected to a 100-watt transmitter can dissipate 50 watts—quite a lot of heat. Cavities are designed to dissipate about 30 mW per square centimeter.

Physically there are many variations of the form that cavities can take, but certainly one of the more interesting is the beer barrel cavity shown in Figure 10.10. The cavity not only looks like, but is in fact constructed from a beer barrel. Figure 10.11 shows the inner conductor from one of these cavities. Not only do these barrels work, but they produced *Q*s well in excess of 10,000. And yes, they really are made from beer barrels! These cavities are used for trunk radio and not for cellular, but the basic operation is the same.

Commercial cavities are mostly made of copper, a non-magnetic material with good conductivity. To further enhance the conductivity, the inside surface is often coated with silver. Temperature compensation is added, and also a tuning mechanism. The tuning consists of a plunger on a screw thread that can be screwed to change the electrical length of the cavity.

To see how closely a resonant cavity is related to a tuned circuit (or lumped elements), an analogy that was presented in Richard Feynman's *Lectures on Physics* (Volume II, Chapter 23) leads us to consider the evolution of the tuned circuit as shown in Figure 10.12.

Starting with a simple tuned circuit in Figure 10.12*a* we can increase the frequency by making the coil have a few less turns (Figure 10.12*b*). This process can be continued until the coil has been minimized to half a turn (Figure 10.12*c*). In each of these iterations the resonant frequency will rise. To bring the frequency down we could increase the inductance by using two half coils in parallel (Figure 10.12*d*) and continue this process with a number of half coils around the capacitor plates. Taking this process to its limit we have a drum (Figure 10.12*f*). In fact currents do flow in the walls of the cavity just as they would in coil, and an electric field exists inside the cavity much as it would in the capacitor (of course, magnetic fields exist as well, because a changing electric field always is accompanied by a changing magnetic field).

By suitably arranging the physical characteristics of the cavities a variety of different filters can be devised. The most common are the band-pass, notch, and pass

Figure 10.10   Cavities constructed from beer barrels; product of the UK.

reject filers. The band-pass filter takes three forms of the cavities discussed earlier and have a characteristic curve as seen in Figure 10.13a. A notch filter (Figure 10.13b) is sometimes added in series to stop a powerful nearby transmitter from intermodulating with the transmitters. It is commonly seen as a harmonic filter where it is used to attenuate the second harmonics of the transmitters. Where a notch is required to be close in frequency to the transmitters (again generally because of some co-sited interferer) the best option might be a pass reject filter, which is essentially a series bandpass and notch filter (Figure 10.13c). The pass reject filter will most commonly be used in a mobile radio duplexer where the notch is used in the receiver path to take out the transmitter RF.

The following is the specification for a typical cavity:

· Resonant circuit
· Impedance: 50 ohms
· Insertion loss: 2 dB
· Input isolation: 14 dB
· Maximum input power: 50 watts
· Minimum frequency separation: 630 kHz

This is the specification for a typical isolator:

· Impedance: 50 ohms
· Insertion loss: 0.6 dB
· Reverse isolation: 50 dB or more
· Maximum input power: 60 watts

This is the specification for a typical junction:

· Impedance: 50 ohms
· Maximum input power: 600 watts
· Intermodulation of 16 TX carriers of +45 dBm shall not produce more than −105 dBm of fifth-order intermodulation products

Figure 10.11   A beer barrel cavity tuner.

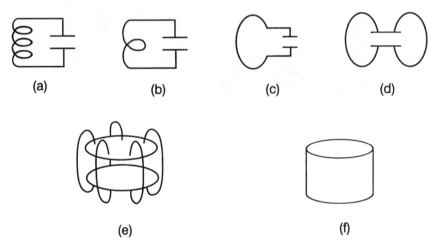

Figure 10.12   The evolution of a cavity from a tuned circuit.

HARMONIC FILTERS

No matter how carefully the combining is done there will be some intermodulation, which will cause the generation of the harmonics of the carriers. Some manufacturers include a harmonic filter, which is connected as shown in Figure 10.14, after the final isolator. The filter may be either a band-pass or notch filter, in which case it will be tuned to notch the second harmonic.

HOW MUCH POWER LOSS IS ACCEPTABLE?

All transmitter combiners will have some power loss, and some, in particular the ferrite hybrid, will have substantial loss. Although not as critical as receiver losses, transmitter losses do have a cost. The first cost may be the cost of the transmitter itself. If high-loss combiners are used, it may be necessary to use a higher power (and hence higher cost) transmitter power amplifier just to

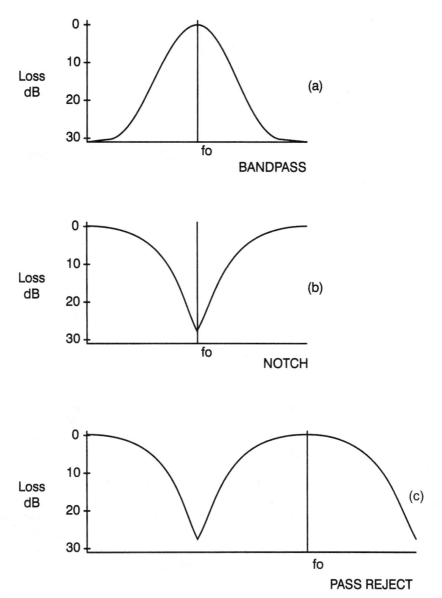

Figure 10.13 The characteristics of various filters.

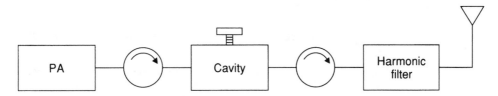

Figure 10.14 A harmonic filter is sometimes part of the combining network.

ensure that the effective radiated power (ERP) is sufficient for the system balance between transmit and receive paths. Typically the power amplifier stage of a transmitter is around 30 percent efficient, and it will be responsible for the bulk of the power consumed by the transceiver. So a 50-watt transmitter will probably require 150 watts of mains power to drive it. When there are a large number of transmitters on site, the cumulative power can be substantial. For example, for a 40-channel site operating at 50 watts per channel the

CAVITIES

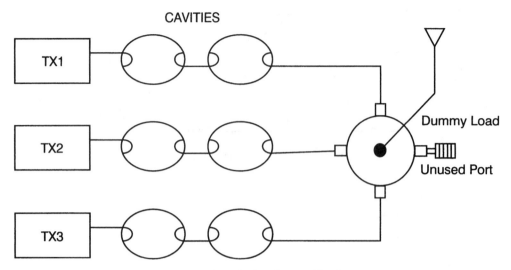

Figure 10.15   A typical cavity combined system.

power consumption would be 2 kW (at the going rate of 10 cents per kilowatt hour, this is 20 cents an hour for power, which depending on usage patterns, could amount to $700 per year). Generally digital base stations will be more power efficient, but a maximally equipped Global System for Mobile Communications (GSM) site with three sectors and two carriers per sector can consume as much as 3 kW of power.

Another cost associated with high transmitter power is the cost of battery backup. Ordinarily the size of the batteries is calculated on the basis of a pre-determined outage time, based on peak hour usage. The size and cost of the battery backup will increase in almost direct proportion to the power consumption.

Intermodulation

Intermodulation can never be totally eliminated, but it can be controlled. In the transmitter path, isolators control intermodulation by acting as a one-way valve (much as a diode does). The degree of isolation (measured in dB) that is obtained is very nearly a measure of the reduction of the intermodulation level.

PRACTICAL COMBINERS

The simplest practical combiner is electrically equal to the *LC* combiner discussed previously. It consists of one or more bandpass cavity filters connected in series with the transmitter and joined together via a star harness (which is a multiport electrical equivalent to the quarter-wave transformer discussed before). This simple com-

biner is cheap, has low insertion loss, and is readily reconfigured. Its disadvantages are that it is physically bulky and is suitable only when the frequency spacing between channels is large. A typical cavity combiner is shown in Figure 10.15; notice that unused terminations on the star combiner are connected to dummy 50-ohm loads. This simple combiner is good for channel spacings down to about 0.5 percent of the center frequency.

When the loss in the combining is high, that loss is dissipated as heat. The heat itself can be a problem, and in air-conditioned rooms it will add to the cooling costs, while in un-air-conditioned rooms it may lead to premature aging of the base-station hardware with consequent maintenance bills.

So there is no clear-cut answer to the question of how much loss is acceptable, but as a general rule 3 dB is a good target, unless other factors (like the need to combine very closely spaced channels) dominate.

ANTENNA COMBINERS AND SPLITTERS

Because many channels ordinarily operate at one base, it is necessary to use antenna combiners. Antenna combiners combine transmitter outputs into a small number of antennas. The frequencies used at any one base have been chosen so that their spacing ensures good isolation between the channels.

Figure 10.16 shows a typical antenna combiner. The whole combining system is known as a multicoupler or multiplexer. Because of confusion that can arise in discussing the RF coupling device and the MUX (also known as a multiplexer), the term *multicoupler* is preferred.

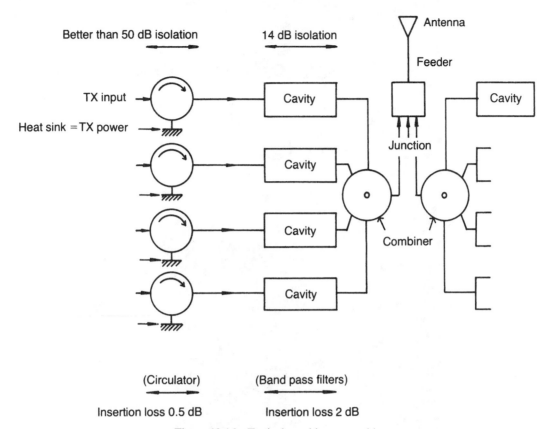

(Isolation is from output to input)

Figure 10.16 Typical combiner assembly.

Isolators allow the TX signal to pass with only 0.5 dB loss in the forward direction, but attenuate any reflected signal by at least 50 dB. These devices are usually circulators and are generally wide-band devices.

The cavities are resonant filters. They have insertion losses of about 2 dB and an input isolation of 14 dB, resulting in about 2.5 dB overall loss of the combiners. Ordinarily, up to 16 transmitters are combined into one antenna.

For cellular systems, 630-kHz separation (minimum) is generally allowed between channels being coupled together. Channel separations less than this are possible only if greater insertion losses are tolerated.

RECEIVER COMBINER

Receivers are multicoupled to one antenna, usually in multiples of up to 64. The most important part of the receiver multicoupler is a good low-noise preamplifier that will determine the ultimate signal-to-noise (S/N) ratios and hence the base-station receiver performance.

Figure 10.17 shows a six-antenna multicoupler. Figure 10.18 shows a typical splitter.

In most cellular bases, diversity reception is used. This means that two antennas are used for each receiver that has two RF inputs and a diversity combiner. These antennas are usually physically separated by about 3–4 meters so that their received signals are not correlated. When one antenna receives a multipath fade, the other antenna probably will not.

ANTENNA DUPLEXERS

Antenna duplexers are sometimes used to connect transmitter-and-receiver ports to the antenna, as shown in Figure 10.19. Antenna duplexers are used mainly when antenna space is at a premium. The duplexer consists of a number of resonant cavities that isolate the transmitter output from the receiver multicoupler input by at least 80 dB. Duplexers are used extensively in mobile equipment (allowing the use of one antenna only), but are only occasionally employed in cellular base stations be-

Figure 10.17 This Motorola receiver multicoupler is contained in its own housing. The six "black boxes" at the top are the low-noise preamplifiers with preselectors for a 3-sector site (with diversity). Each of the preamplifiers then feeds a four-way splitter (the six light-colored boxes below). Additional splitters are found in the receiver racks.

cause they are quite lossy (2 to 3 dB) and impair the overall sensitivity of the base-station receiver.

Problems with intermodulation are often associated with this configuration. These usually arise with time as the duplexer becomes detuned or the antenna/feeder develop minor faults (corrosion water ingress), which unbalances the impedances, and consequently detunes the duplexer.

RECEIVER COMBINING

Unlike transmitter losses, which while inconvenient, need not be performance affecting, any receiver losses or

impairment can be translated directly to reduced handheld coverage. It is usual practice to combine all receivers onto a single antenna, because for reasons that will become clear later, this can be done without any significant losses at the channel level.

A typical receiver combiner is shown in Figure 10.20. The essential elements are the preselector filter, which is a bandpass filter, passing only the receive channels, a low-noise amplifier (LNA) and the distribution amplifier.

The preselector filter typically passes the whole of the cellular radio system receiver bandwidth in a single window as per figure 10.21a and will have a loss of less than 3 dB. Sometimes the receiver channels will not all be contiguous and they may be split into a number of

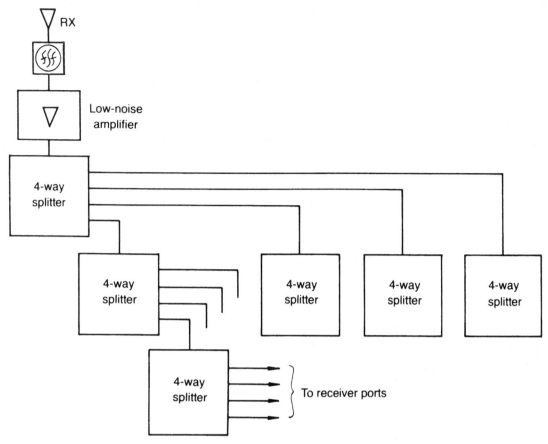

Figure 10.18   A total of 64 receivers can share a single antenna by multiple splitting of the RF path. Some of these splitters may be active but the S/N performance is largely determined by the first low-noise amplifier.

different groups. This can be accommodated by multi-window preselectors (Figure 10.21b) that will come at a somewhat higher cost and with additional losses. This happens in the Advanced Mobile Phone Service (AMPS) and CDMA band in the United States and in most systems in Japan.

In most systems, following the preselector will be an LNA. It pays to use a good LNA. For practical purposes the noise figure (and so performance) of everything that follows is determined by this device. The contribution to the overall noise figure of the system by devices after the LNA is reduced by a factor that is inversely proportional to the gain of the LNA, and so the gain of the LNA is also important. Offsetting the positive contribution of the LNA gain is the intermodulation distortion produced by it. An LNA that is subjected to high levels of spurious signals, or high levels of wanted signals, will produce intermodulation signals (the most troublesome being second-order products) that will de-

grade the receiver performance. Ideal devices do not produce intermodulation distortion, but virtually all real devices do. An ideal device is one where the output level is directly proportional to the input. For most real devices the relative amount of distortion produced increases as the input power level is increased. This occurs because of inherent non-linearites. For a typical RF amplifier the relationship between the power in and power out follows a simple linear relationship, while the third-order intermodulation level follows a log linear relationship. Figure 10.22 shows that for a low level of input the distortion products are relatively low, but as the level rises the distortion levels rise more rapidly, until at 27-dBm input the distortion power is equal to the fundamental (or wanted signal). Obviously nobody is going to operate an LNA at such a level, but this level, called the *third-order intercept point*, is a good measure of the power-handling capabilities of the amplifier. Its value is significant when determining the preselector filter char-

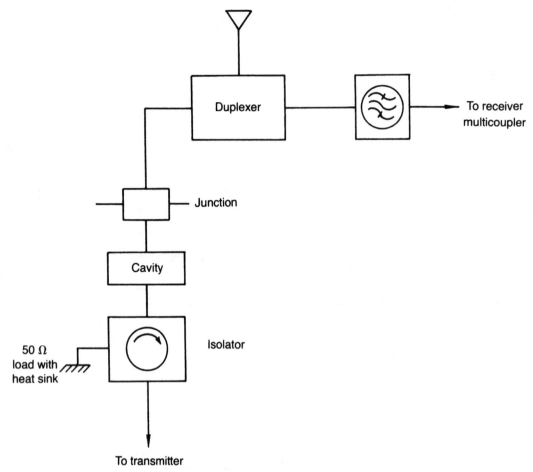

Figure 10.19 When receivers and transmitters are coupled into one antenna, a duplexer is needed to isolate the transmitter output from the receiver input.

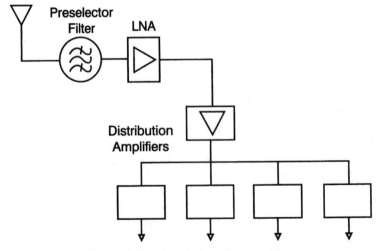

Figure 10.20 A typical receiver combiner.

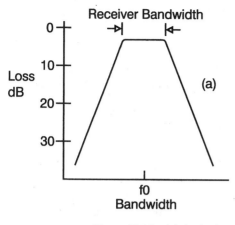
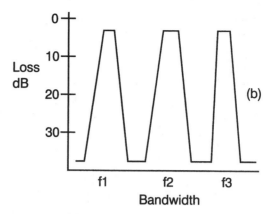

Figure 10.21   (*a*) A single preselector window, and (*b*) a multiple window.

acteristics. Fortunately for system operators, most LNAs have a third-order intercept of around 35 dBm, and so this is one less variable to worry about.

It is not only active components like amplifiers that contribute to intermodulation distortion, because to some extent all real-world devices are non-linear. Even components like cavity filters and connectors will produce some third-order products, and in most cases they will have a measurable third-order intercept.

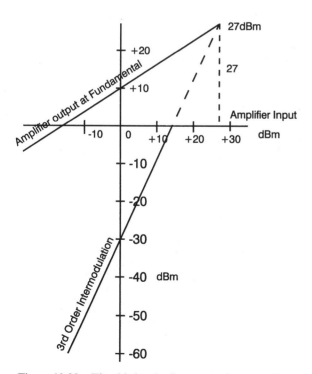

Figure 10.22   The third-order intercept of an amplifier.

NOISE FIGURE

The other parameter that measures the performance of an LNA is its noise figure (NF). This is usually expressed in dBs, and it is the additional noise that is injected into the system by the amplifier itself. A high-quality LNA will have a noise figure of around 0.1–0.5 dB, while an average one may be more like 3–4 dB.

Passive components like cavities and cables introduce additional noise that is equivalent to their loss. So a 3-dB attenuator (or a length of cable with a 3-dB loss) connected in series with the antenna will add 3 dB to the total noise figure of the receiver. So if this were followed by an LNA with a noise figure of 3 dB, the overall noise figure would be 6 dB. However, the importance of the LNA can be seen by comparing the loss if the same 3-dB attenuator were connected after the LNA. Assuming the LNA had a gain of 10 dB (equals 10×), the total noise figure contributed by the cable is only 0.962 dB. For a more detailed explanation of this, see Chapter 35. The important thing to realize is that anything that follows the LNA contributes only marginally to the overall noise factor (and hence the sensitivity) of the receiver.

RECEIVER MULTIPLEXER

Figure 10.20 shows a series of distribution amplifiers (also known as receiver multiplexers) immediately after the LNA. Because, in the way that has already been explained, these amplifiers contribute little to the overall noise performance, a good number of them can be cascaded in series. It is usual that they branch off in multiples of 4, 5, or 8, with the number being determined by the rack layout expected. In total there may be 60 or more receivers fed from a single antenna. The gain of these amplifiers is generally low, but sufficient to overcome the distribution losses. For example, if the amplifiers are branched in groups of four, a minimum gain of $10 \times \log 4$ (6 dB) will be necessary to maintain the LNA levels. Too much gain here can be a problem, because it could overload the first RF amplifier of the receiver causing intermodulation distortion.

CUSTOMIZED RF

If you need to customize your RF, there are a few things that need to be quantified before contacting a supplier.

1. The actual frequencies to be used need to be known, and if these are not available, then at least the band needs to be identified

2. The maximum transmitter power level and the maximum acceptable combining loss

3. Where space is at a premium, the limitations on size need to be specified

4. The number of channels to be combined as well as the long-term expansion plan (the maximum number of channels on any one site)

5. The frequency separation between transmit and receive

6. Any channels that need to be notched out (co-sited potential interferes)

7. The maximum preselector filter loss

8. If an LNA is to be used, whether it will be tower or rack mounted

9. Preferred connector types (N-type are the most common)

10. Do you want the supplier to do an intermodulation study on the chosen frequency plan? (There may be extra charges for this)

ADVANCED CAVITY TECHNOLOGY

There are two accepted ways of producing high-performance cavities. The first is to use conventional technology with advanced designs, and the second is to use superconductors to make up the cavity body.

Cavity design techniques now allow for elaborate cross-coupling between cavities, which results in higher Q. A low-tech way to produce a high Q is simply to make the cavity bigger. High Q preselector filters will permit closer spacing of channels with minimal loss.

Using superconductor technology, a number of new companies have been able to produce some spectacular filters. In principle all that is needed to improve the sharpness of the filter is to cascade more cavities. The downside of this is that it also introduces more loss. By using superconductors, the losses can be minimized even when a large number of cavities are used.

FILTRONICS

Filtronics makes a range of conventional and ceramic filters for the cellular industry, and have achieved Qs of 6000 with their comb-line technology (see Figure 10.23). The combine filters have one tuning screw and one coupling adjustment per resonator. Depending on the complexity of the filters, tuning, which is done manually, can take from 5 minutes to 60 minutes.

The filters are tunable over a few percent of their

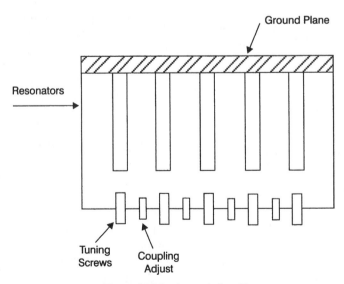

Figure 10.23 A comb-line filter.

center frequency, and a retuning service is available. Although these filters can in theory be retuned in the field, in practice this has not been successful. It takes about six months for a factory worker to become reasonably skilled at tuning, and it is regarded as more of an art than a science, with non-technical people often excelling at it. The best tuners usually have years of experience.

At the heart of this type of filter is the cross-coupling between the resonators, which if controlled can insert nulls at frequencies close to the passband. With careful control this enables filters with relatively small numbers of resonators to have sharp band-pass responses. The importance of this is that more resonators mean more loss, and using this technique the total number of resonators can be kept small.

Ceramic Filters

Using ceramics filtronics has been able to achieve Qs of 50,000 on 5 and 6 pole filters, and can typically achieve a Q of 20,000 using 8 poles with these filters. The Q of a filter is independent of the number of poles, but the loss through the filter increases with the number of poles (resonators).

Despite the high Qs and small size, there are several disadvantages. Ceramic filters are more expensive than comb-line filters. They have spurious responses/spikes (at typically 1.2× center frequency), whereas comb-line filters can have stop bands beyond 5× center frequency. Their power handling is lower because of the intrinsic problem of removing heat from the insulating mounting materials.

There is a wide range of suitable ceramics, and these range from relatively cheap to very expensive (e.g., from $15 to $250+ per kg of raw powder). Typical examples are zirconium tin titanate, barium zinc tantalate, and calcium titanate (see *Dielectric Resonators*, by D. Kajfez and P. Guillon, Artech House, Dedham, MA 1986, for more details).

CO-SITED GSM AND AMPS/CDMA

Increasingly there are many sites where different technologies share the same base station site and tower, and this can pose some challenges for filtering. Consider the case of cosited GSM/AMPS and/or code division multiple access (CDMA), a situation that is common throughout Asia and Australasia. In fact, except for the United States and Europe, it is common that both the GSM 900 and AMPS/CDMA bands are both in use and that the systems compete directly in the marketplace.

The GSM uplink [base transceiver station receiver (BTS RX)] and the AMPS/CDMA downlink [BTS transmitter (TX)] are adjacent. In most countries there will be a guard band between the two systems of 2–4 MHz. In some countries, however, the guard band is missing altogether, and ad hoc arrangements must be made for separation. The situation is seen graphically in Figure 10.24.

Solving the GSM Co-Sited with AMPS/CDMA Problem

The preceding general problem of co-siting GSM and AMPS is now looked at in more detail. This exercise is a real one involving the development of a filter by Filtronics to meet the needs of Telstra, a cellular operator in Australia with both GSM 900/GSM 1800 and AMPS/CDMA systems co-sited. While it is not unusual for the GSM and AMPS to be co-sited, it is unusual that both systems belong to the same operator. In many ways this makes it easier, because both ends of the problem are under the control of the one operator.

For this study I am indebted to Greg Delforce of Filtronics and Alan Hogg of Telstra who put together this problem and its solution. Although this is a specific problem the principles can be applied to most filtering problems.

The basic problem is seen in Figure 10.25, which shows that the co-sited AMPS TX antenna is only a few

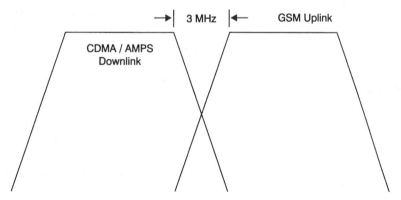

Figure 10.24   The spectrum of GSM RX and AMPS/CDMA TX.

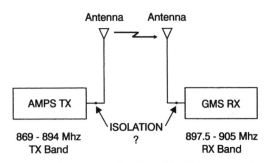

Figure 10.25 How a nearby AMPS/CDMA transmitter can desensitize a GSM base-station receiver.

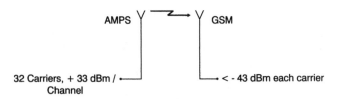

∴ ISOLATION   33 - 43 = 76 dBm

If 32 Carriers ∴ ADD 10 log $\frac{32}{2}$ = 12 dB (Worst Case)

∴ ISOLATION   > 76 + 12 = 88 dB

Figure 10.27 A total of 88 dB isolation is required.

MHz away from the base-station RX band of the GSM system. This situation can cause five identifiable mechanisms for interference.

- AMPS causing GSM RX blocking
- GSM RX intermodulation of AMPS TX signals causing interference to GSM receivers, especially in the LNA and multicouplers if tower mounted
- AMPS TX intermodulation causing GSM interference (from antennas, combiners, isolators, connectors, and cables)
- AMPS TX spurious causing interference to GSM
- AMPS TX noise causing interference to GSM

Notice that in all of the preceding instances the objective is to protect the GSM system from the AMPS TX, through one of the many modes by which the transmitters can cause interference. Looking at each of these problems separately:

1. *GSM Blocking by AMPS TX* High levels of adjacent channel interference will cause desensitization. GSM specification 5.05 requires the base station RX to operate normally with a blocking carrier at −13 dBm, which is more than 800 kHz away, and still experience a sensitivity decrease of no more than 3 dB. The AMPS TX is 3.5 MHz away and this should provide an additional advantage over the specification. The power level of the AMPS TX channels is +33 dBm/channel (on

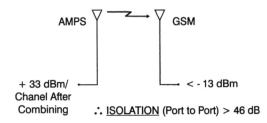

Figure 10.26 AMPS TX channels as seen by the GSM RX.

Telstra's systems after combining) so the situation is as depicted in Figure 10.26. Here we see that it will be necessary to provide at least 46 dB isolation port to port to protect from this mode.

2. *Intermodulation Within the GSM RX, Induced by AMPS Carriers* GSM specification 5.05 states that the GSM BTS will operate normally with a desensitization of no more that 3 dB in the presence of co-channel third-order products produced by carriers with a level of −43 dBm.

As there are a number of AMPS carriers, the net effect of these has to be considered, and in this instance Telstra allows for a maximum of 32 carriers; in the most unlikely worst case, the total power of all the AMPS carriers might be contributing to the intermodulation and so, as seen in Figure 10.27, in this mode we are seeking a net isolation of 88 dB.

3. *AMPS TX Intermodulation Occurring in the TX Stage* All transmitters produce intermodulation, and it is generated to some extent in all components of the system (both passive and active), so we can expect to see some intermodulation at the preamp (PA), the combiners, isolators, cables, harmonic filters, connectors, cables, and antennas.

The GSM-specified susceptibility to interference is 12 dB C/I, so relative to the GSM nominal sensitivity of −104 dBm, the interference "I" must be less than −104 dBm−12 dBm = −116 dBm. At this point it is worth noting that most modern BTSs have a sensitivity of at least −107 dBm, and so this figure is not conservative.

Again each and every carrier *may* be involved in producing an in-band intermodulation product, and so conservatively the 32 AMPS carriers can be added. The AMPS specification calls for the TX IM level to be −100 dBc, and so as can be seen in Figure 10.28, the net isolation required here is a minimum of 61 dB.

4. *AMPS TX Spurious Products that Fall into the GSM TX Band* As in case 3, the spurious products will be seen as GSM in-band interference, and so the

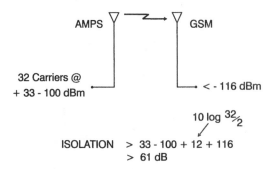

Figure 10.28 The AMPS intermodulation products as seen by the GSM RX.

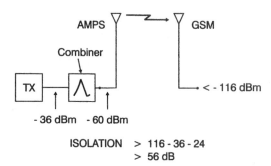

Figure 10.29 AMPS spurious seen by the GSM RX.

level from the GSM specification should be less that -116 dBm, as before at the GSM RX.

A typical AMPS TX has spurious levels of <-36 dBm, and the combiner will attenuate the levels at the GSM frequencies by a further 24 dB. Thus, the situation as depicted in Figure 10.29 is that a net isolation of 56 dB is required.

5. *AMPS Thermal Noise* AMPS thermal noise is similar to case 4, as the thermal noise to consider is that in the GSM band; therefore, at the GSM RX port the level should be below -116 dBm.

Telstra's AMPS TX system was measured to have a typical noise level of -95 dBc over a 250-kHz bandwidth (which is comparable to the GSM 200-kHz chan-

nel width). Again in the worst case each of the 32 possible channels could be a contributor to this noise source, and so they can be considered to be added. Hence from Figure 10.30, the net isolation needed here is 42 dB.

If we now summarize these conditions, we find (the cases in boldface represent the worst situations):

	Isolation (dB)	Solution
1. GSM RX blocking	46	GSM RX filter
2. GSM RX intermodulation	**88**	**GSM RX filter**
3. AMPS TX intermodulation	**61**	**AMPS TX filter**
4. AMPS TX spurious	56	AMPS TX filter
5. AMPS TX noise	42	AMPS TX filter

The isolation will be provided by a combination of methods

- Site separation
- Antenna orientation
- RX and TX filtering

Applying the Results

To see how this result is applied to an actual installation, let's look at an installation where the AMPS and GSM systems are situated at sites that are 50 meters apart (see Figure 10.31). For simplicity we will assume both sites are using 12-dB omni antennas.

Here the isolation provided by the separation is 63.5 dB (using near-field loss calculations, you will get 65.6 dB if you assume far-field path loss). Gain, however, is added at both antennas and there will be some cabling losses. Assume the cabling losses are 3 dB, then the isolation provided is

$$\text{Isolation} = 12 - 3 - 63.5 + 12 - 3 = 45.5 \text{ dB}$$

Comparing this result with the requirement for condition 2 (GSM RX intermodulation), we find that the GSM

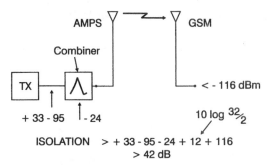

Figure 10.30 AMPS noise seen by the GSM RX.

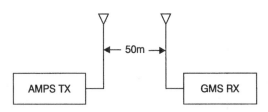

Figure 10.31 The spatial separation of an AMPS and GSM system.

EXAMPLE CALCULATION

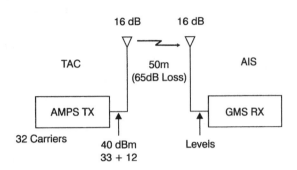

Mechanism # 2: (GSM RX IM)
    Level = 40 + 16 - 65 + 16 = + 7 dBm (Versus - 43 dBm)
    ∴ Use filter at GSM RX (CB519 or CB479)

Mechanism # 3: AMPS TX IM @ - 100 DB (Versus - 116 dBm
    ∴ Use AMPS TX Filter (CB425)

Figure 10.32   Example calculation.

RX filter must provide $88 - 45.5$ dB or 42.5 dB attenuation of the AMPS TX frequencies. Figure 10.32 shows the system gains. Figure 10.32 AMPS TX noise into GSM RX.

The third case, AMPS TX intermodulation, requires only a modest 61 dB of isolation, and in this case, 45.5 dB of it is provided spatially, so an additional $61 - 45.5$ or 15.5 dB is required on the AMPS TX path to produce a result like that seen in Figure 10.33.

Antenna Separation

The calculation of coupling losses due to antenna separation is rather complex. It can easily be shown that at distances greater than 250 meters, in the preceding example there will be no need for the AMPS TX filter and a 30-dB GSM filter will suffice. (Incidentally, this result implies that *all* GSM 900 systems operating in the same

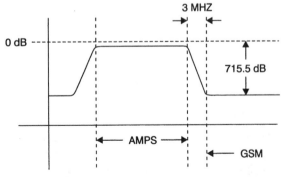

Figure 10.33   The required AMPS TX filter.

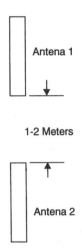

Figure 10.34   The vertical spacing of antennas mounted for isolation.

locality as an AMPS system should have a basic 30–40-dB filter at the AMPS TX frequency.)

At distances closer than 250 meters, near-field effects come into play and the coupling losses will be less than the far-field (inverse square law) losses.

Often at co-sited locations the separation will be provided by vertical separation of the antennas. The problem with this is that the coupling is very difficult (perhaps impossible) to calculate. Free-space near-field calculations for single dipoles can be done with reasonable accuracy, but the problem lies with the complex radiation patterns of gain antennas and the distortion of the pattern, particularly in the near field by the tower. Reflections off the tower may well be the main mode of coupling between the two antennas.

For vertical separation, as seen in Figure 10.34, the isolation as measured will be around 20–30 dB for omni antennas and 40–60 dB for sectored ones. In free space two dipoles separated by only 2 meters and *exactly* in the same plane will have an isolation of 85 dB. However, move one of them horizontally by 0.2 meter and the isolation will drop to 76 dB, at 0.5 meter horizontally, it will be 65 dB. This might seem to imply that it is vital to get the antennas *exactly* in the same plane, but in fact the main coupling mechanism on a tower and between gain antennas depends more on other factors, as explained earlier.

PCS/PCN Bands

As the personal communication service/personal communications network (PCS/PCN) offers become more an off-the-shelf item, so it is increasingly likely that they will come with built-in 75-MHz preselector filters (the whole of the band). It is very difficult to design filters for such a wide passband (relative to the center frequency), which also have a sharp cutoff in the 20-MHz region between the TX and RX bands, so, often the units as shipped leave a lot to be desired. Most PCS/PCN operators have 25 MHz of spectrum or less, and in many cases it is desirable to add additional sharper preselector filters for the actual spectrum in use.

Purpose-Specified Filters

Increasingly as the spectrum becomes more crowded there is a need for purpose-designed filters. Filtronics make many filters to the users specification and claim a turnaround time of 6 weeks. What follows is a checklist of things that should be specified for built-to-order filters.

- Passband frequency
- Stop band (rejection frequencies)
- Passband loss
- Return loss (usually around 20 dB)
- Connector type-N type and 7/16 the most common
- Mounting details, indoor/outdoor
- If outdoor, mounting bracket description
- Intermodulation levels, if an issue
- Power handling
- Number of carriers and the power of each
- Size
- Temperature range
- Filter height. It is cheaper if they can be kept at least 80 mm for 800 MHz and 40 mm for 1800 MHz.

What Is Still a Problem in Filter Designs?

Precise filter designs are still problematic. Designing a filter that is very close to specifications, that can be tweaked readily to meet the requirements, is routine, but designing a filter "ready to go" still, in 2000, requires computing power beyond even a top-of-the-line PC.

Although in principle any temperature changes can be compensated for, in practice it is difficult and costly to cater for temperature extremes and still keep the filter within specification.

Most filters will produce spikes at multiples of the passband, and this should be considered when co-siting 900 MHz and 1800-MHz systems.

Masthead LNAs

Nearly all PCN/PCS base stations can benefit from a masthead LNA. This is because, in general, PCS/PCN is competing directly with conventional cellular systems, which have the dual advantage of higher operating power and lower frequency, giving them a net effective gain of at least 6 dB. Even at 900 MHz GSM often has to compete with CDMA or AMPS, both of which have a significant advantage in coverage.

CDMA systems themselves have a breathing problem, which effectively limits their range under load, and in many instances these systems can also benefit from and LNA installation.

Filtronics offer the following LNA's: for CDMA and LNA with a noise figure less than 1 dB, and a net gain of 15 dB; for PCS/PCN, a masthead LNA with a 1.5–2-dB noise figure for a 25-MHz bandwidth.

SUPERCONDUCTORS

Superconductors were first discovered in 1911, when it was found that the resistance of a mercury wire suddenly plunged to zero at about 4 K (0 K is −273°C). Until 1986 all known superconductors were metals. In fact 27 of the elements are superconductors at low temperatures and pressures and they are all metals. An additional 11 elements are superconductors at high pressures and low temperatures; included among these are some common metals such as aluminum, tin, lead, and mercury. Conduction in metals is accomplished by a sea of free electrons that move easily throughout the metal structure. This movement is hampered to some extent by impurities, dislocations, grain boundaries, and lattice vibrations (phonons); collectively these impediments account for the resistance of the metal. It was long known that at reduced temperatures resistance decreased (as order increased), so it was a fairly obvious step to follow this down to very low temperatures to see what might happen. It was found that for direct current (DC) currents superconductors ceased to have resistance at temperatures approaching absolute zero. At the time this was a bit of a surprise, and something that remained unexplained for a long time.

In superconductors at low temperatures, a bonding occurs between electron pairs, called *Cooper pairing* (the pairing is a phenomenon of momenta more than position). Today there are hundreds of known superconductors and most are compounds or alloys. Most of these are low-temperature (below 10 K) superconductors. All superconductors also have a critical magnetic-field level (which is dependent on the superconductive material), above which they revert to normal conductivity behavior. This magnetic property is most important in engineering applications, as it places an upper limit on the current that can be carried without destroying the superconductivity.

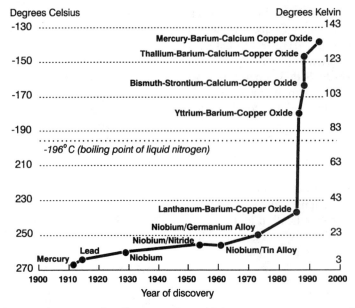

Figure 10.35 The progressive discovery of ever higher temperature superconductors.

Low-temperature superconductors are usually kept in liquid helium and have limited engineering applications because of the high cost and difficulty of maintaining those temperatures. High-temperature superconductors can operate in liquid nitrogen and are ceramic. The highest known temperature that a superconductor will operate at is 134 K, and this is a complex of copper oxide. High-temperature superconductors are lanthanum, yttrium, or another of the rare-earth elements, or bismuth or thallium; usually barium or strontium (both alkaline-earth elements); copper; and oxygen. The progressive discovery of ever higher temperature superconductors is shown in Figure 10.35.

It is traditional in the industry to abbreviate the full chemical symbols for the elements by using only the first letter used, so that

$$\text{yttrium (Y)} = Y$$
$$\text{Barium (Ba)} = B$$
$$\text{Copper (Cu)} = C$$
$$\text{Calcium (Ca)} = C$$
$$\text{Oxygen (O)} = O$$

SUPERCONDUCTING FILTERS

High-temperature superconductors will work in liquid nitrogen, a relatively inexpensive gas to freeze. It does mean, however, that along with a cavity it will be necessary to have a refrigeration plant. Adding to the bulk

of the arrangement is the fact that the cavity must be able to continue to work even in the event of a power failure; this means building in thermal inertia and very good insulation. Cavities with 6 hours power-off capability have been demonstrated. Naturally, this means that normal operations are at temperatures somewhat lower that the minimum required for superconductivity.

The advantage of superconductors for filter construction is that elaborate filters can be constructed to meet exacting designs without the attendant introduction of high ohmic loss. To obtain sharper cutoffs, or more complex filter masks (like in-band notches), it is necessary to add more poles, and more poles mean more loss, which is minimized with superconducting construction.

Just as conventional filters are often silver plated to lower the cavity resistance, superconducting filters are "plated" with superconducting material. There are two commercial approaches being taken with superconducting filters. These are known, respectively, as thin film and thick film. Thin-film fabrication is done by "sputtering" the film onto the cavity walls, using a vacuum deposition process. Thin films are of the order of 1 micron, while thick films are about 100 microns (as a comparison, economy cars today are regularly painted with 5-micron-thick paint). Thick films are painted on and then kiln fired.

Thick-Film Superconducting Cavities

The thick-film approach is taken by Illinois Superconductor Corporation (ISC), which has used the technol-

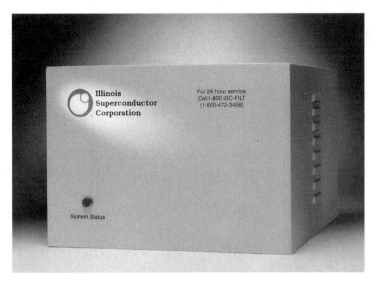

Figure 10.36   The ILS Omega 150 series. (Courtesy of Illinois Superconductor Corporation.)

ogy to produce not only receiver filters but also transmitter and combiner units. The essence of thick-film technology is to use the superconductor as a surface coating on an otherwise conventional filter to reduce losses and so improve the Q. Because the losses are lower, more complex filters can be made and higher Qs follow.

A filter for CDMA/time division multiplex access (TDMA)/AMPS, which has a hot standby backup (a regular filter that is switched into service), in the event of coolant failure is seen in Figure 10.36 and has the performance curves shown in Figure 10.37. The 300 A covers the A-band, and the 300 B the B-band.

The characteristics of this filter are listed in the following table:

	Normal (cool)	Backup (warm)
In-band noise figure	0.5 dB	3 dB
Gain	15–25 dB	3 dB below set point
IP3	15 dBm	15 dBm
Return loss	>13 dB	>10 dB

It is interesting to note that superconductors have zero resistance to DC currents, but they offer some resistance to alternating current (AC) (and so they offer resistance to the RF currents that are circulating in a cavity). Of course, the resistance to the current will result in heat dissipation, and this heat increases the cooling load. Another disadvantage of superconductors in cavities is that they produce intermodulation distortion in the superconducting material itself.

Superconductor filters have been demonstrated in practice in cellular and PCS preselector filter applications, and in the laboratory transmitter combiner cavities with Qs of 10,000+. However, such Qs are not much higher than the best conventional filters, some of which have Qs of the order of 11,000, but they are capable of much more elaborate filtering where complex passbands are needed.

Increasingly in filter technology the resonators themselves are taking on new and variant forms. ISC uses doughnut-shaped resonators, as seen in Figure 10.38. These rings are coated with a black layer of thick-film superconducting material, and they are mounted so that they are clear of the cavity walls.

Thin-Film Superconductors

The other approach is to use thin-film technology. The nominal operating temperature is 77 K.

Thin films allow designs with up to 19 poles and a net filter loss of less than 1 dB. Conductus (California) has produced a commercial thin-film filter that has an insertion loss of only 0.4 dB and a band-pass characteristic that allows a 25-MHz total bandwidth as, seen in Figure 10.38.

Thin-film technology permits the construction of very small filters using small wafers of materials and construction techniques similar to semiconductors. An 800 MHz, thin-film filters can be as small as a few centimeters across.

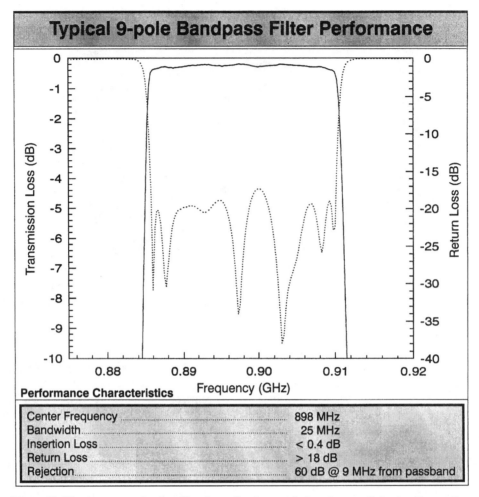

Figure 10.37 A superconducting filter from Conductus, designed as a cellular band-pass filter.

Figure 10.38 The resonators of a thick-film superconductor filter. (Courtesy Illinois Superconductor Corporation.)

The filter itself is a lumped-element filter, which means that it is essentially made up of distributed capacitance and inductance. The distribution of these elements is so thorough that the term "quasi-lumped" elements is often used.

The filters are made in a manner similar to a microchip by placing the superconductor on a substrate. The superconductor is grown epitaxially (single crystal) on a single-crystal oxide substate; typically MgO or LaAlO$_3$ is used for the substrate. The crystals are individually grown and are chosen for their low loss at microwave frequencies and for high superconductor yields, and are from 0.375 mm to 0.5 mm thick.

In 1999, these MgO crystals were grown in wafers 2 inches square, and because a practical filter needs to have many individual filters combined to achieve the desired response, dimensions around 1 cm are about the maximum for one resonator. However, because of the lumped elements, a resonator can be very small compared to a wavelength.

The High-Temperature Superconductor (HTS) can be of TBCCO or YBCO.

The combined crystal substrate and the filter, complete but unpackaged, is known as a die, and is the equivalent of a semiconductor chip.

Conductus is working on a three-dimensional structure that will allow more larger filters to be built around such a structure.

As of 1999, Qs of up to 500,000 and powers of up to 100 watts have been achieved. Operational frequencies from 18 MHz to 90 GHz have been demonstrated. Improvements in the technology have seen Qs increase by around 50 percent per year.

When to Use Superconducting Filters

Superconducting filters are undoubtedly the *best* filters, but they are also the most expensive. They are justified under certain conditions.

When spectrum is tight and there are multiple operators using the same bands, guard bands that are put in place to minimize adjacent channel interference are a costly waste of spectrum. Superconducting filters can permit the recovery of much of the guard band by providing filters with sharper cutoffs.

For situations where coverage is a prime consideration (in rural locations and along highways), the use of a premium filter can reduce losses by 2–4 dB, which in turn can increase the range by about 20 percent. This in turn can increase the area covered by around 45 percent, which in many instances could translate directly to a 45 percent decrease in the number of base sites. GSM systems in general have a path budget disadvantage over analog and CDMA, and this is even more pronounced with GSM 1800 and GSM 1900. For GSM 900 there are applications for recovery of guard band and for mainly rural improvements in path budget. For GSM 1800 and GSM 1900, capacity is rarely a problem, but range in low-density areas is.

For CDMA the main application would be to reduce the guard bands to the adjacent analog (or sometimes digital) networks.

Analog applications of the technology abound. Adjacent channel problems between A and B carriers are well known, and the complex structure of the B-band represents a real challenge for conventional filters that is easily tackled by superconducting filters.

11

CELLULAR REPEATERS

Because cellular base stations are expensive, repeaters are sometimes used to fill in small areas not properly covered by the base-station network. Cellular repeaters are cheaper than base stations and can be very useful, but their limitations must be understood. A number of cellular-radio repeater systems are now available. They are useful in isolated, low-density areas. This chapter discusses typical applications for repeater systems.

In analog networks, repeaters are essentially only used to fill in coverage areas that were awkward or not profitable enough to be covered by a base-station site. Digital systems, and more particularly personal communication service/digital communications system (PCS/PCN) have changed all that. A digital PCS system (and to a lesser extent a CDMA 800 system) that is designed to cover a reasonably wide area will almost certainly have excess capacity, because the small cells and large carrier capacities dictate this.

It is the general rule, particularly in outer suburban and rural areas, that PCS systems [whether Global System for Mobile Communications (GSM) or code division multiple access (CDMA) based] can be installed at lower cost by making repeaters an integral part of the design; better still, this can be done without in any way compromising the quality of service. In fact, in CDMA systems advantage can be taken of the massive built-in capacity at the rollout, by making extensive use of repeaters (they may exceed the number of base stations by a factor of 4 or more). Repeaters are significantly smaller than a base station and can ordinarily be mounted on a utility pole. Because of this they have less environmental impact and can be deployed very rapidly. Modern repeaters offer network management facilities and regular status reporting.

DIGITAL REPEATERS

Digital repeaters can use channel-selective repeaters where Surface Acoustic Wave (SAW) filters are used to ensure that only the required donor base-station frequency is repeated. The carrier bandwidth of 200 kHz for GSM or 1.25 kHz for CDMA can be programmed in for any of the operator's frequencies.

Both the European Telecommunications Standards Institute (ETSI) (for GSM) and the American National Standards Institute (ANSI) (for CDMA) have recognized the important role that repeaters are playing in the rollout of the new digital networks and have released standards GSM 05.05 and J STD 008, respectively, to cover them.

There are basically two types of repeater: the broad-band repeater (also called the cell-extender repeater) and the cell-replacement repeater. These repeaters are shown in Figure 11.1. The broad-band repeater amplifies all the channels in the operator's spectrum. To have reasonable total power consumption and isolation, this repeater must have low power per channel.

The high-power cell-replacement repeater amplifies a few channels only. Frequency translation occurs so that the repeated cells are on a different frequency from the main base. Cell-replacement repeaters are used along highways where the extended coverage is not expected to attract high traffic volumes.

CELL-EXTENDER REPEATERS

The cell-extender repeater allows the coverage of a cell to be extended cheaply in areas adjacent to the main

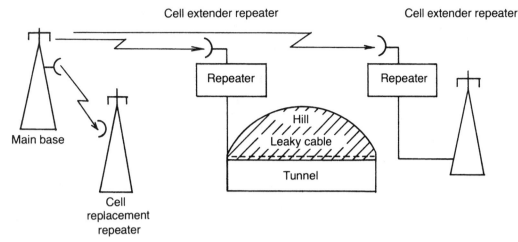

Figure 11.1 A cell with a cell-replacement repeater and two fill-in cell extenders, one behind a hill and another feeding a leaky cable in a tunnel.

cell that may not otherwise be covered. No modification is required at the original cell site. This extender is designed to cover relatively small areas, typically about 0.5–2 km.

As shown in Figure 11.2, this type of repeater is simply a wide-band amplifier that amplifies and repeats the host base-station channels. The cell extender typically costs only about 10 percent of the cost of a new cell site, but it has some serious limitations.

Using a wide-band amplifier makes it difficult to have

a really sharp cutoff at the edge of the band. Figure 11.3 shows that the decision to place the control channels for the A- and B-bands adjacent to each other clearly was done without regard for this type of repeater limitation.

In Figure 11.3, you can see that with practical filters, either the gain of the system over the control-channel band is not consistent or the adjacent control channels are also amplified.

A filter that causes differential control-channel amplification severely restricts the cell-pattern flexibility

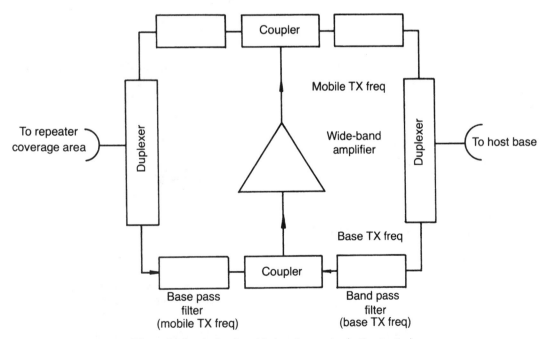

Figure 11.2 A simple wide-band repeater (cell extender).

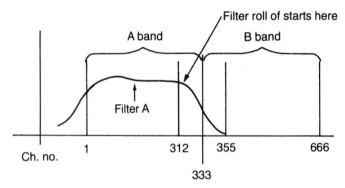

Figure 11.3 Repeater filters have cutoffs asymmetrically across the control channels. This band-pass filter cuts off at the edge of the speech channels.

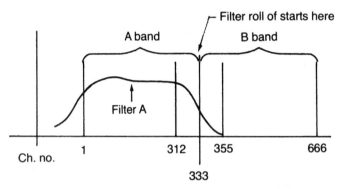

Figure 11.4 Filters can give a flat response over the desired control channels. This band-pass filter is flat over the band of operation, but significantly amplifies the adjacent control channels.

because only cells with control channels, the frequencies of which are close to the voice channels, can be repeated.

If a filter that amplifies all the control channels equally is used, then problems can occur in the adjacent band. For example, if such a filter is used on the Advanced Mobile Phone Service (AMPS) A-band, as shown in Figure 11.4, then the adjacent B-band control channels are amplified to some extent. Amplifying the A-band control channels causes many unsuccessful call attempts to occur on the A-band system because the repeaters repeat the control channels but not the voice channels. (This applies only to incoming calls that use the control channels for paging.)

Another problem in a simple repeater system is multipath cancellation on the control channels. Figure 11.5 illustrates this problem. Destructive interference that renders the area unserviceable can occur over the area where the direct path from the host base and the re-

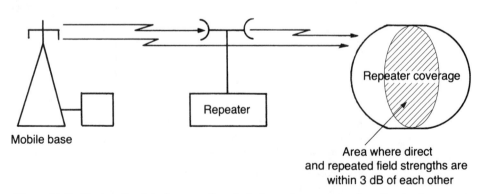

Figure 11.5 Destructive interference of control channels can occur in cell-extender repeater systems.

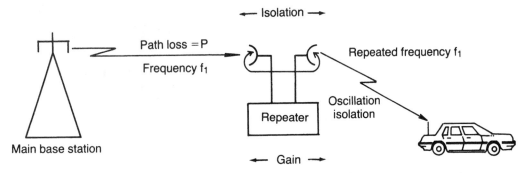

Figure 11.6 A repeater is a potential oscillator. Cell extenders are potential oscillators as the input signals are amplified and retransmitted without frequency translation.

peated transmission have field strengths within ± 3 dB. This is a problem mainly on the data channels; it is recognized by the existence of areas where calls cannot be reliably established or received, but where voice communications, set up elsewhere, are available.

Because the repeater amplifies its own input, it is a potential oscillator, as shown in Figure 11.6. In practice, the isolation between the receive and transmit antennas must exceed the gain by a margin of at least 10 dB, to ensure stable operation. That is, the total isolation between the input and output must be 10 dB greater than the gain. Unity gain or 0-dB isolation is needed for oscillation.

This isolation is usually provided by high-gain antennas with good front-to-back ratios, plus vertical and horizontal antenna separation. Repeaters usually have a variable gain, adjustable over a wide range. Purpose designed antennas for repeaters have recently become available, which offer high front to back ratios (and hence high isolation).

In order to get reasonable power levels from the repeater, it is necessary to operate fairly close to the maximum-gain margin (10 dB). Gains of about 60 dB are typically used. Instability can be caused by the external environment, however, when such things as passing aircraft or wind in the trees alter the feedback levels.

You can calculate the antenna isolation using the following formula:

$$\text{Vertical isolation} = 25 + 40 \log(2.8d) \text{ dB}$$

$$\text{Horizontal isolation} = 22 + 20 \log(2.8d) \text{ dB}$$

where $d =$ spacing in meters.

Figure 11.7 shows the antenna locations and the way in which the spacing (d) is measured.

Usually, although not necessarily, the repeated cell covers a relatively small region that can be served by a directional antenna. Where an omni-directional antenna must be used, the isolation is decreased and the maximum power at the repeated site is limited (due to the oscillation margin).

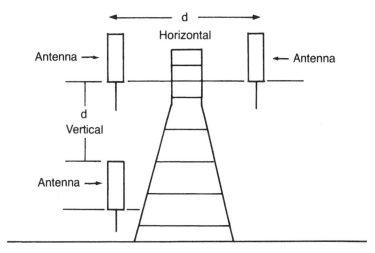

Figure 11.7 Vertical isolation is measured as the smallest distance between the two antennas. The distances are measured as indicated.

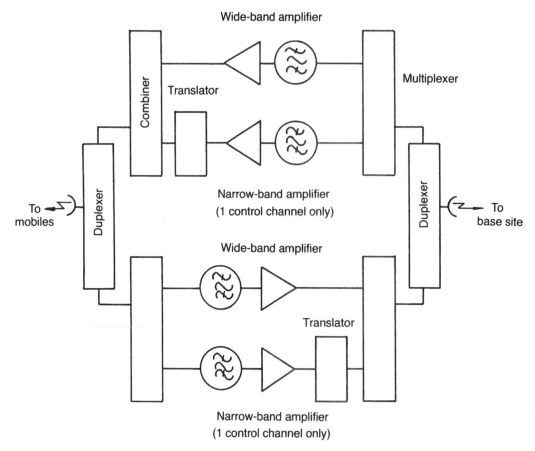

Figure 11.8 In an enhanced repeater, the control channel can be filtered and translated in frequency. The send-and-receive channels are independently translated.

The maximum gain that can be used at the repeater is as follows:

$$\text{Maximum gain} = \text{isolation} - \text{oscillation margin}$$

As mentioned earlier, 10 dB is a practical value for oscillation margin.

To calculate the radiated power, start with the Effective Radiated Power (ERP) at the base; for example,

$$50 \text{ watts} = 47 \text{ dBm}$$

$$\text{Subtract path loss } P = -P$$

$$\text{Received signal} = 47 - P$$

$$\text{Maximum repeater gain} = \text{isolation} - 10 \text{ dBm}$$

$$\begin{aligned}\text{Maximum ERP at repeater} \\ = 47 - P + \text{isolation} - 10 \text{ dBm}\end{aligned}$$

This isolation is mainly determined by the antenna selection and location.

Intermodulation is always a problem with wide-band amplifiers, and a well-designed repeater minimizes intermodulation products. Because of the wide separation between the send-and-receive frequencies of cellular systems (usually 45 MHz or more), using separate amplifiers in the send-and-receive directions can contribute significantly to lower intermodulation.

Because of the bandwidth, gains, and number of channels, the output power of a cellular repeater is typically limited to about 0.25 watt per channel. This is adequate for many repeater applications.

ENHANCED-CELL EXTENDERS

Because most of the limitations of the simple repeater in analog systems adversely affect the control-channel operation more than the voice channels, an obvious refinement is to separate out the control channel and, using a narrow-band amplifier, process it separately. The enhanced-cell repeater is a refinement of the simple repeater. Using frequency translation on the control channel only can eliminate the interference problem previously discussed. This is done in modern repeaters.

Figure 11.8 shows an enhanced-cell repeater.

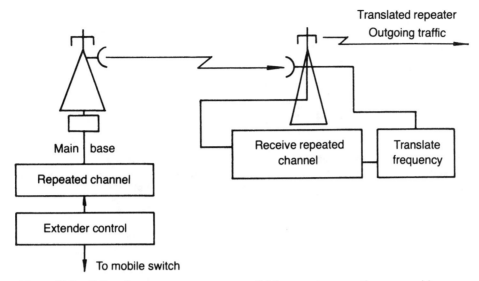

Figure 11.9 Cell-replacement repeaters are useful for repeater operations over wide areas.

TRAFFIC CAPACITY OF THE SIMPLE REPEATERS

Because this type of repeater repeats only those channels required for calls, the base station cannot distinguish between direct traffic and repeater traffic. Thus, the marginal traffic capacity of the repeated cell is added arithmetically to the local traffic. In this way, the base station is equipped with enough channels to carry the cell traffic only. It is not so simple with extended cells.

CELL-REPLACEMENT REPEATERS

The cell-replacement repeater operates quite differently from a simple repeater and requires a hardware and software interface with the main base and switch.

Figure 11.9 shows the cell-replacement repeater. In this instance, the main base can act as a normal base or, under the control of the extender controller, a particular channel can be used as a link channel to the repeater base. Seen from the view point of the mobile switch, the repeated channel is frequency agile and can change from the normal frequency to the repeated frequency.

This system avoids interference from the main base to the extended base because different channels (frequency translation) are used at each base. The repeater can therefore use high-power omni-directional antennas. Typically, the repeater channels are 10 watts ERP, as shown in Figure 11.9.

The cell-replacement repeater is useful when additional coverage of a large area with few subscribers is anticipated.

The repeater base consists of two back-to-back repeaters; it performs much the same as a normal base station, but it needs no link system and no controller. This is cheaper than a full base station only for small cells, as you can seen from comparing the equipment required in Figures 11.10 and 11.11.

For small numbers of channels to be repeated, the cost of an extra transceiver is less than the cost of a link-

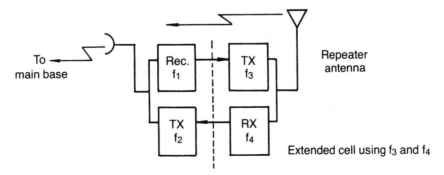

Figure 11.10 A cell-replacement repeater is basically two back-to-back transceivers that allow frequency translation. No site controller or link system is necessary.

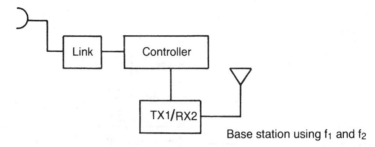

Base station using f_1 and f_2

Figure 11.11   A base station consists of a transceiver, a site controller, and links to the switch. The controller sends different channel assignment instructions to the mobile, depending on whether the channel f_1/f_2 is directly accessed or is accessed through the repeater.

plus controller. However, the cost per channel of the repeater (requiring two transceivers/channel) rises faster with additional channels than does the cost of adding channels at a base station.

For example, suppose you have the following costs:

- A transceiver costs $8000 (including combiners and antennas)
- A link costs $9000
- A controller costs $30,000
- An extender-controller costs $10,000
- Assume other infrastructure costs are equal (batteries, building, towers, and so on)

The cost of the base station is

$$90 + 30 + 8 \times N = 120 + 8 \times N + \$1000$$

where N = number of channels. The cost of a repeater is

$$(8 + 16) \times N(2 \text{ repeaters/channel} + \text{original channel}) + \$10,000$$

The costs are equal when

$$120 + 8N = 24N + 10 \text{ (that is, } N = 7 \text{ channels)}$$

Because dedicated repeaters are usually available in modules of seven carriers (at about 10 watts each), it is reasonable to assume that this is the upper limit of the economic viability of this type of repeater.

TRAFFIC CAPACITY OF CELL-REPLACEMENT REPEATERS

The traffic capacity of the cell-replacement system presents some interesting problems. Imagine channels that are extended as last-choice routes (channels that are extended and used only for local traffic when all non-extended channels are fully occupied). Figure 11.12 illustrates this situation.

The optimum number of channels, K, can now be found using the conventional traffic theory for a main

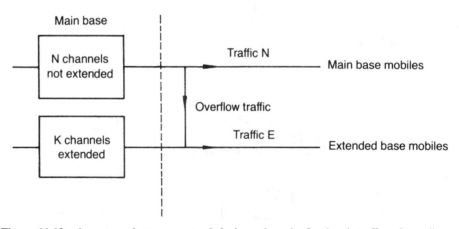

Figure 11.12   A system that uses extended channels only for local traffic when all non-extended channels are used.

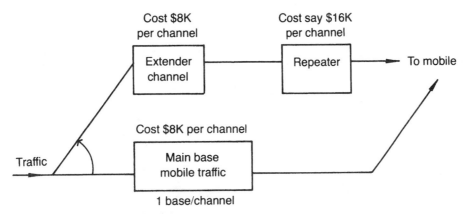

Figure 11.13   A cell extender as an alternate routing system.

route with one overflow path. The optimum number of channels, K, may exceed the traffic capacity required just to serve the extended base mobiles. This depends strongly on the relative costs of the alternative routes, but at current prices, little or no overflow could be justified. Be sure to calculate this extra capacity when comparing the cost of a repeater with the cost of an additional base.

Figure 11.13 shows a cell extender as an alternate routing system. (For more information about routing, see Chapter 20, "Traffic Engineering Concepts.") From Figure 11.13, it can be seen that:

Cost of the main route $= \$8000/\text{channel} = C_1$

Cost of the overflow route $= 24,000/\text{channel} = C_2$

Occupancy of the main circuits ≈ 0.7
$= B_1$ Erlangs/circuit

Occupancy of the overflow circuits ≈ 0.5
$= B_2$ Erlangs/circuit

$$\text{Cost factor } H \approx C_1 \left[\frac{B_2}{C_1 + C_2} \right] = \frac{8 \times 0.5}{(8 + 24)} = 0.125$$

A cost factor of 0.125 suggests that almost no overflow is justified, so it would be just as economical to dedicate the repeater channels at the base to repeater use only (see Figure 20.13, which shows that this cost factor results in circuits being provisioned at a rate about equivalent to a Grade of Service (GOS) of 0.05, i.e., there will be no overflow). In fact the number of circuits for $H = 0.125$ and, for example, for 15 Erlangs is 22 (from Figure 20.13) and this compares to 21 circuits for 15 Erlangs carried at a GOS of 0.05, using Erlang B tables. If the cost per channel of the repeater were lower, however, this might change.

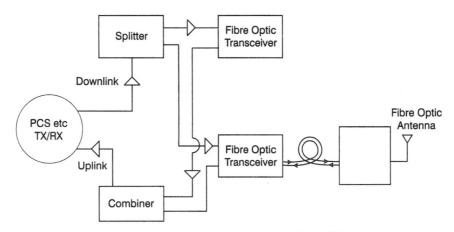

Figure 11.14   The universal repeater MirrorCell.

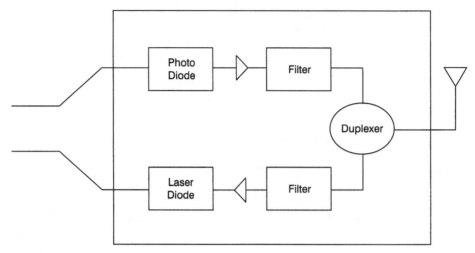

Figure 11.15 The antenna of the MirrorCell.

TUNNELS

In the case of a leaky cable feeding a tunnel or building, the isolation is very high, and therefore very high-gain (high-power) repeaters can be used. Because the loss in the leaky cable occurs exponentially, it will probably be necessary to cascade amplifiers in the cable. Sometimes it is necessary to use cable sections that taper, being lowest loss nearest to the first repeater and then increasing in loss toward the end.

UNIVERSAL REPEATERS

Using the capabilities of fiber-optic systems, it is possible to repeat any radio frequency (RF) signal regardless of its format or protocol. Ortel makes such a repeater, which is called MirrorCell, and while most of its applications are for PCS, it can be used for CDMA, GSM, cordless phones, or any other protocol. The way the repeater is connected to the radio transceiver is shown in Figure 11.14. The system is available for just a few distributed antennas up to much larger systems being served by a distribution hub that can support 16 remote antennas.

The basis of this system, as shown in Figure 11.15, is the photodiode to convert the light signal to RF and the laser diode to convert the RF to an optical signal. This diagram is a little oversimplified, as on the uplink there is an amplifier before the laser diode.

12

ANTENNAS

Antennas play an important role in shaping cell patterns and in determining reuse patterns. There are two basic antenna types. The first type has an omni-directional pattern; the second has a sector, or unidirectional, pattern.

GAIN

Because most antennas are passive devices, they can achieve gain in one direction only at the expense of gain in another. Just as a car headlight concentrates the radiated power of the filament in one direction, so gain antennas cause the signal to be relatively stronger in one direction than another. For most mobile applications upward radiation (into space) and downward radiation (into the ground) are undesirable, so minimizing radiation in these directions while concentrating it in the forward direction is advantageous.

VSWR

An antenna with a voltage standing-wave ratio (VSWR) of 1.0 will transmit all of the power presented to it. As the VSWR rises, an increasing amount of power will be reflected. It is generally accepted that a VSWR of 1.5–1.7 is the highest acceptable value for an antenna. An antenna with a VSWR of 1.5 will reflect 4 percent of the total power.

BANDWIDTH

The bandwidth of an antenna is the range of frequencies over which the VSWR remains below 1.5–1.7 or some other defined VSWR usually less than 2. The VSWR will vary as a function of frequency in a manner shown in 12.1. Occasionally the bandwidth will be specified for a VSWR of 2.0, and such an antenna will not be as good a match as one specified to a VSWR of 1.5.

BEAMWIDTH

Because the gain of an antenna is a result of pattern compression, there will generally be a direction in which there is maximum gain, as seen in Figure 12.2. The beamwidth is defined by the two points that define the half-power levels (down 3 dB).

FRONT-TO-BACK RATIO

The front-to-back ratio is ordinarily measured as the ratio of the gain of the maximum lobe compared to the gain at 180 degrees to that direction. As can be seen in Figure 12.2, that number may not give a true impression of the actual power levels that are scattered in the backward direction.

ANTENNA CONSTRUCTION

The earliest cellular antennas were simple dipoles and sector antennas derived from dipoles (effectively dipoles surrounded by a reflector). These devices served the industry well and are still the mainstay of small and rural networks.

Later came the antennas with mechanical downtilt. These antennas, while simple in concept were the source

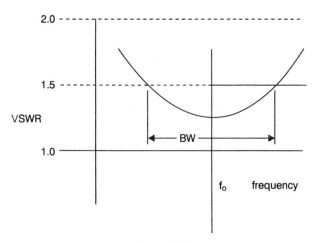

Figure 12.1

of many network problems. The mechanical downtilt distorted the azimuth beam pattern and often gave unpredictable results. A further improvement in the early 1990s was electrical downtilt, which relied on phasing of the antennas and produced an undistorted downtilted pattern.

Experiments showed (and theory predicts) that the polarization of a signal that had traveled extensively in a mobile environment was no longer vertical. Further studies showed that two cross-polarized antennas received signals that were sufficiently uncorrelated, and that they provided diversity similar to that of two spatially separated vertically polarized antennas. These cross-polarized antennas could be mounted into a single radome, thus giving diversity with half the number of antennas.

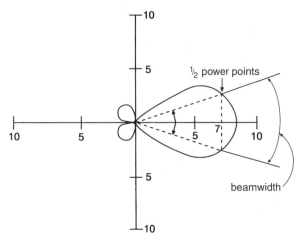

Figure 12.2 The beamwidth of antenna is defined by the 1/2 power points.

The most recent development in antennas has been cross-polarized antennas with downtilt.

Increasingly there is a need for dual-band antennas, and one elegant solution to this need comes from Radio Frequency Systems (RFS), which has a dual-band antenna (900/1800 MHz) that features a single feeder cable and comes with a splitter that is accessed at the base station to separate the two frequencies.

OMNIDIRECTIONAL ANTENNAS

An omnidirectional antenna is usually a collinear dipole (a number of dipoles in a line with a phasing harness). Figure 12.3 shows an omni-directional antenna.

In an omni-directional antenna, a power divider may be required to phase a number of dipoles within the one gain antenna or to connect two antennas to the one feedline. The quarter-wave transformer shown in Figure 12.4 is a simple power divider. Because the power divider is fed inside the antenna, the internal wiring harness is quite complex. A cellular omni-directional antenna is usually of this kind, and as it has a good wide-band performance. The matching section of the quarter-wave transformer has a characteristic impedance of $Z_o = \sqrt{Z_{IN} \cdot Z_i}$. (See Figure 12.4)

The unit in Figure 12.3 is an 8.5-dB collinear antenna constructed with a fiberglass radome. Extreme care must be taken with the choice of materials used in the antenna, as dissimilar metals can cause corrosion, which in turn leads to intermodulation problems.

Simpler, more compact collinear antennas can use passive radiators, as shown in Figure 12.5. Such antennas are widely used in two-way radios because they are much cheaper to construct than the type illustrated in Figure 12.3. However, they operate only on a narrow bandwidth.

OTHER TYPES OF OMNI-ANTENNA

Another version of the collinear antenna that is used for cellular applications is illustrated in Figure 12.6. The concept is to use series instead of parallel-phasing elements between the radiators. This type of antenna is limited to around 6-dB gain because the losses are cumulative, and the ever-decreasing power that is applied to each successive radiator will result in a decreasing contribution to the overall radiation pattern.

Commonly encountered, but rarely used for cellular, the folded dipole antenna, shown in Figure 12.7 is particularly rugged. Its elements are at direct current (DC) ground potential, and this makes it particularly suitable for use in lightning-prone areas. It also is a wide-band

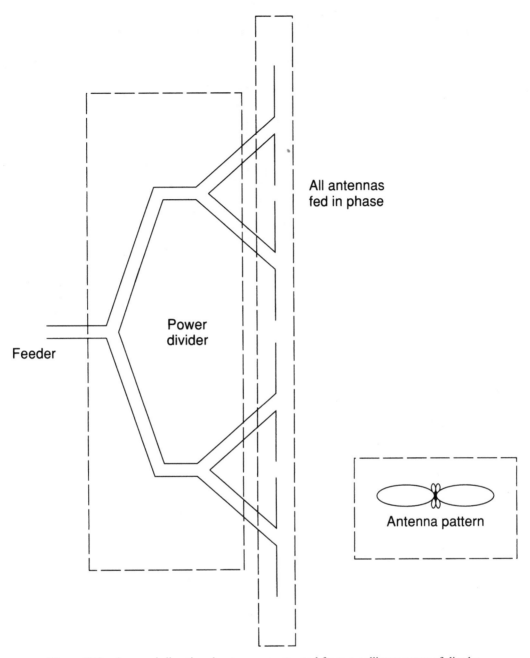

Figure 12.3   An omni-directional antenna constructed from a collinear array of dipoles.

device. Versions exist with the dipoles all offset to one side, and with the dipoles mounted with a 90-degree or 180-degree rotation around the central pole.

The common fiberglass-enclosed collinear antenna (also shown in Figure 12.7) is often used for cellular, but other versions, which can usually be identified by the fact that they are thinner and/or tapered, are used in public mobile radio (PMR) and paging. They generally are somewhat cheaper and less rugged that the folded dipole.

SECTOR ANTENNAS

Sector antennas usually have higher gain than omni-directional antennas and are typically 14 dBd or 17 dBd. Sector antennas may combine the power gains obtained by using phased arrays with the additional gains obtained by using reflectors. Figure 12.8 shows a sector antenna with 17-dBd gain. Such an antenna is, in fact, constructed with the omnidirectional radiator (see Figure 12.3) mounted in front of a reflector. This configu-

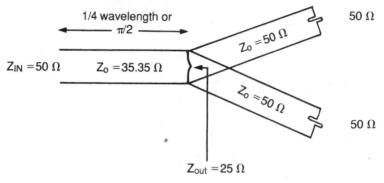

Figure 12.4   A 50-ohm power divider.

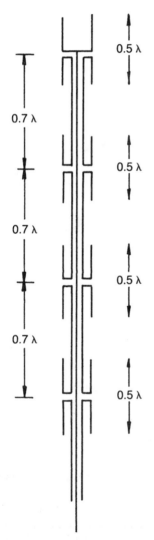

ration has the advantage of needing only one set of spares (an omni-directional radiator) to service both omni-directional and sector antennas. Table 12.1 lists the specification for the antenna shown in Figure 12.8.

Figure 12.9 shows that the VSWR varies considerably over the rated bandwidth, but is always less than 1.5. The maximum gain, however, is virtually flat at 17 dBd over the rated frequency ranges. The radiation pattern

Figure 12.5   A collinear dipole with passive radiators. It is possible to drive only the top radiator and to use induced currents to derive the required pattern. This principle is similar to that of the familiar Yagi.

Figure 12.6   Series-fed radiators have been used to make cellular collinear antennas.

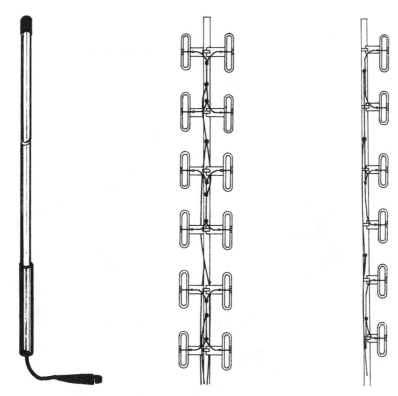

Figure 12.7   Omni-antennas can take all these forms.

Figure 12.8   This a sector antenna has 17-dBd gain and an overall length of 3.76 meters. (Photo courtesy of Deltec.)

TABLE 12.1   Specification for 17-dB Sector Antenna

Type number	MTA860-8-UN
Frequency	825–896 MHz (to 860 MHz)
Bandwidth	71 MHz
Input impedance	50 ohms
VSWR	Less than 1.5:1 over 71 MHz
Gain	17.0 dBd
Vertical beam width	6.5 degrees
Horizontal beam width	60 degrees
Maximum power	500 watts
Polarization	Vertical
Termination	"N" type plug fitted to 50 cm of PTFE coaxial cable
Reflector screen	All-welded alloy finished Alocrom 100 pretreated and coated in white polyester
Mounting section	300 mm × 48.5 mm × 7 swg aluminium grade 6082 finished Alocrom 100
Overall length	3.76 meters
Weight (incl. mounting clamps)	18.2 kg
Wind area	0.734 square meters
Wind loading	107.5 kgf at 160 kph
Mounting clamps	2 × 9099 galvanized steel parallel clamps
Packing	Case with timber ends and sides; hardboard top and bottom panels

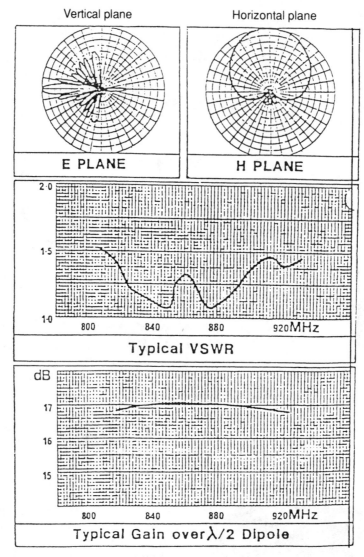

Figure 12.9 Radiation patterns, VSWR, and gain of the antenna shown in Figure 12.8.

illustrated in Figure 12.9 was derived from a computer simulation. When measured on the test range, real antennas show anomalies in their patterns that are not suggested by the simulations.

There seems to be a rather poor understanding of antenna patterns within the industry, and frequently articles appear that deal with nulls beneath high-gain antennas. These nulls do not exist, as an examination of the E- and H-plane diagrams in Figure 12.5 will reveal.

Often antenna patterns are simplified to show only the major lobe and the minor lobes are forgotten. A low-gain antenna will have a much broader major lobe and far fewer minor lobes than its high-gain counterpart. This is because the lobes are the product's interference between the radiation patterns of the various elements that make up the antenna. Because high-gain antennas

have more elements, they generate more interactions. The result is that high-gain antennas, far from having nulls below them, are likely to have quite significant downward radiation. In fact as a "go–no-go" test when surveying a high-gain antenna for cellular radio, the field strength immediately below it (for example, under the tower) should be around 90 dBμV/m (or around 1,000,000 times more than the minimum power needed for a cellular phone to work).

The polar diagrams that depict the vertical radiation pattern usually have a linear scale so that any lobes that appear significant on the diagram will be around 10–20 dB down from the main lobe. At close range (like immediately under the antenna) this will result in very significant local field strength. Referring again to Figure 12.9, notice the significant lobe at the back of the E-

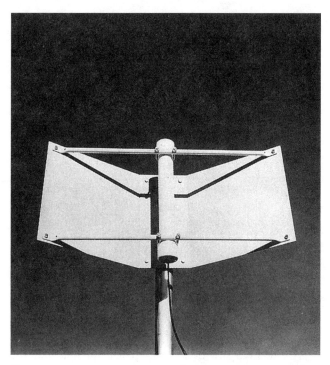

Figure 12.10 A simple corner reflector can provide gains of about 8 dBd in a very compact unit. (Photo courtesy of Deltec.)

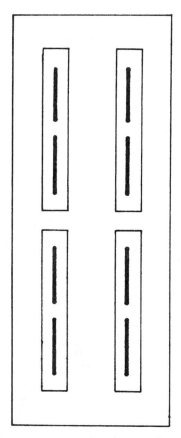

Figure 12.11 A panel array has two parallel arrays of collinear dipoles. These are usually enclosed in a radome.

plane, which is about 20 dB down on the main lobe. This level is known as the front-to-back ratio, and the level is quoted for the antenna in free space. In real-world situations, additional interference paths will reduce that ratio by 6–10 dB. Ongoing efforts by some manufacturers to improve the front-to-back ratio seem to be somewhat misplaced, as there is little prospect of gaining an advantage from this except in rural environments, where it is generally not needed.

It is these back lobes in particular that make even high-gain antennas susceptible to interference for distant mobiles. Knowing that these lobes exist, you can improve the interference immunity by mounting the antenna so that power in the backward direction is decreased—for example, by mounting the antenna against the side of a wall. There are limits to improvements that can be made, however, since a significant amount of unwanted signal from behind the antenna will be a result of reflections from structures in front of it.

It must be kept in mind that the antenna polar diagram is a free-space characteristic, but that the antenna must operate in a cluttered, multipath environment that approximates free space very poorly.

In places where high gain is not necessary, a smaller and simpler reflector can be used. Figure 12.10 shows a simple corner reflector. Such reflectors are made from

unity gain radiators mounted in front of a radiator. The main application for this type of antenna is covering small localized regions such as tunnels and parking garages.

PANEL ARRAYS

Panel arrays have long been popular in TV applications, but have recently found applications in cellular radio. As shown in Figure 12.11, they consist of a number of dipoles stacked horizontally as well as vertically.

As can be seen in Figure 12.12, modern panel antennas are physically smaller and less obtrusive than the conventional corner reflectors, and as such are most suitable for mounting on buildings and in places where aesthetics are important. This 13-dB Deltec antenna features both electrical and mechanical downtilt, operable over the range of 2–15 degrees.

Panel antennas are becoming increasingly more popular. Although they have somewhat higher wind loadings than the conventional corner reflectors, they are less

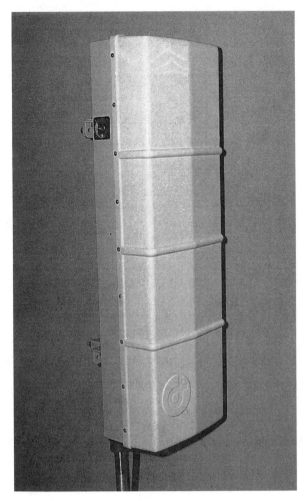

Figure 12.12 A panel antenna operable from 820 MHz to 960 MHz. (Photo courtesy of Deltec.)

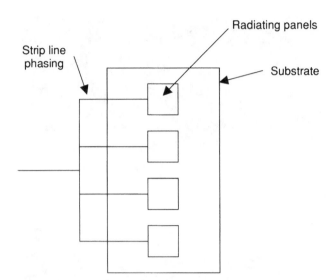

Figure 12.13 Modern panel arrays are sometimes etched in a process much like that used on printed circuit boards.

conspicuous (a major factor when it comes to convincing the building owner to allow the rooftop to be used for an installation) and they are believed to be more reliable. The construction of a typical panel antenna is shown in Figure 12.13. Note that the dipoles are plates and so will have a wide bandwidth compared to a collinear device. Antennas with a bandwidth sufficiently wide to permit the same unit to cover both the code division multiple access (CDMA) and the Global System for Mobile Communications (GSM) band are now commercially available.

If the antenna is filled with a suitable dielectric foam, the ingress of water (the major culprit in corrosion) can be greatly inhibited and so reliability will be increased.

By the use of appropriate phasing these antennas can be designed to have half-power bandwidths from 60 degrees to 180 degrees and so meet all the requirements of cellular sector antennas.

The relative size of a small panel antenna and the conventional corner reflector can be seen in Figure 12.14.

The patch antenna is a microstrip that is generally either square or round in shape and is mounted onto a dielectric substrate, with a conducting ground-plane backing, as seen in Figure 12.15. A resonant square is one half a wavelength on a side, while a round patch is resonant when the radius is about 0.3 wavelength. The patches can be made easily from double-sided printed circuit boards and the unwanted material is etched away in the regular manner. The feed point either will be at the edge or via a hole in the ground plane. The ground plane forms part of the antenna, much like it does for a monopole, and so the radiation analysis is usually done in the half-plane (as it is with a monopole).

POLAR DIAGRAMS

Some considerable care should be taken when using polar diagrams to draw conclusions about the performance of an antenna in the multipath environment that is usually encountered in mobile radio. When manufacturers measure the polar pattern, they will usually do it in a purpose-designed antenna range, which has been carefully set up to give the true *free space* performance of the antenna. This means that the tower will be designed so that it will not interfere with the radiation pattern (it may in fact be made of wood); there will be no obstructions such as other antennas or lightning rods and the measurements will be conducted in the near-field completely free of multipath.

The situation is even more pronounced for mobile antennas where, in the real world, the vehicle (or person) forms an imperfect ground plane that is moving in three dimensions. In particular the antenna is rarely ever vertical, and it is rarely receiving a signal that is incident to it at right angles. Because of this, the carefully drawn polar plots of mobile antennas mounted on a vehicle that is being rotated on a turntable with the intention of showing the relative merits of one mobile antenna over another or one mounting position over another should be treated with skepticism. The effectiveness of an antenna destined to operate in the multipath environment *can only be measured in that environment*.

VOLTAGE AND POWER LIMITATIONS

It is widely appreciated that the number of transmit channels that can be combined into one antenna is limited by the channel spacing required to keep losses and intermodulation in the combiners at acceptable levels. UHF high-gain antennas consist of a number of radiators connected together with a harness for phasing and power splitting. This harness has to be relatively low-loss and at the same time be small enough to fit inside the collinear harness. The loss of the harness will be about 1 to 2 dB. This may seem a small loss, but in terms of power dissipation it means that between 20 percent and 33 percent of the power applied to the antenna will be lost in the harness. When radio frequency (RF) power on the order of 500 watts is driven into the antenna, the result can be from 100 watts to 165 watts to be dissipated by the harness. For this reason, on the specification sheets of most antennas, there will be a power rating. This power rating is a function of the power-dissipation capabilities of the phasing harness.

In cellular applications, it is not, however, the power rating that is the main stressor of the antenna, but rather

Figure 12.14   The relative size of panel antennas and the older corner reflectors.

The nice, clean pattern with its distinctive major and minor lobes applies only to the free space environment in which the measurement took place. In the real world, the pattern will be more complex, with multipath effects significantly reducing the back-to-front ratio, and generally distorting the free space pattern.

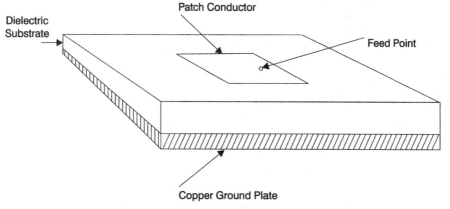

Figure 12.15   A patch antenna.

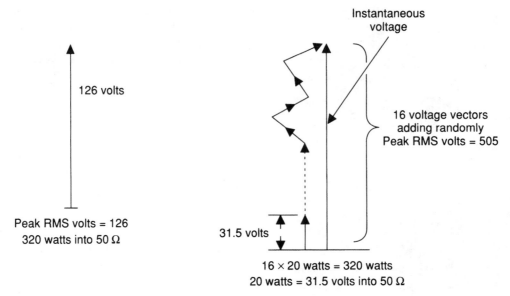

Figure 12.16 The voltage vectors in an antenna are such that the instantaneous peak voltage of a number of small RF sources is higher than the peak voltage of a single source with the same total power.

the voltage. The voltage-to-power relationship is

$$\text{Voltage} = \sqrt{(\text{power} \cdot \text{radiation resistance})}$$

If this relationship is used to determine the voltage of a 20-watt channel, the voltage is found to be 31.6 volts. If 16 such channels are combined into a single antenna, their individual phases will be random and the voltage vectors will combine in the manner shown in Figure 12.16.

It can be seen, however, that at some time all the voltage vectors will be aligned to produce a peak voltage of 16×31.6 or 505 volts. By comparing this voltage peak to that of a single RF source of the same total power, namely 20×16 or 320 watts, you will find that the peak voltage is a mere 126 volts. It should be kept in mind that these high voltages are in the feeder cables, across all connectors, as well as in the antenna and that they can readily lead to an arc-induced breakdown at points where there is moisture ingress or corrosion.

A problem, which can frequently occur because of this phenomenon, is that the antenna surge arresters, which are often discharge devices designed for single-source loads, will break down sporadically under peak voltage conditions at a cell site. Unless the arrester has been specifically designed for cellular, it should be noted that a power rating based on 505 volts peak root mean square (RMS) (that is $505 \times 505/50$ or 5000 watts) should be used. In practice this usually means using the highest-rated device you can find. Failure to take heed will result in spurious intermodulation.

ANTENNA IMPEDANCES

A quarter-wave antenna, as depicted in Figure 12.17, has an effective radiation resistance of 37 ohms when provided with a good ground plane. In this case the radiation pattern will be the same as a dipole. However, as the ground plane is reduced in size to dimensions below a quarter wavelength, the radiation pattern tilts upward (as shown in Figure 12.18) and the radiation resistance drops.

It will be remembered that cables are usually 50-ohm impedance, and this can be seen to be a compromise be-

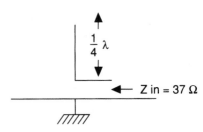

Figure 12.17 A quarter-wave antenna.

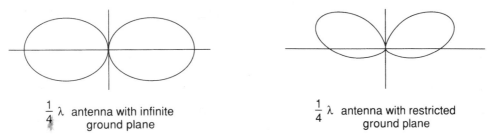

$\frac{1}{4} \lambda$ antenna with infinite
ground plane

$\frac{1}{4} \lambda$ antenna with restricted
ground plane

Figure 12.18 The effect of the ground plane on the radiation pattern of a quarter-wave antenna.

tween the 73 ohms of a dipole and the 37-ohm quarter-wave. In practice, most antennas are designed for a minimum VSWR into a 50-ohm load. This means that the antenna will include an impedance-matching device and that for simple antennas (like whips) the elements will not be exactly electrical multiples of a quarter-wavelength long.

ANTENNAS WITH DOWNTILT

Antennas with built-in downtilt are becoming widely available. Some panel antennas have combined electrical and mechanical downtilt. Typically there will be 5 degrees of mechanical tilt, which can be read off from a calibrated scale, and 5 degrees of electrical tilt, which will also have an indicator of the level of tilt. It is important to realize that the difference between mechanical and electrical tilt is that electrical tilt shifts the whole transmission uniformly, whereas mechanical tilt will change the pattern in different ways for different directions

Electrical tilt is accomplished by phasing the feeds to the dipoles that make up the antenna. Often it is as simple as placing different lengths of coax in series with the feeder to the top pair of antennas, so that the feed path is lengthened (or shortened). There will be some small power losses in the phasing harness, particularly when this is done by adding extra connectors.

Antennas Specialists have an omni-antenna that has a calibrated electrical downtilt of 3 to 8 degrees. The use of downtilt for omni-antennas is *not*, as is often portrayed in magazines, to eliminate dead spots underneath the antenna nulls. Where this popular piece of mythology came from is unknown, but it won't go away. US magazines regularly run stories by some well-meaning writers explaining that high antennas will have dead spots of nulls directly underneath, and describing all sorts of cures for this problem, usually including the use of low-gain antennas and downtilt. The myth is not, however, confined to the US—it seems to be quite universal.

Anyone who has done a field-strength survey will know that the highest field strength recorded is usually almost directly below the antenna, and not, as these stories would imply, at some distance from it. Referring to the radiation pattern in Figure 12.7, you will see that there are indeed nulls directly below the antenna. However, these nulls are not areas of zero field strength but rather areas that are about 20 dB down from the main lobes. Measurements made directly below the antennas will still be high because the distance is so small that even given the reduced radiation in that direction, the actual field strength measured will usually be *higher* than the nearest point in the far field that is receiving power directly from the main lobe.

The real value of downtilting omni-antennas is exactly the same as it is for sector antennas. *Downtilt reduces coverage, and so reduces potential interference with distant cells.* It categorically will not improve handheld coverage, and only in very rare instances will it help to fill in dead spots that are below the lobe of the main antenna.

Downtilted antennas can be used to increase the frequency-reuse factor, and for this they are highly effective. The rest is myth.

POLARIZATION

Radio propagation is characterized by its polarization, which refers to the relationship between the plane of the electric vector (the E-plane) and its angle with respect to the vertical. The H-plane or the magnetic plane is at right angles to the E-plane and in the direction of the maximum radiation, as shown in Figure 12.19.

It has been traditionally assumed that for mobile services the E-plane is vertically polarized and antenna design has been based on this assumption. This assumption is based on the fact that both the receive and transmit antennas are vertically polarized. The effects of multipath on the polarization, is significant and in

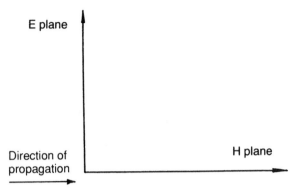

Figure 12.19 Polarization of cellular systems.

particular reflections can cause up to 180-degree rotations of the E-plane vector.

When a vertically polarized wave is incident on a perfect reflector, as can be seen in Figure 12.20 the component parallel to the reflector undergoes a 180-degree phase shift, while the component that is normal undergoes no phase shift. Where the reflector has more complex characteristics (for instance, resistance), the reflected phase changes are more complex.

Refraction is a little more complex, as there will always be a boundary condition that will include a reflection. Continuity at that boundary will require that the vector sums of the incident and reflected waves equal the transmitted wave vector.

In short this means that virtually every reflection and refraction will result in a polarity rotation so that a mobile radio wave in the far field can be expected to have undergone significant E-plane twisting so that its polarity will be indeterminate. This twisting has been shown to produce almost uncorrelated signals at widely different polarizations.

DIVERSITY

Antennas are typically either 9-dB omni-directional, or 17 dB or 14 dB, 120-degree or 60-degree, sector antennas.

Because diversity reception is frequently used, antennas should be mounted as shown in Figure 12.21. Diversity results in an effective 6-dB improvement in the receive path where diversity combiners are used and 3 dB where switching diversity is used. The mounting arrangement shown in Figure 12.21 ensures acceptable isolation and diversity reception.

However, it should be noted that the separation for effective diversity performance is related to base station height above ground and that a separation $= 1/10$ antenna height can be shown to be about optimum. Where this results in a separation much greater than 3 meters, it may not be practical to implement and consideration should be given to saving the cost of the second antenna, as the diversity may not be very effective.

Diversity works best at right angles to the plane of the antennas. There is virtually no diversity effect in the plane of the antennas, as seen in Figure 12.22. This is because when the antennas are in line they receive signals for the same path, and so the multipath effect advantage cannot be taken. If switching diversity is used for two antennas in the plane of the received signal, the second antenna will contribute virtually zero gain, whereas for combining diversity the gain will be 3 dB (the power in the two antennas will be added). The mounting of omni-antennas can therefore affect the system performance. They should be mounted to achieve best diversity reception from the direction required (alternatively, the direction of the plane of the antennas should be one where minimal coverage is required). This problem does not occur with sector antennas.

Vertical diversity may be used, but it generally requires greater physical separation to achieve the same

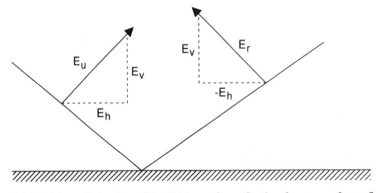

Figure 12.20 The polarization of the E-plane after reflection from a perfect reflector.

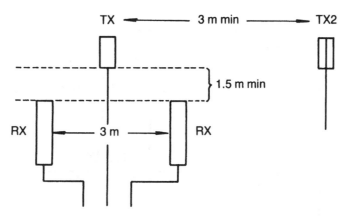

Figure 12.21   Antenna mounting for a 30-meter tower.

results. It does not work well for sites where the base station height is not large compared to the separation. At low antenna heights, the lower height gain of the vertically lower antenna will significantly reduce its effectiveness. For diversity to work well it is necessary that the two antennas have the same gain and thus contribute equally to the received signal. If one has a higher gain, it will simply dominate the level and make the low-gain unit virtually redundant. Vertically separated antennas, unlike their horizontal counterparts, work equally well from all directions.

The diversity receivers come in two types: the diversity-combining receiver and the switched diversity receiver. The diversity-combining receiver aligns the phases of the incoming signals and then adds them. The switched diversity receiver chooses the best of the two signal paths and switches to that path. A gain of 6 dB can be obtained in the first instance, and a gain of 3 dB in the second. Figures 12.23 and 12.24 illustrate these configurations.

LEAKY CABLES

Leaky cables, which are effectively inefficient antennas, can be used to good effect in tunnels and buildings. In fact, it is becoming a common practice to install leaky cables in new railway tunnels as a matter of course in anticipation of their future use by mobile services.

A leaky cable is one that is designed to radiate a portion of the signal carried along its length. Of course to some extent all cables are leaky, since they will radiate some RF along their length. The main mode of radiation at high frequencies is due to magnetic flux leakage through the shield. Braided cables are far more leaky than solid cables, as the "open" areas between the braids are the launching points for magnetic leakage. Leakage can be provided by cutting small slots in a coaxial cable, as shown in Figure 12.25a. The amount of loss can be controlled by the choice of slot size and regularity along the cable. Leaky cables can also be constructed from twin-lead, loosely braided cable (see Figure 12.25d), and continuous slot cable. This same technique can be used in high-frequency waveguide.

There are many other ways of producing a leaky cable. Figure 12.25b shows a coaxial cable where the screen is loosely wound around the center conductor. Figure 12.25c shows a triaxial cable in which the outer conductor is in two halves. The triaxial cable is more predictable when mounted near other conducting surfaces.

Cables also leak power by dissipation of the carried

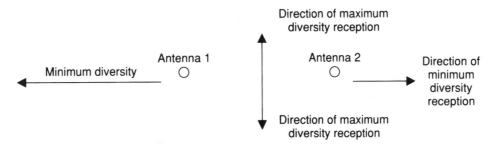

Figure 12.22   Diversity effectiveness depends on the plane of reception.

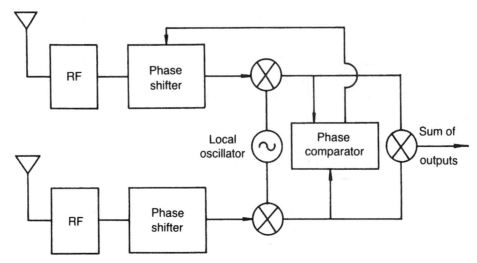

Figure 12.23 A diversity-combining receiver continuously adjusts the phase of one of the received signals to produce a phase-consistent output that can then be added. Such a receiver configuration can improve S/N by about 6 dB.

current, as heat due mainly to the conductor resistance and dielectric losses. This loss still has to be minimized as there is no RF radiation associated with this loss, and as a consequence larger leaky cables will be more efficient than smaller ones.

When using these cables one should be aware that the leakage will not be uniform—for example, if a cable has a loss of 10 dB per 100 meters and is fed by a 100-watt Preamp (PA), then 90 watts will be radiated in the first 100 meters, but only 9 watts in the next 100 meters and 0.9 watt in the next, and so on. Consequently, if long

cable runs are planned, it will be necessary to use not a single cable but a number of cables connected so that the first span is relatively low-loss, and with each successive span being slightly higher loss so that the average power leakage per meter is approximately maintained at a constant.

Because leaky cables are deliberately very lossy, there will be a limit on how long a cable can be and still be effective unless some amplification is used. Cable repeaters work much the same as off-air repeaters, and amplify the signal from both directions equally. They

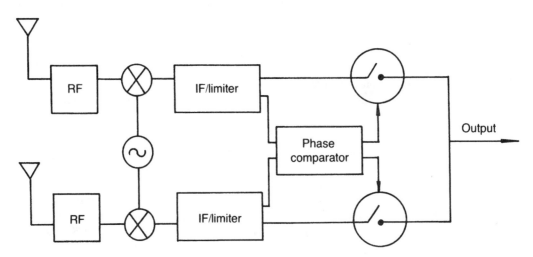

Figure 12.24 A switched diversity (also known as selection diversity) receiver chooses the best of the two signal paths.

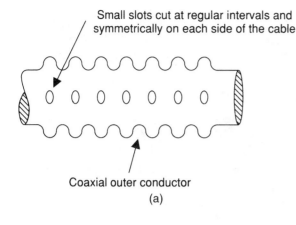

Small slots cut at regular intervals and symmetrically on each side of the cable

Coaxial outer conductor

(a)

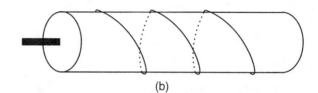

(b)

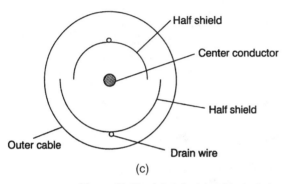

Half shield

Center conductor

Half shield

Outer cable

Drain wire

(c)

Figure 12.25 (*a*) A leaky cable is designed so that the shield has wide gaps to permit a regulated RF leakage. (*b*) A loosely wound shield leaky cable. (*c*) The cross section of a triaxial leaky cable.

will consist of back-to-back amplifiers and duplexers to separate the two paths. The output levels are typically 0–10 dBm.

If repeaters are used, one should keep in mind that they will be linear amplifiers and as such very prone to intermodulation distortion. To avoid this they need to have the power backed off to the linear region (which is usually about 3 dB below the maximum gain).

Often it will be necessary to share a leaky cable with other users. There may, for example, be a trunked radio or other RF facility operating in the same region as the cellular system. Where this is the case, some specialized repeater modules, like the one shown in Figure 12.26, that house repeaters and filters for two systems are intended to work from the one leaky cable.

In tall buildings leaky cables will usually be installed vertically in non-metallic service shafts, whereas in places like shopping centers and malls they will run along the ceiling. Often the cable will be terminated with an antenna, which will provide some local coverage. In general, it is not absolutely necessary to terminate a lossy cable if the overall loss exceeds 10 dB. If the loss is

less than that level, then a 50-ohm termination should be used to ensure a reasonable VSWR.

TUNNELS

While leaky cables are effective in tunnels, advantage can be taken of the fact that losses in straight (or nearly straight) tunnels are linear with distance, so that a Yagi placed at the mouth of the tunnel can effectively service a much greater distance than it would in free space. It is often possible to use a simple low-powered repeater and Yagi combination to service a small, straight tunnel. If this technique is used, it will often be the case that a handoff will nearly always be initiated at one end of the tunnel, and this may itself cause problems.

USE OF YAGI ANTENNAS

Yagis are not widely used in cellular, except in special applications such as repeaters, fixed subscriber units, and sometimes as base-station antennas in Nordic Mobile

Figure 12.26 A two-frequency leaky cable repeater designed to operate simultaneously at 450 MHz and 900 MHz. (Photo courtesy of Andrews Australia.)

Figure 12.27 Vertically polarized Yagis mounted like this from the side of a tower will suffer pattern distortion.

Telephone (NMT) systems. Yagis produce high gain at low cost, but they are subject to considerable pattern distortion when mounted near other conducting surfaces. In particular, vertically polarized Yagis such as those shown in Figure 12.27 will suffer pattern distortion from the tower legs. Horizontally polarized antennas mounted the same way would be less affected. For this reason grid-pack antennas are often preferred, even though they cost a little more, as they are virtually immune to pattern distortion.

INDOOR COVERAGE WITH LOSSY CABLES/ MICROCELL ANTENNAS

While small antennas can provide satisfactory coverage indoors of large open spaces such as halls and shopping centers, better coverage of cluttered environments such as offices, is generally obtained with lossy cables.

The loss from a radiating cable will depend on frequency, and this dependence can be difficult to estimate in a cluttered environment. While generally, the net path loss will increase with frequency, this may not always prove to be the case, and sometimes 1800 MHz systems may propagate with a lower path loss than a 900 MHz system.

The loss will also depend significantly on the mounting of the cable, and in particular how close it is to the walls (or ceilings and floors). In a regular office environment it should be possible to cover three floors with one cable (one floor above and one below the floor that the cable is installed in).

Figure 12.28 An omni-antenna, if it is to be side mounted, should be offset from the tower as is the case for this omni. Caution should be exercized with this type of mounting because it is prone to movement in high winds and may need regular maintenance to keep the antenna vertical.

ANTENNA MATERIALS

Because antennas are exposed to the elements (sun, rain, ice, smog), the choice of materials is critical. All metals must be electrolytically compatible or else local corrosion cells will form. Joints with corrosion are potential intermodulation sites. Cellular omni-directional antennas usually have about 9-dBd gain and are of collinear construction, encapsulated in a fiberglass radome.

The radome must also be of high-grade fiberglass because water leakage can lead to corrosion of the elements and the ultimate failure of the antenna. Many fiberglass products, however, have metallic additives that make them unsuitable for antenna construction.

MOUNTING

Cellular antennas must be properly mounted to function effectively. Omni-directional antennas should be mounted on top of the tower or building, or well offset from the tower as shown in Figure 12.28, because side mounting can seriously distort the original omni pattern. Roof tops are effective platforms for most antennas (see Figure 12.29).

Often existing tower structures, which were not originally designed to accommodate cellular antennas, can be modified with the addition of suitable platforms. The example in Figure 12.30 shows one tower in Henley, England, that has undergone extensive modification to allow usage by both Racal-Vodaphone (the triangular platform) and British Telecom (just below the offset omnis) as well as a host of other users. In rural areas, an old microwave tower can be simply modified by building a small platform on the top, as seen in Figure 12.31.

Sometimes the existing towers can be suitable for cellular mounting without modification as seems to be the case in Ireland where the existing Telefon Eireann microwave towers have very wide tops. As can be seen in Figure 12.32 the omni cellular antennas are mounted around the corners of the tower. Vertical separation between transmit and receive is obtained by raising the transmit antenna to about the same height as the centrally mounted lightning rod. Perhaps the most prominent of all rooftop mounting structures is the platform on top of the Singapore Telecoms building in the heart of downtown Singapore (Figure 12.33). What looks from a distance to be a massive scaffolding is in fact an elaborate antenna-mounting frame that dominates the top of the 30-floor building.

Figure 12.29   Rooftop mounting can permit the antennas to be located at various positions around the building in order to get the best near-end clearance, as seen in this British Telecom installation.

Figure 12.30   With a little innovation many users can be accommodated on a single tower, as this example shows.

Figure 12.31   An old microwave tower modified simply and cheaply for cellular in a rural area.

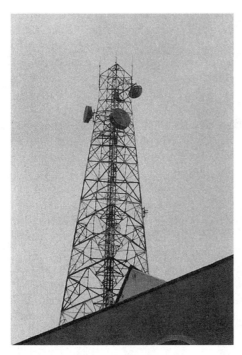

Figure 12.32 Mounting is simplified when the towers have wide tops.

Figure 12.33 Singapore Telecoms Headquarters building has a massive antenna mounting structure on the rooftop.

Because cellular antennas are high-gain devices with a very narrow E-plane pattern, even small deviations from the vertical can very seriously effect coverage. The antennas should be vertical to ± 1 degree or downtilted (if intended) to the specified angle ± 1 degree. This variance is usually much less critical in land mobiles, where base-station antennas rarely have gains greater than 6 dB.

Unidirectional antennas can be side mounted. Side mounting can improve the front-to-back ratio by using the mounting structure to decrease sensitivity to interference from the backward direction. In most big cities it is a good idea to mount unidirectional antennas on a mounting that allows for future downtilt of the antenna. Figure 12.34 shows such a mounting.

The connection between the main feeder and the antenna should include a tail of 2–3 meters, as shown in Figure 12.35. The tail makes installation easier and is important for future maintenance, when it may be essential to replace an antenna with one of a different type.

Table 12.2 lists the typical specifications of an omnidirectional cellular antenna.

Mounting on a tower without a special platform can be easily done with a custom mount, as shown in Figure 12.36.

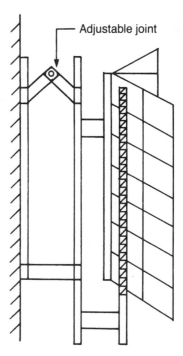

Figure 12.34 A mounting that allows for antenna downtilt.

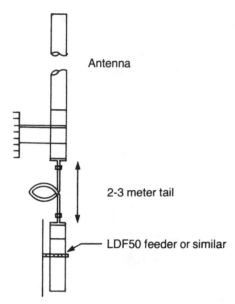

Figure 12.35 The connection between the feeder and antenna should be by a 2–3-meter tail.

Figure 12.36 A simple adaptation to an existing tower makes mounting easy.

DRAINAGE

Drain holes must be clearly identified and correctly oriented. Because even the best-constructed antennas will suffer some water leakage, antennas are usually fitted with small drain holes. These holes are located at the lowest points of the antennas. This is an important factor when considering mounting the antenna upside down (as is often done in cellular installations). If the antenna is not designed to be mounted this way, there will be a drain hole only in what was the bottom (and is now the top). In these cases, it is necessary to seal the existing drain hole and carefully drill a new hole in what was the top (and is now the bottom).

INTERMODULATION

When two signals of a different frequency mix in a non-linear device (for example, a rusty wire, loose joints on

TABLE 12.2 Specifications for an Omni-Directional Antenna

Frequency	825–896 MHz
Bandwidth	71 MHz
Input impedance	50 ohms
Gain	8.5 dBd
Maximum power	500 watts
Half-power bandwidth	6.5° (E-plane)
Wind area	0.175 square meters

towers, and rusty fences), the result is intermodulation. Intermodulation can be a problem at any site that has two or more transmitters. Cell sites, which often have co-sited paging, trunked radio, and other mobile services, are likely to experience intermodulation. This problem appears as interference to other (nearby) mobile service users, or it can introduce interference into the cellular operator's equipment and cause blocking.

Finding the intermodulation source can be very time-consuming and usually proceeds by eliminating likely offenders by trial and error. Because the problem is often intermittent, it can be very frustrating to track down.

Rusty bolts have long been blamed as the major culprits, but recent tests have revealed that the worst offenders are rusted galvanized mild-steel rope, mild-steel chains (the very worst offenders), and mild-steel wire fences.

Loose joints with small areas of contact are the main intermodulation points. Large areas of corrosion, such as on decking or galvanized iron sheets, may not necessarily produce high levels of intermodulation.

The intermodulation products are ranked according to their order, which defines how far removed they are from the fundamental frequencies. The original frequencies are regarded as first order, and so for two frequencies, F_1 and F_2, the intermodulation products are

Second Order

$$2 \times F_1$$

$$2 \times F_2$$

$$F_1 + F_2$$

$$F_1 - F_2$$

Third Order

$$2 \times F_2 + F_1$$

$$2 \times F_2 - F_1$$

$$2 \times F_1 + F_2$$

$$2 \times F_1 - F_2$$

etc.

In the cellular environment, much more complex products will occur, as there will be more than two frequencies. Intermodulation will then occur as above, and the general expression for the products for first-order frequencies $F_1, F_2, \ldots, F_n$ is

$$m\text{th-order product} = (a \times F_1 \pm b \times F_2 \pm \cdots \pm k \times F_n)$$

where

$$a + b + \cdots + k = m$$

and any of these coefficients can take values from 0 to m.

Notice that although the intermodulation products are infinite in number (as there is no limit to m), their magnitude decrease rapidly and it is unlikely that products beyond the third order will cause operational problems.

A number of commercial computer programs are available that can take the first-order (RF transmitter) frequencies and do a search for intermodulation products that fall in the receive band.

MEASURING VSWR

Antenna VSWRs should be in the range of about 1 to 1.7, and each antenna should be checked upon installation. Most base stations have a built-in VSWR meter that can be switched to each channel.

To measure VSWR, it is necessary to distinguish between forward signals and reflected signals. In a VSWR meter, a directional coupler is used to selectively read the signal in either direction. If two couplers are used, it will be possible to get a direct reading of VSWR by using some simple circuitry to derive the ratio of the forward and reflected signals. Figure 12.37 shows a simple directional coupler.

A directional coupler relies on placing the sensing loop so that the induced currents from the electric and magnetic fields are equal. The induced currents resulting from the electric field are indifferent to the direction of that field. The magnetic fields, however, are of opposite phase in the forward and reflected waves, and will thus cancel the electric field in one direction. By specifying port 1 or port 2 as the sensor, a directional coupler can be made to read either the forward current only or the reflected current only.

REMOTE ANTENNA MONITORING

While there is nothing complex about measuring the VSWR of an antenna, it can be a tedious business because of the need to handle the antennas very carefully on a live system. The transmit antennas often have some kind of VSWR monitor built in, but it is rare that the receive antennas do.

In very remote sites, and ones where for various reasons access may be difficult, it is possible to install a device like the Loral Cats (continuous antenna test system), which will continuously monitor the VSWR of the antenna and report the results back to a remote mon-

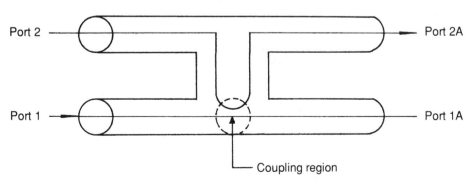

Figure 12.37 A simple directional coupler.

itoring site. Model 72000 can simultaneously monitor six transmitter (TX) and six receiver (RX) antennas. These remote monitors generally work by injecting small out-of-band signals and measuring the return loss.

SMART ANTENNAS

Increasingly cellular and satellite systems are coming equipped with "smart" antennas. Surprisingly, these smart antennas not only have more gain than fixed ones, but can be expected to cost less (in the near future) and to include interference canceling. These antenna systems are so good, that it is hard to see how conventional antennas can survive in the long term except for the most basic of applications. To date usage on cellular systems has been limited, probably because of the cost and complexity of the commercially available systems. However, both of these impediments are being minimized.

To understand how a smart antenna works, it is necessary first to look at the phased array, around which a smart antenna is built. All multielement antennas are phased arrays (that includes the common co-linear dipole). Consisting of a number of radiating elements connected by a phasing harness, the desired pattern is achieved by adding the received signals from each of the dipoles in a way to produce the desired antenna lobe pattern.

To get a feel for how a phase array works consider first an omni-directional antenna pattern as seen in Figure 12.38. Now look at the pattern formed by two omni-antennas separated by 0.25 wavelength and phased an additional 0.25 wavelength, as seen in Figure 12.39.

This will give you the football-shaped pattern. What you are seeing is reinforcement in some directions and cancellation in others.

Next look at Figure 12.38c where we see an antenna separation of 0.5 wavelength and a phase delay of zero.

Now, you have a coverage that is decidedly directional, with peaks toward the top and bottom of the page and nulls at right angle to the peak direction.

Let's assume that this is just the kind of pattern we are looking for, but we would like to increase the directivity. This is easily done simply by increasing the number of antennas in the (linear) array. Increase the number to nine and you see the result is similar to Figure 12.38d. You will find that, in general, increasing the number of antennas while keeping the other parameters fixed will sharpen the pattern.

For many applications a directional array like the one you have just looked at may be all that is needed. However, sometimes something more sophisticated is needed. Another variant that could be used is a two-dimensional

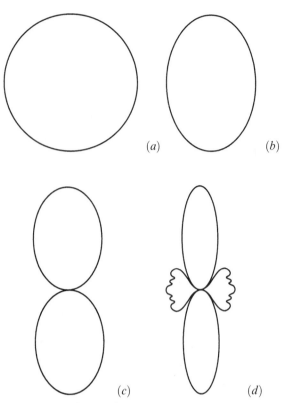

Figure 12.38 (a) The uniform pattern of a single omni-directional antenna. (b) Two antennas separated by 0.25 wavelength. (c) Two antennas separated by 0.5 wavelength. (d) Nine antennas separated by 0.5 wavelength.

array of antennas, to give even more patterns, and this is what is used in practice.

You will have noticed that there are two ways to vary the pattern: you can either change the antenna spacing, or you can change the phasing between the antennas. Changing the phasing is as simple as changing the length of the cables connecting the two antennas.

Suppose it was required to use a fixed array of antennas to provide a number of different patterns, for example, to direct broadcasts in different directions at different times. This could be done by having a number of phasing harnesses and switching in the required harness using positive intrinsic negative (PIN) diodes, which can be biased to be open circuit or low-impedance short circuit. There is no real limit to the number of patterns that can be generated in this way, but a large number of harnesses can get messy.

The Adaptive Array

Adaptive arrays are the heart of the smart antenna. What is needed is a way to vary the relative phase of the antennas *continuously*. While there are a number of

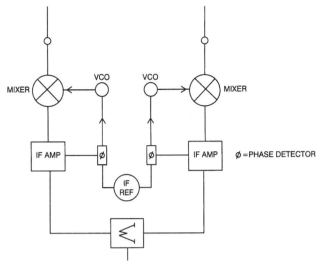

Figure 12.39

ϕ = PHASE DETECTOR

mechanical devices that could be used to vary the phase, there is a neat way to do it within a phase-locked loop.

Consider the two-antenna array in Figure 12.39. Here the intermediate frequency (IF) reference oscillator determines the angle at which the signals incident on the two antennas are phase locked. We now have the basis of an antenna array that can be steered continuously by phasing the wanted signals. When the reference oscillator is controlled by a processor, automatic adaption to maximize a signal or to minimize interference becomes possible. Even if there is more than one interfering signal, more nulls can be generated to counteract the problem. Additionally the performance of the systems can be enhanced by *simulated patterns*, which are generated as part of the signal processing.

Smart antennas are finding applications mainly in high-density central business district (CBD) areas; to date, rural applications are limited.

Space Division Multiple Access

Smart antennas can be used as a multiple-access device, when an array of antennas is operated so that individual mobiles are catered to by their own synthesized beams. If these beams are sufficiently narrow, then it will be possible to reuse the channel even from the same site for communications with a mobile outside that beam. Hence the term space division multiple access (SDMA).

SDMA antennas ordinarily will have very narrow beams and very high gains (around 24 dB). Part of the gain comes from the beam compression, which is what allows reuse, but at the same time it decreases unwanted signal pollution. Higher gains also mean that lower

TX power can be used and/or more range is available. ArrayComm's Intellicell is an example of this technology. This antenna allows for spatial channels, which means that the same frequency can be reused from the one site because the beams are spatially separated.

Nulling Interference with Adaptive Arrays

One of the most complex things that a smart antenna can do is to null out interference. This means that the antenna must distinguish between the wanted signal and the interference. The way this is done is to maximize the antenna gain in the direction of the wanted signal, and to minimize it in the direction of the interference (which is assumed to be coming from a different direction). This is equivalent to minimizing the unwanted lobes in the direction of the interference.

For cellular applications we have the advantage that most wanted signals carry identifiers such as a color code, and most interference is from distant co-channels that carry with them clear markers. This makes the job of designing the null array just a little less difficult.

There are three "classic" ways of interference nulling, to be used in the cases where the *direction* of the interference is not known [in the case where the interference is from a known fixed location, two of these methods (the last two), could be used with the software discussed earlier to form phased arrays, with you the user, controlling the algorithm].

The array can have a bank of receivers for each array element, and a correlation matrix can be formed for each receiver. This matrix is then optimized for the best signal-to-noise ratio (SNR) for the desired signal. This is effective for relatively small arrays, but does require a high degree of receiver calibration.

Another way is the Monte Carlo approach, which literally guesses the weightings for the nulling until the desired output (SNR) is achieved. This is generally too slow for cellular applications.

The third method is to minimize the power of the interference by examining the gradient vector when small changes are made in the phase and or amplitude of the elements, and seeing what effect this has on the overall S/N.

A fourth method has recently been reported in the *Microwave Journal* (January 1999) by Randy Haupt and Hugh Southhall, which uses a genetic algorithm to minimize the received power from the interference.

This method uses only one receiver, and phase shifters and attenuators at the antenna, as seen in Figure 12.40. The algorithm starts without any preconceived assumptions (that is, the chromosomes are randomly set), and

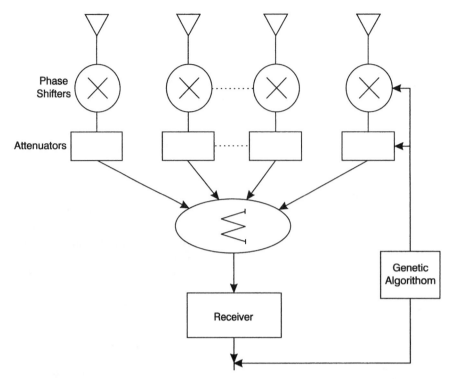

Figure 12.40   The phasing between the two antennas can be changed by lengthening one of the connecting cables. An adaptive phased array controlled by a genetic algorithm is shown here.

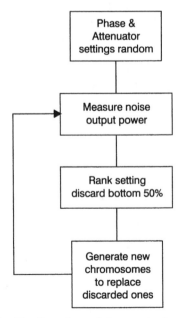

Figure 12.41   The flow chart of the genetic algorithm for control of the nulling antenna.

the algorithm is optimized through a series of measurements. After each measurement 50 percent (the worst 50 percent) of the chromosomes are discarded and a 0.1 percent mutation rate is introduced. The process is seen in the flow chart in Figure 12.41.

The convergence for the 5-GHz array under test occurred with an average of around eight iterations, which is quite fast.

A variant of this is the Trailblazer by ANC of California. This is a nulling technology for adjacent interfering sites. It works by pointing an antenna at an interfering source (which may be its own network or a foreign network), and the device then specifically nulls that interferer.

Beam Switching

Mainly because it is easier to do, switched-beam mode is probably the most common way that smart antennas are implemented. This can be done either by having passive, narrow-beam sector antennas (virtually a narrow-beam version of the common 120-degree sector antenna), or

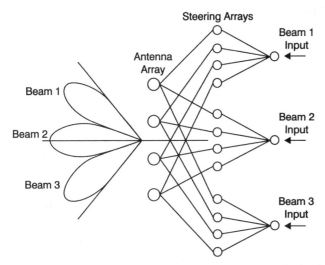

Figure 12.42 Standard array of anttennas of a synthesized beam antenna.

Figure 12.43 The Metawave smart antenna.

alternately the beams can be synthesized. In a synthesized beam antenna the array can consist of a standard array of antennas seen in Figure 12.42, with the sectors being steered by phase shifters.

In any case, each beam is monitored by separate receivers (or sometimes a bank of scanning receivers) and according to the algorithms in place, switching to the most appropriate beam will occur.

What Smart Antennas Offer

A properly functioning smart antenna will improve capacity and net carried traffic, improve SNR, increase spectrum efficiency, control interference better, reduce dropped calls, and probably increase range. Given the wide spectrum of service-affecting capabilities, it can be expected that the algorithm that drives the antenna (the software that makes it work) will play a critically important role in just how effective the smart antenna actually is.

Smart-Antenna Implementation by Metawave

Metawave has taken a practical approach to smart-antenna implementation, which simplifies the hardware with a minimum of performance compromise. While the phasing of the antenna arrays can be controlled by the smart-antenna system, the Metawave approach is to *fix* the phasing in what amounts to conventional antenna technology, which produces 30-degree beam-width antennas. The antennas themselves are entirely passive, with all the electronics being in the base-station enclosure. Gains of 15 dBd on the standard model and 16.4 dBd on the high-gain model are available.

The hardware rack used to implement this system is seen in Figure 12.43. The technology is equally applicable to PCS and WLL.

The Metawave approach is to integrate the smart-antenna system transparently into the base station, at the antenna interface; then using the intelligent algorithms in the software, to optimize the sector configuration both in beam width and gain. For any given channel it may well be that the sector with the strongest signal is not the best selection, as it may not have the best C/I. The choice of the best sectors is made by the embedded algorithms.

The approach is to replace the "conventional" 3-sector, 120-degree beamwidth antenna system (Figure

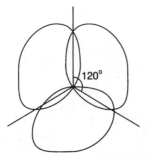

Figure 12.44 A conventional 120-degree antenna pattern.

12.44) with a 12-sector 30-degree beamwidth one (Figure 12.45). Four 30-degree antennas are mounted on a single 120-degree panel. These antennas form software-definable phased arrays. This antenna can directly substitute for the conventional pattern and offer immediate benefits by limiting the area over which the receive antenna can detect interferers. Because the beam width has been reduced to one-quarter of its original value, the net interference energy will be reduced by the same amount (6 dB). This will improve the C/I ratio accordingly. An additional improvement will come from a net reduction in the contribution to interference to the rest of the network that will come from the tight clusters that result from the narrow beam width. The evidence is that this last factor will improve the C/I by an additional 1.5 dB.

Capacity Improvements

With the improved C/I alone, it is evident that for an analog system, capacity gains can be achieved by migrating from an $N = 7$ reuse to $N = 4$. It is worthwhile looking at the capacity gains from this. Assuming there are 300 channels available and the $N = 7$ pattern results in 21 distinct frequency clusters, then each sector has 300/21 or 14 channels. Similarly, for an $N = 4$ pattern, there will be 24 channels in 12 clusters. Using the Erlang C table for a grade of service (GOS) of 0.05, we find that

the sector capacity of the $N = 7$ sector is 8.27 Erlangs, while for $N = 4$ it is 17.08. So the net traffic gain *per site* is 107 percent.

Additionally, the larger sectors are more efficient carriers of traffic, so that while the 14, $N = 7$ channels will carry 0.59 Erlang per channel, the 24, $N = 4$ channels carry 0.71 Erlang per channel; an improvement of 20 percent in traffic efficiency per channel. While it is possible to use $N = 4$ patterns without a smart antenna, it is inadvisable to do so in most cases because of the difficulty in controlling interference.

For CDMA the capacity improvements occur through three mechanisms. First, the antenna pattern is sharper and the rolloff at the edges is also sharper, which decreases the soft handoff between sectors. This effect can account for a 5 percent to 10 percent increase in capacity.

Next, traffic load balancing can be used (as explained in the next section) to increase capacity. Finally, the CDMA patterns can be finely sculpted, using the 16 dB of power control (done with attenuators on the transmit side) available at each of the 30-degree sectors to reduce pilot pollution, improve load balance between the cells, and control soft handoff.

Traffic Load Balancing For any cellular network, there will be time-of-day and day-of-week shifts in the traffic distribution among cells and among base stations. This often means that the peak traffic load on one cell at a base station does not coincide (is not time-consistent), with the peak load of another cell. This might mean that while one cell is working at capacity, other cells at the same site have unused channels. Because the traffic patterns vary in time, the situation may be reversed later in the day, so that a different cell is at capacity while others are idle. The Metawave smart antenna can compensate for this situation by changing the effective coverage area of any sector. The way that this is done is shown in Figure 12.46, where an underutilized cell expands to accept more traffic, while an adjacent busy one con-

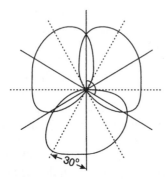

Figure 12.45 A 120-degree sector system replaced by a 12-sector 30-degree antenna system.

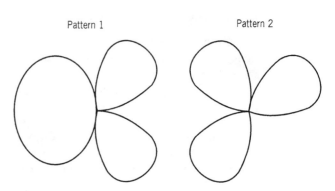

Figure 12.46 Cell grooming to accommodate traffic changes.

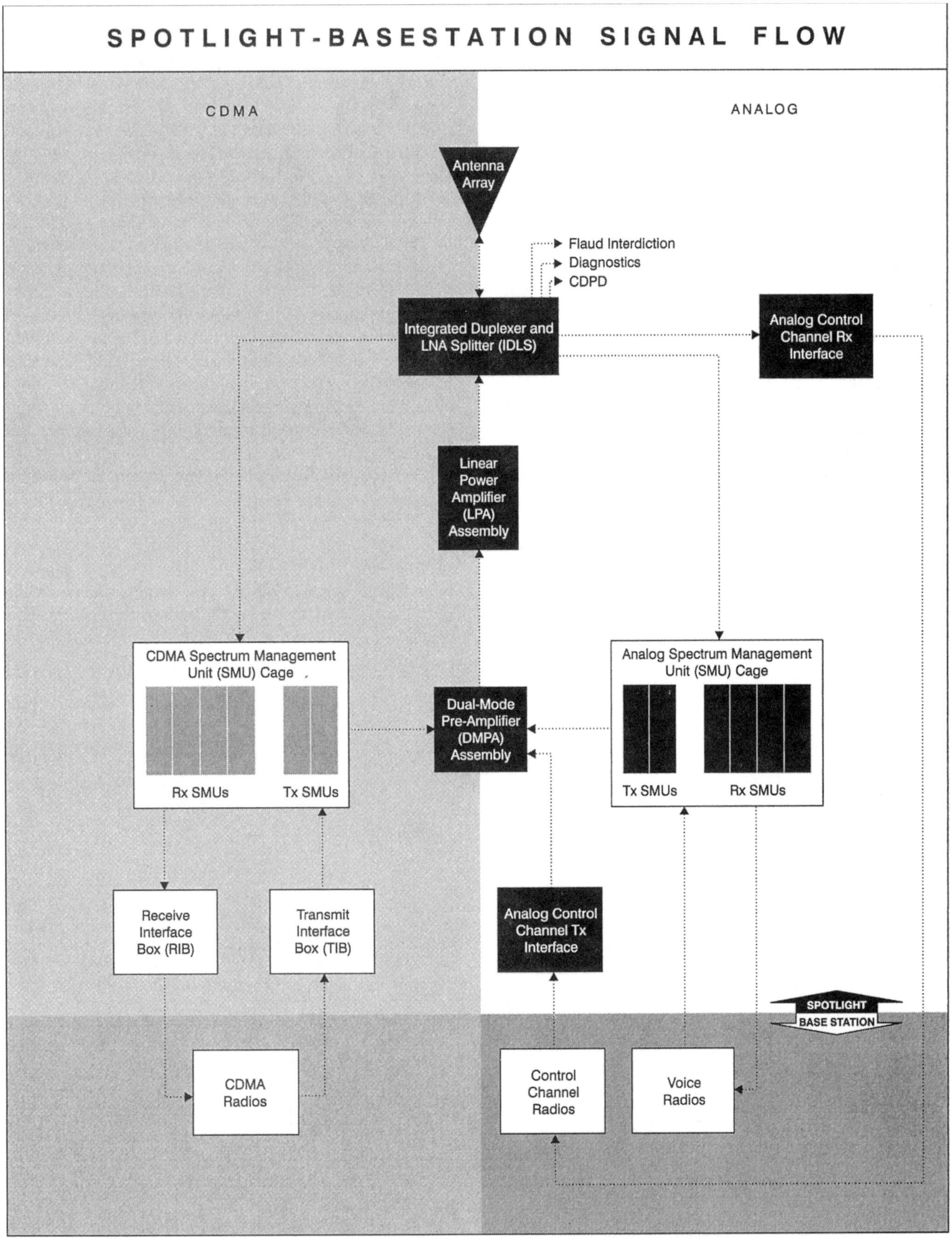

SPOTLIGHT-BASESTATION SIGNAL FLOW

CDMA

ANALOG

Antenna Array

Flaud Interdiction
Diagnostics
CDPD

Integrated Duplexer and LNA Splitter (IDLS)

Analog Control Channel Rx Interface

Linear Power Amplifier (LPA) Assembly

CDMA Spectrum Management Unit (SMU) Cage

Rx SMUs   Tx SMUs

Dual-Mode Pre-Amplifier (DMPA) Assembly

Analog Spectrum Management Unit (SMU) Cage

Tx SMUs   Rx SMUs

Receive Interface Box (RIB)

Transmit Interface Box (TIB)

Analog Control Channel Tx Interface

SPOTLIGHT
BASE STATION

CDMA Radios

Control Channel Radios

Voice Radios

Figure 12.47   The Metawave smart-antenna structure.

tracts. The actual pattern can be programmed to change with time-of-day traffic changes so that sectors can be expanded or contracted to cater for peak hours. It is planned that there will soon be a dynamic allocation capability, so that the base station can sense the traffic conditions in real time and reconfigure accordingly.

The effect of the cell grooming is to make capacity available as needed, and were this to be allocated dynamically, in effect any of the site channels could be made available to any mobile. To a very good approximation (where an optimal algorithm is used; real systems may be less efficient, particularly while the smart antenna is not integral with the base station), this will turn a sectored site into an omni for traffic efficiency purposes. So for the $N = 7$ case, the 14 channels allocated to each sector become a pool of 3×14 or 42 channels. The 42 channels can carry 31.5 Erlangs (using Erlang C and 0.05 GOS), compared to a combined traffic of $8.27 \times 3 = 24.81$ Erlangs for the separate sectors. Thus, the increased traffic efficiency is 27 percent. In fact, the gains can be expected to be higher in practice, because the situation where all three separate sectors are equally fully loaded in a time-consistent way would be rare.

Using the same reasoning for an $N = 4$ design, an increase in circuit efficiency of 14 percent could be expected.

Handoff Management in CDMA   Handoff management can be a significant factor for an operator trying to get the most out of a CDMA network. While proper operation of the soft and softer handoffs is essential for good CDMA performance, excessive use can tie up valuable resources. In a typical CDMA network the *handoff overhead factor* (the average number of handoff links that are active per subscriber) is in the range 1.65 to 2.0.

These handoffs consume system resources, such as computational time and forward transmit power levels (which increases forward link noise, and so reduces forward link capacity). It is worth noting here that most CDMA systems are capacity limited in the forward link. By using a smart antenna to tailor the coverage and load balance, the handoff overhead can be reduced by about 10 percent. So for CDMA networks, traffic efficiency is improved by sector grooming *and* reducing handoff overheads.

The net gain from all of these factors is seen as a decreased net noise level that reportedly gives about a 20–30 percent increase in traffic capacity.

CDMA with AMPS   The Metawave system supports both AMPS and CDMA into the same antennas, while allowing different sector configurations for each. This is effectively done, by utilizing separate processing (called

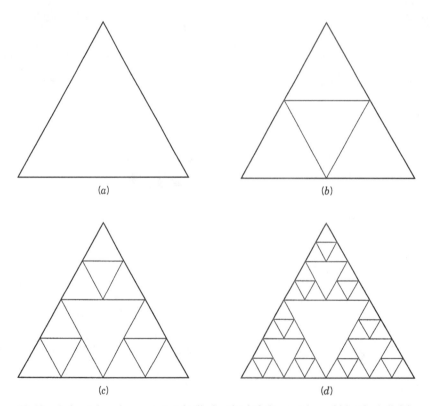

(a)   *(b)*   *(c)*   *(d)*

Figure 12.48   A fractal evolves symmetrically by the infinite number of identical divisions.

Spectrum Management Units) for the CDMA and AMPS signals as seen in Figure 12.47. While AMPS systems are somewhat constrained in sector orientation (due to frequency reuse considerations), a co-sited CDMA system, can have any orientation due to the unity reuse of spectrum. Ordinarily, this might mean that the CDMA system will need to use its own antenna system, if the optimum CDMA pattern is required. The Metawave smart antenna allows sharing of antennas while still allowing the CDMA and AMPS patterns to be independent.

The GSM implementation works a little differently, with the main objective being to improve the C/I ratio. This is done by dividing the cell into twelve 30-degree sectors and optimizing the cell pattern on a time-slot-by-time-slot basis. This results in the sector pattern following the user through the coverage area. An improvement of 5–6 dB in C/I has been noted in early trials.

THE FUTURE

In the future it can be expected that the "intelligence" of the smart-antenna systems will be more widely distributed, so that a site that is experiencing excessive interference from another can request that site to power down or change the sectorization in a way that will reduce the unwanted interference.

Future 3G systems will have support for smart antennas incorporated into their specifications. This applies to the CDMA-2000 and IMT-2000 systems now being specified, and is likely to apply to any future system.

WIDE-BAND FRACTAL ANTENNAS

In the March 1999 issue of the magazine *Fractals*, Nathan Cohan and Robert Hohlfeld showed that (mathematically) for an antenna to be extremely wide band, it must have symmetry about a point and have the same basic appearance on every scale. In short, it must be a fractal. A fractal has symmetry down to the smallest scale, as can be seen in Figure 12.48. At all scales the symmetry is maintained.

Putting these principles to practical use, fractal antenna systems are developing smaller, more efficient arrays. Additionally it has been found that the shapes of the fractal elements can be used to generate capacitance and inductance as required for making smaller narrow-band antennas.

13

CELLULAR LINKS

MICROWAVE

In its simplest form, a cellular system can be thought of as a switch connected to a number of base stations. The connections between the cellular radio bases and the switch [or the switch and the Public Switched Telephone Network (PSTN)] are an essential part of the network. Cellular operators usually think of microwave, fiber optic, or lines rented from the wireline service as the means of linking up the network. Often a mix of these options is the best choice.

The short lead times, low installation costs, and often the availability of existing towers (particularly for non-wireline operators) make microwave links particularly attractive for cellular operators. Base-station towers can also be used for microwave links.

Microwave systems are so called because they operate in the "microwave band," which was originally the ultra-high frequency (UHF) band (300–3,000 MHz) but today extends to 300 GHz. Typically, they have capacities ranging from a few voice channels to many thousands.

A number of software programs are available that can quickly and reliably calculate microwave path performances. These programs can also give the path performance as a function of tower height and determine the effect of potential interferers. Cellular operators who use microwave extensively should obtain this software and become familiar with it. This chapter looks mainly at the concepts involved in the selection and design of microwave links.

Links and transmission equipment account for 15–25 percent of a cellular operator's infrastructure cost. All too often they are an afterthought, and this can result in a future maintenance nightmare if no consideration is given to the functionality of the link network and its future management.

In this chapter we consider all the things that a link designer should look at. Even if a software package is used for the design (and this is recommended), better designs will come from a better understanding of the underlying principles.

When it comes to designing links, there are four basic steps:

- Define the operational frequency with regard to availability of bands, terrain, and rainfall data for the region. In tropical areas, high frequencies (>11 GHz) lead to poor reliability in tropical rain; higher frequencies should be avoided where possible. If higher frequencies must be used, keep the hops short as they will be more reliable than long ones.
- Define the link performance [as signal to noise (S/N) for analog or bit error rate (BER)] for digital systems
- Determine the radio frequency (RF) signal (that is, the receiver sensitivity) that will be needed to meet the performance criteria based on the selected microwave equipment
- Select the other hardware, such as antennas and feeders, that will assure that the required RF level is available.

Microwave systems suitable for cellular radio will generally be in the 2–38-GHz band. Frequencies lower than 2 GHz have been used in the past, but are not likely to be generally available for future cellular applications. Table 13.1 shows the bands and their suitability.

Modern microwave links use digital modulation

TABLE 13.1   Microwave Frequencies

Frequency (GHz)	Applications
2	Widely used for cellular, low and medium capacity
6	Widely used for cellular, medium and high capacity
10	Only recently utilized for small capacity
11	Medium capacity, medium-haul
18	Short-haul, moderate to low rainfall only
23	Short-haul, moderate to low rainfall only

TABLE 13.2   BER as a Function of Carrier S/N and the Number of States in the Digital Code of a Typical Link

	Carrier S/N		
BER	2 Level	4 Level	16 Level
10^{-3}	11	17	28
10^{-4}	12	18	29
10^{-5}	13	19	30
10^{-6}	14	20	31

TABLE 13.3   BER Threshold Levels for Typical 2, 4, and 8 Mbit/s Systems

	System Size		
	2 Mbit/s	2 + 2 Mbit/s	8 Mbit/s
10^{-3} BER threshold	−95 dBm	−92 dBm	−89 dBm

microwave (earlier systems used analog links). Figure 13.1 shows a typical microwave link. The performance of a digital link is measured in BER. The threshold level for a digital link is often defined to be the level at which the BER equals 10^{-3}, although other BERs can be used. The threshold level is approached very rapidly from much higher BERs as the S/N level of the carrier approaches the threshold. This can be seen in Table 13.2. Terminal equipment often has the threshold level defined for a 10^{-3} BER (or similar BER) as a specified parameter. Table 13.3 lists typical threshold values.

The BER is usually measured by looping back a E1/T1 channel (or single channel) and measuring errors over a period of 1–12 hours.

The BER is a function of the input signal level, which in turn depends on the transmission loss or the loss between the two ends of the link. You can calculate the transmission loss for Figure 13.1 from the following formula:

Transmission loss

= Free-space path loss + Terrain losses

+ Antenna feeder losses

+ Antenna branching loss − Antenna gains

The free space path loss is calculated as:

Equation 13.1

$$L_f = 32.5 + 20 \times \log d + 20 \times \log f$$

where

$$d = \text{hop length in km}$$

$$f = \text{frequency in GHz}$$

$$L_f = \text{path loss in dB}$$

MARGINS

Microwave systems are always designed with margins to ensure reliable working, even in adverse conditions. At higher frequencies (above 7 GHz) rain attenuation will

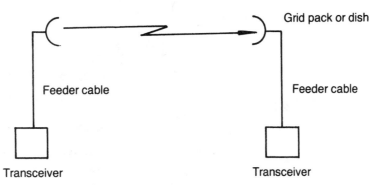

Figure 13.1   A typical microwave link.

TABLE 13.4 Fade Margin Versus System Reliability

Margin (dB)	Typical Reliability (%)
0	50
20	90
30	99.9
40	99.99

be the most significant performance-affecting factor, and this can readily be allowed for. The system reliability can be directly related to the fade margin, as shown in Table 13.4. However that relationship will be distance dependent.

FRESNEL ZONE

In a microwave link, the radio transmission exhibits wavelike characteristics, and the zone where wavelike interference can affect the propagation path can be approximated by the Fresnel Zone. (This is really a quantum-mechanical consequence of the way the signal propagates but the "wave" theory approximation has served the industry well for many decades.) The Fresnel Zone is widest in the middle of the link and can be calculated from the formula:

Equation 13.2

$$R_{FZ} = 17.3 \times \sqrt{(d_1 \times d_2)/(d \times F)}$$

where

R_{FZ} = Fresnel Zone radius (meters)

d_1 = distance zone base 1 (km)

d_2 = distance zone base 2 (km)

$d = d_1 + d_2$ or the length of the hop

f = frequency in GHz

Figure 13.2 shows the calculation of the first Fresnel Zone radius.

Microwaves do not normally propagate within the atmosphere in straight lines; they ordinarily travel in curved paths (usually curved downward) due to atmospheric refraction. The amount of curvature is usually defined with respect to the Earth's curvature, which is designated as K, where $K \times R$ (R = the Earth's actual radius) gives the effective radius of the Earth as seen by the microwave path.

If the Fresnel Zone is obstructed, some additional path losses will occur. When there are no obstacles within 50 percent of the Fresnel Zone radius for $K = 4/3$ (the most usual value that approaches a "flat Earth"), then the obstacle generally causes negligible loss. When, however, an obstacle protrudes into the path of the link by more than 50 percent of the first Fresnel Zone, an adjustment must be made for the additional losses incurred.

The loss due to path obstacles is a function not only of their intrusion into the Fresnel zone, but also of the actual shape of the obstruction and its reflectivity. Because of this, no simple relationship exists between the loss caused by the intrusion and the degree of intrusion. Some general statements can however be made about the minimum clearances that are acceptable.

For high reliability routes:

Equation 13.3a

$$\text{Clearance} > 0.3 \ F_1 \text{ at } K = 2/3 \quad \text{and}$$
$$> 1.0 \times F_1 \text{ at } K = 4/3 \qquad (13.3a)$$

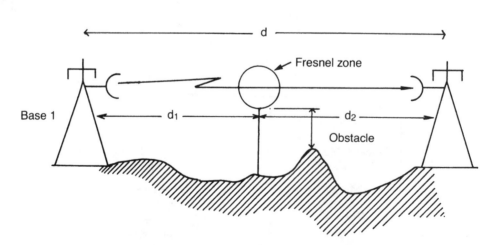

Figure 13.2 Calculation of the first Fresnel Zone radius.

For medium to thin line routes

$$\text{Clearance} > 0.6\, F_1 + 3 \text{ meters at } K = 1.0 \quad (13.3b)$$

Where $F_1 = $ the first Fresnel zone.

Because of changes in the refractive index of the atmosphere, the effective value of K varies with time. Smaller values of K increase the attenuation due to obstructions, particularly on longer path lengths. You should check to ensure that potential variations in K will not degrade the service.

The change in clearance (C_C) where the clearance was first calculated for $K = 4/3$ can be approximated by

Equation 13.4

$$C_C = 0.078 \times d_1 \times d_2 \times \{0.75 - (1/K)\} \text{ meters}$$

The limiting values of K are

$K = 1$ for wet climates

$K = 0.8$ for temperate climates

$K = 0.6$ for desert climates

It is normal to check the path profile for the extremes of $K = 4/3$ to $K = 0.8$.

When Line of Sight Is Not Possible

It is not actually necessary to achieve a line-of-sight path. This is because the RF energy, as it travels disperses, and that energy traveling just a little skyward of the line of sight encounters thinner air, and hence a lower refractive index. The effect of this is to cause this energy to refract (or curve) downward. This ducting allows the RF to propagate further than the line of sight. To approximate this effect, a traditional way of accounting for it has been to imagine the radius of the Earth as being larger than it actually is, hence yielding a larger value for the line-of-sight distance. Commonly a curvature of 4/3 (meaning line of sight based on an Earth radius $4/3\times$ bigger than it actually is) is used.

The refractive index of the air at any time is a function of the water-vapor pressure. This pressure varies with time, and for this reason the lower values of the refractive index more accurately depict the worse-case senarios.

If the path is further than line of sight, a penalty will be paid in terms of reliability. In abnormal atmospheric conditions, such as temperature inversion, the propagation may be such that the effective earth radius decreases even below line of sight, and so the path will fail. Traditionally designers of links have taken into account var-

ious possible values of effective curvature (designated K factor: where $K = $ the ratio by which the effective Earth radius is increased) from $K = $ infinity, $K = 4/3$, $K = 2/3$. $K = $ Infinity corresponds to perfect ducting, where the radio path follows the Earth's curvature, as in Figure 13.3a. The effect of other values of K can be seen in figure 13.3b.

Free-Space Conditions

Free-space conditions are said to exist to a good approximation when the ground clearance is 0.6 of the first Fresnel Zone. A conservative link design approach is to specify the 0.6 F_1 clearance for $K = 2/3$.

DESIGN SOFTWARE

A wide variety of commercial microwave design software tools is available that will automatically look at all these parameters and more. If a number of links are to be designed, or if there is any reason to consider a link marginal, then these tools *must* be used. The software comes in two basic forms. The first is one that requires the manual entry of a selected number of reference points and their elevations along the proposed path. These are then used to calculate the path loss and the existence of any obstructions. Notice here that "line of sight" does not ensure that there will not be obstructions in adverse atmospheric conditions, and the software will look at the worst-case scenario. Alternatively, fully automated versions exist that contain digitized regional maps, and all that is necessary to input is the coordinates and the height of the link ends.

Digitized maps are today widely available, but care is needed when using them. In particular, for short hops the number of positional data points may be small and good resolution may not be possible.

No matter what the computer prediction, there must still be a visual inspection (often called an *ocular inspection*) to ensure that the path is truly free of obstacles and that the site itself is suitable for a microwave site.

Ocular Inspection

A check must be made to ensure that there are no near-end obstructions, and as a rule of thumb the clearance immediately around the microwave dish should be at least twice the dish diameter, although if possible three diameters should be the target. This clearance requirement is shown in Figure 13.4. Also check that there are no obvious near-end buildings in existence or likely to be built in the near future that obstruct the path.

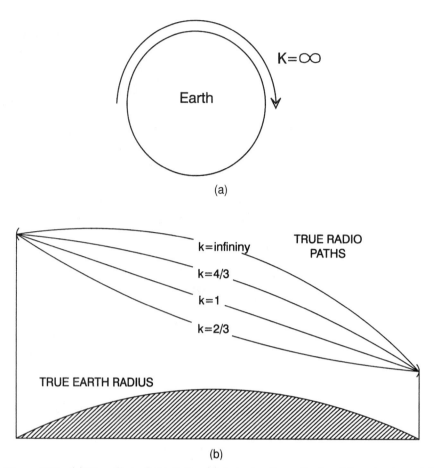

Figure 13.3 (*a*) The effect of *K*-infinity. (*b*) Actual radio paths for different values at *K*.

Ensure that if public access to the site is permitted that the antenna assembly is well out of reach. Also ensure that access ladders are available and secure.

Check for other installations that may obstruct the path, particularly in the case of guywires. Badly installed or maintained third-party installations can cause intermodulation problems, or in extreme cases, can be a physical hazard, if the structure is not secure.

If it is proposed to use an existing tower, make sure it is in good condition (see Chapter 22, "Towers and Masts," for an inspection check list).

Fading

All links will at some time suffer fading, and there are a number of mechanisms that can cause this.

Diffraction Fading Diffraction fading is caused by the radio path grazing the ground or other obstacles. As the effective value of *K* decreases, so the ground clearance decreases, and in bad atmospheric conditions this fading will cause increased path losses.

Surface Reflection Fading Reflected signals that bounce of the Earth's surface may interfere with the signal from the direct path. As the effective *K* value changes, so the point of reflection moves, and the degree of interference will vary.

Atmospheric Multipath Fading For well-engineered systems, the first two kinds of fading should not present a serious problem. Multipath fading, however, is more difficult to predict and to control. It comes about when the atmospheric conditions are such that there is more than one radio path between the links, and these paths combine in a random phase. This can cause rapid changes in signal level.

FADING DEPTH

The correlation between BER and fading on a digital link is not very high, and the fade margin has less meaning than it does in analog systems. The flat-fade margin can be calculated from the empirical formula. The formulas used are tailored to geographic regions, so

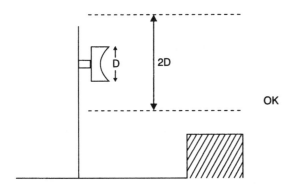

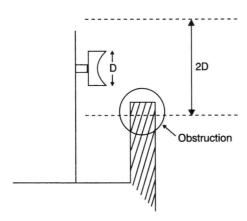

Figure 13.4   Microwave dish clearances from obstruction.

the constants vary a little from place to place. The main variables are temperature, humidity, and rainfall. For Western Europe, Equation 13.5 applies.

Equation 13.5

$$F_D = 35 \times \log d + 10 \times \log f - 10 \times \log p$$
$$- 58.5 + 10 \times \log C_f + 10 \times \log Q_f$$

where

p = the percentage of time of outage

C_f = the climatic factor

Q_f = the terrain factor

F_D = fading depth

and

$Q_f = 0.25$ for hilly or mountainous paths

$= 1$ for rolling terrain

$= 4$ for very smooth terrain (or over water)

TABLE 13.5   Typical Coupling Losses

System	Coupling Losses at a 2 GHz dB
Unprotected	5
Hot standby	10
Polarization protection	5
Hot standby with space diversity	6

and

$C_f = \frac{1}{8}$ for dry climate

$= \frac{1}{4}$ for temperate climate

$= \frac{1}{2}$ for humid climate or coastal area

$= 1$ for coastal area in hot humid climate

LOSSES IN ANTENNA COUPLING

Table 13.5 shows typical coupling losses. The antenna coupling losses are dependent on the actual coupler used. Use Table 13.5 as a guideline only; consult the manufacturer's specifications.

The multicoupling used on a microwave system varies according to configuration. A typical protected system is illustrated in Figure 13.5.

CALCULATION OF OUTAGE TIME

The percentage of outage time, due to multipath fade only, can be calculated by Equation 13.6. The probability of outage time P is

Equation 13.6

$$P = C_f \times Q_f \times 3 \times 10^{-7} \times F^{1.5} \times d^3 \times 10^{(-M_f)/10}$$

where

P = probability of outage with no diversity

M_f = fade margin

F = frequency in GHz

or for a diversity system

$$P_{od} = P/I$$

where

$$I = 0.0012 \times S^2 \times (F/d) \times 10^{M_f/10}$$

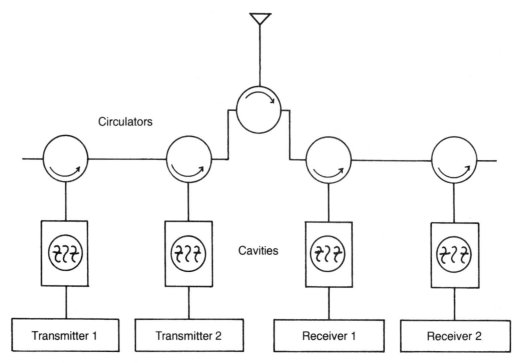

Figure 13.5 Branching and coupling of a singularly polarized protected system.

P_{od} = probability of outage with diversity

S = spacing between the antenna 5 m < S < 15 m

M_f = fade margin of the first antenna

SYSTEM GAINS

The smallest antennas used on microwave links are usually grid packs with about 25-dB gain. This relatively high minimum gain is used in order to ensure narrow beam width and hence good frequency reuse potential.

Table 13.6 shows the antenna gains available at 2 GHz.

At these frequencies it is possible to use grid packs that are lighter, cheaper, have lower wind-loading, and are almost as efficient as solid dishes (within 0.2 dB).

TABLE 13.6 Antenna Gain with Diameter at 2 GHz

Dish Diameter (m)	Gain (dBi)
1.2	26.4
1.8	30.5
2.4	32.8
3.0	34.6
3.7	35.5
4.6	37.5

Microwave antennas come in many forms, as can be seen in Figure 13.6, in which the tower is covered in various sized dishes and horns.

GAIN MEASUREMENTS

In microwave radio it is normal to express gains in dBi (or gain over an isotropic antenna). Mobile radio gains are nearly always quoted as dBd (or gain over a dipole). These two measurements are directly related by

Equation 13.7

$$\text{dBi} = \text{dBd} + 2.15 \text{ dB}$$

FEEDER LOSSES

Feeder losses are most significant at microwave frequencies and must be taken into account when determining system gains. Table 13.7 shows typical feeder losses at 2 GHz. You can obtain actual feeder losses from the manufacturers' catalogs. Some variations occur with the dielectric type and manufacturers. Connector losses of 0.5–1 dB can be expected at these frequencies.

At higher frequencies, waveguides are generally used, with EW-77 elliptical waveguide being most common

Figure 13.6 A microwave tower.

above 7 GHz. Elliptical waveguides are semi-flexible and can support only a single polarization mode, but the ease of handling, particularly the ability to install without joins, makes the elliptical waveguide far more popular than the square or round sections.

INTERFERENCE

Microwave systems, such as those used in cellular systems, are expected to operate in an environment of frequency reuse and have evolved a number of techniques

TABLE 13.7   Feeder Losses at 2 GHz

Foam insulated 1/2″	11.3 dB/100 meters
Foam insulated 7/8″	6.4 dB/100 meters
Foam insulated 1¼″	4.7 dB/100 meters
Foam insulated 1⅝″	4.1 dB/100 meters

to limit interference. The interference takes two main forms: intersystem (correlated) and intrasystem (uncorrelated). In both forms, co-channel (same channel) and adjacent-channel interference can occur. Figure 13.7 shows these interference modes.

Unlike cellular radio, microwave links have no inherent immunity to interference. Microwave links rely on careful planning and usually field survey to ensure adequate interference immunity. There are three major types of interference: carrier-beat interference, threshold degradation, and sideband-noise interference.

Carrier-beat interference is caused by small differences in frequency between the interfering carriers. This causes a beat frequency that may lie within one or more of the demodulated baseband channels. This is mainly a problem with analog systems.

Threshold degradation occurs when an interfering carrier, offset in frequency from the system frequency, is of such a level that even after passing through the system filters, a significant signal level persists. The interfering carrier effectively increases the thermal noise and consequently degrades the system noise factor. This results in an effective decrease in receiver sensitivity.

Sideband noise occurs when the sideband spectral power densities and the inband power densities overlap. Upon demodulation, this interference appears as noise and degrades the received S/N ratio. This is why it is important to see that channel overmodulation does not occur.

MARGINS

Digital systems require a margin of 20 dB ± 5 dB for correlated interferences. To include the fade margin, 25 dB is normally allowed between the interference level and the desired signal level. Cross-polarized signals have an inherent immunity of about 35 dB and, in well-designed systems, do not ordinarily present problems. Non-correlated interference can only be determined by measurement. A margin of 30 dB should be allowed for such interference. Often the directivity of the antenna system can be used to provide the bulk of this margin.

SYSTEM CAPACITIES

Systems are divided into three capacity groups:

- Low capacity 2, 4, and 8 Mbits/s
- Medium capacity 17, 34, and 70 Mbits/s
- High capacity 102 (3 × 34) Mbits/s and above, with 140 Mbits being the most common

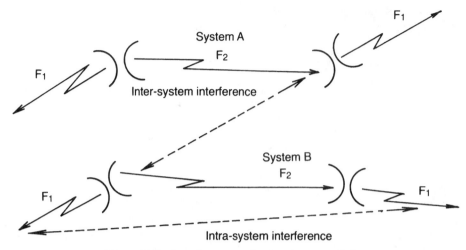

Figure 13.7   Interference modes in microwave links.

The channel capacity for North America and Japan is 24/1.5 Mbit/s, but most of the rest of the world (Europe and Australia-Asia in particular) uses 34/2 Mbit/s.

The 32 channels are based on 64 kbit/s per channel, with 30 channels available as speech channels. The 24-channel (or T1) systems are also 64 kbit/s and have all 24 channels available for speech.

The US system has a structure as indicated below

DS-0   (64 kbits)
|
DS-1   (T1)
|
DS-2   ($4 \times$ T1)
|
DS-3   ($24 \times$ T1)

For cellular links the usual practice is to use either low-capacity links, which come packaged as 2-Mbit (E1) units expandable to 8-Mbit (or DS1, DS2, US), or medium-capacity links, which are 8-Mbit units expandable to 34-Mbit (or DS3, US). The connection to the cellular base or switch is usually in 24- or 30-channel streams (depending on the system) and multiplexers are available to allow this connection.

ADVANTAGES OF DIGITAL SYSTEMS

Digital radio systems have a number of advantages over analog systems, including higher S/N and interference immunity, and the ability to regenerate traffic at each repeater without adding additional noise. Higher S/N and interference immunity results in greater spectral efficiency. The ability to regenerate traffic at each repeater allows for very long hops without serious noise degradation.

The main disadvantage of digital systems is the lower immunity to frequency-selective fading, which can result in the need for an additional 18-dB flat-fade margin above that required for a comparable analog system.

RACK SPACE

Microwave equipment generally requires only one 600-mm rack spaces, which includes space for the multiplexer (MUX) and inverters (which are generally needed for the MUX and may be needed for the RF hardware). These racks should also have space for an intermediate distribution frame (IDF) and alarm panels.

MICROWAVE LINKS IN CELLULAR SYSTEMS

Every base station needs to link each of its channels back to the switch. There are three ways of doing this: by conventional land-line facilities, with fiber optics, or with microwave links.

Many operators prefer microwave links because they have full control over these links and often they are cheaper. Microwave links can also be readily redeployed if base stations must be moved.

The frequencies used for microwave links range from the long-haul 2- and 6-GHz bands to the relatively short-haul 11-, 18-, and 23-GHz bands. The availability of spectrum varies and some countries, notably the United States, have complicated rules about which frequencies

TABLE 13.8   Minimum Path Length for Certain Bands in the United States

Frequency (MHz)	Minimum Path Distance (km)
2110 to 2130	5
2160 to 2180	5
3700 to 4200	17
5925 to 6425	17
10,700 to 11,700	5

TABLE 13.9   Hierarchical Level for Data Transmission

Level	Bit Rate (Mb/s)	Number of Voice Channels	T1s
DS1	1.544	24	1
DS2	6.312	96	4
DS3	44.736	672	28

can be used under certain conditions (for example, the United States has a minimum-distance rule for some bands, as shown in Table 13.8). Other rules cover minimum capacity on links in certain bands.

Increasingly, the short-haul, high-frequency systems are becoming more economically attractive. The frequencies above 11 GHz are approximately three times more susceptible to weather-related outages such as rain and snow. The susceptibility decreases rapidly at frequencies below 11 GHz.

For economic reasons microwave units operating above 26 GHz rarely have an low noise amplifier (LNA). As a result, their noise figures are 7 to 10 dB down from those operating at lower frequencies. This puts an additional limitation on the range that can be expected from these links. An LNA will reduce the noise figure to below 5 dB.

UNITED STATES AND JAPAN

The 24-channel units used in the United States and Japan are known as T1s; each T1 bit stream occupies 1.544 Mbits/s. Figure 13.8 shows a simple T1 circuit. Table 13.9 shows the hierarchy of transmission that has been established. High-capacity systems comprising multiples of 56 and 84 T1s are also available.

REST OF THE WORLD

Microwave elsewhere in the world is 2 Mbit or E1 based. The systems of most interest in cellular radio are the 2/8 (2-Mbit expandable to 8) and 8/34 (8-Mbit expandable to 34) -Mbit systems. There is usually only a small difference in price between a 2-Mbit system expandable to 8-Mbit and a 2-Mbit system (not expandable). For this reason, if expansion is anticipated, a 2/8-Mbit system is usually chosen even though a 2-Mbit unit would suffice initially. A price break usually occurs at 8 Mbit, but an 8-Mbit unit expandable to 34-Mbit is similar in price to a non-expandable 8-Mbit system.

These microwave units may come in racks or may be intended to be wall mounted (see Figure 13.9).

Even in the case of analog mobile systems, the switches are digital and so it is logical for the links to the base station to also be digital. This requires an Analog-to-Digital (A/D) converter (also known as a multiplexer or MUX) to be placed at the base station. Figure 13.10 shows multiplexer equipment at the cellular base. Some manufacturers provide MUX equipment as an integral part of their base stations, but others do not.

The switch can accept 2-Mbit data streams. Each 2-Mbit stream has 30×64 kbits/s voice-channel capacity plus two control channels.

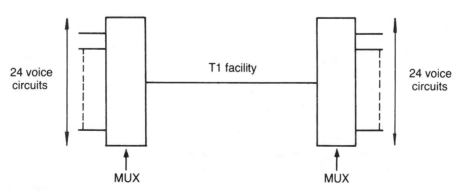

Figure 13.8   A T1 facility equipped with voice channel MUX.

Figure 13.9 Four racks of microwave equipment. The first two racks on the left contain the MUX equipment. The next rack contains two 2-Mbit microwave transceivers, and the end rack contains a 34-Mbit transceiver. Microwave equipment is usually designed to be wall mounted. (Photo courtesy of Extelcom.)

DROP AND INSERT

The MUX equipment occupies a full 2-Mbit or T1 data stream each time it is accessed. Drop and insert facilities, however, can take a single timeslot and drop or insert that timeslot at any location along the microwave link. These facilities might be used for a thin (low-density) route covering a highway, as shown in Figure 13.11.

The power consumption of a microwave link in the hot standby mode is from 60 to 200 watts, making it a relatively low-power device when compared to the rest of the base station.

SYNCHRONOUS DIGITAL HIERARCHY (SDH)

By the early 1990s most of the world's trunked networks (outside the United States) was based on the European plesiochronous digital hierarchy (PDH), with its familiar but complex interleaving of 64-kBit channels into 2-, 8-, 34-, 140-, and 565-mBit/s bearers. The very complexity meant that individual timeslots could only be accessed by demultiplexing all the way down to the 2-mBit/s level. Synchronous digital hierarchy (SDH) is essentially the same thing without the complex and somewhat ad hoc structure, which allows direct access to individual 2 mBit/s from the higher-order bearers.

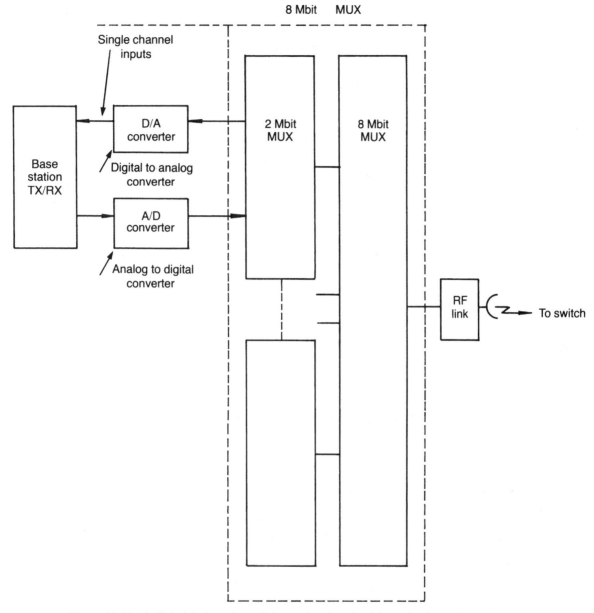

Figure 13.10 A digital link to the switch. Notice that the A/D and D/A and 2-Mbit MUX conversion may or may not be part of the base-station equipment.

The SDH structure has now become an International Telecommunications Union (ITU) standard and covers layers from 64 kBit/s to 622 mBits and beyond, even to 2.5 GBits/s and 10 GBits/s. Interworking with the US SONET structure can most effectively be achieved at 155 mBit/s. It is specified in some detail at the network management level, and this allows a single-ended maintenance, if required, or structured regional and centralized management.

FACTORS IN CHOOSING MICROWAVE

The installation time for a microwave system (after obtaining approval and assuming equipment is available) can be less than one week. Approval times and equipment delivery times, though, can add up to months or, in the worst cases, years.

The reliability of modern microwave equipment is very high and the mean time between failures (MTBF)

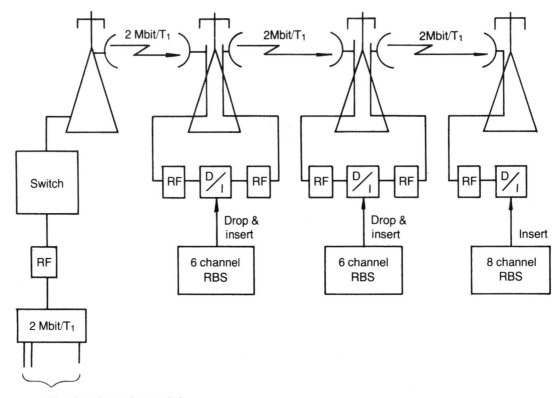

Figure 13.11 Drop-and-insert facilities used to fully utilize bearer capacity.

can be expected to run to many years. There are many different ways of calculating MTBF, and since there is no standard, one should not place too much weight on figures supplied by manufacturers.

Although heavy rain and snow can cause outages, the probability of a complete outage on a well-designed system is very low. Weather restrictions do limit 10-GHz systems to 50 km and 18- and 23-GHz systems to 15 km (and even less in tropical regions where a few kilometers may be the usable range), but these are large distances by cellular standards.

A microwave system used for cellular should not require extensive additional sophisticated (expensive) test equipment to maintain. Consider this when ordering test equipment for the cellular network. Microwave equipment, being significantly higher in frequency than cellular bases, requires more sophisticated and hence more expensive test equipment. Where possible, test equipment for the microwave network should also be usable at cellular frequencies.

In order to have effective system management, extensive diagnostics and alarms should be built in. Because space is almost always a problem in cellular installations, choose a system that can be wall-rack mounted without consuming too much equipment space. Some-

times microwave racks, as sold to wireline companies, are too high for cellular base stations. The rack height should not be higher than the base-station rack height. This may require a reconfiguration of the link equipment.

SURVEY

Because most cellular base stations are either in line of sight to the switch or to a suitable branch point to the switch (clearly) or obstructed (clearly), the process of a survey is somewhat simplified. Because of the relatively high cost of microwave links (compared to a base station, microwave links represent about 10–30 percent of the total costs), a trade-off may occur between good cellular design and good link design when base stations are deliberately placed behind obstacles (to improve frequency reuse). In these instances, either an alternative to microwave or the use of a relay site present themselves as options to compromising the cellular design.

Where any doubt exists, a proper detailed microwave survey should be done but, in a modern city, line of sight usually guarantees an adequate path over the relatively short hauls required for cellular. Long hauls (greater

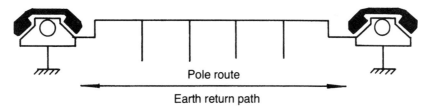

Figure 13.12 The earliest (and cheapest) transmission systems used the Earth as the return circuit. This method generated considerable noise and transmission-level uncertainties.

Figure 13.13 A negative-impedance repeater produces gain in a two-wire cable and provides a moderately effective means of extending transmission distances in two-wire cables.

than 20 km) should be surveyed at RF frequencies similar to those of the proposed links.

WIRELINE—ITS DEVELOPMENT

It is worthwhile taking a brief look at the history of wireline, as much of the the terminology and even practice of today derive from the need for backward compatibility. Telephone traffic can be carried by one-, two-, four-, or six-wire circuits. Each of these types of circuits has its place, and it is worth considering each separately.

Figure 13.12 shows a single-wire circuit. A one-wire circuit is an Earth-return system (the Earth provides the second wire). Early telephone links were mainly single-wire, or single-wire Earth return (SWER), because this type of circuit was cheap. The disadvantages of such circuits are many and include variable performance due to soil resistivity, low noise immunity, and safety hazards in lightning-prone areas. Despite these problems, some single-wire circuits exist even today in rural areas.

The two-wire line is an obvious improvement from the SWER line, and it does eliminate a lot of the SWER line problems. These phone lines were heavy-gauge single-stranded copper wires that were hung on insulators like those seen in Figure 13.14. The insulators have today become collector's items (Who knows why?) and the copper has long since been sold as scrap metal. This survivor of the past was seen by the author in the United Kingdom in 1991.

When longer routes were considered, a two-wire system has limitations when amplification is needed. Some ingenious amplifiers known as negative-impedance repeaters (NIR) were developed to provide some amplifi-

Figure 13.14 A copper wireline power pole complete with insulators and wires.

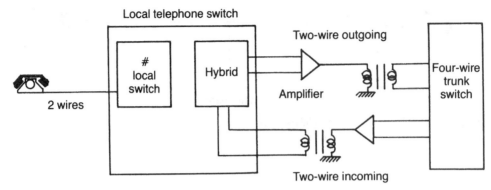

Figure 13.15 The four-wire transmission (sometimes called six-wire or four-wire E&M) enables independent amplification of the send-and-receive directions. E&M are signaling media derived over the four-wire channels.

cation over long two-wire system routes. A conventional telephone is an example of a two-wire circuit.

Figure 13.13 shows an NIR between the two switches. The NIR provides some gain to compensate for line losses between the switches. The operation of an NIR is such that the maximum theoretical gain that can be provided on the route (end-to-end) is 0 dB (a gain of 1) before instability occurs. In practice, gains of −6 dB were more common.

The circuit shown in Figure 13.15 shows a subscriber telephone connected by a twisted pair to a telephone switch. Two-wire connections from subscribers' units to the telephone switch are usual.

On longer routes it was necessary to provide a considerable amount of gain to make up for system losses. The only practical way of doing this is to use four-wire transmission (that is, to separate the send-and-receive paths and amplify each). High-level (trunk-level) exchanges are usually four-wire switches, as shown in Figure 13.15. With four-wire switching, transmission without loss can be achieved in practice, with noise considerations and hybrid leakage limiting the total amount of gain achieved.

Six-wire switching can be regarded as four-wire with two "wires" used for simple signaling purposes. (Four-wire links can derive the signaling purposes. (Four-wire links can derive the signaling channels in a number of other ways.) The two signaling wires are known as the M lead (the "M" was for "mouth") which transmits the outgoing signal, and the E lead (the "E" was for "ear"), which carries the incoming signal.

Wireline links will not often be used for cellular systems, but when they are, the most likely transmission mode will be pulse code modulation (PCM). This will provide individual 64-kbit channels over copper routes. Occasionally physical copper pairs are used.

ECHO SUPPRESSION

Echo has always been a problem in telephony. In short-wire systems, the hybrid will leak some of the signal from the incoming line back to the return wire. In short systems where the total propagation delay is less than 30 ms, the effect is to produce the familiar side tone in the distant receiver. This tone has proved beneficial, as it acts as a confidence tone for the speaker. Once the transmission line is long enough that the delay exceeds 30 ms, the feedback signal is heard as an annoying echo.

In digital transmission systems, the digital multiplexer process and signal regeneration introduce new delays, and because even today most links eventually connect to a two-wire telephone (and hence a hybrid) the potential for echo arises.

Although echo-reducing devices are commonly called "echo suppresses," this name really refers to the first-generation devices, which were designed to attenuate or suppress the echoes. Modern devices are digital and are cancellation devices that sample the forward signal, process it (using a convolution process), and then by looking at the signal on the return leg the echo, identify it and cancel it out.

In cellular radio applications the handsfree mobile phone has become an everyday item, and to a large extent it relies on an efficient echo cancellation process to work. When talking handsfree, the signal from the speaker will reflect off solid surfaces (such as car windows) back to the microphone, and thus there is a potential echo. Left unchecked, this echo will result in positive feedback and ultimately oscillation. The early model handsfree mobile phones used echo supressors and worked by allowing only half-duplex speech (so if the user was speaking, the microphone was on but the speaker was blanked out and vice versa). Current

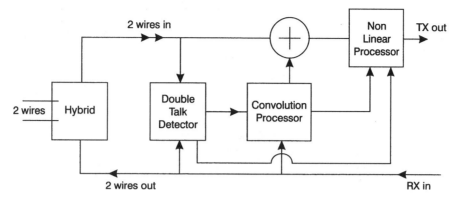

Figure 13.16 An echo canceller.

models use echo cancelers, and thus allow full duplex operation.

In cellular systems sources of echo are either in the 2- to 4-wire conversion points (as in the hybrids shown in Figure 13.15), which will be present in most land-line to mobile calls, or from the mobile handsets and handsfree devices. The acoustically coupled echos are mostly only a problem with digital phones. Some digital phones have the ability to remove any acoustic echos, but most do not have this capability. Digital phones that have low bit-rate voice coding also have considerable processing delays, which can add significantly to the echo problem.

To minimize echos at the hybrid it is important that the loads all be well balanced, so that an isolation of around 15 dB can be achieved.

Echo cancelers have three main components: a double-talk detector, a convolution processor, and a non-linear processor, as seen in Figure 13.16. The convolution processor (also known as an *adaptive filter*) makes a mathematical model of the circuit's impulse response and subtracts the unwanted echo from the returned signal, which consists of both the wanted speech path and the echo. The non-linear processor then removes any echo components missed by the adaptive filter. The double-talk detector is there to ensure that in the instances when both the send and receive parties are talking at the same time the convolution processor is informed of this, otherwise the signal may be wrongly processed.

For digital services, echo cancelers will generally be installed on digital lines connecting the base stations to the switch. For some systems where digital and analog systems co-exist, pooled echo suppressers can be used at the switch and put into service only for digital cellular calls. The pooled method will consume two switch ports for each echo canceler, but this is still cost-effective.

A modified pooled system that routes the mobile to

PSTN calls directly to inline PSTN trunk echo cancelers (which will generally be provided as a matter of course), are worth considering.

MINILINKS

Originally minilinks were meant for short hops (from 1 to 3 kilometers) and for applications where cost was more important than system reliability. To meet these constraints, the links used cheap RF devices, which were necessarily low power and of doubtful reliability. The links were meant to be used only in single hops and so that sophisticated monitoring and order wire facilities could be dispensed with. These links are less than one-half the cost of a conventional microwave unit, and some were so low powered that they could be used without a license. Mean time between failures (MTBFs) of around a year were all that could be expected.

Over the years RF device technology improved, and it is now possible to buy medium-power, high-reliability, GHz RF amplifiers at reasonable prices. Seeing this, the developers of minilinks realized that they could go up-market at minimal cost.

Until around 1992 minilinks would not have been considered by serious cellular system designers, but things have changed and it may be worthwhile to re-think this option. Consider the model 7014, 2 Mbit/s minilink from Codan, Australia. At $17,500 (current price), it is almost half the price of a conventional link, and it comes complete with a 600-mm dish included in the price. It has an MTBF of 32,000 hours, or 3.7 years, and will run on any voltage from 12 to 48 volts.

The link operates in the 10.5–10.68 GHz range and the power output is 26 dBm (or 0.4 watts). An order-wire can be provided, and local or remote system mon-

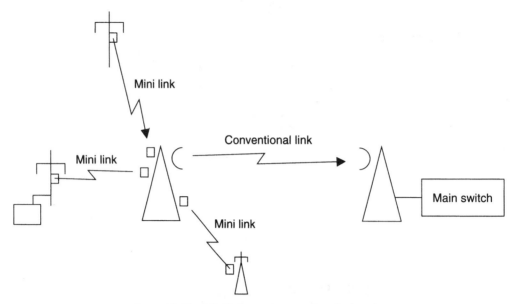

Figure 13.17 Minilinks are best used to feed nodes.

itoring facilities are available. When used in a chain the individual repeaters can be monitored in turn from a centralized monitor, which costs only $3400.

Of course, the MTBF becomes a problem if too many of these links are chained in series, so the 7014 can be provided with redundancy (which will increase the MTBF to almost 14 years and, of course, significantly increase the price) and automatic change-over. The link can even be configured to carry traffic on both 2-Mbit/s channels during normal operation, and then to switch whichever of the two 2-Mbit/s streams is priority into service in the event of a partial failure.

Because the MTBF is somewhat lower than conventional microwave, it would make good economic sense to consider using minilinks to feed nodes, which would be brought back by conventional microwave, as seen in Figure 13.17. This will avoid having too many minilinks in series, which seriously reduces the system MTBF.

It is interesting to speculate on what the MTBF of minilink equipment would be if it were built to operate in the same way as conventional units. If the equipment can operate in the hostile, lightning-prone environment at the top of a tower and still have an MTBF of 3.7 years, how much longer would it survive in an air-conditioned room, effectively isolated from the environment by cables and surge arrestors?

Minilinks differ from conventional microwave in that the electronics are located in a module, either on or near the antenna (see Figure 13.18, which shows a typical minilink system). This has the advantage that expensive RF feeders are not needed to feed the antenna because

Figure 13.18 A minilink system, showing three minilinks fitted with radomes. (Photo Courtesy of Codan).

the link output is at the 2-Mbit/s level rather than at the carrier frequency.

However, not all minilinks are necessarily small, and some are fitted with microwave dishes up to 1.8 meters in diameter. When it comes to mounting these larger units, the "mini" seems somewhat out of place.

A recent trend is to have higher capacity minilinks, and 34-Mbit systems are widely available. In general, these higher-capacity links are relatively low powered and thus are short-range devices.

SATELLITE LINKS

Satellite links have been used successfully in all types of cellular systems. If the implementation is done properly, they can be quite effective. This is particularly true when interconnection between distant sites is required, or when remote rural services are being considered. Poor-quality links with excessive delays or echo will not be tolerated today, and care must be taken with echo canceling and to ensure that the overall delay is kept within acceptable limits (around 600 ms).

Satellite capacity can be rented on a demand-assignment basis (meaning that it is paid for only when in use) or as dedicated channels. For cellular purposes, dedicated channels are the more practical, given that delays and blocking may occur with demand-assignment channels.

The major problem with all satellite links is the time delay in the transmission, which also causes echoes. A delay of around 0.5 second can be expected for each satellite link (although the actual delay will depend on the relative location of the send and receive stations with respect to the satellite). This delay means that only one satellite link can be used within the network and care must be taken to ensure that the switching is configured to prevent interconnection using a second link. Usually this will be possible except on international calls where the cellular operator will have little control over the trunking of the call in foreign countries.

There will ordinarily be a need to make timing changes at the switch to accommodate the longer processing delays that can be expected with satellites. Consideration needs to be given to any inband signaling, timing constraints on all data exchanges, handoff timing, and other constraints. Most switches today can accommodate this.

So if it is desired to connect three or more switches together using satellites, then *each* switch must be separately connected to each other switch in order to avoid more than one satellite link. For major links like switch-to-switch, it may be necessary to install an earth station at the switch site. Modern earth stations are not partic-

ularly big and could occupy as little as 10 square meters with an integrated transceiver and antenna system. In general the power consumption will be relatively high particularly for the up-link. The earth station can be expected to require around 20 kW of power (which includes air-conditioning). It is possible to trade off transmitter power (and hence consumption) against the antenna size (and hence its cost). Narrow-band earth stations (such as would be suitable for a base-station link) can have final RF power levels as low as 10 watts and use dishes only a meter or two in diameter.

Satellites are getting bigger and more powerful. A large satellite like INTELSAT VI has a 120,000 voice channel capacity. In 1985 most weighed less than 1 ton, in 1991 50 percent were bigger than 2 tons, and some are as big as 4 tons. As the number and size of satellites increase, so the cost of individual circuits will decrease, and this mode will become more attractive, particularly in isolated areas.

Today most satellite links are digital and so transcoders can be used to expand the voice capacity of a single channel to at least two and possibly up to five voice channels. Cellular systems can do all their signaling in a relatively narrow analog bandwidth, so the degradation due to the channel compression is minimal. This is most fortunate, as satellite channels are not cheap.

Compression

Care should be taken when considering voice-compression techniques, as these add to the overall delay experienced by the users. If compression is to be used, it should be ensured that the net additional delay is less than 80 ms.

VSAT

Very small aperture terminal (VSAT) may have some limited applications for cellular operation. VSAT systems are physically small, low-capacity earth stations that may have capacities from a few hundred baud (bits/second) up to a few digital channels (64/56 kbit). More recently systems with capacities of 6, 12, 24, and 30 channels have become available.

VSAT terminals are much cheaper than a full earth station and are much more portable. As the power and space requirements are minimal, in theory a VSAT terminal could be put in place in a day.

VSATs use parts of transponders from a geostationary satellite, which are linked to a centralized hub [or master Earth station (MES)]. There will be significant delays in a VSAT voice link, and care must be taken not to use more than one such link, where possible, to avoid echoes and annoying speech delays. The hub antenna is

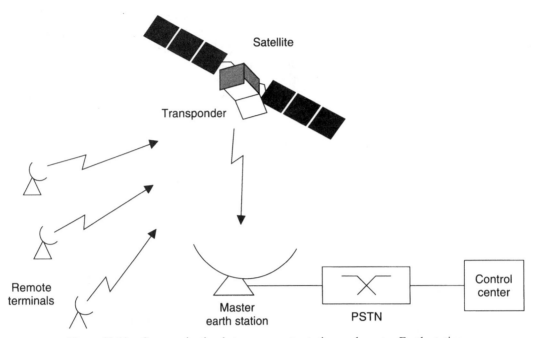

Figure 13.19   Communication between remote station and master Earth station.

necessarily large (around 12 meters) because of the remoteness of the satellite and the wide bandwidth it needs. Communication between the small remote stations and the MES is as seen in Figure 13.19.

VSATs operate mainly in the C- and Ku-bands using dishes that vary in size from 2.4 meters to 3 meters in the C-band to 0.6 meter to 1.8 meters in the Ku-band. The actual size will depend on the footprint field strength and the bandwidth. Narrow bandwidth equipment, such as slow-speed data, can use very small dishes. By VSAT standards links to cellular base stations are "wide bandwidth" systems.

The network configuration is controlled by a small control station, which can reconfigure the system in real time or according to some time-dependent program. For a small remote base station, the configuration may, for example, have the control channel and three voice channels permanently linked through to the switch during the day, plus four channels on demand assignment (being called in as they are needed). The nighttime configuration, which could be programmed to automatically, run from 6:00 A.M. to 8:00 P.M. may have just the control channel directly linked with all speech channels on demand assignment. An additional program could be to download the processed billing records once a week to a local computer. This could involve, for example, opening up further 4 × 64-kbit channels for a half hour every Saturday evening from 10:00 P.M. to 10:30 P.M. Reconfiguring is cheap and can take account of regional demands as required.

Special Considerations

If satellite links are to be used, then it is essential to fit echo suppressors to each channel. It also will be necessary to alter some system parameters so that delay is accounted for in the signaling. As an example, handoff instructions from the switch usually expect confirmation of success within a few tens of milliseconds after the instruction. Where a base station is connected by satellite, the transmission delays (which are almost a full second) will need to be allowed for. This can normally be done without any custom software.

Frequencies

Satellites are allocated spectrum in the four main bands listed in Table 13.10.

TABLE 13.10   Spectrum for Four Main Bands

Band	Up-link (GHz)	Down-link (GHz)	Bandwidth (MHz)
C-band	5.9–6.4	3.7–4.2	500
X-band	7.9–8.4	7.25–7.74	500
K-band	14–14.5	11.7–12.2	500
V-band	50–51	40–41	1000

FIBER OPTICS

During the 1980s the ever-decreasing cost of optical fibers, together with the improved performance in terms of loss, has made optical fiber a competitive alternative to other modes of transmission. Losses as low as 0.1 dB/km mean that long-haul systems without repeaters are practical.

One of the main attractions of this transmission mode is the high capacity, with modulation bandwidths of more than 10 GHz being possible. Naturally the cellular operator is unlikely to have a need for this type of capacity, but even simple systems offer very high channel capacity, which means that once installed a fiber cable can be expected to serve easily both present and future capacity requirements.

Optical fibers have very high immunity to electric, magnetic, and electromagnetic interference.

Increasingly, cellular operators are using fiber-optic links to connect base stations and microcells. While microwave equipment is similar in nature to the cellular infrastructure, fiber optic is a new field for the cellular operator. New skills and equipment are required to install and maintain the cable.

Problems can occur with obtaining rights-of-way to run the cable, and the authorities may be reluctant to allow new access. When this happens the operator can go to the local exchange carrier (LEC), which will generally be using fiber-optic cable routinely. When dealing with an LEC there are a number of optional approaches that can be taken. The LEC could be asked to provide and install a working link on a turnkey basis. This solves the installation problems, but it still leaves the cellular operator with the maintenance responsibilities. The equipment to maintain the cable, together with the expertise, can be expensive, and this option should be considered only by operators who will have enough cable to justify the infrastructure expense. An alternative would be to contract out the maintenance to the LEC.

Alternatively, the LEC could be asked to provide the cable only. Often this will be a problem for the LEC, some of which object to the "cable installer role." It does, however, give the cellular operator more control over the link service and hardware. Costs can be expected to be in the vicinity of $30,000 per kilometer.

For base-station links, where a few E1 or T1 circuits are all that is required, there should be no problem using a general traffic cable. For microcells that retransmit the base-station carrier and need a wide bandwidth, it would be best to lease a whole cable. This is sometimes known as leasing a "dark or dry" cable, indicating that as leased there are no other users (and so the cable is "dark"). In fact, microcells that use fiber-optic cables are designed for exclusive use.

Reasonably priced cables that have moderate losses (around 0.4 dB) can be used for distances of around 25 km without introducing excessive losses that would require repeaters. Repeaters increase the cost and are an additional maintenance item. A loss of 10 dB between links should be acceptable for most cellular applications.

It should be noted that when the broad-band mode is used to transmit sets of RF channels, the modulation technique is amplitude modulation (AM), and so the cable can be expected to contribute to the overall noise level of the system. Also although cables, unlike microwaves, are not subject to atmospheric fades, they are highly susceptible to "backhoe" fades, which can put them out of service for hours or sometimes even days.

Fiber-optic systems mostly use wavelengths of 0.8 μm or 1.3 μm, because of the ready availability of suitable lasers at these frequencies. These wavelengths (which translate to 300,000 GHz) are in fact in the infrared band rather than visible light, but the propagation characteristics are similar.

An optical fiber consists of a glass core surrounded by glass cladding. The loss is largely a function of the purity of the glass because impurities cause scattering. Common glass-based fibers are made from silica oxides or sodium borosilicate; more recent fibers are not glass at all, but are made of plastics with good optical qualities.

SONET

SONET is the synchronous optical network standard that allows for a fiber-optic system with a great deal of survivability. The system provides for line and path switched rings with a network healing time of around 60 ms. Mostly SONET comes with 100 percent redundancy, with the main and standby bearer both sharing traffic.

INFRARED LINKS

A recent development has been the availability of infrared links. Like microwave, these links are transmitted through the air, but they have the advantage that they require no licensing and so can be deployed very rapidly. This makes them especially suitable for emergency links, unplanned installations, and short hops where cable is not an option.

The light is in the 860–920-nm band and is generated by lasers. The whole unit is about the size of a shoe box and can handle data rates up to 622 Mbps. The cost is a function of the capacity, but broadly the range is $7000 for a low-power, low-capacity link to $50,000 for the highest-capacity high-power links. The links can serve as a direct replacement for fiber optic. Naturally the link requires a clear line of sight to operate at all.

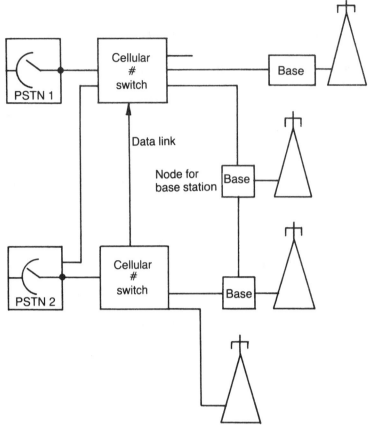

Figure 13.20 The cellular switch can be interconnected to a number of PSTN points. PSTN 2, for example, could be in a different town.

Reliability figures in the range of 99.1 percent to 99.9 percent are quoted by one supplier, Infrared Communications Systems, Inc. (ICS). These figures are relatively low compared to most telecommunications links, but would certainly be adequate for temporary and emergency links.

A specification for a link supplied by ICS for a 4 × T1 link (E1 links are also available) is:

Range	1000 meters
TX power	30 mW
RX sensitivity	−45 dBm
Capacity	4 × T1
Power consumption	50 watts

The link is transparent to the rest of the system, and "looks" like a part of the transmission cable, and so no special interfaces to it are required.

The links come in a number of models with different outputs that give different ranges. These models include those for 200, 500, 1200, 2000, 4000, and 6000 meters.

TRUNKING

The call-routing-and-system-link configuration is known as the trunking of the network.

Cellular systems can be linked by copper pairs, fiber-optic cables, or microwave (analog or digital). Any combination of these systems can be used. A fully developed cellular network can have a complex trunking scheme involving multiple switches and route diversity. Some operators like to spread circuits to and from the PSTN over different routes and even over different systems. For example, half the circuits from the switch may go via microwave and half by fiber-optics cable. Figure 13.20 illustrates the trunking of a multiswitch system.

Today, unless the cellular operator is a wireline company with a large surplus of copper pairs, some form of digital link is likely to be the most economical way to connect the cellular switch to the base stations and the PSTN. Even where copper pairs are available, it is generally economical to use PCM to increase the trunking efficiency.

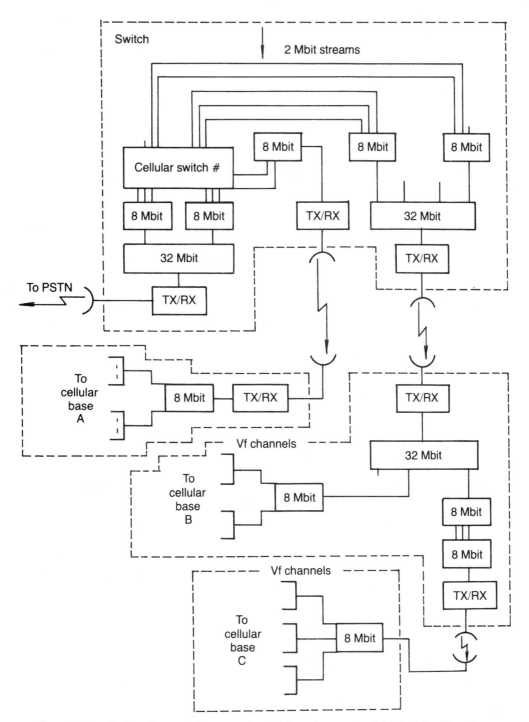

Figure 13.21 A three-base cellular network trunking scheme using 2-Mbit links. Three bases are connected to a switch via a microwave link. Base B picks up the link from base C and then through connects to the switch.

For most cellular operators, the choice is between optical fiber, digital microwave, or PCM.

The links for cellular systems are often group connected in multiples of 24 or 30 channels. The switch is usually configured to accept incoming and outgoing traffic in these multiples, and it is usually necessary to use multiplex equipment to access traffic at the individual channel level.

In the trunking plan, some branching may be necessary to increase the total capacity of the links in the network. Branching involves using one or more base-station links as nodes (or connection points) in the linked network.

Figure 13.21 shows the connection of three radio bases to a cellular switch, and the connection of the switch to the PSTN. The figure assumes that the largest group that can be separated out from the main stream of data is 2 Mbits. In some systems, it is possible to connect at the 8-Mbit level.

Figure 13.21 also shows base B being used as a node for the 8-Mbit link from base C. One half of the 34-Mbit capacity of the link between base B and the switch is tied up and is not available for any purpose on other routes. The need and opportunity to use some base stations as nodes (or collection points) for data streams grows as the system capacity grows, regardless of whether you use microwave or fiber-optic systems.

Frequency congestion on microwave systems will often encourage the use of local nodes, as will route availability when using fiber optics. In order to attain maximum circuit efficiency, it may be necessary at some nodes to go down to the voice frequency (VF) level to avoid too much use of low-occupancy 24- or 30-channel streams.

MUX equipment frequently employs a 48-volt power-positive ground; this equipment usually requires the use of inverters at 24-volt cell sites if the base-station batteries are to be used to provide backup power for the microwave.

STANDBY

Hot standby—that is, a fully active parallel system powered up and on standby—is usually used for main routes and the PSTN route. A changeover switch is provided (automatic) to allow the traffic to be switched from one link system to the other when a failure occurs.

Cold standby is much the same as hot standby, but the second redundant system is not powered up. Cold-standby systems usually have manual changeover. The reliability of modern microwave systems is such that it is practical in thin-route (light-traffic routes) situations to dispense with standby altogether and have only a single link.

For fiber-optic systems, the risk of loss of service is increased by the chance of having the cable cut by excavation. Fiber-optic systems, for this reason, often employ route diversity (that is, a link via two different paths).

A simple system with standby is called a $1 + 1$ (one active plus one standby); an unprotected route is designated a $1 + 0$.

Larger systems can provide a good degree of protection at increasingly lower costs. For example, a large system might consist of 3×34-Mbit bearers in service with a 1×34-Mbit hot standby. In this configuration, any failure of one of the main bearers can be taken up by the standby bearer. Such a system would be designated a $3 + 1$.

SPLIT ROUTES

Sometimes, in order to further decrease the trunking costs, it is worthwhile to spread traffic over the main and protection bearer. As shown in Figure 13.21, when the PSTN route (34 Mbits) becomes full, it is possible to abandon the hot-standby configuration in favor of two parallel links with the cellular switch to PSTN traffic spread between them.

In this instance, failure of one half of the system results in a loss of half the traffic capacity of the route. This may not be acceptable on the main PSTN route, but it would generally be acceptable on a route to a node.

Where parallel systems are used, it is necessary to have two active systems employing two separate frequencies.

14

BASE-STATION MAINTENANCE

A fully developed cellular system will have a very large number of channels and radio bases. Fortunately, most cellular systems have good housekeeping software; problems such as voltage standing-wave ratio (VSWR) occupancy, transmitter (TX) failure, channel-blocking, power and entry alarms, and speech path conditions are constantly monitored and are automatically reported to the maintenance-control center.

Maintenance usually consists of board or panel replacement, and this replacement is directed by the base-station controller, the switch, or the operations and maintenance center (OMC) (that is, apart from reporting the fault, diagnostics suggest the maintenance procedure).

Routine maintenance is usually handled by one or two technicians who respond to a fault by replacing the suggested boards or panels. Ongoing maintenance in a growing cellular system also includes channel re-arrangements to shift channels, as required, from base to base to match traffic capacity to demand. Staff should be capable of retuning the base transceivers and multicouplers.

There is a fine line between extending the capacity of a base (installation) and rearranging channels to suit traffic conditions (maintenance). In those companies where these two functions are performed by separate groups, it is necessary to clearly define the role of each group in each instance.

Routine maintenance should include approximately monthly visits to check on the condition of the batteries, air-conditioning, and site upkeep. In particular, the grass and foliage around the site should be neatly kept to avoid fire hazards. This visit can be handled by non-technical staff. Audio levels, radio frequency (RF) power

receiver sensitivities TX levels and frequency should be confirmed every 6–12 months.

Because cellular bases are often on remote sites, careful attention should be paid to vandalism and attempted vandalism. Having a good security fence and some means of preventing intruders from climbing towers and structures is essential.

If this routine maintenance was all that was required, then base-station maintenance would indeed be easy. But there are other considerations. The rest of this chapter discusses these considerations in detail.

BASE-STATION MAINTENANCE

The base station is subject to the extremes of nature. Even high-density installations (microcells excluded) are usually situated so that the antennas are in the most prominent position in the locality. They are subject to lightning, and the external plant is constantly subjected to the weather.

Personal communication service/personal communications network (PCS/PCN) has brought new demands for small, light, and cheap construction. Often (but not always), the cost-cutting measures compromise the survivability of the hardware. Economic lifetimes of equipment are rapidly falling, both due to pressures to "keep up to date" and because of lower equipment costs.

There is a false sense of security provided by the elaborate OMCs that can lead operators to think that the whole network is being continuously and adequately monitored. There are many faults the OMC will not see! I recently inspected a system on which I found more than 2000 faults including about 25 percent of channel resources not available and the OMC saw nothing.

What Goes Wrong

The base station comprises a number of modules, each of which is connected to the power grid through the power supply, and some of which are connected to the outside world via the antenna structure.

The connection to the outside world makes the equipment vulnerable to power surges and lightning strikes. Of course, there is nearly always some protection built in to the hardware, but even the original designers did not anticipate that the protection they put in would withstand a direct lightning strike or a major power surge. To make matters worse very few installers understand the need for proper grounding; and without this, even the protectors that are installed will not function properly.

The base-station modules are made of components that themselves have a finite life, and will deteriorate over time. Some fail gradually, some catastrophically. As a rule, gradual failures are harder to detect. Frequency references drift, connector contacts wear, improperly mating connectors gradually fail, receiver sensitivities drop, and transmitter outputs fall.

External Plant

The external plant, which includes antennas, cables, towers, and power plant, survives in a difficult environment and has accelerated failure rates. Antennas in particular are not only subject to lightning but can be moved by strong winds and damaged by flying debris.

External Problems

Base-station problems may not necessarily be internal. Other sites that are malfunctioning can cause excessive interference into a base station. Installations by competitors (particularly where they use different systems with adjacent frequency assignments) can be a problem.

Signal blockages can occur because trees have grown in the line of sight, new buildings have sprung up, or other obstacles have appeared. This can be particularly threatening in the case of PCS/PCN or microwave systems for which propagation is largely line of sight.

Base-Station Testing

Despite the cost cutting, base stations of today do require less maintenance than those of the past, but this does not mean that they are maintenance free. The monitoring of a base station is fourfold.

- Monitor the network diagnostics and in particular watch for trends
- Do routine drive tests
- Have a routine external plant inspection (approximately yearly)
- Check the overall performance of the base station with a good service monitor (at least every 18 months).

Service Monitor Testing

Whatever the system there are some basic tests that can be done effectively with a service monitor:

- Check TX power
- Check receiver (RX) sensitivity
- Check system clocks
- Check VSWR
- Check cable losses

MAINTENANCE WORK LOAD

In the United States in 1999, the average technician was responsible for 15 cell sites, and this is expected to increase to 25 by the year 2002, due largely to increased automation and improved hardware and software.

Maintaining Quality of Coverage

A VSWR alarm and low TX power alarm will generally suffice to detect most RF failures or partial failures, but will not find all problems.

Without detection, the following problems can cause reduced radiated power:

- Water in the antenna
- Partial lightning damage to antenna/cable
- New (since the cellular installation) buildings, foliage growth, tower extensions, and other obstructions
- Damaged feeder
- Damaged, faulty, or waterlogged connectors

The frequency of these problems increases as the number of channels increases. If each cell has two RX and two TX antennas, then there are four antennas per cell. The number of other passive components is proportional to the number of channels.

A modern city will have at least 50 cell sites (at least 200 antennas and probably many more); PCS systems will have hundreds of sites. If the mean time between failures (MTBF) of the antennas is 20 years, then in an average year, 10 (200/20) antenna failures can be expected. Catastrophic and severe failures will, of course, be detected by the base-station controller; less severe failures may not be detected. Most of these failures will result in reduced coverage, but they may escape detection by the alarm system. This is particularly true in the case of installations with high feeder losses.

VSWR Testing

VSWR specifications for cable and antenna systems generally specify a maximum VSWR of 1.5. At higher levels the reflected signal can become a network problem, as the reflected signal reflects again from the antenna isolator and reappears at the antenna as a delayed version of the original signal, causing an interference similar to multipath fading.

High VSWR can be due to many things, including cable discontinuities, damaged or cut grounding shields, connector problems, poor-quality connectors, connector pin offset, antenna problems, and lightning damage.

Many of these things can be hard to locate from a visual inspection, and a new range of VSWR meters that cannot only measure the VSWR value, but by using fast Fourier transform (FFT) techniques and sweeping the cable with a wide-band signal, can locate the discontinuity, often to a degree of accuracy of 1 meter or better.

Because many of the causes of bad VSWR develop slowly over time, it is good practice to measure and record the VSWR of all cables at the installation. Future checks will show a degradation in this reading if a faulty condition exists.

The VSWR is measured by comparing the level of the forward RF signal to the reflected signal, and if measured from the TX output, only the forward component is accurately measured. If the fault is at, or near, the antenna, the reflected component has effectively traveled the distance to the antenna and back, and hence has suffered attenuation $= 2 \times$ feeder loss. This is how serious antenna problems can escape detection.

The VSWR is most easily measured by measuring the actual reflected power through a directional coupler. Once this value has been measured, the VSWR can be calculated as follows:

Equation 14.1

$$\text{VSWR} = \frac{1 + \sqrt{\dfrac{\text{Reflected Power}}{\text{Forward Power}}}}{1 - \sqrt{\dfrac{\text{Reflected Power}}{\text{Forward Power}}}}$$

An inline RF power meter can measure both of these

components. A base-station antenna should have a VSWR of no greater than 1 to 1.7.

If the reflected power is measured at the transmitter terminals, it is necessary to increase the value of the measured reflected power by the cable loss $\times 2$ to convert the reading to the value at the antenna input. Figure 14.1 shows this VSWR measurement at the transmitter terminals.

The effect of cable loss on the measured VSWR at the bottom of the cable compared to the antenna VSWR increases dramatically with increased values of measured VSWR, as shown in Figure 14.2. As most cables together with the connectors will contribute a loss of at least 2 dB, it can be seen that to ensure that the antenna is below a VSWR of 1.7 the measurement at the cable termination always should be below 1.5.

The accepted nominal value of VSWR at 1.7 maximum is based mainly on the capabilities of the antenna manufacturers rather than the need to keep the VSWR in that range for purposes of efficiency. Broad-band antennas that are used in cellular will have a VSWR range that may vary from 1.1 to 1.5 over the frequencies used. An antenna with a VSWR of 1.5 will have a loss of about 0.4 dB (compare this to its gain, which may be 13 dB). An out-of-spec VSWR tells you that the antenna is not performing as it did when it left the manufacturer and that most probably something is amiss.

Be aware, however, that simply confirming that the antenna VSWR is in-spec does not *absolutely* confirm that it is radiating correctly. Twice I have encountered antennas with VSWRs very close to 1.0 that barely radiated at all (that is, the gains were worse than -20 dB). One was full of water and the other had a design fault from the factory.

Continuous VSWR Monitoring

Most base stations have a VSWR monitor on each TX port. These ports continuously monitor VSWR and activate remote alarms when a preset threshold is exceeded; some in fact shut down the channel at a sufficiently high VSWR. Rarely, however, are the receiver antennas so monitored. After-market monitors are available that connect to every antenna (including receive ports). Devices like Loral's communications antenna systems (CATS), can monitor power, VSWR, and do trend analysis and diagnostics, holding up to six months of data.

Visual Inspections

About 10 percent of base stations can be seen by visual inspection to have at least one antenna mounted incorrectly with respect to the design downtilt. Omnis are

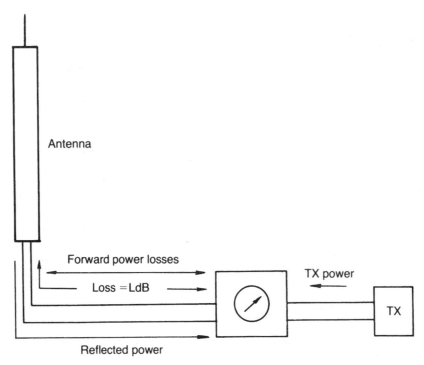

Figure 14.1   VSWR measurement at the transmitter output terminals.

often mounted so that with time they will slip to off-vertical. This will seriously reduce the range of the system. A simple visual inspection from the ground will often reveal the most serious cases like the one shown in Figure 14.3, where the two top-mounted omnis provide an easy visual reference that at least one of them is leaning. The visual inspection can be made more effective by using a pair of binoculars (about 8×20 will do). Such inspection should be a routine part of site maintenance. The alignment of the antennas should be checked by a rigger at least once a year, and as a matter of course during routine tower inspections.

Sector antennas are often incorrectly aligned both in the plane of the sector and vertically (downtilt). It is not unusual to see antennas serving the same sector but pointing as much as 20 degrees the wrong way (and sometimes even with respect to each other). Sectors are sometimes connected to the wrong set of antennas.

LOCATING A CABLE FAULT

A VSWR problem may signify a cable fault, but it cannot locate it. A time domain reflectometer (TDR) will give the fault location as a readout in meters from the injection point. It works by sending a pulse down the cable and then, like radar, measuring any reflected pulses and the time for those reflections to return. Any discon-

tinuity in the cable, which may be caused by physical damage, water, or faulty connectors, will cause the forward signal to have a discontinuity that will result in some of the energy being propagated forward while some is reflected, as shown in Equation 14.1. Because of the way it works, the TDR will locate only the nearest fault if there are more than one in the same cable, and so the test should be repeated after each fault has been located.

Recently TDRs have become affordable as maintenance tools for cellular applications. A TDR will cost around $1500 for a digital device, which may have an outlet to provide a time domain waveform of the reflection (a screen image of the reflection from the cable as a function of time) using an oscilloscope. A more upscale version with a built-in printer and screen will cost about $6000.

Measurements are best done with the cable disconnected at both ends from the termination equipment, as this connection will tend to absorb the reflections. In the case of antenna feeders, it can be difficult to disconnect the antenna, and fortunately this may not be necessary. The antenna will be designed to present a 50-ohm termination at the center frequency to which it is tuned. At other frequencies it will have a much higher, complex impedance. So unless you are unfortunate and have a TDR that operates at or near the antenna frequency, the antenna can be left in place.

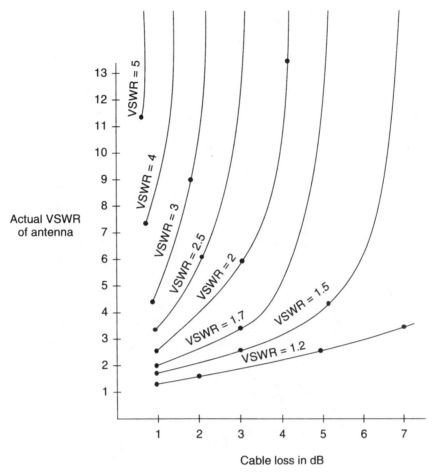

Figure 14.2 The actual antenna VSWR from the value measured at the end of the feeder cable.

AIR-CONDITIONING

The integrity and proper functioning of the air-conditioning is most important to ensure reliability of the base-station equipment. It is well-established that temperature cycles contribute significantly to the fault rates of electronic equipment. It is a good idea to provide dual air conditioners so one unit is available if one should fail.

The trade-off of having the air conditioners out of service during power failures and the cost of providing emergency power should be considered. Equipment that has a low thermal operating range may not be able to function for any significant period without air-conditioning, but some more robust equipment can. The fact that some equipment can operate at high temperatures, however, does not mean that it is immune to the deleterious effects of temperature cycling.

Routine maintenance consists essentially of changing the filters at regular intervals. These intervals are largely determined by the airborne particle concentrations in the area and can vary from weeks to months.

Maintenance and operations staff in warm climates should also watch for the formation of condensation, which can be indicative of excessive ingress of warm moisture-laden air. This can be readily rectified by blocking the air leaks.

However, the most likely source of moisture remains insufficient waterproofing.

Condensation on the outside of the building indicates excessive heat loss through the shelter and can be rectified by improved insulation.

Failure to rectify these problems can lead to base-station failure due to moisture on complementary metal oxide semiconductor (CMOS) boards or corrosion of boards and connectors. Also, inefficiently operating air conditioners will consume excessive power.

A humidity meter should be placed in each base station and the levels recorded periodically. In general the humidity should be kept in the 35–65 percent range.

Figure 14.3   Off-vertical omnis, like those at the top of this tower, often account for poor coverage.

MEAN TIME BETWEEN FAILURES

Mean time between failures (MTBF) is defined as the mean time to the *first* failure. For this reason it cannot be used to reliably say anything about a unit once it has been repaired, because the failure can occur as a consequence of a fault in any of a large number of components. Apart from those parts replaced by the repair, all other components are now farther down the failure path than they would be on a new item.

MTBF is also strongly influenced by a high "infant mortality rate" of new equipment, which occurs because a number of manufacturing or burn-in failures are likely to occur in the first few months. After the settling-in period, the failure rate settles down usually for some years, until old age begins to take a toll and the failure rate once again climbs.

Figure 14.4 shows the way in which a power amp (PA) can fail as a function of temperature. Notice that at high ambient temperatures the failure rate increases dramatically. The high temperatures that cause that failure

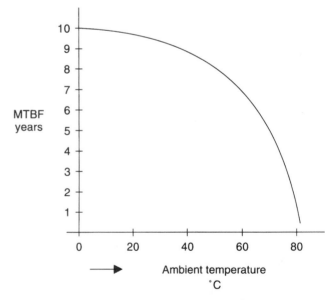

Figure 14.4   The MTBF of a PA versus temperature.

rate can be induced by inadequate ventilation. Even more damaging to reliability is temperature cycling, and in any environment except a fully temperature controlled one, this may be the biggest cause of failure.

The most important conclusion to draw from all of this is that MTBFs mean very little because the conditions under which they are measured largely determines the result.

TRANSCEIVERS

Without a doubt, transceivers require the most consistent maintenance attention. About 10 years MTBF can be expected. Published figures are invariably based on different measurement criteria of MTBF, which reflects sadly on efforts to standardize MTBF measurement.

The least reliable part of the transceiver is the RF power stage. Because the loss of the control channel puts the base off the air, it is almost always recommended that redundancy be provided for this channel. For the same reason, when a control channel has been taken out of service, it should be attended to quickly.

In many cases the RF receive stage will give a lot of problems. If this is happening, it is likely to be due to improper surge protection on the RF cables and/or grounding inadequacy. RX problems, and in particular low sensitivity, can go undetected for years unless a rigorous maintenance program is in place that includes scheduled full base-station testing.

BASE-STATION CONTROLLER USING STATISTICS

The switch statistics (and sometimes the billing system) produce information that can be used to diagnose base-station faults. The following sections discuss two examples.

Blocking

Blocking (also known as sealing) is the term used for a channel that is temporarily taken out of service by the base-station controller due to interference. This can occur while the channel is in use or when it is on standby.

The most probable cause of blocking is interference from another mobile; provided that blocking is kept relatively low (less than 2 percent), it can be considered normal. Blocking above 5 percent, however, generally indicates that there are problems, including system design errors or intermodulation problems or RF interference from other equipment. Check that new cell sites

have not been added or that an existing cell on the same or adjacent channel has not been modified.

By looking at the individual distribution of channel blocking, it may be possible to identify the frequency and nature of the interferer.

Cell Occupancy

Cell occupancy statistics tell a good deal about how the system is performing. A sudden drop in cell occupancy can indicate problems:

- Antenna/feeder problems may cause reduced coverage (in particular, antenna downtilt or obstructions caused by new work on the tower).
- The signal-strength receiver may be out of calibration.
- The control channel may be low in RF level or off frequency.
- The receiver multicoupler and/or connecting cables may have introduced losses.
- Antennas may be on the wrong sectors.
- RX sensitivity may be down.
- TX could be low.

Overloaded cells can in some cases manifest themselves as increased blocking or cross talk, as the mobile is handed off to a more distant cell. Consider the situation of a cell in a local valley, as seen in Figure 14.5. If a call is attempted and the local cell is overloaded, the call will be sent on retry to a neighboring cell. Because of the valley, the neighboring cells, even though they are close, will not provide good coverage. As the mobile moves in the valley it will soon seek a handoff. The local cell, which is congested, will be the "best choice," but the next choice could well be a distant high cell that was never intended to serve this area.

Because of the distance, the mobile will be operating at a high power level. If it happens that the distant cell is on the same frequency as the valley cell, the mobile will cause strong interference to it.

To make matters worse, the valley cell base station will in turn interfere with the mobile. The net result is that many of the calls in the valley will suffer interference, and others will have both poor signal-to-noise (S/N) performance *and* interference.

It is also worth noting that each mobile that is forced onto the distant co-channel station by the congestion is likely both to be interfered with and to cause interference to a valley station (because, being congested, it is reasonable to expect that the same channel will be in use locally), as seen in Figure 14.6. In turn, the local station will attempt to handoff because of the interference and

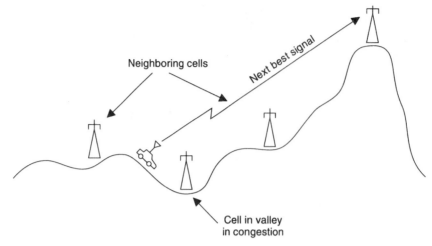

Figure 14.5   A call in a valley may be sent to a distant cell, which will cause local interference.

may itself end up on the same distant base. *Overloaded systems deteriorate rapidly for more reasons than just a lack of traffic capacity.*

CUSTOMER COMPLAINTS

Some operators rely on customer complaints to inform them of Effective Radiated Power (ERP) problems of reduced coverage. This detection system, apart from causing poor public relations, will not work if there is significant base-station overlap and other bases take over from the faulty one, or if the fault is gradual (an example, water accumulation) and subscribers come to accept the "new" coverage as normal.

To make matters worse, many customer complaints are ill-informed. "The service doesn't work at my house" may mean "It doesn't work in my underground garage," but of course the service is fine outside the garage.

There can be a number of reasons why a customer reports less-than-satisfactory service.

- The mobile is faulty.
- The customer has incorrectly operated the mobile equipment.
- The customer is outside the normal coverage area.
- The system is congested.
- A base-station fault is evident.
- The customer attempted calls during the down time of the system/base.
- The customer likes to complain.

At this point there is a need for clear communication between the person recording the fault, the mobile maintenance staff, and the base-station and switching staff, to avoid duplication.

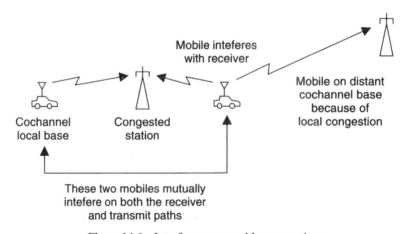

Figure 14.6   Interference caused by congestion.

This is an easy way to proceed:

1. The fault recorder has a coverage map, with good street-level accuracy, and can ascertain whether the complaint originates from within the normal service area. The fault recorder should ask sufficient questions to ensure that the customer is using appropriate operating procedures.

2. If the complaint is outside the service area, the customer should be advised accordingly.

3. Checks should be made to see if there are other recent reports from this area

4. If the complaint is inside the service area, then an appointment to check the mobile should be arranged with the mobile maintenance staff. At the service center it may be possible to determine absolutely whether the fault is mobile or not, but, due to lack of adequate test equipment, it may also be that trying a replacement mobile is the only solution.

5. Unless the mobile or its associated hardware has been clearly implicated, the fault report should now be passed on as a possible base-station fault. The fault could also have been a temporary switch or base-station problem (for example, outage due to maintenance). Because of this, these reports should be handled centrally and forwarded to the switching or base-station staff only when a clear pattern has emerged.

6. Analysis of consistent fault reports in areas thought to be well-covered may reveal local "dead spots." This information is most useful to system planners and designers, so it is important that reports are carefully processed and the results forwarded to the relevant personnel.

There is a tendency to "do maintenance at the lowest possible level of competence." This philosophy comes from the public switched telephone network (PSTN) operators; most readers who have had a faulty telephone will likely have experienced the frustration that this policy causes.

In my own case, I have had to resort to threats of media complaints on two occasions to get relatively simple switching faults on my line repaired. The PSTN operator insisted on sending "the lowest-level staff," who understood only telephone apparatus faults and who therefore wanted to replace or repair the telephone. On both occasions, I clearly defined the problem (and its precise origin, which was in the switch), but it took about six months of insistence to have the fault cleared.

On one occasion, the fault occurred in a rented apartment; after the problem was finally rectified, it was confirmed that the fault dated back six years and three tenants. The second problem also involved a switch-related fault resulting in crossed signaling wires. It was accurately identified, but the repair personnel spent months trying to tie it down to a phone or cable fault. This problem was fixed by a visit to the switch to point out the offending equipment.

To work effectively, maintenance staff must have a reasonable overview of the system. It is false economy to adopt the PSTN operators' approach. The maintenance staff must understand the network beyond simply replacing specific pieces of hardware.

LINE-UP LEVELS

Most PSTNs have a series of test numbers that will return certain test tones. These numbers can be used to evaluate the quality of the PSTN interface and line-up levels.

The objective of the cellular operator is to make the handoff almost imperceptible. Modern cellular systems have handoff times that meet this standard, but handoffs can become conspicuous if a level change occurs during handoff. For this reason, it is necessary to accurately set all system levels. When the cellular operator uses leased lines (which may be rerouted), checks on levels should be made regularly. This is particularly true if copper cables are used for links.

After replacing or repairing any multiplexer (MUX) equipment or other line equipment, it is essential to check all affected line-up levels. Line-up levels and deviation should be routinely confirmed every six to nine months.

TEST MOBILE

Nearly all system manufacturers offer a test mobile. This unit is virtually identical to a normal mobile but has incorporated an automatic answer and the ability to loop back the line to the switch.

Under the control of the switch, this mobile can automatically confirm the proper operation of each channel sequentially. It can also "check" RF propagation, because the test receiver measures the base-station RF off-air. If the test mobile is deliberately given a poor antenna (for example, a dummy load) it may be possible to also use S/N/bit error rate (BER) measurements to detect degraded propagation. The test mobile can loop back a test tone for S/N (or BER) measurements.

Because the mobile is controlled by the switch, it can be programmed for extensive after-hours testing of the channels.

Note that a single, successfully answered call to a test mobile only confirms that the base station is at least partially on air. This does not confirm satisfactory operation of the base.

To progressively test each channel, a call to a test mobile, once established, can be forced to handoff to other channels at the base station. To be effective, this should be done late at night when most channels are likely to be free.

The test mobile can be connected individually at each site so that it is local to a base.

Often, automatic test routines are available that can run either periodically (once every five minutes or so) or at certain times late at night.

Test mobiles and other "in-house" engineering services should have easily identifiable numbers so that they can be distinguished for billing purposes. Some methods reserve all numbers where the last three digits are consecutive (for example, ABC789), or all numbers where the second-to-last last digit is 9 (for example, ABCD9X), or a specially defined group (such as ABC1XX or ABC2XX). Similarly, "in-house" non-engineering services could also be identified like this to ultimately save the accountants or fraud controllers many headaches.

SITE AUDIO TEST LOOPS

Some suppliers include a site audio test loop to distinguish remotely between noisy channels that are caused by the switch to base link from those that are caused by the base RF equipment. This test tells staff located at the switch if the fault is a link or a base-station RF problem. Because each of these problems is likely to be serviced by different personnel, a good deal of diagnostic time can be saved by early identification.

INTERACTION WITH THE SWITCH/BSC

Base-station maintenance requires close cooperation with the technical people in charge of the switch/BSC, because it is at the switch that most of the alarm information is available. It is important that the lines of communication between the switch technician and the RF technician are good.

A great deal of time can be saved by having an accurate diagnosis of the problem before staff are dispatched, particularly to a remote base. There are many diagnostics available at the switch that can be used to determine the nature of a fault.

Since any service affecting alarms will prompt a response from the switch/BSC staff, they should be notified before any channels are taken out of service. In some cases it will be necessary for the switch to perform a station reload if significant parts of it have been put out of service.

In particular, it is most important to diagnose whether the fault is from the switch, the transmission link, the site controller, or the RF. The test mobile is an important tool to differentiate between these categories.

Some manufacturers have remote-alarm monitoring equipment that can be located in the RF repair center. This equipment can reduce the dependence on switching staff to diagnose base-station faults.

System designers can assist by co-locating a switch/BSC with a base station so that the switch operators become more familiar with RF hardware. There is a tendency in cellular (as in all other enterprises) to form the "us and them syndrome" (that is, the switch operator always tends to assume that the fault is in the domain of the RF staff and vice versa). (Of course, co-siting should be done only if it does not compromise the system design, but it is generally possible to achieve such a design.)

The switch can provide details on service affecting conditions such as channels out due to interference, low power, data errors on the link, timing and synchronization problems, as well as faults detected while routinely polling the test mobile.

The switch can also readily provide call tracing, which can be most useful in identifying handoff and other problems.

SITE LOG BOOKS

All sites should have locally maintained log books. In these log books are recorded all visits (entries) to the site and the purpose of each visit. Each entry should cause an alarm at the switch that must be "canceled" by the person entering the base station. The entry is recorded by both the switch attendant and the person at the base station. Table 14.1 shows a typical log book.

CALL-OUT PROCEDURES

After hours, it is usual to have only a skeleton staff available. The staff may amount to only one person directly on call at the switch or even one person on call-out via alarm rerouting. In either case, it is unlikely that the person on call-out will be an overall system expert, so procedures that determine the priority of call-out must be established.

In any large city there will be considerable overlap in base-station coverage, so that the outage of any one station, particularly after hours, would not be service

TABLE 14.1   Typical Entries in a Log Book

Date	Arrival Time	Name	Designation	Department	Purpose of Visit	Departure Time	SIG
09.10.89	08.30	S. Davey	Technician	Maint.	Replace chn. 215	10.30	...
09.10.89	13.00	N. Kelly	Cleaner	Cleaning	Polish floors	14.00	...

affecting. This means it is not noticed by customers. In even larger sites, two, three, or even more non-adjacent bases may be out of service without affecting service.

Rules must be drawn up to enable the call-out staff to decide on a course of action. For example, consider a medium-sized city with 100 base stations. After discussions between switching RF and transmission staff, the following rules may be decided upon:

1. In normal working hours up to 4:30 P.M., any base-station outage is treated as urgent and must be attended to immediately.
2. Between 4:30 P.M. and 7:00 A.M., and from 4:30 P.M. Friday to 7:00 A.M. Monday, any two base stations can be out of service without call-out.
3. All urgent alarms are to be attended to by the switch operator (either an on-site operator or one with a remote terminal if provided). This operator will decide on any subsequent call-out procedures.
4. In both the switching and radio areas, at least three personnel, with a predetermined call-out priority, will be available. If called, the first party to answer will attend to the fault and determine what other call-outs are needed.
5. Each of the call-out personnel will be supplied with a pager and mobile telephone.
6. Special call-out procedures will be defined from time to time for special holiday periods.

EQUIPMENT

To service the base stations, it is necessary to have at least one dedicated vehicle and two technical staff. A station wagon or a small van is a suitable vehicle. The vehicle should be permanently outfitted with the necessary test equipment and tools.

For most installations, base-station servicing will be ongoing, on a daily basis. Spare cards and parts will normally be stored at a central area. This should probably be the same storeroom as the one in which switch spare parts are kept (because some of the parts are the same). Maintenance staff need frequent access to this storeroom, so a central location is most important.

The following is the minimum equipment necessary:

1. Two VSWR meters to 50 watts.
2. One general-purpose mobile test set incorporating at least the following to 2 GHz:
   a. RF signal generator
   b. RF level measurement
   c. RF deviation measurement (analog only)
   d. RF frequency measurement (accurate to ± 1 kHz)
   e. Audio modulation frequency modulated (FM) and amplitude modulated (AM) (analog only)
   f. Preferably a spectrum analyzer
   g. SINAD/BER measurement
3. Three digital voltmeters.
4. One spectrum analyzer to 1-GHz RF and 10-Hz intermediate frequency (IF) resolution.
5. Specific equipment for servicing the particular manufacturers' bases [for example, printed circuit boards (PCBs), PCs, special cables and plugs]. Note that most special cables and plugs should be stored at each base station.

Because the time to travel to and from a base station is usually significant, the maintenance van should carry a good range of tools and miscellaneous parts (such as connectors, transition connectors, cables, and lugs). It is not usual to leave test equipment permanently on site at a base station.

The maintenance staff should also have some independent means of communicating with the switch, such as via the PSTN or a land mobile. This is important in the event of link loss or complete system failure.

QUALITY AND CALIBRATION OF TEST EQUIPMENT

Cellular radio equipment represents state-of-the-art hardware that is beginning to approach the theoretical limits of performance at room temperature. For this reason, it is essential that only high-quality and well-calibrated equipment be used when servicing or adjusting the hardware. The old service monitor of the early 1980s is probably only accurate to ±5 or 6 dB, and even most modern ones are struggling to achieve ±1 dB.

The equipment used to service base-station equipment should be accurate to at least ±2 dB. This rules out most old equipment and even new equipment that has not been calibrated in the past 12 months. Errors of frequency, signal level, modulation level, or VSWR can lead to poor system performance. Additional sources of error include poor-quality connecting cables, connectors, and the use of multiple adapters.

TEST SETS

A cellular test set has many features specific to cellular radio that can contribute significantly to cellular maintenance. A good test set can completely simulate either a base station or a mobile for mobile testing. An operator with anything but the smallest of systems needs at least two test sets.

A good test set should include most of the following:

- SINAD or BER measurement with the appropriate weighting
- SINAD measurement of out-of-band test tones [for example, supervisory audio tones (SAT) in Total Access Communications System (TACS) and Advanced Mobile Phone Service (AMPS), or phitone in Nordic Mobile Telephone (NMT)]
- Simulation of signaling tones
- A simple spectrum analyzer
- A frequency counter (to 2 GHz)
- A signal generator (to 2 GHz)
- Deviation measurement (analog only)
- RF power measurement
- RF millivoltmeter

In general, it is a good policy to have at least two of each test equipment item so that the calibration of one can be checked against the other. Down time with test equipment can be excessive (repair times of 3–6 months are not unusual), and the second piece of equipment can be extremely valuable at this time.

A good-quality spectrum analyzer is necessary to adequately assess the transmitter performance. The analyzers normally found on service monitors do not have adequate resolution for this purpose and cannot be used. The spectrum analyzer should have a tracking capability so that it can be used to sweep filter and cavity assemblies. The analyzer needs an IF filter resolution of 10 Hz and should be able to display spurious and harmonic generation up to 4 GHz (when accurate measurement above 1 GHz is not necessary).

A spectrum analyser that can simulate a mobile, and so "talk" to the base station is essential for digital site maintenance and is highly recommended for analog systems. Alternatively a purpose-built test set like the Rhode and Schwarz Global System for Mobile Communications (GSM) test set CMD 57 might be used.

The CMD 57 can do many useful test functions, the first of which is a basic check on the transmitter. Figure 14.7 shows the test set looking at the control channel and reporting the frequency error, phase error, and base transceiver station (BTS) power (the power for this test, was picked up "off-air" and about 80 meters from the BTS antennas).

When the CMD 57 test set records errors greater than a preset level, it highlights the values in order to make them easy to spot. Ordinarily a phase-and-frequency error test would be done at the BTS test point, and the BTS measurement as shown in Figure 14.8 passed when the test was done that way. However, when the same parameter was looked at a few meters after the antenna, it was well out of specification. This indicates that additional phase errors are being induced in the RF transmission path.

In digital systems it is important to ramp the power levels up in a controlled manner, to limit RF interference. Even if it were possible to have an infinitely short ramp time (a perfect ramp), such a ramp would develop so many unintended RF by-products, that it would not be desirable. What is done instead is to ramp the level up, so that the ramp remains within certain levels defined by a mask, as seen in Figure 14.9. The signal in this example stays within the mask and so satisfies the requirement. Notice that the mask permits a rather large overshoot on the initial ramp-up; this is mainly an artifact of the early days of GSM, when the overshoot was inevitable.

Figure 14.10 shows a BTS not meeting the power ramp requirements. The BTS appeared to have some clocking instability that caused this.

On all BTS tests, a critical performance figure is the RX sensitivity, not only because it is generally the uplink that is the weakest but also because the RX front end is a delicate thing, and through the antenna and cables it is highly exposed to the elements.

The CDM 57 does a sensitivity measurement by first establishing a connection (basically it sets up a call) and then by gradually reducing the RF power level and monitoring the BER as it goes, it can find the power level corresponding to a pre-defined BER. In the example in Figure 14.11, it was seeking a 2 percent BER on the class II bits, and it has found the corresponding level to be −110.2 dBm.

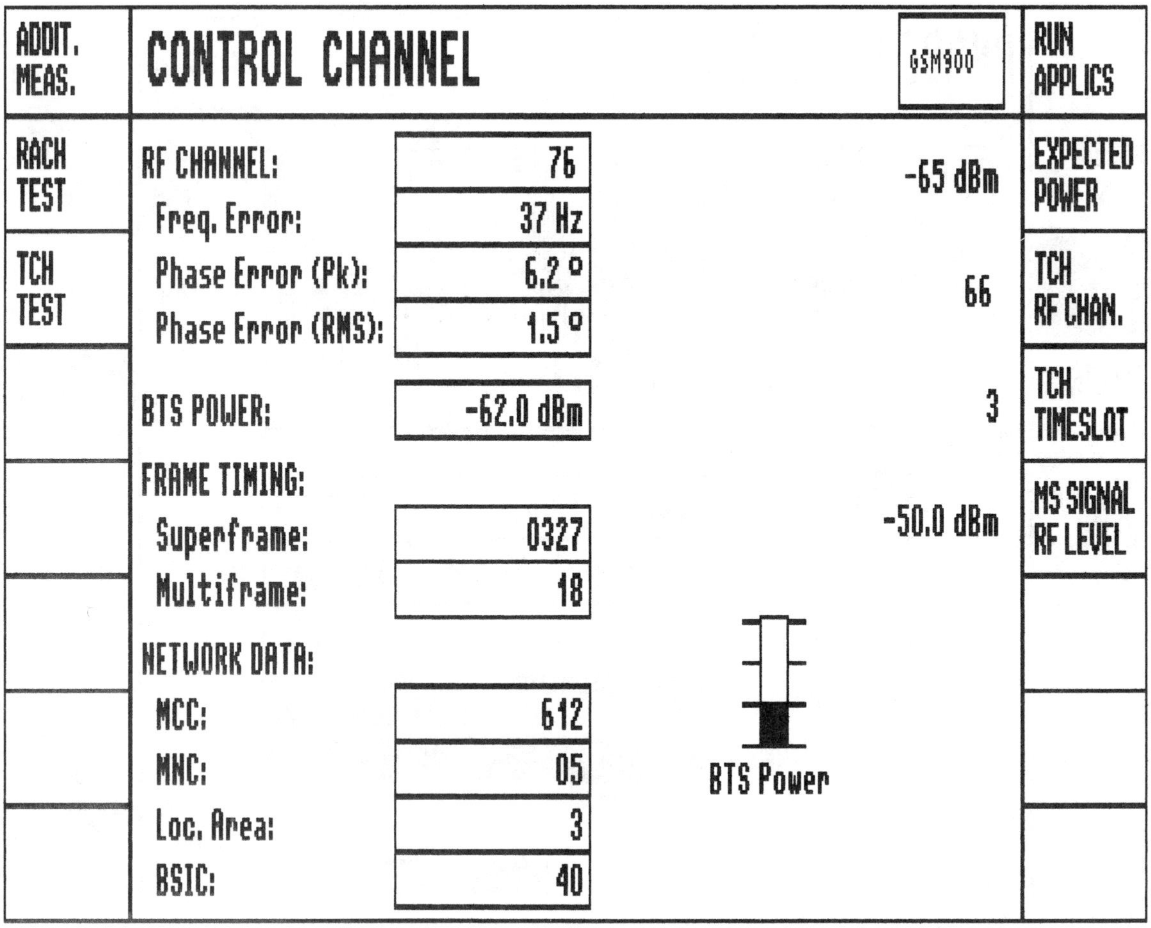

ADDIT. MEAS.	**CONTROL CHANNEL**		GSM900	RUN APPLICS
RACH TEST	RF CHANNEL:	76	-65 dBm	EXPECTED POWER
TCH TEST	Freq. Error:	37 Hz		
	Phase Error (Pk):	6.2 °	66	TCH RF CHAN.
	Phase Error (RMS):	1.5 °		
	BTS POWER:	-62.0 dBm	3	TCH TIMESLOT
	FRAME TIMING:			
	Superframe:	0327	-50.0 dBm	MS SIGNAL RF LEVEL
	Multiframe:	18		
	NETWORK DATA:			
	MCC:	612		
	MNC:	05	**BTS Power**	
	Loc. Area:	3		
	BSIC:	40		

Figure 14.7 An off-air look at a GSM BTS control channel.

The ability to simulate a base station is usually an add-on feature and is necessary only if the unit is to be used to test mobiles.

Notice that mobile-specific wiring harnesses are necessary to connect the test set with each model of mobile to be tested. A typical analog service monitor is shown in Figure 14.12.

SINAD/BER measurements are some of the most fundamental measurements; the method is best illustrated using a separate RF generator and SINAD meter. Figure 14.13 shows a SINAD/BER meter.

ANALOG

The RF generator sends a signal at a specified test frequency and deviation (usually about 1-kHz frequency an 1-kHz deviation) to the mobile. The SINAD/BER meter has internal filters that separate the test tone (wanted signal) from the other components (assumed to be noise and distortion from the receiver).

A meter reads the SINAD directly and the RF level is adjusted until the desired SINAD is obtained. Various weighting filters are used to limit the bandwidth and pass characteristics of the noise and distortion components. This is because the mobile channel itself is filtered and the transducer (earpiece) and the ear combined do not respond to all frequencies equally. Thus, the "weighting" networks attempt to account for the subjective noise-level rather than the actual S/N levels.

The distortion function of the SINAD meter can be used in this same test to measure the receiver distortion. If the RF signal level is high enough, the processing gain of the receiver will reduce the noise to around 70 dB below the signal, and at this level it is negligible. If the 1-kHz tone level is now raised so that the signal generator is at 25 percent of the maximum system deviation (for AMPS that means 25 percent of 12 kHz or 3 kHz), it will be at an energy level that corresponds to the average of a voice signal. The SINAD meter will now be reading the level of distortion in dB relative to the signal. This

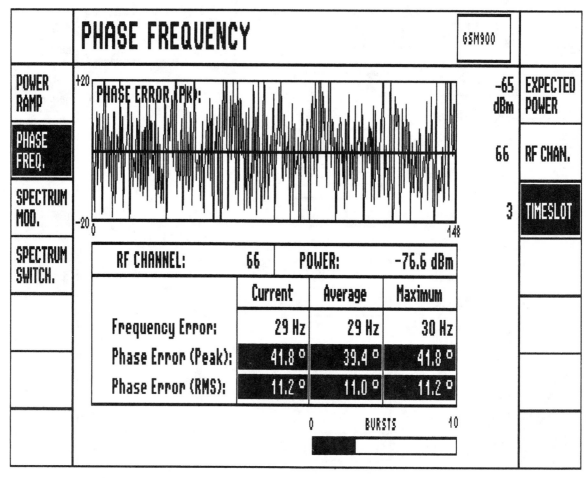

Figure 14.8 When the CMD 57 detects errors that are out of specification, it highlights them as seen here.

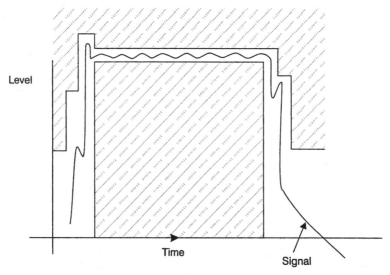

Figure 14.9 The power amp of a GSM signal must remain inside the mask.

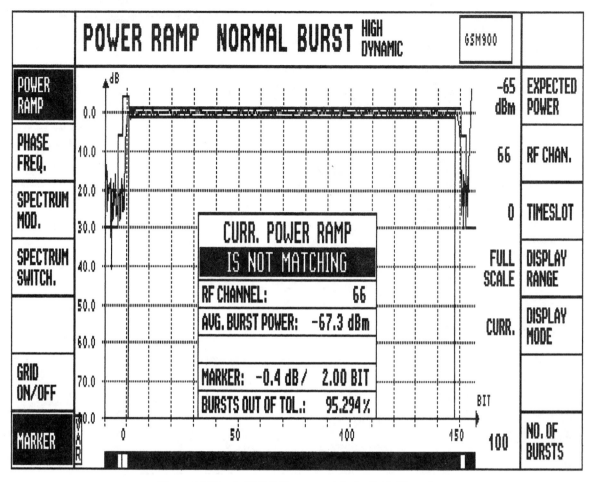

Figure 14.10 A BTS failing to match the RAMP mask.

should be below 5 percent (although this specification may vary a little from system to system) or $10 \log 0.05 = -13$ dB relative to the tone

Adjacent Channel Rejection

Although only one service monitor is needed for in-band tests, at least two are needed for out-of-band testing; the second monitor can also be used as a reference (that is, discrepancies between them will alert the user to a calibration error). As frequency reuse increases, adjacent-channel interference increases and the need for out-of-band measurements increases.

Measuring adjacent-channel rejection requires that one signal generator be set to measure 12-dB SINAD in-band, and the second signal generated be coupled to the receiver input and set to the adjacent channel. The second generator is adjusted in level until the SINAD falls by 6 dB, as shown in Figure 14.14.

Alternate Channel Rejection

Alternate channel rejection is essentially the same measurement as the adjacent-channel rejection, except that the second signal generator is tuned two channels away from the wanted frequency.

Co-Channel Rejection

Co-channel rejection is a measure of the receiver's ability to receive a wanted signal in the presence of an in-band interferer. The equipment is set up as shown in Figure 14.14, except that both signal generators are set at the same frequency. With only signal generator A turned on, set up the conditions as described for the SINAD measurement to give 12-dB SINAD, using a 1-kHz modulation. Now connect the second generator with its output set to minimum and with a 400-kHz tone modulating to 25 percent of the maximum deviation. Now increase the

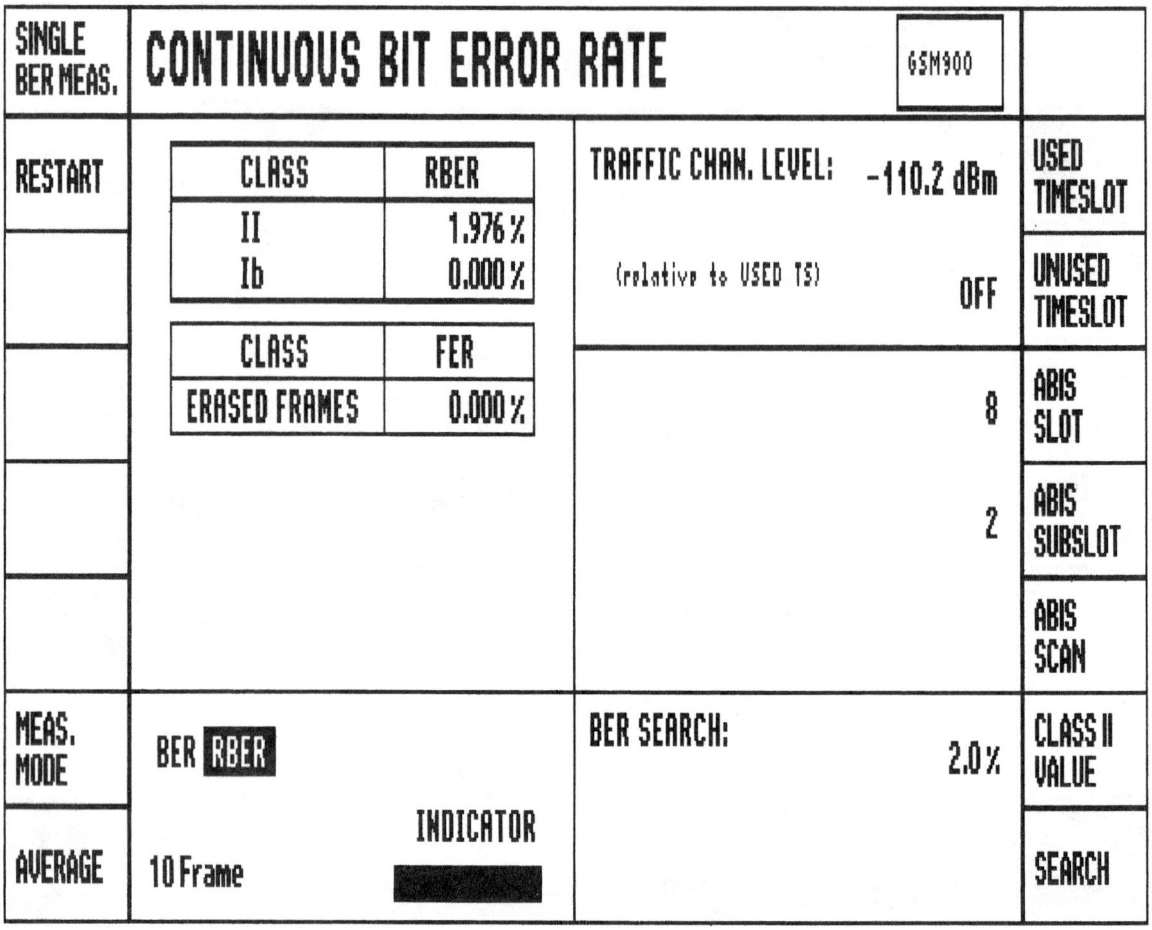
SINGLE BER MEAS.
CONTINUOUS BIT ERROR RATE
GSM900
RESTART
CLASS
RBER
II
1.976 %
Ib
0.000 %
CLASS
FER
ERASED FRAMES
0.000 %
TRAFFIC CHAN. LEVEL:
-110.2 dBm
(relative to USED TS)
OFF
USED TIMESLOT
UNUSED TIMESLOT
8
ABIS SLOT
2
ABIS SUBSLOT
ABIS SCAN
MEAS. MODE
BER RBER
BER SEARCH:
2.0 %
CLASS II VALUE
INDICATOR
AVERAGE
10 Frame
SEARCH

Figure 14.11 A BER test on the CMD 57.

level of the second generator until the SINAD is reduced to 6 dB. Because the receivers may not be exactly on frequency, it will then be necessary to sweep one of the generators over a bandwidth equal to 30 percent of the maximum deviation while watching the output to ensure that the "on frequency" reading produces the lowest SINAD. If another minimum is found, then the test should be repeated at those frequencies and this result used. The difference in levels of the two generators is the co-channel rejection.

Unwanted Spurious Response

In the case of an unwanted spurious response, proceed exactly as for the co-channel rejection, setting signal generator A to read 12-dB SINAD at 1 kHz while setting signal generator B for 25 percent deviation at 400 Hz.

Now set signal generator B to 85 dB *above* the level of generator A and scan manually from 100 kHz to 2 GHz, watching for a dip in the SINAD measurement. When one is noted determine the level of generator B that reduces the SINAD to 6 dB and record the difference in levels of A and B (in dB) and the frequency.

Results at frequencies that are submultiples of the receiver frequency should be discounted, as they are probably harmonics of the signal generator. Also results between adjacent channels should be ignored.

Image Rejection

In superheterodyne receivers (which includes most cellular phones) there will be a response to signals that are twice the first IF frequency away from the tuned frequency. To check for this, set the first generator as before to give 12-dB SINAD at 1-kHz modulation and 25 percent deviation. Set the second generator to twice the IF frequency above the tuned frequency modulated at 400 Hz and 25 percent deviation. Adjust the level for 6-dB SINAD and the image rejection will be the difference

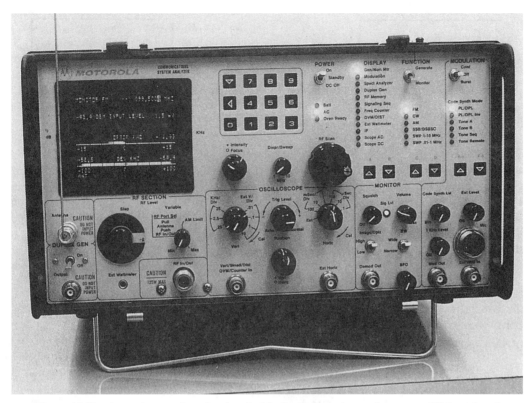

Figure 14.12 A service monitor that is suitable for base-station maintenance. With optional additional equipment, this monitor can be used for mobile unit service.

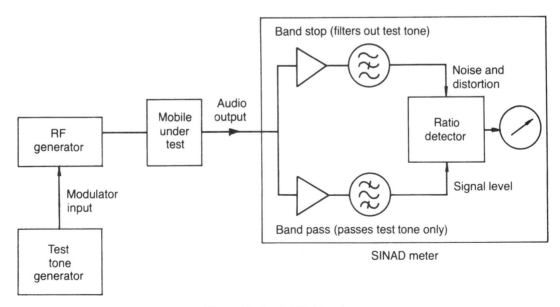

Figure 14.13 A SINAD meter.

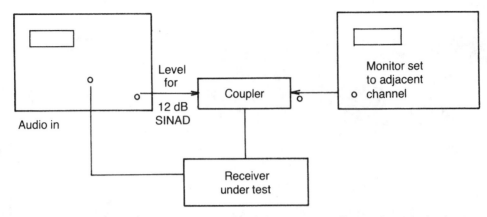

Figure 14.14 Two signal generators are needed to measure adjacent channel rejection.

in level of the two generators. Repeat the above tests for the frequency that is twice the IF *below* the center frequency.

Blocking (Desensitization)

When operated in the vicinity of other transmitters the receivers will experience some desensitization. To measure this, proceed as if the measurement were for adjacent channel sensitivity, except that generator B is set for zero modulation at 85 dB *above* the level of generator A (which, of course, is set up to read the 12-dB SINAD point). Now manually scan the frequency on generator B over the range ±100 MHz of the center frequency, noting dips in the SINAD reading and determining the level of generator B that produces 6-dB SINAD. The actual level of generator B (not the difference) is the blocking level of the receiver.

CDMA Network Problems

CDMA networks have problems that are similar to other networks, but they manifest themselves differently. Pilot pollution is similar to co-channel interference and results when, in a given area there is no dominant pilot. This manifests itself as poor Ec/Io and poor frame error rate (FER) in an area of otherwise good signal strength. The remedy is simply to reorganize the network to make one of the pilots dominant in that region.

Search window minimization is rather like housekeeping. The search window is the number of chip delays that the mobile needs to search to look for pilot PNs for a handoff. If this is made too large, then all possible candidates will be checked, but this can take so long that a moving vehicle may not have time to complete the search. By narrowing the search window, faster handoff

is assured, but with the trade-off that some handoff candidates may be overlooked.

A measurement of the quality of a code division multiple access (CDMA) transmitter is its ability to assign power to each of the orthogonal codes. Termed *code domain analysis*, this test measures the capability of the modulator.

BER Testing

BER testing will usually be done as follows. The test set will establish a call to the base station. It will do this at a high RF level. Once the call is established, the test set will slowly reduce the RF level sent, monitoring the BER the whole time. Once the BER has reached a predefined level (typically 2 parts in 100 for a bit stream without error correction), then it will report back that RF level as the sensitivity.

QUANTIFYING COVERAGE PROBLEMS

Good baseline records of the original coverage must be available in order to determine if a problem has occurred since installation.

The GSM ABIS

In GSM the network structure is as shown in Figure 14.15. The radio base station is linked to the switch by the Abis, which is a 2-Mbit link that transmits the voice and control data to the mobile. The voice speech path is available here (unencrypted), as is a lot of link data. Abis interface testers can be used to find out a lot about the air interface.

The Abis will generally be accessible from a socket on the BTS and at the BSC.

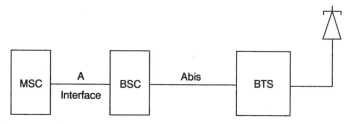

Figure 14.15   GSM system interfaces.

What Is on the Abis?

There are alarms that indicate a loss of signal [alarm indication signal, (AIS)] and synchronization alarms. These alarms are found on the Abis and time-stamped.

BER testing of the 2-Mbit Abis line between the BTS and the point of measurement can be done from here.

A check on the speech quality of a channel at the audio level can be done by listening, and the frame erasure rate (FER). A high FER indicates a poor radio path. On the call path it is also possible to read the 64-bit RxLev (in dBm, if the test set can translate it) and the 8-bit RxQual.

LOSSY T

Sometimes you will be confronted with a receiver that has symptoms of lack of sensitivity but when measured the BER or SINAD is within specification. The problem could be external interference and noise. To measure this, you will need a lossy T.

A lossy T is just a T section with a capacitance coupling to the third port, as seen in Figure 14.16. To measure the effect of the antenna, first measure the sensitivity of the receiver at the antenna port. Next use the lossy T to connect the input port to the antenna and the output to the receiver. Now connect the test set to the lossy T port (see Figure 14.17) and measure the sensitivity. Let us assume that the measurement was on a GSM base station and the sensitivity without the lossy T was

−108 dBm. The measurement through the lossy T port was then −85 dBm.

To calculate the noise input of the antenna, we need to know the loss of the lossy T. (This may be known, and if not a measurement of the receiver sensitivity measured from the input port, compared to the same measurement made through the lossy port [both with no antenna attached], will give the loss).

By adding the loss of the lossy T (say it is 20 dB) to the lossy port reading to get −105 dBm, we can see that the contribution of the antenna systems to the overall noise figure is 3 dB.

OFF-AIR MONITORING

It has long been the complaint of maintenance engineers that no matter how many alarms are used for base-station monitoring, what really counts is the signal going to air and that just is not monitored. Smith Myers Communications, a small UK-based company, has developed a device called a CSM8000 that can resolve this problem for AMPS and TDMA.

The CSM8000 was designed to monitor the progress of mobiles off-air, while at the same time recording their compliance with the specifications, and of course their mobile identification number (MIN). Mobiles that were found to be out-of-spec could be called in for service. During the early trials it was soon realized that the intelligence in the monitor could easily handle base stations as well if only the receiver bandwidth was widened.

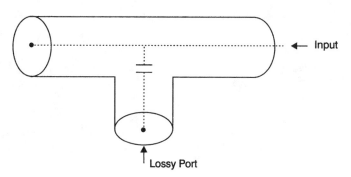

Figure 14.16   A lossy T connector.

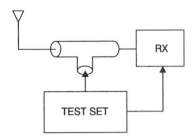

Figure 14.17 Measuring sensitivity through a lossy T.

The results of monitoring with the CSM8000 device reveal that about 1 percent of all mobiles are out-of-spec. For a network of 100,000 mobiles, this means that around 6000 are probably performing below "par." One of the most frequently encountered problems is over-deviation. This problem will cause a little distortion for the user, but could also cause sputter in adjacent channels, as the power bandwidth of the mobile will exceed 30 kHz. Some have carrier-frequency errors of up to 8 kHz and still seem to work. Again these will cause adjacent-channel interference.

The CSM is 19-inch, rack mounted (see Figure 14.18), and is meant to be remotely controlled via a modem from a compatible PC. It has six antenna inputs and can scan for the best channel, or in the manual mode can be directed to monitor a particular antenna. It can be controlled from a cellular mobile fitted with a modem.

As a base-station monitor it can monitor all channels in use by scanning or it can monitor particular channels. Because the whole system is controlled by a central PC, it is possible to program a monitoring activity that is time dependent. For example, a few hours a day can be devoted to mobile monitoring, followed by a base-station scan for the rest of the day. Because it can measure received field strength with an accuracy of around 2 dB, the device can monitor effectively the base-station ERP. In practice a few dB difference in ERP can be expected from channel to channel, but larger discrepancies will be indicative of a problem.

ROGUE MOBILES

Problems have been found with mobiles that, instead of simply listening on the control channel, transmit continuously on that channel and so block it out of service. These are known as *rogue mobiles*. This problem has been found in a number of different mobiles and is usually related to poor construction or poor repair or modification, which results in the incorrect processor command to switch the transmitter on. Tracing such a mobile in the conventional manner (a transmitter hunt) is nearly impossible in a high-density cellular network. Another job for a cellular monitor!

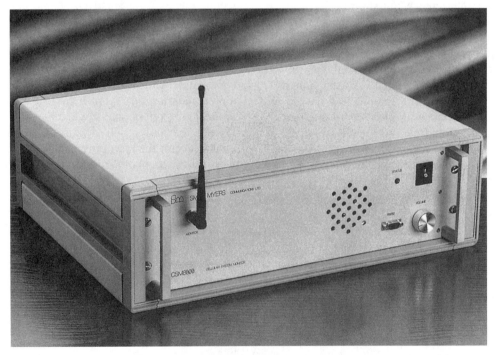

Figure 14.18 A cellular system off-air monitor. (Photo courtesy of Smith-Myer Communications.)

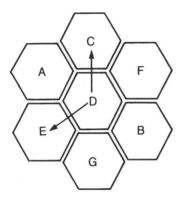

Figure 14.19 Channel adjacency in a 7-cell plan. Cells D and E and cells D and C are adjacent.

CO-CHANNEL AND ADJACENT-CHANNEL INTERFERENCE

As stated earlier, interference occurs mainly from the mobile to the base-station receiver. Sometimes, however, interference occurs between two base stations.

All systems are prone to data corruption from co-channel interference. (A co-channel is a channel that has the same frequency but operates from a different site; an adjacent channel is one channel width away.) To some extent, these kinds of interference are potential problems in all cellular systems (both digital and analog).

A mobile can fail to receive an instruction if a data clash occurs from two different bases. For problems to arise, the two sources must be within 3-dB signal levels of each other. If the difference is greater than this, the dominant signal prevails.

Frequency plans usually will take adjacent-channel interference into account by arranging the cells for maximum distance between adjacent channels. Figure 14.19 shows a typical frequency plan. Channels A and B, B and C, and so on, are adjacent blocks of frequency. This plan separates all cells that are co-channel by at least one cell (except D and E, and D and C), so the problem has been reduced, not eliminated.

Co-channel interference can be more severe and can cause whole base stations to fail if it occurs on the control RX channels; however, it is usually limited to a few high areas that have equal access to the two sites. Sectoring reduces this problem. A sure sign of co-channel interference is the identification of certain local areas where initiating and receiving calls is not reliable, but where reliable service is normal outside those areas.

In these instances, a frequency change is usually the best short-term solution, but the problem should be addressed as a design problem. It may be necessary to sector the site, reduce power, lower antennas, or even relocate.

THIRD-PARTY INTERFERENCE

When cellular radio was first switched on in most countries, it was common to find existing point-to-point links in the band. When the cellular-radio operator was also the wireline operator, it was not unusual to find that some of these offending links were the operator's. Most of these links are easily tracked down, and after negotiation, the fixed link can be moved. In some instances, retuning to a frequency outside the cellular band is all that is necessary.

When interference persists, it manifests itself in a number of ways. The most easily recognized is interference that brings the base-station receivers out of mute (blocks channels and reduces base capacity). Another sign is calls lost during handoff. In this instance, the radio base station (RBS) requests a handoff attempt and all other neighboring bases are asked to measure the field strength of the mobile and determine a base suitable for handoff.

If there is a foreign carrier with the same TX frequency as the mobile (at least in the AMPS system), it will be measured and assumed to be the mobile. If the foreign carrier records the highest field strength, the mobile is instructed to go to the base that measured the foreign carrier. Often, the mobile cannot access that base and so the handoff fails.

Although it is relatively easy to find point-to-point links that are causing interference, tracking ultrahigh frequency (UHF) mobiles is almost impossible. A number of mobile-locating systems exist, and although they all work fairly well when a fixed mobile repeater is the source, none is really satisfactory.

The problem with mobile interferers is the intermittent operation and multipath that causes false bearings to be read; this makes it very difficult to get accurate fixes.

Because of all these difficulties, it is perhaps inevitable that some of the more persistent non-cellular users of the band will stay on, undetected, until such time as the interference to them, from the cellular system, becomes unacceptable.

T1/E1 Interference

The T1 carrier lies in the 540-kHz to 1700-kHz broadcast band (the 2 Mbits for the E1 also overlaps this band) and so can be susceptible to interference from high-level AM radio station transmissions.

Testing in the Switch and Transmission Network

The switch and transmission part of a network is much the same as a land-line network; it has most of the same elements and problems.

Transmission systems have long been tested with BER testers that also test for bipolar violations (BPVs). Unless these are within specification, there will be dropped and corrupted calls. The same test set can confirm that the transmission at the E1/T1 levels (and higher) are being clocked at the correct frequency, and that transmission is occurring at the correct levels.

Transmission tests are best performed in the out-of-service mode, operating in full duplex, with the test set synchronized to the system clock.

Jitter testing confirms that the phase error is within tolerance. Excessive jitter will cause lost calls.

Switches have a translation table that translates the number dialed to a particular route. Any error in this table will result in calls being wrongly forwarded and probably in a dropped call (although in some cases a wrong number is possible). Calls should be made to all valid codes through an automated call test set. This will verify that the translation table is correct. When calls are made through the PSTN, this test will also confirm that wire-line section of the call routing.

The switching systems use SS7, IS41/MAP, and occasionally other protocols for signaling. Mostly SS7 is used for interswitch signaling, and the IS41/MAP runs on top of the SS7 providing the wireless signaling between the cellular switches and other wireless hardware. To test these protocols the appropriate test set is required, but it is important that the relevant links be tested first to ensure that they are within specification.

SPARE PARTS

For cellular systems, by far the most unreliable component is the RF channel equipment; adequate spare parts should be kept to allow for a one-year requirement.

Be sure to test all spare boards soon after delivery by placing them into service. A non-functional spare is of no use and a significant number of boards are delivered from the manufacturer as dead on arrivals (DOAs).

At least five spare antennas of each type should be on hand. These often have fairly long lead-times on delivery and are liable to fail in considerable number during the electrical storm season.

Disposable items, such as fuses and light bulbs, should be purchased in quantities adequate for two years. The frequency of failure, coupled with the cost of processing orders, makes small holdings economically unattractive. Good quantities of spare connectors should also be stored. It is false economy to save on spares, and adequate quantities should be held to ensure that they will be available when needed.

SYSTEMS IN CHAOS

Systems are sometimes implemented with no regard to sound engineering practice. These systems usually have managers whose background has not been telecommunications, and so they fail to understand that the inherent robustness of a cellular system to a little overload belies its almost complete lack of defense against chronic overload.

When a cellular system is overloaded by no more than 10 percent, it will mostly perform reasonably well, as the adjacent cells pick up traffic that was destined to be served locally. The only thing an astute customer will notice is a slight drop in "line quality" due to the fact that more distant cells, which occasionally are used, will have a lower S/N performance.

As the system is loaded more, the probability of getting the best (nearest) cell decreases rapidly, so that virtually all calls are connected to a distant cell. This means that the path will be lossy and the call noisy. Because the call proceeds through a non-optimum cell, it will soon need to hand off. The handoff in really bad systems probably again will be to something other than the nearest cell. The switch consumes most processing time handling handoffs. The situation soon occurs where the processor itself overloads and dropped calls are frequent during handoff.

A chaotic failure of the type described above can occur locally in even the best designed system where for some reason one cell becomes overloaded. Unfortunately, it also occurs chronically in badly designed systems, and I have used systems where virtually every call is a bad one. The users make multiple attempts to get connected and when they get through they talk quickly to get the message through before the inevitable dropout occurs.

Such systems are very difficult to fix. The actual degree of congestion is difficult to estimate, as callers, resigned to the fact that the call probably will be unsatisfactory, refrain from making calls. The system statistics can also be unreliable, as frequent lost handoffs and dropped calls can be due in part to hardware and in part to congestion. It is often difficult to distinguish.

Technical Support

Increasingly as operators become more business oriented they are becoming less engineering oriented. Increasingly new telecommunications companies are needing to rely more heavily on the equipment vendors for support. According to some of the major vendors, the support role is currently the fastest growing segment of the industry.

OPTIMIZATION

Once a system has become established, and a number of ongoing extensions have been made, it will eventually become evident that the system is drifting away from its optimum configuration and effectively needs to be replanned.

System optimization is often done as a result of customer complaints that have nothing to do with the network configuration. While typically it may be necessary to reoptimize the system approximately ever two years, it is necessary to optimize the hardware once every year. If the hardware optimization has not been done, then system optimization is a total waste of time.

Optimization of the hardware consists of a series of checks on the hardware that are not dissimilar to a full acceptance test. The object to is confirm that the hardware, including the base station, power supplies, antennas, cables, links, and switching, is performing as installed. It is insufficient to rely on the network maintenance system, which will ordinarily detect serious network shortcomings, like failed hardware, but often fails to detect more subtle problems like high cable losses, combiner and cavity faults, transceivers out of specification (especially sensitivity and transmit power), antennas pointing the wrong way (they were often either installed that way or have moved in the wind since), cables connected to the wrong antenna (it happens during maintenance), towers and support structures that are rigid, the erection of new buildings or other infrastructure that interfere with cellular or microwave paths, voice and data path lineup levels that are wrong, and a great many other problems that can evade the system's inbuilt fault detection.

Prior to doing a complete network hardware optimization, it is a very good idea to look at the current complaint statistics and base-station performance statistics. Any problem areas should be highlighted as the information is made available to the optimization crew, in order that they can look for the possible causes at the sites visited.

Once the hardware optimization has been done, a settling period of a few weeks should be allowed and then the network statistics should be reviewed. Typically, some sites will now be carrying significantly more traffic, have alleviated some congestion, and possibly even new sites are congested (due to their improved performance, they may now be attracting more traffic).

Having reviewed the network, it is now time to start an overall optimization that will likely involve a major retune (frequency allocation), a revision of the downtilting, system parameters (handoff levels, hysteresis, neighboring channel lists, etc.). For practical purposes this part of the optimization is a complete redesign of the network and its parameters.

When the optimization has been completed, then it is important once again to review the network statistics, to confirm that changes were for the better, and that no new problems have been introduced. Drive tests should follow to confirm coverage, that cell boundaries are as expected, and that audio and link quality is within expectations.

Getting It Right First

Increasingly base stations are provided as "black boxes" that are installed in a few days and then left. If these base stations were not installed correctly in the first place, then they will never perform properly. From firsthand experience here are a few of the performance-affecting problems that can already be in the network from a poor installation:

- Sectors connected to the wrong antenna (causing handoff failures)
- Antenna downtilts wrong (coverage shortfalls or interference)
- Antennas on the one sector pointing in different directions (bad site performance and dropped calls)
- Faulty equipment installed, in particular low-sensitivity RX and low TX power (site performs badly and may drop calls on handovers)
- Base-station parameters incorrect (poor performance of site)
- Insufficient direct current (DC) power supplies for fully loaded base (dropped calls)
- Antenna support structures not rigid (dropped calls)
- Antennas obstructed (poor performance and dropped calls)
- Insufficient battery backup (poor survivability)
- Alarms not connected (site can be out and no one knows)

In addition to these faults that cause poor performance on day one, there are others that will induce poor performance in the future, including the following:

- Poor grounding practice (will cause equipment failures intermodulation problems, or impairment)
- No labeling of cables (makes future maintenance difficult)
- Connectors not tightened properly (will lead to premature equipment failure)

- Racks not secured properly (premature equipment failure)
- No station log books (routine maintenance goes unrecorded and eventually no one is sure exactly how the station is configured)
- Base station not waterproof (premature equipment failure)
- Switches not labeled (equipment will get turned of accidentally, or may inadvertently be worked on live)
- Cables and antennas not secured firmly (can lead to poor quality calls, dropped, calls or site failure)

If it is not right in the first place, there is no point in looking at optimizing.

Network Performance

The real objective of network performance is for the operator to get a measure of the way the customer sees the network. As multiple competitive networks spring up it is also important for the operator to obtain relative performance measures against the other networks.

Not very many network performance evaluations are very revealing, and it pays to look first at how easy it is to get it wrong.

While it should be easy to evaluate a network, because after all from the customer's viewpoint it comes down to simply this: making a call successfully, holding the call, and ensuring that the quality is good for the duration. It seems simple, but as we will see, that assessment is simplistic.

Installer/Vendor Evaluation

It is not unusual that it is the installer, or worse still, the vendor, who is responsible for evaluating the network performance. Evaluations should be unbiased and independent. How can someone who is relying on a good report to get the last check turn in a report that is bad?

Often installers (who may be part of the operator company or may be vendor employees) are the ones to evaluate the network. These people have a vested interest in a good report. If they find fault with the network, they are in effect finding fault with their own work.

Keep the CEO Happy

Very often the CEO takes the position that what really matters is his or her personal evaluation. While the interest might be commendable, it is clearly not objective. Once the staff becomes aware of this "measure," they will optimize the network for the CEO. This means that every effort will be made to ensure that the office area, the CEO's drive route, and house are well covered. Soon the CEO will get the false impression that all is well!

Trusting Switch/Base-Station Statistics

Switch statistics can reveal a lot about the network, but what they reveal best is trends. If these statistics show a change, then it is usually indicative that something is happening. However, these figures can be biased by the users themselves. If a base station in a certain area is performing badly and dropping calls, the users will get to know this and avoid making calls in that area.

Certain call statistics are difficult to interpret. Congestion should mean that no circuits are available, but from a customer's point of view it may mean that a call has failed. While it is possible to estimate the number of failed attempts that result from congestion, that number is itself something that the customers can influence. If the system is basically running well, a dropped call will result in an immediate retry and the number of these retries can be estimated from the carried traffic (however, it is very common that these calculations are erroneously done in the statistics processing package using Erlang B or other inappropriate models, and so even in this case the estimate is wrong). However, as the system becomes more congested, and the customers learn that a retry is likely to be futile, they tend not to retry. This means that once again the statistics will underestimate the problem.

Switch and cell statistics can be a problem from the sheer number of them. There can be several hundred parameters that are reported, and to make something that resembles a cohesive report from these can be daunting.

Do It Right: Drive Test

A drive test over an unbiased route is the only real way to determine how well the system is working. Traditionally this has been a labor-intensive task that is done both by technical staff, who collect detailed records, and nontechnical staff, who record subjective opinion scores.

While this evaluation remains the best, recently it has been possible to automate it to a considerable extent. A number of vendors have packages that will automatically collect and evaluate the drive-test data. Mostly these can monitor more than one service at a time so that comparative network figures are also available.

If the drive test is done properly, it will be the nearest thing to the experience of the customer, and will probably reveal the network to be worse than any other measure of the performance.

Recently, fully automatic testing has become possible and to ensure unbiased testing, the test equipment can be placed in a local taxi, courier truck or the like, so that an unbiased route is selected. This is the ultimate in network testing.

Voice Quality Measurement

Traditionally voice quality has been assessed by a person who rates the quality subjectively. The problem with this is that different people will rate things differently, and after a days "rating" there is no real assurance that the rating is to a consistent standard.

A device from Ascom called Qvoice, solves the "human" problem by using a neural network to rate the voice quality. The system has two mobile phones (that can monitor two networks at the same time). The unit not only sets up and monitors the calls automatically, but it logs the call's progress, and any overhead messages. It has built-in Global Positioning System (GPS) tracking. It can be used both in a vehicle and indoors, and comes complete with analysis and presentation packages.

The only downside to this system is the reliance on mobile phones as test instruments when they are not designed for that purpose.

Measuring Call Quality

There are numerous ways of quantifying network performance; most involve collecting a lot of performance statistics. However, it often happens that so many statistics are collected that it takes an expert to make any overall judgment from them. The following method attempts to simplify the appraisal of a network or part of a network from the statistics collected.

The figures that are of most immediate interest to the user include the following:

1. Call success rates (Cs)
2. Handover failure rate (Hf)
3. Dropped-call rate (Dc)
4. Call quality (Qc)

Depending on the method of statistic collection, there may be other parameters that might be considered, for example:

5. Congestion (C)
6. Blocked channels (B)

Considering parameters 1–4 first, all of which can be obtained from a relatively straightforward drive test,

let's devise a set of benchmarks (these are to some extent arbitrary).

Parameter	Benchmark Value (%)
Cs	>95
Hf	<1.5
Dc	<2
Qc	>85

From these benchmarks we can derive a network performance index by weighting the parameters in a way that the weights add up to 1.0. We need to select the parameters so that the relative weighting of these performance figures are similar, then proceed as follows. If we let the network weighting factors be A, B, C, D, then the performance is

$$A \times (100 - Cs) + B \times Hf + C \times Dc + D \times (100 - Qc)$$

Assuming for the present that we agree that all of these factors are important, but we rank their importance in the order shown in the table below, then at the point where each of them reach their benchmark levels, the equation should reflect this relative importance. If more than four performance parameters are used, proceed in the same way, but make the sum of the weights still be 1.0.

Parameter	Weighting
Cs	0.25
Hf	0.35
Dc	0.25
Qc	0.15
Total	1.0

The relatively high weighting for the Hf is due to the fact that multiple handovers may be encountered during the average call. For a system in a rural area where handovers would be less common, a lower weighting for Hf could be used.

Nominating the performance of a marginal system where all parameters are just within the benchmark of 100 (again this figure is nominal), we have

$$A \times (100 - Cs) = A \times (100 - 95)$$

$$= 0.25 \times 100 \text{ or } A = 5$$

What we have done here is make the value of the first parameter, at the benchmark point, equal to its rating $\times$ 100, then

$$B \times \text{Hf} = B \times 1.5 = 0.35 \times 100 \quad \text{or} \quad B = 23.3$$

$$C \times \text{Dc} = C \times 2 = 0.25 \times 100 \quad \text{or} \quad C = 12.5$$

$$D \times (100 - 85) = 0.15 \times 100 \quad \text{or} \quad D = 1$$

We now have a formula for the system rating:

$$\text{Network rating} = 5 \times (100 - \text{Cs}) + 23.3 \times \text{Hf}$$
$$+ 12.5 \times \text{Dc} + (100 - \text{Qc})$$

This network rating will be lower for better overall quality of service, so we can invert it by subtracting it from 100. The formula now is

$$\text{Network rating} = 100 - 5 \times (100 - \text{Cs}) - 23.3 \times \text{Hf}$$
$$- 12.5 \times \text{Dc} - 100 + \text{Qc}$$

or

$$\text{Network rating} = 5 \times \text{Cs} - 23.3 \times \text{Hf}$$
$$- 12.5 \times \text{Dc} + \text{Qc} - 500$$

Thus, for a system in which the network rating is running rather well, such as the one below

$$\text{Cs} = 97\%$$

$$\text{Hf} = 1\%$$

$$\text{Dc} = 1\%$$

$$\text{Qc} = 91\%$$

we have Network rating $= 485 - 23.3 - 12.5 + 91 - 500 = 40.2$.

It might be better if such a network, which is obviously rating rather well, achieved a nominal score of 100, something that can be done by multiplying all the parameters by $100/40.2$ (lets say $100/40 = 2.5$). The equation then becomes

$$\text{Network rating} = 12.5 \times \text{Cs} - 58.3 \times \text{Hf} - 31.25$$
$$\times \text{Dc} + 2.5 \times \text{Qc} - 1250$$

With this rating formula, we still have a system that is just marginal for all parameters rating zero.

To see how this formula would rate another reasonable network, which is better in some respects but not as good in others, let's assume

$$\text{Cs} = 95\%$$

$$\text{Hf} = 0.85$$

$$\text{Dc} = 1.3\%$$

$$\text{Qc} = 95\%$$

We then find that the network factor is 84.82.

Now let's apply this to a truly bad network, where

$$\text{Cs} = 50\%$$

$$\text{Hf} = 7\%$$

$$\text{Dc} = 5\%$$

$$\text{Qc} = 81$$

The network rating of this network is -986.8.

Thus, we have a simple tool for comparing the performance of the network over time and comparing parts of a large network with each other. For small systems, and locally for large systems, this rating can be used to compare the performance of base stations. The results can easily be used by non-technical management to get an overview of the network performance.

If more parameters are measured, then they can be factored in, following the procedure outline above.

Given that a computer, and perhaps a spreadsheet, will be used to calculate this network rating, there is no reason why many more parameters cannot be similarly included. Also, the operator may like to give a greater weighting to any parameters that exceed the benchmarks for performance. For example, one might choose to double the rating of that part of any parameter that exceeds the benchmark.

Rating the Network

An extension of this method is to look at the whole network the same way. One method would be to average the measured values over the whole network, and then using the same rating method, give a rating to the network as a whole. Additionally it would be useful to report for each parameter how many sites were less that the benchmark values.

It is apparent that an easy way to handle all of this is through a spreadsheet. That way changes can be easily made and the methods refined as the network evolves. Be careful, however, because once the rating system is changed, it will be difficult to get a clear idea of the relative performance of the network over time.

MONTHLY ROUTINES

The sheets that follow can serve as a guideline for routine testing of base stations. Some additional tests may be necessary as a consequence of the particular hardware installed. These tests should only take a few hours (at most), and will ensure that the network is functioning correctly.

Although most systems today come with impressive looking control centers, and these centers do perform many valuable functions, they still have not reached the stage of development where old-fashioned site visits and testing can be bypassed.

I have seen many systems where this kind of testing was not done, and while the OMC was reporting nil failures, a closer look at the base stations revealed up to 50% of the resources (channels) not functioning correctly.

MONTHLY ROUTINE BASE-STATION MAINTENANCE

On a monthly basis the following checklist should be completed.

BASE-STATION SITE NAME

_____

Year _____ Month _____

Tested by _____

INITIALS

_____ All fans (usually mounted on PAs and racks).

_____ Air-conditioning units (in particular where two or more units are provided, run them one at a time for at least 15 minutes to test working under full load Monitor room temperature).

_____ All rack alarms.

_____ Building lights and emergency lighting.

_____ Cabling and connections.

_____ Tower lighting and grounding.

_____ Test all rectifiers by switching them off one at a time. When turned back on they should resume proper load sharing.

_____ Fire extinguisher condition.

_____ Inspect building walls and ceiling for condensation.

_____ All grounding points for corrosion and firmness of connections.

_____ Combiner tuning.

_____ Control channel output level.

_____ Using a spectrum analyzer check all carrier channel levels against previously tested control channel level. Attend to low RF levels.

_____ All redundancies should be tested.

_____ Measure and record all battery cell voltages.

_____ Check generator fuel level and run the generator for 20 minutes on load.

_____ The −48 &/or −24 volt supply rails should be redundant. Check the voltage levels, load sharing, and redundancy.

SIX MONTHLY ROUTINE MAINTENANCE–DIGITAL

BASE-STATION NAME

_____

Year _____ Month _____

Tested by _____

INITIALS

_____ Check control channels(s), TX power.

_____ Check control channels(s) redundancy.

_____ Check control channels RX BER.

_____ Look in log book for new work, if modifications or retuning have occurred, then check sectors properly allocated.

_____ Check antenna orientation and tilt.

_____ Check rectifier redundancy and load sharing.

_____ Check battery date (not past serviceable life).

_____ Once a year check all TX power and RX sensitivity.

_____ Place a few calls on each sector.

_____ Complete the full monthly check.

SIX MONTHLY ROUTINE MAINTENANCE—ANALOG

BASE-STATION SITE NAME

_____

Year _____ Month _____

Tested by _____

INITIALS

_____ Transmitted and received audio levels checked on all channels.

_____ Check SAT data and voice deviation.

_____ VSWR on all antennas (including receive antennas).

_____ Voice and scanner receiver sensitivity (12-dB SINAD/or BER).

_____ Channel carrier frequencies.

_____ Check linearity of SSI receiver.

_____ Complete a full monthly check.

_____ Upon leaving the cell site place some test calls and have them traced and monitored by the switch.

15

BASE-STATION CONTROL AND SIGNALING

The air interface (or the interface between the base station and the mobile) is specified precisely for all systems. This precise specification allows different manufacturers to produce compatible systems. All systems use "handshake signaling," meaning that all instructions must be acknowledged. Most systems use data blocks and some error-correcting code.

Base stations have a control, or signaling, channel for signaling purposes. In some instances, the control channel is also available as a speech channel (when not used for signaling). Two such instances are Nordic Mobile Telephone (NMT) systems, which use a voice channel for signaling (control), and some Advanced Mobile Phone Service/Total Access Communications System (AMPS/TACS) systems, which use the redundant control channel as a voice channel when it is not required for control. Although some limited signaling is done on the speech channel, this is usually by out-of-band (or inaudible) tones. Control channels usually send instructions in a digital format (even in analog systems), because this is faster than using analog tones and is more immune to simulcast-type interference.

All signaling on all systems is fundamentally the same, but the details may vary and the digital systems are somewhat more complex, with more detailed authentication procedures. What follows is the signaling in an AMPS system, which can be used to understand the basis of signaling in any trunked radio or cellular system.

CALL TO MOBILE STATION

A call to a mobile station requires sending a page to the mobile, a response from the mobile, sending a ringing tone, and finally, connecting the voice circuit. This process involves a number of distinct steps, as illustrated in Figures 15.1. through 15.9.

First, the land party dials the mobile number. The public switched telephone network (PSTN) forwards the number to the cellular switch for verification and forwarding to the mobile. The PSTN switch generates call progress tones. All mobiles receive the page call (see Figure 15.1).

Only the desired mobile responds by sending an identification that acknowledges it received the page (see Figure 15.2). Next, the base station determines an appropriate free channel, turns on that channel, and transmits its supervisory audio tone (SAT) tone (see Figure 15.3).

The mobile now has to be directed to the appropriate channel and told which SAT tone to expect (see Figure 15.4). When the mobile switches to the voice channel it automatically loops the SAT tone, which informs the base that the connection is complete (see Figure 15.5). An alert order (a 10-kHz tone) tells the mobile that a call is on line (see Figure 15.6). This 10-kHz signaling is also known as *blank and burst*. It gets this name from the fact that to transmit the instruction the other signals currently being sent out (usually voice) are first blanked out, then the signaling is accomplished.

The ringing tone is then generated locally at the mobile and sent back to the network by the mobile switch (see Figure 15.7). When the handset is lifted, the mobile loops the SAT tone, telling the base to stop the ringing tone and that a normal call is in progress (see Figure 15.8). During the conversation, call supervision occurs for the entire duration of the connection to ensure quality and check the continuity of the call (see Figure 15.9).

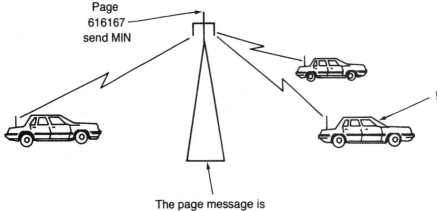

Page
616167
send MIN

The page message is
transmitted over all control channels
in the service area.

Figure 15.1   A paging call is sent.

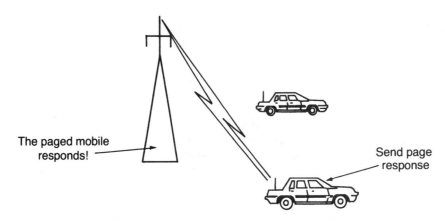

The paged mobile
responds!

Send page
response

Figure 15.2   The mobile acknowledges receiving the call.

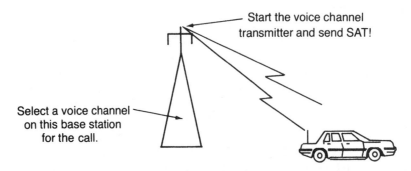

Start the voice channel
transmitter and send SAT!

Select a voice channel
on this base station
for the call.

Figure 15.3   The base station selects a call (voice) channel.

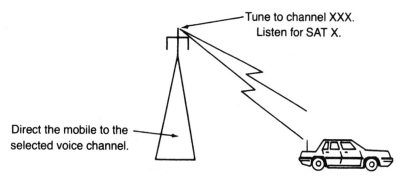

Figure 15.4   The mobile switches to the appropriate voice channel. The base informs the switch of the selected channel.

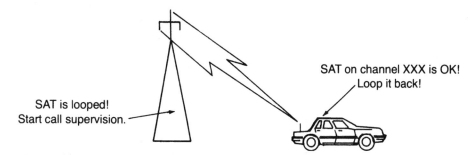

Figure 15.5   The looping of the SAT tone, which is generated by the base, received by the mobile, then sent back to the base, confirms that the mobile has arrived on the call channel. This process is time-supervised.

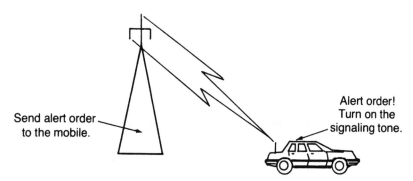

Figure 15.6   The ringing (call ready to proceed) tone is generated for the land subscriber by the switch and locally in the mobile for the vehicle unit.

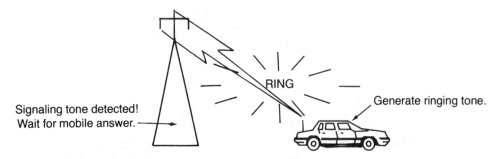

Figure 15.7 Signaling is done via a 10-kHz tone burst from the mobile to signify that its ring signal has been started.

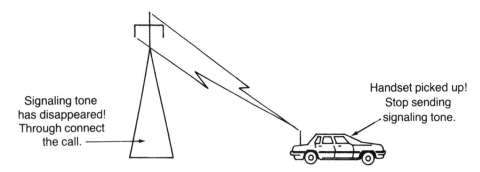

Figure 15.8 The looping of the SAT tone indicates that the mobile has answered.

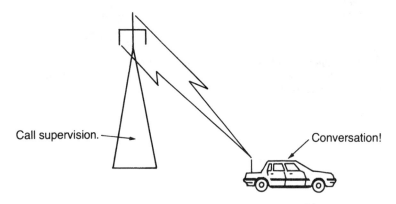

Figure 15.9 The call undergoes constant supervision.

MOBILE-ORIGINATED CALL

A mobile-originated call proceeds much the same as a terminated call, except that this time the mobile calls the base. The mobile knows if it is out of range and does not attempt to send if there is no control channel present, although it continues to scan for the presence of a control channel.

Periodically, the base station transmits overhead information that, among other things, defines what the mobile should send for identification. Particularly where encryption is used, it may not be desirable to send all the identification information because some of it is used as part of the encryption "seed."

Because all mobiles use the same control channel to request a call, the possibility of two mobiles attempting

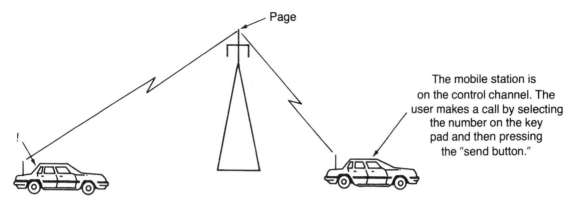

Figure 15.10 The mobile is in "listen mode" when idle; it listens out on the best control channel. The subscriber initiates a call by pressing the "send button."

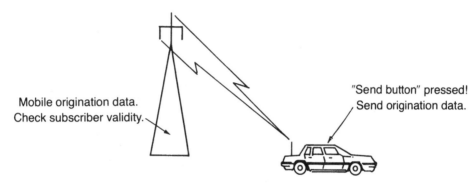

Figure 15.11 The base station communicates the subscriber's information to the switch, which checks the validity of the subscriber. The base station assigns a free voice channel.

to call simultaneously must be allowed for. Should an originated-call attempt clash with another mobile (and cause data corruption), a built-in algorithm ensures a retry attempt at a time that (hopefully) is different from that of the conflicting mobile.

The steps involved in a mobile-originated call are as follows:

The mobile subscriber enters the desired number and then presses the "send button." The call can then be sent almost instantly on the correct channel because, in idle mode, the mobile regularly scans for the best control channel (see Figure 15.10).

At first, the mobile sends its identity [mobile identification number (MIN2) plus, if requested, MIN1 and its serial number]. The MIN2 is associated with the subscriber's telephone number, and it may, in fact, be the subscriber's number (but not necessarily). This information is checked for validity at the switch. The called number is also sent (see Figure 15.11). Next, the base selects a voice channel and sends a SAT tone, as in the case of a terminated call (see Figure 15.12). The mobile is then instructed to switch to the designated voice channel (see Figure 15.13). The looping of the SAT tone confirms that the mobile is on frequency (see Figure 15.14). The call is connected and the network ringing tone is received until the call is answered. Then call supervision takes place (see Figure 15.15).

CALL SUPERVISION

Call supervision lasts for the full duration of all calls. The call quality is "sampled" by a scanning-signal measuring receiver, which samples each of the active channels. This process takes about 50 milliseconds per channel. The channel itself is monitored for interference by constantly monitoring the presence of foreign SAT tones.

If necessary, a handoff request is made. Figure 15.16 shows the call-supervision process.

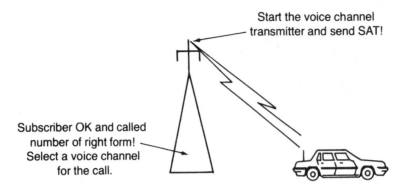

Figure 15.12 The SAT tone is generated locally at the base station to distinguish the base from others using the same frequencies.

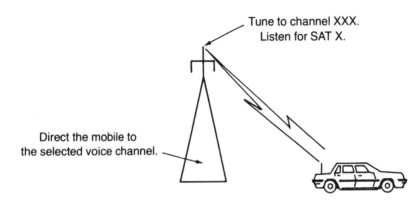

Figure 15.13 The mobile switches to the speech channel.

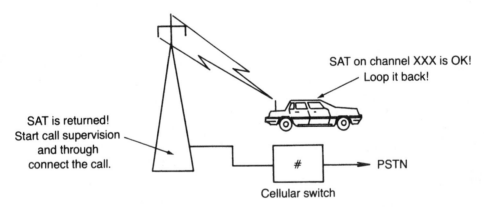

Figure 15.14 The mobile loops the SAT tone, confirming its presence.

SETTING UP A CALL BETWEEN TWO CARS

The call-setup procedure for calls between cars is very similar to that for mobile-to-land-line subscribers where an external telephone exchange is involved, except that the connections are made via the incoming/outgoing interface. That is, the call is completely handled by the mobile switch. Note that the switching process described in this section is for an AMPS system, but the basic concepts are the same for all systems.

A call request from the car in cell A is sent via the access channel (a radio channel dedicated to setting up

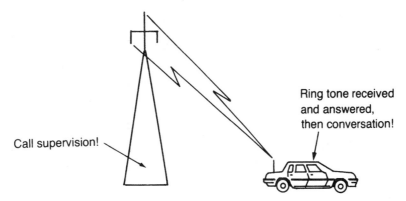

Figure 15.15   Tones to the mobile are generated by the switch to indicate call progress (for example, ringing tone, busy tone, and number unobtainable).

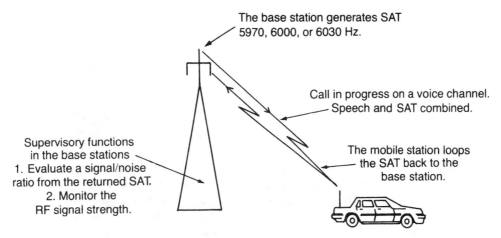

Figure 15.16   Active subscribers are logged on to a lookup table at the base station. A dedicated scanning receiver monitors for S/N performance, RF field strength, and the presence of the correct SAT tone. When the subscriber hangs up, a 10-KHz tone burst from the mobile signifies release of the call.

calls from mobiles) to the terminating circuit. This circuit separates out control signals (which are sent to the controller) from voice signals (which are handled by the switch).

The terminating circuit sends the request for a channel to the controller, which examines the A number (the calling party's telephone number) to determine the nature of the caller.

The controller checks the following:

• Is the caller local or a roamer?

• What is the category (barred international calls, barred outgoing calls, and so on)?

• Are there any special facilities (call forwarding, call answering, and so on)?

Assuming the call is legitimate, it is marked as busy to prevent conflict with other calls that might be attempted to the A party.

The B party (or number called) is then analyzed, usually only to the first three or four digits to determine the charging, number length, and routing. The B party category is also analyzed to confirm that the called party is currently a valid subscriber.

Next, a check is made to see if the B party is free; if it is, the line is marked as busy.

The B party is then called over the control channels (from all bases, because the location has not yet been determined) via the terminating circuit.

When the B party mobile acknowledges, the most appropriate cell is selected. (The B party has already

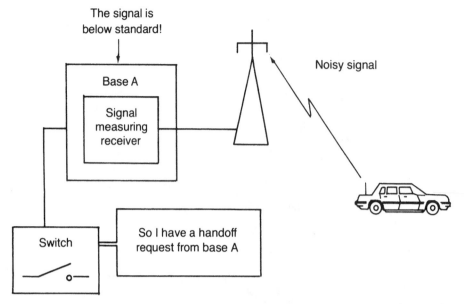

Figure 15.17 A noisy path is detected by the scanning-signal measuring receiver and the switch is informed. A handoff is requested.

designated the best cell by choosing the strongest control channel.)

The B party is then assigned a voice channel and instructed to switch to it.

The B mobile switches to the voice channel and acknowledges its presence. This acknowledgment is sent to the controllers.

Similarly, the A party is assigned a voice channel. The controller then instructs the switch to connect the two voice channels.

Call supervision now takes place via the terminating circuit to the controller. Call-quality assessment, as judged by signal to noise (S/N), absolute signal strength, and detection of interference, occurs throughout the call; necessary handoffs to adjacent cells are attempted if problems occur.

At the same time, the system is monitoring for possible cleardown signals (end-of-call indications) from the A or B party (which, of course, terminates the connection).

HANDOFFS

Handoffs are initiated when the signal, as monitored at the base station by the signal-level scanning receiver, is deemed to be below a certain quality. This quality can be judged simply on signal strength or it may involve S/N measurements, or both.

At the base station, in all systems, there is a receiver,

the sole function of which is to scan all mobiles in use and to monitor the signal from each. The scanning time is usually about 50 milliseconds per channel. This scanning receiver can be a standard voice channel transceiver or a dedicated scanning receiver. A dedicated unit is usually cheaper than a transceiver (which will have an unused transmit section), but it also required dedicated spare parts for backup.

When a mobile is judged to be below a pre-defined standard (and this is a user-definable parameter), the base-station controller requests that the switch attempt a handoff.

The switch refers to its lookup table of sites that are adjacent (or neighboring) to the site requesting the handoff. The switch then asks each of these sites to scan the transmit frequency of the mobile in question and report its field strength (or S/N, or other quality). The returned parameters are then compared to those of the existing base; if they are better by a defined amount (usually 3 dB), then a handoff to a free channel on the preferred base is initiated.

The mobile then has a set period (typically 50 milliseconds) to report on the new channel.

The scanning-channel measuring receiver first detects a substandard signal path and then requests a handoff from the switch, as shown in Figure 15.17.

The switch now refers to its lookup table to determine which base stations are adjacent to base A and therefore candidates for a handoff. Because this process consumes considerable processor time, it is typical to limit

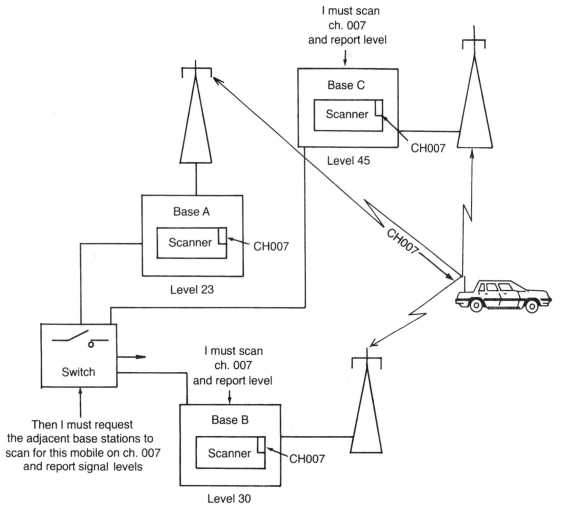

Figure 15.18 Neighboring bases are requested to tune to the mobile frequency and report on the received signal level.

the "neighboring" bases to only immediately adjacent ones.

The neighboring bases are then asked to tune their measuring receivers (which can scan any channel) to the channel used by the mobile, and then to report back on the signal level. Figure 15.18 shows neighboring bases reporting signal levels.

Each base station then sends back a report on the received level; the switch attempts a handoff if the next best base is better than the original signal by a predefined level (a system parameter). Figure 15.19 illustrates this process.

If one of the bases reports a sufficiently high level, then a handoff is attempted and the mobile is requested to report to the new base on a new channel. In Figure

15.20, base C has the highest field strength and so is the station chosen for handoff.

AMPS SIGNALING FORMAT

The signaling on the air interface is achieved by sending binary-coded-hexadecimal code in bit streams called *words*. A word sent to the mobile consists of 28 bits; a word from the mobile contains 36 bits. An additional 12 bits of error correction code is added, resulting in word lengths of 40 and 48 bits, respectively. For security, each word is repeated a number of times.

A word train begins with dotting, a term used to describe the signal sent to synchronize the mobiles. Dotting consists of the digits 1010 . . . 10. This is followed by a

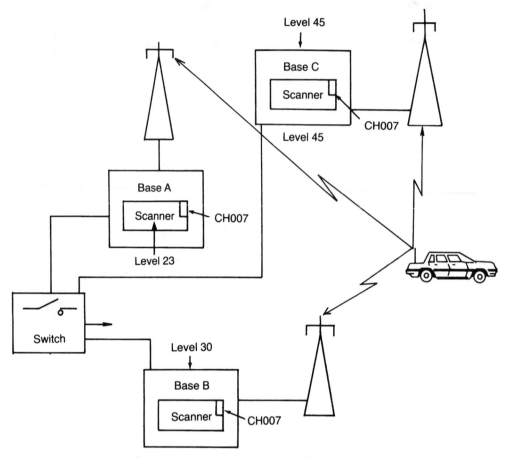

Figure 15.19   The bases send back a report on the received signal level from Channel 007.

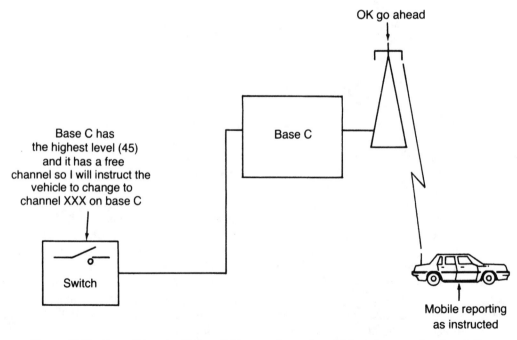

Figure 15.20   Base C has the highest field strength, so the mobile is instructed to handoff to that base. After reporting, the conversation continues as before.

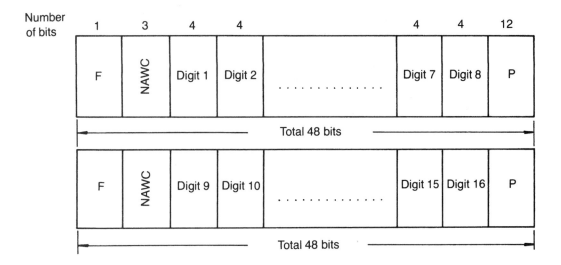

where   F = first word in the field
NAWC = number of additional words coming
P = parity bits

Figure 15.21   The format of the message from a mobile requesting a network number is two words of 48 digits each, describing a 16-digit number. This information is repeated five times.

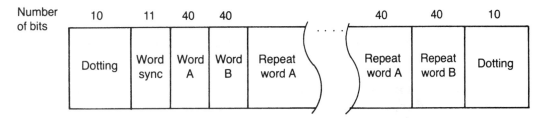

Figure 15.22   The general form of a forward message includes the dotting and word synchronization followed by the word repeated five times.

word synchronization code of 11100010010, and then by two words.

A mobile can request a connection to a given number by sending the information shown in Figure 15.21.

The message is preceded by the dotting and word synchronization and the whole message is repeated five times. A 16-digit number is assumed.

A forward control-channel stream (a message from the base station to the mobile) takes much the same form. The general form of such a message is shown in Figure 15.22. The word is repeated five times to ensure the correct transfer of information that might otherwise be lost in a multipath environment.

Figure 15.23 shows a typical word sent by a base station for the case of a voice-channel assignment.

SIGNAL STRENGTH PARAMETERS

There are number of user-defined system parameters that determine the behavior of a base station. Table 15.1

TABLE 15.1   A Number of User-defined Levels are Available to Customize the Coverage and Behavior of a Base Station

Signal Strength (dBm)	Parameter
−72	SSD—mobile power decrease
−82	SSI—mobile power increase
−95	SSH—mobile handoff level
−110	SSB—block channel level

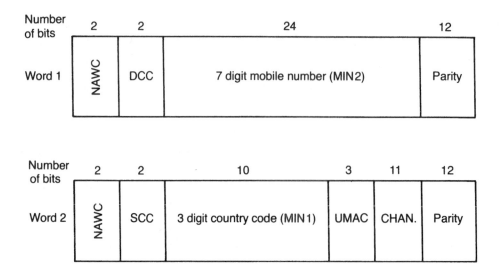

UMAC = Voice Channel Mobile Attenuation Code
CHAN = Channel *Number*

DCC = Digital Color Code
SCC = SAT

Figure 15.23 A typical forward control-channel message.

shows some of the most important system parameters and typical values for them.

Other parameters that can be set include SSHY, which sets the hysteresis for a handoff attempt (usually about 3 dB, which means the alternate base must be at least 3 dB better before a handoff is attempted), and SUH, which is the supervision time for a call before the handoff is declared unsuccessful.

Many systems also measure S/N. Two other parameters SNH (signal-to-noise before attempting a handoff, usually around 25–30 dB) and SNR (signal-to-noise before the call is released, usually around 12–15 dB) are also very useful.

16

POWER AND DISTRIBUTION

AIR-CONDITIONERS

Air-conditioners are an essential part of most mobile-telephone base-station installations. The heat output from the equipment may be such that, without air-conditioning, the equipment would soon overheat in all but the coldest climates. In very cold climates, reverse cycle units may be needed.

An air conditioner in its simplest form is a heat pump, that pumps heat energy from one region (the equipment rooms) to another region (the outside). Figure 16.1 shows the basic operation of an air conditioner.

In planning for reserve battery power, you should note that unless the air-conditioners can also be run off the emergency power, the base-station power amplifiers may exceed the base's operating temperature (usually around 60°C) and protection circuits will close down the base during power failures.

Table 16.1 shows the typical heat loads for 25-watts per-channel bases.

The air-conditioning load is the sum of the components shown in Table 16.2. Notice that the use of two air-conditioners (in case one fails) is a good idea.

Domestic air-conditioners are ordinarily too small for cellular base-station operation and units typically two to four times more powerful are required. Because of the high power consumption (typically 4 kilowatts), the air-conditioner motors are often three phase. This can be a problem in some areas, so a single-phase alternative should be available.

To conserve space, wall-mounted air-conditioner units are preferred, but these can be difficult to find, particularly if a single-phase motor is required.

CALCULATION OF HEAT LOADS AND LOSSES

In a well-designed shelter the heat loads will be largely equal to the cooling load presented by the equipment power consumption. There are, however, other losses that should be considered if the air-conditioning is to be optimized.

The following assumes that the equipment needs cooling to bring it to a desirable operating temperature, although the same principles can be applied to a cooling load.

A most useful concept for calculation of the heat flow through the walls is the thermal resistance, R, or the thermal resistivity per meter R_m. The heat loss through a conductor with a thermal resistance of R and applied temperature difference T_d is T_d/R watts. This is illustrated in Figure 16.2.

The resistance of slabs in series can be visualized as being equivalent to the resistance of a number of electrical resistances in series and can be added arithmetically. Where thermal resistances are in parallel this analogy also applies.

In a real situation it may be that a number of insulating materials are used in series and so their resistivities should be added. Consider the construction in Figure 16.3, which could be typical of a switch room building.

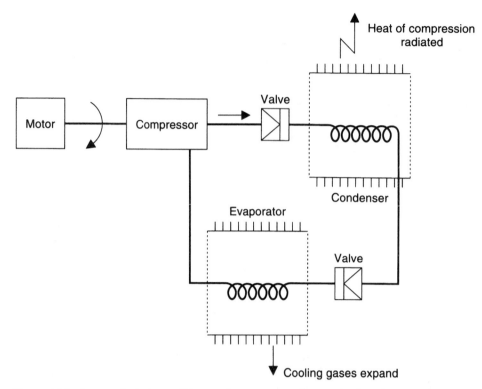

Figure 16.1 In a typical air-conditioner, the compressor pumps coolant into a compression chamber, where the generated heat is radiated to the atmosphere. Upon release from the chamber, the coolant expands to produce cooling.

TABLE 16.1   Heat Loads for 25-watt Bases*

Base Type	Heat Load
Radio channels*	180 watts/channel
Radio controllers	400 watts each
Rectifiers (100 A)	500 watts each
Exhaust fan	40 watts each
Microwave link	200 watts

* Increase radio-channel load proportionally for higher outputs.

Table 16.2 shows values for R and R_m for some commonly encountered building materials. From Table 16.2 the total resistance of the wall in Figure 16.3 is

Outside air film	0.03
Brick face	0.07
Air space × 2	0.3
Fiberboard	0.23
Mineral wool	1.9
Plasterboard	0.08
Inside air film	0.03
Total Resistance	2.64

per square meter

The outside dry-bulb temperature can be estimated from experience or can be obtained from the weather authorities. The temperature that should be used is the temperature that is not exceeded on 95 percent (or sometimes 97.5 percent) of the days.

The inside temperature is the comfortable operating temperature for the equipment and is usually taken to be about 22°C where heating is needed and 24–26°C if cooling is needed.

As an example consider a switch room with dimensions of 10 meters by 6 meters which is constructed with a wall as illustrated in Figure 16.3. The walls will be taken to be 3 meters high and the roof is lined with 75 mm of mineral wool.

Assume that the outside 95 percent air temperature is 28 degrees centigrade and the inside is to be maintained at 22 degrees.

The total wall area is $2 \times (10 + 6) \times 3$ or 96 square meters. The roof area is 10×6 or 60 square meters. We will ignore the floor loss for this calculation. So, the wall heat load is

$$96/2.64 \times (28 - 22) = 218 \text{ watts}$$

TABLE 16.2 Thermal Resistance

Material	$R_m{}^a$	R^b
Insulating		
Mineral fiber, batts 25 mm		0.6
Mineral fiber, batts 50 mm		1.25
Mineral fiber, batts 75 mm		1.9
Mineral fiber, batts 150 mm		3.8
Loose fill mineral fiber	23	
Interior materials		
Gypsum/plasterboard 15 mm		0.08
Soft wood	8.5	
Hardwood	6.0	
Plywood	8.6	
Particleboard 7.4		
Fiberboard 13 mm		0.23
Exterior walls		
Face brick	0.8	
House brick	1.4	
Stone	0.55	
Concrete	2.0	
Metal surface with 10-mm insulating backing	0.32	
Roofing Materials (except metal)	0.07	
Air Interface		
Surface air, still	0.13	
Surface air, moving	0.03	
Air space	0.15	

a Thermal resistivity/meter.

b Thermal resistance.

and the roof load is

$$60/(1.9 + 0.07) \times (28 - 22) = 182 \text{ watts}$$

Hence the total heat loss for this building is 400 watts. You will see later that this calculation would only be valid if the room in question were surrounded by air-conditioned rooms. Where this is not the case, an additional factor that depends on the floor length will be the most significant loss. Notice that if the roof were not insulated the heat load would be $60/0.07 \times (28 - 22) = 5000$ watts!

HEAT LOADS

Heat loads are the radiated effective powers of the various sources of heat. When the air-conditioning load is considered, however, some derating can occur for heat loads that are not radiating 24 hours a day. This may include most of the base-station load, lights, and human heat. The derating occurs because most of the radiation given off is first absorbed by the surrounding walls, floors, and hardware, and is only slowly reradiated. In this way the shelter will exhibit a thermal inertia. A derating factor for partial occupancy can be applied as follows:

Derating of Heat Loads

1. Base stations 75 percent of peak
2. People on site for 8 hours a day, 85 percent of peak

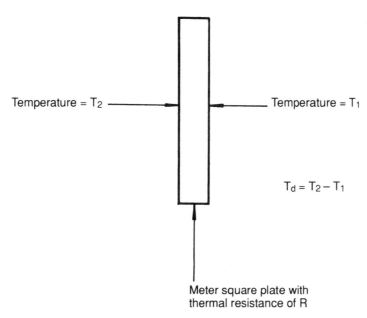

Figure 16.2 The heat flow through a conducting slab is proportional to the temperature difference.

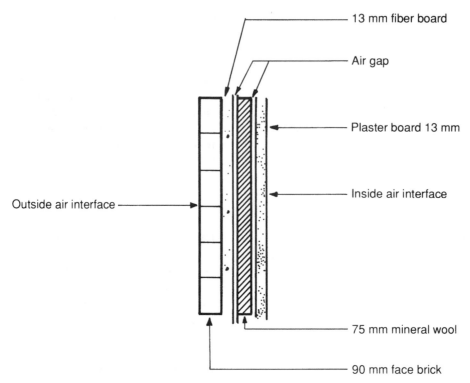

13 mm fiber board

Air gap

Plaster board 13 mm

Inside air interface

Outside air interface

75 mm mineral wool

90 mm face brick

Figure 16.3   The construction of a typical switch room wall.

Heat Load Sources

1. Switching heat losses = power consumption of the switch
2. Lighting losses = rated power of the light if incandescent or 1.2 × rated power if fluorescent (ballast power loss).
3. Rectifier loss = power delivered × (1 − efficiency). Assume an efficiency of 75 percent if unknown.
4. People, 150 watts
5. Other electrical equipment = rated power consumption.

The appropriate derating figures should be applied.

SOLAR HEAT

Most equipment shelters are exposed to the sun's radiation and will gain heat from that source. It is usual that the windows contribute most to solar heating and most switch rooms, and that base stations are designed to minimize this source of heat by having a minimum of windows.

SIMPLIFIED CALCULATIONS FOR OFFICES AND OTHER BUILDINGS

As an approximation Table 16.3 can be used to calculate the cooling loads for offices and other buildings in temperate climates. Notice that the heat losses here are greater than that of a purpose-designed switch room, and the main difference is the extensive insulation that is used in the latter.

Length-Dependent Losses

Where the floor is raised, as in the case of a hut on stilts, the floor losses are calculated in the same manner as the walls. For the more usual cellular floor construction of slab concrete on gravel, the floor area losses are more nearly proportional to the total perimeter than to the total area.

$$\text{The loss} = L \times (\text{outside temperature} - \text{inside temperature}) \times F$$

where

TABLE 16.3 Heat Losses for Offices and Other Buildings

Surface	Watts/Square Meter
Ceiling	20
Insulated ceiling	10
Floor to a non-air-conditioned space below	40
Slab floor or enclosed low-level space below	0
Windows	
External screens or vegetation	120
Double glazing or heavy drapes	120
Awnings	160
Internal blinds or drapes only	300
Plain window	400

If the window faces the setting sun, multiply above by 2.

If the window faces south in the southern hemisphere or north in the northern hemisphere, divide by 2.

For skylights multiply by 2.

L = the perimeter length in meters

F = the slab factor

The temperture is in degrees centigrade

Factor F varies from 1.4 watts/meter/degree for an uninsulated edge to 0.9 watt/meter/degree for an edge with 2.5 centimeters of insulation. Taking account of the slab loss, if the 6×10 room previously considered were on an uninsulated slab, then the slab losses would be $32 \times 1.4 \times (28 - 22)$ or 269 watts. So the total heat load for the air conditioner, being the sum of the wall, ceiling, and slab losses, is $400 + 358$ or 748 watts.

Remember to add heat generated by any sources in the building. For electrical appliances add the rated power of the device and allow 150 watts per person.

UNITS OF HEAT

Throughout this book the metric system is used and units of heat are measured in watts. In many texts and equipment specifications other units are used, and so it may be necessary to convert. Some conversions are

- 1 joule = 1 watt for 1 second = 1 watt second
- 1 kCal (1000 calories) = 1000 watt seconds
- 1 Btu = 0.252 watt
- 1 ton = 3024 watts

POWER CONDITIONING

There are many times when it is necessary to consider power conditioning in the cellular environment. The billing system must have a reliable and continuous power supply, as must other computers that are handling databases such as subscriber lists and customer data. A power failure that occurs while data are being manipulated can cause corruption of the database and perhaps even the irretrievable loss of valuable information.

Computers are notorious for their unreliability in areas where the consistency of the power service is poor. This is largely because most computers use switch-mode power supplies, which offer the circuitry that they power very little protection against power surges.

The most common form of backup for a computer is the uninterruptable power supply (UPS). This is usually a device consisting of a rectifier that charges a small sealed battery, which in turn drives an invertor. These devices usually have ratings of around 900 watts to 2000 watts, and are rated for approximately 30 minutes. Larger cabinet-sized units may have capacities of up to 15 KVA for 4 hours. As is illustrated in Figure 16.4, the conventional UPS is a simple double-conversion device where the incoming alternating current (AC) is converted to direct current (DC) which in turn is converted to AC.

Such a device is free-running so that in the event of a power failure the battery continues to provide the power to run the computer or switch. For typical computer usage, a backup time sufficient to finish the current job in such a way that data will not be lost as a result of the

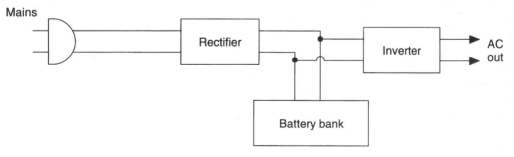

Figure 16.4 A typical UPS configuration.

power failure is all that is needed. Other equipment and the billing computer may be required to operate even during sustained power breaks and may therefore require more substantial backup times.

Since the efficiency of the typical double-conversion UPS is not high, (being typically 60 percent at full load), the efficiency is very load-dependent, decreasing rapidly at lesser loads. You should not rely on the implied ampere-hour capacity and so assume that for lighter loads the UPS will run longer. In practice, for most UPSs when run at 50 percent load you cannot expect much more than a 10 percent increase in backup time.

Big double-conversion units (more than 5 KVA) also tend to be noisy, to the extent that they cannot be placed in the vicinity of the workplace. Also, the low efficiency means high heat losses, which can be a strain on the air-conditioning. The heat load can be calculated directly from the efficiency, because the losses are all dissipated as heat.

FERRO-RESONANT UPS

A more efficient, but more costly solution, is the ferro-resonant UPS, which uses a saturated resonant transformer to clean up the incoming power line. It uses a flywheel effect to overcome short-duration voltage sags, and inherently dampens spikes, surges, and noise.

The ferro-resonant transformer can supply clean power for around 15 ms after loss of the primary power. In a properly designed system this will allow time for the standby inverter to start up and take over the supply. This is all accomplished without any noticeable loss of power, but it means that the inverter is only running when it is needed. Because of this, ferro-resonant UPS is much more efficient than the double-conversion system. Efficiencies of 90 percent can be expected, and so the heat load is much reduced. Noise is a lesser problem, and the inherent filter characteristic of the transformer eliminates the need for additional filters. The inverter, which is only run on a needs basis, will have a longer life and increased reliability.

Ferro-resonant conditioners are based on a constant voltage transformer principle. They are simple, reliable, and perform well in practice. They are (as the name implies) resonant, and so are dependent on the line frequency. Exporters should note that both 50-Hz and 60-Hz main power supplies are commonly used throughout the world and that a ferro-resonant conditioner will not work (without modification) at the wrong frequency.

An interesting aside here; the choice of 60 Hz had nothing to do with getting a convenient ratio for the second (3600 cycles of 60 Hz = 1 min but 3000 cycles at 50 Hz also equals 1 minute). The choice was based mainly on the fact that for technology limitations at the time low frequencies were preferred but anything much lower than 50 Hz produced a noticeable flicker incandescence lights.

UNINTERRUPTABLE BATTERY SUPPLIES

Uninterruptable battery supply (UBS) is a relatively new concept based on a combination of a ferro-resonant UPS, with the batteries backed up by a DC generator.

POWER STANDBY UNITS

Power standby units are similar to the UPS, but with one very important difference, which is that they operate on standby and only come into service when an actual power failure occurs. The time to come on-line is usually hundreds of milliseconds, and so this type of equipment is not suitable for protecting operational computers.

Power standby units come in sizes up to 15 KVA. They are often powered by standard vented (wet) lead acid cells and so require monthly maintenance. The maintenance must include inspection of the acid levels and the terminals, which need to be checked and greased.

The power output from these units often deviates significantly from an ideal sine wave, and so if the intention is to power computers or switches it will often be necessary to insert a line conditioner between the computer and the standby unit. This is necessary to prevent damage to the computer's power supply particularly during the changeover operation.

Sometimes power standby units can be reconfigured for continuous operation and so can operate as a UPS.

POWER CONDITIONERS

These devices are used to clean up the waveform and to limit the power excursions that can occur. There are basically two types. The electronic conditioners sense the line voltage and respond to fluctuations. As a result they do not have instantaneous response times, and they usually respond with step-function corrections.

THE DC DISTRIBUTION PANEL

The distribution panel incorporates safety, isolation, and battery protection features. The primary function of the

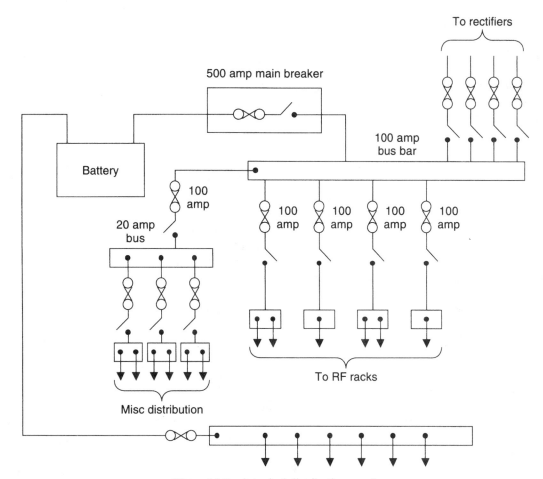

Figure 16.5 A typical distribution panel.

distribution panel is to distribute the DC supply to each of the bays in a way that enables the operator to isolate individual bays for servicing as required. A typical distribution panel is shown in Figure 16.5. Although this panel shows only the active battery lead being switched by the main circuitbreaker, it is usual to switch both the active and ground by a double pole-breaker. The fuses ordinarily would be circuit breakers.

DC–DC CONVERTER

Often it will be necessary to provide DC–DC conversion to allow 24-volt batteries to power 48-volt equipment and *vice versa*. Suitable converters are readily available in both 19-inch and 600-mm rack mounting. Depending on the equipment being powered and the consequences of its failure, 100 percent redundancy, with suitable isolation between the converters may be needed. In general, converters are considered to be reasonably reliable but

not sufficiently so that they can be used to provide power, without battery backup to service affecting modules like links or multiplexer (MUX) on a stand-alone basis.

RECTIFIERS AND BATTERIES

Rectifiers are available in the conventional transformer-coupled style and as switch-mode devices.

The conventional rectifier, while at least two to three times the size and weight of the switch-mode device is more reliable, particularly in areas where power conditioning is poor. Although in recent years the reliability of switch-mode rectifiers has increased significantly, a conventional rectifier can be more easily serviced, as it contains no really complex components and its components can readily be sourced. The transformer itself provides good isolation from power surges and transients. The mean time between failures (MTBF) of a well-engineered rectifier of this type is around 10 years.

Switched-mode supplies are generally serviced only by board replacement.

Increasingly the switched-mode rectifier is becoming popular for telecommunications equipment, and it has the advantage of being so small and light that a rectifier bay, complete with sealed batteries, can be positioned in the equipment suites in standard 600-mm bays. Not only does this enhance the appearance of the equipment and save space, but with attention to the placement of the rectifier bays, it is possible to make the DC power distribution losses very small indeed. Particularly when compared to the old-style battery room with its bus-bar distribution network, the savings in copper alone can amount to 20 percent of the total DC power costs.

On the down side, the complexity of switch-mode equipment is greatly increased by the fact that the supplies are virtually DC coupled. This means that in the event of failure of part of the system, repair is nearly impossible without access to dedicated repair facilities. Because of this, manufacturers of this type of equipment usually limit the internal construction to two or three boards at most and recommend that local repair be on a board-replacement basis. At least one manufacturer has recently produced a single-board rectifier so that repair simply amounts to replacing the board. When purchasing switch-mode rectifiers it is essential to order adequate supplies of spares and to be comfortable with the supplier's arrangements for component-level repair.

Whichever type of rectifier is used, it should have load-sharing capability so that the load is automatically shared between the available rectifiers.

The rectifiers have alarms for such conditions as power failure, low output, and rectifier failure. These outputs can be wired to the spare alarm positions in the base station so that they can be remotely monitored.

AC power can be provided either as single phase or three phase. If a three-phase supply is used, the load can be distributed (as equally as possible over the three phases). This must be done carefully because most base-station hardware is single phase.

Rectifiers can be supplied in rack sizes compatible with the cellular equipment. They usually come in modules of 25, 50, 100, or 200 watts (single phase), and somewhat larger in three phase.

Synchronous Rectifiers

As power supplies become increasing efficient, it has been determined that a good deal of the inefficiency of a modern rectifier is in the heat losses across the junction diode. A more efficient way to address this problem is to use switched field-effect transfer (FETs) as the rectifying device. Because the FETs may have a forward resistance as low as one-tenth that of a diode, heat losses are minimized. Naturally, the FETs will need to be driven by an appropriate switched circuitry that ensures that they only conduct current in one direction.

POWER RATING

Rectifiers will be available up to about 150 amps DC in single phase, but units from 150 amps to around 1000 amps will be available in three phase. The DC output can usually be increased to any desired level by using parallel rectifiers. If this is done, automatic load sharing is essential.

Ordinarily, two or more rectifiers are provided on a load-sharing basis, with provision to load share in the event of a rectifier failure (see Figure 16.6).

BATTERIES

Batteries are needed to keep the equipment functional during power failures. Sealed batteries are popular because of their low maintenance cost and flexibility in mounting arrangement.

The only sure way to determine the capacity of a battery is through a properly conducted discharge test. These tests involve discharging the battery at a standard rate (it could be, for example, 3, 5, 10, or 20 hours) into a dummy load until the battery voltage reaches some defined level (typically 1.88, 1.84, 1.80, or 1.75 volts per cell) the actual values may depend on the manufacturer's specifications or company policy. However the choice of any one of these levels will not greatly affect the derived ampere-hour rating, because once the battery gets near 1.9 V/cell, it will discharge to any nominal lower value very quickly. While the figures may vary, it is important to note that the battery testing should not be taken too far or permanent damage can occur. It is also important that the battery be recharged promptly after the testing, as the lead sulphate in discharged batteries can harden to the point where recharging no longer allows the reverse chemical reactions, which are

$$\{\text{Charge}\}\ \text{Pb} + \text{PbO}_2 + 2\text{H}_2\text{SO}_4$$
$$\Leftrightarrow 2\text{PbSO}_4 + 2\text{H}_2\text{O}\ \{\text{discharge}\}$$

These tests were once done religiously by PSTN operators, who had skilled personnel to carry them out. Apart from the skills needed, these tests were time-consuming and generally assumed that the batteries were in redundant banks, so that one bank could be taken out of service, while the other was tested.

Today's cellular operator probably has neither the

Figure 16.6   Three 100-amp conventional rectifiers working in a load-sharing mode.

skills or the time to undertake such tests. Nevertheless, a typical operation may have hundreds or even thousands of cell sites, all with battery backup. There is little point having batteries if it cannot be ascertained with a reasonable degree of surety that they will perform when called upon.

A traditional test for a lead-acid battery is to measure the specific gravity (SG), a test that can't be done with a sealed battery. The SG reading tells the state of *charge*, but not the capacity. In fact, a series of tests done on banks of telecommunications batteries and reported to the ITELEC conferences, have revealed that there is virtually no correlation between the SG of a battery and its capacity (measured in ampere-hours).

The state of charge as a function of SG is also a function of temperature. The following table gives approximate relationships:

Hydrometer Reading	State of Charge[a]
1.270–1.290	Full
1.215–1.230	Half
1.110–1.130	Discharged

[a]"State of Charge means relative charge given the actual ampere-hour value of battery.

The other common test is to measure the open-circuit voltage of the cells. This also has been shown to be uncorrelated with the capacity of the cell.

LOAD TESTING

The batteries should be load tested at least once a year. This involves connecting a dummy load to the batteries to discharge them at the 10-hour rate. Before the test all terminals should be checked and cleaned if necessary. The batteries should maintain at least 80 percent of their rated ampere-hour rating.

The conventional way to test batteries is to discharge them under heavy loads and to plot the cell voltages as the battery discharges. Generally this can be done without completely discharging the batteries, since to take them below 20 percent (approximately 1.8 volts/cell) of capacity can seriously decrease their net life.

When the batteries have been discharged it is necessary to recharge then separately (not in a series connection) before they are placed back on line.

If the battery is a wet cell, its (SG) should be checked prior to load testing. An SG of 1.265 and a battery voltage of 2.1 (with only small variations from cell to cell) can be considered normal, but does not really indicate the condition of the battery. The electrolyte level should be checked and topped up if necessary. Note that some loss of water is normal for a wet cell. Only distilled water should be used to top up the cells. Tap water contains impurities that will reduce the battery life.

Similarly, the generators should be load tested by being run at their full rated power into a dummy load for 4 hours at least once a year. These tests will ensure that when emergency power is needed it will be available.

Both battery and generator load tests should be logged in a permanent log book, which will enable trends to be established.

CONDUCTIVITY TESTING

To the rescue of the busy operator comes a simple test that with some degree of surety will identify faulty batteries and can be done by staff with limited training. Conductivity (the inverse of resistance) is a measure of the capacity of a battery.

A conductivity test is done in a few minutes using a device like the Midtronics tester seen in Figure 16.7. The conductance measured in mhos (Siemens in some countries) is correlated to the cell capacity strongly enough to be able to identify around 80 percent of faulty batteries. The tester measures the conductance at frequencies of around 25 Hz for telecommunications batteries and around 100 Hz for automobile batteries.

When using a conductivity tester it is preferable that the battery history be complete, in that conductance measurements were made on installation, and then periodic checks reveal loss of capacity as a decrease in conductance. When the historical records are not available a method that has been shown to have a good deal of validity by Midtronics is to measure a pool of batteries and tag the 20 percent with the lowest conductivity as potentially due for replacement. These replaced batteries can be removed and re-tested by the discharge method before discarding them all.

Conductance testing is a process that has been known since the 1970s and is still surprisingly little known today. This involves sending a pulse (or an AC current) through the battery to measure its conductance (see Figure 16.8). As the battery ages, the conductance, which is a measure of the efficiency of the plate surface, will decrease. However the actual decrease will not be the same for all batteries, which can make this test somewhat uncertain. The tester will work on batteries in service, as floating conditions in the telecommunications environment generally produces noise levels of less than 3 amps peak to peak, which the device has been designed to cope with.

Midtronics is a company that manufactures battery conductance testers for batteries from 10 to 2000 A · hs. A plot of the capacity of a battery bank, as shown in Figure 16.9, shows that measuring the voltage of each cell can fail to detect a few very weak cells, which have very low capacity. The plot of the conductance against the capacity in Figure 16.10, however, shows a strong correlation between cell capacity and conductance.

Batteries should be tested about once every 18 months; this applies equally to new and old batteries, as the infant mortality rate of batteries is rather high.

Another Look at Conductance Can the impedance of a battery really tell the condition of a battery? Knowing

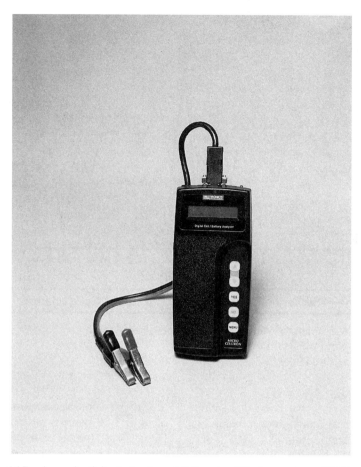

Figure 16.7   A conductivity tester from Midtronics. (Photo courtesy of Midtronics.)

that there are probably millions of batteries in service that have not received any on-going attention, it seems obvious that many of them are faulty. Will a simple conductance test find faulty batteries?

To test this proposition I arranged for a number of lead-acid car batteries that were known or suspected of being past their service life to be tested. The test was twofold. First a discharge at the 10-hour rate was done. All of these batteries were rated at over 40 ampere-hours, and all failed the long-term discharge test.

Next the batteries were checked with an auto-electrician's battery tester. The tester places a load equal to the rated cranking current (around 300–400 amperes) and passes or fails the battery depending on the terminal voltage while supplying the cranking voltage (see Figure 16.8). The results are seen in Table 16.4. Of four batteries tested that were known to be faulty, only one was correctly so identified. The conclusion to be drawn is that this test is not definitive.

It was suggested by the auto-electrician that the tests were done when the batteries had just been freshly charged and that if tested a few days later, many of those that had previously passed would fail. This was subsequently confirmed.

It is interesting to note that the battery in Table 16.4 is in my 4.1 liter Ford and was known to be low in capacity because leaving the lights on one day over lunch meant a dead battery. That battery also had the lowest

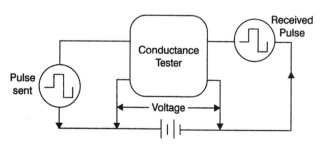

Figure 16.8   A conductance tester.

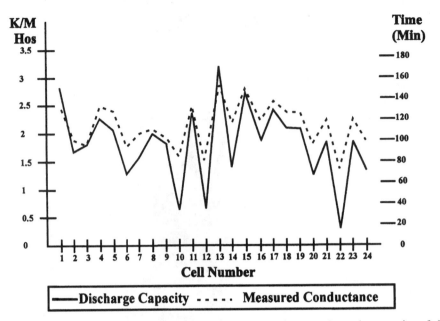

Figure 16.9 Measuring the voltage per cell does not reveal much about the capacity of that cell. (Graph courtesy of Midtronics.)

measured ampere-hour capacity. A year later, that same battery is still starting the car daily and it recently passed, once again, the conductance test.

Record Keeping Although this testing method is relatively simple, it really requires that a complete battery history be kept. A variation in the battery impedance of + or −20 percent indicates that something is wrong (a negative change in impedance probably indicates a shorted cell). Unless good records are kept of the batteries from new, then there is not a lot that can be concluded from conductance testing.

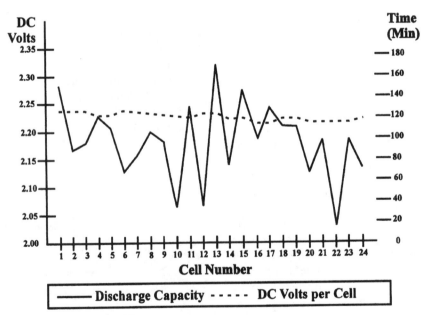

Figure 16.10 The conductance measurement corresponds well with the cell capacity. (Graph courtesy of Midtronics.)

Figure 16.11 A conventional auto-electrician's load tester, being used on my classic car.

Sealed Batteries

Sealed batteries are more properly known as valve-regulated lead-acid batteries (VRLA), and rely on stabilizing the electrolyte in a non-liquid form and pressuring the casing. The water decomposition that ordinarily accompanies battery charging and results in low water still

TABLE 16.4

Battery	Rated (peak current)	Voltage	Pass/Fail	A · h (measured)
Battery 1	300	10.2	P	15
Battery 2	400	8.0	F	26
Battery 3	280	10.2	P	20
Battery 4	430	10.5	P	Not measured, known faulty cell

occurs, but the battery is so arranged that the oxygen and hydrogen are recombined to maintain a constant water (electrolyte) volume. These batteries then fall into two main kinds: gelled electrolyte or absorbed glass matt (AGM). Gel batteries have a significantly longer life than the cheaper AGMs. Each has its place, but for critical applications, the gel battery is to be preferred.

A little-appreciated shortcoming of sealed batteries is their intolerance to high temperatures, which can see their service life halved for every 10 degrees rise in ambient temperature above 20°C. To some extent temperature rises can be mitigated by compensation of the float voltage, but temperature control is the most effective way. Conventional battery enclosures usually have ventilation to allow the gases to escape, while enclosures for sealed batteries rarely have such vents. Venting the gases may no longer be a problem, but heat accumulation is, and it is important to ensure that the enclosure has adequate ventilation, and that a 10-mm space (as specified in BS6133) is provided between cells and monocells.

Another problem that is exacerbated by heat is case expansion, in which the case enlarges with time, thus causing reduced capacity. This can cause loss of plate or separator contact with the electrolyte. Batteries designed for remote locations should therefore be steel cased (or reinforced).

Generally, VLRAs are not as reliable as the flooded cells they replace. VLRAs can fail in a number of ways, including drying out, grid corrosion, and loss of contact to the active material or internal corrosion. Any of these problems will cause loss of capacity. Additionally, value failures can hasten failure.

The nominal life of a telecommunications-grade sealed battery is 10 years, but general-purpose batteries are available that have a life of only 5 to 8 years. Regular commercial-grade batteries have a design life of only 3 years. It is important to be aware that all batteries are not created equal. The personal communication service (PCS) industry in particular, probably because of the large number of sites involved, seems to have been tempted to buy more on the basis of price than quality. However, even good-quality batteries often fail to meet their nominal lives.

BATTERY LIFE

The life of a sealed lead-acid battery (more correctly "sealed" batteries should be called value-regulated lead-acid batteries (VRLA), but "sealed" is the more commonly used term) can be as much as 10 years or as little as one. The factors that most affect the life are the temperature of operation and discharge cycling. Lead-acid batteries will achieve the longest life if kept at the correct float voltage at a constant temperature. Ironically, this means they last longest if they are never used.

Some batteries have to perform deep discharge cycles frequently. This may be the case in areas with unreliable power and in the case of solar-powered operation, where discharge starts as soon as the sun sets. Solar battery capacities are usually dimensioned for around five days of overcast sky, which means that the batteries on most days will discharge only partially. Unreliable power grids are becoming more common in developing countries where economic success has led to a ravenous demand for power, which cannot be met. The result is load-shedding, where parts of the city (or country) lose power for a few hours a day on a rotating basis, so that the generation capacity can be matched to the demand. This usually occurs during week days in business hours when the demand is at a peak. Unfortunately, this is also the peak demand period for mobile systems and so the battery discharge will be at a high level.

Battery life is defined as the time for the usable storage capacity to be reduced to 80 percent of the rated value. A typical battery may be able to withstand 3000 discharge cycles if the discharge is only 5 percent of capacity. If discharged to 100 percent, the same battery may achieve only 150 cycles. As a good approximation, these two points can be plotted on a linear graph to reveal the expected battery life for other discharge cycles.

Where regular and deep discharge is to be expected, it may be worthwhile to consider using tubular batteries. These are high maintenance, and require regular attention to the fluid levels, but being designed for frequent discharge applications (such as for electric forklifts), they will outlast conventional batteries very significantly. An alternative approach is to provide larger batteries than would be needed just to meet the load, so that the discharge will be relatively lower.

The correct float voltage varies with temperature, and if the batteries are stored in an unregulated temperature environment, they should be charged with a temperature-regulated charger. This will have a temperature probe, which is used to determine the float voltage placed adjacent to the batteries.

It should be noted here that premature failure of sealed batteries is rather common. This problem is especially prevalent in batteries that offer smaller size and weight for a given rated capacity. It is worthwhile asking for references before choosing a battery, as it is true to say that not all batteries were created equal.

Fuel Cells Fuel cells were first demonstrated by the British physicist William Grove in 1839. For more than a century they remained laboratory curiosities, largely because of technical difficulties and the costs involved in the technology. They were revived in the 1960s by NASA as light-weight but efficient power sources for the space program.

The most immediate problem was that hydrogen, the most common fuel for fuel cells, is difficult to store. Liquid hydrogen, which has a very high energy density, needs to be stored at a few degrees above absolute zero, which makes it impractical for portable use. Methanol, a liquid at room temperature, is a rich source of hydrogen, but the extraction of the hydrogen is a cumbersome and difficult process.

Although great advances have been made in recent times, the cost of a fuel cell is currently around $3000 to $4000 per kilowatt, which is high compared to the gas-fired turbines used by industry, which cost $500 to $1000 per kilowatt. The relatively high cost of the fuel cell is all in the manufacture, as the raw materials used in its construction are cheap, even compared to lead-acid and nickel cadmium (NiCad) batteries. Once the manufacturing complexities are sufficiently refined, the fuel cell may well be a cheap source of power for the future.

In its simplest form a fuel cell passes hydrogen through an anode, where the hydrogen atom is stripped of its electron. The hydrogen ion then passes through an electrolyte, to the cathode, leaving the electron to make its way to the cathode via an external circuit. At the cathode, oxygen that is diffused into it combines with the hydrogen ion to produce water as a by-product. This process is seen in Figure 16.12.

In the future fuel cells may well feature mobile phone applications running on methanol and producing about 20 times more power per "charge" than NiCad batteries of the same weight, leading to about one month between "charges." Additionally, the charging process would consist of nothing more than topping up the methanol level.

Fuel cells and conventional batteries both rely on chemical reactions, the main difference being that in a fuel cell, the active chemistry is provided by the fuels, while in conventional batteries, reactions of the electrode materials provide the power.

REDUNDANT BATTERIES

Some operators insist on having two battery banks in parallel (with half the reserve capacity each) to ensure against base-station failure in the event of the failure of one battery bank. A single cell or even a circuit breaker can cause battery-bank failure.

BATTERY AND RECTIFIER LOADING

You can calculate the rectifier load and battery load from heat loads

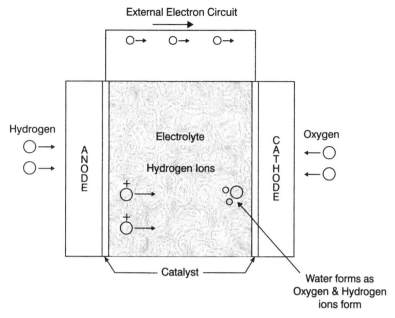

Figure 16.12   A hydrogen-powered fuel cell.

Rectifier load = Sum of heat loads

Battery load = Rectifier load + Air-conditioning
load (if included)

where the air-conditioners also run from a battery and
the approximate battery load = 2 × rectifier load (that
is, most electrical energy is dissipated as heat so the air-
conditioning load is approximately equal to the rectifier
load).

Some manufacturers produce equipment to temperature specifications that are sufficiently high so that air-
conditioning is not essential during power failures. This
results in significant savings in both batteries and emergency plant costs. It, however, results in a substantial
temperature cycle for the equipment, which has been
shown convincingly to increase MTBF.

Battery ampere-hour ratings are usually quoted for a
10-hour discharge. A mobile base is ordinarily equipped
with 2–3 hours battery reserve; at these higher discharge
rates, the battery ampere-hour capacity is reduced by 10
to 20 percent.

EMERGENCY PLANT

It is often necessary to provide a diesel generator set to
back up the base station. When such a generator plant is
provided, it is possible to reduce the battery capacity to
the point where it merely ensures continuous operation
from the time of power failure until the generator starts.
Ordinarily, the generator starts automatically when a
power failure occurs.

Generators for cellular base stations can be either
single phase or three phase, depending on the main
power supply. The generator is usually a diesel, and with
proper maintenance can be expected to give 10,000
hours of use if it is low revving (below 1800 rpm) and
about 5000 hours if it is a higher speed unit. In many
applications, the cellular operator can rightly consider
that the generator will receive only very light use, and
the prospect of installing a used unit may arise. Generally there are few disadvantages to doing this, and the
savings can be considerable. A used unit with less than
1000 hours of service, which has been properly maintained, may cost half as much as a new unit and give
virtually the same service life in the cellular environment.
Naturally maintenance costs will be a little higher than
could be expected for new machines, and there will be no
warranty, and these factors must be offset against the
capital savings.

Maintenance should include running the generator
once a month, ensuring that the fuel is not more than 12
months old, and following the procedures recommended
by the manufacturer. Full load tests, which may require
the use of a dummy load, should be conducted annually.

Gravity feed tanks can be very dangerous in the event
of a fuel-line failure, which could flood the equipment
room, and are therefore not recommended. The preferred delivery method is by a pump. To ensure that
sufficient fuel is available for starting the generator,

however, place a small gravity feed tank in series with the main fuel tank. The main fuel tank is best placed underground with a pump for fuel delivery. Dual pumps with manual changeover are a good idea.

Because it is costly to move a generator plant, it is best to purchase a unit that can run a fully equipped base station from the start. However, diesel plants do not perform well at partial load, and, unless expansions are foreseen within a few years, it may be necessary to plan to upgrade the generator plant at a future time.

Generators are usually rated in KVA. Unless the power factor is known, a figure of 0.7 should be used. Thus the generator rating, in KVA, is

$$\frac{\text{Watts}}{1000} \times PF \times E_f$$

where

Watts = the total power consumed by the base station

PF = power factor

E_f = rectifier efficiency (typically 70 to 80 percent)

Such a generator consumes about 0.3 liter fuel/KVA/h.

You should provide a fuel tank sufficiently large to provide a one-week backup (this can be tailored according to the reliability of the local power supply).

Diesel fuel does not keep indefinitely; do not plan to store it longer than six months.

DC DISTRIBUTION

Because the base stations may consume 10–100+ amps of current, the DC distribution system must be properly designed. In particular, it is important to provide switches that can isolate each piece of equipment used. This isolation of the batteries, rectifiers, and equipment bays is imperative. Heavy-duty copper cables, carrying around 60–100 amps each, are usually used for power distribution, with each radio frequency (RF) rack being individually supplied via a separate fused path equipped with a circuit breaker.

Use cables for rack wiring that are sufficient to carry the current safely. Table 16.5 shows the current capacity of various wire gauges. As the ambient temperature increases, a derating factor must be applied to the cable. Table 16.6 gives the appropriate derating. Table 16.6 assumes that not more than three separate conductors are placed in one cable or raceway. When more than three cables are bundled together, a further reduction in capacity occurs, as shown in Table 16.7.

TABLE 16.5 Copper Cables and Their Dimensions and Current Carrying Capacity at DC Continuous Rating at Room Temperature (30°C)

AWG	Metric	Maximum Current in amps	DIA (MM) 19-Strand Cable
26		1	
22	1	5	
18	1	10	1.16
14	1.5	17	1.84
12	2.5	23	2.32
10	4	33	2.95
8	6	45	3.7
6	10	60	4.67
4	16	80	5.9
2	25	100	7.42
1	35	125	8.43
0	50	150	9.47
00	70	175	10.06
000	70	200	11.9
0000	95	225	13.4

TABLE 16.6 Correction Factors for Higher Temperatures

Temperature (°C)	Derating Factor
40	0.82
45	0.71
50	0.58
55	0.41

TABLE 16.7 Derating Factor for Multiple Bunched Cables

Conductors in one Cable or Raceway	Derating Factor
4–6	0.8
7–24	0.7

CABLES

Cables used for carrying the DC supply should be adequate for the peak current expected and the maximum expected operational temperature. Table 16.5 shows the current-carrying capacity of various wire gauges at 30°C.

As the temperature increases, a derating factor should be applied as per Table 16.6.

A typical RF bay at a base station will require around 100 amps. If this is derated to 45°C (to allow for air-conditioning failure) then a derating factor of 0.71 would apply. That is, the cable should be able to carry 100/0.71, or 140 amps. This would make an American wire gauge (AWG) "0" suitable for wiring these bays. If a

Figure 16.13   Remote solar-powered base station.

slightly higher temperature is anticipated, an AWG "00" would be needed.

Since the main factor limiting a conductor's current-carrying capacity is heat dissipation, bunching conductors together will lead to mutual heating and so to effective derating. The derating factor to be applied to bunched cables is given in Table 16.7.

SOLAR-POWERED BASE STATIONS

Solar power is available in abundance. At noon the incident power is around 1 kW/m^2. Most solar cells are designed to deliver their rated peak voltage with an incident power level of around 200 watts/m^2, and so

can be used to charge storage batteries for most of the day.

It is possible and sometimes desirable to power remote sites with solar power. The base station shown in Figure 16.13 is a small omni rural base station on a hilltop, which is accessible only by a chair lift (or a long climb), and on which there is no main power. Generally, remote bases will be small, which makes solar more viable. In conventional solar design it is recommended that the battery reserve be five days. This is to allow for extended periods of cloudy days.

A typical solar panel that might be used in such an installation will be about 1000 mm × 500 mm × 40 mm and will be capable of producing a peak power of 75 watts at a nominal 12 volts (actually about 17 volts).

Two panels will need to be connected in series to obtain the 24 volts needed for a base station.

The most critical parameter for a solar panel is the watts produced per dollar. In 1999, panels were available for around $5.00 per watt. Prices are falling, but slowly. Care must be taken when selecting solar cells to buy units only from reputable manufacturers, as there are still some panels on the market that are very poorly made. When the storage battery price is added (noting that it needs to be good for five days), in most instances solar will not compare very favorably with other sources of power. The exception is in remote locations, where transporting generator fuel is a dominant cost.

Some cells claim to be self-regulating. This claim is based on the fact that the voltage never gets high enough to damage the battery, and thus it does not become overcharged. These cells generally do not perform very well in either very strong or very weak sunlight. They are not recommended for telecommunications applications and are mostly useful for trickle charging applications.

For cellular it will be necessary to use a voltage regulator. This regulator must be specifically designed for use with solar cells; otherwise, the efficiency will be severely compromised. The angle at which the cell is mounted is critical for high efficiency, and this will depend on the latitude at which it is being used. Contact the supplier for the correct angle for your site.

Silicon solar cells are made of a layer of N-type and a layer of P-type silicon, which when exposed to light produce a potential difference of around half a volt. Most solar panels are 12 volts. This voltage is obtained by stringing around 30 cells in series. In many panels the cells are physically different and easy to distinguish.

Given that in 1995 a typical commercial solar panel was around 8 percent to 10 percent efficient and maybe 12 percent efficient in 2000, but that laboratory panels with efficiences higher than 30 percent have been achieved, there is room for speculation that a new generation of high-efficiency, low-cost panels is not too far away. Analysts suggest that solar plants that can achieve a cost efficiency of 10 cents per kilowatt hour (approximately the cost of commercail electricity today) are about a decade away.

There are a few places where solar cells are just not practical. In areas exposed directly to strong winds from typhoons or cyclones, solar panels tend to get blown away. On remote islands inhabited by large numbers of sea birds, bird droppings can cover the panel very quickly, rendering it ineffective.

WIND GENERATORS

Wind generators are only rarely used for cellular power supplies, but they can be considered in high latitudes where strong steady winds are available. They also have been used on islands (even in the subtropics), where reliable sea breezes make them viable.

The theoretical output of a wind generator varies as the cube of the wind speed, but only as the height to the power of 0.14. From this it is evident that wind speed is the critical factor in the choice of this technology, and that high towers contribute little. Of course, the wind speed varies with the terrain and is higher on hilltops than on the flats.

THREE-PHASE POWER

The power supply may be single or three phase. If a three-phase supply is used to power the base station, it

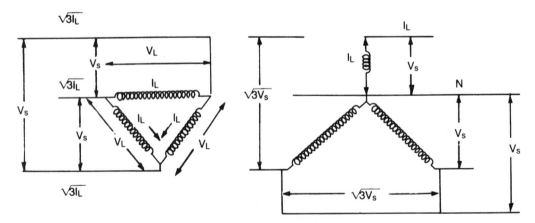

Figure 16.14 Power-grid three-phase supply configurations. Star- and delta-connected main power transformers showing the voltages from line-to-line and line-to-neutral as a function of V_S (the single-phase voltage, which is usually 110/220/240 volts) and the current relationship for three balanced (equal) loads.

can be used directly only if the rectifiers and air conditioners are three-phase units. Even in this case it is necessary to derive some single-phase supplies for auxiliary items such as the lights and power outlets.

Because three-phase power is not available at all locations, the decision to use three-phase equipment where appropriate means that the network will probably end up being a mix of three-phase and single-phase hardware. Consequently, spare rectifiers and air conditioners of both three-phase and single-phase-type must be kept.

Single-phase voltage feeds can be derived from three-phase supplies and the load shared between the phases. Figure 16.14 shows the relationship between the line and phase voltages of a three-phase supply from both star and delta transformers.

Providing a power inlet on the outside of the building so an emergency generator can be plugged in if required is a good idea. If an emergency generator is used, a suitable isolation switch should be provided at the switch board to enable the generator to be engaged and disengaged without danger to the operator or the equipment; a three-position switch that includes a neutral position will suffice.

Power companies generally supply bulk power in a three-phase form because in this form transmission costs are lower than for a single supply. Consider the delta connected supply in Figure 16.11. If all the loads are equal, then the power supplied to the three loads =

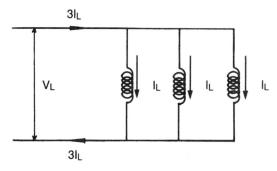

Figure 16.15   The single-phase equivalent of Figure 16.11.

$3 \times V_L \times I_L \times Pf$ (where Pf = power factor). The corresponding current carried by each of the three feeders is $\sqrt{3I_L}$.

If these loads were supplied by a single-phase line, the line current would be $3 \times I_L$, as shown in Figure 16.15. Hence, two conductors carrying $3I_L$ each would be necessary. The conductor size is a function of the current carried. If $I_L = 10$ amps, then the single-phase conductors would need to be AWG 6 or 4.13 mm. A 1-km feed length would require 237 kg of copper (two wires).

A three-phase conductor feeding the same load would need to carry $\sqrt{3} \times 20$, or 34 amps, and would be AWG 10, or 2.59 mm. A 1-km feed length would require only 141 kg of copper.

17

PROTECTION AND GROUNDING

LIGHTNING PROTECTION

Lightning strikes the earth at a rate of about 1000 times per second. It kills five people per year in the UK, 14 in Japan, and 90 in the United States. About 30 percent of strikes to humans are fatal, and death usually results from failure to breathe. On being struck, both the heart and lungs are affected; the heart recovers but breathing doesn't, and thus the heart stops again due to lack of oxygen. To assist a victim, immediately begin cardiac pulmonary resuscitation. Those struck but not killed by a strike usually report effects ranging from "being thumped in the back" to "no memory of anything." Survivors usually have no serious long-term problems; only about 5 percent do. During the last 15 years it has been responsible for the loss of nine aircraft in the United States and has caused damage to aircraft ranging from minor pitting to punching 20-centimeter holes in thousands of others. Lightning most frequently strikes as a result of storms but has been recorded in clear, blue sky conditions. Damage is most prevalent when the soil is of low conductivity. Granite hills are particularly vulnerable locations. A radio antenna on top of a tower is especially at risk for anything from minor pitting to total vaporization.

The rate at which lightning strikes regionally is strongly correlated to the average temperature, which is why tropical regions are far more prone to strikes than temperate areas. The sensitivity to temperature is very high, as testing in Darwin, a tropical city in northern Australia, showed. During a period when the average temperature was only 2 degrees higher than normal, the rate of lightning strikes went up 100-fold. Global warming therefore also means an increasing incidence of lightning activity worldwide.

For many centuries the most likely target for lightning was the village church, which was usually the tallest building in the town. Long after Franklin had clearly established that effective protection could be made available with the installation of a simple lightning rod, many churches refused to install the devices, claiming that lightning strikes were the will of God and nothing should be done to interfere. Many churches, like the one shown in Figure 17.1, were needlessly destroyed before this policy was abandoned. This spire carries a plaque reading "The spire of St. Bride's church, Fleet St. London. Removed in 1764 due to lightning and erected here, Park Place Estate in 1837." It now stands in an open paddock, only a few hundred meters from a tower jointly used for base stations of British Telecom and Racal-Vodaphone.

LIGHTNING CHARGES

The mechanism of lightning charge generation has recently come to light. At the base of a cloud small droplets of ice crystals form, which rise upward in the drafts. As they rise, they collide with larger particles of hail, and in the ensuing collision, the smaller particles acquire a positive charge (leaving the hail with a negative charge). The light particles continue to rise after the collision. Recent studies indicate that the charge is transferred by a thin layer of "melted ice", that is a few hundred atoms thick. This "surface melting" occurs on virtually all solids and persists at temperatures below the freezing point of the solid. As a result of this the cloud builds up a positive charge near the top and a negative one on the bottom.

Figure 17.1 The remains of a London church, struck by lightning in 1764 and reassembled in a field near a modern base-station site.

A fully developed thundercloud will generally be positively charged at the top and negatively in the middle. Thunderclouds or cells generally occur in clusters, arising independently of each other. Cloud-to-cloud discharge is common, but it is the cloud-to-earth discharges that are of most interest to us. A local region of positive charge at the bottom of the cloud is often noted, but it is not particularly important for this consideration of the subject.

The bottom of the cloud will typically be at a potential of around 20–100 MeV relative to the ground. A strike happens when a small initial discharge makes its way along a "leader" that conducts some negative charge to the ground at around one-sixth of the speed of light, in a stepwise fashion moving about 50 meters on each step. As the leader leaves behind it an ionized high-conductivity path, it prepares the way for the main strike, which starts when the leader initially contacts the ground, and travels *upwards* from the ground toward the cloud.

Once the first stroke has occurred others may follow at intervals of around a few hundredths of a second.

The strike will typically carry around 10,000 amperes and deliver to the ground a total charge of 20 coulombs. About 90 percent of strikes deliver a net negative charge. The cloud having discharged, can build up for a new strike in as little as 5 seconds.

A lightning strike can consist of a current pulse of more than 200 kA, with a pulse width from 10 to 350 microseconds. In order to sink the current from the strike, it is essential that a good ground be provided for the lightning rod, and that the connecting cable be adequate to carry the surge current. The waveform of the pulse is such that even the inductance of a straight cable will be significant, and any bends or kinks may present themselves as high impedances.

STATIC AIR CHARGES

The air has a nearly constant field potential of around 100 volts per meter (measured vertically). This potential gradient is continuous up to about 50 kilometers, where a region of high conductivity prevents a further buildup of voltage. At this altitude the voltage is around 400,000 volts with respect to the earth. The natural conductivity of the air, which is brought about largely by particles ionized by cosmic rays, ensures that this voltage difference causes a current of the order of 1800 amperes to be flowing continuously between the air and the ground. This is the equivalent of a 700-megawatt generator!

The source of power for this field-strength generator is the lightning discharges delivered continuously over the surface of the earth. The field peaks (by about 15 percent) at 7.00 P.M. universal time (UT) and is at its minimum at 4.00 A.M. UT. Notice that the field follows the intensity of the thunderstorm activity worldwide, and that the peak occurs everywhere at the same time (local time is not a factor).

When the air is dry the conductivity goes down and the potentials build up. These gradients can destroy high-impedance, modern integrated circuits (ICs) in fractions of a second. All printed circuit (PC) boards should be treated as static sensitive, and the proper handling procedures should be followed. Damage to ICs can be immediate, but sometimes internal damage can occur that may take months or even years to lead to ultimate failure.

Precautions that must be observed are:

- Transport all PC boards in conductive plastic bags.
- When handling PC boards ensure that a wrist strap is used and that conductive mats are used on the benches
- Ensure that wrist straps and connection points are fitted to all equipment racks
- Keep all spare ICs on conductive foam-rubber sheets. Aluminum foil is not as effective, as some leads may not be contacting the metal and the foil will transfer charges very efficiently during wrapping and unwrapping.

Figure 17.2   A 1850s gunpowder magazine in Beechworth, Australia, with the original lightning protection.

Dry conditions also allow local charges to build up on insulated objects like carpets and clothing. Discharges from these sources can induce very high instantaneous currents, which can permanently damage ICs by puncturing semiconductor surfaces.

Basic Strategy

The basic strategy for handling lightning, as expounded by Erico, is

- Capture the strike at the preferred point using purpose-designed air terminals
- Conduct the strike to ground using a purpose-designed down conductor that minimizes side-flashing
- Dissipate the energy into the ground with a minimal rise in ground voltage through a low-impedance grounding system
- Eliminate ground loops and differentials by creating an equipotential grounding plane under transient conditions. Ensure all grounds are tied together!
- Protect the equipment from any induced lightning surges as well as power-line-induced surges and transients

- Protect the equipment from surges and transients on communications and signal lines

Lightning Rods

An early example of the use of lightning rods is seen in Figure 17.2, where a pair of lightning rods protrude above the roof of an 1850s government gunpowder store located in Victoria, Australia. The rods and the bar connecting them are made of mild steel, and there are quite a number of joins and sharp turns in the bars. Like too many installations today, they are also bonded to independent grounding points.

Lightning protection works by providing a path of low resistance for a lightning strike. For this reason, the lightning conductor must be the highest point on the structure and have a good path to the ground. That path is best provided by copper straps.

Since the days of Benjamin Franklin (who invented the lightning rod), some people have argued that the protection of a lightning rod comes from "discharging" the atmosphere around the tower and preventing strikes by preventing static buildup. This belief has been soundly disproved by every generation since Franklin, but it still persists. Consequently there is no evidence that

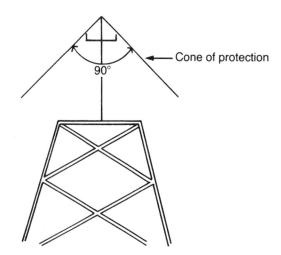

Figure 17.4 Lightning rods can be envisioned as providing a "cone" of protection to an area beneath them.

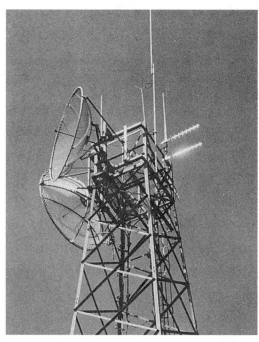

Figure 17.3 An omni mounted well above the lightning rod is tempting fate.

pointed lighting rods work better than rounded ones. Even today some manufacturers claim to produce such devices.

A very dangerous variant of the "discharge" type of lightning rod, which was popular a few decades ago, was the *radioactive* prong rod. In this instance a small capsule of radioactive material was embedded in the end of the prongs. The theory was that the radiation thus produced would ionize the air adjacent to it and so provide a low-resistance path for the discharge. Although the amount of radiation from such a rod was minimal most manufacturers have discontinued the production of such devices. Today where the old rods are rapidly decaying to the point where the radioactive capsules are exposed (or may even fall to the ground), the hazard is greatest.

Lightning rods are the first line of defense. They should be placed to minimize potential interference with radio frequency (RF) propagation and to maximize protection. No matter how often the warning is repeated, there is always somebody willing to tempt fate—as in the installation in Figure 17.3.

The "zone of protection" can be defined as the 90-degree cone around the antenna, as shown in Figure 17.4. The protected antennas should be inside the "cone" of protection described by the lightning rod.

With the number of antennas on even a medium-sized cellular installation, it is often difficult to find a place to put a lightning rod where it will not cause significant pattern distortion. For this reason, many operators dis-

pense with the lightning rods and rely on direct-current (DC) ground-potential antennas (antennas designed to withstand lightning discharges). This approach seems to work reasonably well.

Because the antenna feeders have large-diameter copper shields, they can make very attractive lightning paths. To reduce possible equipment damage, it is good practice to ground the feeders at the top and bottom of the tower and every 20 meters, as well as at the entry point to the building structure.

Lightning protection is an essential consideration and all antenna-mounting structures should at least be fitted with a lightning rod that is well grounded (directly to a proper ground via a copper strap). The tower itself, as a minimum requirement, should be grounded at each leg to separate grounding rods. The rods should be tied together with a buried busbar, and the earth-ring should be tied to the building earth, as shown in Figure 17.5.

GROUNDING

In order to protect personnel and equipment it is essential that all installations are properly grounded. A good ground will minimize damage from power surges, lightning strikes, as well as from noise and interference.

The golden rule of grounding is to avoid ground or earth loops. To do so it is essential that all grounds are firmly bonded together by straps of adequate current-carrying capacity. A typical ground loop, and one that can be quite dangerous, occurs between an unprotected telephone and its remote connection to the public switched telephone network (PSTN).

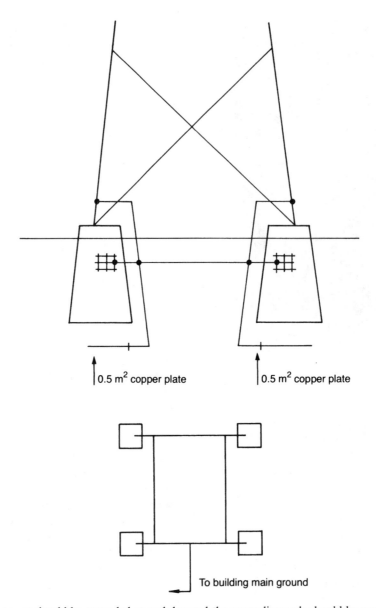

Figure 17.5 A tower should be grounded at each leg and the grounding rods should be connected together.

As seen in Figure 17.6 a lightning strike to a distant switch will cause a local potential rise that can amount to several thousand volts. Provided all the hardware in the distant switch is properly grounded and those grounds are bonded, there should be no damage done at the switch. However, the user of an unprotected telephone line can provide an unintended ground loop with disastrous consequences. Statistics from Australia, with 10 million telephone lines, indicate that 80 people per year receive shocks in the manner indicated in Figure 17.6. To date, the results have been burns, jolts, and hearing problems, but no deaths.

Good practice at all switch and base sites is to install a ground ring, which is bonded to the ground by grounding rods, around each structure, spaced at intervals of 2–4 meters (depending on soil resistivity). A typical ring grounding is depicted in Figure 17.7.

To ensure a low-resistance bonding, all joins should be exothermically welded, and tinned copper wire of gauge AWG 2 or larger should be used. Mechanically bonded joints should not be used for below-ground bonding.

Notice that it is very important to avoid direct connection of bare copper wire with galvanized steel, as the

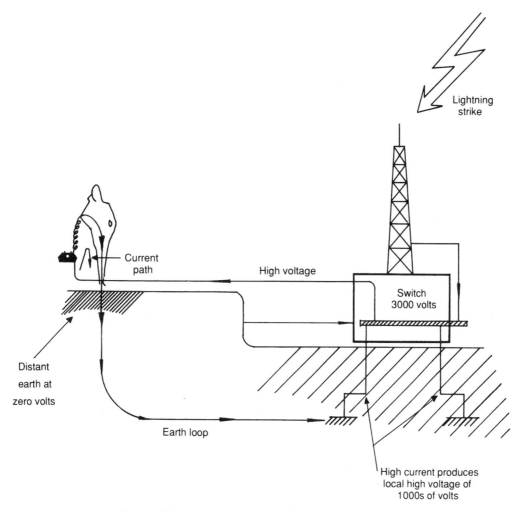

Figure 17.6   An unintended and dangerous ground loop.

combination will cause a serious corrosive, potential difference.

The 2-meter radials are there to assist the path to ground of large surges. The radials should be connected to 2-meter grounding rods at each end. The rods should consist of 2-meter, 15-mm-diameter copper-clad steel rods. Stainless steel can be substituted if the rods are to be used near large mild-steel structures with which copper may case corrosion problems.

The ring should be connected to the base station at 2-meter intervals by a gauge AWG 2 copper cable that is placed in a polyvinyl chloride (PVC) pipe from the connection to the ring to a point at least 150 mm above the soil surface.

The tower should be connected to the ring at each leg in the manner described for the base stations. If a monopole is used, it should be connected to the ring at four different points around its circumference.

Avoid sharp bends, since these can present a high inductance to surges. All bends should have a radius of at least 300 mm.

INTERNAL GROUNDING

An internal ground ring that completely loops around the inside walls of the base station at a height of 2 to 2.3 meters should be provided for internal grounding. Ideally this will consist of a flat copper busbar of cross section 15 mm × 5 mm, otherwise a bare wire of gauge AWG 2 can be used.

This internal ring can be connected to a number of smaller busbars, which are designed as ground terminating points. They consist of a flat copper plate of approximate dimensions 250 mm × 100 mm × 8 mm, and

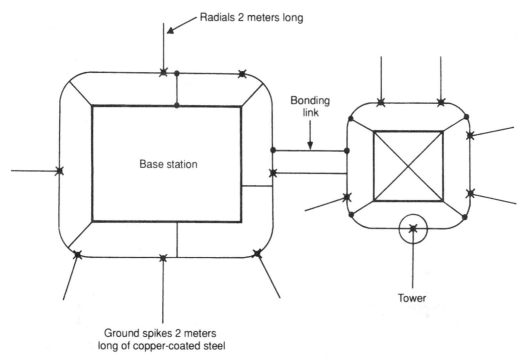

Figure 17.7 A grounding ring. Note that the tower and base-station rings are bonded together.

with holes drilled in them to make convenient connection points for straps to the equipment to be grounded.

A typical ground bar is depicted in Figure 17.8.

All of the equipment racks should be connected to one of these ground bars using stranded 25 mm (AWG 6) copper wire. The racks should also be connected to each other by strapping each rack to the next. In order to avoid ground loops, it is advisable to preclude other grounds by insulating the racks from the floor. The cable tray should also be connected, preferably at a number of different points.

Grounding bars should be placed as needed and are usually drilled with spare holes for additional future connections, as seen in Figure 17.9.

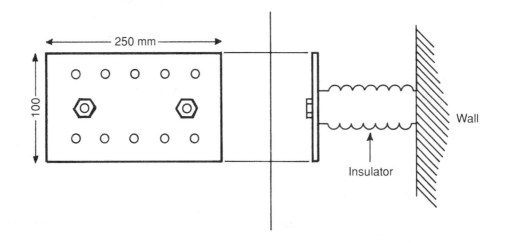

Figure 17.8 An internal ground bar.

Figure 17.9 A grounding plate with holes for future connections.

Cable Grounding

All RF cables should be well grounded at the top of the tower, the bottom of the tower, and at the point of entry to the building.

The cable window should have a grounding plate on either side so the cable sheath can be grounded at the external point of entry, and on the internal side there will be a good direct ground for the gas arresters.

All other cables, such as telephone cables, inter-building cables, and power cables, need gas-discharge protection.

For telephone lines, the usual solid-state protectors should not be regarded as sufficient, and three-leg gas-discharge devices should be used.

Avoid Induced Currents

When a surge occurs on a grounding cable, it will induce high currents into any nearby cables. Always run grounding cables as far as possible away from any other cables. This also applies to power cables, which can carry large surges; isolate them form any other cables. Power cables should also be separated from grounding cables. *Never tie grounding, power, or signal cables into a bundle! Keep these three types of cables at least 200 mm away from each other.*

High-Resistivity Areas

It is possible to get special grounding compounds that can be used to reduce the ground resistance by as much as 50 percent. These take the form of nonsoluble gels, which surround the grounding rods and straps.

Grounding Base Stations on High Buildings

It is not adequate for an installation on a tall building to simply run a long grounding cable to an external ground (of basement). The impedance of that long cable at the transient pulse rate of a lightning pulse will be high. It is essential that the high-rise installations have their grounding firmly bonded to all other building and utilities grounding systems.

If a 10 mm round wire is run 20 floors, each 3 meters (total 60 m), then we find it will have an inductance of 103 mH. If we assume a strike of 200 kA in 100 microseconds, then the voltage developed across this wire is

$$L \times di/dt = 103 \times 10^{-6} \times 200 \times 10^{3}/10^{-4}$$
$$= 206,000 \text{ volts!}$$

Where a good low-resistance ground proves difficult to achieve a "Ufer" ground can be considered. This type of ground consists of a cable embedded in concrete. It works on the principle that concrete is highly porous and will readily retain the moisture that it can absorb from the surrounding soil. An improvised Ufer ground can be obtained in low-rise buildings by attaching to the reinforcing bars of the concrete. To be effective a number of different connection points are usually necessary. This technique is of little use in buildings more than a few storys high.

Acceptable Ground Resistance

The resistance of the grounding should be less than 10 ohms (although 5 ohms would be preferable), when measured with a null-balance earth-resistance test set.

It should be noted that the conventional methods of measuring ground resistance measure only the low-frequency impedance. Lightning strikes are high-frequency pulses, and the impedance that they see may be very different than the "ground resistance" as measured by conventional methods. Most transients are high-frequency signals, and will encounter the reactance of the ground circuit, which will generally be much larger than the low-frequency resistance. Additionally, there are capacitive and inductive parts of the grounding circuitry.

Take the example of a copper grounding circuit made of 4-mm diameter copper wire. A 3-meter straight length of this wire encountering a current with a rise time of 10,000 amps/microsecond will drop 40 volts due to its resistance and 40,000 volts because of its inductance. The overriding importance of the inductive part suggests that every effort should be made to minimize turns and loops in the wire, and runs should be kept as short as possible. Splices should avoid sharp intersections, as shown in Figure 17.10.

Discontinuities in the cable run caused by joints, bends, or even proximity to alternative ground paths can cause reflections, which may exacerbate the damage.

Figure 17.10 Avoid abrupt intersections on grounding runs.

It should be noted that for RF applications the same inductive problems occur. The difference is that transient currents are typically around 1 MHz and high current, whereas RF signals are much higher frequency but of lower current, but nevertheless, RF ground path impedances may be much higher than the DC (or low frequency) resistance used to measure grounding paths. RF grounding problems will often manifest themselves as noise and interference.

Metal clamps and steel supports will increase the inductance of the wire and should therefore be avoided. Plastic clamps and cable ties are preferred. If the conductor must pass through a metal conduit, ensure that it is connected to the grounding cable at both ends.

Low-frequency devices measure the impedance not only of the point being measured, but of all the other grounds that are bonded to it, no matter how distant. In fact, the whole Earth is part of a grounding network at low frequencies. Because the duration of a strike may be very brief, in practice the bonded grounds may play no part in dissipating the energy. High-frequency pulsed ground impedance testers, measure the impedance seen by a typical strike. This is often 2 to 6 times greater than the low-frequency reading.

MEASURING GROUND RESISTANCE

The ground resistance can vary from a few ohms to hundreds of ohms. to ensure that the grounding resistance meets minimum standards (often a minimum of 10 ohms), it is necessary to undertake a series of measurements. Ground resistance cannot be measured easily with DC, because the soil will generally contain some electrolytes that will cause a DC voltage to arise between grounding rods.

Purpose-built AC ground measuring devices, such as the one shown in Figure 17.11, can generally measure down to small fractions of an ohm (in the case of the device pictured down to 0.02 ohms) and up to several thousand ohms. The device depicted generates a 108 Hz voltage for the measurement.

Clearly there is no way a single ground-point resis-

tance can be measured without reference to some other point. Even with two grounding points, only their sum can be obtained. However if three points are measured, it is possible to get a set of simultaneous equations that can be solved to give the individual ground impedances.

If the three grounding points are designated A, B and C respectively, we can get the sum of the resistance between any two as:

$$R_a + R_b = R_{ab}$$
$$R_a + R_c = R_{ac}$$
$$R_b + R_c = R_{bc}$$

Then solving this as a set of simultaneous equations, we find:

$$RA = 1/2(R_{ac} + R_{ab} - R_{cb})$$
$$RB = 1/2(R_{ab} + R_{bc} - R_{ab})$$
$$RC = 1/2(R_{ac} + R_{bc} - R_{ab})$$

If three grounding points are not available, then the additional measuring points can be made by temporarily inserting rods to a depth of about 0.5 meters at a distance of 5 m or more from existing grounding points. It should also be noted that once the measurement of any three points has been completed, the known resistance of any one of those points may then be used to measure any others.

GROUND-LOOP CURRENTS

Tests show that even an apparently well engineered grounding system often has significant circulating currents within the grounding system. As the grounding system is made of heavy gauge copper, all points connected to it should be at the same potential and therefore there should be no circulating currents. Any loop currents are indicative of high resistance joints or induced currents. These are the potential weak spots when a strike occurs. Accu-Scan Corporation has a device to monitor these currents. It has 16 clamp-on Hall-effect devices, which can do a detailed long-term monitoring of the circulating currents. Those due to high resistance joints will show up readily, whereas some will be intermittent and occur only when the offending devices (such as generators and standby power supplies) are operated.

Transient and Surge Protection

Irregular line voltage conditions can be caused by lightning, power-system faults, electrostatic discharge, or

Figure 17.11   AC ground measuring device.

radio-frequency interference. Modern semiconductor components are extremely vulnerable to damage from two of these effects.

The most severe and most probable source of irregular line voltages in cellular systems is lightning, which can easily find its way to the sensitive hardware by way of the antennas, power cables, or land-line links.

The next most likely problem is electrostatic discharges. These can be the most insidious since the damage may not actually cause immediate device failure, but rather may lead to drastically reduced device lifetimes.

The magnitude of the problem for telecommunications companies has been increasing with the sophistication of the hardware used as can be seen from Table 17.1.

Transient Surge Detection

Surge diverters are a must in any telecommunications network, but the diverter itself can be a hazard under some power-line faults and operating conditions. In a case experienced by the author, a base station was found to have the diverter units melted and obvious evidence of intense heat behind the diverter unit. However, there was no evidence of a strike, and it seemed more likely that the surge protection had been activated by excessive supply-line voltage, which had caused the transorbs to conduct and overheat.

The diverter was replaced with one rated at a 30 percent higher voltage, and no further problems were noted.

In rural areas and areas where power regulation is poor, this problem can occur. One solution is provided by Erico in the form of a transient surge discrimination protector. This device switches in the protection only when a surge is detected. It will ignore the slower voltage swings caused by the power grid, allowing supply swings of 48 volts.

Basic Protection

Clamping devices offer a cheap way to provide a reasonably good level of protection. There are two main types of clamping device: those that contain a spark gap and divert the surge directly to ground, and varistors and such devices, which absorb a good deal of the surge.

Spark-gap devices act quickly (within the order of 100 billionths of a second), but they ordinarily have rather high clamping voltages (of the order of thousands of volts). This means they offer a good first-order protect,

TABLE 17.1   Energy to Damage

Compenent	Damage Energy in Joules for a 1-Microsecond Pulse	Technology Timeframe
VLSI/ASIC	10^{-6} to 10^{-8}	1980 to present
Low-noise transistors and FETs	10^{-7} to 10^{-8}	1980 to present
Digital ICs	10^{-6} to 10^{-3}	1970–1980s
High-power transistor	10^{-3} to 10^{-1}	1960s to present
Wire-wound resistor	10^{-1} to 10	current
Valves	10^{-1} to 10	1900–1960s

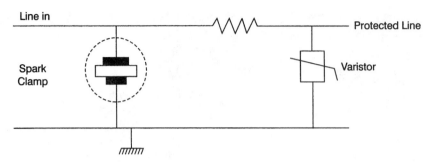

Figure 17.12   A basic two stage surge protector.

but they can still pass voltages that would be harmful to most equipment. Because of this they are best used in combination with varistors or transorbs, as seen in Figure 17.12, because these devices will clamp at much lower voltages. The π circuit shows that the varistor is isolated from the spark-gap clamp by a resistor.

Power-Line Protection Scheme

- Provide adequate lightning protection and good grounding
- Bond all grounding systems together
- Protect all power lines connected to your equipment
- Protect all RF cables at the point of entry
- Protect all data cables and land-lines entering the building
- Eliminate ground loops

There are a variety of protection devices available:

Power-Line Surge-Reduction Filters   These are placed in series with the lines and offer common and differential mode protection. They typically are rated at 100 kA per phase and are available in load ratings from around 10 amps to 1000 amps. The performance of the filter is virtually independent of the actual load. It is essential that good grounding practices be followed.

In principle these filters are similar to the familiar π RF filter; a circuit diagram for one is seen in Figure 17.13.

In practice the problem is to have capacitors and inductors that can stand the high voltage and current surges. The inductor in particular has a tendency to fly apart under the stress of very high currents. Figure 17.14 shows a three-phase surge filter suitable for the protection of a base-station power feed.

Power-Line Shunt Protection   The protection offered by shunt devices is limited and requires very good installation practices to be effective at all. The clamping can leave relatively high residual voltages.

Power-Line Filters   Power-line filters are installed in series with the equipment to be protected. Being similar in principle to power-line surge protectors, both common-mode and differential-mode protection is provided. These devices are often small self-contained plastic or metal boxes that rely directly on the ground from the power outlet. Typical power ratings are 1 to 15 amps, with surge ratings of around 5 kA.

True Transient Protection   Most surge protectors are designed to switch in when the line voltage exceeds about 115 percent of its nominal value. While this is adequate in most locations, there are rural areas where transition beyond 115 percent are common. When this happens, failure of the protection devices can occur, because the frequent transitions cause overheating, and finally meltdown of the protection device; sometimes there may even be a fire if the overvoltage is sufficiently high.

A solution to this problem is to cause the surge protection to *switch* in on a voltage transient rather than an absolute voltage. Swings in the power-line voltage will

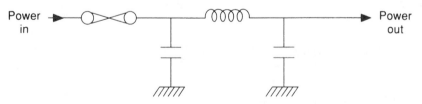

Figure 17.13   The circuit of a single-phase surge filter.

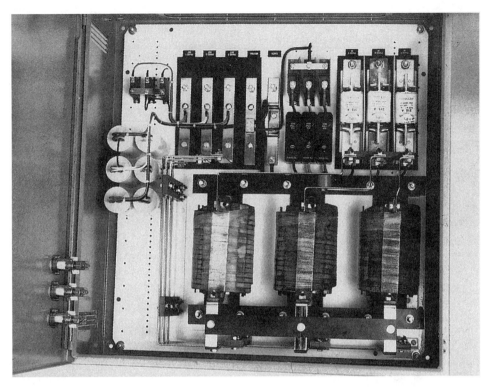

Figure 17.14   A three-phase surge filter. (Photo courtesy of Erico Lightning Technologies.)

be slow, so they will not be clamped, but real surges are fast and so can be easily discriminated. Erico Lightning Technologies has such a product, which they call "Quick Switch."

Line Conditioners   The main purpose of a line conditioner is to regulate the supply voltage to the protected equipment. A common version uses a transformer with a number of taps that can be connected as required to obtain the desired power output.

Communications Lines Protection

First Level   Simple shunt devices such as gas discharge tubes can provide some protection. The devices are slow-acting but cheap. They are usually found on the subscriber's termination point of the MDF at land-line telephone switches.

Second Level   A combination of series transorbers and shunt gas arresters can significantly improve the line protection. These are still relatively cheap but not completely reliable.

Third Level   Complex combinations of transorbers, gas arresters, and filters are used when maximum protection is required.

Working Voltages

Telephone lines are usually clamped at 200 volts. This is largely because of the ringing voltage, which can reach 130 volts. Other clamping levels are 7.5, 15 (for modems), 30 (RS232), 68, and 135 volts.

Coaxial-Line Protection   Coaxial-line protectors are available with clamping voltages from 90 to 1000 volts and can be provided with N, BNC, UHF, and IBM connectors.

Transient Ground Clamps   Where the local regulations provide for isolated grounding, an extra degree of security can be provided by placing earth clamps between the earth systems, which are normally open-circuit (and so preserve the grounding isolation), but which will break down under surge conditions to clamp the grounds together.

RFI Filters   Where the source of interference is a radio-frequencies series, RF high frequency or (HF) filters can be placed in the lines.

Fiber-Optic Cables   These naturally provide good immunity against virtually all surges.

18

TRUNKING

Trunking is the routing of the links between the cellular switch and the public switched telephone network (PSTN) or to other switches in the cellular network. The trunks will generally be digital links, but analog and even wire-cable systems are sometimes used.

There are a number of factors that will determine the trunking configuration, and to some extent the optimum trunking will be determined by the PSTN configuration.

MAINTENANCE CONSIDERATIONS

From a traffic point of view, the larger the trunk group the more efficient the circuit provisioning, but for ease of maintenance the trunks should be divided into functional groups.

Trunks should be separated into incoming and outgoing groups. Generally circuit groups smaller than 15 circuits can be left as both-way (B/W) groups, but bigger groups should be split. Both-way circuits can be of limited use because of signaling clashes that can occur when the both-way trunk is seized simultaneously from both ends. The alternative is to split the route into two unidirectional circuits. The problem of clashed seizures does not occur when a common channel signaling system such as SS7 is used. In this case there is no limit on the size of both-way circuits. The fact that the both-way circuits carry larger volumes of traffic than the two unidirectional circuits means that they will be more efficient.

The second group of trunks that should be separated comprises

- Local
- National long distance
- International

If these groups are separated into identifiable groups, then it will be easy for maintenance personnel to identify PSTN or transmission problems as they occur.

To see how this could be useful imagine the instance when the above three traffic groups are combined, a fault occurs that causes all international calls to fail. As international traffic represents only a few percent of all calls, the failed-call records would not give the maintenance staff any indication of a serious problem. Of course, ultimately the statistics package will reveal the problem, but usually these are processed only once a month and sometimes even less frequently.

ROUTE DIVERSITY

It is a good idea to plan to have two physically separate routes (as shown in Figure 18.1) to the PSTN because a failure of this route will cause severe service disruption.

It is not necessary (usually) to use a different transmission medium, but in the case of a cable it is best to use two different routes, since cables are subject to damage by excavation. Where possible the two routes should also terminate in different tandem switches so that complete diversity is achieved.

CIRCUIT SPREADING

Within a given digital transmission system the circuits will be subdivided into 2-Mbit groups (or T1 spans). For the purposes of simplicity, the term "group" will apply to 2-Mbit groups or T1 spans. Each of these groups will be associated with some dedicated multiplexer hardware, which like all hardware can fail.

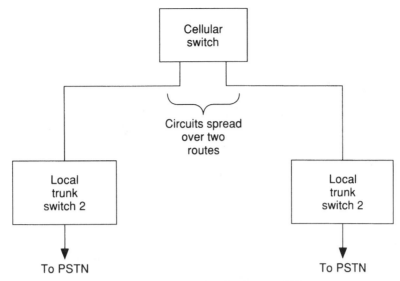

Figure 18.1 Route diversity is a good idea.

A number of routes can be spread physically over different groups, as illustrated in Figure 18.2.

ROUTE CAPACITY

A route is a number of circuits to a single destination. A cellular switch will have a number of outgoing routes and these routes will also be of limited size, which will be dependent on the switch architecture. The route capacities will be in multiples of one group. Each group is accessed via a port (or group inlet).

A typical switch may have a route capacity of 256 circuits (that is, there may be many routes but none can individually have more than 256 circuits). Although this means that circuit groups must be kept smaller than 256 circuits, groups this big have already reached a circuit efficiency sufficiently high so that there is no disadvantage in this restriction.

The switch will consume a number of its ports on housekeeping functions such as recorded announcements, storage drives, and internal switching.

The maximum number of routes is also limited and may be around 200 total routes.

There may be other limitations within the total number of routes such as the maximum number of incoming (I/C), outgoing (O/G) and B/W circuits. In some cases B/W circuits are provided only as an option.

REDUNDANCY

Redundancy in trunking is usually used on major routes such as the cellular-to-PSTN link. This involves duplication of the link hardware such as the microwave or fiber-optic system. In the event of a system failure (as detected by high error rates or loss-of-received-signal level) the redundant system will be brought into service.

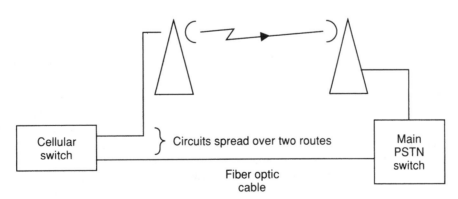

Figure 18.2 A number of different routes can be spread over the groups for added security.

There are three types of redundancy:

1. Manual changeover
2. Automatic changeover from cold standby
3. Automatic changeover from hot standby

The first type is self-explanatory, and the main difference between hot and cold standby is that in hot standby the redundant equipment is normally fully operational and the changeover involves no start-up time. In the case of cold standby the redundant equipment is only turned on as required and information may be lost during the changeover.

For microwave systems hot standby can be very wasteful of spectrum and may not be permitted. With fiber optics, hot standby is a very practical option.

USING REDUNDANT EQUIPMENT

As an economy measure, sometimes the redundant hardware can be used to carry live traffic by spreading the traffic equally between the two systems such that a failure of one system will cause a 50 percent drop in circuit capacity. While this will still cause serious degradation at peak hours, it will probably provide a reasonable grade of service (GOS) at off-peak times and will at least provide some service in the peak hours.

EFFECTIVE USE OF THE SWITCH-PORT CAPACITY

As mentioned before, the switch will be equipped with ports each with the capacity of one group. Generally it will be convenient to connect to other switches on the group level (E1/T1). When this is done some caution should be exercised to see that the port capacity is not unduly tied up with lightly used ports.

Switches will usually have sufficient port capacity to allow for some lightly loaded ports, but some operators have found that they have trunked the network in such a way that while the switch still has spare capacity otherwise, all the ports are exhausted.

It should be noted that it is possible to spread different routes across the same port, so that as long as they are bound for the same destination, it should be possible to effectively use most of the ports. For example, a PSTN route may have separate I/C routes from the local PSTN, trunk, and international switches. Provided these switches are co-located, it is possible to mix all three routes on the one group. However, to do this it may be necessary to demultiplex down below the group level and multiplex back up to meet the switch interface.

A compromise must be made between conserving ports and the cost (and maintenance liability) of splitting groups. In general splitting groups would only be done where the ports are becoming a limiting factor on the switch.

TIME-DEPENDENT ROUTING

In countries such as the United States where the cellular operator can choose from a number of long-distance carriers, there may be charging structures that make it cheaper to use particular routes at certain times of the day. Usually the cellular switch can be programmed to route the traffic onto the most economical route for the time of day. Note, however, that this means that each of the alternate routes must be provisioned to carry the peak traffic expected at the times of service, and this can lead to poor port utilization.

MULTIPLE-SWITCH OPERATION

When multiple switches are used they will need to be interconnected, and if a large number of switches are interconnected, a good number of ports will be consumed.

It is good practice to locate multiple switches physically in at least two different locations, whenever this can be done at a reasonable cost. The reason is the added security of spreading the switches so that in the event of catastrophic switch failure, at least half the network can still function.

Interswitch handoffs consume a good deal of processor time and efforts must be made to reduce the incidence of it. In particular the incidence of different switches handling adjacent base stations should be minimized. Unfortunately this will also reduce the network's ability for one switch to take over capacity from the other.

All networked switches will need interswitch data links to provide roaming capability, which needs to be dedicated.

USING BASE STATIONS AS NODES

A well-developed cellular network will have a reasonably complex distribution of base stations and it may well be that the base stations can themselves be used as nodes (or transit routing points) for future base stations.

When designing the links to base stations give some thought to this aspect, and particularly where microwave or fiber-optic cable is used, provision for some extra capacity should be considered.

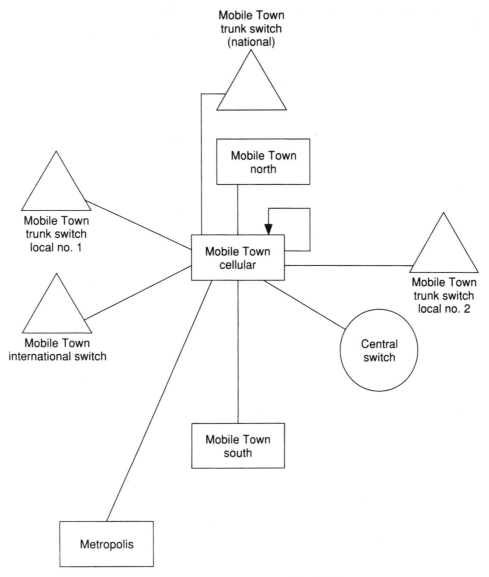

Figure 18.3   The trunk network of Mobile Town.

A TYPICAL TRUNKING OPTIMIZATION

Once you have a traffic dispersion as discussed in Chapter 20, it is then possible to proceed with the trunk-network optimization.

Assume, for a hypothetical small town of Mobile Town (as depicted in Figure 18.3), that the traffic dispersion in Table 18.1 has been obtained.

For this optimization we will assume that it is possible to get trunks from the Mobile Town cellular switch to each of the destinations in Figure 18.3. The availability of trunks is, of course, a major limitation in real life.

To optimally dimension these routes it will be necessary to calculate the cost factor for each route, as explained in Chapter 20. When this has been done Table 18.2 can be constructed. There are some interesting implications of the results in Table 18.2.

The traffic to North Mobile Town justifies only two direct circuits. Most operators have set a minimum number of circuits required before a direct route can be justified. Typically this minimum varies from 2 to 10 circuits. This minimum is justified by concerns about the accuracy of the traffic forecasts, the cost of establishing a route, and the variability in the traffic dispersion with time, which may see some small routes shrink to zero in a timeframe of a few years. In the case of a small cellular operator in Mobile Town it probably would not be worth considering establishing the North Mobile Town route.

The minimum number of circuits necessary before a

TABLE 18.1   Traffic Dispersion, Mobile Town Cellular Switch, Forecast for December 1992

Traffic Sink	Erlangs
Local network	
Mobile Town cellular switch	50.7
Central switch	125
North Mobile Town	8.8
South Mobile Town	11.8
Other Mobile Town codes	132
National network	
Metropolis	55.9
Other national codes	44.9
International	
International	9.76

direct route is provided will vary depending on whether the operator leases lines (in which case circuits of small quantities may be justified) or whether the trunk network is provided by the operator. In the latter case the decision to provide a direct route will largely depend on the cost of the channels. Thus the addition of a new 2-Mbit/T1 to an existing fiber-optic cable may be very cheap, but the provision of a new cable route may be prohibitive.

The traffic that overflows from the local direct routes is accumulated by the addition of the mean and variance, and this total traffic of 15.5 Erlangs with a variance of 52.2 then overflows to the local trunk switch.

It is a characteristic of such traffic that the measured traffic on the overflow route will be less than the calculated traffic. The reason for this is that the mean traffic on the various direct routes will not have the same busy hours and so the sums of the traffic on the various routes will be greater than the actual traffic at any particular time. Typically the discrepancy will be about 10 percent.

A lower grade of service (0.001) is used on the inter-switch link, because this is a low-cost route that can be provided by simply linking the switch ports to each other.

TRANSCODERS

Digital circuits are 64 kbits/56 kbits (depending on whether the system is 2 Mbit/T1). These data rates are more than adequate to satisfy the Nyquist criteria of sampling at twice the frequency of the signal for voice circuits. Devices known as transcoders take advantage of this fact to compress two or more voice channels onto a single digital voice channel. Most typically channel ratios of 2 and 5 are found.

Originally transcoders were designed for long-distance circuits such as satellites and seabed cables. In recent times, however, they have become inexpensive enough to be considered for cellular applications even over small hops.

Pacific Communications Sciences, Inc. (PCSI) has produced an 8 kbit/s multiplexer that allows 8 channels to be derived from 64 kbit/s with quite acceptable voice quality. Although this unit will support voice at 5.6 kbits, the quality at this speed is not adequate for cellular radio purposes.

Care should be exercised when using these devices in tandem, as the voice will degrade with each link. A general rule is that no more than two such units should be used in series. This is sometimes difficult to control, as the PSTN operator may, on some routes, also be using compression devices.

TABLE 18.2   Circuits and Overflow for the Dispersion

Traffic Sink	Erlangs	Cost Factor	CCTS	Mean	VAR	GOS
Local Network						
Mobile Town Cellular Switch	50.7	N/A	72			.001
Central switch	125.0	.55	123	7.2	41.0	N/A
North Mobile Town	8.8	.85	2	7.0	8.2	N/A
South Mobile Town	11.8	.30	14	1.3	3.0	N/A
Net local overflow traffic	196.3			15.5	52.2	
Other Mobile Town codes	132.0		160			.002
National Network						
Metropolis	55.9	.45	59	3.9	16.0	N/A
Net national overflow				3.9	16.0	
Other national codes	44.9	N/A	63			.002
Net traffic on national trunk route						
International						
Net traffic on national trunk route						
International	9.76		20			.002

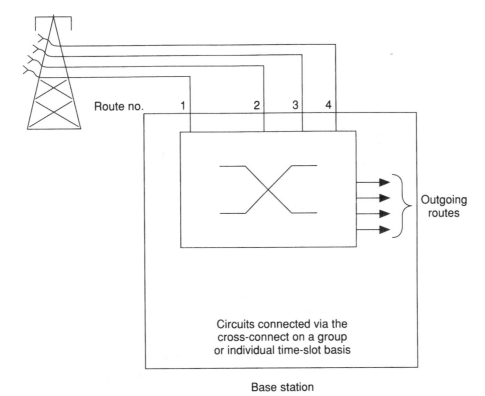

Route no.   1   2   3   4

Outgoing routes

Circuits connected via the
cross-connect on a group
or individual time-slot basis

Base station

Figure 18.4   Shows a nodal digital cross-connect.

Transcoders for cellular applications that use a factor of 2:1 (these are ADPCM and convert the 64 kbit/s to 2×32 kbit/s streams) are most suitable for urban applications. If higher factors are used, problems can be expected when the cellular link is connected to a long-distance circuit that also uses high levels of circuit multiplication.

INCREASING TRUNKING EFFICIENCY

In digital trunk networks it is usual that the trunking is mainly done at the T1/E2 level. This can result in a network consisting of a large number of partially full trunk groups, the efficiency of which may be very low. In order to improve trunk occupancy it is worth considering a "grooming-and-filling" process at the major trunk nodes.

There are two basic ways of grooming and filling. One is to select some of the lowest occupancy trunk groups and combine them into a number of full trunk groups by using voice-level multiplexer (MUX) equipment configured back to back. While this may be a little messy, it is practical for small circuit groups, and in particular for trunk groups that are not subject to high growth rates, since reconfiguration is a manual affair.

The other (and operationally by far the best) way of grooming and filling for larger nodes is to use a digital cross-connect (Figure 18.4), which allows the redirection of any of the trunk groups or any individual circuits into any other group under the control of software.

The advantages of these techniques are most evident to the operator using leased circuits. A leased T1/E1 link may cost around $2000 per month. A digital cross-connect will cost around $100,000 to $200,000. At these prices it is easy to see that a few T1/E1 circuits saved can pay for a cross-connect over a few years. Where the cellular operator uses in-house trunking, the savings will not be so evident (but operational expediency will be enhanced).

For switches with a large number of small circuit groups there may well be advantages in "grooming and filling" in order to save the investment of a new switch for switches where the capacity has been limited by availability of inlets/outlets.

ADVANCED GROOMING

Commercial pressures and poor planning have resulted in many trunk networks today becoming virtually unmanageable and highly inefficient. In the rush to connect base stations, often little thought is given to the link

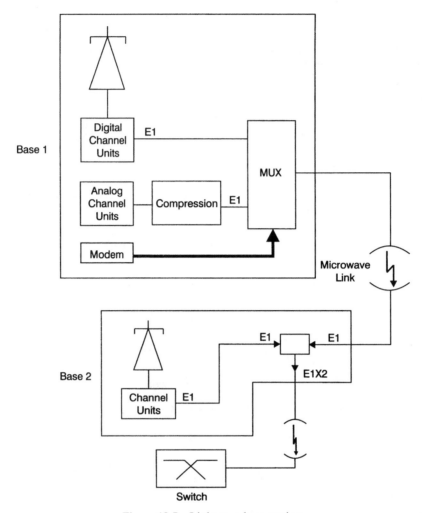

Figure 18.5   Links to a base station.

network that can become unwieldy. Trunk networks can also be expensive, costing as much as 10 percent of the network investment.

When the networks were mostly analog, it was often sufficient to use a MUX at each base station to convert the T1/E1 link to analog and to bring back the link digitally. This often meant wasted switch ports, but these could be groomed out. Today's rapidly expanding networks with enhanced capabilities make this far more difficult. Even an analog station can have CDPD (Cellular Digital Packet Data), security, emergency call, and voice fingerprinting data links in addition to the cell-site transmission requirements.

Many site links have a good deal of excess capacity that could be put to better use. This applies particularly to digital systems, which often have built in quite efficient compression on the base transceiver station (BTS) interface. Because microwave/fiber-optic suppliers often change from contract to contract, the network monitoring and grooming capabilities built into the microwave

are often difficult to access, particularly at the interfaces between incompatible hardware. Some operators in their rush to become operational even forget to order trunk network monitoring software/hardware, or wishfully suppose that the base-station monitoring will be sufficient.

Many operators will be using third-party (usually PSTN) backhaul routes, which are just leased lines. Usually this option is more expensive than the links that the operator owns (when the present value of both options is considered), and so in this case it is even more important to optimize circuit usage.

While it is possible to find a box full of devices that can do the network grooming and monitoring, unless they are all compatible, then there is little chance that they can be configured as a group. A software-controlled transmission equipment platform is what is needed. Fortunately these are available off-the-shelf.

Consider the trunking of two base-station sites as shown in Figure 18.5. Here with a lot of hardware it is

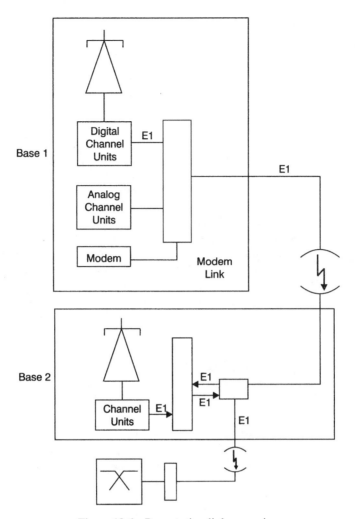

Figure 18.6 Base station link grooming.

possible to connect up a number of different types of equipment from base site 1, and to send that information along a single E1. It would even be possible to put a second MUX at the second base-station site and send all of the information to the switch on a single E1. The problem is that this is complex, requires numerous boxes, and is difficult to monitor in the event of a network fault.

A better approach, seen in Figure 18.6, is to use a single interface at each cell site that does all the grooming and compression in a network-friendly way. Because the interfaces can communicate with each other, it is easy from a central location to monitor and track faults. Just as important, the resources can be audited and controlled centrally so that it is possible to get the most out of the transmission network.

One such solution is provided by Paragon Networks International, which supplies a complete product line of back-haul network equipment, including their Web-browser-based graphical interface to the network, for system management.

Digital Cross-Connect

Digital cross-connect systems (DCS) provide more advantages than just flexibility. They generally come with a good-quality monitoring system and can provide advance real-time monitoring. In the advent of a partial failure of the transmission network, the DCS allows rapid keyboard reconfiguration of the network.

THE 2 MBIT/S CCITT DIGITAL STANDARD

The standard 2-Mbit circuits widely used throughout the world are actually 2.048 Mbits and comply with Consultative Committee on International Telegraphy and Telephony (CCITT) specifications G.703, G732 and

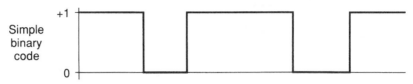

Figure 18.7   A simple binary signaling scheme.

G704. The transmission code is known as high-density bipolar three (HDB3), and a full definition of this code is found in Annex A of G.703.

The simplest binary code is one in which a "1" is represented by voltage and a zero by zero volts, as seen in Figure 18.7. This system requires a direct-current (DC) transmission medium. To send the signal over a system that cannot pass DC, the voltage transition can be turned into a pulse. However, if all the pulses are the same polarity, the net signal will have a high DC component and so will suffer significant distortion when passing through a bandpass filter (which includes transmission lines).

An improvement on the unipolar signaling is to use a bipolar code such as alternate mark inversion (AMI). This code sends marks as pulses, which invert their polarity after each mark, as seen in Figure 18.8. HDB3 is a derivative of this code. It needs modification, however, because the 2-Mbit/s signaling system derives its timing from the signaling pulses, and a long string of zeros could cause a loss of synchronization. The "3" in HDB3 signifies a scheme that ensures that no more than three consecutive zeros are sent. In their place is a sequence that the system recognizes as zero. It does this by violation of the inversion sequence. Because AMI requires each alternate pulse to be inverted, the presence of a pulse that violates this inversion signifies a zero (see Figure 18.9).

FRAMING

HDB3 has 32 time slots (channels), and the frame format is defined in CCITT recommendation G.704. The

frame format is shown in Figure 18.10. Time slot zero is reserved entirely for frame alignment bits and frame alarms, as shown in Table 18.3. The recommendation also allocated time slot 16 for channel-associated signaling, and so there are 30 channels in total available for voice. The 30 voice channels are numbered 1–30, with the first 15 voice channels being time slots 1–15, and voice channels 16–30 being time slots 17–31.

Each time slot consists of 8 bits, so at the frame rate of 8000/sec there are 8×8000 or 64 kbits/s per channel.

THE MULTIFRAME FORMAT

Each voice channel has a 4-bit signaling word associated with it, which it transmitted in time slot 16. The 30 channels can send the signaling for only two channels in one frame, and thus 15 frames are needed to send all the signaling information. An additional channel is used for multiframe alignment signal (MFAS), making the multiframe 16 frames long.

Time slot 16 can also be used for common channel signaling, such as SS7.

REDUNDANCY

A cyclic redundancy check (CRC) is inserted into the multiframe structure. The CRC word is 4 bits long and is inserted into each half multiframe. The frame alignment signal (FAS) is sent in time slot 0 of the even-numbered frames.

FAULTS IN THE 2-MBIT/S STREAM

Any part of the link may develop faults, which will show up as code errors, bit errors, frame errors or jitter, and slips. Poor connections can be a major source of faults, with faulty connectors and imperfect shielding responsible for timing errors. Storms, power surges, and interference can cause frame, code, and CRC errors. These errors will be intermittent and thus may be difficult to track down. Fading in radio systems can be responsible for high bit error rates.

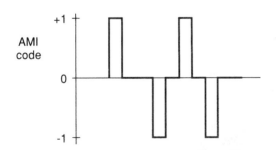

Figure 18.8   The AMI code.

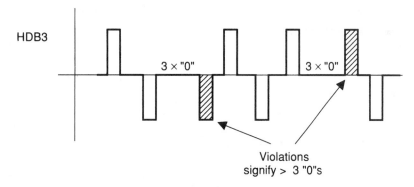

Figure 18.9   HDB3 uses AMI violation to avoid more than three zeros in a row.

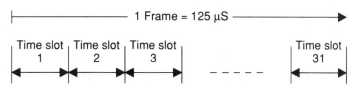

Figure 18.10   The HDB3 frame structure.

MODEM ACCESS

If the base station is analog, it will often be necessary to access the base-station controller, control channel, and maintenance functions by means of a modem. Modems are used to send data over analog lines. These may be used for control, interrogation, or sometimes even intersite system communications. Modems generally have a wide range of speeds, modulation techniques, and error correction codes. This means that you need to be very careful about compatibility. Although most modems have baud-rate (transmitted change-of-state rate) compatibility with early versions from the same stable, they are not necessarily compatible with each other. Historically, relatively slow data rates around 300 bauds were the norm because the data were destined for mechanical printers.

Once electronic storage became available, the demand for faster transfer rates spurred development. This de-

velopment was hampered by the cost of producing high-speed data systems. Because of this slow development, manufacturers had to produce new modems that could still communicate with the slower previous generation ones. To do this a negotiation protocol was established to enable the modems, on first contact, to automatically communicate to each other their baud capabilities in the same way as group 3 and group 2 fax machines can synchronize data rates.

Error correction is vital in most data communication schemes, and very much so in high-speed modems. Although early modems largely used propriety correction codes most manufacturers today adopt the CCITT standards.

Modems can be divided into duplex and half-duplex units. Duplex modems (sometimes called full duplex) can transmit and receive simultaneously. The connection between the modems is 4-wire (two send, two receive) or over a conventional duplex radio link. This full-duplex transmission is the most efficient way to communicate.

Where the connection is made by a conventional 2-wire telephone line, it is necessary for the modems to transmit one at a time (otherwise there would be data clashes on the line). This type of transmission is known as half-duplex.

BITS/SEC VERSUS BAUD RATE

There is often a lot of confusion over the difference between baud rate and bit rate, so it is worth considering

TABLE 18.3   Time-Slot Zero Structure

BIT NO.	Time Slot Zero							
	1	2	3	4	5	6	7	8
Alignment frame	S	0	0	1	1	0	1	1
Non-alignment frame	S	1	A	X	X	X	X	X

S-reserved for international use
X-reserved for national use
A-Remote Alarm set to "1" for alarm

why these two are not necessarily the same. In the case of a simple modulation scheme, such as two-tone frequency shift keying (FSK) (i.e., one tone for logical "1" and another for logical "0"), the baud and data rate are the same. This is so because one tone equals one bit.

More complicated modulation schemes such as quadrature amplitude modulation (QAM) allow each symbol sent to be modulated in both phase and amplitude so that at the receive end the demodulation takes account of both the amplitude and the phase, with each amplitude and phase corresponding to a bit. A typical QAM scheme can have 16 states, which means that each piece of information sent (or each baud) contains 16 bits of information. Therefore, if the baud rate of this system is 300, the bit rate will be 4800 bits/sec (bps).

THE 300/1200/2400 STANDARDS

These standards are now obsolescent but will still be found in some older equipment. The 300-baud standard mostly uses the AT&T Bell 103 (U.S.) or the CCITT V.21 (rest of the world) standard, and the 1200 bps uses V.22 and Bell 212A.

The 2400-baud rate proved difficult to implement, and a number of proprietary systems came onto the market. The current standard for this speed is the V.22bis.

Like the early 2400-baud modems produced up to 1992, there were a number of proprietary 9600-baud units produced. These were completely incompatible with each other. Today there is now a strong tendency to stay with the V.32 standard (full duplex) and V.29 (half-duplex) for this speed.

Modern modems often have a dozen or more inbuilt protocols to ensure backwards compatibility to those earlier standards. The 56 K Bits/second standards of today's modems come close to the theoretical limits of the band-limited wireline medium.

Although there were a lot of players in the modem market, a few were so successful that they should be considered separately. An early entry into the market was made by Hayes, which produced a wide range of competitively priced effective modems. Telbit introduced the Packetized Ensemble Protocol (PEP) trailblazer, and Robotics produced the Courier High Speed Technology (HST) 14,400-baud modem (with standard 300/1200/2400 compatibility). The tendency today is for all of these manufacturers to adopt CCITT standards.

ASYNCHRONOUS MODEMS

It is usually more efficient to transmit data synchronously (in fixed time slots), but sometimes it is important to transmit data as they are generated (asynchronously).

THE X.25 PROTOCOL

The CCITT-defined X.25 has become a protocol of choice for interconnection of data networks, particularly those with multiple data processors. X.25 is defined within the three communications layers of the Open Systems Interconnection (OSI) model to be able to connect any type of computer or terminal equipment.

The X.25 protocol was released in 1970 and was the first data protocol meant to be used for both point-to-point and switched data circuits. X.25 offers high security and quality even over analog lines. It was rapidly adopted by telephone companies around the world, and soon became the protocol of choice for many data applications.

High-speed applications of X.25 allow data rates of 2 Mbps and beyond. X.25 is still evolving.

Typical applications in private networks would be to connect anything from a few to about 100 terminals through multiprotocol ports to a common network. X.25 is used for data links in the Global System for Mobile Communications (GSM) and digital amps (DAMPS) systems to connect the data processors [such as Home Location Registers (HLRs) and Visitor Location Registers (VLRs)] together.

A derivative of X.25, Frame Relay is a faster but less robust, high-speed protocol, which is a subset of X.25, defined on level 2 of the OSI.

Frame Relay

Frame relay is a derivative of X.25 that reduced the end-to-end controls on error correction and traffic regulation to take advantage of the more reliable digital networks and to increase the data throughput. However, the downside is that congestion and loss of data can occur, so that X.25 remains the protocol of choice for applications like banking, airline bookings, and data transport within the PSTN.

ATM

Asynchronous transfer mode (ATM) is a packet-switched system that allows prioritizing of voice and video data that are intolerant of the delays that would occur within a conventional packet-switched network.

Conventional switched circuits provide a link between two points that stays up for the duration of the call. Because of this it is ideal for voice and video, which require an unbroken flow of information. Packet switching, on the other hand, is fine for pure data transmission and is more economical with bandwidth, as it does not tie up circuits if there are no data to send. The fact that the packets arrive at irregular intervals presents no problem, once the original message is reassembled.

The irregular arrival of the packets can cause what is known as *jitter* on transmissions that need to be continuous. Because ATM identifies the packets and mini-mizes jitter on critical circuits, it can be seen to have the circuit efficiency of a packet network combined with an almost jitter-free switched-like network when necessary.

ATM is a protocol that requires dedicated switches, but can use any transmission medium, such as E1/T1, SDH/SONET. Data rates to 622 Mbps are available.

19

SWITCHING

Until the advent of cellular radio, it was relatively easy for a switching engineer to have little or no knowledge of radio systems and for a radio engineer to know nothing of switching. In cellular radio, however, the two technologies interact inseparably.

In fact, the BSC, which is basically a switch, is often regarded as part of the radio network. The basic concept of a switch is to connect one line (usually a subscriber) to another in such a way that any subscriber (or line) can eventually connect to any other. When the number of connections is small, this can easily be done manually. Consider the situation of the four subscribers shown in Figure 19.1. By placing the link between any two subscribers (as shown between subscribers 1 and 2 in Figure 19.1) the operator can connect them in any order.

As the number of subscribers grows, the operator's task becomes increasingly difficult; automatic telephone exchanges are needed to cope with the number of potential links. In fact, the complexity of the switching rises rapidly as the number of subscribers increases, and calculations show that it would be literally impossible to switch the traffic of a modern city manually. The number of operators needed tends to be equal to the population, so unless everybody was involved working as an operator, the traffic could not be carried; however, if everybody is an operator, then no one is generating traffic, and so the scenario is impossible. It is interesting, however, that the driving force behind automation of switching was not economizing on operator costs or numbers, but rather was instigated by an undertaker who, believing the telephone operators were unfairly directing business to his competitor, set about to design the first electromechanical switch. His suspicions could have had some substance, as the operator in question was also his competitor's wife. The undertaker was Almon Strowger, and until quite recently there were millions of lines of electromechanical switches in service around the world that still bear his name.

The first automatic telephone exchange was produced in 1892 at La Porte, Indiana. It was electromechanical—electric switches driven by electromagnets, which by mechanical movement performed the switching function. Switching was accomplished by sending pulses (dialing) to indicate the number required. In this sense, these early switches were digital.

Signaling was done by the still familiar rotary dial. The dial is spring driven, and when rotated to, say, 5, the rotation winds up the spring. Releasing the dial causes the spring to return the dial back to the resting point at a speed regulated by a governor. On the way back it makes and breaks the circuit five times, sending out a train of five pulses. The pulse-train frequency is usually either 10 or 20 per second, as seen in Figure 19.2. Because only two wires were available to connect the telephone to the switch, this pulse train was sent by alternatively making and breaking the loop between the phone and the switch, as seen in Figure 19.3

SWITCH CONCENTRATORS

Figure 19.4 shows a simple concentrator switch. The switch is non-blocking because every inlet has potential access to every outlet. Call blocking can still occur, however, because the number of simultaneous calls permitted is limited by the number of outgoing routes. In the switch shown in Figure 19.4, three simultaneous calls are allowed.

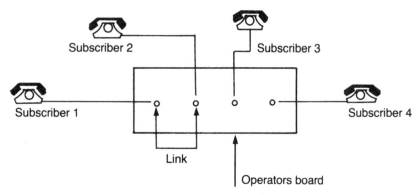

Figure 19.1 With this simple operator-controlled switch, a manual connection between subscriber 1 and subscriber 2 is made by the link.

The eight inlets A–H in Figure 19.4 are concentrated into the three outgoing routes (1–3) by activation of the cross-points; only one switch in any row or column can be closed at one time. This principle, called *line concentration*, is used extensively in all telephone switching. A radio base station acts as a line concentrator because it connects the mobile subscribers to the cellular switch in such a way that approximately 20 mobile subscribers can be connected to the switch by one channel.

These switches have one very significant practical disadvantage: The connection between any inlet and any outlet occurs by only one path. The failure of any switch cross-point means that certain paths are no longer available. This limitation can be overcome, however, by introducing a second row of switches, as shown in Figure 19.5.

In Figure 19.5, you can see that the path between two ports (for example, B and D) can be connected by engaging the B and corresponding D row bars on any of bars 1–3. For example, B can be connected by engaging the cross-connection B-1 and then 1-D or alternatively B-2 and then 2-D, and so on. This results in three paths connecting B to D. Although this configuration doubles the number of switches, it provides a very valuable redundancy in internal paths.

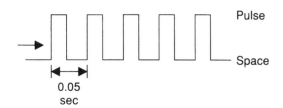

Figure 19.2 The pulse train sent out when a "5" is dialed.

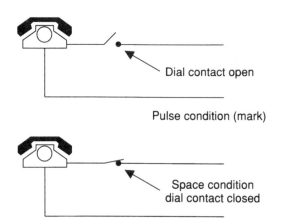

Figure 19.3 The dialed pulse is accomplished by opening and closing a switch across the line.

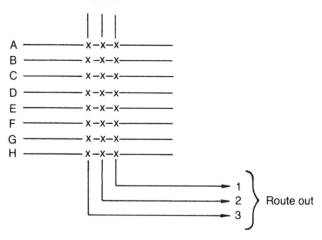

Figure 19.4 It is easiest to think of a switch as a crossbar where connections are made by activating (connecting) the cross-points. This limited-availability concentrator has eight inlets and three outlets.

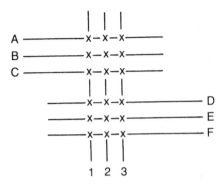

Figure 19.5 This switch has multiple internal paths between inlets and outlets.

THE TELEPHONE

The humble plain old telephone (POT) has changed very little since it was first introduced. It is important to understand the functioning of the telephone because the way network switches have evolved, even to this day, is greatly influenced by the terminating equipment they originally had to serve. The earliest phones were meant to be connected manually and so had no use for a dial; when automatic switching was introduced, the dial was literally added on. Today's phones have Dual Tone Multifrequency (DTMF) dialing (tone dialing), but that is not a big difference.

A block diagram of a modern phone is shown in Figure 19.6. One of the essential parts is the line switch, which either connects or isolates the receiver apparatus from the line. This switch is in parallel with the hook switch. When on-hook it presents a capacitive load to the line. When off-hook, it presents a low impedance to direct current (DC) and a 600-ohm load to voice frequencies.

Because the microphone and earpiece operate independently, it is necessary to connect the 2-wire line to a 2-wire-to-4-wire converter. In the earlier phones this would have been a hybrid transformer; today it will be an integrated circuit (IC).

The keypad is simply in parallel with the microphone (usually with the facility to cut the microphone out of the circuit while tones are being transmitted, so that background noise is not detected at the DTMF decoder).

The ringer is simply a bell, or in more modern phones, a piezoelectric buzzer, which is connected to a sensor that detects the low-frequency (around 17 Hz) ring tone.

Often the parts of the phone will be controlled by a microprocessor, which will usually also have some memory for redial facilities.

STEP-BY-STEP SWITCHES

Step-by-step (SXS) switches take the dialed pulses and convert them into switch positions by arranging a rotary switch like the one seen in Figure 19.7, to step around one position for each pulse. This switch can select any one of 10 outlets. This type of switch is known as a *uni-*

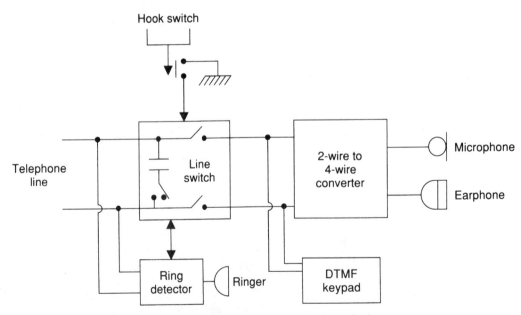

Figure 19.6 Block diagram of a modern telephone.

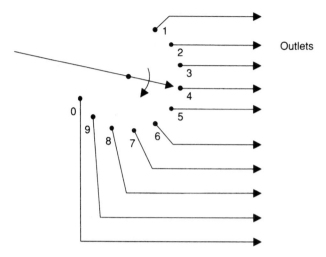

Figure 19.7 The uniselector can be pulsed around to any one of 10 positions.

selector because the mechanical movement (of the wiper) is in one plane only. Later switches were able to move the wiper along a shaft to a selected group of 10 outlets; it then rotates to select an outlet. In this way the bi-motional selector can select one of 100 outlets.

By connecting one or more uniselectors in series 100, 1000, or even more distinct circuits can be selected, as seen in Figure 19.8. Often uniselectors are used with bi-motion selectors to drive 1000 line groups.

The cellular operator will not have an SXS switch, but it will sometimes be necessary to interconnect with a SXS switch or perhaps even private automatic branch exchange (PABX) with SXS switching. However, even in Third World countries SXS switching is becoming rare.

Later systems used various forms of memory so that the exchange switching could be done asynchronously (at a different speed) to the dialing pulses. With refinements, this type of switching and dialing remained the standard until very recently.

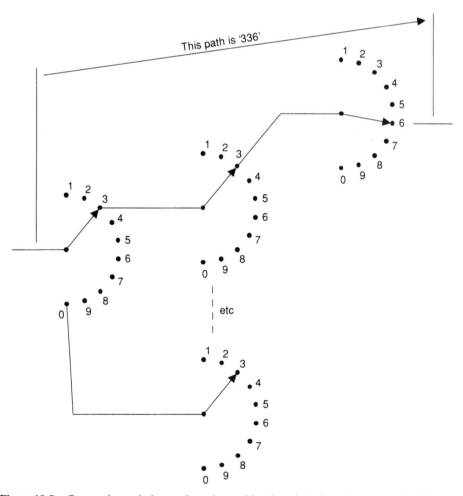

Figure 19.8 Connecting uniselectors in series enables the selected number to grow indefinitely.

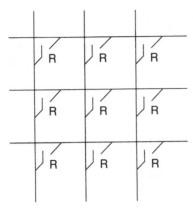

R = Reed relay

Figure 19.9   A reed-relay-based crossbar matrix for a 3 × 3 switch.

CROSSBAR SWITCHES

The SXS switches have the "intelligence" of the switching system arranged so that it is part of the switch. The mechanical movement of the switch uses the pulse train to move the wiper and so select the outlet. This is wasteful of resources because the control is used only to set up and clear down calls. Because setup and clear-down take only about one second, the "intelligence" is largely doing nothing during the progress of a normal call.

Crossbar switches (often abbreviated as Xbar), which came into service in the 1960s, use switches that have no local intelligence, but rather operate under the control of some external circuitry. Because the control equipment is pooled, it is called *common control*. Most crossbar

switches take this concept further and pool other resources, such as tone decoders and registers. The logic is usually relay based, but later versions will have microprocessor-based controllers. The switches are still activated electromechanically.

Reed relay switches, which, as the name implies, use reed relays for switching, have banks of reed relays in a crossbar-type configuration. The relays are operated by coils of wire wrapped around the reeds, and they cause the reeds to close when a current passes through them. They are configured as shown in Figure 19.9, where it can be seen that for practical purposes they form a crossbar matrix.

Figure 19.10 shows how the common control of a typical crossbar works. The switch will have two crossbar switches, one that switches telephone traffic and one that switches in tone senders, receivers, and storage registers as they are needed.

DTMF DIALING (TONE DIALING)

Pulse dialing was universal until around the early 1970s, when network operators began to introduce tone dialing. Even in the 1990s, a significant proportion of the world's phones are still pulse dialed, but virtually all new installations are tone, and most operators have an ongoing program to convert the remaining pulse phones to DTMF dialing phones.

Interestingly most modern processor-controlled switches retain the capability of decoding either DTMF or pulse dialling which is why the vintage phones still work mostly on PSTN networks.

In this system, two tones are sent simultaneously to a line to indicate the desired number. Figure 19.11 shows

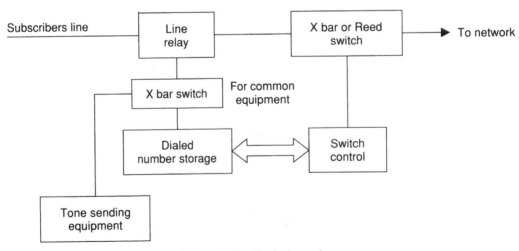

Figure 19.10   Typical crossbar.

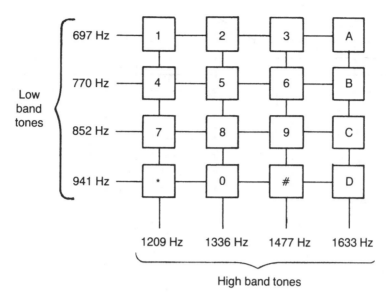

Figure 19.11 DTMF, the standard for tone dialing, consists of two tones—one "low-band" and one "high-band"—generated from the matrix shown.

the tone pairs and their associated numbers. The A, B, C, and D keys are not usually provided for POTS applications; they are reserved for special purposes.

DTMF dialing has been available for a number of decades. It was designed to take advantage of the potentially higher dialing speeds that were obtained by using code receivers with memory to store the digits and forward them as required by the switches.

In SXS systems, each number dialed represents a train of pulses that cause mechanical switching in real time, making them necessarily slow. These systems are sometimes referred to as "stagger by stagger" systems by those who have seen the switches in operation.

With crossbar systems (and with modification, some SXS systems), code receivers were provided as exchange-common equipment. They were switched across the subscriber's line for the duration of dialing and could decode and ultimately store the DTMF pulses, which could then be sent further on in the network at any desired speed. Although DTMF dialing is a feature of almost every cellular telephone, it is by no means a new idea.

Each pair of tones in the DTMF scheme consists of one high-band tone and one low-band tone. This increases immunity to false decoding from voice or noise, as do other requirements, such as a minimum signal-to-noise ratio and a correspondence in-level (known as *twist*) between tone levels for a successful decode. The tones are structured so that false decoding due to voice or noise is unlikely.

SPACE SWITCHES

Telephone switches were originally all "space" switches; that is, the switches physically connected one circuit to another with a connection in space. SXS, crossbar, and reed switches are all examples of space switching. In order to make a call between two telephones, it was necessary first to physically connect the two telephones with a wire. Space switches were used to connect subscribers for the duration of the call only, so that the links could be used by other subscribers after the call was completed. Today, although a physical connection usually does not occur (because the digital switches are time-multiplex-devices), two telephones are connected by a dedicated route to each other for the duration of a call.

Figure 19.12 shows a switch in which each inlet can connect to each outlet, as well as being able to park in a neutral position. This is known as a full-availability switch, since each inlet has a path to each outlet. The number of possible paths is 16 ($4 \times 4 = 16$).

As the switches get larger, the total number of possible paths rapidly increases. Consider a 400-inlet switch with 400 outlets; the total number of paths is 160,000 (400×400). Because of these huge numbers (and hence the massive amount of hardware), early switches were limited-availability switches, which means that each inlet could access only a limited number of outlets. For example, if the outlets per inlet are limited to 20, the total number of possible paths is 8000 ($400 \times 20 = 8000$), which is considerably more manageable.

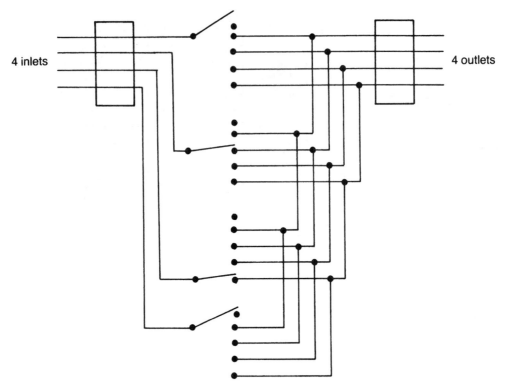

Figure 19.12 A four-inlet, four-outlet switch. Inlets 1 and 4 are in the parked or neutral position, inlet 2 is connected to outlet 2, and inlet 3 is connected to outlet 4.

TIME SWITCHES

Time switches became available with digital techniques. These switches work on the principle of switching a particular inlet to a particular outlet at a certain point in time. Figure 19.13 shows how inlets are assigned their respective timeslots. The input data are then rearranged (switched) under the direction of the control store so that each incoming timeslot is connected to the desired outgoing timeslot.

In Figure 19.13, each telephone line is sampled in its respective timeslot. The telephone in timeslot 1 on the A side is connected to the telephone in timeslot 4 on the B side. Notice that the switching is done by rearranging the timeslots, not by physical wires, so that the information can be carried by a single path between switch A and switch B.

Modern telephone exchanges, such as shown in Figure 19.14, and cellular radio switches generally use a combination of time and space switches to minimize the total hardware needed.

SPC SWITCHES

Modern switches operate under a central program, much like a computer. The system configuration, subscriber base, and routing are all held in the central database, which is loaded as required from a tape or CD. Stored program control (SPC) switching, as is known, is the basis of all modern switching. Most switches have purpose-designed processors because the requirements of a switch processor differ from a computer in that a computer is required to perform serial operations at high speed with very high accuracy, and a switch processor, on the other hand, needs to perform thousands of parallel processes at moderate speed in an environment where the occasional error would not be a big problem.

Recent advances in computer parallel-processing technology have allowed the merging of the two technologies, and many of the latest designs for switch processors are based on common computer processing chips.

SPC switches can use both space and time switching or a combination of both. Figure 19.14 shows a modern SPC switch housed in a neat metal cabinet and designed to be used in an air-conditioned room. The cabinets house the processors, the switches, and often the power supplies, which are switching regulators.

Intelligent Networks

Intelligent networks (INs) take this whole process one step further by taking the processing power away from

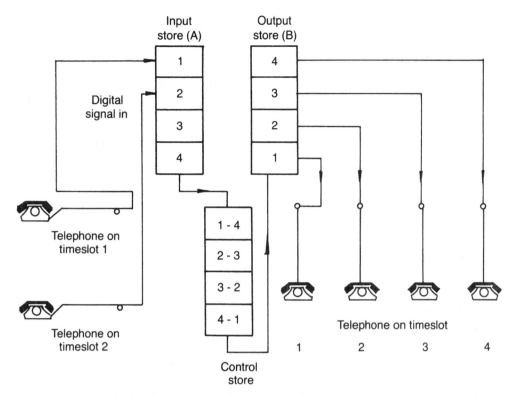

Figure 19.13 A full-availability time switch uses timeslots to connect any inlet to any outlet. The timeslot translation is shown in the control store. Inlet 1 is connected to outlet 4, inlet 2 to outlet 3, and so on.

Figure 19.14 An AXE10 from Ericsson—an SPC switch that can be used for conventional telephone or mobile telephone switching. (Photo courtesy of L.M. Ericsson.)

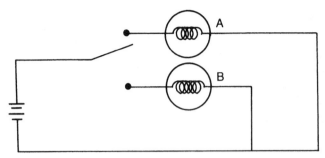

Figure 19.15 Blocking occurs in limited-availability switches, and congestion or information loss can occur in the switch. The simplest blocking switch, shown here, enables one light or the other, but not both, to be turned on at the same time.

the switch to a high-power centralized computer network that makes all the hard decisions. The switch processors become subordinate and mainly carry out basic switching instructions. Things like the Home Location Register (HLR), "follow me" call instructions, and validation are assigned to the IN.

LIMITED-AVAILABILITY (BLOCKING) SWITCHES

Figure 19.15 shows the simplest limited-availability, or blocking, switch. In this example, either light A or light B can be on, but not both. Because the switch can transmit information to only one of the outputs, it can lose information (for example, if a condition indicates that both A and B should be turned on simultaneously, some information is "lost" in the switch because the switch can only indicate the first state. A switch that can lose information is called a *limited-availability* or *blocking switch*.

The simplest telephone switch is an extension of the limited-availability switch, as shown in Figure 19.16. This simple uniselector switch allows a number of telephones to share a common outgoing line, but it has the disadvantage that only one telephone can use the line at any one time.

Subscribers' telephone switching stages will always be limited-availability switches. This is because one of the main functions of the subscriber's switch is to concentrate a large number of individual low-traffic telephone lines into a smaller number of high-usage lines that can be used to distribute the traffic efficiently. However, once the traffic is concentrated into parcels of about 0.5 Erlang per circuit, it is efficient to use full-availability switches for onward trunking. Trunk switches are usually full-availability.

FULL-AVAILABILITY (NON-BLOCKING) SWITCHES

A full-availability or non-blocking switch is one through which it is possible to connect any idle outlet to any idle inlet, regardless of how many other connections have been made. Figure 19.17 shows the simplest full-availability switch. In this example, you can see that the switching of states A and B are independent and that information will not be lost through any limitation in the switch.

The switch concentrator shown earlier in Figure 19.4 is an example of a limited-availability or blocking switch. For example, if three of the inlets, A, B, and C, are connected to three outlets, 1, 2, and 3, respectively, then no other inlet can be connected until one of the established connections is dropped.

The switch shown earlier in Figure 19.5 is an example of a full-availability or non-blocking switch. The number of inlets and outlets must be equal in a full-availability switch.

A simple one-stage switch can easily and economically be made non-blocking for small-sized switches. Such a switch must have links from every inlet to every outlet, so the number of links increases as the square of the number of inlets. For large switches, this soon becomes prohibitive.

In 1953, Mr. C. Clos of Bell Laboratories published an analysis of three-stage switches, showing the relationship between the switch configuration and the num-

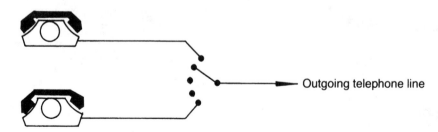

Outgoing telephone line

Figure 19.16 A simple uniselector line switch (which is a line concentrator with limited-availability) can be used to concentrate telephones into a limited number of lines.

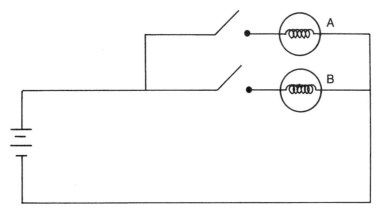

Figure 19.17   The non-blocking switch shown here allows either or both lights to be in either the "on" or "off" state. No information loss occurs due to limitations within the switch.

ber of links used. He demonstrated that for non-blocking it is necessary that each stage be non-blocking and that the number of center stages be:

Number of center switch points

$= 2n - 1$

$= 2 \times$ (the number of inlets/outlets per group) $- 1$

Figure 19.18 shows a three-stage switch.

The path of any call can route from any inlet group to any center group by one link and from any center group to any outlet group by one link. Thus, there are K-paths through the switch from any inlet to any outlet.

It can be shown that the total number of cross-points for the switch system in Figure 19.18 is

$$T = 2NK + K \left[\frac{N^2}{n} \right]^2$$

where

$N =$ the number of inlets/outlets

$n =$ the size of each inlet/outlet group

$K =$ the number of center arrays

The number of center arrays of switches can be determined by imagining a switch that has all circuits busy

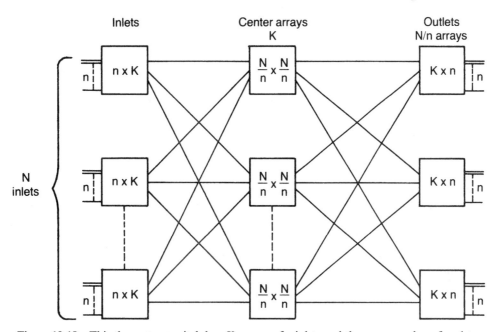

Figure 19.18   This three-stage switch has K groups of n inlets and the same number of outlets. To ensure full-availability, there must be at least two $(n - 1)$ switch cross-points at the center.

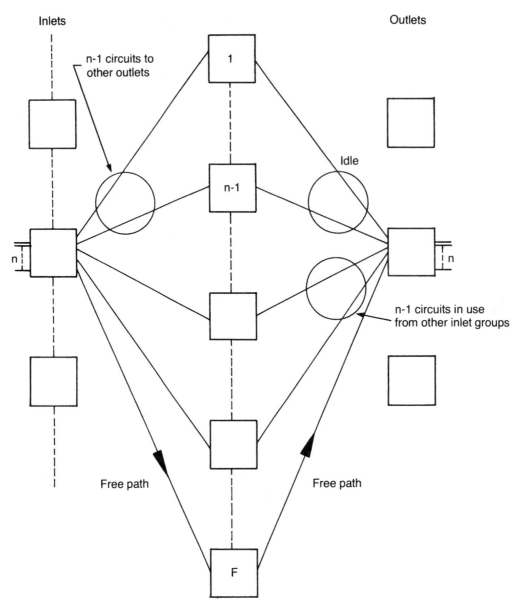

Inlets   Outlets

Figure 19.19 This figure shows n inlets in one group that have $n - 1$ circuits busy to other outlet groups. The required outlet group also has $n - 1$ circuits occupied. To avoid blocking, at least one free path (via center switch F) must be available.

except for one inlet and one outlet. In this instance, the worst case is an inlet group that has $n - 1$ active outlets and attempts to connect to an outlet group that also has only one free outlet (the one sought), but that is accessed from a different group of center switches. Figure 19.19 illustrates this. To be full-availability, the switch must still be able to switch the path between the inlet and the desired outlet, so at least one other free path must exist.

The minimum number of center switch points is

$$(n - 1) + (n - 1) + 1 = 2n - 1$$

A PSTN SWITCH

A public switched telephone network (PSTN) switch is made up of a number of switching stages and contains both blocking and non-blocking switches, as seen in Figure 19.20. The subscribers' telephone lines (which are still predominantly copper wires) come into the switch and terminate at the main distribution frame (MDF). The access to the PSTN is then via a subscriber's switching state, which is a blocking switch. This state then connects to a group switch, which also picks up private automatic branch exchanges (PABXs).

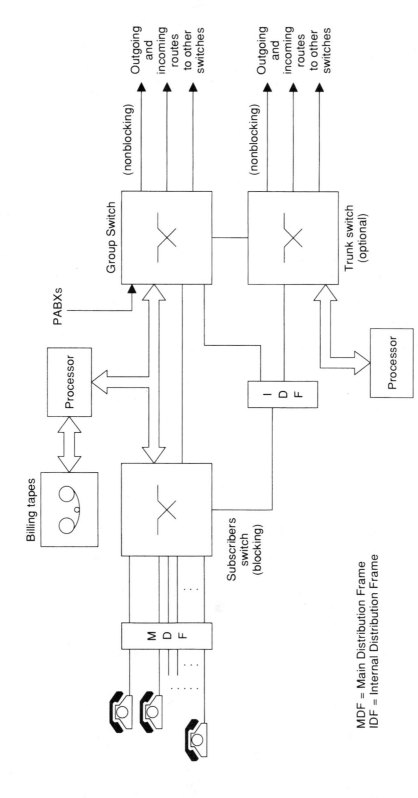

Figure 19.20 Block diagram of a PSTN switch.

MDF = Main Distribution Frame
IDF = Internal Distribution Frame

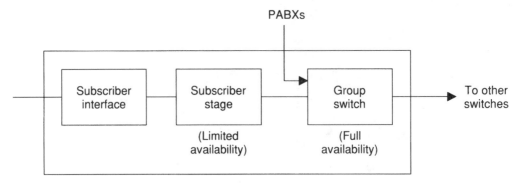

Figure 19.21 Functional diagram of wireline switch.

An internal distribution frame (IDF) is used to connect the various stages together. It may be that the switch will include a trunk stage, which will be used to route some of the local calls and act as a transit for other switches.

In most cases the whole system will be controlled by a few (usually redundant) processors, although some manufacturers have switches in which the processing function is totally decentralized. Notable among this type of switch is the Alcatel 1000 S12, in which every switch block consisting of 2 Mbits has its own microprocessor.

Most important for the operator are the billing records, which record every detail of the call for charging purposes.

WIRELINE TELEPHONE SWITCHES

Wireline switches use a mixture of full- and limited-availability switches, as seen in Figure 19.21. Switching systems are designed on probabilities, and it is most unlikely that more than 10 percent of a large population of domestic subscribers would be using the system at any one time. For that reason it is not necessary that each subscriber be able to connect to an outgoing junction simultaneously. Because limited-availability switches are significantly cheaper than their full-availability counterparts, they are used for the subscribers' stage.

The most expensive part of a modern wireline switch is the subscribers' equipment. Each subscriber must be provided with an individual inlet and the line interface equipment to make the phone work. The interface performs what is sometimes called the "BORSCHT" functions, an acronym from the key letters of the functions:

"B"   Battery—the interface powers the telephone
"O"   Overvoltage—surge arrestors and protection
"R"   Ring current and detection

"S"   Supervision—is the phone on-hook, dial pulses, DC conditions?
"C"   Codec—converts the incoming analog signal to pulse duration modulation (PCM) and the outgoing signal back to analog
"H"   Hybrid—converts the telephone 2-wire link to 4-wire for the switch, and vice versa
"T"   Testing—of the subscriber's line and equipment

Most of these functions are done by pooled equipment, but they must be made available to every line.

PABXs will have a much higher traffic density because they have already concentrated the traffic from a large pool of users before connecting to the land-line switch. The traffic per circuit from a PABX is usually around 0.6 Erlangs/circuit. For this reason PABXs will be connected to the switch at the group stage, which is designed to handle higher traffic.

In cellular switching the role of the subscribers' stage is performed by the base stations, which pool a thousand or more subscribers into a concentrated stream.

CELLULAR SWITCHING

A cellular switch can be thought of as a trunk switch that performs a large number of non-switching functions. For this reason, a cellular switch is usually considerably larger (physically) than a trunk switch of the same capacity. Figure 19.22 shows a cellular switch, and Figure 19.23 shows the wiring of a cellular switch.

The mobile switch is essentially a stored program control (SPC) switch that enables connections to be made between the mobile bases and the rest of the telephone network.

Smaller switches may consist of a simple single-stage time switch with full redundancy (or $n+1$ redundancy) of the switching stage for reliability. Bigger multistage

Figure 19.22 A complete Motorola EMX500 with a subscriber capacity of 15,000 consists of eight racks of equipment. The racks are usually wired as a single suite. Tape drives and digital voice announcement equipment are included. (Photo courtesy of Motorola Communications and Electronics, Inc.)

switches use a combination of space and time switching. Figure 19.24 shows the structure of a mobile switch.

The switch is composed of four main parts: the terminating circuitry, the switch, incoming and outgoing interfaces, and controllers. These elements are discussed in the following sections.

Terminating Circuitry

This circuitry connects the switch with the outside world. It transmits instructions (data) and voice (usually coded as data). The terminating circuits, like all other parts of the switch, are controlled by the processors and the stored programs.

Alternate routing of trunk groups is often provided, and four alternate routes are typically available. Trunk routes from the switch can be either unidirectional or bidirectional; the numbers and size of each type of route is usually limited. At least 50 such routes are ordinarily

available from a medium-size switch. This allows traffic to be distributed in a cost-efficient way and can be used to avoid wireline toll charges when the system operates over a number of charge zones.

The Switch In cellular radio systems, the switch is normally a non-blocking (full-availability) switch. (Non-blocking means that every inlet has a path to every outlet.) Earlier switches and some simpler switches are limited-availability types, which means that each inlet has only a fixed number of paths (often in multiples of 10) to any outgoing route. In practice, this arrangement means that the switch can block a call, even though there are free outgoing circuits, because all internal paths in the switch are in use.

The cellular switch is characterized by these basic parameters: total number of ports, total PCM ports, and busy hour call attempts (BHCA). These parameters are discussed next.

Figure 19.23 A cellular switch is quite complex internally, as can be seen from the mass of interconnecting cables behind the neat cover panels of a Motorola EMX500 (circa 1990).

Total Number of Ports   The total number of ports is the sum of the inlets and outlets (whether to the switched network, the base stations, or peripheral equipment such as recorded announcement machines). These inlets/outlets are in multiples of 24 or 30 channels.

Note that small switches using digital links to the bases consume port capacity as a function of link size rather than channels in use (that is, a 2-Mbit, 32-channel link uses 30 ports regardless of the number of active channels). Figure 19.25 illustrates using multiplexer (MUX) equipment to save inlet ports. The link between the MUX and the cellular switch should be less than 250 meters unless repeaters are used. Back-to-back MUXs can be used as repeaters (but this is expensive).

Total PCM Ports   The total number of PCM ports equals the total number of PSTN and radio voice chan-

nels. There may, however, be other ports in addition to the PCM ports.

Busy Hour Call Attempts   Varying from about 1000 to more than 1,000,000, BHCA measures the maximum number of call attempts that can be handled during a busy hour. This maximum, reflecting the processing power of the switch, will limit the number of customers that can be connected to the switch.

Incoming and Outgoing Interfaces

Incoming and outgoing interfaces connect the switch to the outside world and provide the necessary signaling and signaling translation so the switch can communicate with other switches.

The signaling between the mobile switch and the PSTN can take many forms. These are the forms most often encountered:

- Dial pulses
- Multifrequency (MF)
- MFC R2
- DTMF
- Consultative Committee on International Telegraphy and Telephony (CCITT) System No. 7 (SS7)

To make matters even more complicated, the "standard" signaling systems can have many variations, and it is usual that the signaling at the cellular switch must be tailored to the local PSTN version of the standard signaling format. This usually involves considerable software costs.

The cellular switch will have limits on the usage of the outlets/inlets for their various functions. These functions provide for limited (but efficient) numbers of

- Both-way junctions
- Unidirectional junctions
- Total number of separate trunk groups
- Maximum trunks in a group
- Radio frequency (RF) cell groups
- Channels in each RF cell group
- Intersystem data links
- Printers
- Tape decks/CD drives/Hard drives
- Voice recordings
- Three-party conference lines
- Tone receivers
- Alternate routing patterns
- Voice mail

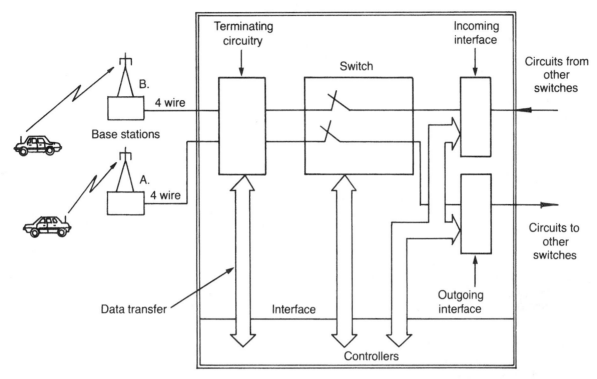

Figure 19.24   Simplified block diagram of a mobile switch.

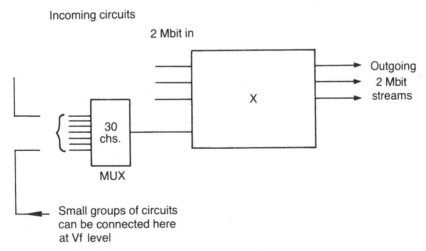

Figure 19.25   Small groups of circuits can be VF-connected to save inlets. Note that every 2-Mbit stream occupies 30 inlets regardless of how many inlets are actually active.

Any or all of these functions may have inherent limitations on the number of outlets available for each function; these limitations do not usually present problems, but they do restrict some fully loaded switch configurations.

Controllers

The controllers consist of processors, memory, software, and hardware that enable the switch to perform the functions required of it. The processors in a cellular switch normally control many non-switching functions such as system monitoring, diagnostics, bill record-keeping, and alarm monitoring.

Notice that although billing records are kept, most cellular switches cannot process these records. Billing records are usually transferred to tapes CDs or removeable hard drives that are then processed using external equipment. Housekeeping routines ensure that the use

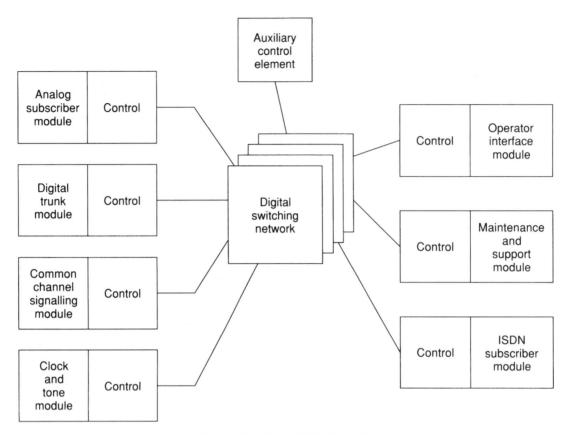

Figure 19.26 The Alcatel 1000 S12 architecture.

and availability of resources (including trunks, receiver channels, and scratch-pad memory) are continually checked.

Global System for Mobile Communications (GSM) switches tend to be very big, like the AT&T 5ESS, which can handle 500,000 BCHA, even with an average handover rate of 0.7 per call. This switch is capable of handling up to 300,000 mobile users, at a calling rate of 0.06 Erlang in the busy hour. This calling rate is high, and for the more usual rates of around 0.03 Erlang/hr, the busy hour call attempts would increase inversely and the capacity would be lowered to around 150,000.

THE ALCATEL S12

The Alcatel S12 is a switch designed as a "special networks" device for private networks, but it is finding many applications in cellular systems. Unlike most other digital switches, which are built around one or more central processors, the S12 has no central processor and instead has distributed processing based on the common microprocessor. A minimally equipped switch consists

simply of two hard drives (which provide data and software backup) and a few cards that can be directly identified with their switching function. This switch can be used for wireline subscribers as a trunk switch or as a high-powered PABX, with virtually unlimited routing capability or in various cellular configurations. Hughes uses this switch in its E-TDMA system, as does Alcatel in its GSM networks.

Functionally, this switch consists of three parts. The terminal module (which determines the terminal function), the terminal control element (TCE), and the digital switch network. It is the TCE that controls the switching, as seen in Figure 19.26.

Although the control for each function is from the TCE, important control functionality resides in the auxiliary control element to provide redundancy. Conventional redundancy is not needed, as each terminal module is independently controlled, and the failure of one card does not affect the rest of the system.

The switch itself is also modular, being made up of independent elements of 16 identical ports. Each switching element has all the logic needed to establish a switching path for voice, data, or control. The TCE has

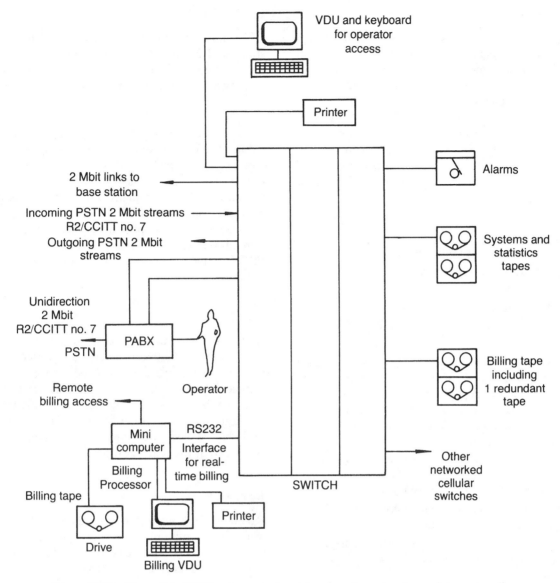

Figure 19.27 The switch CPU has to control peripheral equipment such as printers, recording devices, visual display unit (VDUs), and perhaps a PABX interface.

access to multiple switching elements, so the failure of one element simply means that calls are rerouted.

ROAMING

For automatic roaming, it is necessary that the switches communicate with each other and exchange information. This exchange is usually done with a dedicated digital link that ties the switches together. Such a setup is practical for Advanced Mobile Phone Service (AMPS), CDMA and TDMA using the IS-41 standard and GSM through the open interface.

Nordic Mobile Telephone (NMT) switches have long had an interconnect specification and can all be connected together regardless of the source of manufacture.

SWITCH PERIPHERALS

Switch peripherals are pieces of equipment associated with the cellular switch and under the control of the switch CPU. This equipment includes system statistics and billing records, PSTN links, PABX (optional), and other items detailed in Figure 19.27 and described in the following sections.

System Records

The system record loads the system software and system parameters that enable the cellular network to operate in its current configuration. Additional network equipment will be associated with a system update to inform the switch of the new configuration.

The system record contains details of both the subscriber's file and the system files. The subscriber's file capacity can vary from a few hundred to more than 200,000. The subscriber's file contains operational data about the subscriber in the processor memory. The stored subscriber data are as follows:

- Home area
- Service status (for example, non-payment)
- Serial number
- Class of service
- Last known location
- Call forwarding and no-answer transfer information

When subscriber information is updated, it is added to the processor memory immediately unless a call is in progress, in which case it is temporarily delayed. The updated subscriber data are also written onto the system tape. The updated information is usually entered using a local or remote keyboard.

The system file contains the following information:

- Routing information on all trunks
- Cell parameters (most of which are downloaded to the base station in an operational system)
- Cell numbers, channels, and channel addresses
- Handoff information (neighboring and directed retry calls)
- Hardware configuration, including peripherals and internal equipment
- System parameters

This information can be modified in real time from a keyboard.

Statistics Records

The statistics records are manufacture-specific and store details about the system performance, including system statistics, traffic and equipment statistics, and subscriber statistics. The statistics records hold such things as system outages, channel outages, channel blocking, congestion, channel usage, handoffs, call attempts, call completions (cellular and land line), traffic, and call-holding times.

PSTN Links

PSTN links are usually provided in multiples of digital streams of 24- or 30-channel unidirectional traffic (that is, the streams are either incoming or outgoing rather than two-way, although two-way steams may also be permitted).

Typically, the traffic will be about 70 percent outgoing and 30 percent incoming, depending on local calling habits. The cellular system must be designed to be compatible with the PSTN and so must use the same signaling format. The two most common standards are R2 (which has almost as many variant forms as there are countries using it) and CCITT signaling system number 7. The equipment supplier must have a detailed description of the actual signaling system used in the local PSTN so that the hardware and software can be tailored accordingly.

PABX

A PABX is generally not required but can be provided as an additional subscriber service to allow a personalized answering service. In some locations (for political or technical reasons), having a PABX as the PSTN interface may be necessary. Note that unless the PABX uses the same signaling system as the PSTN, additional and probably expensive decoding hardware and software will be needed at the cellular switch.

Roamers Files

Roamers files contain block-validation information (for example, all subscribers in a certain foreign group are permitted access) and negative-file information (for example, stolen units and bad debts).

Billing System

The billing system is a separate entity from the switch and may have no physical connection because the billing can be done by processing the raw data on the billing tape at a remote location. The billing computer for an average-sized cellular operator is a minicomputer. If the operator is a wireline provider, however, the billing may well be integral with the wireline billing and use a mainframe computer. For real-time billing (that is, billing available instantly when requested), a data interface between the computer and the switch is necessary. Most billing systems will have limited real-time capacity, so real-time billing is usually reserved for a subgroup of flagged customers. These customers are usually short-term renters.

The link may be RS232 or X.25; like other such world

"standards," RS232 has many non-standard forms, so the type of RS232 should be ascertained from the switch provider. The "standard RS232" is EIA-232D, published in January 1987, which conforms with CCITT V.24 and V.28 and International Organization for Standardization (ISO) IS 2110. These three references give functional, electrical, and physical standards, respectively.

The billing system may or may not also have an integral Management Information System (MIS) that also logs and analyzes system functions, such as channel usage, outages, traffic, and other housekeeping. The MIS may be interactive (that is, system commands such as subscriber validation can be input from the billing/MIS computer). Most billing systems expect a number of remote terminals to be operating simultaneously off the host computer.

Alarms

All alarms, both local and from the base stations, need to be reported centrally. Often the cellular switch features remote access to the alarm status to allow remote monitoring. Some systems also have remote access to each base station.

Alarms are usually divided into two or more categories, including major alarms (which affect service to a degree likely to be noticed by the user) and minor alarms (which may lead to partial reduction in capacity). The alarms are classified by the operation depending on the severity of the disruption to service of a particular fault. In a large city, the complete loss of one base station could be a minor problem, whereas in a small city that has only one base station the same loss would be a major problem. Lower levels of severity exist when, for example, a redundant unit is faulty and is switched out of service, or a simple channel is blocked in a large base station.

Routing

Dialed digits can be deleted or prefixed as required to ensure correct onward-routing. Alternate routing may be specified for various trunk groups, enabling the most economical route to be selected. This is particularly important when the switch operates over a number of PSTN charge zones.

CALLS TO/FROM MOBILES TO PSTN

For incoming calls (from the switched network to a mobile subscriber), the significant digits of the called number must be communicated to the control circuits. For example, if the mobile subscriber's number is OAB-

CDEFGH and OAB signifies a mobile number, only the last six digits are required. The mobile switch is thus structured as a group-switching stage.

HANDOFFS

Handoffs involve switching the call from a channel on one cell to a channel on another. The procedure is internally quite complex and sometimes results in a small number of calls lost within the switch. Typically, the switch handoff success rates are about 98 percent and losses attributable to other factors are dominant.

A handoff is initiated by the base station when the base scanning receiver detects low signal-to-noise ratio, low signal power, or a foreign carrier. The base station controller then requests a handoff from the switch.

CALL SUCCESS RATES

As in all telephone switching systems, many call attempts do not result in completed calls—the called party is busy or not in attendance (the main causes of incomplete calls) or because the call failed for system reasons. In a cellular switch, completed calls can be expected to be distributed as approximately follows:

- 30 percent land to mobile
- 65 percent mobile to land
- 5 percent mobile to mobile

These have success rates of

- 50 percent for land-to-mobile calls
- 85 percent for mobile-to-land cells
- 50 percent for mobile-to-mobile calls

The low success rate of attempted calls to mobile units largely accounts for the smaller traffic in that direction. This is mainly because the mobile is unattended, out of range, or switched off at the time of the call attempt.

INTERSWITCH OPERATIONS

The earliest cellular switches were often designed to operate as stand-alone devices and had no facilities for interswitch handoff or roaming. This meant that once a small switch was full, it had to be replaced by a bigger one. This problem was soon resolved by most cellular suppliers, and large nationwide networks like the Australian, Canadian, and Scandinavian soon evolved. However, these early networks (with the exception of Scandinavia) relied on the implementation with switches of one manufacturer only.

Some interesting early attempts at interswitch networking were made, particularly in the case where one manufacturer managed to edge out an established competitor. In this case, it would prove necessary to operate the two switches in some sort of coexistence, at least during the changeover.

An instance of long-term coexistence occurred in 1990 in Seoul, South Korea, where the operator placed 30,000 lines of AT&T AMPS equipment in parallel with a similar number of lines of Motorola. The solution chosen was to relegate one switch to service mobile-originated calls only, while the other handled mobile-terminating calls.

IS-41 HISTORY

IS-41 is a standard for intersystem handoff as developed by the TIA tr-45.2 committee. It was developed in two versions: IS-41-0, which allows interswitch handoff and precall validation, and IS-41-A, which additionally provides call delivery, subscriber data transfer, feature control denotation, roamer validation, and registration. It is important to note that even though IS-41-A was developed from IS-41-0, the two systems are incompatible. One of the main features of IS-41-A is that it uses the No. 7 signaling system and can thus be connected into an intelligent signaling network.

The next standard, IS-41-B, takes into consideration dual-mode operations, path optimization, faster call handoff, conference features, and other enhancements.

The path optimization of RevB was sorely needed, as the earlier versions allowed a condition known as "shoelacing" to occur. Shoelacing is the result of multiple interswitch handoffs and occurs because each handoff requires the temporary establishment of a link between the two switches. This link is not cleared down if a new handoff is undertaken but is held up as a series connection in the new handoff. In this way it is possible for one roamer to tie up all the links between two switches.

While there is no question that each IS-41 revision was a big improvement on the earlier ones, a situation arose in that carriers throughout the United States might have been equipped with no IS-41 or any of revisions 0, A, B, or C. Revision 0 is not compatible with any other revision, and although revision A and revision B equipment can communicate with and usually understand each other, they do not always have the ability to respond.

The cost of an upgrade to IS-41 was around $500,000 (depending on the switch).

Most manufacturers have introduced IS-41 as a translation device that accepts the regular interswitch signaling of the switch and converts it for external communications, as seen in Figure 19.28.

SS7 enables direct communication between switches on dedicated channels. In this way, the subscribers' database can be directly interrogated by a remote switch and the call can be routed through the network utilizing PSTN switches. While this gives seamless roaming, it does generally mean bypassing the tried and proven third-party negative files. As the cellular switches lack the powerful processing power of the clearinghouses, some important fraud indications may be missed. This can in turn be overcome by locking the clearinghouse into the SS7 network.

A lot of cellular switches are now communicating using the far less powerful X.25 protocol. Although it is effective for transferring data, X.25 does not allow the interaction that SS7 does.

Although clearinghouses in the United States have long provided roamer validation, the advantage of the IS-41 standard is that the validation is done in real-time during a call setup. This will decrease the potential for roamer fraud and give the subscriber a more efficient service.

ANSI-41

IS-41 was the interim standard-41 for mobile communications. IS-41C has been raised to a national standard and is now know as ANSI/TIA/EIA-41, but more commonly as ANSI-41. An excellent reference on this standard is *Mobile Telecommunications Networking with IS-41* by Michael Gallagher and Randell Snyder, McGraw-Hill. 1997.

Cloning

Mobile cloning is the most common kind of fraud in analog networks. The cloning consists simply of obtaining the MIN and ESN numbers of a legitimate mobile and copying these into a clone. MIN and ESN numbers in an unprotected network are easily obtained, either off-air or from data bases.

Part of the U.S. response to a tighter authentication routine was ANSI-41. This authentication applies today to suitably equipped AMPS, NAMPS, TDMA, and CDMA mobiles and is not unlike the GSM authentication procedures; to date GSM appears to be clone-free.

The ANSI-41 standard requires that the MIN and ESN form part of the ANSI-41 encryption code. The MIN is a 10-digit decimal number, which essentially identifies the mobile subscriber; it often (but not always) is the same as the subscribers phone number. The ESN is

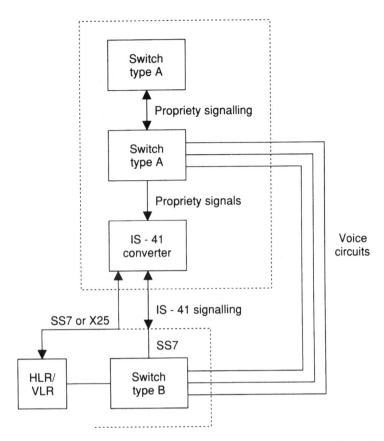

Figure 19.28 Protocol convertors are used to convert propietary signaling to IS-41.

a 32-bit serial number which is the phone's identification number. Although the ANSI-41 procedure uses both MIN and ESN, some operators are only using the MIN to identify the subscriber which compromises security.

THE A-KEY

The A-Key is a 20-digit number with an additional 6-digit check bit that is held both in the mobile (MS) and the authentication center (AC). This information is securely enbedded in the mobile and in the AC. The A-Key is never transmitted but is used together with the ESN, MIN, and a 56-bit random number, to generate the shared secret data (SSD). The SSD is used to authenticate calls, and is a semipermanent number that can be changed at the request of either the mobile or the AC. The generation of the 128-bit SSD is seen in Figure 19.29. The SSD is, in fact, in two parts, each 64 bits; the SSD_A, which is used for authentication and the SSD_B, which is used for encryption. The SSD ensures security by using three kinds of challenges. A challenge is a call from either a base station or a mobile to the other part of the network to send the SSD, in order to prove authenticity.

Global Challenge

Global challenges are issued on all registrations, call origination, and call terminations (i.e., whenever the mobile receives a call). Because they occur frequently global challenges use a lot of system resources.

Unique Challenge

A unique challenge can occur as the result of any event as specified within the AC. This may include the same events as the global challenge, or random or periodic time intervals, during calls or some other specified event. Most unique challenges occur over the voice channel and so mostly are specific to one mobile.

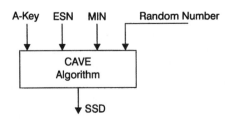

Figure 19.29 The generation of the ANSI-41 SSD.

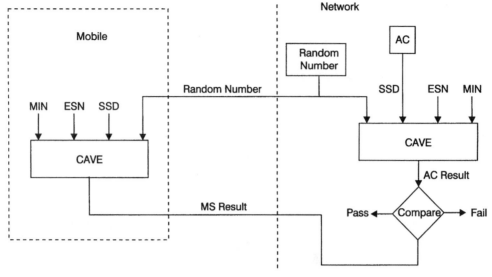

Figure 19.30

Base-Station Challenge

A base station challenge is the only challenge originated by a mobile to the network. Mobile cloning, surveillance, and tracking, are often done by using a "false" base station, that forces the mobile to re-register. In so doing, it will identify its MIN and ESN, and if and ANSI-41 code is used, it will reveal its SSD. However if the mobile initiates a challenge, then a new SSD will be generated, so that only if the base station's AC has the correct A-Key can further transactions take place. The basic authentication process is seen in Figure 19.30. A fraudulent user who has a valid MIN, ESN, and SSD can make phone calls until detected. The detection is mostly likely going to occur as a result of an SSD update, which can be brought on by a routine procedure. In the event of a verification procedure, when a verification fails, the network will usually respond with an update on the SSD.

The A-Key is the most critical part of the authentication. With the A-Key, a fraudulent user can make a perfect clone—one indistinguishable from the genuine mobile. Equipped with the known A-Key, it is easy enter it into the mobile because provision has been made to directly enter it from the keyboard. In fact, in TIA/EIA TSB50 "User Interface for Authentication" the method is spelled out. Just press 2-5-3-9 (A-KEY) then "function" + "function" + "store" and then enter the A-Key 26-digit number. Then a clone phone is easy to arrange once the A-Key is known. To use the clone phone however, the mobile will also need to know the current SSD. When the SSD is unknown the authentication will fail,

and as a result a new SSD will be generated. The perfect clone will then have a connection. The clue that there is a clone, will be that many "first attempts" at authentication will fail. In the case where a fraudulent user has obtained access, the SSD will be changed and so when the genuine user comes on line, it to will be denied access initially. Once a clone has been detected the genuine user will have to be given a new A-Key.

To secure the authentication process, it is desirable that the A-Key be a number that is secret to all parties including the user, the network personel, mobile phone manufacturers, and distributors. This is a rather difficult ask, and in many networks, the A-Key is far from secure. In fact there are a lot of systems where the manufacturers default key (which may be something as simple as a string of zeros) is used. It is relatively easy to determine the check bits once the A-Key is known (from the CAVE algorithm); so perfect clones will be easy enough to make. Although the authentication center will eventually catch the clone, the fix, which involves a new algorithm for the legitimate user, may cause considerable inconvenience.

There are a number of ways to store and enter the A-Keys. They can be provided by the manufacturer, together with a database to be used on the provider's network; alternatively the number can be provided by the AC and distributed to the mobiles as required. However they are stored, security is enhanced if access to the numbers is difficult. Surprisingly since in ANSI-41 encryption is option, many operators have chosen not to implement it, probably due to the mistaken belief that digital modulation is the equivalent of encryption.

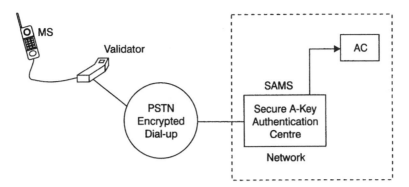

Figure 19.31 Synacom's secure A-Key center.

Over the Air Activation (OTASP)

OTASP is destined to become the main way in which mobiles are activated. Analog phones generally are not capable of this form of activation, but it is a standard feature of digital phones. There are still some security issues to be dealt with and some providers are still uncomfortable with sending this kind of information to air.

An Implementation by Synacom

A secure way of managing the A-Key is provided by the Synacom "Secure Key Management System" (SAMS) which generates random A-Keys, stores them in an encrypted form, and directly validates a mobile over the PSTN through a validator that reads the encrypted A-Key. The validator reads the mobile's ESN and sends it to the SAMS, which in turn returns an A-Key based on the ESN. Also system information such as SIDs and other NAM parameters, can be automatically loaded by the validator, saving the mobile provider at lot of programming time. In this way it is not necessary for human intervention and so security is significantly enhanced. The SAMS is shown in Figure 19.31. The SAMS system may be centralised, or it may be distributed over different parts of the network.

As an adjunct to the SAMS, Synacom provides a secure authentication center (SAC) and provides the AC functions together with a secure interface into the SAMS.

DISCONNECTION

Disconnection ordinarily occurs when either party hangs up, but the release can be missed if the subscriber drives out of range (for example, into a garage) before con-cluding the call. Because it is possible to use voice-operated switching (VOX) as a power-saving feature, loss of carrier alone does not necessarily mean that the mobile is out of range. For this reason, it is necessary to have an audit instruction that instructs the mobile transmitter to key up. If it fails to do so, the call is then disconnected.

UNCHARGED LOCAL CALLS

Many networks worldwide have uncharged local-call access. This approach presents no problems except in those countries, all of them developing countries, that have taken it further and offer local-access (uncharged) calls only from PSTN numbers as an option, and have subsequently installed telephones without charging or metering equipment.

These numbers are normally barred access to any trunk numbers and thus cannot reach mobile telephone numbers unless the mobile telephone system has a local telephone number. In either case, they cannot be charged for local calls originating from local-call-only numbers. This problem is usually overcome by adding a PABX with access permitted to "local" subscribers, who then book a manual call to the operator, as illustrated in Figure 19.32. Because this method is clumsy, it is a good idea in this situation to establish a new category of mobile subscribers who can (optionally) elect to receive local calls from non-metered numbers at their own expense.

In order to allow access by uncharged subscribers, a special access code is needed to trunk the barred subscriber directly to the mobile switch. The cell would be marked from the barred terminal exchange as a non-metering call. It would be diverted automatically to the mobile switch, where it would be connected normally

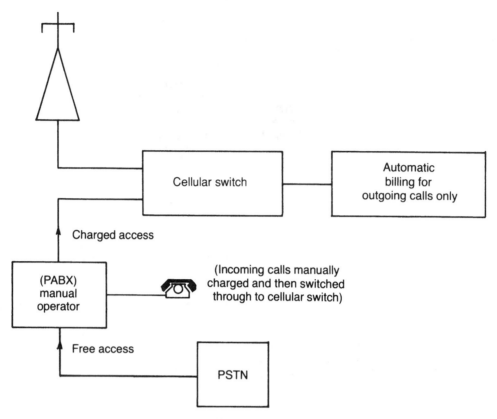

Figure 19.32 A PABX can be attached to the cellular switch. Manual operation may be necessary when local calls are (generally) uncharged and automatic line identification is not available.

(except that a tone would be sent to indicate to the mobile subscriber that the call is a reverse-charge call).

The worst aspect of using operators to manually connect calls is the same problem that led to a strong drive from the 1950s through the 1970s for automation—the cost escalates linearly as the system grows. Because cellular switches can be as large as 1,000,000+ lines, the number of operators needed can be staggering.

During the normal day shift (and dependent somewhat on caller patterns and habits), operators are needed at the rate of 1 per 100 customers manually connected. The operators, of course, handle only incoming calls from the PSTN to the cellular switch. Outgoing calls from a mobile are fully automatic.

DTMF overdialing could be offered as a solution to this situation and would require the land-line user to dial the PABX (which automatically answers) and then dial further digits to indicate the mobile number sought. This solution has two drawbacks: the first is that a network without international direct distance dialing (IDD) access is unlikely to have DTMF telephones; the second is that, without automatic-line identification (again unlikely to

be available), only reverse-charge (to the mobile) calls can be made.

SWITCH CONFIGURATIONS

A telephone exchange consists of a switch to which 50 to 1,000,000+ customers are typically connected. These subscribers normally have direct dialing access worldwide, which means the calls have a very wide dispersion. Consider a hypothetical town that has three subscribers' telephone switches, as shown in Figure 19.33.

If the town is relatively isolated, a fairly high percentage of the total traffic can be carried by the inter-exchange routes, so it would be justified to have direct trunks between those exchanges. When traffic to another area is considered (for example, to a town 200 km away), the traffic will be relatively light and the economics of providing three separate routes to the distant town may be rather poor. In this instance, a hierarchical switch, called a trunk switch, can be used effectively to

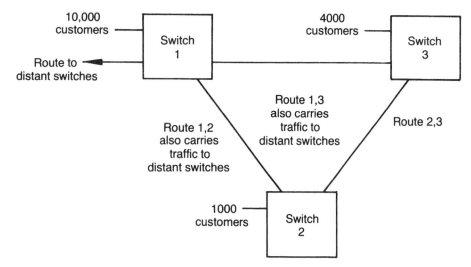

Figure 19.33 Although it may be economical to directly connect local switches in a small town, it is often more economical to route all distant traffic via one (usually the largest) local switch.

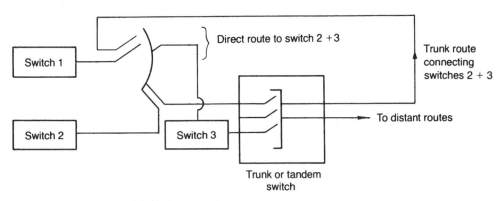

Figure 19.34 A simple trunk switch configuration.

concentrate the three traffic streams into one. Switch 3 in Figure 19.34 is a trunk switch.

All traffic routes from any switch in the town that are too small to justify a direct circuit can be switched through the trunk switch. In practice, the trunk switch may physically be one of the three subscribers' switches, but the trunk portion of it functions as the trunk switch shown in Figure 19.34.

This principle is also applied to international calls; a few switches collect all international traffic and disperse it to distant destinations. At the distant end, calls are routed to their destination through successively lower-ranking trunk exchanges until they finally arrive at the desired terminal exchange.

Modern digital switches do not employ a rigid hierarchical structure, but rather are able to reconfigure their routing to take advantage of the best available route for any call. Thus, a local exchange can be both a terminal

and a trunk exchange and can handle transit traffic like a tandem. Exchanges that can perform this function are called *nodes*.

There are many possible trunking schemes that can be used for cellular radio; the most common schemes view the cellular switch as either a terminal or a trunk switch. The type of trunking scheme and the type of access code are related in cellular systems. Those that use access codes similar to long-distance codes (that is, they begin with a 0 or a 1) will switch as though they were in fact distant switches. Systems that use local telephone numbers will be connected to the PSTN in the same manner as a local telephone switch.

Because of the wide community of interest in both schemes, the cellular switch is best connected at a high level in the switch hierarchy. Regardless of which trunking scheme is used, there are some common considerations of network security.

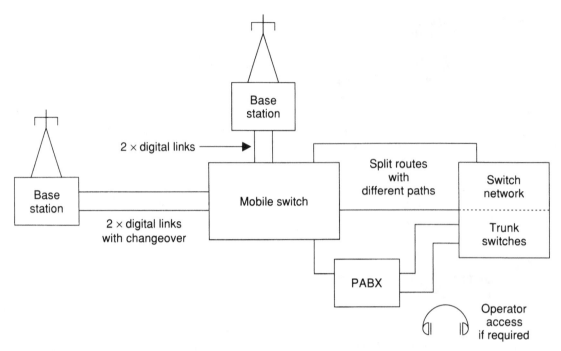

Figure 19.35 A simplified cellular network consisting of switches, base stations, links, and PSTN.

Figure 19.35 shows a typical trunking scheme of a mobile switch, with dual routes to each base and to each trunk switch. It is a good idea to use route diversity with half the circuits on each route and with automatic changeover in the event of failure of one route. This can be achieved economically using modern digital links. Similarly, when an expansion to more than one switch is contemplated, it is worth considering placing the second switch in a different physical location.

SWITCH HIERARCHY

It is recommended that the mobile switch be placed high in the network hierarchy. The advantage of high-level trunking is that, in general, fewer switches are required for the average call, so trunking costs and transmission losses are minimized.

To visualize the relative position of a mobile switch, consider a typical city network, as depicted in Figure 19.36. The mobile switch ordinarily draws its customers from all over the city, and therefore has no geographical center of interest [except perhaps the central business district (CBD)]. If the mobile switch is connected low in the hierarchy—for example, at a primary center—then most calls must be routed through a number of switches before reaching their destinations. This switching delay is true for both mobile-originated and mobile-terminated calls.

The worst possible connection level from the viewpoint of network efficiency is at the local switch level, where, the 2-wire transmission is the first problem. All mobile circuits are 4-wire, and so a 2- to 4-wire hybrid must be used on all routes resulting in unnecessary losses. Further, as Figure 19.36 shows, the community of interest of a mobile customer is the whole city, and most calls must be routed to a different area using the network hierarchy. This involves more switching paths, and hence more loss than would be the case for higher-level switching. The mobile switch ordinarily signals other exchanges in the standard format of the country (for example, CCITT's signaling system number 7 or R2).

Switch Location

Wireline operators have traditionally placed their main switches at what is known as the "copper center." The cheapest location for a land-line switch is the one that minimizes the total length of cable to the subscribers (and other exchanges). The location that minimizes the amount of copper (or total cable length) is known as the copper center.

Using a strictly mathematically accurate configuration to minimize the total length of copper is complex. However, the copper-center concept is relatively simple, as you can see in Figure 19.37. Here, the switch is located at the "center of gravity" of the copper mass feeding A, B, and C. For this reason, conventional

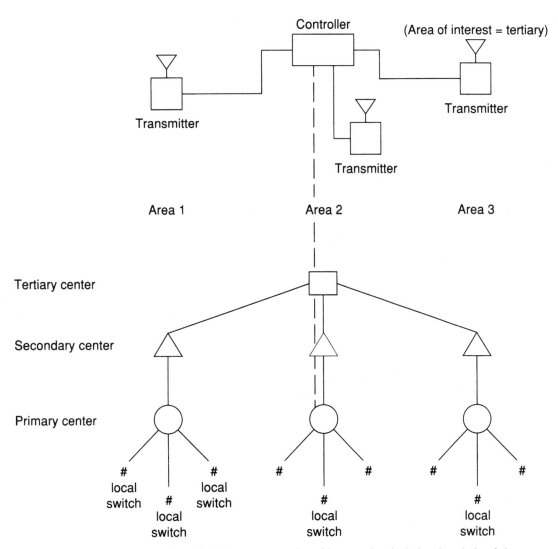

Figure 19.36 The mobile switch has a community of interest that includes the whole of the local service area of the PSTN. The switch should therefore connect at a high level in the trunk network.

wireline switches are usually found at the population centers.

It should be noted that cellular switches do not have subscriber links. Instead they have "trunk" links to the bases and the PSTN. Because only a relatively few links are involved—and with modern technology the cost of these links is not strongly distance-dependent—the concept of "copper center" is no longer dominant in choosing a switch site.

A wireline operator almost invariably chooses to place the cellular switch in an existing high-security building, which for the wireline operator is the cheapest location. Such a site will be even more economical if the chosen building has a trunk switch (that is, the trunk link costs will be low).

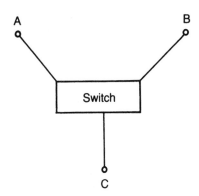

Figure 19.37 The copper-center concept. A PSTN subscriber switch is placed at the copper center to minimize the total length of cable used.

Also consider that the links to the base stations and the switch site should be well-located with respect to the bases. Base-station links are generally microwave or optical fiber. When microwave links are used, ease of link access to the switch site is a primary consideration. Whether microwave or fiber-optic links are used, the link cost is not strongly dependent on the link length.

NON-WIRELINE SWITCH LOCATIONS

When non-wireline switches are used, there is a wide choice of switch locations. Rental properties can be used for switch rooms, but are not recommended for the following reasons:

- Insecurity of tenure
- Lack of security where other tenants are involved
- Cost of moving the switch should it ever be necessary
- Unscrupulous landlords, who know the cost of relocation, could demand unreasonable rents
- Lack of control over the building, its expansion, and use

Purchased properties are much more practical for switch rooms. Because there is no real need for a central location, costs can be controlled by locating the switch outside the expensive inner-city area. A site that minimizes the total number of links will probably also minimize the cost.

SWITCH TO BASE-STATION LINKS

In large systems with multiple base stations it will be necessary to have a number of nodes for the microwave networks that ultimately converge back to the switch, as seen in Figure 19.38. Note that a scheme that links all base stations back to the switch directly will cause problems with finding enough space to mount the microwave dishes at the switch, as well as placing severe restraints on the base-station locations, as they will need to be line of sight to the switch. Additionally, link frequency congestion can be expected if this is done. In all but the smallest systems, multiple node links will have very great advantages over a single node at the switch.

In large systems the planning of the link networks should be done to optimize the frequency reuse of the microwave links. This involves identifying suitable nodes, which will be sites that have good line of sight to a number of base stations, arriving at a frequency assignment plan that maximizes reuse (see Chapter 13, "Cellular Links").

Consideration should be given to providing redundancy on the routes that support two or more base stations.

SIGNALING

Signaling is the means by which information about the dialed digits, and the line condition and other network information is passed around the network. The early signaling systems were based on DC pulses; around 1950

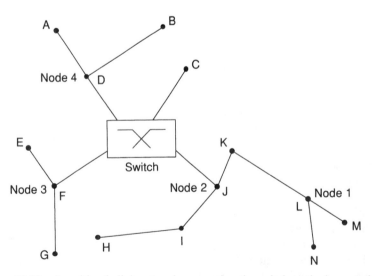

Figure 19.38   A multinode link network connecting the switch to the base stations.

a series of systems based on inband and out-of-band tones were developed. These tone-based signaling systems form the basis of the modern signaling formats although they now are more widely used in digital versions.

The most common form of signaling over trunks today is common-channel signaling, although voice-channel signaling is still widely used. In common-channel signaling, a different circuit is used for inter-switch communications from that used for speech. Common-channel signaling enables switching to occur at very high speeds and means that voice channels are not tied up with signaling. The standard 2-Mbit, 32-channel system uses two of those channels exclusively for signaling.

Inband and out-of-band signaling use the same channels as those that are used for voice. Voice frequencies are assumed to be those in the bandwidth 300 to 3400 Hz. An early inband tone is the CCITT R1 signaling system, which uses a 2600-Hz continuous signal tone (with a notch filter to line for voice). Cellular radio AMPS systems use 6-kHz out-of-band SAT tones to identify and control interference.

Although there are a number of standardized signaling formats, it must be appreciated that within the standards there are user-definable spare codes. Almost every operator worldwide has managed to find some use for these spare codes and the interconnecting cellular operator needs to carefully examine the local version of the signaling system, since often the absence of the optional code bits will cause signaling alarm errors. The cellular software should be able to be set to accommodate any of the usual variants.

DC or Loop Signaling

Necessarily, the early signaling systems were very simple and were based on line pulses that alternately placed a "1" or "0" state on the line. In its simplest form, as seen in a rotary-dial telephone, this is achieved by simply placing a short-circuit or open-circuit condition on the line. In the early step-by-step switches these pulses were translated directly into the physical movement of switch wiper blades, which in turn switched the call.

Use is made of pauses to indicate the end of a signaling system. For example, the pause between dialing one digit and the next on a rotary telephone is used by the system to determine the end of one string of pulses and the beginning of a new one.

Line-polarity reversal is used to indicate certain states, including that the called party has hung up.

A major disadvantage of loop signaling is that the pulse waveforms deteriorate badly over long lines and that line pulses had to be separately reconstituted where amplification was used. This was usually done by converting the pulses to tones, which were ultimately converted back to pulses.

R1 Signaling

R1 signaling is composed of a line-signaling part and a register-signaling part. This is illustrated in Figure 19.39.

Line signaling is accomplished on the speech path either during the presence or absence of speech. The line signaling is done using a single 2600-Hz tone and its presence or absence is the basis for the signaling. The line contains filters at the subscriber's end to filter out the tone.

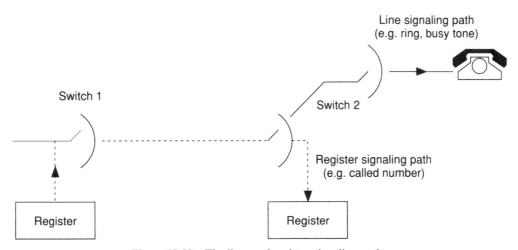

Figure 19.39   The line- and register-signaling paths.

The registers form the "memory" of the switch and are used to store and forward the dialed digits. When a subscriber dials a number it is stored in the register, which can begin to switch the number as soon as sufficient digits have been dialed to determine which is the next switch.

Register signaling is done using two simultaneous tones chosen form a total of six. This is called *multifrequency code* (MFC), sometimes known as "two out of six MFC."

This signaling is fully specified in the *CCITT Blue Book Fascicle vi.4*.

R2 Signaling

Analog R2 uses an out-of-band signal of 3825 Hz for line signaling and a total of 15 forward and 15 backward inter-register signals. It is frequently used for both national and international signaling.

R2, like most other signaling systems, comes in both an analog and a digital version. The analog version is listed in Table 19.1.

The signal type can have two meanings, in the forward direction, as defined under group I (selection) and group II (service class of calling party). Table 19.2 lists these meanings.

The backward signal direction also has two signal groups, called group A (certification and control) and group B (status of calling party). These are detailed in Table 19.3.

R2 signaling is fully described in the *CCITT Blue Book Fascicle vi.4*.

TABLE 19.1   R2 Signaling Tones

		Frequencies Hz					
	Forward	1380	1500	1620	1740	1860	1980
Signal	Backward	1140	1020	900	780	660	540
1		×	×				
2		×		×			
3			×	×			
4		×			×		
5			×		×		
6				×	×		
7		×				×	
8			×			×	
9				×		×	
10					×	×	
11		×					×
12			×				×
13				×			×
14					×		×
15						×	×

TABLE 19.2   R2 Signaling Group A

Signal Number	Group I Signal	Group II Signal
1	Digit 1	Subscriber without priority
2	Digit 2	Subscriber without priority
3	Digit 3	Maintenance equipment
4	Digit 4	Spare
5	Digit 5	Operator
6	Digit 6	Data transmission
7	Digit 7	International use
8	Digit 8	International use
9	Digit 9	International use
10	Digit 0	International use
11	Spare	Spare
12	Request not accepted	Spare
13	Access to test equipment	Spare
14	Spare	Spare
15	End of sending digits	Spare

CCITT No. 5

The CCITT No. 5 signaling system is meant for interswitch signaling and has largely been superseded by CCITT No. 7. It consists of one or two inband line signals and two out of six MFC inter-register signals.

A full description can be found in *CCITT Blue Book Fascicle vi.2*.

CCITT No. 7

CCITT No. 7, known as SS7 in the United States is the most widely recommended interswitch signaling system,

TABLE 19.3   R2 Signaling Group B

Signal Number	Meaning of Group A	Meaning of Group B
1	Send next digit	Subscriber's line free
2	Send digit before last	Subscriber has changed number
3	Change to reception of group B signals	Subscriber line busy
4	B signals congested	Trunk congestion
5	Send calling party's category	Unallocated number
6	Set up speech path	Metering line
7	Send second last digit	Non-metering line
8	Send third last digit	Called line out of order
9	Resend whole number	Spare
10	Spare	Spare

and it uses common-channel signaling (that is, the signaling is done on dedicated channels).

For economical operation of a common-channel signaling system, signaling transfer points (STPs) can be used to provide dynamic routing of the signaling traffic.

SS7 signaling is one of the most recent of the signaling systems, and has been designed to allow interactive networking of switches. The information exchange using this format can be quite complex, and interacting switches can be made to operate in a cooperative way that is not possible with the earlier signaling formats.

Increasingly, SS7 (sometimes referred to as intelligent signaling or intelligent network) is becoming the signaling system of choice of the cellular operator. GSM uses it between switches and base stations as well as between switches. Even analog systems can use it for interswitch communication and more particularly for communication between PSTN and cellular switches. The successful implementation of IS-41 in the United States depends on this signaling. It can virtually eliminate tumbling fraud, as it will allow interrogation of the home subscriber database in real time.

The digital form is a full-duplex signaling system operating at 64 kbits/s (56 kbits in the SS7 version), an analog version that can operate over 3-kbit channels is available.

SS7 operates only on dedicated lines and other line conditioning such as echo suppressors, and A or mu Law equipment must not be used.

When performing high-level control it is similar to high-level data-link control (HLDC). Full details can be found in *CCITT Blue Book Fascicle vi.7, vi.8,* and *vi.9.* CCITT publications can be obtained from:

I.T.U.

General Secretariat—Sales Section

Places de Nations, CH1211

Geneva 20 Switzerland.

Typical costs will be $20–$70 per Fascicle.

INTERFACING SWITCHES

The signaling used between two switches will usually follow one or more of the above standards. For two switches to be interconnected, they must have a compatible signaling system.

In modern switches it is usual that all of the above signaling types would be available, but apart from E&M most of the others will be options that will cost around $40,000 to $100,000 (US$).

When purchasing a switch, it is essential to ensure that it will be compatible with the PSTN to which it will connect.

Devices do exist that can interconnect two dissimilar signaling systems, and these can be purchased as stand-alone devices.

It should be noted that the optional parts of the various signaling systems are widely used for custom purposes, and it will usually be necessary for the interconnecting cellular operator to match the "modified" version used by the local PSTN.

SYNCHRONIZATION

Virtually all modern digital systems require synchronization in order to interwork. The cellular switch and the digital transmission systems will need to be synchronized to the networks to which they interconnect.

The cellular system can be either slaved to the PSTN, which presumably has a high-stability reference clock such as a cesium beam, or it may be synchronized to its own high-stability clock. If it is slaved, then it should be connected to two separate clock sources to ensure redundancy.

Large network operators will usually have a number (2–20 depending on the network size) of high-stability clocks distributed throughout the network to which the rest of the network can slave.

The slave system will usually use the master clock to synchronize its internal clock, so that should the source of the master fail (as, for example, during transmission fades), its internal reference can keep the system running.

Where the clocks fail to synchronize there will be data loss as a frame slip, which can be either chronic, a prolonged frequency offset (PFO), or a transient synchronization loss (TSL). The latter is more likely to occur when the slave clock overreacts to a noise pulse or fails to stabilize when a changeover to a redundant clock source of arbitrary phase produces frequency swings. Between nodes the frequency accuracy should be 1×10^{-11} or better.

INTELLIGENT NETWORKS

A conventional telephone network is a simple transport system. It connects one fixed service to another (so transporting the call). Business people typically have a fixed phone, a fax, and a mobile phone. All of these are independent and all of them are basically used to establish or terminate calls. While this has been adequate for almost a century, the shortcomings are starting to become obvious.

INs have long been whatever the vendor says they are. INs separate functionally the service and call processing logic. This means that new services and facilities can be grafted onto the network with the minimum of fuss.

The IN concept has been evolving since the 1960s. Originally all of the processing power of the switched network was integrated with the switch. As more and more functionality evolved, it became common practice to make the processing part of the switch more stand-alone (if for no other reason than the ease of a processor upgrade).

Soon the concept of a switch that followed the commands of an external processor, evolved into a switch plus a powerful computer that controlled the switch, but also was capable of doing a lot more service functions.

By standardising software modules, additional functionality can easily be added, and by standardizing hardware interfaces, multivendor supplies can be assured.

At the risk of oversimplifying, the conventional network is essentially a purpose-designed hardware/software implementation that has a range of pre-defined capabilities and a set of options. Beyond its designed-in capabilities, it can do no more without a major and expensive upgrade that must be done in the manufacturer's laboratories. A seemingly simple additional piece of functionality may require a massive design effort input only because there was no allowance made for this change in the original design.

An IN network, on the other hand, is built to be flexible. Much like today's PCs, it comes with a certain amount of built-in ability, but is very modular in the design of both its hardware and software so that changes and enhancements can easily be made. These latter include things like virtual private networks (VPNs), universal personal telecommunications (UPT), and toll-free phones.

Intelligent Cellular Interfaces

Perhaps one of the first applications for a cellular operator to use an IN solution is the HLR. This database can be moved off the switch and onto a service control point (SCP), as seen in Figure 19.40. While the SCP communicates directly with the switch via SS7 or some other protocol, new services can be provided by the SCP using software and computing hardware that is familiar to the programmers, who need not necessarily have any detailed knowledge of switching.

The SCP can use commercial high-end computers that have more memory and faster access times than switch-based processors, and so can serve very large customer bases. In mission-critical applications, it is possible to operate two SCPs redundantly in different locations to ensure the survivability of the network.

The SCP solution is particularly valuable for carriers that use switches from multiple vendors, as these

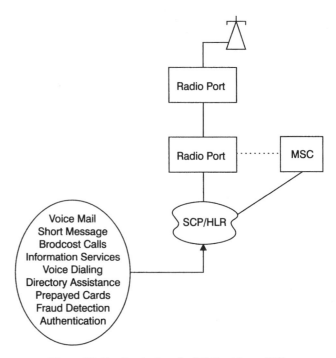

Figure 19.40 Replacing the HLR with an SCP.

switches will require expensive equipment-supplier software upgrades for new services, the price of which may be even higher if the features and services across the network must be identical.

Because the SCP HLR has a global view of the network (basically, it knows who has a phone switched on and the status of every phone in the network), the SCP solution is ideal for Centrex applications and message forwarding.

Travelers may find that they can move throughout the country and receive calls on their mobile phones without any action on their part to redirect the calls. This is done by the network keeping track of the user's location and rerouting the calls accordingly. This may, however, not be enough. When the traveler goes to an area without mobile coverage, the "follow-me" facility is lost. While traveling it may be desired that only urgent calls be diverted, so a call-screening facility would be needed.

An intelligent network is one that provides a complete communications facility to the user rather than simply transporting calls. All of the features of follow-me interaction between the landline network and cellular, call screening, and other special features would be available to the user on a universal number, regardless of location. So while the conventional network switches calls, the intelligent network knows where the user is and how the call should be treated, and it will screen the call in much the same way as a secretary.

To enable intelligent interaction of the switching network it is necessary to divide the functions into switching

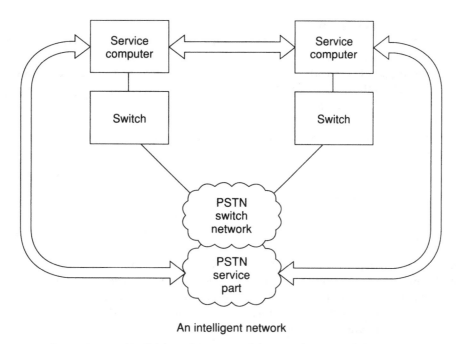

An intelligent network

Figure 19.41   The division of the network into service and switching parts.

and service, as shown in Figure 19.41. In this structure all calls are processed by the service computer (which is physically separate from the switch) before switching occurs. The service computer will check on the current status of the call (location, screening instructions, validity, authentication of the caller) before passing on the switching instructions to the switch. To do this, it may be necessary for the service computer to interrogate one or more other services before the call can be set up. The switch network will continue to use much the same conventional hardware and software, but instead of switching strictly according to the dialed number, it will switch per the instructions of the service computer.

The link between the intelligent processor (from the SCP) and the switch part of the network [service switching point (SSP)] will be by some protocol such as SS7, as shown in Figure 19.42. There is no need for the SCP to be co-located with the switch, and it can serve any

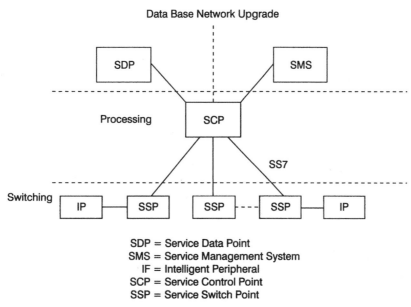

SDP = Service Data Point
SMS = Service Management System
IF = Intelligent Peripheral
SCP = Service Control Point
SSP = Service Switch Point

Figure 19.42   How the SCP connects to the networks.

number of switches; there may also be any number of SCPs in the network. Every switch must have some access to an SCP to function, and the SCP is the real "intelligence point" behind the whole network.

The service data point (SDP) is the network database, and it may contain a great deal of data, especially for things like UPT services (where perhaps hundreds of thousands of numbers are cross-referenced) or HLR and VLRs in cellular services.

The intelligent peripherals (IPs) are things like interactive voice services and voice recognition services.

The service management system (SMS) is the point where new software is introduced and distributed into the network.

Of course, with increasing capabilities there is a corresponding increase in the complexity of the billing services. In fact, simply sorting out how the IN services are charged, especially when (as is often the case today) the service may well involve transport across more than one network, and how costs are apportioned, are extremely difficult.

Cellular Implementation

These functions are performed in the cellular environment to a somewhat limited extent by the HLR/VLR.

The service computer has been named the SCP, and it is the point at which the HLR/VLR data are entered. Communications between SCPs will permit the transfer of service data as required. Once all switches (both cellular and wireline) have an associated, compatible SCP, it will be practical to issue universal numbers. Now communications are person-to-person rather than telephone-to-telephone.

Other examples of limited intelligent networking are nationwide centralized calling (like the 800 numbers in the United States); time of day switching, with which local telephone numbers, for example, for airlines, are automatically diverted to a single (or sometimes multiple) after-hours location; and call diversion on busy or no answer. A truly intelligent network will have all of these features (and more) available nationwide and with the same access format so that the user does not have to account for location.

In the United States, the ANSI-41 interswitch links are a step toward an intelligent network. Lack of coordination in the early days of cellular has led to different access codes for different features and non-uniform feature sets, which needs to be resolved for the future. This has occurred because individual operators were able to define their own feature sets, and they did. Most other countries with nationwide networks will already have uniform procedures in place.

20

TRAFFIC ENGINEERING CONCEPTS

The measurement of telephone or circuit traffic and its application to circuit dimensioning (provisioning) is fundamental to all large-scale communications systems. Traffic is measured in Erlangs (most simply, an Erlang is one circuit in use for one hour). The traffic on one telephone line can be measured with an ammeter or voltmeter, as illustrated in Figure 20.1. Indeed, this was how traffic was measured in days gone by.

When a line is looped (that is, the handset is off the hook), a direct current (DC) current of about 50 mA flows. This current can be detected by an ammeter, and the DC voltage drops from the open-circuit value of about 50 volts to about 5 volts. The actual loop current depends on the loop (line resistance), which can vary from 0 to 1500 Ω. When the handset is replaced, the current flow is zero and the line voltage returns to 50 volts.

A more modern traffic meter uses a microprocessor connected to an interface that monitors many lines simultaneously, and it constantly scans each line to determine whether the line is in use. Figure 20.2 shows an example of a traffic measurement system.

Modern switches collect traffic statistics in real time by writing all call details to memory.

Traffic measurements in cellular systems should be examined on a monthly basis to ensure adequate network provisioning. Traffic can be represented by a number (of Erlangs), but remember that traffic varies with time and any practical representation is a compromise.

Fortunately, not much traffic engineering is involved in cellular radio, but you should understand some basic concepts.

A cellular system will have a traffic distribution that varies with time somewhat like the one shown in Figure 20.3. The total volume V of traffic carried, measured in Erlang hours, is

$$V = \int I(t) \cdot dt$$

where $I(t)$ = Instantaneous traffic at time, t

This integral is a crude measure of the cellular operator's revenue, particularly since the call charge is largely an air-time charge. Thus, $V \times$ air-time charge $\simeq$ air-time revenue.

This leads to the concept of call-holding time (the time that a call is held up). From the previous equation you can see that a call lasting one hour will generate the same traffic volume as 20 three-minute calls over the same period. The call-holding time of a typical cellular subscriber is 120–180 seconds.

The average calling rate per subscriber is the total traffic at the measured time divided by the number of subscribers. Ordinarily the rate is quoted in Erlangs per subscriber per hour. This figure determines the number of subscribers who can be placed on any given system, and it varies immensely. Average calling rates from 0.005 to 0.045 Erlang/subscriber have been recorded and it would appear that the calling habits of different countries are very diverse. A calling rate of 0.005–0.015 could be regarded as normal.

These rates are also quoted in milli-Erlangs (0.001 Erlang = 1 mE). To calculate the total system traffic, it is merely necessary to multiply the average calling rate by the number of subscribers.

Notice that traffic varies in many ways with time. These are the most significant variants:

- Instantaneous variation of call arrivals.
- Hourly variations that depend on demand; cellular

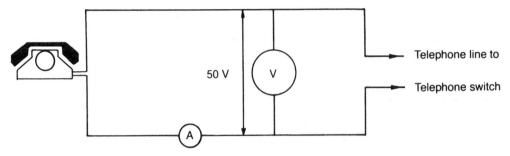

Figure 20.1 A looped (in use) telephone drops the circuit voltage from 50 volts (open circuit condition).

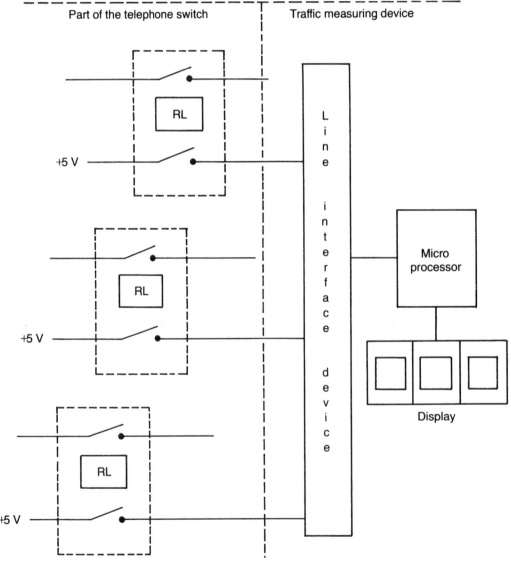

Figure 20.2 The line relay RL, which activates when the telephone circuit is seized, can be used to measure traffic. A simple integrating device is used to measure the total traffic (in Erlangs).

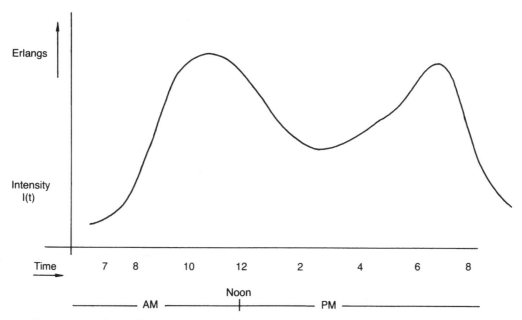

Figure 20.3   The traffic density is shown here as a function of time of day. The traffic carried by a cellular system shows two peaks characteristic of the prime "drive times." The peaks occur early in the morning and late in the afternoon.

radio usually has an early-morning and late-afternoon peak with a lunchtime minimum. The cellular peak is different from the public switched telephone network (PSTN) peak, which generally occurs either side of a lunchtime minimum. The cellular peak occurs during drive times; that is, the cellular peak occurs during transit to and from work. The PSTN peak occurs after arrival at work.

- A daily pattern is usually distinguishable. There is a marked difference between weekdays and weekends, and there may be significant day-to-day variations during weekdays.

- Seasonal peaks that occur in the PSTN (for example, Christmas, Easter, Mother's Day) are likely to be reflected in the cellular network, although that reflection is sometimes negative (lower traffic as a result of holidays).

- Tariff variations can cause significant (but usually temporary) variations in the call rate. These effects, for cellular radio, usually last only three to six weeks before the original calling pattern returns.

- Long-term variations in traffic can occur over periods of months or years.

Because it is not usually economical to provide circuits (or for cellular base stations, channels) to cater to the peak demand, a compromise based on the provision of an acceptable grade of service in most instances is usually adopted.

Although traffic is measured in Erlangs (recall that an Erlang is defined most simply as one circuit in use for one hour), in practice many different interpretations of the unit of traffic are used. The two main ones are instantaneous Erlang and busy-hour traffic (in Erlangs). Instantaneous Erlang means the number of circuits in use at the instant in question. Busy-hour traffic means the average number of circuits per unit time in use over the busiest hour. The traffic readings are usually taken in half-hour periods over the day, and the sum of the two adjacent half-hour blocks with the greatest total is defined to be the busy-hour traffic. This is why the busy hour usually starts either on the hour or the half hour. Table 20.1 shows the busy-hour calculation.

TIME-CONSISTENT BUSY-HOUR TRAFFIC

The concept of time-consistent busy-hour traffic (TCBH) is another measure used for dimensioning telephone cir-

TABLE 20.1   Busy-Hour Calculation

Time	Traffic (Erlangs)
09.00	4.3
09.30	4.8
10.00	7.2) Largest adjacent readings – Busy-hour traffic =
10.30	6.1) $(7.2 + 6.1)/2$ 6.65 Erlangs
11.00	5.1
11.30	6.2

TABLE 20.2 Time-Consistent Busy-Hour Averages the Busy Hour Over a One-week Period

Time	Traffic Mon.	Traffic Tue.	Traffic Wed.	Traffic Thu.	Traffic Fri.	Average of 5 Days
09.00	4.3	2.1	4.0	3.9	2.9	3.44
09.30	4.8	4.0	5.1	3.9	4.1	4.38
10.00	7.2) 6.65	6.1	3.0	4.7	4.7	5.14
10.30	6.1)	5.4	4.5	4.8	5.7	5.3)
11.00	5.1	6.4	6.1	4.8	5.6	5.6)*
11.30	6.2	3.9	3.1	4.8	4.0	4.4

* Highest adjacent sum equals TCBH traffic.

cuits. It is only a concept and has no real physical meaning. Despite that, however, circuits are provided on this basis. TCBH attempts to obtain a weekly average traffic measurement. One should be aware that TCBH is variously defined and because of this different administrations will obtain different values of TCBH from the *same* set of data.

A possible implementation of this concept is best illustrated by an example. Table 20.2 shows example base-station readings taken over a week. In this example, the traffic is measured as before, for every working day. Then, for each half-hour period over the five days, the average traffic measurement is found and put in the average traffic column. From the average traffic column, the busy-hour traffic is found as before; in this case, it is $(5.3 + 5.6)/2$ or 5.45 Erlangs. Notice that Monday's busy hour is the highest (6.65 Erlangs). The circuits are provided only for 5.45 Erlangs, so the nominated grade of service (probability of congestion) applies only to this theoretical TCBH traffic and not to any particular day.

Another definition of TCBH is to *define* the time at which the traffic is to be measured and measure all traffic at that hour. Typically peak-hour times such as 10 A.M. to 11 A.M. or 3 P.M. to 4 P.M. may be used.

MEASUREMENT OF CONGESTED CIRCUITS

In congested circuits, it is likely that for significant periods of time all circuits will be busy. Because of the random nature of telephone traffic, even congested systems occasionally have free circuits, and so the traffic carried is always less than the number of circuits.

Methods exist to determine the offered traffic on such circuits from the measured average traffic, but these techniques produce uncertainties that increase rapidly as the grade of service increases (that is, as it gets more congested). Once the network becomes congested, it is difficult to measure traffic accurately enough to determine the offered traffic (and hence number of circuits

needed) with any certainty. This is even more difficult in cellular radio, because traffic that cannot be directed to the nearest and best cell often "overflows" to the next choice cell. Thus traffic measurements from a congested system should be used with considerable caution.

GRADE OF SERVICE

A grade-of-service (GOS) figure is sometimes used to express the probability that a call will be lost due to switching or transmission congestion. Because it is a probability, the highest value it can have is 1.0; all calls will fail on any system that has this GOS. In cellular systems that are not faulty, few calls are lost. Calls that encounter congestion simply resend. The GOS in this case is a measure of the probability of a resend.

Like banks, which rely on the fact that it is unlikely that all their customers will require their money at the same time, telephone companies rely on the improbability that all their customers will attempt to place a call at one time. Banks will start a panic run if they are unable to meet the demand for funds at any time, so they must operate at a very low GOS (that is, the probability of not having funds available to meet the demand at any time must be very low). For any one bank to do this alone would mean that it would need to keep a very large proportion of its funds available at any time in cash form. This would be very expensive. For this reason, vehicles such as the short-term money market exist to ensure that each bank has access to large sums at short notice without tying up too much of its own money.

Fortunately, telephone providers do not have quite the same problem. Usually, customers will just try again if their first call fails. But like the bank, these repeated attempts place a strain on the network, and too many repeated attempts can lead to total system collapse. The system must therefore be designed to minimize the possibility of such problems.

Telephone companies use a typical GOS range from

0.002 to 0.05. The acceptable range of call fail rates (due to equipment availability) is from 2 per 1000 to 5 per 100 attempts. For cellular purposes, the GOS varies from operator to operator, but 0.01–0.05 for base-station links and 0.002–0.001 for the switch to PSTN link are reasonable values.

DIMENSIONING BASE-STATION AND SWITCH CIRCUITS

It is normal to use a fairly high GOS for the radio path because of the high cost of the base-station channels and because an effective overflow regime exists whereby traffic offered to one base station can, if no circuits are available, be carried by a neighboring one. A GOS between 0.05 and 0.01 is ordinarily used for base stations. Appendix H can be used directly to dimension bases from measured traffic. A short BASIC computer program that can be used to calculate traffic for any GOS is found within this chapter. This program is not copyrighted and can be freely used.

The path from the switch to the PSTN, which carries all the mobile network traffic, is usually dimensioned at either 0.002 or 0.001 GOS, because any traffic lost on this path causes calls to fail.

Traffic Capacity of a Base Station

The traffic capacity of a cellular base station can be determined from the Erlang B table in Appendix D. (The Erlang B table is also known as Erlang's formula of the first kind, and Erlang's loss formula.)

These are the assumptions of the Erlang B table:

- That the switch is a full-availability switch.
- Subscribers generate calls individually and collectively at random.
- Lost calls are cleared with zero holding time; that is, upon striking congestion, the subscriber hangs up and does not immediately attempt to redial.

The first assumption, that every channel at the base station is available for use by any incoming or outgoing caller on every attempt, is true for cellular systems.

In a traditional small telephone exchange, a good percentage of the calls will be in the local community of interest. It is more likely that the call will be local rather than distant. This means that there is a high probability that subscribers will call someone on their own switch. If the total population of local callers is small (traditionally less than 200), then each local call reduces the probability that another local call will be made. In the extreme case of two subscribers on the switch, if a local call is made to the other subscriber, then the probability of another local call must be zero.

A small cellular base station may well have a capacity of less than 200 subscribers, but because most of these subscribers are roaming, they are individually and collectively independent. There is no reason to think that a cellular subscriber is any more likely to call another subscriber in the same cell than to call a subscriber in a different cell. Consequently, the second assumption, that subscribers generate calls individually and collectively at random, is true.

The third assumption, that repeated attempts are unlikely, is not true, but it may be less obvious why. Because of the redial facility, a subscriber striking congestion is most likely to retry. This retry will occur on the access channel and not on a voice channel. Congestion causes the call attempt to fail, and the caller is dropped off the system very promptly.

The failure of the third assumption means that Erlang C or Erlang B extended is more appropriate.

ERLANG B TABLE

The Erlang loss formula is based on the probability of congestion B being

$$B = \frac{P(N)}{P(0) + P(1) + P(2) + \cdots + P(N)}$$

where $P(N)$ is the probability that the Nth circuit is busy when offered traffic A, such that

$$P(N) = \frac{A^N e^{-A}}{N!}$$

This simplifies to

$$B = \frac{\dfrac{A^N}{N!}}{1 + A + \dfrac{A^2}{2!} + \cdots + \dfrac{A^N}{N!}}$$

It is common experience that most users of a cellular telephone will immediately hit the redial button should a call fail for any reason. The repeated attempt corresponds to the extended Erlang B method, which contains the assumptions of the Erlang loss formula modified by the assumption that a percentage of callers who encounter congestion will redial.

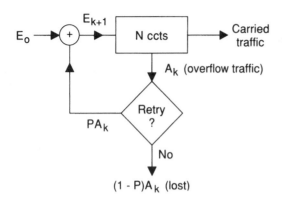

Figure 20.4 The extended Erlang B model.

EXTENDED ERLANG B

Because the Erlang B formula assumes that blocked calls are lost, the effect of redialing is to present additional traffic to the circuits (relative to the Erlang B case), which in turn will increase the blocking by these circuits. To estimate the value of blocking, B, under redialing, where the proportion of calls that encounter blocking are redialed equals p, consider Figure 20.4.

First assume the Erlang B traffic E_0 is offered to the N circuits. Then

$$A_0 = E_0 B(E_0, N)$$

where

$$A_0 = \text{carried traffic.}$$

Now and pA_0 to E_0 to give the first iteration of the total traffic offered to the N circuits and repeat the process.

$$E_1 = pA_0 + E_0$$

and

$$A_1 = E_1 B(E_1, N)$$

This can be repeated using the general form

$$Ak = E_k B(E_k, N)$$

and

$$E_k + 1 = E_0 + pA_k$$

Having thus calculated the probability of blocking in the N circuits, it can be shown that the probability of a call being blocked in the redial situation is

$$L = B(1 - p)/(1 - Bp)$$

Notice that if $p = 1$ (everybody retries), then $L = 0$, meaning that no calls are blocked and that all calls will eventually get through. From experience this is true. Even in the worst networks, if time doesn't matter, the call will eventually be carried through repeated tries. This method makes no attempt to quantify the delay, or the number of repeated tries that may be necessary to get the call through.

ERLANG C

The Erlang C formula assumes that all calls retry, and this is probably the case in the cellular environment (which is equivalent to the extended Erlang situation where $p = 1$). The blocking for the Erlang C case can be derived directly from the Erlang B calculation. If the blocking experienced by a traffic of A Erlangs over N circuits is B, then the blocking under Erlang C conditions is

$$Bc = B/(1 - (A/N) \times (1 - B))$$

For realistic evaluation of cellular base-station circuits it is necessary to think in Erlang C terms, or in terms of redialing. The industry traditionally uses Erlang B dimensioning, which for good grades of service is a reasonable approximation, as can be seen from Table 20.3. It can be seen that the Erlang C table is the most conservative and will always require more circuits than the Erlang B approach. Conversely, the widespread use of the Erlang B table consistently overestimates the GOS. Appendix D tabulates the Erlang C function.

As previously mentioned, the Erlang B table (see Appendix D) can conveniently be put into a simple computer program to allow ready calculation of any grade of service. The following BASIC program can be run on any personal computer.

TABLE 20.3 Comparative Erlang B and Erlang C Traffic Capacities

	GOS			
	0.002		0.05	
Traffic	ERL B	ERL C	ERL B	ERL C
10	15	17	20	21
20	26	29	34	35
40	46	52	58	61
80	86	97	103	108

```
    REM program to produce Erlang B circuits
    PRINT "Erlang B table"
30  INPUT "offered traffic"; A
    INPUT "GOS"; G.
    C = 1
    N = 0
70  N = N + 1
    C = 1 + N * C/A
    B = 1/C
    IF B > G THEN GOTO 70
    PRINT "No of circuits"; N
    GOTO 30
    END
```

This program can be used to calculate the number of circuits for any traffic and grade of service. It can be useful for calculating the number of circuits needed between the switch and the PSTN or between switching centers, as well as for calculating the number of base-station channels needed for a given traffic.

The GOS used will vary from administration to administration, but as a guide, GOS = 0.002 is recommended between switches and GOS = 0.05 for base-station channels.

If your interest is in programming, you will soon find that this program cannot produce the Erlang B table in Appendix D. While a relatively simple iterative relationship exists between traffic, GOS, and circuits, no such simple relationship is available between circuits, GOS, and traffic. However, the program was used as the basis to produce the Erlang B table of circuits for a given GOS. It is derived from an iterative method, outlined below.

1. Write an algorithm to estimate (guess) the traffic for the given GOS and number of circuits. A first approximation could be traffic = 0.6 × circuits. By developing more elaborate approximations, you can improve the starting point and so speed up the calculation, but this will get you started.

2. Using the approximation for traffic (A) for the given number of circuits (N), calculate the actual GOS using the Erlang B calculation.

3. Since you know that the relationship between traffic offered and circuits for a given GOS is monotonically increasing (that is, if one goes up so does the other), it can easily be seen that if the guessed traffic is too high, then the resultant GOS will also be high (and conversely for low values of traffic, the GOS will be low). So you know which way to adjust the estimated traffic.

4. Devise an iterative routine that gives the correct traffic to a given precision.

Use a binary chop, which involves using the guess first to determine simply whether the guess is too high or too low. Next use another algorithm to force the guess to go the other way (i.e., if the original guess is too high, then eventually find a guess that is too low). Now that the true answer is "cornered" between a too-high and too-low value, use a method that halves the interval between the two values successively until the answer is found to a prescribed accuracy. The method will be found in any good mathematical text on numerical methods of calculation.

If your BASIC is an interpreted BASIC, then the program could take a very long time to execute. In fact, even the simple Erlang B program above will run slowly in interpreter BASIC, and even slower on a handheld calculator. Please be assured that it runs quite quickly on a PC. Interpreter BASIC reads and acts on one line of instruction code at a time. I use Power BASIC 3.0, which is a compiled BASIC and runs very much faster than interpreter BASIC. Visual Basic running on a Pentium PC is also fast enough.

OVERLOADED CIRCUITS

Good engineering means that circuits will not normally be overloaded, but there are times when overloads occur, such as when partial trunk route failure occurs or equipment deliveries are delayed and a temporary overload occurs on a route. There are two things that you need to know about overloading circuits. The first is that the GOS deteriorates very rapidly; the second is that by its nature any measure of the overload becomes increasingly *less accurate as the load increases*, so you soon reach the situation where you know it is bad but you can't accurately quantify either how bad or what the magnitude of the lost traffic is. Figures 20.5 and 20.6 are

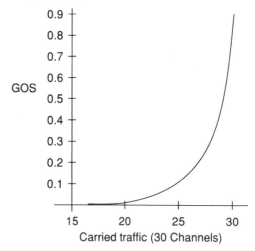

Figure 20.5 GOS versus carried traffic for an E1 2-Mbit route.

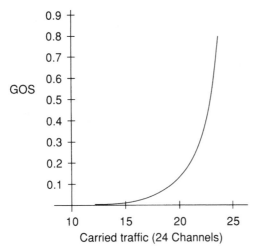

Figure 20.6 GOS versus carried traffic for a 24-channel T1.

plots of carried (or measured) traffic on an E1 or T1 circuit versus actual GOS. From these graphs you can see that very small increases in measured traffic on an overloaded system are associated with large increases in GOS. It is the difficulty of measuring the carried traffic precisely enough that makes good estimates of GOS difficult on the overloaded circuit. Notice that when the GOS is large there is a very big difference between carried and offered traffic, and care must be taken not to confuse the two.

ESTIMATING CARRIED TRAFFIC

Often, for various reasons, a base station or traffic route will become congested and it will be necessary to estimate the degree of overload. The traffic statistics will tell you how much traffic is being carried, but to correct the situation it is necessary to know how much was offered. There is no simple relationship between offered and carried traffic, but starting from offered traffic and the number of circuits, it is relatively easy to calculate the carried traffic.

The BASIC subroutine below will calculate the carried traffic for any offered traffic to a given number of circuits. Notice that this version was based on power BASIC and so has no line numbers. If you are using interpreter BASIC, then simply add these as required.

```
REM This subroutine calculates
SUB Oflow(A, N, M, V, C)
REM N Circuits offered A Erlangs
REM A=TRAFFIC
REM N=CIRCUITS
REM M=OVERFLOW MEAN
```

```
REM V=OVERFLOW VARIANCE
REM C=Carried traffic
IF A=0 THEN
    V=0
        M=0
        RETURN
    END IF
 K%=INT (N)

REM calculates traffic carried by K% circuits and A
        erlangs
        I%=0
        Cx=1
FOR I%=1 TO K%
            Cx=1+I%*Cx/A
        NEXT I%
M=A/Cx   'the carried traffic

REM for one additional circuit
Cx=1+(K%+1)*Cx/A
Mx=A/Cx
C=A-M   'the carried traffic
' calculates the variance
V=(1-M+A/(N+M+1-A))*M

END SUB
```

To run this subroutine use this simple program

```
INPUT "Offered Traffic";A
INPUT "Circuits";N
CALL Oflow(A, N, M, V, C)
PRINT "Carried traffic is";C
```

By proceeding this way you can determine the actual offered traffic by trial and error. However, you will find that you can cause it to converge fairly quickly with a bit of practice.

If you have a frequent need for this calculation you can write a routine to make it converge automatically. It is not too difficult to write one that inputs carried traffic and the number of circuits and returns the offered traffic.

It should be noted that in cellular systems that are not congested, the carried traffic is the throughput of traffic, and the overflow is basically redialed traffic. This is because those who fail to get a circuit will generally attempt redials until the traffic gets through and so the traffic is essentially re-offered, as seen in Figure 20.7. This redialing means that the GOS is no longer a measure of lost calls, as essentially all calls are carried eventually. The GOS is more a measure of the probability that the caller will need to redial, except in the case of very severe congestion where the offered traffic exceeds the number of circuits available. In this case some traffic

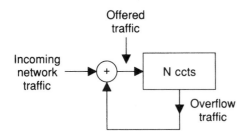

Figure 20.7 Traffic that doesn't get through on the first attempt will redial.

must be lost. It should be noted that there will always be lost calls if the offered traffic exceeds the number of circuits available.

USING ERLANG C AND GETTING IT RIGHT!!!

Traffic can be processed as Erlang B (the most commonly, but incorrectly, applied table for cellular networks), Erlang C (the relationship most applicable to cellular networks) or as a Queued traffic system (such as might be seen in trunked radio networks).

Erlang C is the correct table to use for mobile subscribers because if they strike network congestion they are most likely to press the resend button (which is consistent with Erlang C) and not permanently cancel the call (as is assumed by Erlang B).

You might be surprised to know that most network designers erroneously use Erlang B, and they get away with it (more or less) because they also use Erlang B assumptions to calculate the offered traffic. It's a case of two wrongs (almost) making a right. The Erlang B offered traffic calculation *overestimates* the traffic, while the Erlang B circuit calculation *underestimate* the circuit requirements. By good luck, rather than good engineering, in many (but not all) situations the designers get away with it.

Let's assume there is some traffic data from a small system of five base stations and 2700 subscribers. The names of the bases are City Central, City South, Northern Suburbs, Great Escape Highway, and Western Ranges. A recent traffic reading gave the following:

SITE NAME	TRAFFIC (E)	CIRCUITS IN SITU
City Central	34.6	40
City South	11.5	12
Northern Suburbs	12.3	18
Great Escape Highway	5.1	12
Western Ranges	2.1	12
Total	65.6 E	

Using Erlang B and doing a conversion from offered to carried traffic (based on Erlang B) we find:

SITE NAME	TRAFFIC (E)	CIRCUITS IN SITU	OFFERED TRAFFIC	CIRCUITS REQUIRED
City Central	34.6	40	38.02	44
City South	11.5	12	33.02	39
Northern Suburbs	12.3	18	12.8	18
Great Escape Highway	5.1	12	5.12	9
Western Ranges	2.1	12	2.1	5
Total	65.6 E		91 E	115

Notice that the traffic listed as total traffic is the total *offered* traffic (as calculated assuming Erlang B) and not the total of the measured traffic. Now, see what happens if we redesign the network using Erlang C.

SITE NAME	TRAFFIC (E)	CIRCUITS IN SITU	OFFERED TRAFFIC	CIRCUITS REQUIRED
City Central	34.6	40	34.6	46
City South	11.5	12	11.5	19
Northern Suburbs	12.3	18	12.8	20
Great Escape Highway	5.1	12	5.12	10
Western Ranges	2.1	12	2.1	6
Total	65.6 E		65.6	101

If you now look at the network using Erlang B, you find that the Erlang B network is somewhat over-designed. But there is a reason for this! The figures chosen for traffic have one of the stations in serious congestion, making the erroneous calculation (Erlang B) more vulnerable. Let's make it easier for the bad designers and remove the congestion at City South. Take the number of circuits from 12 to 20 (the site is no longer in congestion). Now run a forecast at 2700 customers. For Erlang B we find:

Total Traffic   70
Total Channels   93

For Erlang C we find as before:

Total Traffic   65.6
Total Channels   101

So now there is not so much difference in the two network designs; but now the Erlang B network is *under* designed. In fact as a general rule, we can say the designs using Erlang B will invariably under provide routes that have adequate circuits (when measured) and over provide on those in congestion.

So the fact that most network designers "get away" with the use of Erlang B calculations when they should be using Erlang C is because it is true indeed that here we have a case where often two wrongs do make a right.

The Erlang C Code

In case you would like to run you own Erlang C code, here is a subroutine that can be used;

```
'Erlang C tables
Sub ErlangC(a, G, N)
'a is traffic in Erlangs
'G is GOS
'N is circuits to be provided

Dim b As Double
Dim C As Double
Dim Cx As Double
Dim Cc As Double
'Does Erlang C table for fixed circuits and GOS
If a = 0 Then
      N = 0
      Exit Sub
End If
C = 1
N = 0
b = 2

Do While b > G
N = N + 1
C = 1 + N*C/a
b = 1/C
Loop
```

```
'does erlang C correction

Cx = 1 - (a/N)*(1 - b)
Cc = b/Cx

Do While Cc > G
N = N + 1
C = 1 + N*C/a
b = 1/C
Cx = 1 - (a/N)*(1 - b)
Cc = b/Cx
Loop
End Sub
```

DUAL-MODE BASE-STATION CHANNEL DIMENSIONING

In order to maximize the efficiency of a dual-mode base station (whether it be code division multiple access (CDMA), narrow-band analog mobile phone service (NAMPS), Digital AMPS, ETDMA, etc.), it is necessary to have the correct mix of channels. There are a number of factors that affect the channel mix. Consideration will have to be taken of the relative number of dual-mode mobiles, the relative cost of Advanced Mobile Phone Service (AMPS) and the other mode channels, and the total number of channels available. The problem can be visualized as shown in Figure 20.8.

This relatively complex dimensioning problem has two optimum solutions. One will be the configuration that carries the greatest amount of traffic, and the other will be the one that has the least cost of base-station

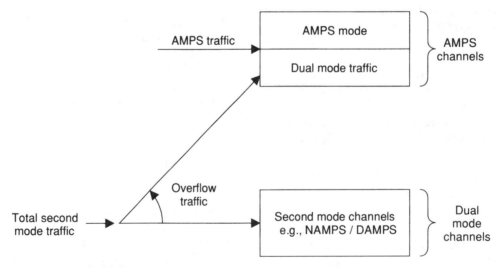

Figure 20.8 Dual-mode base stations can be visualized as offering the second-mode traffic to a group of channels that overflow to the AMPS channels.

TABLE 20.4 Traffic Tables for a Dual-mode NAMPS System

Total Traffic	Cost	NAMPS Channels	Cost/ Erlang	NAMPS GOS
63.6	70.2	9	1.103	0.029
65.2	71.6	12	1.099	0.023
66.6	73.0	15	1.096	0.017
67.9	74.4	18	1.096	0.013
69.0	75.8	21	1.098	0.008
70.0	77.2	24	1.103	0.005
69.7	78.6	27	1.128	0.002
69.2	80.0	30	1.156	0.001
68.5	81.4	33	1.188	0.0002
66.8	82.8	36	1.239	0.00003
66.1	84.2	39	1.273	0.00000
64.5	85.6	42	1.327	0.00000

hardware per Erlang of traffic carried. Either solution may be "correct," depending on the goals of the operator.

To see how these various solutions may be arrived at, consider an example for NAMPS as shown in Table 20.4, which lists the total traffic carried (NAMPS + AMPS), the total cost (in arbitrary units where one AMPS channel has one unit of cost), the number of NAMPS channels, the cost per Erlang, and the effective GOS offered to the NAMPS mobiles.

Assumptions

1. Original number of AMPS channels 66
2. Cost of an AMPS channel 1 unit
3. GOS of AMPS channels 0.05
4. Proportion of NAMPS phones 0.3
5. Cost of NAMPS channels = 1 unit

There are quite a few salient points that can be derived from Table 20.4. Notice that the GOS of the NAMPS is always better than that of the AMPS channels. The GOS experienced by a NAMPS mobile is in fact the product of the GOS on the NAMPS and the GOS on the AMPS channel.

It can also be seen that the maximum traffic carried is 70 Erlangs, which corresponds to 70 NAMPS channels. The cheapest configuration can be seen to correspond to 1.096 units of cost per Erlang and occurs when there are either 15 or 18 NAMPS channels.

The derivation of these tables is by iteration:

1. First estimate how much traffic will be carried by a particular configuration. (For example, from an original pool of 66 AMPS channels, a composite of 9 NAMPS and 63 AMPS channels = total

of 72 circuits. An estimate might be based on a circuit occupancy of 0.7, the total traffic = $0.75 \times 72 = 54$ Erlangs.)

2. From the portion of traffic that is NAMPS (0.3 in the example), derive the corresponding NAMPS traffic of $0.3 \times 54 = 16.2$ Erlangs.

3. Using the subroutine Oflow (listed earlier in this chapter), calculate the traffic carried on the nine NAMPS channels and the amount that overflows to the AMPS channels.

4. Add the AMPS traffic (0.7×54 Erlangs) and the NAMPS overflow traffic together to get the total traffic on the AMPS channels.

5. Using the GOS routine, estimate the number of circuits needed to carry the combined AMPS traffic.

6a. If the number of AMPS circuits available (63 in this example) is greater than the number required, then increment the traffic estimate upward and repeat steps 1 to 6.

6b. If the number of circuits available is less than the number required, decrement the traffic estimate and repeat steps 1 to 6.

6c. If the number of circuits available on the AMPS route equals the number required, then the solution has been found.

To speed up the process, more efficient iterative methods such as the binary chop can be used to hasten convergence.

CIRCUIT EFFICIENCY

It is important to note that when a small number of channels are used at a base station, the circuit efficiency (which can be defined as subscribers per circuit) is very low.

Converting traffic (in Erlangs) to subscribers is relatively straightforward. First, it is necessary to know the average calling rate of a subscriber. This is typically six calls/day of 150 seconds duration, which translates to 0.03 Erlang/subscriber (given suitable assumptions about the busy-hour pattern). Hence the number of subscribers per base is

$$\frac{\text{Traffic capacity of the base}}{\text{Calling rate of subscriber}}$$

To show how the circuit efficiency varies with total circuits (channels), Figure 20.9 shows subscribers/channel versus number of channels for a 0.01 grade of service, assuming a calling rate of 0.03 Erlang/subscriber (note that 0.03 Erlang is often written as 30 mE).

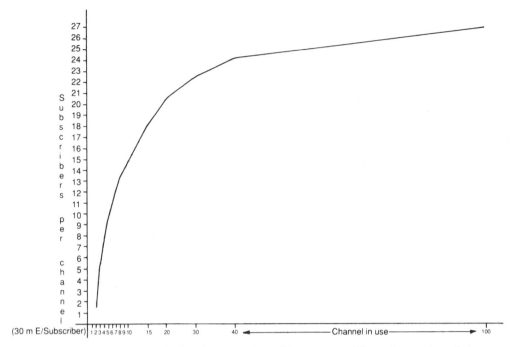

Figure 20.9 The number of subscribers per channel increases rapidly as the number of channels in use increases.

DISPERSION

Dispersion measurements are used to indicate the sources and sinks (destinations) of traffic. Although direct occupancy measurements can give the total volume of traffic, they do not give any information about the direction or origin of that traffic unless all traffic on the route has only one sink. For example, all switches to the south may have their traffic combined on one route, so

Clearly small cells (less than 10 channels) are very inefficient in terms of subscribers/channel (and hence in dollars/channel) and should not be used unless it is unavoidable. For this reason, most operators, even of analog systems which can be expanded by one-channel increments, use 7 channels (one rack of equipment) as the minimum installed capacity of a base station. At this size, the channel efficiency of 11.9 subscribers is still only about one half that of a fully equipped base.

Sectoring divides the total number of channels per cell by the number of sectors. A 21-channel, three-sector cell is, in fact, three times a 7-channel cell and therefore carries 250 subscribers ($3 \times 7 \times 11.9 = 250$). Compare this to a 21-channel omni site that has two less control channels and can carry 430 subscribers ($21 \times 20.5 = 430$), and it is easy to see that sectoring reduces circuit efficiency.

that a knowledge of the total traffic only says nothing about the traffic distribution.

Dispersion measurements involve analyzing the called number to typically six or more digits and recording the total holding time.

The switch will record the dispersion information on its statistics records. From this a matrix of the originated traffic is obtained.

TRAFFIC FORECASTING

The dispersion matrix is the basis of traffic forecasting. To see how to use it consider the simplified matrix in Table 20.5.

To use the table to forecast future traffic, it will be necessary to make some assumptions.

TABLE 20.5 A Simplified Dispersion

Traffic from Cellular Switch One		
Destination	Size	Traffic (Erlangs)
Self-traffic	10,000	20
Local	1 million lines	290
Cell switch two	6000 lines	12
Long distance		15
Total		337

TABLE 20.6 New Traffic Distribution, Two Years After Table 20.5 Data

Destination	Size	Traffic
Self-traffic	10,000	15.8
Local	1.21 million	267.7
Switch two	20,000	31.5
Long distance		13.0
Total		337.0

The main assumption, which holds fairly well for short time frames, is that the proportion of traffic to each destination is directly proportional to the switch size and the traffic measured at the time of the dispersion measurement. It will also be assumed that the total traffic generated by switch one will remain constant.

For a future time, assume two years, we may find for example that

- Switch one remains the same size as it is at capacity
- The local network is expected to grow at 10 percent yearly and so will be 1.21 million lines
- Switch two may be 20,000 lines
- Long-distance traffic grows at 5 percent yearly

Using these assumptions it is found that the new traffic distribution in two years will be as shown in Table 20.6.

In this example all routes are expected to receive a reduction in traffic except the route to the second cellular switch, which is attracting traffic at a greater rate due to the rapid growth of the cellular network.

In this way dispersion can be used to forecast future trunk and junction requirements.

Where multiple switches are used for a cellular network, it is important to know the interswitch traffic so that appropriately dimensioned routes can be provided. A phenomenon of local traffic applies: the percentage of local traffic (mobile-to-mobile) remains approximately constant. As the total number of switches increases, the traffic between any two switches may increase or decrease as seen in the previous example. This is best illustrated by Figures 20.10 and 20.11.

If the local mobile-to-mobile traffic is, say 10 percent, then the dispersion to any particular switch decreases as the total number of switches increases. Figure 20.10

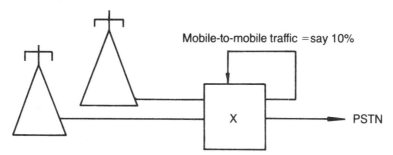

Figure 20.10 Local cellular traffic is usually small. Typically, about 10 percent of local traffic (mobile-to-mobile) is generated by a cellular network.

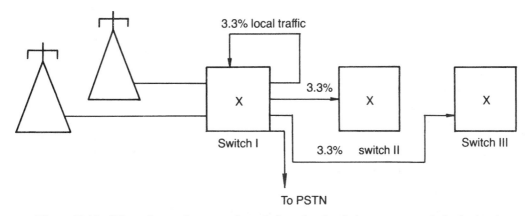

Figure 20.11 When the service expands and three local switches are networked, the local traffic remains a (nearly) constant percentage of the total traffic (all switches are assumed to be of equal size), and the gross internal traffic on any particular route decreases.

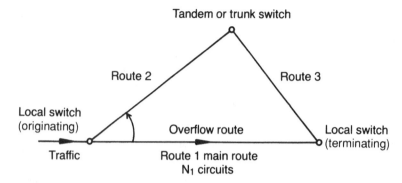

Figure 20.12 Alternate routing involves collecting small parcels of traffic into a combined trunk route, and distributing traffic between a direct route (route 1) and an alternative overflow route (route 2).

shows a cellular network that begins as a single switch with 10 percent local traffic. As the system grows and more switches are added, the local traffic on the original switch, and to other switches, decreases. In Figure 20.11, the three switches are assumed to be of equal size, so the traffic is distributed (approximately) evenly between switches.

ALTERNATE ROUTING

Like the banker's short-term money market, there are ways of improving the grade of service without undue investment in circuits. The most common means is alternate routing, whereby traffic carried on a number of small routes can overflow to a common alternative path. This means of gaining extra circuit-efficiency is often accidentally incorporated into cellular systems via the overlapping coverage provided between cells. If the most-favored (highest field-strength) base is not available, the call can be carried (overflowed) by an adjacent cell. The umbrella cell concept is an example of where this happens deliberately.

The cellular-radio provider must consider alternate routing, particularly when it is necessary to interconnect with a number of other carriers that can provide circuits and switching using a variety of different means. The simplest case of alternate routing has only one overflow route, as shown in Figure 20.12. Usually, a number of main routes (called direct routes) overflow to the same alternate route (route 2–3).

A parameter defined as cost factor is used to determine the optimum configuration where the cost of $route_n = C_n$ and the traffic carried per circuit on $route_n = B_n$. (More strictly, B_n equals the marginal capacity, which is the marginal additional traffic carried by one additional circuit. For all but very small routes, this is virtually the same as the average traffic per circuit.)

$$\text{Cost factor } H_1 = C_1 \left[\frac{C_2}{B_2} + \frac{C_3}{B_3} \right]^{-1}$$

There are various tables and computer programs that enable the cost factor to be used to determine the number of main (direct) circuits and the value of the traffic overflowed. The cost factor H_1 is known as the marginal occupancy of route 1 and is the marginal increase in carried traffic per added circuit when the offered traffic is held constant.

OPTIMIZING CIRCUITS

The cost factor can be used to calculate the optimal number of direct circuits by proceeding as follows:

1. Estimate an approximate value for n, the number of direct circuits.
2. Calculate from the offered traffic the mean of the overflowed traffic with n circuits.
3. Repeat 2 for $n + 1$ circuits.
4. Compare the marginal traffic carried by the extra circuit with the cost factor.
5. If the marginal traffic is less than the cost factor, then reduce n by one circuit and repeat the procedure from point 2. Or if the marginal traffic is greater than the cost factor, add one circuit and return to point 2.
6. Once the best match between the marginal traffic and the cost factor has been found, the number of direct circuits is determined.

Obviously the best way to handle this calculation is with a computer, but a graphical method can be useful for individual calculations. Figure 20.13 shows a plot of the economic circuit provisioning for a number of cost

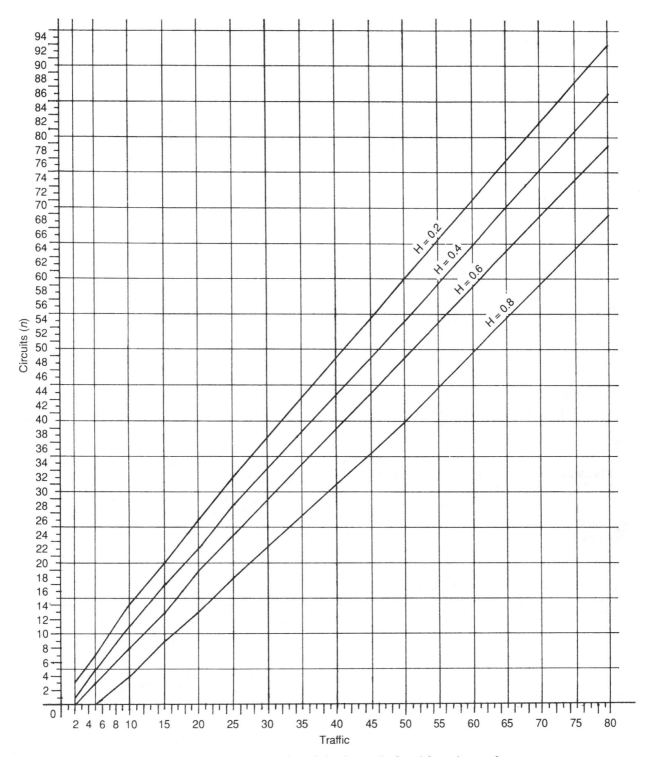

Figure 20.13 The optimal number of circuits can be found from the cost factor.

factors. For cost factors other than those shown it is possible to interpolate.

Where alternate routing is used, the traffic that overflows from the direct route is "rougher" than the pure-chance traffic from which it was derived. This roughness means that the predictability of the traffic is low and so more circuits are needed to carry this rough traffic than would be needed for smooth traffic with the same mean.

The parameter most used to characterize this roughness is the *variance*. Variance is a measurement of the

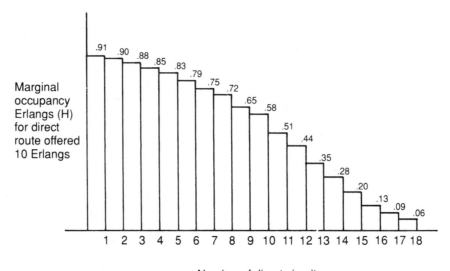

Figure 20.14 The marginal capacity of a given number of circuits offered a fixed traffic decreases rapidly with increasing circuits.

deviation of the traffic from the mean value. For smooth (or pure-chance) traffic the variance is equal to the mean. For any overflow traffic the variance will be greater than the mean.

The traffic carried on the direct route is super-smooth, and the variance of this traffic is less than the mean.

Thus when dimensioning an alternate route the parameters that are needed are the offered traffic and the cost factor. From these you can derive the optimal number of circuits, the mean of the traffic overflowed, and the variance of the overflow traffic.

The overflowed traffic can be calculated from the Erlang loss formula as before. From the overflow traffic the variance can be determined as it is related to the mean by the formula:

$$V = m\left(1 - m + \frac{A}{n + 1 + m - A}\right)$$

where

A is the offered traffic

n is the number of circuits on the direct route

m is the mean of the overflow traffic

The number of circuits required to optimize a route such as shown in Figure 20.12 is usually calculated by an iterative procedure as follows:

1. Estimate B_2 and B_3 (typical values, 0.5–0.8).

2. Calculate H_1 using the previous formula.
3. Find N_1 for the marginal occupancy from Figure 20.13.
4. Calculate the overflow traffic from N_1 to N_2 and determine the number of circuits required on route 2 and route 3.
5. Find B_1 and B_2 as determined from step 4, and if very different from the original values, insert the new values in step 1 and repeat.

On a computer, this is easy to do; it is quite a task if done manually.

If the marginal occupancy H_1 of a route is known and the cost of a circuit on that route is C_1, then the marginal cost per Erlang is C_1/H_1. For any direct route offered a fixed level of traffic, the marginal occupancy will decrease as the number of circuits increases. This is shown below in Figure 20.14, which shows the margin occupancy for a direct route offered 10 Erlangs.

The cost of carrying an Erlang on the direct route is $C_2/B_2 + C_3/B_3$. If it is assumed that if $C_2 = C_3 = \$1,000$, and that $B_2 = B_3 = 0.75$, then the cost of carrying an Erlang on the alternate route is \$2666. By plotting this value on the graph in Figure 20.15 the intersection can be seen to be the point where the cost of carrying the traffic on the direct route just equals the cost of carrying the traffic on the alternate route.

From this you can see that

$$C_1/H_1 = C_2/B_2 + C_3/B_3$$

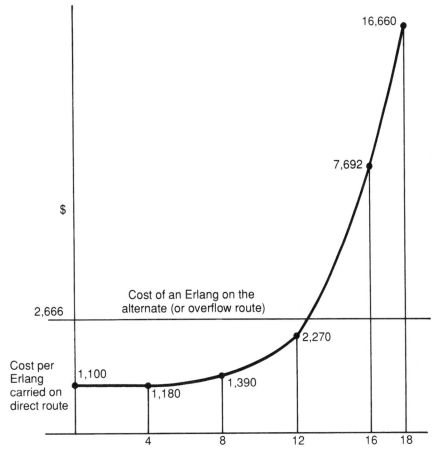

Figure 20.15   The cost per Erlang carried on the direct route if a direct circuit costs $1000.

or

$$H_1 = C_1/(C_2/B_2 + C_3/B_3)$$

EFFECT OF ALTERNATE ROUTING ON GOS

Where large volumes of traffic exist between the cellular network and other switches it may well be that some direct routes are justified. Particularly this may apply to a second cellular operator in the same city or to one or more telephone switches in smaller cities that have a small number of relatively large switches.

Where alternate routing has been used, a call has two separate possible paths from the cellular switch. For traffic parcels, which have both a direct and an overflow route, a higher GOS can be used on the link to the overflow route.

For example, considering Figure 20.13 the two possible paths are direct to the local switch or via the tandem switch.

Assume that 50 Erlang of traffic is offered from the originating switch to the terminating switch. Typically, the direct route may have 56 circuits (as calculated using the cost factor), which would carry 46.8 Erlangs and overflow 3.2 Erlangs. It can now be seen that the probability of a call failing to get access to the direct route is 3.2/46.8 or 0.068.

If the overflow route has a GOS of 0.01, the probability of the call failing totally is $P = .068 \times .01$ or .00068. That is, a very good GOS overall is available.

Software packages that can do all the mathematics of circuit provisioning are available. All of the calculations described here, and many more, are in my software package called Mobile Engineer.

21

MOBILES

Until recently consumers were not too demanding of cellular phones; most were more than delighted that the technology even worked. Today's consumer, however, expects a lightweight device, convenient to operate, and that has long standby and talk times. Increasingly, these expectations are being met by ever-improving technology.

EARLY MOBILE PHONES

The design of early mobile phones were necessarily large and heavy. When the first AMPS phones came onto the market, the portables were a whopping 8 liters in volume. The development of the mobile phone as it has reduced in size over the years is seen in Figure 21.1, which shows the size of phones from the first AMPS phone to today's small handhelds.

Even-earlier phones were bigger. Figure 21.2 shows the body of a Japanese NAMTS phone from 1981. These units weighed in at 7 kg, and additionally came with a cradle and locking system, that was of course vehicle-mounted. This unit was then cabled to a fixed handset as seen in Figure 21.3. However even in 1981, portability was possible and the equivalent transportable, weighing about 8.5 kg is shown in Figure 21.4. These phones were not light on power consumption either and required 20 watts on standby and 80 watts when transmitting (with a 10-watt output). The phones cost $2500 in 1984 and had a network connection fee in Australia of $88.00 per week. These phones were fully cellular and featured a 120 channel capacity with handover. Handover times were quoted as "up to 3 seconds," with the actual time really depending on how busy the processor was at the time the handover was requested.

CDMA One

In 1999 Samsung introduced a code division multiple access (CDMA) phone/watch. The whole unit weighs 39 grams without the battery and 50 grams with it. It allows 90 minutes of continuous talk-time and 60 hours standby. Naturally it has voice-activated dialing, a phone directory, is handsfree, and has vibration alert.

GSM

Other lightweights include the Mitsubishi Global System for Mobile Communications (GSM) 900/1800 at 69 grams, Motorola's v3688 GSM 900/1800 at 83 grams, and the Mitsubishi Arai GSM 900/1800 at 85 grams.

The demand for dual-band phones is driven more by capacity demands than by roaming. This is because many GSM 900 operators have acquired a GSM 1800 spectrum for relief in traffic "hot spots."

MULTIBAND

Ericsson's trimode time division multiplex access (TDMA) phone, R250d, works on 800-MHz analog and 800-MHz and 1900-MHz digital. Ericsson is also about to release a trimode GSM phone. Most manufacturers now offer one or more dual-band mobile.

Roaming accounts for some of the dual-band offerings, as for example, Ericsson's 1888 World Phone, which has both GSM 900 and GSM 1900 capability.

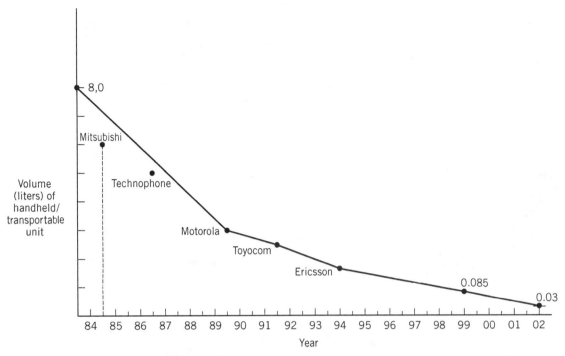

Figure 21.1   The volume of a portable/transportable cellular phone over time.

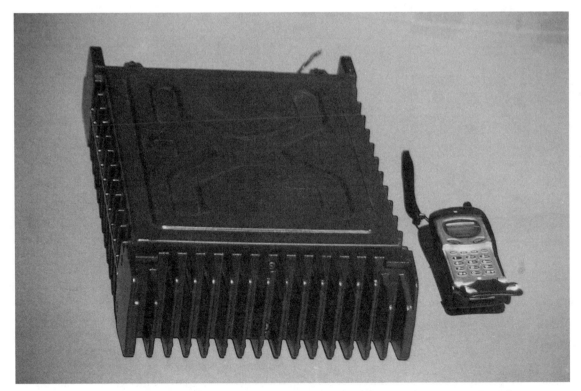

Figure 21.2   The 7 kg NAMTS phone (shown without the handset unit) in 1981; beside it is a Hyundai CDMA phone (1999).

Figure 21.3 The NAMTS handset unit mounted in a vehicle 1983. Phone courtesy Lynne Martin (seen in the photo).

Figure 21.4 The NAMTS transportable 1984.

THE TRENDS

Mobile phones are getting smaller (with new ones typically being around 80 grams), the screens are getting bigger and battery life longer. A typical design is this dual-mode GSM phone from Voxson, released in 2000, and seen in figure 21.5. It has a larger screen for short messages, and few buttons. Figure 21.6 shows the board that drives the phone. There is in fact only one board inside the phone and the two boards shown are actually the same board seen from either side.

Nextel Communications, the trunked radio operator, is about to introduce a dual-mode GSM integrated dispatch enhanced network (iDEN) phone.

CELLULAR TEST AND MEASUREMENT SET

It is necessary for every well-equipped mobile phone center to have a cellular test set. The simple go/no-go test done by attempting a call will verify the validation and little else. About 2 percent of phones will come from the manufacturer out of specification. This does not mean that the phone will not work, in fact it may perform completely normally, but if it starts in service this way, it probably will not be long until the fault develops into a malfunction.

There are many different test sets on the market and they vary in price from around $5000 to $40,000. Although the rule that "you get what you pay for" certainly applies to the test sets, there is no need to pay for

Figure 21.5 A Voxson GSM phone.

what you cannot use. The top-of-the-range instruments are high-precision test equipment that are versatile enough to be used for mobile testing but can also be used for base-station, public mobile radio (PMR), and paging maintenance. If all of this is not needed, then why pay for it?

Ideally a mobile test set should be simple to operate, menu-driven, and should guide the user through the test procedure.

SENSITIVITY AND PERFORMANCE

There is very little difference between the mobile telephone models in terms of specified sensitivity or on-air performance (at least while we consider phones working within any given protocol). This uniformity exists because the technology used in virtually all mobile phones approaches state of the art.

What does vary in mobile telephones is quality and consistency; some manufacturers meet their specifications more often than others, and so their telephones may perform better on average.

Most phones meet their specifications, but some don't. To get type approval the units sent must at least meet the specifications, but this is not an assurance that future production runs will necessarily also meet those standards.

Particularly in the digital phones (where there is more complexity), there can be a wide variation in sensitivity, from well below specification to well above it. It is the

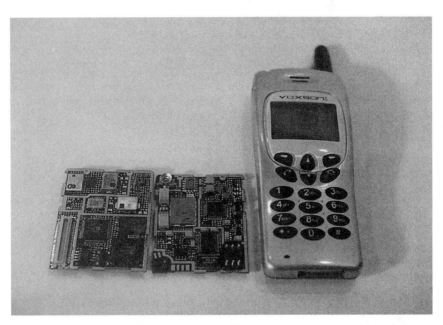

Figure 21.6 The fully assembled board of the Voxson phone (two identical boards shown; there is only one per phone but they are populated on both sides).

phones that are well below specifications that make it hard to design networks. Just how bad do you let the mobiles get?

Remember that in fringe areas the propagation is time- and space-dependent, so a particular phone that performs well on a particular day in such an area may not necessarily perform as well on the following day.

Even the most reputable manufacturers will admit that not all their mobile phones are created equal; unscrupulous suppliers can sometimes even use one of their better mobile phones (that is, the best from a production spread) against a bad example of the competitor's to "prove" superiority. Rented mobile phones will frequently have short talk times (battery life) due to indiscriminate battery charging; this should not be used to evaluate the talk time of the model rented.

GSM PHONE TESTING

As mobile phones get smarter and more complex, they are increasingly fault-prone. However, network problems are also ubiquitous, and it may be that up to half of the phones returned for service are not actually faulty. For most large mobile-phone retailers it is desirable to have a means of testing the product sold.

There are a number of mobile-phone testers on the market, and one of the most unusual is the Wavetek Console tester, which is designed to allow the users to test their own phones. The unit has a touch screen and guides the user through the tests with screen prompts. The phone can be tested with the network subscriber identity module (SIM) card, and the report is given to the user in plain language, with messages such as *bad microphone or speaker, faulty key pad, bad receiver or bad transmitter*. The unit is available for GSM 900, 1800, and 1900 phones. For the technician, Wavetech also has the 4100 series of handheld GSM testers, which can give detailed diagnoses of faults and can also log the faults for later analysis.

ANTENNA TYPES

Antennas are an important part of a cellular system and they can have a significant effect on the performance of a mobile. They come in four basic types and there will probably be a fairly wide variation in price and appearance. The four basic types are roof-mount, through-the-glass mount, trunk-mount (an elevated feed type), and disguise antennas.

Car Kits

The fall of the transportable means that for many car-mount kits today the radio frequency (RF) part simply consists of a cable and external antenna. Nevertheless, this will give a net gain of around 10 dB over a handheld used inside the car, and this represents a considerable gain for someone who relies on a mobile phone for business.

Dual-Mode Antennas

A helical antenna can form the basis of a dual-mode mobile antenna. In theory it is possible to arrange that a helical antenna is resonant at two frequencies, especially when they are almost in a two-to-one ratio, as is the case with GSM 900/PCN or AMPS/CDMA. All that needs to be done is to make the antenna a quarter wavelength at the higher frequency and a half wavelength at the lower frequency. It has been found by the ACE Antenna company that having a descending-pitch helical winding, seen in Figure 21.7, offsets some of the more difficult capacitance effects that might otherwise have to be dealt with.

The dual-pitch antenna in the retractable mode has a gain of 0 ± 1 dB, but the gain can be increased to 1 ± 1 dB, with the addition of a coupled resonant extendable whip, as seen is Figure 21.8. The radiation pattern is omni-directional and the polarization is vertical.

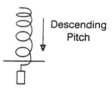

Figure 21.7 A helical antenna for dual-mode mobiles resonant at two frequencies.

Figure 21.8 A helical antenna supplemented by a retractable whip.

TABLE 21.1   Antenna Gains

Gain	Improvement Factor
0 dB or Unity	1×
3 dB	2×
4.5 dB	2.8×*

Antenna Gain

The difference in performance is measured by the *antenna gain*, which is usually quoted in decibels (dB). The gains normally encountered are shown in Table 21.1.

An improvement of 3 dB (or 2×) will definitely be significant in fringe areas, and for this reason unity gain external antennas are not to be recommended. As a rule of thumb, a 3-dB gain will give an improvement of about 30 percent in range provided there are no natural boundaries such as hills.

Because higher gains are associated with narrower beam widths, antennas with gains higher than 3 dB do not, in practice, improve the performance of the mobile. An exception to this is the case where the installation is fixed (say to a house); then the full benefits of the higher gains are realized.

Antenna Types

1. *Roof-Mount*   A typical antenna intended for roof mounting is the 3-dB unit. The best location for these antennas is at the center of the vehicle roof. It should be noted that these antennas are not suitable for mounting elsewhere on the vehicle. However, it appears that many installers do not understand this and so it is common to see these antennas mounted on trunks and even side-mounted over the driver's window. They will not perform adequately when so mounted.

Some difficulties can be expected with this type of antenna if the vehicle has a fiberglass roof. This is because a metallic ground-plane afforded by most vehicle rooftops is in fact part of the antenna system. (This can be overcome by placing a metallic tape on the underside).

The rooftop is the best location to receive a good signal and so it is possible to use simple and cheap antennas in this location.

Some people will have trouble with roof clearance (particularly on vans) if a roof mount is used. A possible solution is to use a unity-gain antenna. A unity-gain antenna will be distinguished by the fact that it lacks the "coil," which is usually located on the bottom half of the antenna.

It is frequently stated, and correctly, that the best performer of all the antennas is the center-roof mount.

However, most people are not too happy about getting a hole drilled in the middle of the roof. Fortunately for them there are other alternatives, which for practical purposes are just as good.

2. *Through-the-Glass Mount*   The through-the-glass mount is deservedly the most popular antenna type. It avoids the need to drill a hole, it is fairly inconspicuous, and it works so well that for most users it is as good as the roof-mount.

Salespeople have usually been told to inform the clients that the roof-mount works best, and although this is true, the difference in performance is so small that it takes expensive high-tech equipment to measure it. For practical purposes, through-the-glass and roof-mounted antennas should be regarded as equal except where continuous operation in a fringe area is anticipated.

Many people are mystified about how the through-the-glass works. But they shouldn't be, because all that happens is that the signal passes through the glass between two metal plates on either side of the glass.

Considering that the signal has already propagated through the atmosphere from a base station that is anywhere from 1 to 20 km away, it is not such a difficult thing for it to pass through an extra half centimeter of glass. In fact, advantage is taken of this gap to help tune the antenna. The two plates on either side of the glass can be considered as a simple capacitor—a device often used in radio circuits (see Figure 21.9).

There can be some problems with through-the-glass antennas, however, including the following.

- They often do not survive the car wash too well.
- They may not work well on some cars with tinted windows.

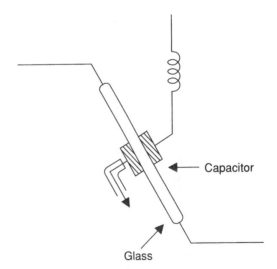

Figure 21.9 A through-the-glass connection can be regarded as a capacitance antenna feed.

Figure 21.10 An elevated-feed antenna.

• Sometimes the rear demister wires can interfere with the performance (however, this is rarely a problem if the unit is correctly installed).

• Extremes of temperature can cause the bonding to come unstuck and the antenna to fall off. It can usually be rebonded by the original installer, however.

3. *Trunk Mount (Elevated Feed)* The trunk-mounted antenna differs from the other types of antennas in that it has an elevated feed. This feed is usually a black section about 30–40 cm long, which is not a radiating part of the antenna but simply a stalk to elevate the antenna clear of interference from the top portion of the vehicle. Again, the performance of this type of antenna is much the same as the above-mentioned types. Figure 21.10 shows an elevated feed antenna.

4. *Disguise Antennas* If you happen to work for the KGB, CIA, or Control and you can't quite afford a shoe-phone but you still don't want anyone to know that you have a car phone, then a disguise antenna may be what you are looking for, as these are much cheaper than shoe-phones. A disguise antenna, however, normally will not perform as well as a garden variety 3-dB antenna and probably will cost a good deal more.

Perhaps not properly classified as a disguise antenna is the Harada CX-400, an antenna that looks like a conventional AM/FM auto-retract antenna, and also functions as a cellular antenna. This means no extra holes and no obvious extra antenna.

5. *Fake Antennas* Some of the most prominent "cellular" antennas come in kits costing from $10–30, which include a fake plastic handset.

Handsfree

The handsfree option allows the telephone user to talk without picking up a handset. It was developed quite early in cellular mobile history but was not perfected until around 1987. Earlier versions used voice-operated switching (VOX) operations, which were often less than satisfactory, particularly in the often noisy vehicle environment. Modern handsfree options are full-duplex (that is, the send-and-receive functions work simultaneously) and are "compulsory" in some countries where using a handset while driving is not permitted.

Studies have shown that drivers using a conventional mobile handset are under demonstrable induced stress that is not present when they use a handsfree option. Whether for reasons of image or perhaps because of possible legal liability, most manufacturers are moving toward including the handsfree option as a standard feature.

Vehicle Mounts

Although purpose-designed vehicular mobiles are rather rare these days, vehicle-mounting kits for handhelds are very popular. The reasons are that the range will be significantly increased by the use of an external antenna (a net gain of better than 6 dB can be expected) and the desirability (and in many cases the legal requirement) of having a handsfree kit installed for use while mobile.

MOBILE ANTENNA INSTALLATION

The mobile antenna is a most important link between the mobile unit and the rest of the telephone network. Pay special care to the installation of the antenna. Approximately 50 percent of vehicle-phone faults are antenna related. Antennas are subjected to extremes of temperature, physical stress, vandalism, corrosion, and sometimes poor installation.

Increasingly, the on-glass, or through-the-glass, antenna is becoming the customer's first choice. Cellular mobile antennas normally come pretuned and should not be trimmed unless instructions are provided otherwise. This precaution holds true as well for some on-glass types, even though the match provided will be a function of the glass thickness.

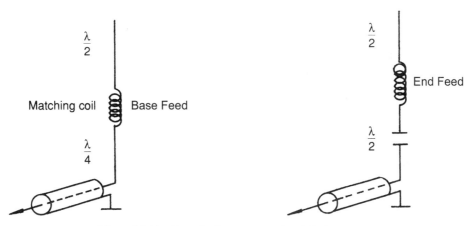

Figure 21.11   Base-feed antenna and end-feed antenna.

Base-feed antennas are commonly used in cellular radio. Most roof-mount antennas are of this type. End-feed antennas are the basis of through-the-glass antennas, where capacitance is formed by two plates on either side of the window. Figure 21.11 shows typical configurations for both base-feed and end-feed antennas.

The base-feed antenna has an input impedance of approximately 50 ohms and is the most common antenna used in cellular radio. The matching coil is sometimes formed by coiling the antenna wire around a former about 30 mm in diameter. The base-feed antenna is durable and inexpensive. This type of antenna is best mounted either on the center of the vehicle roof or centrally on the trunk.

The end-feed antenna is the basis of the "through-the-glass" antenna; the capacitance is provided in part by two metal plates mounted either side of a glass window. Additional series capacitance is usually provided to enable fine-tuning of the input impedance [to 50 ohms or voltage standing-wave ratio (VSWR) = 1.0]. The biggest problem with on-glass antennas is that they sometimes simply fall off due to the effects of heat, moisture, pollution, temperature, and (not infrequently) the automatic car wash.

Through-the-glass antennas are fixed in place either with double-sided tape or with acrylic bonding material. Double-sided tape can be most effective, but remember that it can be used only once (so care must be taken in positioning). To ensure a good bond, the area around the antenna fixing point should first be carefully cleaned with a commercial window cleaning product or other cleaning materials recommended by the antenna supplier. It is a good idea to seal around the edges of the double-sided tape with a suitable sealant. This type of fixture is removed using a razor blade or similar cutting tool.

Acrylic bonding is a very powerful bond whose application should be strictly to the manufacturer's specifications or it may not be possible to remove the antenna later. It is definitely not advisable to substitute other bonding materials because of incompatibility of their coefficients of expansion. Carefully follow the curing instructions, especially the amount of time needed before the joint becomes load bearing.

Glass-mount antennas must avoid the defroster wires and can be rendered useless when the vehicle has metallic window tints. It is possible to buy test kits for use with standard digital voltmeters that detect metallic components by their effective capacitance.

The antenna cable will frequently be run via the ceiling or floor of the vehicle. It is important to ensure that the cable is well hidden. If this is not possible, ensure that the cables are well restrained and will not catch on luggage or be easily tampered with.

When running the cable along the floor, be careful of the screws which are sometimes used to hold down the trim. These can puncture a carelessly laid cable. If the cable runs via the ceiling, firmly attach it using a suitable tape to ensure rigidity so that the cable will not move because of the vibrations. When the cable is fully run, cut off any excess cable because it is quite lossy at 800 MHz.

Attach the coaxial connector using a suitable crimping tool and stripping tool—it is not acceptable to improvise with pliers.

The position of the center conductor is important if a good connection is to be made; take care to ensure it is fitted properly and check again after installation.

Drilling Holes

Pay special attention when drilling holes in vehicles. Always ensure that the location is a safe one and that the drill will not puncture the gas tank or other functional parts. When the location has been determined, measure the placement of the antenna hole—don't guess at it. Then, use a spring-loaded center punch to start the hole. Drill a pilot hole and then use a hole punch to punch out the antenna hole. Remove burrs from the hole with a paint brush or vacuum cleaner.

When fitting the antenna, make sure that a good water-seal is achieved and that there is a good electrical contact between the antenna base and the vehicle body.

ANTENNA MOUNTING

Cellular antennas suffer considerably from vandalism; it has been found that the less conspicuous the antenna, the less likely it is to be vandalized. Moreover, smaller antennas also seem to survive a car wash a little better.

The mounting and antenna type are largely determined by customer preference, but the installer should be aware that a poor selection can seriously degrade the performance.

Take particular care with fiberglass bodies or panels where no ground plane is readily available. In this instance, use ground-plane-independent antennas, such as a coaxial dipole, or use a standard antenna with a ground plane formed by a layer of conductive tape.

The best place to mount a mobile antenna (for reception) is in the center of the vehicle roof. Most customers object to this, however, and so less optimum locations need to be used. Almost as good as the roof-mount are the on-glass, rear-window, elevated-feed center-trunk, and fender mounts.

Rooftop antennas are often very difficult to mount, and so it is probably fortunate that most subscribers do not want this mounting. Some vehicles have a double skin on the roofing that can buckle under the tension of an antenna and result in water leakage.

If rooftop mounting is requested, however, it is essential to make sure that a center-mounted antenna is actually centered. Take care to double-check the measurement of the proposed hole. Patching a hole drilled in the wrong area is very expensive.

Drilling a hole through the roof requires some care not to catch the headliner underneath. It is advisable to use two people for this operation—one to drill and one to ensure the integrity of the headliner. Using an area of the roof close to the interior light (if it is in the center) can simplify installation by making internal access easier.

An elevated-feed antenna raises the height of the antenna to a level comparable with that of a roof-mounted antenna. Because these antennas are, in effect, dipole-fed antennas, they are ground-plane independent and so can

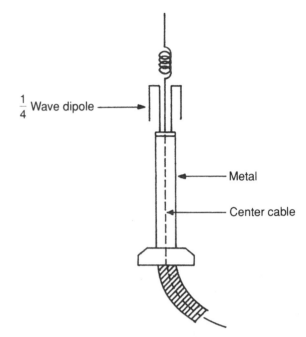

Figure 21.12 An elevated-feed antenna showing that the elevation rod is an extension of the coaxial feeder. It is ground-plane independent.

be mounted almost anywhere on the vehicle. Figures 21.10 and 21.12 show elevated-feed antennas.

NOISY ANTENNAS

Sometimes an antenna will develop a whistle at speed. Usually this is due to a resonance in the loading coil. A simple and often effective way to eliminate this is to place a piece of heatshrink over the coil.

ANTENNA GAIN

Antenna gain cannot be adequately tested except by using digital sampling techniques that collect large amounts of data on antenna performance over closed loops. Comparing two antennas in the same general location using just listening or S-meter tests is not sufficient. Because of the standing-wave pattern, the same antenna will give very different results when moved as little as a few centimeters. Some manufacturers have exploited this situation to claim that they have produced "high-gain antennas" or "fringe-area antennas," knowing that their claims are difficult to check.

Any claims of more than 4.5-dB gain for cellular antennas should be treated with a good deal of suspicion,

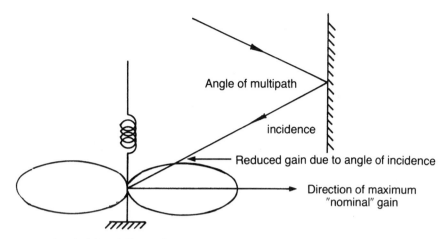

Figure 21.13 "High-gain" mobile antennas are limited in effectiveness. The "nominal" antenna gain is the maximum gain along the major axis. Other angles of incidence reduce this gain.

because such antennas are most difficult to manufacture. True 4.5+ dB antennas will normally have at least three segments and are somewhat unwieldy.

"True" 4.5+ dB gain means only gain measured normal to the antenna and along its most sensitive axis. Due to multipath, much of the RF energy will impinge on the antenna off-axis, and hence the net gain will be less than the "nominal" gain.

Figure 21.13 illustrates this effect. The higher the gain of the receive antenna, the more narrow will be the lobes of the field pattern, and so the more multipath will reduce the effective gain.

To successfully test the gain of a mobile antenna, it is necessary to sample the field strength along a closed loop of roads at least 6 km in diameter. It is preferable to average the results in both the clockwise and counterclockwise directions because, apart from a center-roof-mounted antenna, almost all other mounting positions are highly asymmetric. The gain will be very much dependent on the relative positions of the transmit antenna and the mobile receive antenna in relation to the body of the car.

Sampling must be done at a rate at least equal to the standing-wave pattern wavelength (approximately one sample every 0.2 meter at 800 MHz). Such sampling requires a good deal of specialized equipment that is ordinarily available to cell site designers only.

Based on a study done by the author in Australia in 1987, the relative performance of antennas mounted elsewhere than the center of the roof have been determined. Table 21.2 shows these results.

All of the measurements in Table 21.2 involve sample sizes of about 60,000 obtained by driving around closed loops of about 6 km.

TABLE 21.2   Effects of Mounting Position on Antenna Gain

Gain of Antenna Relative to Center Roof Mount	Mounting Position
−1 dB to 0 dB	Magnetic base (various bases), roof-mount
−6 dB to 0 dB	On fender with magnetic bases (various locations)
−5 dB	Gutter-grip-mount above driver's door (not ground-plane independent type)
−1.5 dB	Fender-mount*
−1.5 dB	Through-the-glass mount**

*Although the average loss was low, the losses recorded traveling directly toward and away from the base are interesting.

**Results from similar tests in the United States.

Table 21.2 shows that gutter-grip mounts (and for the same reason other similar mounts that use brackets that lift the antenna away from the ground plane) and magnetic mounts vary widely in efficiency and should be thoroughly tested before being recommended for customer use.

For the fender-mount, the gain is very directional. To ascertain the degree of directionality, measurements were made in the two extreme directions of directly toward and directly away from the transmitter. Figure 21.14 indicates how these measurements were taken.

The results of the measurements were as follows:

$$G_{\text{FORWARD}} = -0.5 \text{ dB (forward gain)}$$

$$G_{\text{BACKWARD}} = -5.5 \text{ dB (backward gain)}$$

Although losses due to fender-mount are not high, in the

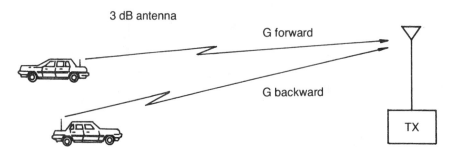

Figure 21.14   Fender-mount antennas are least effective when the vehicle body is between the antenna and the base station.

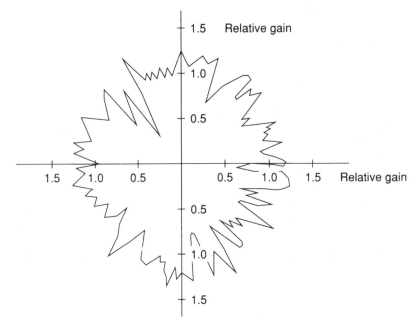

Figure 21.15   A polar plot of a typical mobile antenna.

particular case where the body of the vehicle is between the base and the mobile antenna, significant losses occur. Despite this potential disadvantage, however, the fender-mount is still a good and very popular option.

Through-the-glass antennas are popular and have a performance somewhere between the fender-mount and gutter-grip-mount options. Through-the-glass antennas do tend to be susceptible to car wash and heat stress. Take care to avoid mounting them near rear-window-demister elements.

The gutter-grip, a relatively common mounting method, has a loss of 5 dB and is not recommended unless used with ground-plane-independent antennas. It is also advisable to avoid the "easy" option of running the antenna feeder between the body and the car door, because feeder damage usually happens eventually.

The more conventional method of measuring field strength of an antenna in polar form is also possible. This method is the most frequently used, probably because it is the simplest. In its basic form it involves connecting the transmitter to a vehicle-mounted antenna and placing the vehicle on a turntable. The vehicle is then rotated and the far-field strength is recorded at various angles of turntable rotation. A polar plot as shown in Figure 21.15 is then made.

Although this method is adequate for base-station antennas, it is nearly meaningless in the mobile environment. The very complex radiation pattern caused by the interaction of the vehicle and the antenna is not only difficult to interpret but also varies considerably with the angle of incidence of the received signal.

An approximation that the signal always arrives in

TABLE 21.3   3-dB Roof-mounted Antenna Gain Degradation when Mounted Off-vertical

Angle Off Vertical	Loss Compared to Vertical Antenna (dB)
10	0
20	2
30	2.5
40	3.5

the plane perpendicular to the receive antenna is usually made when the polar diagrams are drawn. Of course, in the real world of multipath propagation, this assumption is not necessarily valid. Assumptions made as a result of these diagrams, particularly about gain and directivity, are highly suspect.

Antenna Rake

Many customers are very concerned with appearances and like to use adjustable-slope antennas that conform with the lines of the vehicle. The general feeling in the industry is that this will result in significant gain losses compared to a conventional vertical mount. To determine the magnitude of this gain loss, a study of raked 3-dB antennas was undertaken by the author in 1988. Some measurements were taken that indicate that small angles from the vertical can be tolerated without significant loss.

Table 21.3 shows the results of a number of measurements on sloping antennas. The method used in Table 21.3 gave readings to an accuracy of ± 0.5 dB, which accounts for the 0-dB reading for 10-degree angle of rake. These tests were conducted at the same time as the antenna location tests. The tests were conducted over the same closed routes, and the sample size was large. It is clear from these results that the small angles of rake (less than 10 degrees) sought by most customers do not seriously reduce or impair gain.

Figure 21.16 shows a vehicle with antenna rake.

DECIBEL'S MOBILCELL

In 1992 Decibel released a product called Mobilcell, which is really a miniature repeater meant to enhance the operation of a handheld used in a vehicle. The device is a bi-directional 38-dB linear amplifier. It uses an external antenna mounted on the vehicle as the primary receiver, which connects to the amplifier. The amplifier, in turn, has a small rubber duck antenna, which forms the link to the mobile, as illustrated in Figure 21.17.

The unit has an internal gain regulator, which limits the output power on the up-link (mobile to base station) so that the maximum output is 0.6 watt. The downlink is likewise amplified and fed via the rubber duck antenna to the mobile.

A minor limitation of this unit is that it is designed to be used by only one handheld at a time. If two or more users attempt to use it, intermodulation can be expected. It also requires a hefty 1 amp of power (even in the standby mode) to power the linear amplifier. For this reason it is not a good idea to connect it directly to the battery.

Figure 21.17 shows that, like all such repeaters, this unit is a potential oscillator if care is not taken to prevent excessive feedback between the antennas. An isolation of 45 dB is required between the two antennas. This is best achieved by roof mounting the external antenna and allowing the roof to act as a shield. Mounting the rubber duck low in the vehicle will also help. There is a light-emitting diode (LED) warning light to indicate too-close coupling between the antennas.

The theory is sound, and reportedly this unit will provide approximately a 10-dB improvement in field strength (as seen by the mobile) in low signal strength

Figure 21.16   Some owners like the trendy look of a raked antenna. Small angles of rake result in only small losses.

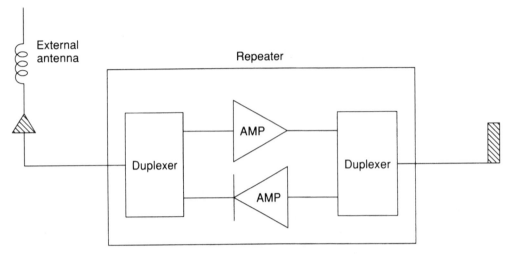

Figure 21.17 Block diagram of Decibel's Mobilcell.

environments, enough to substantially improve the range of a handheld.

PASSIVE REPEATERS

In 1990 a controversial mobile product—the passive repeater—reached the market. Based on an earlier product, the "beeper booster" (an external antenna for boosting the range of a pager), the passive repeater consists of a cellular antenna mounted on the outside of the window and a "radiator" on the inside, which is supposed to re-radiate the antenna power to a handheld in the vehicle. Now, there is a major difference in concept between the beeper booster and the passive repeater, and that is in the degree of coupling. The beeper booster has a small cradle that holds the pager up against the radiator, and it has been proved effective in quite rigorous tests. Were the handheld physically wired to the external antenna (as is the case for some external antenna systems), then like the pager with a beeper booster it would undoubtedly work.

The passive repeater, although it has its supporters, has never been rigorously tested, and such tests as have been done are clearly not done to standards sufficient to be conclusive. From a mathematical point of view the passive repeater will not work. The coupling between the antenna and the "radiator" would typically be around 20+ dB down, which means that the energy coupling between the handheld and the repeater would be negligible. The windows, being glass and many wavelengths wide, offer almost no loss to an incoming 900-MHz signal. Thus the only advantages that the repeater antenna has is that it is mounted a little higher and it does not

have the losses associated with blocking by the metallic parts of the vehicle.

Some limited testing has been done by the author and *no* improvement of any kind was found. However, the testing was not exhaustive, perhaps because there was no reason to expect it to work.

BATTERIES AND TALK TIME

A major limitation of handheld and transportable equipment is the battery "talk time" (transmission time). The first handhelds (see Figures 21.18 and 21.19) had a talk time of just over half an hour and standby times of about 8 hours. Figure 21.20 shows a typical Advanced Mobile Phone Service (AMPS) transportable from the early days of cellular. These batteries were mostly lead–acid and weighed about half a kilogram.

Substantial improvements have been made in battery technology over the last decade. Analog phones can be fitted with batteries that provide more than 120 hours of standby time, and digital phones (which inherently consume less power) can have standby times of more than 300 hours. Although nickel cadmium batteries (NiCads) are still very common, lithium ion batteries are becoming more widely available; their acceptance is even greater because of the lack of the memory effect.

Nickel Cadmium Batteries

Nickel cadmium batteries (more properly known as cadmium nickel batteries) have been around for a century, having first been patented in 1899; then in 1912, Edison patented an alkaline iron-nickel-oxide sealed-cell

Figure 21.18   A Motorola handheld circa 1984.

system. Commercial production came about 40 years later.

NiCads use an alkaline electrolyte, which is usually potassium hydroxide, with a concentration of about 25 percent by weight. They are the least expensive of the rechargeable batteries, with an estimated cost of 0.03 cent per cycle.

One major shortcoming of the NiCad is that cadmium is a toxic metal that has been cited as a cancer causing agent, and is implicated in kidney disease and osteoporosis. A major concern is that improperly disposed of batteries could leach cadmium into the groundwater.

Unlike lead-acid batteries (for example, car batteries),

Figure 21.19 A Toyocom handheld circa 1989.

which have a prolonged life if kept fully charged, Ni-Cads exhibit a memory phenomenon. That is, if a NiCad battery is lightly used and recharged frequently, it tends to remember this usage pattern and will later be unable to deliver its full ampere-hour capacity. Therefore, these batteries should be fully charged and then used until fully discharged whenever possible, particularly when the battery is new. A few deep recharging cycles will contribute to a long battery life.

Batteries that show evidence of "memory," discharge very quickly but can often be restored by two or three repeated deep discharge cycles.

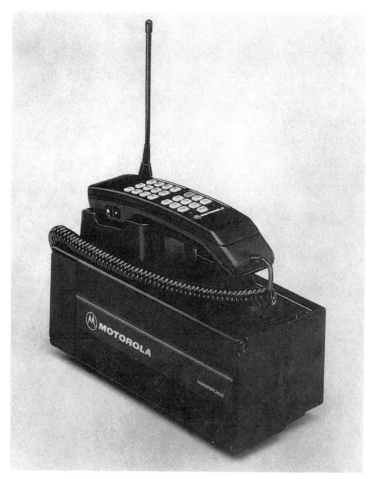

Figure 21.20 A Motorola AMPS transportable, circa 1989. (Photo courtesy of Motorola Communications and Electronics, Inc.)

NiCad batteries have a shelf life of around three years and an in-service life of about half that. The number of recharge cycles depends on the duty cycle. Fully discharged/recharged batteries will, as the manufacturers recommend, hold a higher charge, but the heavy cycling will reduce their life span to around 500 discharges. A battery that is regularly cycled to only 50 percent of its capacity, will begin to suffer the memory effect, but its life will be increased to between 3000 and 5000 cycles. Under well-controlled laboratory conditions with ideal charge/discharge cycles, NiCads have been found to last in excess of 30,000 charge/discharge cycles. Their maximum life is increased by a monthly "exercise" cycle in which the battery is discharged to one volt per cell. This cycle can also be used to rejuvenate low-capacity batteries, provided they have not deteriorated too much. The cycling process mainly causes the unwanted crystalline structures to dissolve.

The discharge to one volt per cell is a compromise. Batteries that are discharged to higher voltages (e.g., 1.2 volts) will have a longer life, but will begin to suffer the memory effect. Batteries discharged to 0 volts will have no memory effect, but will have their life shortened by perhaps a factor of 10.

One other (but not recommended way) of countering the memory effect is to freeze the batteries to break up the crystals. This works, but it can cause mechanical damage to the battery and so shorten its life. An additional shortcoming of this method is the time it takes, and the necessity to defrost the battery back to 10°C before charging can take place.

Reflex Charging

Reflex charging allows partially discharged batteries to be recharged without the memory effect. The technique uses pulse charging, followed by a short-duration discharge pulse on each cycle, as seen in Figure 21.21. The technique prevents the buildup of gases at the battery cell plates, which in turn leads to crystalline growths.

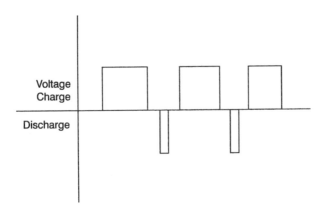

Figure 21.21 A reflex charge cycle.

In practice most people will randomly cycle their batteries, sometimes recharging them after only light use and at other times fully discharging them. Such batteries will have an intermediate life of around 1000 cycles and will exhibit only a minimal memory effect.

Batteries should never be charged if their temperature is below 0°C.

Keeping a battery that is not used often in a state of readiness is a problem. A fully charged NiCad does not have a long shelf life, as it has a significant self-discharge of about 10 percent per month. Trickle charging a NiCad is a sure way to induce memory-effect losses. A compromise is to charge the battery and cycle it every two to three weeks.

NiCads, with very long life cycles have been made, and in the space industry for satellite applications NiCads deliver a 70,000 cycle life over a period of 15–20 years. However, batteries for mobile phones have a life of about 900 cycles. A battery is said to have reached the end of its life cycle when the fully charged capacity falls to 80 percent of the rated capacity.

Some consumer items carry labels recommending that the NiCad batteries should be *fully discharged*. This is a misinterpretation of the discharge cycling and can cause serious damage to the battery.

Several repeated discharge cycles can often be used to rejuvenate low-capacity batteries.

A common problem with batteries is that the external charging contacts become dirty, and so effective charging (and sometimes discharging) cannot be achieved. The terminal should be cleaned periodically by swabbing with alcohol (or methylated spirits).

The cause of the memory effect is complex. However, the mechanism of memory is known. In a normal full charge/discharge cycle, the charge causes the negative plate to lose oxygen and convert from cadmium hydroxide to metallized cadmium, while the positive plate goes into a higher form of oxidation, changing nickel hydroxide to nickel oxyhydroxide.

If a complete discharge occurs, then the battery reverts to its original state. However, a partially discharged battery, when charged, will, in the presence of the negative plate's metallized cadmium, cause some of the cadmium hydroxide crystals to clump together and form large crystalline structures. These crystalline structures reduce the reactive surface area and so reduce the charge capacity of the battery.

Another effect that can occur is "voltage depression." This results from overcharging and results in a lower terminal voltage. Again this can be restored by a full charge/discharge cycle. This voltage depression, which results in a lower terminal voltage when a state of overcharge is reached, is used by some chargers to signify that the battery is fully charged. These chargers are known as $-\Delta V$ chargers.

Life Cycles

A progressive and irreversible capacity loss will occur; this is mainly the result of a reduction in the electrolyte volume due to evaporation or loss through the pressure vents.

Crystalline structures of nickel and cadmium can grow to the point where they cause internal short circuits. In some cases, these structures can be reduced by a conditioning process that aims to fuse the shorted links. Permanent loss of capacity occurs by degradation of the separator and by internal short circuits that build up over time.

Overcharging is a major culprit, causing capacity loss, and can be minimized by the use of "smart" chargers that assess the condition of the battery before charging begins, and regulate the charging accordingly.

Batteries with less that 80 percent of rated capacity are in need of rejuvenation. If this fails to restore the capacity to above 80 percent, then the battery should be discarded.

NiCad Battery Chargers

A lot of damage can be done to NiCad batteries by incorrect charging procedures. Many of the chargers supplied with mobile phones are budget models and can easily cause damage to the battery. The expected battery life is at least 500 cycles.

Perhaps the most popular and simplest charger is the trickle charger. It delivers a constant current of about 1/10 of the ampere-hour rate of the battery. Often this charger will have no cutoff mechanism and so it can easily overcharge.

A simple cutoff arrangement uses a timer. This can overcharge batteries that are not fully discharged before recharging.

Fast chargers, by their nature, need some limitation on the charge rate. One common version is the negative delta charger, which senses a decrease in battery voltage as overcharging commences and then switches to trickle charging. The counterpart for the NiMH battery is the zero delta, which senses the end of the rise in battery voltage, which for a NiMH battery indicates the fully charged state.

Nickel Metal Hydride Batteries

Nickel metal hydride (NiMH) batteries offer around 30–40 percent more capacity than NiCads, with the added bonus of being free from the environmentally damaging cadmium. They do, however, have a high self-discharge rate of around 35 percent per month. The cell voltage of an NiMH battery is 1.2 volts, and so they can directly replace NiCads.

The NiMH battery is composed of a positive nickel electrode and a negative one made of a hydrogen-absorbing rare earth alloy. The electrolyte is essentially potassium hydroxide. During the charging cycle, the current breaks down the water in the electrolyte, generating hydrogen. The hydrogen, as a gas, is absorbed by the negative electrode. Discharge is the reverse of this.

NiMH batteries are similar to NiCads, except that in the NiCad the mechanism involves the absorption of hydrogen into a metal alloy (which replaces the cadmium as the active negative material). During charging, the metal hydride breaks down the water, releasing hydrogen, which is absorbed into the hydride alloy. The discharge process releases the hydrogen, which in turn is oxidized.

Hydrogen absorption alloys were discovered in the 1960s, and it was found that they could absorb over a thousand times their own volume of hydrogen. They usually consist of two metals, one of which absorbs hydrogen exothermically, and a second one that is endothermic and acts as a catalyst. These alloys can be of many different compositions, but the AB_2 (e.g., $ZrNi_2$) and AB_5 (e.g., $LaNi_5$) are among the most important. More complex alloys with up to eight metallic components have been used.

NiMH batteries became commercially available only in the early 1990s and are costly relative to NiCads, with an estimated cost of 0.10 cent per cycle. These batteries should be cycled to a fully discharged state (that is, to a cell voltage of 1 volt) about once every three months.

The downside of the NiMH is its low number of recharge cycles, which is ordinarily only about 400, although it seems that in the commercial market, people are prepared to accept a short lifetime in return for a long talk-time. Additionally, like the NiCad, it has a significant memory effect. They cost about 50 percent more than NiCads, and the extra capacity may be largely illusionary, as the newer high-capacity NiCads can provide similar capacity.

Lithium Ion Batteries

Lithium ion (Li-Ion) batteries were first used in hearing aids because of their relatively high storage capacity, being about 50 percent higher than NiCads. They have no memory effect, which makes them very user-friendly. The life of 800 cycles for a phone battery nearly matches the NiCad, as does the effective operational temperature range of -20 to $+60°C$.

Discovered in 1912 by G. N. Lewis, it was not until the 1970s that commercial applications arose for the non-rechargeable Li-Ion battery. In the 1990s the first commercial rechargeable batteries appeared, but not without problems. In 1993, a large batch of batteries had to be recalled after a battery in a cellular phone in Japan exploded in a man's face causing burns (thermal runaway was the problem). By that same year, however, the Li-Ion battery was established commercially.

Li-Ion batteries have a linear discharge voltage, which makes it relatively easy to determine the state of charge of the battery, especially when it is compared to NiCad and NiMH batteries, which have very flat discharge curves.

They have a terminal voltage per cell of 3.6 volts, and a self-discharge rate of only 10 percent per month (which compares favorably with the 20 percent rate for a NiCad). Since most phones today operate at 3.6 volts, a single cell is all that is needed.

The relatively high internal impedance of a Li-Ion battery, which is around 300–400 milliohms (3–4 times higher than a NiCad), can be a challenge with digital phones, which can require up to 1.7 amperes of pulsed current.

The Downside Li-Ion batteries are about five times more expensive than NiCads on a per kilowatt hour basis, although ongoing development is reducing that price differential. They are relatively fragile and require elaborate charging techniques. In operation they need protective circuitry to keep them within safe operating limits. The batteries contain a peak charge and voltage-limiting circuit in each pack. The protection device typically operates if the cell temperature approaches 100°C, the cell voltage reaches 4.3 volts, or the cell voltage drops to 2.5 volts. Additionally, protection is provided against short circuits and in some batteries against high pressure.

Most manufacturers do not sell Li-Ion cells, but only complete battery packs.

The chargers for Li-Ion batteries must meet tight

specifications and trickle charging is not an option, as there is no mechanism to absorb overcharging.

Li-Ion batteries begin aging even when not used at a rate much higher than NiCads, and experience a noticeable fall in capacity after one year and probable failure in two.

Lithium Polymer Batteries Currently under development, lithium polymer technology may be the next advance. The shortcomings, however, are still a limited recharge cycle time and high internal resistance.

Alkaline Cartridges

Alkaline cartridges that use standard AA or AAA cells to power a mobile phone are available for some phones. Although they will never replace the rechargeable battery (who wants to have to buy a set of batteries everyday?), they do have some appeal. They last about twice as long as a NiCad and have a much greater shelf life. They are therefore ideal to be kept as backups. Travelers will also find them useful.

Rechargeable Alkaline Manganese Batteries

Rechargeable alkaline manganese batteries have recently been released, initially mainly as a direct replacement for disposable zinc-carbon cells. They have a cell voltage of 1.5 volts and can be recharged 600 times. The cells are made of non-toxic materials.

ZINC-AIR FUEL CELLS

Zinc-air fuel cells are now readily available as standby batteries for most popular cellular phones. The technology provides disposable batteries, with a long shelf life that have capacities up top 4500 mA·h; about twice the capacity of a good-quality rechargeable battery.

These batteries are not meant to be a replacement for the rechargeable battery, but their role is more as a supplement, particularly in remote areas or in an emergency situation. Electric Fuel is one company that makes a range of zinc-air batteries that use no environmentally hazardous materials and are available for a wide range of phone models.

Known as air-depolarizing cells, zinc-air batteries use oxygen in air as the cathode material and zinc as the anode. One of the difficulties in constructing such a battery is that the oxygen must be kept away from the zinc anode, as it will attack and corrode it. To this end, the oxygen usually has a waterproof polymer-bonded porous carbon layer around it. Zinc-air cells provide relatively high watt-hour ratings at reasonable cost.

BATTERY RECYCLING AND DISPOSAL

Worldwide, about 90 percent of lead-acid batteries are recycled. Typically the batteries are physically separated into their components and then recovery of the lead is done in a smelter.

Cadmium-based batteries are more problematic. Cadmium is a very toxic element, and worldwide little has been done to recover the cadmium. Most exhausted NiCads end their life in local dumps. There are a few plants operating in the United States and Europe that recover the cadmium in a smelter.

Lithium is not toxic, but the element is highly reactive, especially with water. Additionally lithium batteries contain some toxic electrolyte solvents, which constitute a hazard.

AMPS NUMBER ASSIGNMENT MODULE (NAM)

The number assignment module (NAM) can be a programmable read-only memory (PROM), an erasable programmable read-only memory (EPROM), or an electrically erasable programmable read-only memory (E^2PROM) that contains details about the customer, the system, and the options chosen.

The PROM is a 32×8-bit memory that is non-volatile and cannot be erased. The NAM programmer is simply a PROM burner. It is called a burner because the links inside the PROM are programmed by applying a current of sufficiently high level to burn out link structures internally, to encode the required information. Today most mobiles have their NAM programmed density from the telephone keyboard, but most mobiles require a NAM programmer. In this case an E^2PROM is used.

Programming from a NAM programmer is made easier by having a master NAM (one that contains all the standard default values, which can be copied so that new NAMs need only have their specific parameters added).

The code is stored in either decimal or hexadecimal (a number system based on 16) and the information is usually input in hexadecimal. The following list shows typical NAM data.

- SIDH. System identification of home mobile service area (5 digits HEX); defines the cellular operator's system and is used by the mobile to distinguish roamer areas

- MIN. Either a 10-digit mobile number (US) or a 3-digit country code plus a 7-digit mobile number
- Telephone number of the mobile unit
- SCM. Station class mark (two digits); indicates if the mobile has VOX and its power output
- IPCH. Initial paging channels in home switch (four digits) usually 0333 or 0334; defines the starting point of the mobile unit's control-channel search
- ACCOLC. Access overload class (two digits); divides the subscribers into 16 subgroups with the possibility of giving each subgroup a different priority under overload conditions
- PS. Preferred system mark (1 digit); used when this function is not keyboard-accessible, which is rarely on modem phones
- GIM. Group identification mark (two digits)

(Additional information on manufacturers' options can also be placed in the NAM.) Most mobiles today have E^2PROM and NAMS and are programmed from the keyboard.

22

TOWERS AND MASTS

Towers (self-supporting structures) and masts (guyed structures) cost about the same if they are both short (that is, at heights ordinarily encountered in cellular radio). As the structures get higher, the costs of guyed masts tend to increase linearly (for the same cross section); the costs of self-supporting towers increase exponentially. Because guyed masts require a good deal of land, they are used mainly in rural areas.

Towers and masts have an important advantage over poles and buildings in cellular applications: They can be used initially with omni-directional antennas mounted on top for maximum coverage, and then later, when cell sectoring is used, the sectors can be mounted at lower levels on the towers or masts according to the coverage sought (that is, each sector can be mounted at a different height to give independent control over the coverage of each sector).

In cellular installations, tall towers are not usually needed and, except for rural areas, poles should be adequate for heights up to 30 meters. A number of imaginative designs have evolved; most use a triangle on the top for mounting.

If a triangle with 3.5-meter sides is mounted on the top of the pole or tower, it is possible to attach up to 3-transmitter/6-receiver (3 TX/6 RX) antennas to it. To improve isolation, transmit-and-receive antennas are often mounted so that the transmit antenna is vertical and the receive antenna is upside down. You must always ensure that water drainage is adequate on the inverted antenna, because antennas usually have drain holes only at the bottom. Sometimes you must drill additional holes and plug the original holes.

The triangle mounting bars should be about 1.5-meters high to vertically separate two antennas. The

triangle configuration is often chosen because of its simplicity for construction purposes.

When a square-section tower is used, the platform can also be square. Figure 22.1 shows a 40-meter tower, specified by the author for Extelcom in the Philippines. This tower has a 3.5×3.5-meter-square platform on top.

Figure 22.2 shows an antenna installation in Indonesia that has a triangular mast section with three triangular platforms. Notice the directional, radome-enclosed Yagi antennas. This is part of the Jakarta NMT450 network.

As mentioned earlier, towers and masts require different amounts of land. Figures 22.3 and 22.4 show the area needed for a mast.

Figure 22.5 shows the amount of land needed for towers of different sizes. Table 22.1 shows the amount of land needed specifically for three-leg and four-leg towers. In this table, T and W are the land dimensions used in Figure 22.5.

No structures of any kind should be built closer to the tower than the edge of the boundaries defined in Table 22.1, because the support of the surrounding soil against turning moments may be diminished. These dimensions are a guide only; the design of a tower or mast depends on such factors as wind loading, local building codes, and local planning-authority regulations.

The choice of monopole, mast, or tower for cellular radio is often made for the operator by the local government rules or environmental considerations. Sometimes, however, there is a choice, so it is worthwhile exploring the alternatives.

The location of the tower needs some careful consideration. In most countries, it will be sufficient that the

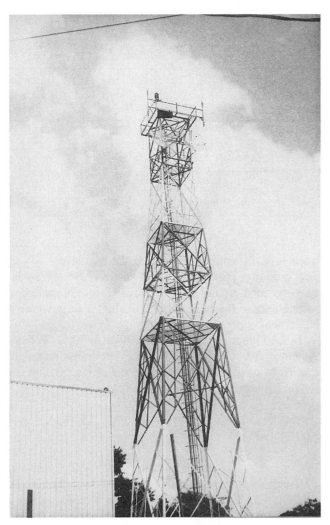

Figure 22.1 A cellular/microwave tower with a 3.5 × 3.5-meter square platform for cellular antennas, designed by the author.

Figure 22.2 These are the triangular mounting platforms used for the NMT450 system in Jakarta. This system uses the "Stockholm Ring" six-cell configuration. The radomes contain Yagis for the NMT450 system. (Photo courtesy of L.M. Ericsson.)

tower win local zoning approval (from the local authorities) and does not constitute a hazard to air navigation. Usually, unless the tower is particularly large or the proposed location is zoned residential, there will not be too many objections from the local authorities. However, they may require the planting of trees around the structure (and in some cases even a painting scheme that is more environmentally sympathetic, such as sky-blue or green).

Conflicting with the requirements of being inconspicuous will be the requirements of the aviation authorities that the structure be visible to aircraft on Visual Flight Rules (VFR) and that it not be an obstruction to any existing or future flight paths.

Generally, if the structure is smaller than 50 meters (in the United States special conditions apply above

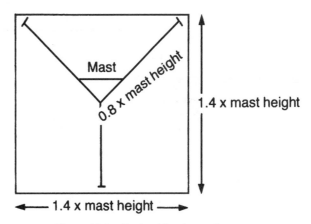

Figure 22.3 Optimal land area for mast.

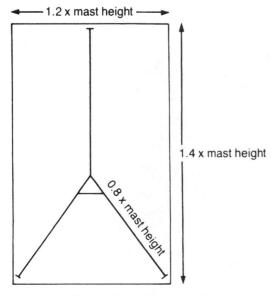

Figure 22.4   Minimal land area for mast.

TABLE 22.1   **Land Usage and Weight for Three- and Four-Leg Towers**

	3 Legs			4 Legs		
Tower Height (meters)	T	W	Approx. Weight (tons)	Tower Height	T	Approx. Weight (tons)
10	7	7	0.7	20	7	1
20	8	7	1.7	30	9	2.2
30	10.2	9	3	40	10	4
40	11.5	10	6	50	12	8
50	13.8	12	10	60	13	12
60	15.5	14	14	70	14.4	16

200 feet) and more than 10 km from any airport, it is unlikely to be a problem. In any case it is a good idea to get a ruling from the local aviation authority on both the location and warning markings/beacons that are required.

Information required by the aviation authorities will include

- Accurate coordinates of the tower location
- Structure height
- Structure type (tower/mast/pole)
- Proposed warning beacons and hazard warning paintwork.
- Location of the nearest airport and other airports within 10 km

Also it may be required in some countries (as it is in the United States) that a full inventory of the radio fre-

quency (RF) facilities—including the frequency, power, and radiation patterns—be provided. These are sometimes needed to assist in the evaluation of potential interference to Instrument Flight Rules (IFR) navigation equipment.

In the United States it is *compulsory* to get an Federal Aviation Administration (FAA) "determination of no hazard to navigation." This can take about 60 days, or longer if the submission is not complete.

MONOPOLES

In general, a monopole is more asthetically pleasing, although very few neighbors are likely to welcome any support structure. The monopole, like a building, has a fixed platform height and usually comes in a very limited range of sizes (typically, 15–50 meters). It may have an internal ladder and cable tray. Its main structural advantage is the small land area required, typically 3–4 meters square.

Monopoles can be erected in about one day provided the base has been poured and cured. Because they are available in a wide range of sizes, they often can be

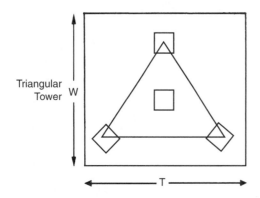
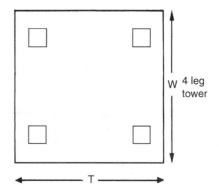

Figure 22.5   Dimensions of land for towers of different sizes.

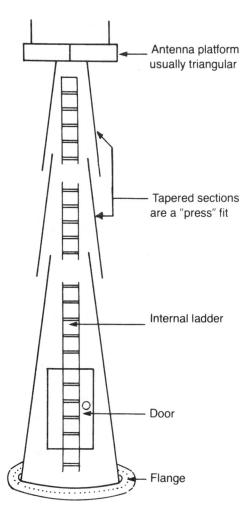

Antenna platform
usually triangular

Tapered sections
are a "press" fit

Internal ladder

Door

Flange

Figure 22.6 The construction of a monopole with an internal ladder as used by Telstra.

Figure 22.7 A mounting platform on top of a monopole. Notice that the bottoms of the sector antennas are not restrained. This can result in very uneven performances in windy conditions.

ordered almost off the shelf and have much shorter delivery times than towers.

Monopoles are usually fabricated in tapered sections of about 10 meters apiece and fit together simply by stacking the sections (see Figures 22.6 and 22.7).

The support is simply a cage, designed to withstand large turning-moments, embedded in concrete. Shafts typically 8 meters deep and tapering from 2 meters at the bottom to 3 meters at the top form the foundation. Other monopoles can have much wider bases with correspondingly smaller shafts. Bolts 2 meters long and 57 millimeters in diameter, embedded in the concrete, attach to a flange at the bottom of the shaft. Up to 50 bolts can be used to hold the flange.

These structures can be designed to give the torsional stability required for low-frequency microwave bearers (maximum half-degree twist).

An alternative construction, shown in Figure 22.8 is used by CENTEL in Las Vegas. This monopole has a much more slender cross section and has climbing pegs external to the pole. Figure 22.9 shows how the rigidity of the antenna structure can be improved significantly by the use of a pair of mounting platforms.

In recent times the antennas have gotten smaller and so have the structures supporting them. Figure 22.10 shows a monopole with three radial arms in place of a platform. The arms each have two vertical rods attached to them for the antenna mounts. Climbing pegs are on the outside. The pegs start about 10 meters up the pole so that a ladder is needed to access them, making public

Figure 22.8 A monopole with external climbing pegs in Las Vegas.

Figure 22.9 Dual platforms significantly improve the rigidity of the antenna assembly.

access a little harder. The cables are fed down through the pole.

GUYED MASTS

Guyed masts are practical only where land is inexpensive. They often prove to be the cheapest solution in rural environments.

Guyed masts are usually constructed of sections of triangular cross section about 6 meters long. The sections are typically 0.5 to 1 meter wide per side and are designed to be bolted together. The strength of a mast is essentially in the guying, so proper tensioning of the support cables is vital. Concrete anchors hold the guy wires. For cellular applications, the standard cross sections should prove adequate to accommodate the cellular and microwave antennas. Figure 22.11 shows a 50-meter mast in Asia.

Hybrid Structures

Sometimes it is not easy to decide if a structure is a mast or a tower. Figure 22.12 shows a structure that started as a tower and later sprouted a mast on top. This structure is on a rooftop in Manila, the Philippines. The mast may have been built to reduce the total loading on the roof.

Fabrication

The cost of a mast is related more closely to the weight of steel than to the height. Table 22.2 shows that the weight of a mast is almost linearly dependent on its height. It can be seen by comparing Table 22.2 with Table 22.1 that self-supporting masts increase more rapidly in weight than guyed masts, particularly for structures higher than 30 meters.

Figure 22.10   Monopole with three radial arms in place of a platform.

Figure 22.11   A 50-meter mast in Asia. Notice the long grass (a fire hazard) in the foreground and recent evidence of fire in the very near foreground.

Figure 22.12   A hybrid mast/tower constructed on a rooftop.

TABLE 22.2 Mast Height and Weight

Mast Height (meters)	Approx. Weight (tons)
30	1.5 to 3
50	3 to 5
70	4 to 7
100	10 to 12

Figure 22.13 Some towers are designed for high-density dish loading. Notice that each dish has a cover for weather protection.

TOWERS

Towers are self-supporting structures that are most practical when land is expensive. Figure 22.13 shows a tower that supports a number of microwave dishes (the solid dishes) and gridpacks (the wire-formed dishes).

Towers require less land than guyed masts and are capable of supporting a large number of antennas, a factor where you plan to rent tower space to other users.

When other microwave facilities are planned (for example, for a wireline carrier), a self-supporting tower is probably the best choice.

A three-sided tower is usually the best value (load-carrying ability per dollar). For the same strength, however, it has a wider base and requires more land than a four-leg tower. A four-sided tower has an extra face and can carry more antennas. For a given strength, it is also smaller.

Tower members can be of various types, including solid, tubular, and channel sections. Tubing is the cheapest material for tower construction; it is available in a large number of sizes and needs little work to make it suitable for towers. Tubing does, however, have a long-term maintenance liability. Moisture can build up inside the tubing and cause corrosion and, in extreme environments, can freeze thereby splitting the tubing. In coastal areas or in areas near heavy industry, this type of construction could prove to be a liability. Such towers need to be designed with weep holes, and the holes need periodic cleaning and unblocking.

Round-bar members can be made of round solid sections; they do not have the corrosion problem of tubular sections. They do, however, require substantially more steel for the same strength and thus weigh and cost more.

The most common material used in tower structures is the channel section, which can be made from formed-plate or angle sections. Formed plate is cut from rolled plate; it is cut to length with the bolt-holes punched while still in the plate form. It is then cold-formed into 60-degree or 90-degree channels along the center axis.

Deformed plate is made from milled 90-degree sections. For 60-degree sections, the plate is bent another 15 degrees on each flange. This plate is cheaper than formed plate, but often it is not precision-formed, which can lead to problems with bowing.

Sites on Transmission Line Towers

As sites become more sought after, it is increasingly necessary to be more inventive to find new locations for base stations. One interesting solution from FWT, of Texas, is the PowerMount. This solution co-sites the base station with an existing transmission tower. The pole supporting the antennas goes in the middle of the existing tower structure, and the antennas themselves appear just above the top of the tower. Said to cost from $25,000 to $30,000 for a 30-meter-high installation, this could be a cost-effective way of extending coverage, particularly in rural areas. Solutions for both cellular and personal communication service (PCS) are offered.

actually what is being seen is only local. A number of samples at different locations will resolve this problem.

OTHER USERS

If the structure is in a particularly prominent position, you should consider, before the tower is designed, the prospect of obtaining additional revenue from leasing tower space to other users. A modest increase in cost at the design stage can significantly improve the load-carrying ability of the structure.

When planning for other users, include them in the overall design by assigning their number, antenna type, and positions on the tower at the design stage. The structural design should also include detailed drawings of the proposed positions of other users so that they can be allocated at a future date without the need for new load calculations.

In general, cellular operators need not fear that including other users will cause interference, provided that they operate outside the cellular band and do not transmit very high power, as is the case with ultrahigh frequency (UHF) TV, for example.

Other users' services sometimes sprout up almost spontaneously in certain prominent areas and are known in the trade as "antenna farms." These "farms" can appear almost anywhere; Figure 22.15 shows one such farm thriving on a rock face in Baguio, in the Philippines.

ANTENNA PLATFORMS

Often in cellular applications, and particularly where sectored cells are employed, it is convenient to provide a platform for the antennas. This platform provides a safe working place and should have hand-railings, grated floors, and kickplates (to prevent tools from falling over the edge). The platform will normally be the same shape as the tower cross section (three or four sides) and should have sides of around 3.5 meters (whether triangular or rectangular).

Alternatively, extension arms can be used. Extension arms are suitable for small base stations, particularly those with few antennas (for example, omni sites with less than 16 channels).

Transmission Cable Support

The coaxial cables need to be securely supported along the length of the structure. For this purpose, the original cable hoisting, which is designed to be used to pull the cable into position during installation, is best left *in situ* after installation. The cable is then supported by cable

Figure 22.14 Taking a soil sample.

SOIL TESTS

Before a tower, mast, or pole can be erected, it is necessary to conduct a soil test. This involves taking core samples of the ground on which the structure is to be built and then having the samples analyzed. The soil samples are taken using a mechanical device like that seen in Figure 22.14. Using the results, the design engineer can determine the load-bearing capacity of the soil and its ability to resist the turning moments of the footings. Only after this test is complete (from one to four weeks) can design of the structure foundations begin.

The cost of foundations, particularly for a large tower, can be very significant. It can even be the major cost in areas of high wind loading. For this reason it is not possible to give a meaningful quote for a tower until soil tests have been done. Particular care should be taken if rock is found, to ensure that the sample is representative. Sometimes a soil test will reveal rock when

Figure 22.15 An "antenna farm" on a rocky outcrop. These antennas are mainly TV antennas mixed with a few links. Notice that the high-gain antennas are often mounted off-vertical, where they won't work well.

hangers, which are spaced about every 1 to 2 meters (the spacing depends on the type of hanger). Plastic cable ties (even the UV proof ones) are not adequate for this job.

For large installations (with many coaxial cables) it is best to install a cable tray, but for smaller applications, there are cable hangers that are designed to bolt onto the tower legs.

Accessories

All the accessories that you need for successful installation of the cables and tower grounding will be available from your cable supplier. Most suppliers are keen to assist, and some, like Andrews in the United States, even offer free training courses.

TOWER DESIGN

The antenna structure must be designed by a structural engineer, but it is worthwhile to consider the design parameters. The structure must account for gravity loads (dead loads) that include structure weight, antennas, and ice, as well as live loads, such as those caused by wind and seismic activity. Invariably, wind, ice, and tower fittings will provide the dominant loads on the tower.

Wind Loads

Until very recently, the dynamic load caused by wind was not fully understood, and towers were designed to withstand a known static load, which was increased by a safety factor (often doubled) to account for dynamic effects. In the light of recent studies, it is clear that early designs tended to overdesign the bases and underdesign the top portions of the structures. Particularly in typhoon and hurricane areas, the top portions of old designs are now being strengthened.

In general, as more collapsed structures are studied and more detailed information of long-term wind peaks becomes available, the minimum requirements of codes for determining wind loads have consistently increased. Old structures should therefore be used only after a thorough survey and inspection.

Wind speeds, recorded by national authorities, are of interest to a tower designer. The designer should know the peak gusts (instantaneous readings) and fastest-minute-wind (the highest velocity sustained for one minute). These two figures are connected by a ratio of approximately 1.3 : 1.

Fifty-year peak wind velocities are sometimes interpreted as ones that are expected to occur 50 years apart. This is not an accurate interpretation. A better interpretation is that 50-year peaks are ones that occur with a probability of 2 percent each year. Therefore, the fact that an old tower is still standing may simply be good luck!

Typical Specifications for a 40-Meter Tower

The tower designer must know a number of things about a tower before beginning the design process. The following list contains the considerations required for a typical 40-meter tower:

- Four-sided (or three-sided).
- 40 meters.
- Designed to Electronic Industries Alliance (EIA) RS222D (the US standard), Australian Design Standards, or other preferred standard.
- Stress factor (that is, suburban or rural safety factor). For example, in the Australian design code for suburban areas, this stress factor is:

  1.7 × factor on steel.

  1.75 × factor on foundations (this factor can be found from the relevant design code).
- Zone specifications and wind loading, depending on location.
- Maximum allowable twist (0.25 degree for 7-GHz microwave or 0.15 degree for 10 GHz).
- Maximum allowable tilt (1 percent for 7-GHz microwave or 0.5 percent for 10 GHz).
- Platforms and walkways at the levels where access to microwave dishes will be required.
- A platform of about 3.5 × 3.5 meters at the top, with guard rails 1.5 meters high and suitable for mounting cellular antennas at the edges. The mounts will be used to attach antennas with tubular supports up to 70 mm in diameter using three heavy-duty clamps.

  Up to 10 antennas can be mounted at the top with an equivalent flat-plate area of 0.23 m^2, weighing 50 lbs for 60-degree, 17-dB sector antennas. (See manufacturers' catalogs for particular antenna types.)
- Cable tray will be accessible from the ladder and will be 0.6 meter wide.
- Safety guard around the ladder, which will be internal with respect to the tower. (See Figure 22.16.)

Figure 22.16 When towers are to be climbed by staff other than riggers, safety guards should be provided.

- IAO standard paintwork and an aircraft warning beacon at the top.
- Tower orientation.
- Specify the microwave dishes, type (solid or grid pack) and mounting level. Allow for future expansion (even if expansion is not planned, it will probably be required; a good rule is to estimate the future requirement and then double it).
- Tower footings should be confined within a square plot of land (as specified earlier in this chapter).

SECURITY

Towers are attractive to youngsters, who see them as a challenge to climb. If towers are climbed untold damage may be caused to the antennas and cables, and even worse, the youngsters may suffer serious injury or death as the result of a fall. In order to discourage unauthorized access, a humanproof fence should be installed around the tower (as shown in Figure 22.17) or access can be barred by the attachment of spikes around

Figure 22.17   A human-proof fence (with spiked steel posts) around a rural cellular site.

the legs (as seen in Figure 22.18). In all cases it is advisable to place a notice, similar to the one shown in Figure 22.19, on the base of the tower to deter trespassing.

HOW STRUCTURES FAIL

A free-standing structure such as a tower is most vulnerable in the compression leg (the side away from the direction of the wind). A mast is similarly subject to compression failure, but because of the multiple guying points, has a more complex failure mode. Failure in both instances will probably be due to buckling.

The stress is very sensitive to wind velocity. It varies as the square of the velocity for static loads and as the velocity to the power of approximately 2.5 for dynamic loads. Wind speed varies more or less regularly with height and has an approximately parabolic gradient from ground level to 400 meters.

A less predictable factor is turbulence, although this is probably the major factor in structural failure. Turbulence is poorly correlated along the length of the structure (it is randomly distributed) and varies rapidly with time. In modern studies, the very unpredictable nature of turbulence is taken into account, and it has been found

that some turbulence patterns are significantly worse than others.

Topology plays a part, and many large towers will be situated on hilltops to gain additional elevation. Hilltops unfortunately produce increased airspeeds over their crests and a 10 percent hill slope can produce a 20 percent increase in airspeed or a 40 percent increase in wind-loading. This is the reason that windmills and wind generators are usually placed on hilltops.

Stiffness (the ability to resist deflection) is a sought-after characteristic in structures and an important factor for reliable microwave operation. Stiffness is often obtained, however, only by using more metal, which increases the cost and weight. For economic reasons, modern structures are designed to minimize the amount of materials used, so a trade-off occurs. Adding extra dead loads (for example, equipment and antennas) reduces stiffness.

Guyed tower anchor corrosion was responsible for eight tower failures in the United States in the years between 1990 and 1995. The collapse of a tower in Sioux Falls, South Dakota, in 1990 (caused by anchor corrosion) that seriously injured two workers, sparked an investigation into this failure mode.

It was found in a subsequent investigation that more

Figure 22.18 Access can be restricted by the use of spikes on each leg and on the access ladder.

Figure 22.19 A warning sign at the base of a tower in England (this is on the same tower as shown in Figure 22.18).

than 50 percent of the towers inspected had some form of anchor corrosion, and that in 10 percent of these cases the corrosion was sufficient to suggest a structural deficiency.

Towers that have collapsed from this problem typically have been found to have corrosion that has removed about 50 percent of the metal. Regular maintenance should have found the problem when less than 5 percent of the metal was gone, and at that stage, prevention methods would have stopped further corrosion. If more than 5 percent of the metal is gone, replacement is in order.

Significant corrosion will only occur over time, so new towers are relatively safe. However, with time the corrosion will become a problem on most unprotected towers. When it is considered that the design life of a tower is 50–100 years, there is plenty of time for the worst to happen.

Often anchor failure can occur below the soil level and this kind of failure is one of the most difficult to detect. Cathodic protection is probably the best insurance against these failures. Cathodic protection is a sacrificial method of protection, and works in much the same way as the zinc coating on iron sheets in galvanised iron does. Cathodic protection does not plate the metal anchors, but it does direct the corrosion away from them. The cathodic materials will themselves slowly be corroded away and will need to be periodically replaced.

Not all soils are equally corrosive and a pH test will reveal the extent of the problem. As a general rule, the soils will be more corrosive at deeper levels, and so testing should include samples taken at the lowest level that the structure is likely to go. Testing the corrrosivity of the soil should be an integral part of tower design.

One controversial proposal related to the use of copper grounding rods. Copper and steel have significantly potential differences and will set up a corrosive cell. One proposal is that galvanized steel rods rather than copper should be used for grounding.

TOWER, MAST, AND MONOPOLE MAINTENANCE

The unscheduled replacement of the antenna support structure can be both costly and disruptive to the installed service and should be avoided if at all possible. The collapse of a tower or mast, particularly in a populated area, at best can be embarrassing and at worst a catastrophe.

Antenna support structures require regular routine maintenance, which is often neglected on the grounds that the structure has been up for years and has not shown signs of fatigue to date.

Figure 22.20 The dishes with conical radomes (to help water run off) mounted at the base of the tower shown in Figure 22.11. Notice the spalling concrete base.

To appreciate the need for competent inspections, it is necessary to first understand how and why structures fail. These are the major causes of failure:

- Poor design, which inadequately allowed for static or, more frequently, dynamic wind loads
- Overloading of the structures with too many antennas and feeders
- Corrosion, particularly where hollow structural members are used
- Insufficient attention to guy-wire tensions and conditions (corrosion)
- Inattention to the indicators of stress
- Guys corroded or improperly tensioned

As an interesting example of the problems facing masts, the mast shown earlier in Figure 22.11 and again in Figure 22.20 is worthy of a closer look. The long grass

Figure 22.21   Cracking and spalling of the guy-wire anchor blocks.

Figure 22.22   The buckle connecting the guy wire to the anchor block in Figure 22.21. Notice that the bolts were not tightened and that no washers were fitted. The structure was, however, relatively rust-free.

in the foreground of Figure 22.11 represents a fire hazard, and evidence in the extreme foreground indicates a recent fire.

This mast uses passive reflectors (the large plates at the top) to deflect a microwave link to the ground-mounted receiving dishes illustrated in Figure 22.20. These dishes are protected by conical radomes. The cracking and spalling concrete seen in Figure 22.21 at the base of the mast is a sign of excessive stress.

Masts are held up by guy wires that are anchored into concrete blocks. Signs of stress were evident at all of the anchor points at the site in Figure 22.20. Figure 22.21 illustrates cracking and spalling at these points at this site. All of the anchor points inspected on this structure showed signs of spalling. This mast was well painted and relatively rust-free, but as Figure 22.22 illustrates, little attention was given to mechanical details. The buckle linking, the guy wire to the anchor point shows that the bolts were not tightened and washers were not used. The large central bolt is about 40 mm in diameter.

Routine inspections should be carried out about once a year for structures located near the coast, and every two to three years at sites more than 100 km from the sea, as well as after severe storms or periods of prolonged heavy icing.

INSPECTION

Very few cellular companies are large enough to employ a full-time, qualified structures inspector. Those that can will invariably be wireline operators.

Because of the special nature of support-structure maintenance, the cellular operator will generally find that there are few companies with the necessary expertise and that the availability of those companies is limited. Having found a competent operator, it is therefore a good idea to arrange the maintenance on a contract basis. The company should have a good structural engineer and experienced inspectors who can climb and inspect every portion of the structure. The inspection process should begin with a review of the existing documentation about the structure and its fixtures. It should then proceed step be step, using a checklist like the one provided at the end of this chapter.

If only cellular or mobile two-way [public mobile radio (PMR)] antennas and microwave links are mounted on the tower, the inspection can be carried out without disturbing the operation. The inspector should avoid prolonged periods of exposure (more than 10 minutes) within 1 meter of the antenna. The relevant local RF radiation limits should be observed.

STIFFNESS

A structure that is too flexible is subject to excessive stress and is liable to failure. All structures have resonant modes about which they vibrate. The primary mode for a free-standing tower involves its whole length and results in maximum movement at the top. The tower will sway under wind loads and the period of this sway is a measure of its stiffness. This period is the time to complete one full cycle (that is, from the vertical position through to the maximum deflection and back to the vertical is one half a period). This period can be measured by observation (difficult and inaccurate), by a video camera (better), and by an accelerometer (best).

Accelerometers are usually located at three or more positions along the length of the structure; the results are relayed to the ground for later analysis. Equipment records motion in two directions, as well as torsion. The optimum period is a function of the structure height, strength, design, and mass. For a 180-meter tower, a two-second period is good; a four-second period would indicate excessive flexibility.

Because early design codes did not fully appreciate the effects of dynamic wind loading, underdesign of the top portions of the structures was common (together with overdesign of the lower portion). As a result, the flexibility of the top portions often, over time, causes high levels of stress. Strengthening the top portions is thus often required at a cost of approximately 10 percent of the structure cost. Cellular operators will probably encounter this problem only if they use an old, existing tower; design techniques today properly account for the distribution of stress. A good indicator of stress is localized flaking paint and, in some instances, corrosion. Flaking paint is best detected soon after a storm when the recent stress highlights the problem.

REPAIR

Any repair and maintenance indicated by inspection should be undertaken as soon as practical. Finding suitable contractors to do the work may be difficult.

Towers should be painted once every five to seven years, depending on the environment. Painting and touch up for corrosion can be done by many contractors, particularly by those who specialize in heavy industry or bridges.

Replacing bolts, adjusting antennas and low-stress members can be done by a suitably qualified rigger.

Stress problems are more serious, however, and require the intervention of a structural engineer. The stress problems could be due to weakened members, but are more likely design-related. After analysis, the structural engineer can recommend the necessary modifications. The replacement of high-stress members requires the services of a specialist structures contractor.

Stress can be reduced by lowering the wind-loading of attachments, but it more often involves adding structural members. The structural engineer usually considers various alternatives to reduce stress and recommends the most cost-effective one.

Welding of strengthening members often destroys galvanizing and other protective coatings, so protective coats will be needed.

MAKING THE TOWER A FEATURE

There have been numerous cases in the past when a communications facility had to be integrated with other functions in order to get approval. These high-budget towers are usually on prominent mountain tops and include viewing areas and restaurants.

One such tower, seen in Figure 22.23, is the Black Mountain tower on a hilltop overlooking Canberra (the Australian National Capital city). It is interesting that no attempt seems to have been made to place the microwave dishes behind a radome, and they are a bit ugly.

From a distance, though, it looks fine, as seen in Figure 22.24. Inside the tower there are:

- Museum of communications
- Lecture room
- Viewing area
- Restaurant
- Sales center for Telstra
- TV transmitters
- FM transmitters
- Various UHF/VHF PMR equipment
- Lots of microwave repeaters
- Microwave distribution services (MDS; i.e., mostly pay TV)

A lower-budget version of this concept, also in Australia, is seen in Figure 22.25 where a large tower has been built in the middle of the town, with an observation platform on the lower level and a large clock just above it.

Figure 22.23 Canberra's Black Mountain tower. (Photo courtesy of Greg Hutchinson.)

Figure 22.24 Canberra's Black Mountain tower from a distance. (Photo courtesy of Greg Hutchinson.)

Figure 22.25   A tower built in the middle of the town with an observation tower and clock as features.

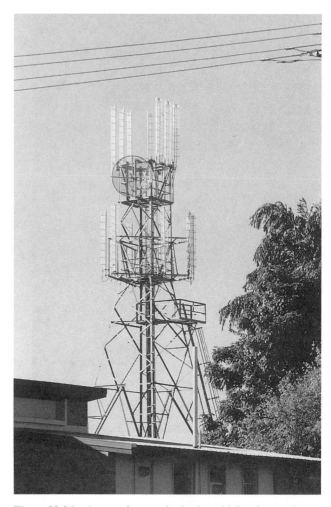

Figure 22.26   A conspicuous site in the middle of a small town in Australia.

WHEN IT GETS TOO HARD TO GET APPROVAL

The difficulty of getting approval for base-station sites is growing in magnitude daily. This is largely a direct consequence of the success of cellular itself. In the United States in 1983 there were only 350 base stations nationwide. By 1993 this number had grown to over 10,000, and now in 2000, there are in excess of 100,000, most of which will be for PCS services.

In the early days of cellular the novelty of the service was enough to ensure a high level of cooperation from local authorities when site approvals were sought. Today, a growing movement opposing towers and other base-station infrastructure on either supposed "health" grounds or the more credible grounds of visual pollution has made obtaining approval a much more difficult process.

What the antenna installation looks like can be highly variable. In a small Australian town the installation shown in Figure 22.26 is right in the middle of the town, on the corner of the main intersection with the through road. About 500 meters away the water tower installation shown in Figure 22.27 is almost inconspicuous, even though no special effort has been made to hide it.

Because of the zoning problems there has recently been a big increase in the number of "disguise" installations. The most basic of these is to paint the antenna a color that makes it blend in with the background. For example, the antenna may be painted yellow against a sandstone wall, or painted with brick patterns against a brick wall.

Antennas can be mounted inside existing structures like signs for gas stations and hotels at costs that sometimes are even lower than the cost of a monopole. All that really needs to be done is to ensure that suitable radome material is used in front of the antenna's path.

Figure 22.27 A site only a few hundred meters away from the one in Figure 22.26 that is discrete. The strange looking tree to the right with branches growing horizontally and even downwards is a Bunya pine tree, a kind of living fossil that is native to the area, and that every two years drops spiked seeds from a pod that looks like a hand grenade, and splinters like one when it drops, but is the size of a football.

Sometimes this will need to be fabricated, but sometimes the existing materials may be suitable.

The health fears are more difficult to address, but as there is little or no scientific basis for these fears, a disguised antenna that is "out of sight, out of mind" may again be the best solution. In this regard, the field strength at a distance of, say, 100 meters from a 100-watt 800-MHz transmitter is a mere 111 dBμV/m; compare this with the field strength experienced by the user of a mobile phone with a nominal power of 1 watt, and the antenna 5 cm away from the head, and we find that the received field strength using a mobile phone is a whopping 159 dBμV/m, nearly 100,000 times higher!

STEALTH ANTENNAS

A company that specializes in discretely placing antennas is Stealth Network Technologies of North Charlston, South Carolina. The company makes a variety of concealment antennas including replacement chimneys, windows, rooftops, and louvers. They place antennas in unlikely places like the BP sign in Atlanta, Georgia, in Figure 22.28. The two antennas here are mounted onto the internal pole so that illuminating the sign at night does not produce shadows. An access panel for easy maintenance is mounted at the bottom.

In Las Vegas, Stealth placed an antenna next to a

Figure 22.28   A disguise antenna in Atlanta, Georgia. (Photo courtesy of Stealth Network Technologies, Inc.)

Figure 22.29   A Stealth disguise antenna housing in a church sign. (Photo courtesy of Stealth Network Technologies, Inc.)

church on a 20 meter, three-legged support tower (Figure 22.29). At the top is a radome that has panels that are 4 meters tall. The radome is designed to cater for multiple carriers. Care has been taken to match the color scheme to that of the church. Another innovation at a church site is seen in Figure 22.30, where the antennas are located in the upper windows of the steeple. The windows are about 2.5 meters high with a disguised wooden look-alike panel to hide the antennas.

Even harder to pick out is the radome structure of a historic building in Littleton, Colorado, seen in Figure 22.31, where the chimney in the foreground in fact houses two PCS antennas. The radome matches the

existing chimney, and blends magnificently with the house.

Flag poles can be used for hiding antennas, and two applications of this are examined. Figure 22.32 shows a flag pole in Laredo, Texas, that is 40 meters tall. The base at the bottom is a monument *and* an equipment shelter. This combination allows three PCS carriers to mount their antennas at different levels behind the RF transparent radomes in the pole. A smaller, general-purpose radome/flag pole is seen in Figure 22.33. This free-standing pole comes complete with the ball and flag, and houses two sets of PCS antennas.

Figure 22.30 This steeple hides antennas which are located in the steeple behind a wooden look-alike panel. (Photo courtesy of Stealth Network Technologies, Inc.)

Figure 22.31 The chimney in the foreground houses two antennas. (Photo courtesy of Stealth Network Technologies, Inc.)

Figure 22.32   The flag pole is really a multicarrier antenna support and the base is an equipment shelter. (Photo courtesy of Stealth Network Technologies, Inc.)

Figure 22.33   A free-standing disguise antenna. (Photo courtesy of Stealth Network Technologies, Inc.)

Pizza Antennas

At first sight the scene in Figure 22.34 looks like a typical Pizza Hut in subtropical Brisbane, with the usual palm trees outside. On closer inspection, while the three trees in the background are real, the one in the foreground has a rather familiar looking hut at its base. Get out the telephoto lens and you will see the hidden microwave and and cellular antennas as shown in Figure 22.35. The fake tree is a good approximation to the real thing, right down to the trunk, but again a careful look at Figure 22.36, reveals that one of them is bolted in. Figure 22.37 shows a real tree, for comparison.

Figure 22.34   This is not what it seems. One of these trees is a fake.

Figure 22.35   A closeup of the tree in the foreground of figure reveals some hidden extras.

Figure 22.36   The "trunk" of the antenna tree.

Figure 22.37   The trunk of the real palm for comparison.

TOWER INSPECLIST CHECKLIST

A tower inspection should include the following steps:

Tower

1. Check foundations, ground points, and straps.
2. Check for corrosion and condition of painting.
3. Check welds for cracks, using ultrasonic equipment where necessary.
4. Check for signs of stress, particularly flaking of paint or bowing of members.
5. Check all bolts for proper tension and corrosion. (Some may actually be missing.)
6. Check guys for proper tension and possible corrosion. In some areas, anticorrosive agents must be applied.
7. Note the position of all fixtures, and when these positions differ from the records, note the details (including photographs).
8. Check for bent or fractured members.
9. Check for tower twist or distortion (sometimes twist can be detected by checking that the tower lines are true).
10. Check the condition of the galvanizing.
11. Check for corrosion in hollow members; this can sometimes be detected by hitting members with a hammer and listening for falling rust. (In some instances and particularly in corrosive environments, such as along the coast or in heavy-industrial areas, a low-stress member can be removed and replaced by a new one. This member can then be examined in a laboratory for strength and corrosion.)
12. Keep a permanent log of the inspection.

Grounding

13. Check that all clamps and ground straps are secure and in good condition.
14. Check that bolts are covered by an anticorrosive material.
15. Check that the lightning rod is secure and in an effective position relative to all antennas (higher than any antenna and at least three wavelengths away). [Some cellular installations dispense with lightning rods and use direct current (DC) ground antennas instead. This is sometimes essential where space on the tower top will not allow for a reasonable separation between the antennas and the rod.]

Antennas

16. Check that all antennas are vertical or at the correct angle of downtilt.
17. Check the physical condition of the antenna; it should be free of cracks, dents, and burns.
18. Check that all bolts and clamps are secure.
19. Check that the antenna grounding is secure.
20. Check that the feeder grounding is secure.
21. Check that the feeder support is adequate and not causing wear or fatigue.
22. Check for any slippage of the feeder.
23. Listen for any audible signs of gas leakage in pressurized systems.
24. Check that the antenna "tail" connector is properly sealed.

Anchorage and Foundations

25. Check that concrete anchors are free of spalling (flaking) or cracks.
26. Check that anchor bolts are tight.
27. Check that grounding is secure.
28. Check that anchor rods are not rusted or corroded.
29. Check for any signs of anchor slippage or creep.

Guy Wires

30. Check for any signs of rust or broken strands.
31. Check that the connectors to guy wires are in good condition.
32. Check that the turnbuckles are in good condition.

Tower Lighting

33. Check that all beacons are in working order.
34. Check that all beacons are in good condition.
35. Check that beacon drain holes are clear.
36. Check that beacon reflectors are in good condition.
37. Check that beacons are free of signs of moisture.
38. Check that beacon lenses are clean.
39. Check that beacon wiring is in good condition.

Ice Shield

40. Check that the ice shield is secure and undamaged.

23

INSTALLATIONS

Most operators opt for turnkey installation, at least for their first system. A turnkey installation is one where the installer (usually the supplier) contracts to provide, design, and install a complete system and hand it over to the operator when it is fully functional (that is, ready to "turn the key" and start).

With turnkey installations, the operator must rely heavily on the supplier; although this may be necessary for new operators, the operator should take steps to avoid prolonged dependence. The disadvantage of being dependent on a supplier becomes obvious when a contract is prepared. An inexperienced operator must either write a very open-ended contract or employ a consultant. Both approaches have disadvantages, but the first is the most fraught with problems.

Suppliers usually dread an open-ended contract because they must specify their offers without any firm knowledge of what other suppliers might be specifying. A supplier who specifies a well-designed and complete network may lose the contract simply because another supplier cut corners to achieve the lowest bid price. A minimum design may seem more attractive to an inexperienced operator because omitting one base station can reduce the cost presented in a proposal by about $200,000 (including links).

The disadvantages of open-ended contracts become apparent when the operator begins evaluation. Because the suppliers are not contracting for exactly the same thing it becomes very difficult to compare the offers. It takes a very experienced cellular engineer several weeks to effectively compare two dissimilar proposals.

Relying on the supplier often runs smoothly through the design and installation phases, but again becomes awkward during the commissioning and acceptance phases. At these stages it is necessary to certify that the work to date has been done adequately and in accordance with good practices and the terms of the contract.

Unless an independent appraisal of the work is available, the operator will find it very difficult to have faith in the acceptance. Even when a supplier goes to some length to ensure a fair acceptance-test procedure is followed, the operator can never be completely sure of the value of that acceptance.

Depending on a supplier for ongoing expertise to keep the system operational also presents problems. When such expertise comes from the supplier, it is very expensive, and the operator does not have complete control over the availability and selection of expert staff. The alternative, using consultants, also has difficulties. With consultants, the operator also has little control over the availability and selection of expert staff. If the peak demands of the operator and the consultant coincide, the operator may not receive the highest-priority service. Therefore, any arrangement with consultants should include a retainer and a guaranteed response time.

Large consulting firms often assign different experts to different phases of the project. This division of labor causes a discontinuity in direction, and it may be that none of the consultants will have an adequate overview of the operator's system to be fully effective. It is wise to require that at least one specific consultant be available for the duration of the project.

Finally, the operator must assume that the consultants are competent because, almost by definition, the operator is not in a position to determine if a particular consultant is competent or not. For this reason the operator should acquire the expertise needed to design and run a network as early as possible.

TRAINING

Most systems offer a training program. These programs typically cost $500 per day per participant; the total cost for a start-up system would be about $100,000. At this price, make sure to get good value for the money spent.

Begin by ascertaining that personnel sent for training have the educational background appropriate to the courses offered. The training courses usually assume a good technical knowledge of radio transmission for the radio frequency (RF) courses, and good knowledge of switching equipment for the links and switch-training courses. The courses are usually detailed and complex and will soon leave behind those who do not have adequate technical background. Personnel must also have a good command of the instructional language (usually English). Insist too that training instructors be experts with a good command of the instructional language.

Courses should be provided in a timely manner; remember that a course undertaken and then not applied for six or more months will be largely forgotten. Course participants should be able to apply their new knowledge within two months of the course. The very fact that the knowledge soon will be needed increases retention.

By participating in the installation phase, employees can gain valuable insights into the functioning of the system. It is not unusual to specify in a turnkey contract that the operator's staff provide some of the labor for installation, thus ensuring active participation from the start. The areas of particular interest to an operator are the design phase (site selection), survey technique (RF path survey), site preparation, and, finally, installation and testing.

Because it will likely be necessary to constantly expand the network, some knowledge of what is involved in such expansions is also invaluable. As discussed earlier, it is nearly impossible to forecast demand (particularly on a new system) and even more difficult to forecast the load on a particular base station. It will be necessary to move channels from one base to another after the system has been switched on so that the channels are placed where the traffic is directed. The operator should be able to relocate channels as required. This means that suitable test equipment must be provided.

THE OPERATOR'S RESPONSIBILITY

When the operator accepts a turnkey system, payment becomes due and it is usually difficult to get the contractor to return for more than minor adjustments. Because the contractor's view of the project is somewhat different from the operator's, it is a good idea for the operator to be particularly alert during acceptance.

The operator's priorities are as follows:

- A good, efficient system with good coverage
- Ability to meet operational targets
- Ability to meet the market requirements at the least cost
- A competitive system
- Low maintenance costs
- Ability to expand efficiently and at minimum cost

A contractor's priorities, however, are somewhat different, namely, to

- Install the system on budget
- Meet the contract specification
- Perform as a credible contractor and win subsequent expansion contracts
- Meet time constraints

The order of these priorities is not always necessarily the same and some operations may have additional priorities, but it is easy to see that the objectives of the contractor and the operator are somewhat different.

For example, if a radio survey indicates that a particular area has marginal coverage and there is a possible (but not definite) need for an additional base, the operator, considering mainly the cost, may decide to take the risk and save money. The contractor, on the other hand, seeing the potential damage to the firm's reputation should poor coverage result, may decide that the doubtful base should be put in as a precaution. Alternatively, the contractor may ignore the problem in order to produce a lower quote. Whatever the contractor's decision, it will be based on different considerations than the operator's.

Contractors are often stressed by the demands of operators and the inability of manufacturers to supply on time. It is very easy, under such duress, to see the fine details such as labeling and documentation as relatively unimportant. The unwary operator who does not carefully check both the overall performance and the detail ultimately pays the price.

ACCEPTANCE TESTING

The operator should be responsible for acceptance testing because this is the one opportunity to ensure that all is well before paying for the system. At the end of this chapter are checklists that detail the items that must be checked before a base station is accepted into service. This checklist should take about one hour to complete.

Acceptance can be either absolute or conditional. If the base is inspected and found to have only a few minor shortcomings (for example, missing labels on equipment racks, some handbooks missing, and some spare parts not available), then it may be appropriate to issue a conditional acceptance (that is, the work is accepted subject to the shortcomings being cleared up within an agreed period—say, one month).

But it is not appropriate to issue any kind of acceptance if major shortcomings are found. Examples would include the following:

1. The radio link to the switch is not functional
2. There are no commissioning test sheets
3. The installation is untidy
4. Grounding straps are not provided
5. The battery/rectifier is not functioning properly.

Because a poor standard of installation results in high maintenance costs over the life of the system, it is necessary to be very firm about acceptance procedures.

COMMISSIONING

As part of the acceptance, the accepting officer should be involved in the commissioning phase, usually in the last two days for a base station and in the last four to six weeks for a switch of the installation. This phase involves testing and aligning the system to ensure that everything operates within specification. The accepting officer should verify that all tests were properly done and recorded. The best way to do this is to be directly involved in the commissioning. Also, this phase is the most instructive part of the installation, and it is an opportunity for the operator's staff to become familiar with the equipment.

A physical check (using a vehicle and a mobile) that coverage is adequate and that handoff occurs correctly should be done for each base-station site. The coverage of the site should be confirmed manually (using the mobile to see the limits of its range) or preferably as a measured field strength. Serious discrepancies between actual coverage and predicted coverage may well point to some problems with the antenna, feeders, or system parameters. For this reason the operator should have access to a field-strength meter (using high-speed sampling). This is a lot cheaper than having the installer do those measurements, and it reduces the operator's dependence on the supplier.

MOVING AWAY FROM TURNKEY INSTALLATION

As the system evolves, the operator will likely gain confidence and be able to undertake a good deal of the work involved. Moving away from turnkey installation often results in large reductions in installation costs, and on that basis it should be at least considered by all operators.

Providing towers, huts (shelters), power, and clearing and preparing sites is work that can most readily be done first. The staff involved in this work should use the expertise available from the original turnkey project to become familiar with the requirements. The following sections discuss preparing the site for installation and establishing a staging area.

Site Preparation for Installation

It is essential to have a properly prepared site before installation can begin. A well-prepared site will be cleared and sealed in such a way as to ensure that a path to and from the site is free of dirt, dust, or loose particles. Other work that is likely to produce dust (for example, building extensions, preparing the site, and landscaping) should not take place simultaneously with installation. Power (preferably three-phase) should be available on the site. Note that the power requirements may be significantly larger than normal domestic requirements. The electrical grounding should be in place, connected, and tested. There should be a communications link back to the switch and preferably to other places as well. This link could be a telephone, a two-way radio, or even the engineering order wire on the microwave. The tower (support structure), antennas, and feeders should all be in place. Air conditioning should be installed and operational, and all doors should be fitted and functional. The doors should have adequate security locks.

Staging Area

It is also necessary to provide a staging area for the network where equipment can be unloaded, stored safely, checked, and sorted for dispatch to particular sites. The area needed for storage is quite large. Figure 23.1 shows two trucks loaded from a staging area ready for dispatch to a base station. In all, the equipment for a switch and three base stations may require about nine medium-sized flat-bed trucks.

The equipment comes in cartons that weigh from 200 kg to 500 kg. A normal forklift is required. The cable used for the RF feeders is usually LDF50 (7/8″ coaxial cable) or similar. This cable comes in large drums, and it is necessary to provide some means of attaching a spindle through the drum to access the cable.

Figure 23.1 The equipment used for cellular installations is bulky, heavy and needs a staging area for sorting. Loading can be a very tiring affair.

Figure 23.2 A frame is needed to hold the RF feeder cable drum so that the required lengths can be conveniently cut.

Figure 23.2 shows such a frame and a smaller drum of rope used on site to measure the length of the cable run. The required length is marked on the rope, and then an equal length of cable is run off back at the staging point.

Cable Installations

It is good practice to sweep the cable before and after installation to detect any damage that might occur during the handling process. The cable can easily get damaged by hitting the tower during the hoisting process, and this can cause dents and flat spots.

If the cable drum is placed well clear of the tower base during hoisting it is easy to avoid impact damage. While the cable is being hoisted if you have someone place their hands around the cable gently as it rises, that person will be able to feel flat spots and dents that may not be detectable by eye (a panel beater will do a similar thing to a car by running a hand gently over the surfaces to feel for minor dents, which will be more evident on a freshly painted car).

Hoisting must always be done with the proper hoisting straps and the hoist grip should be placed well clear (and below) the top connector. A hoist strap should be used every 60 meters. The cable hoist straps are meant to be left on the cable after installation to support the weight of the cable.

The cable hangers are meant to keep the cable in place, but their function is not to support the weight of the cable. Cable hanger clips should not be overtightened, as they can damage the cable.

Once the cable is in place, it should be supported by good-quality purpose-designed cable brackets, which are usually placed at intervals of 1 or 2 meters, as seen in Figure 23.3.

The cable should enter the building through a properly designed cable window and the entrance should be well grounded, as seen in Figure 23.4. Although this installation of the cables has been well done, the nearby power-cable entry (to the left) has not been sealed properly.

Figure 23.3 The correct way to support a cable on the tower, with support brackets every few meters.

Connectors

Only good-quality cable connectors should be used, and they must be correctly tightened. Too tight and they can damage the connector; too loose and they can be responsible for serious intermodulation problems. Most connectors come with a recommended torque wrench setting and this should be followed.

DIN Connectors

Deutsche Institut fur Normury (German Instituto for Standardization) (DIN) connectors are commonly replacing N-type connectors in cellular and personal communication service (PCS) applications. While these can handle more power, and they do result in a low voltage standing-wave ratio (VSWR), it is even more critical that the tension used to install is correct. A torque wrench must be used and the correct values set (however this is not widely followed by many installers). The correct torque settings are often not found in the cable catalogs, so you should be prepared to request them from the manufacturers.

Checklist for Cable Installation

It is important that the cables be properly installed, and this starts with making sure that all the needed hardware is available. The following list can serve as a guide.

- Antenna
- Antenna mounting brackets
- Jumper cables
- Weatherproofing kit
- Connectors (top)
- Cable (transmission line)
- Hoisting grips (every 60 meters)
- Hanger kits (every meter)
- Cable tray
- Grounding kits (top, bottom, and every 20 meters)
- Grounding bus-bar kits
- Entry-port kit
- Cable window
- Bottom connector
- Lightning protectors

Figure 23.4 A cable window with adequate grounding at the cable.

- Jumper cable in equipment room
- Cable trays for equipment room
- Color coax markers to identify sectors at multi-sector sites.
- Heat or cold-shrink kits

Always observe the recommended minimum bending radius, as tight curves will damage both the inner and outer conductor, and may necessitate the replacement of the damaged section of cable.

Water Migration

Ingress of water into a cable is the end result of poor installation or maintenance practices, and is to be avoided where possible. Water in the cable between the inner and outer conductor will significantly increase the cable loss and increase the VSWR.

Water can get into the cable only in a few ways. First, if the outer polyethylene jacket is damaged, water can get in. However, provided the outer jacket is intact, the only way for water to get in is through the ends. The connectors are designed to be waterproof, and if properly installed they can provide an adequate barrier.

Cables are nominally tested for waterproofing in a way that measures their resistance to water penetration. Typically a cable is tested (see Figure 23.5) under a head of 3 feet of water, and the cable passes the test if water

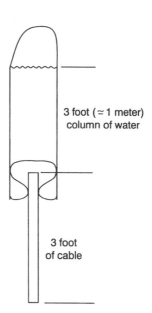

Figure 23.5 Diagram depicting how cables are tested for waterproofing.

has not leaked through after one hour. Passing this test in no way ensures that the cable is actually waterproof.

Once water has penetrated the cable, it can migrate because of pressure differences caused by temperature changes (typically the swings from day to night), or changes in the barometric pressure of the atmosphere.

ACCEPTANCE-TEST SHEETS

The following acceptance-test checklists can be used by acceptance-test personnel for cellular radio base sites.

Site (Name)   _____

Location   _____

Switch (Location)   _____Tel._____

Installed by   _____

Installation supervisor   _____Tel._____

Inspected by   _____

         ☐ on completion   ☐ work still in progress

Acceptance date   _____

Signed   _____

Conditional acceptance date   _____

Signed   _____

(Subject to rectification of items indicated on attached sheets)

Date in service   _____

POWER RECTIFIERS

	OK/NOT OK	COMMENTS
1. Mounting and layout	_____	_____
2. Cabling and terminations	_____	_____
3. Alarm extension	_____	_____
4. Designations	_____	_____
5. Commissioning test results	_____	_____
6. Handbooks	_____	_____
7. Load sharing	_____	_____
8. Safety signs	_____	_____
9. DC distribution	_____	_____

BATTERIES

	OK/NOT OK	COMMENTS
1. Fusing	_____	_____
2. Battery spacing	_____	_____
3. Electrolyte level (unsealed batteries only)	_____	_____
4. Battery vents	_____	_____
5. Battery lead burning/connections	_____	_____
6. Battery cabling	_____	_____
7. Designations	_____	_____
8. Drip trays (unsealed batteries only)	_____	_____
9. Safety equipment	_____	_____
10. Safety signs	_____	_____
11. Water supply (unsealed batteries only)	_____	_____
12. Fuse alarm extension	_____	_____
13. Accessibility for testing	_____	_____
14. Battery function and continuity (test)	_____	_____
15. Floor loading of batteries within limits	_____	_____

EXTERNAL PLANT

	OK/NOT OK	COMMENTS
1. Tower, mast, or pole	_____	_____
2. Gantry	_____	_____
3. Guys and anchors	_____	_____
4. Lightning protection	_____	_____
5. Tower, mast, or pole grounding	_____	_____
6. Mains surge protection	_____	_____
7. Tower lighting	_____	_____
8. Corrosion protection	_____	_____
9. Safety signs	_____	_____
10. Site RF radiation records	_____	_____
11. Equipment shelter	_____	_____
12. Base security, fences, locks, gates, etc.	_____	_____
13. Cable window and seals	_____	_____
14. Grounding of equipment rooms	_____	_____
15. Access to tower restricted	_____	_____
16. Antennas correctly mounted	_____	_____

INTERNAL PLANT

	OK/NOT OK	COMMENTS
1. Mounting and layout	_____	_____
2. Cabling and terminations	_____	_____
3. Alarms and telecontrol	_____	_____
4. IDF cabling	_____	_____
5. IDF labeling	_____	_____
6. System performance	_____	_____
7. Commissioning test results	_____	_____
8. Station logs	_____	_____
9. Designations	_____	_____
10. Handbooks	_____	_____
11. Drawings	_____	_____
12. Test cables and extender cards	_____	_____
13. Spares	_____	_____
14. TRX plug crimps	_____	_____
15. RF N-connectors	_____	_____
16. Independent link to switch (control center) functional (telephone line or radio link)	_____	_____
17. Lightning arresters on all incoming cables	_____	_____
18. Door alarms functional	_____	_____
19. Suitable fire extinguishers	_____	_____
20. Redundant control channel functional	_____	_____

ITEMS TO BE CHECKED

POWER RECTIFIERS

Mounting and layout

☐ Correct positioning of cabinets.
☐ All mains terminals covered.
☐ All cabinet components supplied, including tops and coverplates for unused positions.
☐ No cracked or non-working meters.

Cabling and terminations

☐ Cabling runs are satisfactory, cable ties used where necessary.
☐ Correct size of cable used for current to be carried for maximum size of installation.
☐ Cable crimps not loose.
☐ No undue mechanical stress by heavy cables on circuit breakers.

Alarm extension

☐ Correct settings and operation of all alarms provided on power cabinets (mains fail, float low, high volts, etc.)

Designations

☐ All circuit breakers labeled.
☐ Cabinets (if more than one) are numbered.
☐ Switch plates clearly marked.
☐ Modules are all numbered, if modular-type rectifiers.

Commissioning test results

☐ Should be provided by the installers and be on site.

Handbooks

☐ Should be left on site by the installers (and usually contain the test results).

Load sharing

☐ If power supply is modular, all modules should supply approximately the same current (but not necessarily equal). Turn the rectifiers on one at a time and note reconfigured load sharing and current-limiting functionality.

Safety signs

☐ Mains hazard.
☐ −Ve 24 V ground (where applicable).
☐ Any others that are required by local regulations.

DC distribution

☐ Cabling from power cabinets to radio cabinets for correct current rating.
☐ Trays used where necessary to support the cable correctly between cabinets.
☐ Busbars and battery feeders insulated and suitably protected against accidental short circuits.

BATTERIES AND DISTRIBUTIONS

Circuit Breakers

☐ Current rating of CBs supplied for battery capacity and current drains.
☐ If indicators supplied, these should be clearly visible.
☐ Spare CBs available.

Battery cabling

☐ Cables correctly tied and supported on cable trays.
☐ Correct size cable used.
☐ No loose crimps.

Safety equipment

☐ Check presence of equipment as required by company regulations (can include rubber gloves, rubber apron, face shield, first-aid kit, etc.)

CB alarm extension

☐ Alarm given when fuse OC or CB operated.

Accessibility for testing

☐ Batteries should be placed in such a position to allow cell replacement without impediment.

Battery function and continuity

☐ Gradually reduce the rectifier output voltage and note changeover to battery-powered operation. A voltage drop of not more than 2 volts should occur and current should remain the same. This checks battery and battery feeder continuity as well as the changeover mechanism. In systems not yet commissioned, the rectifier power should be turned off for this test. The battery voltage will initially drop rapidly and will then rise, stabilizing about one volt above the minimum.

Floor loading of batteries within limits

☐ Battery loads (kg/m^2) should be confirmed as being within floor structural limits.

EXTERNAL PLANT

Tower, mast, or pole

☐ State which, and give height. Check all ironwork is galvanized, no evidence of early rust, and all nuts and bolts are in place.

Gantry

☐ Feeders between structure and building are adequately supported by a gantry or tray.

Guys and anchors

☐ Check masts having multiple guys and concrete anchors. Guys should be examined for correct tightness and rusting, and concrete anchors for flaking and cracking.

Lightning protection

☐ Usually provided at top by antenna. Ensure lightning rods will not interfere with antenna pattern.

Tower, mast, or pole grounding

☐ Structure to be strapped to ground at ground level. Feeders to be grounded top and bottom of tower and at cable entry. Tower grounding connected to building ground.

Mains surge protection

☐ Usually provided on mains power line into building.

Tower lighting

☐ Provided on structures near airfields, in accordance with local aviation regulations.

Corrosion protection

☐ All ground strap connections are sealed with anti-corrosion kit.

Safety signs

☐ Relevant safety signs are prominently displayed at foot of structure and on site fence.

Site RF radiation records

☐ When required, this record contains the maximum radiation from each antenna, and the safe working level of radiation on the tower top.

Equipment shelter

☐ The equipment shelter is properly and completely finished, waterproofed, and secure.

Base security, fences, locks, gates, etc.

☐ The site fencing and security are ensured.

Cable window and seals

☐ Cable window and seals are properly fitted to prevent water seepage.

Grounding of equipment rooms

☐ The equipment room is adequately grounded and the ground resistance less than 10 Ω. The test results should be available.

Access to tower restricted

☐ The tower is separately fenced and locked.

Antennas correctly mounted

☐ Antennas are correctly spaced and either vertical or at the correct level of downtilt.

Antenna sector check

☐ Antennas connected to correct sectors.

INTERNAL PLANT

Mounting and layout

- ☐ Positioning of cabinets and supporting framework is as per design drawing.
- ☐ Blank panels, covers, etc., provided as required.
- ☐ Feeder supports provided above cabinets and up to the cable window.
- ☐ Frame racks firm and secure.

Cabling and terminations

- ☐ Neat distribution and positioning of all intercabinet cables; tied down at regular intervals.
- ☐ Cable plugs to be complete, no missing components.
- ☐ Plug labels correctly supplied and marked.
- ☐ Combiner-module and combiner-star connector tails are free and unstrained; N-connectors to be tight.

Alarms and telecontrol

- ☐ All alarms as specified are correctly returned to the mobile switch or monitoring center.

IDF cabling

- ☐ Neatness and tying of cables.
- ☐ Terminations correct for the type of termination applicable at IDF.

IDF labeling

- ☐ All circuits correctly labeled appropriate to the type of IDF system.
- ☐ Record book correctly made out.

System performance

- ☐ All channel modules within specification.
- ☐ All combiners within specification.
- ☐ Feeder-antenna return loss within specification.
- ☐ Base-station controller, redundancy functional.
- ☐ PCM or link system.
- ☐ System lineup levels on each channel or port correct.
- ☐ All alarms functional.
- ☐ Final call-through test on each channel module prior to cutover.

Commissioning test results

- ☐ To be left on site by installers for subsequent use by maintenance staff.

Station logs

- ☐ To be provided at cutover by installation staff for batteries, attendance, or other as locally required.

Designations

- ☐ Cabinet numbering.
- ☐ Module numbering.
- ☐ −Ve ground signs.

Handbooks

- ☐ To be left on site by installers for maintenance use.

Drawings

- ☐ Copy of floor layout and cabling records to be left on site with handbooks.

Test cables and extender cards

- ☐ Available where needed.

Spares

- ☐ Check any spare parts ordered are available before cutover.

TRX plug crimps

- ☐ Check crimping of wire-to-plug connection.

RF N-connectors

- ☐ If available, use N-connector gauge on all feeder connectors and tails. Some center pins may be out of specification and can cause damage to the socket.
- ☐ Check also correct lightness.

Independent link to switch (control center) functional (telephone line or radio link)

- ☐ Check for proper functioning of voice link to switch or control room.

Lightning arresters on all incoming cables

- ☐ All cables entering and leaving the building should have suitable lightning arresters.

Door alarms functional

- ☐ Test all door alarms and confirm the proper working.

Suitable fire extinguishers

- ☐ Suitable non-conductive fire extinguishers are in place.

Redundant control channel functional

- ☐ Turn off the operational control channel and confirm the proper functioning of the changeover.

24

EQUIPMENT SHELTERS

Equipment shelters should be designed to maximize flexibility for expansion, to minimize operational costs, and to provide a clean, safe working environment for the staff. There are many different ways to achieve these objectives. This chapter presents some of the factors to be considered when designing shelters.

BASIC CONSIDERATIONS

Planning for expansion is particularly important because it seems that no matter how large an equipment shelter may be, soon after installation it will be found to be too small. Even when no expansions are foreseen, it is wise to build the shelter so that at least one wall can be removed to extend the building.

When transportable huts are used as equipment shelters, they should be placed on the site in a way that allows for additional huts to be added. Too often transportable huts are placed centrally on the site, mainly for cosmetic reasons. This can make future expansion awkward.

Switch rooms in particular should be built with ease of expansion in mind. It is very expensive to relocate a switch once it is placed in service. As additional switches are added, the need to interconnect and monitor them is best met if the expansion allows good access to the internal distribution frame (IDF) and the control room. High-level digital interconnections are cheaply and effectively achieved if they can be made without repeater equipment; but the switches must then be located within about 100 meters of one another.

There are a number of ways of providing for switch-room expansion. In the example discussed here, a store-room is incorporated in the switch-room building with the plan that it will ultimately become a switch room itself.

Cellular operators, particularly new operators, should not underestimate the amount of space that may be needed for other services. In particular, space must be found for microwave, power and, sometimes, billing equipment. Other services, like paging, trunked radio, and voice messaging, can also consume considerable space.

Shielding

In an ideal world, shielding of the shelter would not be a consideration, but with the proliferation of other radio frequency (RF) services, interference to the equipment through the shelter wall is always a possibility. Because of this, preference should be given to construction techniques that minimize interference. In areas where high-powered RF sources are known to exist, such as amplitude-modulated (AM), frequency-modulated (FM), or short-wave (SW) broadcast transmitters, as well as on shared sites, only metallic-clad buildings should be considered. In other areas a preference for metallic cladding should exist, provided this does not compromise structural integrity or—to a significant extent—cost.

Concrete shelters, with their steel reinforcing, can double as effective shields, particularly if care is taken to bond the reinforcing bars together. This may need to be allowed for at the design stage.

Fiberglass shelters can be lined with conductive tape and filled with metallic-backed insulating material. Lining of the internal walls with metal sheets should be considered if there is any suspicion that interference will be a problem.

SWITCH BUILDING DESCRIPTION

The building will generally have only one floor and include the following areas:

- Equipment room
- Control and (perhaps) billing room
- Battery and power room
- Emergency-plant room
- Uncrating area
- Storeroom
- Air conditioning plant
- Toilets
- Staff facilities
- Cleaner's/janitor's room

Equipment Room

The equipment room houses the switch and microwave equipment. Because the equipment is often bulky, good access for delivery vehicles and movement of equipment is essential. The area must be air conditioned and the humidity regulated.

Because demand is difficult to forecast, the room should be designed to expand in at least one direction (at least one wall should not be structurally supportive). Expansion can be facilitated by the use of steel-framed, non-load-bearing walls, which can readily be removed when additional area is required. The position of the equipment room with respect to the site should allow for this expansion.

An equipment room allowing for three suites of equipment should be about 7.5 × 5 meters. The actual dimensions will depend on the actual equipment purchased.

Floor tolerances for the equipment room should be precise. In the largest dimension of any floor area, the level should not vary by ±12 mm; in any 3-meter length, the floor should not vary by more than ±5 mm; in any 300-mm section, the floor should not vary by more than ±2 mm.

Control and Billing Room

A room about 5 × 10.8 meters could be included to house the control and billing equipment. If necessary, billing functions can be handled remotely, and there are often good reasons for doing so. Real-time billing (hot billing), however, requires a data link between the switch and the billing computer.

Battery and Power Room

The battery and power room should be about 7 × 5 meters—large enough to accommodate two battery stands and the rectifiers. High ohmic distribution should be employed. The room should be fitted with a small washbasin and a handheld hose spray attachment, as well as an exhaust fan.

When unsealed lead acid batteries are used, the battery and power room should be physically isolated from the switch room. Today, however, sealed, rather than unsealed batteries are usually used, and these can be located in the switch room. In fact, with sealed batteries, it is common to place the batteries and rectifiers very close to the equipment to reduce copper costs (smaller bus bars), which can result in overall savings of 30 percent.

Emergency-Plant Room

The emergency-plant room should accommodate a diesel generator adequate for the total load of equipment, air conditioning, and lighting. Typically, the load will be from 30 kVA to 150 kVA (the upper limit being where a co-located base station is included). This room contains an electrical switchboard and must have an exhaust fan. Measures should be taken to contain any fuel spillage in this room, including surrounding the room with curbing 150 mm high (or making the floor 150 mm lower than the rest of building). Figure 24.1 shows a typical emergency-plant room. To allow easy access and safe passage, raised ramps should be provided for doorway access.

The generator room should be soundproofed, as the noise levels can be quite high.

An essential part of the emergency-plant room is to remove the heat generated by the diesel. Most large generators (bigger than 100 kVA) will need to have the

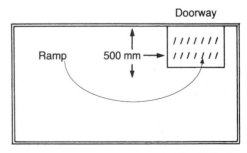

Figure 24.1 The emergency-plant room should be lower than the rest of the building so that fuel spillages will not seep into the main equipment room.

radiator mounted so that it is directly exposed to the outside air. The installation manual for the generator will show the recommended mounting details. As the airflow from a large generator is hot and the volume is large, there may be a need to duct the airflow above the level of passersby. Notice that the emergency plant has its cooling fan arranged so that the airflow blows *out* (the opposite of a motorcar).

The room should be designed to facilitate the airflow. An air-intake of at least 1 square meter is needed and an exhaust fan, placed high in the room in the vicinity of the radiator, can be used to assist the airflow.

The engine exhaust pipes and muffler are dangerously hot. If the generator has auto-start, a sign indicating "Caution this generator may start at anytime" should be posted in a conspicuous position.

Uncrating Area

The uncrating area, a part of the switch room, is deliberately left vacant to provide space for additional equipment. The area serves as a workplace for installation and maintenance staff and is the last area of the switch room to be occupied. Good access should be available for vehicles transporting equipment, and the area should be fitted with doors 1.5 meters wide.

Storeroom

The switch building will probably house its own spare parts. The value of spare parts is usually about 5–10 percent of the switch value, so good, safe storage is essential. A storeroom 3 × 4 meters would be adequate if storing only switch-room spare parts. The switch room shown later in Figure 24.3 has a very substantial storeroom meant to hold base-station spare parts, too.

Walls

The walls should be made of bricks or cavity blocks. Because cavity blocks cannot be cut to size as bricks can, when cavity blocks are used, room dimensions must be exact multiples of the block length. The internal walls to the power room/storeroom, equipment room, and emergency-plant room should be non-load bearing and surfaced with a fire-stop plasterboard or other fire-retardant material with a one-hour rating. The internal ceiling height should be 0.6 (minimum) to 0.9 meter higher than the switch rack or the minimum height specified by the local planning authority, whichever is the highest. Such considerations ordinarily yield a wall height of around 3.4 meters (equipment height 2.5 meters), although suppliers have rack heights from 2.2

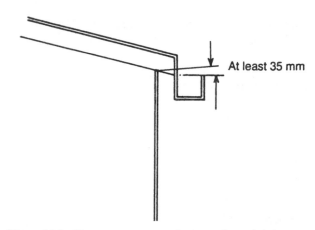

Figure 24.2 Because gutters can become clogged, it is recommended that the outside part of the gutter be lower than the wall side to prevent overflow down the walls and perhaps ultimately to the equipment.

meters to 2.9 meters. Of course, these measurements do not apply if under-floor cabling is used.

When an existing building is used and the ceiling is significantly higher than the optimum, adding a false ceiling is worthwhile to substantially reduce the air conditioning load.

Roofing

The building must be totally waterproof because water on the equipment can cause total malfunctioning of the switch. The roof should have a minimum pitch of 5 degrees (assuming steel decking is to be used). The appropriate local high-wind code should be applied. Particular attention should be paid to the guttering and downpipes, which should be a "leaf-free" type to decrease the likelihood of blockage. The gutter sections should be designed to prevent overflow of blocked systems into the building, as shown in Figure 24.2.

Insulation Cladding

Table 24.1 shows the minimum insulation in fiberglass batting or equivalent that should be provided. Insulation should be used in walls and ceilings to reduce air conditioning costs. All materials should be fire resistant.

Floor Loading and Construction

The floor should be designed to carry live loads of 9.5 kPa throughout the building. Suspended floors should be used only where there is no economic alternative.

TABLE 24.1 Thermal Insulation for Equipment Shelters

Climate Zone	Insulation Thickness (mm)
Temperate	100
Mediterranean	75
Subtropical	50
Tropical	75

Ideally, the whole building should be on level, consolidated ground. A reinforced concrete raft slab, incorporating edge beams to the perimeter and ground beams under the walls, should be used. Newly filled land requires time to compact and is not recommended for the switch room.

The raft slab should be placed on a consolidated base, leveled with sand, and covered with a waterproof membrane.

The floor should be elevated sufficiently above ground level to ensure against the entry of water under the worst flood conditions. (A flood-free site should always be selected).

Ceilings

Ceilings should be of plasterboard or similar material and insulated to the recommended thickness.

Windows

Windows should be of laminated glass. External windows, for security, should be high and covered with a metal security screen that also diffuses sunlight. To reduce air conditioning costs, double glazing or thick glass (60 mm+) should be used for external windows. Adequate provision should be made for cleaning the windows (a sliding construction with a key lock facilitates cleaning).

Appearance

Because the switch room will probably be located in a residential area, building design is important. The building should not look like a residence, but it should blend in with the area and should not be conspicuous. The visual effect of the large equipment-access doors can be reduced by painting the top of the doors in a dark color. Increasing the visual effect is more important to make base-station sites blend in with the surroundings. This may mean that the site has to look like a tropical grass hut, a rock, or even just a grassy hill; sometimes it is even necessary to place the shelter underground.

Tornado/Hurricane/Cyclone Areas

In tornado/hurricane areas, brick or blockwork should be reinforced with galvanized tie rods, from the footings to the roof beams, providing post-stressing.

Structural Steelwork

All structural steelwork should be painted to protect against corrosion. Galvanizing is not necessary except for lintels.

Cable Window

A cable window measuring about 1000 × 1000 mm provides cable access into the equipment room (see Figure 24.3). This window will be positioned so that its bottom edge is rack height above the floor. The window should be made of galvanized, painted mild steel, with one plate on the outside of the wall and one inside. A cable tray (gantry) supports cables from the tower, see Figure 24.4.

INTERNAL FINISHES

A dust-free environment, ease of cleaning, and hard wear are the main factors to consider when choosing internal finishes.

Floors

The toilet and entrance lobby should be floored with ceramic or vinyl tiles. All other areas can be finished with vinyl or similar tile.

Walls

All brick walls should be cement, rendered with a rubber float tool to a fine-sand finish. Partitions to the emergency-plant and battery rooms should have a one-hour fire rating. Steel-studded plasterboard can be used in other areas except the toilets. The surface coating should be a gloss enamel finish to minimize maintenance.

Doors

All internal doors should have a durable gloss enamel finish. Door frames in brickwork should be pressed steel and finished with the same coat as the adjoining walls. The doors to the uncrating area should be 1.5 meters wide and able to swing 180 degrees to be flat against the external walls. All external doors should be faced with sheet metal.

Figure 24.3   A cable window with a number of unused cable boots that line up with the internal cable tray.

Figure 24.4   The gantry supports the cable that runs up from the tower to the building.

EXTERNAL FINISHES

External finishes should be maintenance-free; painting should be avoided. Aluminum frames should be anodized, and steel fixtures should be formed from material with an aluminum/zinc coating and finished with polyester.

All external doors should be fitted with dual heavy-duty mechanical locks vertically separated by 1 meter. An electric combination lock for use during normal hours of operation is also recommended. A combination lock allows the staff to move freely but keeps out others.

EXTERNAL SUPPLY

Three-phase, four-wire (or sometimes three-wire, single-phase for small-switches) power is required. Electrical utilities usually offer a range in price; select the option that is most economical.

The main switchboard should be located in the battery room. The power board should be fully enclosed with hinged doors for access to all components except main switches and changeover switches.

ELECTRICAL POWER OUTLETS

All equipment and control rooms should have adequate power outlets. In most locations, double outlets every two meters is adequate. The outlets should be located at least 150 mm above the floor.

EXTERNAL EMERGENCY PLANT

Because the emergency-plant room requires adequate access to a diesel generator, the doors to the room should be at least 1 meter wide.

Provision should be made to attach an external emergency generator. A manual changeover switch with three positions should be provided: normally On, Off, External Emergency Plant.

A suitable three-phase external socket should be fitted to the building.

ESSENTIAL POWER

When using the emergency plant, only the essential power is provided. This power supports

- Main switch rectifiers
- Power and battery-room ventilation fans
- Control room
- One three-phase future expansion with fuses
- Emergency lights and outlets
- Essential air conditioning

AIR CONDITIONING

The whole building should be air conditioned except for the battery and power room and emergency generator room. Significantly longer life can be expected from batteries when they are sealed and placed inside air-conditioned rooms. Separate air conditioning for the switch room and control room should be powered from the emergency plant. It is probably not necessary to run the rest of the building from the emergency power.

Temperature should be kept in the range 20–30°C, with a relative humidity of 20–60 percent (non-condensing) and dust to a maximum of 60 mg/28 cu meters of air by weight (5 micron diameter).

TYPICAL SWITCH ROOM

The switch room can take many forms, and maximum provision should be made for future expansion of the switch room. Figure 24.5 shows a typical switch room.

The switch room in Figure 24.5 could be used for a cellular switch. A billing computer room has been included. The billing need not be co-sited, but because the switch has emergency power and high security, it may be desirable to do so.

Figure 24.5 also shows a large storeroom. For a new cellular business (non-wireline), safe and secure storage is vital; normal office-type storage probably will not suffice. Locating storage in the switch room at first and then using that space as an expansion of the switch room in the future may prove cost-effective.

BASE-STATION HOUSING

The base-station housing can take many forms—for example, existing buildings, transportable huts, and specially built structures (usually to meet town planning requirements). With the exception of very remote sites, all new structures or acquisitions should be adequate in size for ultimate expansion to full-base size. This full-base size depends on the system type, frequencies allocated, and the cell plan used. Site sharing has become popular and is mandatory in many countries. Access by third parties should be considered. This problem does not occur in low-density regions.

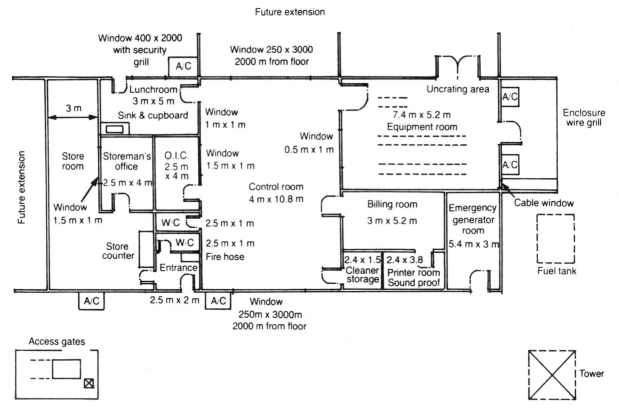

Figure 24.5 This typical switch room has a larger than normal storeroom (to be used also for base-station spare parts) and has the billing computer co-located with the switch.

The housing should be designed for ease of equipment-rack expansion, especially in installations where estimates of channel requirements at particular sites are, at best, guess work.

Base-station housing can present many challenges for the installation engineer. It is possible to use a standard shelter for new sites; for installations in existing buildings, however, it is necessary to conform to the space and layouts available. This constraint often also applies to building rooftop installations, where sufficient space for an optimum layout is not generally available. It is often necessary to use small spaces, often with awkward shapes, in order to conform with rooftops that were never designed to accommodate a base station.

Dead-load limitations can also restrict the type of installation that can be placed on a rooftop; this is particularly true if a tower structure is also required.

Shelters can be made of concrete, metal, or fiberglass. The cost of a well-designed shelter built with any of these materials is similar, and all materials can be expected to be suitable for about 20 years. Locally available materials and expertise may well determine the most cost-effective choice.

Concrete generally provides the most durable shelter and, where necessary, a virtually bulletproof enclosure. The concrete must be sealed about every four years, but otherwise requires very little maintenance. Concrete hardens with age, however, and can suffer cracking in extreme weather conditions.

The incidence of bullet damage in the United States, although not widely reported, is believed to be on the increase. It is generally considered that high-powered rifles represent the biggest threat to the equipment. A number of operators, particularly in rural areas, are turning to bulletproof structures that range from 4-inch concrete-walled buildings to prefabricated "bulletproofed" shelters.

A consideration when "bulletproofing" is how much reinforcing is necessary. More resistance means more cost. In the United States the Underwriters Laboratories code "Standard for Bullet-Resisting Equipment, UL 752" provides a good definitive basis for the level of bullet resistance.

A concrete shelter in the United Kingdom, which was not necessarily chosen for its bulletproof qualities, is shown in Figure 24.6.

Figure 24.6   A concrete base station.

Steel structures require a little more maintenance, but can be expected to last about 20 to 25 years. Steel and fiberglass panels are easily moved and are most suitable for assembly in awkward places like rooftops. A steel base-station in England is shown in Figure 24.7.

Fiberglass, with careful upkeep, can be expected to last 15 to 20 years. Care should be taken to ensure that panel joints and sealing (caulking) are permanent and do not require annual maintenance. Fiberglass and steel are both less vandalproof than concrete.

Transportable huts can be prefabricated and have all hardware installed before being moved onto the site. These huts can also be placed on top of office blocks and other high-rise buildings. When using high-rise buildings that are not the property of the cellular operator, locating equipment in prefabricated huts is recommended because the huts can be moved relatively easily in the event of future leasing disputes. It is often necessary to construct the huts on site because of access problems that make the placement of a complete hut on the rooftop impractical.

Using transportable huts reduces installation costs (because of standardization). It may be possible to assemble the huts and the base-station equipment before transporting them from a central workshop site. Atten-

tion should be paid to adequate thermal insulation to reduce air conditioning costs. The huts should be of a type that is easily dismantled, or they should be designed for transport by helicopter, vehicle, or crane.

It is easier to meet deadlines if prefabricated huts are used, because they can be under construction even before the site is selected. Base stations can be constructed of brick, as shown in Figure 24.8, but it should be remembered that brick will not, by itself, provide adequate insulation.

A different approach to base-station shelters is offered by the AT&T Autoplex Series II Compact Base Station. With a footprint of only 2.75×1.22 m, it is literally a ready-made base station. The cabinet is designed to be bolted onto a concrete slab. Consisting of three racks of hardware (see Figure 24.9) in a weather- and vandal-resistant cabinet, the Autoplex Series II simply requires an antenna, alternating current (AC) power, and a link to the switch and it's ready for service.

Inside the cabinet the installation is complete upon delivery. The air conditioning is built in, and because most links today are digital, so is all the required multiplex equipment. Plug-in modularity with a view to digital compatibility is a feature and up to 28 analog Advanced Mobile Phone Service (AMPS) voice-channel units can be accommodated.

Figure 24.7   A steel transportable base station.

Figure 24.8   A brick base station.

The linear amplifier technology perfected by AT&T enables the channels to be combined at low power levels so that the need for cavity combiners is eliminated. The linear amplifier is distinctive in Figure 24.9 as a large circular construction composed of wedge-shaped amplifier units. Each unit is a 10-watt wide-band amplifier and they are all connected in parallel through a low-loss combiner. Should only a few active amplifiers be needed

Figure 24.9 An AT&T Autoplex Series II Compact base station. (Photo courtesy of AT&T.)

(for small bases), the unused slots are filled with dummy modules to ensure adequate cooling. The black central circle is a radial fan, which forces air outward between the cooling fins.

Floor Loading

A large radio base station will have up to 4000 kg of equipment installed. Table 24.2 details the equipment.

The floor loading should be designed for the actual equipment loading, typically 700–800 kg/m^2.

Additional Area Required

Operators contemplating expanding into paging and possibly public mobile radio (PMR), should provide room for expansion.

TABLE 24.2   Equipment for a 48-Channel Radio Base Station

Equipment	Weight (kg)	
	Analog	Digital
48 fully equipped radio channels	1500	300
Base-station controllers	900	600
2 rectifiers	300	500
Batteries (2-hour backup)	1300	400
Total	4000	1800

Weight

The choice of construction materials will seriously affect the weight of the structure, which can vary from a few tons for wooden and fiberglass, 5–6 tons for aluminum clad, and around 15 tons for a concrete shelter. The weight can be a significant factor in the cost of foundations and transport. Some roads have weight limitations (which may apply year-round or seasonally due to heavy rain or snow/ice) that can make transport in an assembled state impossible. Disassembly for transport is usually a possibility, but not an economical one. Heavy buildings often will require two cranes for site placement.

Lighting

Adequate lighting can be provided by eight (4 × 2 twin units), 40-watt fluorescent lights within the hut. Incandescent lights should not be used because of their low power efficiency and high heating load.

Security

It is usually necessary to build a humanproof fence around the hut/tower installation to prevent vandalism. The fence should clear the structure by 2 meters and should not provide an access to the building roof. The antenna structure should have some means of preventing attempts to climb it. Serious legal liability could result

Figure 24.10   The sign on an unmanned transformer station in the UK.

Figure 24.11   A warning attached to Racal-Vodaphone cellsite in the UK.

should an intruder fall and be injured. All doors should be fitted with entry alarms that are monitored 24 hours a day at some central location.

A number of innovative signs have been attached to unmanned base-station sites to deter trespassers, some of which are shown in Figures 24.10, 24.11, 24.12.

Special attention should be paid to doors, which can be a weak point. Good-quality locks, with at least two latching points, should be provided.

An independent means of communicating to the main operations control center should be provided. A conventional telephone and land line, or a PMR (two-way radio) link, is adequate. This is needed if complete failure of the base station, its link, or the switch, occurs.

Equipment Mounting

The floor of the building should be constructed to allow the racks to be anchored by suitable bolts. The wall construction should be such that equipment (such as power boards and microwave links) can readily be attached. In the case of metal construction, mounting can use the supporting studs and the steel frames between them. For fiberglass construction, it may be necessary to include a plywood layer in the wall that can be used as a structural support. With fiberglass, it is important not to

rupture the outside wall, because a rupture can lead to problems with waterproofing.

Insulation

To minimize the air conditioning load, fiberglass-batting insulation should be placed in the walls and ceilings of the shelters. Table 24.1 gives the proper insulation thickness.

The grade of insulation is a trade-off between cost and energy conservation due to heat losses. A thermal-resistance rating in the range R-12 to R-22 covers the range of insulation normally used in such shelters.

Figure 24.12   A warning sign on a microwave site in Honolulu.

Cable Window

It is necessary to bring RF cables for the base station and perhaps microwave antennas into the base-station housing. Because these ordinarily come from the same antenna support structure, a single cable entry of approximately 800 mm × 800 mm suffices. The cable window should be able to support additional cables as the base station expands.

For transportable shelters, the orientation of the shelter cannot always be determined by the cellular operator before construction (the shelter often must be made to fit into available space), so it is sometimes necessary to bring the cables into the shelter through any of the walls. In these cases, the location of the cable window must be flexible.

Electrical

The shelter should be wired for the following:

- Air conditioners
- Rectifiers
- Eight double, general-purpose power outlets
- An exhaust fan (when lead-acid unsealed batteries are used)
- Lights

- Emergency lights that can run off the equipment batteries
- An external power-inlet socket for an emergency portable generator
- General-purpose power outlets, one double unit every 2 meters, at least 150 mm above the floor level

Notice that the power consumption of a large base station can be quite high (around 30 kW); if possible, it is a good idea to use a three-phase power supply (some authorities may require it). Approximately 200 watts per channel should be allowed for equipment power and a similar amount for air conditioning.

If an external generator is used, it must be housed in a suitable weatherproof shelter. Access for refueling (of diesel, propane, or natural gas) is essential. Above-ground fuel storage is both cheap and allows regular inspection of the tank. Below ground is sometimes used because of local regulations or where space is at a premium.

Access

Good access is essential for any base-station site, both for ease of installation and for maintenance. The access road should be a minimum of 4-meters wide, but preferably 4.5 meters. The access gradient should not exceed 15 percent, and any bends should be at least 20-meters radius. Good drainage around the site, tower, and access roads is necessary to prevent water erosion.

Factors that could affect access at particular times—including fires, heavy rain, winds, snow, and ice—should be considered.

GROUNDS AND PATHS

For both the switch and the base station, it is important that dirt and dust not get into the equipment rooms. The grounds surrounding the shelters should be paved and the rest of the ground should be free of dust, mud, or other potential pollutants.

25

BUDGETS

Before undertaking any project, it is necessary to ensure that sufficient resources are available to complete it. Many cellular projects fail because of poor resource allocation. For budgetary purposes, some broad assumptions about costs and staff hours need to be made. These assumptions should be sufficiently general so that costs can be estimated before getting a detailed engineering study, and equipment orders can be placed before completing a detailed engineering design. Equipment orders generally require from 3 months to 6 months of lead time because manufacturers produce equipment only upon receipt of an order. And, at any given time, equipment manufacturers normally have quite a few unfilled orders on hand. Note that the guidelines presented in this chapter provide a first-order approximation useful only for budgeting studies.

EQUIPMENT REQUIREMENTS

Before determining how much equipment is needed, you must first establish the level of demand and when service will need to be provided. This exercise is relatively simple if the equipment is for ongoing expansion only. Estimating quantities becomes more difficult, however, if the installation is for a new area. The estimates must consider the number of base stations, channels, switches, links, shelters, and towers, as well as the type of billing system to be used.

Base Stations

In most new installations, assuming that high-density coverage is not required initially, the number of base stations depends more on the area to be covered than on any other factor. When the area to be covered is small, say less than 500 km², a first-order engineering approximation of base-station requirements is necessary in order to account for large variations in local terrain or other conditions. In larger areas where this variable tends to average out, the general rules discussed below are more applicable.

Table 25.1 gives the expected range of base stations in a city environment from a 30-meter structure. In rural areas, significantly larger towers and larger ranges can be achieved, so these areas should receive at least a map study before estimating coverage.

The number of bases required for a given installation can be estimated from Table 25.1.

Low-density areas permit the use of prominent sites, and coverage in these regions can extend to 20 km for NMT900 systems, and to 40 km for Advanced Mobile Phone Service (AMPS) systems.

Additional bases may be required, depending on density (traffic-capacity) rather than coverage. The time to make this determination is when estimating the number of channels that will be needed. Also, some isolated, high-density areas may require dedicated bases. Whether your area requires dedicated bases can be determined later, when a detailed coverage study is done.

Base-Station Rentals

Base-station rentals are rising, and $2000 per month for a base-station site is not uncommon in the United States. In developing countries site costs are still relatively low, but they are on the rise. This expense is roughly equivalent to the depreciation costs on a $170,000 base station over 7 years.

TABLE 25.1 Expected Base-Station Coverage for Various Systems from a 30-meter Antenna Elevation

Terrain	Area Covered (km^2)			Equivalent Range (km)		
	AMPS NMT450	TACS	NMT900/GSM*	AMPS/CDMA	TACS	NMT900
High-density urban	12.5	10.2	4.5	2	1.8	1.2
Med.-density urban	50	41	18	4	3.6	2.4
Suburban	120	92	41	6	5.4	3.6
Outer suburban	200	163	72	8	7.2	4.8

High-density urban: Central business district (CBD) of major cities, such as London, Hong Kong, New York, and Sydney
Medium-density urban: CBD of smaller cities, such as Manila, Bangkok, Munich, and Singapore
Suburban: Area with only a small proportion of buildings higher than two stories, mainly residential
Outer suburban: Mixed residential, farm land, and open land

*Global System for Mobile Communications (GSM) because of timing restrictions, has a maximum usable range of 35 km.

SWITCHES

The switch is usually a "stand-alone" unit (that is, it is not used for non-mobile purposes).

There are several factors that may lead to the decision to install a second switch: capacity, link costs, reload time, and security.

- *Capacity.* Smaller switches are available from specialist vendors of 5,000 to 100,000 lines, most major manufacturers have switches from between 100,000 and 1,000,000 lines.
- *Link Costs.* When the base stations become more and more remote, link costs rise. It may be more economical to provide a second switch. Generally, this is not a factor for links shorter than 100 km.
- *Reload Time.* In the event of a system "crash," a reload at the switch occurs, and each base station must be downloaded from the switch sequentially, with its data and software. The time to reload one base is about 2 minutes (but can be much more on some systems), and as the number of bases increases, the total reload time can become prohibitive. The reload time, however, is very dependent on the system architecture and can vary greatly from one manufacturer to another.

- *Security.* In the event of a major disaster, such as fire, earthquake, or war, the system integrity is greatly enhanced if two physically separate switch locations are used.

Channels

This is an area of great uncertainty because the number of channels depends on the forecast demand, which experience shows is never very predictable. If actual traffic figures are available, they should be used, but for new service areas or even modified service areas (for example, a sectored omni cell), estimates must be made.

The sum of the traffic of individual cells (measured as time-consistent busy-hour traffic) will usually be 5–10 percent more than the traffic carried by the switch (measured in the same way).

This difference in traffic occurs because traffic peaks occur at different times for different cells and, in effect, measuring time-consistent busy-hour traffic at different bases is not the same thing as measuring time-consistent busy-hour traffic at the switch.

Table 25.2 can be used to formulate channel-capacity requirements when traffic readings are not available.

When actual traffic figures are available, they should be

TABLE 25.2 Budgetary Estimate of Customers/Channel

Number of Channels/Cell for Sector Sites	Formula
1. Less than 10 channels/cell	$K \times 8$ customers/channel
2. More than 10 and less than 30 channels/cell	$K \times (0.6 N + 2)$
3. Greater than 30 channels/cell	$K \times 20$

where N = number of channels/cell
$K = 30$/calling rate (mE)

Add 1 control channel/sector or 2 if standby exclusive control channel provided.
Increase above by 30 percent for omni sites.

TABLE 25.3   Work-Hour Requirements

Installation Task	Work Hours	
	Analog	Digital
Design radio		
(site selection and radio-path design)		
Up to 4 bases	1000 hr	1500 hr
5 to 15 bases	300 hr/base	300 hr/base
16 or more bases	400 hr/base	400 hr/base
Radio survey		
(includes survey of rejected sites)	200 hr/base	200 hr/base
Switching design	500 hr	1000 hr
Installation of bases		
(includes links and installation design, and commissioning)	800 hr/base	100–400 hr
Installation of switch		
(includes installation design and commissioning)	3000 hr/switch	3000 hr/switch
Acceptance testing		
Per base	40 hr	60 hr
Per switch	400 hr	600 hr
Mobile installation in vehicle	3 hr/unit	3 hr/unit
Handheld sale		
(includes paperwork and customer instruction and validation)	1/2 hr/unit	1/2 hr/unit

processed with the Erlang C table to determine requirements (for more information, see Chapter 20, "Traffic Engineering Concepts").

Install sufficient channel capacity to allow for the lead time for the next expansion (typically one year) and, if the network is a new one, build in some overcapacity (for example, 30 percent) to allow for errors in forecasts.

TYPICAL WORK-HOUR REQUIREMENTS

Table 25.3 gives the approximate work-hour requirements for cellular radio installations.

COSTS

Table 25.4 shows the cost of switches. Table 25.5 shows the cost of switch sites, and Table 25.6 shows typical base-station costs.

Switch Software

Switch software comes in packages that cost from $500,000 to $1,000,000 and includes a guarantee of support and updates for a fixed period (usually about one year). Some manufacturers charge once only for software; others charge on a per-switch basis. Usually, there are additional charges for ongoing software support

TABLE 25.4   Switch Costs

Maximum switch size	Cost (US$)	
	Analog	Digital
500 lines	300,000	3,500,000
2,000 lines	500,000	3,600,000
10,000 lines	1,300,000	4,000,000
50,000 lines	3,000,000	5,000,000

TABLE 25.5   Switch Shelters (including land and site work)

Number of Subscribers	Cost (US$)	
	Analog	Digital
0–5,000	100,000	300,000
10,000–20,000	300,000	300,000
More than 20,000	500,000	500,000

after the first year. The switch costs shown in Table 25.4 include an allowance for the software component.

Cost Models for Line Transmission Systems

Cellular bases are ordinarily connected to the switch by 2-Mbit, 8-Mbit, 34-Mbit, or T1 spans. Table 25.7 gives rules for determining the cost per circuit of the links.

TABLE 25.6 Base-Station Costs

Item	Cost (US$) Analog	Cost (US$) Digital
Base-station shelter	40,000	30,000
Tower/mast	1,000/m	1,000/m
Antennas and feeder (each)	2,000	2,000
30-meter pole (each)	10,000	10,000
Power supply for base stations, incl. batteries	12,000	12,000
Emergency power plant	30,000	15,000
Base-station radio equipment	10,000/channel	3,000/channel
Common equipment (incl. base-station controller)	230,000	50,000
Site and site work	40,000	25,000

TABLE 25.7 Cable Costs as a Function of Route Distance

	$(A + B \times L)$ Cost Coefficients in $1,000 per VF Circuit for Cable Types*					
	Cable Types					
	CPFUT		SQC		SMOF	
System Type	A	B/km	A	B/km	A	B/km
2 Mbits (protected)						
Existing (25 km)	0.19	0.08				
Existing (100 km)	4.63	0.08	0.54	0.08		
New (100 km)					0.55	0.16
2 + 2 Mbits						
Existing (25 km)	0.13	0.062				
Existing (100 km)	2.47	0.062	0.27	0.039		
New (100 km)					0.34	0.081

* Costs of cable routes can be modeled as Cost = $(A + B \times L)$, where A = fixed costs, B = distance-dependent costs, and L = distance.

CPFUT = cellular polyethylene-insulated filled-core unit twin (cable)
SQC = single-quad carrier (cable)
SMOF = single-mode optical fiber

This model is based on a cost model of

$$\text{Cost (\$)} = A + B \times L$$

where

A = fixed costs

B = length-dependent costs

L = length (km)

DIGITAL RADIO SYSTEMS (DRS SINGLE HOPS)

Digital radio systems (DRSs) have a very small marginal cost for an additional circuit and so costs are figured using a different model of

$$\text{Cost (\$)} = C + D \times N$$

where

C = fixed cost

D = cost per channel

N = number of channels

This model assumed that the existing cellular towers, power, and shelters are used for the microwave. These costs include installation. Table 25.8 shows the cost of microwave systems.

Wireline carriers have a good chance of having an existing plant, but if new links are provided over distances greater than 10 km, DRS usually are the least expensive.

TABLE 25.8   Cost of Microwave Systems*

System Type	C	D
2 Mbits	$28,000	$500/channel
8 & 34 Mbits	$35,000	$500/channel

*Microwave systems modeled as Cost $= (C + D \times L)$, where $C =$ fixed costs and $D =$ additional costs/channel. For hot standby, add 40 percent.

BILLING SYSTEM (INCLUDING COMPUTER AND SOFTWARE)

The cost of billing systems has dropped by factors of 3 to 5 over the last few years. It was once cost-effective for the operator to develop software for billing, but the cost (and complexity) of a commercial package is now at a level that this should not be a consideration.

Table 25.9 shows the cost of typical billing systems.

Fiber Optics

The preferred method of linking base stations today in developed cities is via fiber optic cable. The terminal equipment for this costs around $15,000. Fiber-optic links give greater flexibility, noise immunity and the possibility of future slave repeater applications.

COSTS OF A "TYPICAL" ANALOG CELLULAR SYSTEM

For an AMPS system (although the basic assumptions are the same for any system, the differences are only in the details), if you make some basic assumptions about start-up sizes and switch capacities, you can calculate the cost of a "typical" cellular system. Figure 25.1 plots costs against subscribers for a system that includes the entire infrastructure (shelters, links, towers) and also the marginal cost where these are not included (that is, the cellular-specific equipment only—a wireline operator may need to add this cost).

The switch in Figure 25.1 is assumed to be supplemented by a 50,000 line unit at 10,000 subscribers. Start-up costs, which are very operator-dependent and can include license fees, lobbying costs, advertising expenses, and other costs, have not been included here.

The average cost per subscriber has been steadily falling from a high of $2000 in 1986 to around $1200 in 1991, to $600 in 2000.

AN EXERCISE

Estimate the cost of covering a city of a million people with an area of 600 km^2 and two CBDs separated by 5 km. Each CBD is approximately 2 km $\times$ 1 km in area and an AMPS A band (not extended) system is planned. Assume shelters are budgeted at zero (already owned) and all bases are omni-directional. A 7-cell pattern is used, and microwave links must be provided. Note that this system needs some infrastructure (microwave only), and so should fall between the two extremes shown in Figure 25.1. Other infrastructure is assumed to be already owned or leased.

Assumptions

You should base this example on the following assumptions:

TABLE 25.9   Billing System Costs (includes computer and software)

	Cost ($)	
System	Software	Hardware
Simple system for 5000 subscribers, basic billing only	80,000 + 10,000/year maintenance	50,000
Simple system for up to 10,000 subscribers	120,000 + 50,000/year maintenance	100,000
System with on-line billing, direct switch validation for up to 10,000 subscribers	250,000 + 25,000/year maintenance	500,000
Fully expandable system for up to 100,000 subscribers, with full-management information capability	600,000 + 50,000/year maintenance*	800,000+

Test Equipment and Tools	Cost ($)
Analog	200,000
Digital	800,000

*Add $120,000 per additional 100,000 subscribers (mainly computer upgrade costs).

Budget $250,000 for test equipment and tools.

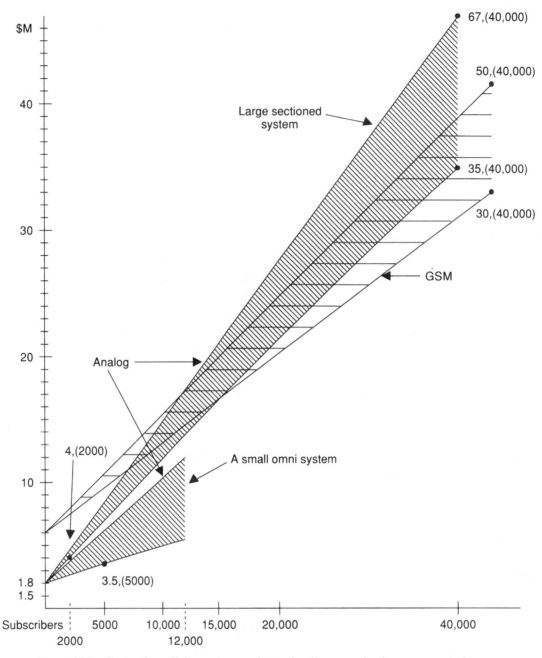

Figure 25.1 Costs of a cellular system against subscriber capacity (assumes most sites are sectored with $N = 7$ pattern).

Switch price $=$ \$2.0 M (50,000 + line switch)

Base stations $=$ \$100,000 and \$11,000/channel

Work-hours $=$ \$60/hour (contract rate)

Calling rate $=$ 30 m Erlangs/subscriber

First year's estimate of customer demand $=$ 5000

The site is configured as a sector site, as rapid expansion is anticipated.

Base-Station Costs

To determine base-station costs, follow these steps:

1. Determine the minimum number of base stations required for coverage only. Each CBD requires

one base (total of two bases). An area of 600 km^2 (assuming it is suburban) requires 600/120, or five, bases, based on Table 25.1. Seven bases are required for coverage.

2. Determine the number of channels required for traffic. Assume that forecast customer demand equals 5000 and that there are 20 subscribers per channel. In this example, 5000/20, or 250 channels, would be required.

Because only the A band is available, it is reasonable to assume that each base has an average of 22 channels (50 percent of the 44 maximum) equipped (assuming a 7-cell pattern). The number of bases required is 250/22, or about 11.

Therefore, the cost of base stations is as follows:

$$(11 \times \$100 + 250 \times \$11)K = \$3850 \text{ K} \quad \text{or} \quad \$3,850,000$$

Base-Link Costs

Assume an average of one link/base (11). For this exercise, all links are considered to be 2-Mbit DRS, except the main link to the public switched telephone network (PSTN). Some sites will have 2 links (as more than 30 channels will be needed at some sites), so say 15 links are needed.

Therefore, the cost of base links (from Table 25.8) is as follows:

$$(15 \times \$28 + \$0.5 \times 250)K = \$545 \text{ K} \quad \text{or} \quad \$545,000$$

Add to this amount $100 K for a 34-Mbit link to PSTN (with standby). Thus, the total link cost is $645,000 K.

Work-hours for this installation are shown in Table 25.10. The cost of work-hours equals $60 \times 18,100 = 1,086,000 K, assuming $60 per workhour. Therefore, the cost of this installation can now be calculated as shown in Table 25.11.

Figure 25.1 leads us to expect that the cost of this system should be around $8.2M.

TABLE 25.10 Work-Hours Calculated from Hours Shown in Table 25.3

Installation Task	Work Hours
Radio design	3,000
Survey	2,000
Switch design	500
Installation of bases = 11 × 800	8,800
Installation of switch	3,000
Acceptance testing of radio bases = 40 × 10	400
Acceptance testing of switch	400
Total	18,100

TABLE 25.11 Cost of Typical Cellular Installation (assuming 5000 subscribers, 11 bases, and 250 channels)

Budget Component	Cost (in millions of dollars)
Work-hours	1.086
Links	0.645
Base stations	3.850
Switch	2.0
Ancillary costs:	
Training	0.100
Tools and test equipment	0.250
Billing system	0.270
For capacity of 5000	
subscribers, the total cost is	8.201

For budgeting purposes, this example gives a good description of the required system. The total work-hour requirements can also be used to estimate involvement by staff and consultants. A detailed engineering design may indicate some deviation from this estimate, but usually the results of this procedure are realistic both in terms of cost and base-station requirements.

REVENUES

In 1985 the average yearly gross revenue per cellular subscriber in the United States was $2064. By 1992 this had dropped to $960; by 2000 it is around $400 and falling. The story is the same elsewhere in the world; cellular revenues are dropping in dollar terms at an alarming rate, especially when it is realized that the $200–$300 per "pop" prices that were paid for cellular licenses in the heyday of the late 1980s were financed by loans that are largely still current. The revenues are forecast to continue to drop at around 10 percent per year. This drop is made up of two main components: a drop in usage rate and to a lesser extent a drop in call charges. The cost of carrying a subscriber can be easily estimated as:

COSTS INCURRED
PER CUSTOMER PER YEAR

	$
1. Depreciation of capital equipment	50
2. System maintenance	50
3. Wages	50
4. PSTN charges	100
5. Repayment of interest on capital equipment	100
6. Administration expenses	50
7. Electricity/water/rates	30
8. Vehicles	20
9. Site rentals	10
10. Interest on license cost ($2000)	140
11. Miscellaneous	15
	605

TABLE 25.12   Historical and Forecast Revenues per Subscriber in the United States

Year	Subscriber Revenues per Year (in 1990 $)
1983	2760
1984	2436
1985	2064
1986	1908
1987	1740
1988	1584
1989	1416
1990	1308
1991	1200
1992	960
1994	629
1998	479
2000	400

These costs are representative only, but they do indicate that the margin between revenues and costs is closing rapidly.

The question must arise, How can the industry pay the massive costs of going digital without a big increase in revenues?

The projection for the future is more of the same, with the return continuing to diminish, as can be seen in Table 25.12.

AIR-TIME CHARGES

Air-time charges vary considerably from country to country, as Tables 25.13A and B show. Countries that have higher air-time charges generally also have higher monthly fees. From this it can be concluded that not all cellular licenses are equal.

Comparing Table 25.13A (1993) with Table 25.13B (2000), it can be seen that call charging has dropped significantly over the past 7 years.

TABLE 25.13A   Air-Time Charges by Country in 1993

Country	Air-Time Charge (US$/min)	
	Peak	Off-Peak
Japan	1.40	0.70
Taiwan	1.00	
UK	0.85	
Australia	0.50	0.25
United States	0.40	0.24
Hong Kong	0.37	0.37
Indonesia	0.30	
Philippines	0.20	
Thailand	0.20	0.16
Singapore	0.18	0.10
Malaysia	0.15	0.08

Note: Air-time charges provided by Teleresources.

TABLE 25.13B   Air-Time Charges by Country in 2000*

Country	$US/min
Japan	0.35
Netherlands	0.33
Belgium	0.28
Spain	0.25
Hong Kong	0.23
Australia	0.22
UK/Sweden	0.22
South Affrica	0.18
Singapore	0.13
Taiwan	0.13
France	0.11
Philippines	0.10
Canada	0.09
USA	0.09
Germany	0.08
Thailand	0.08
Indonesia	0.05
Bangladesh	0.04

*Note: Because of the usually complex tariff schemes, it is not possible to indicate sample Peak and Off-peak rates in 2000.

26

BILLING SYSTEMS

HISTORY

In the early days of public switched telephone network (PSTN) billing all calls were manual and the bills were recorded manually as well. Then came the introduction of automatic switches, and the billing had to be automated. Each time a subscriber made a local call, a meter pulse would be initiated, which in turn would make the subscriber's meter tick over to register one more call. If local calls were timed, then these pulses would occur at each charge interval. Long-distance calls were still mostly manual.

Later, when long-distance calls were automated, the solution was simply to speed up the meter pulses so that these calls were charged as multiple local calls. So, for example, if the local call rate was 10 cents a minute and a long-distance call was $1.80 per minute, the subscriber's meter would tick over at 18 registrations per minute.

At the end of each billing period somebody would come around and read the meters. If the meter had clocked up 670 registrations over the billing period, then the bill was $67.00 (670 × the local call rate of $0.10). As simple as it was, this method had its problems. Reading hundreds of meters can be very tedious, and mistakes are easily made. Some of the meters that were put in before long-distance calls were automated had only four digits; users who made a lot of long-distance calls could run the meter "around the clock" and so get undercharged. This was often detected only after a negative reading was noticed (that is, the meter had run right through to 0000 and then back to a new reading that was less than the previous one).

CELLULAR BILLING

The billing system is an important but independent part of a cellular system. Most cellular system suppliers do not offer a billing system as part of their product line and recommend that this be purchased from an independent supplier. Wireline operators will almost certainly already have a billing system for their other operations, and this can be adapted for the cellular system. Some operators incorporate the billing so that the customer receives the cellular bill in the same format as if it were an additional wireline telephone; others prefer to bill separately.

In principle, the concept of billing is simple. The switch records all the call details on a billing tape, the tape is later read by another computer that analyzes that tape, and from the called number and the length of time that the call was held up, the charges to the calling party are determined and billed.

In practice, getting a billing system fully operational and effective is very time-consuming and requires the understanding on the part of the operator that just because it is all computerized does not mean that it is infallible. In fact, if a newly installed billing system can compute the charges due so that 99 percent of the records are correctly processed, then that would be considered rather good.

Billing systems have human interfaces, and data entries include the tariff table, the charges on a basis of the called number, the charging algorithm (for example, are only whole minutes charged? rounded up minutes? etc.). This interface and the software itself is where problems arise. It is generally considered that computer code that has not more than one line of faulty code per

1000 lines is fully debugged. This one line is the gremlin in the system, and it usually surfaces only occasionally and usually in the most mysterious way. You can expect that the price of accurate billing is eternal vigilance.

The billing computer is physically separate from the switch; frequently, the only link is a tape or data line. The raw billing information is collected by the switch but is not processed. The tape is then removed and sent to be read and analyzed by the billing computer.

This billing computer can be a minicomputer or a mainframe computer that has its own software to interpret the billing tape and then produce the subscriber's bill and the operator's ledger. Because various manufacturers have different proprietary operating systems, the billing-tape format is different for each equipment type. If the billing system provider has not previously worked with a particular switch, some software must be written to enable the billing data to be read.

Charges vary widely from country to country and vary somewhat from operator to operator within the same country.

Charging can involve many parameters. These are some of the main parameters:

- Cost of local calls
- Cost of airtime
- Call rate variation with time of day
- Charges for long-distance calls
- Charges may vary with call length (for example, the first minute may be charged at a higher rate than subsequent minutes)

The following requirements should be built into the software:

- Frequency of billing (monthly, two-monthly, and so on)
- Charges applicable to roamers
- Billing format (what the bill looks like)
- Currency (this can present problems in countries where the exchange rate is thousands of units to the dollar so that existing billing formats may not be able to accommodate a sizable bill)
- Language (sometimes the bill may be required in languages other than English)

Real-time (or hot) billing requires that the billing computer can access the switch in real time and produce a bill on demand. The advantage of this is particularly evident in the case of roamers and rental units, where an instant bill is needed. To keep the demands on the processor within practical limitations, only some of the customers are flagged for real-time billing. Typically, a real-time billing package may allow up to 1000 customers at a time to be marked for real-time access. Such a system could use an RS232 or X.25 link between the billing computer and the switch. Figure 26.1 illustrates this.

Billing systems can perform billing and ledger functions only, or they can perform validation and produce Management Information Systems (MIS). There are considerable advantages in the more capable systems, but they are also considerably more costly.

Billing systems that can perform validation allow customer data to be entered manually only once into the billing system. If validation is not available, the dual entries of customer data into the switch and to the billing system are sure to leave considerable room for false and missed entries. Once the system is bigger than a few

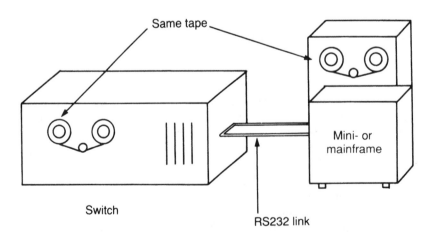

Figure 26.1 The link between the switch and the billing computer for real-time billing may use an RS232 or X.25 link. The bulk of the billing information, however, is transferred via tape.

thousand customers, it is almost impractical to use two parallel data entry systems.

UPWARD COMPATIBILITY

A start-up operation usually tries hard to minimize cost. This can be difficult with billing systems. Most commercial systems are designed for large networks (millions of customers). As the system grows, there is a need to upgrade the computer, making the old one redundant, except in the rare cases where the computer itself is physically upgradeable by adding more processors.

In most cases, the billing system vendor will offer a scaled license fee, which is based either on the actual billing platform (that is, the computer model) or on the subscriber database size. This enables the new or smaller operator at least to obtain full-powered software at an affordable price.

THE DANGERS OF CONTRACTING FOR A BILLING SYSTEM

Every operator must at some time obtain the services of a billing service and regardless of whether an in-house system, or outside billing service, is chosen the operator will be rather heavily dependent on ongoing support for the billing software.

In one case that the author has been involved with, the billing supplier was somewhat less than reliable, and as soon as the payments had been made for the in-house system, the support dried up completely. A few months after the system was installed a severe power failure caused a number of the files to be corrupted, which caused the software to fail. This problem was never rectified by the supplier, and the billing system was never again functional. As a result, the operator was unable to bill the customers. As a precaution it was necessary to cease to connect new customers until the problem could be resolved. In the meantime, the lost revenue amounted to around $100,000 per month. A quick check of the contract revealed that the operator's recourse was to demand that the software be made functional (under the terms of the warranty). This was done, but still no support was forthcoming. The final option available was to terminate the contract. Losses incurred as a result of the incompetence and/or bad faith on the part of the billing-system provider were specifically excluded, as claimable liabilities under the conditions of the contract.

So in essence the billing supplier only guarantees to rectify the hardware/software. If, however, this is not honored, the only recourse is to go to someone else. Doesn't really sound fair, does it?

In subsequent dealings the author learned that all billing-system providers similarly require an indemnity against non-performance of their wares. Although this approach is understandable, it does leave the operator very vulnerable to unscrupulous billing-system providers. There is not a great deal of risk with most of the more reputable suppliers, but it does pay to check the reputation of the people you deal with very carefully.

Choosing a Billing Supplier

Choosing a billing-system supplier can be a daunting task. There are about a dozen major vendors to choose from, and differentiating between them can be taxing. If you require service bureau facilities, then the field is partially cleared, as not all billing-system suppliers have a billing service.

Some billing suppliers specialize in large systems (1,000,000+), and their systems operate only on mainframe computers. This is fine if you are big, but if your system has only a few thousand subscribers, then the mainframe will cost almost as much as the network. An operator with a small system should look for a billing system that can operate on a minicomputer. The upgrade path must be considered if future growth is anticipated.

Pricing packages are available from most vendors. They are usually based on either the platform (you pay for software upgrades as you upgrade the computer) or on the usable subscriber database. Sometimes when a computer is upgraded the software has to use a new operating system, which will be very different, but usually the upgrade is relatively simple. What to do with the secondhand, now useless, computer isn't so simple. Where the system is database limited, software restrictions are placed on the number of users that can be processed, and these are removed only when an upgrade license is acquired. Recently PC-based systems have appeared for smaller cellular operations.

RESELLERS AND THIRD-PARTY VENDORS

The concept of a reseller evolved in countries that originally did not allow the cellular operator also to be a mobile retailer (notably the UK). This concept, however, can apply to all cellular operators. It involves giving limited access to third parties to the validation and/or the billing system. In this way, a reseller can validate and bill a subgroup of customers; in effect, the reseller becomes a de facto cellular operator. Figure 26.2 shows a billing system for resellers.

Most billing systems were developed with the major US or UK markets in mind, so this concept is generally available on most commercial billing systems.

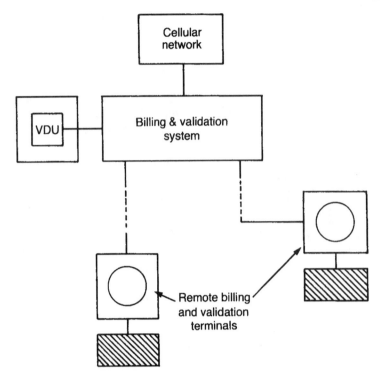

Figure 26.2   The billing system can be configured for resellers so that the reseller effectively becomes a de facto operator.

The billing package can have a retail module, which allows remote modem-connected access for validation of new customers by a network of dealers. A real-time modem access for rental phones also may be required.

FLEXIBILITY OF TARIFFS

Tariff structures are often quite complex and may have built-in day-, time- and distance-dependent rate structures. These rates are not fixed forever, and so it is important that the billing system offers an easy means of altering tariffs as rate changes occur.

Sometimes the billing system may need to analyze call charges that are dependent on the location of the PSTN caller as well as the location of the mobile subscriber. In this case the billing system must be able to access the PSTN calling-party's number

Where the cellular switch is connected to several different PSTN operators, the frequency of tariff changes will be correspondingly higher.

ACCOUNT SETTLEMENT

The billing system must be able to produce a monthly statement of accounts due to the systems interconnected

to the cellular switch. It may be that the settlement with various operators is structured quite differently and a good deal of flexibility is required for this part of the program.

METERED BILLING

A form of billing, called *metered billing*, was introduced in 1993. For carriers who require high deposits from new subscribers, metered billing allows an option of a no-deposit, paid-up-credit facility. The subscriber can use the system to the extent *only* of the paid-up credit. Once that credit is exhausted, the system service is withdrawn. Credit cards can be used to make payments.

This concept developed into the very popular prepaid service.

BILLING CYCLES

The cellular operator will have to decide on a billing cycle that maximizes revenue collection without unduly loading the billing system. Typically, billing cycles range from one month to three months.

When the system is large, it may be necessary to have continuous billing cycles in order to avoid unmanageable volumes of paper, which would occur if all bills were sent on the same day.

ITEMIZED ACCOUNTS

These are readily available from the billing system and will give the customer details of each call, its duration, time of call, called number, and charges. Being detailed, these accounts can be bulky and it is not unusual that they are offered to the customer at an additional charge (usually a few dollars per bill).

REMOTE BILLING

The billing system will normally have the capacity to support remote billing terminals. Apart from use by re-sellers, which has already been discussed, remote terminal access can be of use for customer service staff who handle billing enquiries as well as for the use of remotely based billing staff who may be handling their own local areas.

MULTIPLE SWITCHES

Where more than one switch is in use, the billing system can be configured to handle each switch sequentially. Some difficulties can exist if the switches are not for the same region and roamers appear on the different switches. It is easier to bill roaming as a separate transaction, although the more sophisticated billing system can collate the various call records.

FOLLOW-UP AND ACCOUNT MANAGEMENT

Not all bills are paid on time and delinquent bills need to be chased. A billing system can do this automatically and flag high-dollar bills for special attention. Automatic devalidation for payment default can be built in.

A series of reminder notices (also known as *dunning notices*) are sent to slow-paying customers to coax payment. These notices can be automatically generated by the billing computer. Usually a period of one to two months late payment is permitted before the phone is devalidated.

Since the cost of carrying bad debts can be high, many operators charge interest on late payments. In the interest of good customer relations the interest rates should be reasonable and in line with the charges of other credit organizations.

BILL PREPARATION AND LETTER STUFFING

The billing system will normally prepare the customer's bill in a form ready for mailing. However, simple systems may still require manual folding and insertion into an envelope. Additional hardware is available to automate all of these functions.

VALIDATION

Usually the cellular operator will initially validate subscribers directly through the switch. Apart from the duplication involved in entering the data separately into the billing system, experience shows that the coordination between the two systems is often a problem. Billing systems can be structured so that an entry/cancellation on the billing system automatically updates the switch records.

TRACKING SALES AND INVENTORY

The billing system can be integrated into the sales network so that it can track the sales of mobiles and keep an inventory of stock. In particular, it can flag stock re-orders as well as especially fast- or slow-moving stock.

ON-LINE INQUIRIES

The billing system should have on-line access to the billing records for the last two to three months. A readily accessible operator position should be available so that queries and complaints can be promptly dealt with. It pays to remember that even though the billing system is computerized, it is far from infallible and that many of the discrepancies brought up by the customer will be legitimate. For those customers who make a habit of complaining some billing systems also have on-line access to previous customer complaint records. Typically you can expect around 1–3 percent of bills to result in customer queries. Disputes over charges can be damaging to customer relations and the more promptly and efficiently they can be handled the better.

GSM ROAMING SERVICES

Because Global System for Mobile Communications (GSM) is designed to operate internationally, it comes with a standard roaming procedure, called the transferred account procedure (TAP), which is laid down in the GSM Memorandum of Understanding (MOU). It describes how the billing information is sent from the visited operator back to the home system.

The TAP file contains the information to allow the home operator to bill the customer in local currency, regardless of the call routing. The information is only sent either via an X.25 network or on magnetic tape and is cleared at least once every two weeks. While tape transfer can be efficient for low traffic volumes between a few operators, X.25 delivery on at least a daily basis is the norm.

An alternative approach is to use a common clearing house, known as a Multinational Automated Clearing House (MACH). In Europe, such a clearing house has been set up in Luxembourg.

BILLING HOUSES

Customer care and billing (CCB) can be outsourced in part or whole to a billing house. In fact, in the United States and Europe a majority of operators choose this method of CCB, while in Asia/Pacific it is relatively rare.

Outsourcing CCB might seem to some companies to be a way of losing control over customer records, and this could be a reason for its lack of appeal in Asia/Pacific. However, because a lot of the billing process is labor intensive (stuffing envelopes and attending to customer inquiries), it could simply be a cost-of-labor decision that keeps billing in-house in regions where labor is cheaper.

Outsourcing Options

It is not always the case that the whole of the CCB process is outsourced. There are a number of partial outsourcing options.

Billing outsourcing can be done in a number of ways.

- Outsource all billing and customer-care functions
- Outsource the billing only, but retain customer care
- Do in-house call processing and analysis of call records, and outsource bill printing and mailing
- Outsource customer care only and retain all of the billing functions in-house.

In-house billing systems cost upward of $100,000 (perhaps as much as $1,000,000 plus the cost of a mainframe computer). Because these need to be in place on day one, the billing system represents considerable capital outlay. It also requires space and staff to operate it. Having a third party undertake this responsibility, particularly at start-up, can make good business sense. A medium-sized minicomputer billing system with four VDU terminals is shown in Figure 26.3.

The outsourcing of billing should particularly appeal to new operators in established markets who are introducing technologies, like personal communication service/personal communications network (PCS/PCN), and expect a rapid increase of the customer base. Not only does the operator have a huge network rollout, but the massive job of billing might well be left to an expe-

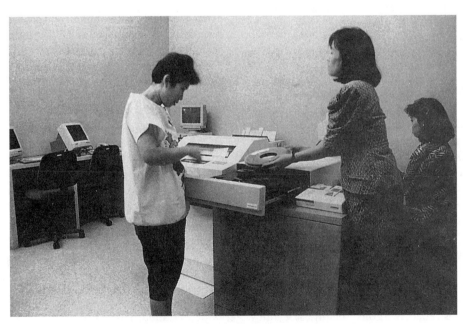

Figure 26.3 The in-house billing room of Extelcom, comprising four workstations, one minicomputer, a magnetic tape drive, and a high-speed line printer. (Photo courtesy of Extelcom.)

rienced billing house that can readily cope with the large number of clients.

Billing houses will ordinarily quote a fee based on the number of customers to be billed. Any new network has a significant margin of uncertainty over the actual (as opposed to the rollout plan) customer base. If an in-house billing system is installed and it's too big, then it will cost too much; if its too small, then it will be a disaster! Allowing a billing house to cater for this contingency not only gives peace of mind but it gives a good level of control over expenses.

Billing outsourcing can be a great boon for a new entrant into the market that may not have the expertise for billing services and may be uncertain about both the size and complexity of the task. Increasing with intelligent network (IN), the network, and therefore the billing, is becoming increasingly complex.

Outsourcing will allow the operator to negotiate a cost for billing services that is directly proportional to the customer base. Costs are fixed and known in advance, and there is no need to invest considerable sums in computer hardware and software that will rapidly become obsolete. Also, as new services like voice mail, prepaid time, and messaging are introduced, taking care of these new services will be part of the billing service vendor's job.

As competition increases and profits fall, there is an ever-increasing need to contain all costs. In many cases, except for the very largest operators, an outsourced billing service will be both cheaper and lower risk than an in-house service.

In 1998 outsourcing of billing services accounted for about 10 percent of the services.

Retaining Control

Billing outsourcing does not mean that the operator has totally relinquished control over this function. It is still necessary to retain sufficient expertise to be able to ensure that the service is provided at the standards expected and that problems are attended to in a timely and efficient manner.

When contracts are being negotiated, it is particularly important that there is sufficient expertise on hand to ensure that the right services have been contracted and that the price is right.

MORE ON BILLING HOUSES

Billing houses have greater access to credit checking, roamer databases and invalid roamer databases, and information on rogue users. They can also provide sophisticated packages for system management to suit a wide variety of user demands.

The billing house can simply process the operator's tapes or, in some instances, can offer on-line facilities via a data link. If a billing house is used, it is always advisable to ensure that at least one data link between the operator and the billing house is established so that new entries and updates can be made in real time.

Although the appeal of using a billing house is strongest for small operators, the cost of a fully operational, large-scale billing system and the mainframe computer needed to run it can make billing houses attractive to customers of all sizes.

Some billing houses have the facility to allow the customer to have full access to the processed billing records on site. Thus a tape image of the processed bills can be used by the cellular operator to provide remote terminal access to customer billing records. In this case the billing house will still do the bill processing at a remote site, but the operator will be able to modify and access the processed data in-house. To do this the operator will need to have access to a minicomputer. If full on-line access to the billing data is required, the computer will need to have a large memory capacity. The computer may also have a direct validation link to the switch.

Finally, it must be remembered that a tape sent away for processing can not only get lost but can suffer partial or complete demagnetization due to stray magnetic fields. Always keep the original in a fireproof safe and send only a copy.

DO-IT-YOURSELF BILLING

It is, of course, possible for a cellular operator to write dedicated software to read and process the billing tape, and a number of operators have chosen to do this. In principle, the concept of reading the billing tape and processing a bill is simple. This can make the "do-it-yourself" approach look attractive, particularly when billing systems can cost hundreds of thousands of dollars (or more). But commercial billing systems do a lot more than just write bills; they offer numerous system-management reports that can be very expensive to incorporate into a "do-it-yourself" system.

Software always takes a lot longer to develop and debug than seems reasonable, and adding on new subsystems can be very costly. Software experts consider five lines of debugged code per day to be a good average rate and 20 lines to be exceptional. Thus, a reasonably complex program of 2000 lines can be expected to take two years for one person to write.

Of course, writing and debugging smaller programs is much easier, and it may be possible to write a 100-line program and debug it in a single day (but not every day).

In general, the "do-it-yourself" approach is really only practical for the very smallest (who don't need sophisticated outputs) and the very largest (who can afford anything) operators.

BILLING SERVICE

The billing system should contain at least the following customer information:

- Age
- Occupation
- Gender
- Business/work address
- Vehicle (for fixed units)
- Average revenue (that can be subdivided into similar revenue groups)
- Areas generating most revenue
- Revenue as a function of time of day (or week)

This information tells a good deal about the types of customers that are connected, but it can never yield the profile of the mythical "typical customer."

A marketing manager can use this information effectively. If it is used intelligently, this information can give a company a competitive edge. The goal, of course, is to get maximum return for minimum investment. In a mature network, achieving this goal means identifying and exploiting unused capacity and using existing capacity more efficiently.

For example, if certain days show underutilization (usually weekends), then perhaps an incentives scheme can be worked out. Similarly, after business hours, the traffic drops dramatically. A two-tier charge rate that offers incentives for nighttime calls could be considered.

Customers who have high calling rates (and hence generate high revenue) can be identified and marked for special attention to ensure that they are retained. To gain customer loyalty, perhaps package deals that offer reduced charges for those users can be arranged. The system could alert customers whose calling rates might entitle them to a better deal if they choose a package. A small drop in revenue is more than compensated for by gaining long-term, satisfied customers.

Incentive packages are often aimed at high-calling-rate customers who, although they generate the most revenue, also use the system most and have a higher system cost per customer. In a fixed network, a customer who is connected but has a low calling rate costs the provider just as much as a high-revenue customer. Such customers are therefore undesirable to the operator. In cellular systems, however, a connected customer who pays the monthly fee and doesn't use the system costs only the price of the billing services. In cellular, therefore, a low-usage-rate customer normally generates the best return/cost ratio to the network operator.

Sensitivity to pricing can be gauged by correlating new customers with price variations. Where the option exists (that is, it is not government controlled), this information could be very useful in determining the revenue mix from monthly access fees versus call rates. It may be found that particular locations are not generating their share of revenue and that a sales promotion in those areas is warranted. Or it may be determined that an area is intrinsically one of low-demand. Knowing why demands are low (or high) can be extremely valuable in future expansion plans. But even if the cause cannot be identified, some generalizations on the structure of the areas with abnormal demand can help to focus expansion.

Information from widely disparate sources reveals at least an order-of-magnitude variation in customer sign-up rates from similar-sized cities with similar per capita incomes in different parts of the world. Within the same country, there are still large variations in what are otherwise similar markets. The best information, therefore, is obtained from the operator's own records; the marketing manager who realizes this is the one who will succeed. In most Western markets, the average monthly bill per customer ranges from $10 to $120.

MANAGEMENT INFORMATION SYSTEMS

Management Information Systems (MIS) are software packages designed to organize customer billing and provide further information about such things as roamers, equipment inventory, traffic distribution and analysis, and customer profiles.

Some billing systems allow, as an input, a field that describes how a potential customer came to inquire about the service. Then these people can be asked how they first heard of the company. This information can then be sorted by the MIS to measure the effectiveness of the marketing strategy.

Advertising can be related directly to sales performance and, by carefully identifying the advertising (for example, TV advertising as opposed to newspapers), the most effective media can be determined.

MIS services are available as specialized software packages with numerous user options. Faced with many choices, it is often very difficult for the cellular operator to define clearly what facilities are needed. Because of this uncertainty, some operators have resorted to using more than one MIS because of their comparative advan-

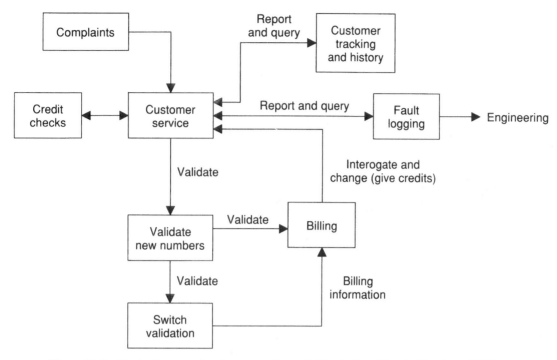

Figure 26.4   Block diagram of customer service activities that could be built into an MIS.

tages in different areas such as customer information, marketing, planning, and, not of least importance, billing. MIS is not cheap, so most operators must settle for a single package. One US company, GTE Data Services' Cellular Business Group, for example, has 150 different reports available from its MIS.

In the United States, it is possible to have either an in-house MIS (that is, one solely operated by the cellular operator) or an MIS provided at various centralized locations. Although there is a definite trend toward in-house service, some operators cite the large database facilities, lower start-up costs, and expertise of the central MIS systems as adequate reasons to use centralized services. This option is generally not available to operators outside the United States.

An MIS will have to be customized to the needs of the user. Shortcomings in the reporting systems associated with the switch, the base stations, and the billing system need to be identified. The interrelation among these areas may not at first seem too obvious, but Figure 26.4 may help to illustrate otherwise. Taking the customer service area as pivotal, it can soon be seen that these people will need to deal daily with customers' queries about their bills, services, faults, and credit-worthiness. Although it is possible to deal with these as separate entities, it is much more convenient to integrate them into a single database network.

When you have an integrated system, you can check immediately, for example: Has this customer had the same complaint before? Is this type of complaint common? What remedies are in hand? When will the problem be fixed?

With such a database it is possible to analyze faults and complaints in a systematic way, which discovers trends in time to rectify them before they become major problems.

Similarly, the engineering department can make effective use of many of the same statistics (see Figure 26.5). If an integrated MIS is not in place, then ad hoc solutions will spring up to solve data-gathering problems as they arise. Almost certainly there will be no coordination among sections, and the communications gap can be expected to be worst between engineering and customer services/marketing. This, in turn, leads to a collection of databases that overlap extensively, do not coordinate, and are invariably incomplete.

When ordering an MIS, it is vital that all sections be consulted about their needs. It is preferable for coordination meetings to be held with all of the proposed users present so that the cross-functional needs can be identified. All MISs will need to be customized to a certain extent, and the best and cheapest time to include a new function is in the initial phase. Whether that function is added on day one or a future provision is made for it, it

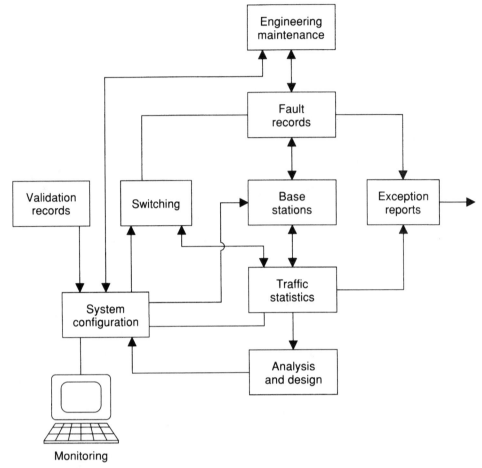

Figure 26.5 Block diagram of some of the engineering functions an MIS may track.

is best that the system designer know about it as early as possible.

Although most cellular systems come with some form of MIS, and all MIS systems come with a wide range of packages, remember that different cellular system providers will require different levels of information on different things. No two cellular systems are *exactly* the same, and a function that may be of minor importance to one operator could be vital to another.

Most MIS systems have been designed primarily for use in the United States or the UK. The relevance of some of their functionality outside these countries may be questionable. However, there is no reason for complacency even within the home countries, as the diversity of operations will still challenge the flexibility of most MISs.

The ultimate in integrated billing/management systems is found in the GSM billing system. GSM is the first system in which the billing is not simply an add-on but is structured into the network. The subscriber identity module (SIM) card production and authentication is central to the GSM concept, and this becomes part of

the network. The customer administration package, which includes lead tracking and fraud control, is directly linked to the mobile switching center (MSC).

The software is built in modules, each of which takes care of certain functions. The modules are usually physically separate code blocks, and it is possible to purchase just the structure required. Usually the code is written in 4GL (fourth-generation language—as opposed to third-generation, such as Pascal or Cobol), and is designed to run on commercially available mainframe/minicomputers. Extensive use is made of relational databases.

WHY THE PROCESSORS NEED TO BE SO BIG

A billing system is essentially a massive relational database. A call record will consist of the calling number, call-length details, and called-party records. On average, each subscriber will make around six calls per day. A record may look like this:

```
786666   11.45.33 097 3334455 11.55.06   45,43,S
^          ^         ^          ^          ^ ^ ^
caller    time     called no. time end   bas stns
```

So if there are one million subscribers, the call records for one day will consist of six million records, each about 100 bytes long. That makes a total data size of 600 MB per day. These can first be read and compressed to maybe one-third that size, so that the record length is 200 MB bytes per day.

It is usual that the records are processed for billing on a monthly basis. Assuming there are 20 working days in the month (weekends do not contribute much to the total call record count), then the billing cycle will be processing 200 MB × 20, or 8000 MB. The only effective way to sort this out is in random-access memory (RAM). If a hard disk is used, it will be very slow and the disk won't last very long, as it will be accessed at a rate that will ensure its early demise.

Of course, the total database size can be reduced by processing less than one month at a time and then merging the results. The general principle remains, however, that as the subscriber base grows, so the record size grows linearly, but the complexity of sorting the larger file grows even faster.

BILLING CHARGES

Towards the end of the 1990s it became clear that long-distance call charges were rapidly becoming cheaper, and the diversity of operators made the resolution of distance based call charges most difficult. As a result by 2000, even physically very large countries like the United States, France, and Australia, had mostly opted for call rates which were fixed nationwide. At the same time, in those countries nationwide long distance wireline rates had fallen to about 15 cents per minute (or less when special deals are on offer).

27

MARKETING

The ultimate goal of a cellular radio system is to sell or rent the end product (a mobile phone) to the customer. Marketing is an integral part of any successful business, and in cellular radio it must be coordinated with all other aspects of the business. Marketing should not be confused with sales. Marketing is the means by which the demand is estimated, the target customer is identified, and the sales strategy (advertising, pricing, and incentives) is determined. The marketing department should also set standards for customer presentation, interaction, and service standards. More than with most products, there is a need in cellular radio for the technical, marketing, and sales staff to establish clear lines of communication.

CUSTOMER DISILLUSIONMENT

Provided the system has been well thought out, most customers are prepared to accept the known limitations of the system. They become very disillusioned, however, if they have been promised more than can be delivered. Some common areas of potential disillusionment are coverage, call charges, contracts, and complex plans.

Coverage should be clearly defined before the sale is made. In particular, any doubt about coverage in areas considered important to the potential purchaser should be clarified. Sometimes a field test is necessary.

Each salesperson should do the following:

- Provide a reasonably definitive coverage map
- Define as accurately as possible the difference between the coverage area for handheld and "in-vehicle" mobiles, especially for rural areas, where this is important

- Indicate that radio systems will always have some areas of poor coverage and that marginal areas may work one day and not the next. This is especially true for CDMA.

Most customers have only the vaguest idea of how a radio-telephone system works. Unfortunately, some salespeople contribute to this lack of information. For example, customers often have difficulty with the concept of a limited coverage area, reasoning that if the area covered is, say, 40 km from the center of the city, then calls can only be made to or from fixed subscribers in the same zone.

Most customers initially expect nationwide coverage and are a little disappointed when they first realize that it is not generally available (a few smaller countries may have this kind of coverage). In addition, call charges are often confusing to the customer (and this is often deliberate on the part of the marketing department). Usually, mobile and rental call charges are much higher than the normal fixed network. Some customers have difficulty accepting these higher rates, and they will have even more difficulty if they first become aware of them with their first bill.

WHAT CUSTOMERS WANT

It has been well established from marketing surveys that customers look for the following five things from the cellular operator (listed in order of importance):

- Coverage reliability
- System reliability
- Price

- Service
- System performance

In certain local areas, the relative importance of these factors can change.

Coverage includes both the extent (what areas are covered) and quality (how well the area is covered). From many surveys, this is clearly the most significant factor. Operators naturally want to cover the greater part of major cities before expanding into more distant rural areas. Operators usually decide (quite correctly) that the demand is mostly in larger cities and expansion brings only marginal returns. What is often not clearly understood (and is difficult to quantify) is that the city customer's decision to buy can be quite strongly influenced by the total service area available (including the area outside the main city), even though in practice the customer may rarely use the extended coverage.

Experience has shown that, when extended coverage is provided (for example, covering the roads between major cities), the local buy rate is generally low, and the transit traffic does not seem to indicate that the extended coverage is heavily used. Despite this, there is considerable evidence that the availability of extended coverage enhances customers' perception of the quality of the service offered even if they rarely use the coverage.

There will, of course, be isolated low-density regions with heavy traffic, particularly when a smaller town is a satellite to a larger one and is interdependent for trade, or where there is a good deal of two-way traffic between the towns and local traffic may be high. Knowing the importance that customers attach to coverage, the operator may be tempted to overstate the coverage, particularly by failing to mention areas of marginal coverage. This "oversight" often leads to disappointed and disillusioned customers.

It is probably better to understate rather than to overstate coverage, so that if marginal coverage is available, customers can "discover it" for themselves; they may then feel as though they've received a "bonus." Calls from elated customers describing good-quality coverage from areas up to 50 km away (on analog and CDMA systems) from published boundaries are often received. These calls are usually made from hilltops or at the coast where an overwater path to the base station is available. Nevertheless, customers seem to feel that they have received more than they paid for and are very satisfied.

Poor and patchy coverage is often a source of customer complaint. This can occur because of poor design (base-station location), failure of the operator to provide an adequate number of base stations, or system overload or excessive blocking. In each case, the result is the same: noisy channels on which it can be difficult or impossible to hold a normal conversation. These areas, when they are within the stated coverage area, should be clearly identified (in the form of a map); and the customer should be aware of these areas before a system is sold.

The price sensitivity of the sales of cellular mobiles is fairly well established and upward discontinuities in sales graphs corresponding to drops in purchase price are a common phenomenon.

When systems approach their capacity, call dropouts, the frequency of cross talk, blocking, and failed handoffs increase, and the result is a degradation of service. Poorly designed and maintained systems suffer the most from this malady, but all systems with frequency reuse will have some problems. Customers will, then, experience a gradual but continuous degradation of the quality of service as the systems grow. As a rule, the bigger the system, the worse the service.

A CUSTOMER PROFILE

In the early days, the cellular customer was typically a high-income earner, who probably worked for a small-to-medium-size company with fewer than 100 employees. In a 1990 survey by *Cellular Marketing*, it was found that 77 percent of US cellular users had an income of $50,000 or more, while 34 percent had incomes above $100,000.

By 1993 things had changed. More than 50 percent of calls were private, but a high 9 percent used a computer interface for laptop access.

In 2000, in most Western countries, cellular penetration has passed 50 percent so that the average cellular customer is precisely the average person in the street.

CUSTOMER CHURN

Customer churn (cancellation of service or migration to a competitive network) has become a serious factor, particularly in countries that offer incentives for users to sign up. Some customers sign up only for the period of the incentive and then look for new offers. Behind this churn there is usually an enthusiastic sales rep who gains yet another activation commission. Throughout the United States the average annual churn rate is about 30 percent (20 percent migration, 10 percent cessation of service), although in some regions it is much higher. With each new customer costing about $300, churn is very expensive and surprisingly little is being done to stem it.

In the United States around 30 percent of customers have switched service providers at least once for reasons other than relocation. The reasons given for the switch are the same as those that influenced the original choice of supplier (see "What Customers Want" earlier in this chapter).

A concept that helps to visualize churn is lifetime value (LTV), which is the measure of the average time that a customer stays on the system. It can be easily calculated from the average churn as:

$$LTV = 1/(\text{monthly churn } (\%) \times 0.12)$$

Using this formula it can be seen that the US churn rate of 2.5 percent per month is 3.3 years LTV.

The concept of LTV is most significant for systems where the connection of a subscriber is subsidized, as is the case in the UK and the United States.

INITIAL MARKETING SURVEYS

Most cellular-radio operators initially commission some type of marketing survey to estimate demand and to identify customer profiles. Nearly all such surveys have been spectacularly unsuccessful. Usually, very little correlation is found between demand for cellular phones and other telecommunications equipment already in service (telephones, private automatic branch exchange (PABXs), facsimile (Fax) machines, and pagers).

Generally, almost no relationship is found between the answers given to market survey questions and public demand for the product. This discrepancy is not surprising when the questions asked are examined (they usually indicate a limited understanding on the part of the market researcher). Furthermore, the answers usually indicate a very vague understanding of the question by the customer.

Today, it is usually unnecessary to spend large amounts of money on futile market research because most big cities have an existing market from which the necessary projections can be determined. If no market exists, cities elsewhere, preferably in the same country or in those with a similar gross domestic product (GDP), can be used as benchmarks. Although this approach may not be very scientific, it is much cheaper and likely to produce results at least as good as market surveys. (Note, however, that expensive, beautifully presented feasibility and marketing studies still tend to impress bankers and financiers, and so may still serve a purpose.)

Cities with populations larger than 100,000 are generally considered capable of supporting multiple operators. Use caution when interpolating the results of big city sales to small ones, however, even when they have the same per capita income. There is evidence that small cities have a lower per capita demand rate than larger ones, and that for relatively small cities (less than 100,000), this demand rate can be very small. As a rough guide, the experience in Australia is that cities above 1 million have approximately a constant density of traffic (in Erlangs) per population, but this traffic decreases rapidly for populations below 1 million, as seen in Figure 27.1 Care should be taken for small cities that are

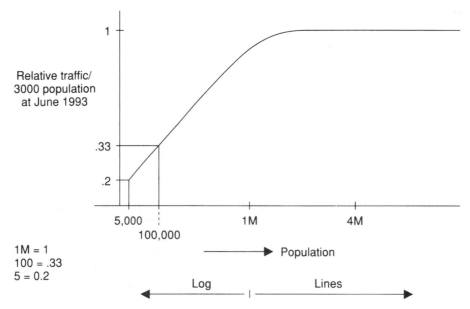

Figure 27.1 The relative traffic in Erlangs/3000 population for Australian cities of various sizes.

tourist centers or junction towns on main highways, which may have exceptionally large roaming traffic. Conversely, towns that are isolated will generally have below-average traffic.

The situation becomes more complex when assessing the potential of developing and underdeveloped countries. The initial reaction was to assume that, because the landline network was (generally) quite bad, business would flock to cellular and that the calling rates would be correspondingly high. This does not appear to be the case, and calling rates in developing countries, at least, are similar to those in developed countries. There is also no strong evidence that there is any trend to regard cellular phones as a serious alternative to the fixed network. It does appear that in developing countries it is still the business/official user who makes up most of the market, and that usage is similar in nature to usage in developed countries, but the profile lags by about 5–10 years.

MOBILE TERMINAL POLICY

The cellular operator can adopt one of these two distinct mobile terminal policies: a monopoly or limited monopoly (the network provider basically controls the mobile terminal market), and an open market.

The operator may not be able to choose which policy to follow, as government legislation may dictate the choice. There are some essential differences between these two policies. These differences are examined in the following sections.

Monopoly

By choosing a monopoly, the following may occur:

- Capital investment is much higher, as the investment in mobiles for a cellular system is about equal to the network investment. Also, all advertising and promotion is now basically with one party.
- The connection rate is lower; regardless of the determination of the supplier, a monopoly can never achieve figures as high as the free market (this also means that network revenue is lower)
- The number of models offered is limited. This is because the operator can get a good deal on the mobiles provided the quantities can be kept up. A reasonable order that can be very cost-effective is about 50,000 units.

- The network supplier can control or tailor the demand as necessary.

Often, in small markets with small sales volumes, a monopoly is the cheapest way to provide a service. Because the cellular market is relatively lucrative, any monopoly will be difficult to keep indefinitely, so marketing plans should include the contingency of ultimately having that monopoly broken.

Open Market

An open-market policy normally reduces the average sales price of a mobile unit and stimulates demand. In small markets, however, the large diversity of models available (there are dozens of manufacturers with an average of three models each) may mean that the benefits of bulk purchase cannot be realized and that costs actually rise as a consequence.

Results of implementing an open-market policy are as follows:

- Greater sales volume and more network revenue due to the cumulative individual efforts of a large number of relatively small, competitive suppliers
- The ability to support a very diverse number of models
- Inability of the network supplier to regulate demand (which may be due to network constraints)
- A large diversity in installation and customer awareness standards
- A need to validate blocks of numbers for each supplier (or alternatively have on-line validation)
- Margins on mobiles are often low

It is generally a good idea for network operators also to have an in-house sales section so that they can get direct customer feedback and also be familiar with installation practices. In some major markets, however, including the United States and UK, network operators are not permitted to be directly involved in the supply of mobiles.

An open market can result in very low profitability, with fierce competition between suppliers. There is little or nothing that the network supplier can do to prevent such competition. The network supplier can contribute to the problem by competing at very low margins, knowing that losses will be recouped on the call charges and rentals or through incentives offered to mobile sales groups. Below-cost mobile phones are now common, as the retailers have become dependent on "term contracts" to be profitable.

SAFETY

In some countries, it is illegal to use a mobile radio or car phone while driving. Only handsfree units can be used legally while driving. In these countries, the driver's insurance can be revoked if an accident occurs while using a handset, regardless of whether the driver was at fault. Studies have shown that using a handset while driving measurably increases stress on the driver, whereas using a handsfree option or a boom microphone causes little stress. (For more details, see Chapter 50 "Safety Issues.")

Dialing requires a high degree of concentration and is probably the one action most likely to cause an accident. The driver normally needs to look away from the road to dial. Dialing is done most safely when the vehicle is stationary or by using the telephone's memory. Voice-activated dialing (using voice recognition to associate a name with the required number) is an elaborate way to overcome the hazard of dialing while in motion. Positioning the keyboard so that it can be seen without looking away from the road can also help. But this is usually difficult to achieve. It is in the interest of the operator to bring these safety matters to the attention of the customer.

DISTRIBUTION OF MOBILE UNITS

The early trend in the cellular market showed that the biggest portion of sales came from the direct-sales sector. As customer awareness grew, so did indirect sales through such outlets as electronics and autosound specialists. This trend is increasing significantly.

Auto accessory dealers, department stores, computer stores, direct-mail sellers, mass merchandisers, and customized sellers have now availed themselves of this opportunity. Together with this diversification, the markets are forecast to stabilize while prices continue to drop.

Special attention should be paid to major customers (corporate and fleet sales), and frequently an "account executive" and experienced salaried staff member will be directly assigned to these customers.

Probably the best source of marketing information available to an operator can be extracted from the company's own files.

Mobile Prices

The wholesale price of mobile phones has been decreasing in real terms over the years, but in recent time seems to be stabilizing around $100.

Sales Promotions

There are in most countries many competing suppliers of mobile telephones. Price cutting is inevitable in such an environment, with many operators even selling below cost in order to get customers into their systems.

Bargain-basement mobiles are sometimes available. These mobiles normally are very basic and offer few extras. Some of these extras, which are not really optional, such as hands-free and a warranty, must be paid for at the time of purchase.

Ultimately there are no beneficiaries of a price war; non-wireline operators, in particular, should be aware that against the resources of the wireline operators, they are particularly vulnerable to chronic cash-flow shortages during such a price war.

End-of-model sales can result in temporary bargain prices, but because the main advantage of the newer models is usually that they are smaller and cheaper to produce, these "sales" are short-lived.

EMERGENCY USERS

Local emergencies and disasters can often be handled more effectively if good communications are in place. By moving quickly at such times and offering an adequate supply of free handhelds, the cellular operator can be seen as a good corporate citizen, while graphically (and publicly) illustrating the value of the cellular product to the community.

Telstra has a service rate that is essentially only for emergency use. The monthly fee is $7.00 ($10.00 Australian), but call charges for airtime are about double the normal rate.

Advertising for the service concentrates on car breakdowns, women traveling late at night, and long-distance travelers, as all the main highways are quite well covered. Figure 27.2 shows a billboard promoting this service.

CHARGES

Charging packages are usually fairly complex and vary considerably from operator to operator both in magnitude and type. From first-hand admissions of major companies, I can affirm the charging scheme complexity is deliberate and designed to be confusing to the customer.

Figure 27.2 A billboard for Telecom Australia's emergency service.

Connection Fee

The connection fee covers the cost of validating a new user on the system and the associated paper work. Typically it is about $50.

Special-Number Charge

Some operators make special numbers available at a premium. Particularly in countries like Hong Kong, where people are somewhat superstitious, numbers like 888 888 can fetch a very high price. The cost and inconvenience to the operator of reserving and providing such numbers is considerable. Most operators either have a policy not to provide such numbers or to do so only at a considerable fee ($100 or more). Special numbers can be considered an additional resource, so it is well worth considering selling them.

Special-Facilities Fees

There are many special facilities available on the average cellular switch, and some operators capitalize on this by charging a fee (either once only or recurring) for fea-

tures such as Call Forwarding, Call Waiting, Conference Calling, Voice Mail, and Call Restriction. The charges levied for these features are usually modest.

Access Fee

The access fee is charged monthly for access to the system. The fee ranges from about $10 to $60 per month.

Connect-Time Charges

Connect-time charges are a gray area, where the number of possibilities is limited only by the imagination of the marketing department. The first major distinction between connect charges is whether the call is charged with airtime added (as in the United States), or only the call is charged but at a premium rate (as in most other countries). The second method is self-explanatory, but the airtime charge needs elaboration.

In general, the airtime charge is based on the user-pays concept. That is, as soon as the mobile begins to occupy a channel, the charging starts. The channel is occupied as soon as the customer presses the send button and continues until the customer hangs up the phone or presses the END button. A variation on this approach is to commence charging when the called party answers. The actual call is charged separately, usually at the same rates as a land-line telephone, so the subscriber is billed separately for call charges and airtime. This scheme is often used when the operator is not a wireline carrier. These two connect-time charging methods can, of course, also be applied when a land subscriber calls a mobile telephone.

A philosophical difference does occur, even here. In the United States, the mobile subscriber pays mostly for a call whether or not the subscriber originates a call or receives one. In many other countries, particularly where only a premium call charge is applied, incoming calls to a mobile are paid (at the premium rate) by the caller. Mobiles using this scheme usually have an area code prefix (a prefix beginning with 0 or 1) so that the caller would normally expect the call to be a "trunk-rate" call.

Special-Package Fees

The line of reasoning discussed in the preceding section suggests that the low-usage customer is the one producing the greatest return on investment and therefore is the one to be encouraged. (This reasoning, of course, does not usually apply to wireline customers.) For example, imagine an operator who has 1000 customers who never use the system. They pay their access fees (the lowest one), for example, $20 per month. The revenue from these customers is thus $240,000 per year and their cost to the operator is billing only; because they do not use the infrastructure (base stations, switch, and so on), these customers represent zero capital costs. These are indeed prime customers! Unfortunately, they are also very rare.

Activation Fees

Where more than one operator is working in a given area, the competition can become fierce. One of the less desirable outcomes of this competition is the introduction of activation fees. The network operators in effect pay the resellers for each customer. The level of these fees, originally set at around $50 is sometimes much higher, which means that mobiles are available at less than cost, and at times are even free (subject to signing up for a specified time of service).

By mid-1990 US activation fees had reached the $200–$250 level. They are the main source of income for most resellers, as the margins on the mobiles have been cut below the level of profitability.

This has led to a high rate of churn, as customer loyalty is overruled by the desire to get the best bargains. Many a customer, who will not buy at $300, but is attracted by $99 or even $100 bargains, will lose interest at the sight of the first bill. Bad debts have risen in indirect proportion to the start-up costs. The bargain-basement customer often turns out to be a low-usage one, which is not necessarily bad (see the preceding section, "Special-Package Fees").

Although most carriers currently rely heavily on indirect activation for the bulk of their customers, the high activation fees are causing many to look more closely at direct sales.

LOOKING AT THE FUTURE

Fixed and mobile convergence (FMC) remains a goal for many operators. Some of the development in 3G is consistent with this aim, but for the most part it ignores the poor net data rate performance of mobile communications, and fails to take account of the enormous progress made on the wireline front.

On the competitive front, things have changed dramatically in the last decade and will continue to do so. The Ernst & Young Communications Competitiveness Indicator shows that the countries that account for 85% of the world's GDP have already moved to competitive market choice for telecommunications services. A quote from Telstra's Chairman, Bob Mansfield, "As the regulatory environment opens up to ensure the growth and development of the industry, investment and service become the critical drivers to delivering customers the service they demand if a company is to compete for market share".

The ten critical challenges that were derived by Ernest&Young interviews with CEOs across the industry and across the globe are:

1. Gaining access to new markets: "Any company will need a global footprint to be successful."
2. Developing successful alliance strategies: "No one has all the skills and capital they need by themselves."
3. Optimizing value extraction from the total customer pool: "We don't have to think in product margins but in customer margins."
4. Redefining brand promise: "The key to success will be whoever creates brand value, like 'Intel Inside'."
5. Driving regulatory reform to promote competition: "Governments have been a barrier to the dawning of the new information age. Instead of constraining the industry, governments need to facilitate the entry of this new age."

6. Creating a truly agile organisation: "The capacity to implement changes in a heavy and complex company evolving in an unstable environment will be our key future success factor."

7. Accelerating new product development: "Success will be based on the ability to continually invent new and innovative competitive products and services."

8. Leveraging OSS for competitive advantage: "When we come with a new idea for an offering, our systems guy says, 'we can't do that'. This is a very big problem for us, because it holds us back."

9. Developing innovative ways to capture and reuse employee knowledge: "To empower employees, we need to empower them with knowledge."

10. Effectively managing assets: "If you are a facilities company, your excess capacity is your enemy."

28

FRAUD

Fraudulent use of mobile networks has become a massive, and growing problem worldwide. Fraud can be defined as access or usage of the network with the intent of not paying for the service accessed. It can be either external or internal to the operator's network, and often involves both.

THE NATURE OF FRAUD

Fraud is something quite different from bad debts. A bad debtor is one who intends to pay, but through financial hardship, is unable to do so. Genuine bad debtors are relatively rare and represent only a small cost to the operator. Fraudulent users, on the other hand, are deliberately out to use the system. Most expect their operations to eventually be detected and disabled, and so they will use airtime frantically. Many have "commercial" operations going, whereby they resell access to third parties; in this case, the phones are running hot. When fraudulent users are detected and disabled, it is usually not long before they are back again with a new access. The cost to the operator of these people is immense, and they should be pursued with vigor.

Often the same groups are perpetrating the fraud, and they get skilled at beating the system. For example, if the operator particularly looks for "irregular use" on the first billing cycle, the fraudulent user may well play the role of a model subscriber for the qualifying period, and then change habits at a later date.

Digital phones have made fraud harder, but they certainly have not eliminated the problem. As the price of phones has fallen—and in virtually every country now they are a common sight—so the range of people involved has widened.

Fraud will increase network costs in a variety of ways. The fraudulent traffic will increase cell loading, switching resources, and trunks. The net increase in network costs will be approximately in proportion to the percentage of fraudulent traffic. As most cellular operators are not long-haul or international carriers, they will bear the full brunt of unpaid toll calls. It should also be noted that fraudulent users access toll routes disproportionately to legitimate users (this is in fact one way that they can be detected). A good deal of operator resources will need to be directed to the control of fraud, and this is another cost that has to be borne.

The problem in the United States is of huge proportions; fraudulent calls make up from 5 percent to 30 percent of all calls made (the actual rate being very much network-dependent). It is not only the magnitude, but the rate of growth of fraud that is worrisome. The actual degree of fraud varies widely from operator to operator. Outside the United States the problem is still worrisome, but the extent in most countries is reduced by the smaller number of operators and the availability of nationwide roaming, which naturally implies a nationwide subscriber database that is instantly accessible. However, in some countries it is a very serious problem. The UK has for many years been struggling with a growing fraud problem. In the Philippines it is estimated that as many as 35 percent of calls have been made fraudulently, and at least one dealer has been arrested and charged with fraudulent usage amounting to $5 million. Pakistan and Indonesia also have pervasive fraudulent problems. Outside the United States, in general, the spread of fraud ranges from around 5 percent to 35 percent.

No operator is immune from fraud, and although it cannot be entirely prevented, it can be controlled. Every operator needs to examine all the ways in which fraud can be perpetrated and to do what is practical to minimize it.

What makes fraud attractive is the fact that it is cheap to do, the difficulty that the operators have in locating a misused mobile, and that the legal system seems to be unable to deal effectively with it. Operators are partly to blame, as they adopt a "head in the sand" attitude to fraud, and try to pretend that it doesn't exist.

It is interesting to speculate about what it is that really makes fraud attractive. Mostly it seems to be the lure of getting "free" international calls. In turn, it is the artificially high tariffs charged by some international gateway operators that is driving the tariff avoidance industry. When international calls are charged on a cost-plus basis there is little incentive to avoid them (as has happened dramatically in the late 1990s).

In the United States the increasing use of IS-41 is stabilizing a situation that was virtually out of control. Once IS-41 is universally available and intelligent fraud control monitoring is in place, much of the problem as it is known today will be solved. Popular electronics magazines carry advertisements for thinly disguised instruction manuals on how to make a clone or tumbler phone. As new controls are put in place, new fraudulent methods appear.

It is believed that fraud in the United States now actively involves organized crime syndicates, which have industry people on the payrolls to keep them ahead of the controls. This "industry" is believed to have its own engineers to design equipment, manufacturers to make it, and a highly organized distribution network. With around half of the world's analog mobile phones, all of which are highly vulnerable to fraud, the United States contends with a massive problem.

CATEGORIES OF FRAUD

Fraud falls into seven categories:

Insider operations

Use of a stolen phone

Roaming

Subscription (fake applications)

Tumbling electronic serial numbers (ESNs)

Cloning

Misuse of operator ESN mobile identification number (MIN) records

The fraudulent user falls into two categories, personal or commercial use.

At the time of writing, analog systems, particularly in the United States and parts of Asia, were experiencing all types of fraud, whereas elsewhere subscription and cloning were the main problems. Digital systems are susceptible primarily to subscription and roaming fraud. They are also very susceptible to the first three categories listed above. In developing countries, subscription fraud is particularly rampant on the Global system for Mobile Communications (GSM).

Subscription fraud is the most straightforward category, but in some ways one of the most difficult to contain. If a user has the intention of running up a massive bill, then a small deposit will not be an impediment. Often competitive pressures reduce the initial deposit to token amounts. The user simply produces false ID, signs up, and then proceeds to make very heavy use of the phone with no intention of paying the bill. Automated credit checks will help reduce this kind of fraud, but will not eliminate it.

Tumbling ESNs work only on analog systems with unchecked roaming. This method involves taking advantage of the "post-first-call" ESN check, which allows the first call of a roamer to be made unhindered, while initiating a check that will decide the fate of future calls. A mobile that has been modified to present a new ESN each time a call is made will always be classified as a first-time user and so the call will be permitted. A little intelligence at the switch could eliminate invalid codes. With more intelligence it would be possible to look for patterns like multiple attempts by "new" roamers on one base station (or a local group of base stations).

Tumbling ESNs have become such a problem that some carriers have chosen to deny access to any unknown roamer. This approach significantly reduces fraud but also disadvantages the customer. GTE has a FraudManager service, which provides positive validation and analyzes for fraudulent traffic. FraudManager can validate a call in seconds and thus prevent illegal calls from maturing. In 1992, GTE estimated that 15 percent of roamer call attempts were fraudulent, with the rate being as high as 25 percent in some markets. However, once FraudManager had been installed, illegal attempts dropped to about one-third in the first six months. By far the majority of fraudulent attempts detected in the United States on analog systems are tumbling ESNs.

Cloning is more universal and very difficult to combat. Cloning can be used to get multiple telephones working on one number to avoid the monthly fee. This is quite common in Asia, particularly in Indonesia where government restrictions (tariffs) on handhelds tempt users to register a vehicle-mounted unit and then connect a cloned handheld or handhelds on the same number.

More sinister cloning occurs when users obtain someone else's ESN and number and proceed to make calls. There are two types of clone phones. The first, known as causal, involves a phone that is a clone of a valid number. It can continue to be used only until that number is disabled by the network. The second kind is used by those who habitually do this. They will have a phone that can be programmed to imitate any ESN and they will have a list of valid ESNs/MINs, which may have come from a supplier, dealer, or off-air. As each number used is identified and disabled, a new one is activated. This type of user is common worldwide.

Clone phones and those with tumbling ESNs first appeared in the late 1980s. At the time, they sold for around $5000, but in recent times they have become available at prices that are only a few dollars above the market price of a regular phone. They are rented out for unlimited international use for a typical charge of $100.00 to $250.00 per night in the United States, and as little as a few dollars per call (any destination) in some Asian countries. This activity is engaged in by some for a living and is called "call-sell." Clone phones are also available on a monthly rental basis with a "free" fix if the number is barred.

Sometimes the source of the cloned ESNs/MINs is the operator's own database. This can be accessed directly by an employee (who may or may not have authorized access) or from computer hackers who can get into the database.

Cloning is not a problem on digital networks, as the networks will not permit multiple users on the same number.

Insider operations can involve the collusion of one or more departments. Where subscriber validation is available from both the switch and the billing system, it is possible to validate users that are not seen by the billing system. This is especially easy when the billing software is an in-house system that is not tamperproof.

The commercial user of the illegal phones is the most sophisticated and by far the most expensive type of fraud. A typical operation involves a fraudulent user in one country working with an associate in a second country (usually a third world country). The accomplice establishes a "calling booth" consisting of a land-line telephone and a sign advising that cheap long-distance calls are available. When a call is required the accomplice rings the cellular phone (which need not even answer, as the ring could signify "ring back"). The mobile phone then contacts the accomplice, who gives his "customers" call details. The mobile phone then puts the accomplice on hold, calls the third party, and establishes a party call (see Figure 28.1). The accomplice collects and so the cycle is complete. This type of operation is thought to net hundreds of thousands of dollars, based on those caught to date.

To perpetrate this fraud the mobile phone requires international dialing and 3-way conferencing. Limiting access to conference calling should be considered, and

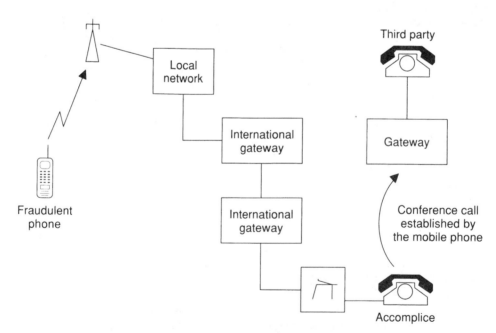

Figure 28.1 The path of a commercial fraudulent call.

calls established via a "post-first call" basis can be restricted, either in access or table time or area. New customers wanting international access plus 3-way conferencing should be suspect.

ROAMING FRAUD

Both within countries (like the United States) and between countries, roaming is becoming increasingly popular as a means of extending the effective coverage of an individual operator. In either case, the total number of potential roaming customers is very large. The size of the customer base presents a major problem of validation.

For simplicity, early roaming agreements between operators allowed all potentially valid subscribers from one operator's system to have access in the other operator's system. This method quickly led to abuses. For example, a mobile dealer could illegally provide a mobile programmed to appear as a valid roaming mobile and could then use the mobile until detected. And detection often took four to seven weeks. Once detected, the ESN could be registered as invalid and details of that number could be entered into a negative file (a "black list" of invalid users).

As a natural progression, particularly in the United States where the opportunity for abuse was greatest, the concept of a negative file was introduced as part of the normal system files. This file contained the serial numbers of all known invalid mobiles from other operators with whom roaming agreements were held.

SPECIAL PRECAUTIONS IN THE UNITED STATES

As the world capital for telecommunications fraud, it is natural that the United States has taken a especially strong stance against this crime. In the United States it is estimated that telecommunications fraud amounts to $1.2 billion per year. Computer fraud heads the list, closely followed by cellular. The Secret Service was originally brought in to combat this problem, but before a case could be brought to the Secret Service the amount of fraud had to be more than $1000. However, the role of the Secret Service effectively ended in 1992, when a district court judge ruled that the Secret Service was not authorized to operate on cases involving changes of ESN. In a case of fine distinction, the judge ruled that although personal identification numbers (PINs) and private access codes fall within the federal arena, ESNs do not.

The Cellular Telecommunications Industry Association (CTIA) has formed a special fraud task force to combat fraud and has been successful to the extent that by 1996 fraud had been contained to about 4–5 percent of calls.

REDUCING FRAUD

Clearly, some form of positive validation is the most effective way of reducing roamer fraud. If negative files are used, they should be regularly updated with some daily analysis. Because fraudulent roamers make about twice as many calls as legitimate users, flagging all high-usage roamers and checking their validity on a daily basis can be worthwhile. Placing limits on toll calls for roamers can also reduce the risk considerably, but at the same time, it reduces the value of roaming to the customer.

PIN NUMBERS

PINs, which work in a way similar to credit card PINs, can be issued to phone users as a fraud-prevention measure that works in two ways. First, it can prevent unauthorized use of a phone; this can be done by making the PIN number integral with the phone's software. Second, and particularly for analog systems, the PIN number can be used to verify the user for access. Having to enter PIN numbers can be annoying for the customers, and in the case of analog networks the PIN can still be read off-air.

Voice Printing

Although PIN numbers have been used extensively for fraud minimization, they have been found to be vulnerable to hackers (on analog networks) and inconvenient for customers. A novel alternative is the SpeakEZ system from T-NETIX Inc., which takes a voiceprint of the registered phone user(s) and relates it to the phone identity. When the user places a call, it will be necessary to give a pre-trained password in order to gain network access. It is claimed that the whole process, including the voice prompts, only takes about 5 seconds, with the authentication process (which is done on a neural network) taking only 300–500 milliseconds.

The neural network will give a weighting for the voiceprint match, and the network operator can set the parameters for how close that match needs to be for the call to be declared valid.

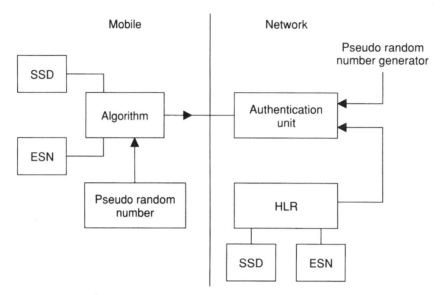

Figure 28.2 The shared secret data concept.

FRAUDBUSTER

In 1993 Computer Science Corporation (CSC) announced its "Fraudbuster," an artificial intelligence system that examines customers' call records in real time, learns the calling patterns, and alerts the operator to unusual activity. A built-in biasing system enables Fraudbuster to be particularly sensitive to certain patterns that the operator may wish to define.

Caller behavior profiling can never be infallible (customers for various reasons may legitimately change their calling patterns, either temporarily or permanently). Detection necessarily occurs after the event, and it can be quite labor intensive, both monitoring the analyzed calls and following up with the customers. Poor customer relations might also occur should a customer be wrongly selected as a fraudulent user and denied access. Effectiveness, however, is often said to be around 90 percent.

SHARED SECRET DATA

The security weakness of analog cellular is due largely to the use of a fixed security number (the ESN), which can be readily copied, read off-air, and so cloned. The shared secret data (SSD) concept increases security around 100-fold. As shown in Figure 28.2, the SSD is a code that is contained in the mobile and is known by the Home Location Register (HLR). These data are never sent to air, but are mixed with the ESN and a pseudorandom number. What is sent to air is an amalgam of these three numbers, so that the actual ESN and SSD are never sent.

The pseudorandom number is one that can be generated by the mobile and authentication unit, but would not be available to an eavesdropper. It could be, for example, that a call counter is installed that counts the calls made by the mobile and uses this number as the randomizer. With this system, a clone would soon get out of phase with the mobile and would cease to work. Other similar randomizers include the "time of the last call in minutes and seconds," "the last call duration," or even last number called. Illegal attempts would be detected by a high error rate in the authentication process.

GSM sends a random key to the mobile on first contact and relies heavily on a complex algorithm for security. With this system, however, once the ESN and encryption key are known (and this is said to be difficult to obtain), a fraudulent user can make calls without ringing alarm bells at the switch because nothing will be abnormal about the call.

Experience to date has shown that the SSD concept applied to a good-quality encryption algorithm will virtually eliminate cloning fraud, as has been demonstrated by the GSM experience to date. A version of this known as A-Key Authentication (part of IS-41) can be used on all cellular systems including AMPS, Narrow-band Analog Mobile Phone Service (NAMPS), time division multiple access (TDMA), and code division multiple access (CDMA).

RADIO FREQUENCY FINGERPRINTING

Yet another tool against fraud is embodied in CELL-SCOPE, an off-air monitor that was originally meant to be used for cell site and mobile maintenance. CELL-SCOPE can examine a call from a mobile and store its profile for future reference. The signal is processed by a digital signal processor (DSP) and stored along with the ESN and MIN. When subsequent calls are made, these three identifiers can be compared. A clone will show up with the same ESN and MIN but a different signature.

Characteristics of a mobile that could be used to "fingerprint" it include Supervisory Audio Tones (SAT) tone frequency, SAT tone deviation, maximum deviation, frequency error, supervisory frequency, and supervisory tone deviation.

Radio frequency (RF) fingerprints are not only unique, but they cannot be replicated, so this technique is also immune to internal fraud.

Corsair's PhonePrint uses signal-processing techniques originally designed for intelligence applications to fingerprint the phones in the network. Cloned phones can be detected quickly and shut down. A roaming facility called PhonePrint Roaming Network allows carriers to share RF fingerprint information, to prevent roaming fraud. It is claimed that this roaming facility has achieved up to a 95 percent reduction in cloning.

On the downside, RF fingerprinting is not interoperable between different vendors, and it is not applicable to digital systems.

FRAUD IN OTHER COUNTRIES

In the Philippines phones are cloned for a few hundred dollars with the price including a free recloning if required (that is, if the network operator detects the clone and invalidates it). Because of the scarcity of landline phones, operators of clone phones can make a living out of making the phone available to the local villagers for outgoing calls in a public call box operation. Calls are typically $1.00 each regardless of the length or destination.

In Indonesia, cloning was originally a way of getting around the artificially high price of phones. Mobile phones cost around $1000 each in Indonesia, the price indirectly subsidizing the air-time charges, which are below cost. Most people who could afford to pay that much for a phone were frequent travelers who soon learned that elsewhere in the region phones were much less expensive. Once a user had paid for a "legitimate" phone and had a number allocated, it seemed reasonable to buy a second or third phone in Singapore cloned onto the same number. That's how it started, but of course it soon got out of hand, with dealers using off-air phone and serial number readers to clone indiscriminately for profit.

In Hong Kong in 1993 a fraud ring was uncovered that had $1 million worth of ESN reading equipment.

FIGHTING BACK AGAINST FRAUD

In most countries, the most difficult problem is that the legal system does not have an adequate way to deal with offenders once they are caught. It may be necessary to engage legal advice and set about protracted lobbying to get the legistators to recognize the problem.

Fraud is a problem for all operators, and no matter how much they compete elsewhere in the market, this is an issue that should bring all operators together. It is not constructive to watch with satisfaction the agonies of a competitor facing heavy fraud losses, because one thing is for sure—it's only a matter of time before the criminal element will turn it attention elsewhere.

29

DATA OVER CELLULAR

Data facilities on cellular inherently will have less potential demand than the equivalent wireline service. With the increasing demand for highly portable handhelds, the idea of an associated, bulky, data-based accessory like a fax machine or PC means forgoing portability.

A major weakness in the 3G proposals for high data rates from mobiles is that along with these high data rates comes high BER. When error correction and resends are allowed for, it is unlikely that any of the current 3G proposals for high speed data over cellular can be viewed as being realistic.

Cellular is not the only medium that can provide mobile data facilities, and in some ways it is disadvantaged against the competition. Trunk radio systems and even some conventional mobile radios can make efficient use of packet switching techniques so that short bursts of data can be sent very efficiently. Most analog mobile systems require a time-consuming voice-channel setup procedure before even a very short message can be sent. Digital cellular and even Narrow-band Analog Mobile Phone Service (NAMPS) offer short message capabilities. Dedicated data networks exist in many countries, but they have not proven to be widely popular.

The advantage that the cellular systems have over their rivals are the universality of the service (note trunk radio systems are usually incompatible with regard to frequency, and rarely are nationwide) and the huge existing potential customer base. As a rule of thumb, information can be transmitted in the data format at 10 times the rate of voice communications, over a given bandwidth.

The efficiency of data transmission can be seen by a few examples. The complete works of Shakespeare amounts to about 5 megabytes, the works of Bach about the same, as is this book. To transmit 5 megabytes of 8-bit data over a standard Integrated Service Digital Network (ISDN) line at 64 kbits would take 10.5 minutes. Over a slower 9.6-kbit line that can be made available from most mobile digital systems, it would take 70–200 minutes depending on the BER. Not long for a lifetime's work!

Because of the harsh environment, transmitting data over a cellular system differs from transmitting data over the conventional telephone network in two ways. First, the transmission medium is much noisier; second, during handoffs, disruption of the data stream can occur for up to hundreds of milliseconds. As a result, a more elaborate code for error detection and correction is needed for cellular systems. This leads to a very significant reduction in net throughput.

ANALOG

Data are typically sent over analog media (such as telephone wires) via a modem. The word modem, comes from the MOdulator/DEModulator, which makes up the device. Modems come in different baud rates (or signaling rates), typically 9600 to 56,000 bits/sec. bauds. The rate at which data can be transmitted over cellular depends on the gross data rate, the error correction used, and the degree of data compression. A modem is simply a device that converts the ones and zeros that form the language of a computer into an analog counterpart that can be transmitted over an analog medium (see Figure 29.1). If the ones and zeros were sent directly down a telephone line or over a cellular phone, the distortion that would occur would make them unreadable at the far end. A simple modem uses a frequency shift keying

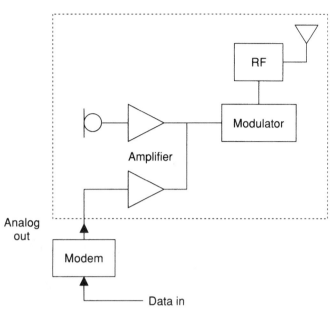

Figure 29.1 A modem.

(FSK) mode to generate two tones, one corresponding to a one and the other to a zero.

A number of special protocols have been developed for cellular radio. They are more robust than those used in land-line environments, reflecting the harsher noise conditions. Noise tolerance is particularly important for computer applications where virtually 100 percent accurate data recovery is essential. Fax applications, on the other hand, can provide acceptable service with much higher error rates. Suitable protocols include the Packetized Ensemble Protocol (PEP), used by the Telebit Cellblazer; Computer Data Link Control (CDLC), used by Millidyne; and the MNP-10 protocol, used by Microcom Microporte.

There is little difference between a wireline modem (of the kind used for internet connection) and a cellular modem. The primary difference is that the cellular modem must have the interfacing cables to the mobile phone, and it needs a protocol, designed to work in the noisy and discontinuous mobile channel environment.

A commonly used protocol is MNP-10 EC (Microcom Networking Protocol), which works on top of the modulation standards like V 22 bis, V 32 bis, and the more common V 42 bis. The MNP-10 protocol is designed to operate under conditions of signal fade, handoffs, interference, and fading that can be expected over a cellular channel. MNP-10 enhanced cellular (EC) is essentially the MNP-10 protocol with some enhancements at the physical layer. Under average conditions these protocols can deliver a 9600-b/s rate, and even 14,400 can sometimes be achieved.

The main cellular modem protocols are MNP-10 EC and enhanced throughput cellular (ETC), for data. Faxes use their own protocol, which cannot be "modified for cellular use."

INTERSTITIAL DATA NETWORKS FOR AMPS

A narrow band of spectrum between the AMPS channels (hence the term *interstitial*) can be used to transmit low-rate data over very wide areas. Using 3-kHz channels, the system was tested by GTE Mobilnet in Houston in 1991. Because the AMPS channels do not have a guardband, it is necessary that the data power be low. A maximum level of 200 milliwatts, which can be dynamically reduced as necessary, has been used.

In theory it is possible to provide one data channel for each voice channel, but initially at least it is proposed to have one data channel for each two 30-kHz channels. The system is packet-switched, with carrier sensing to avoid data collision.

The data network consists of an X.25 packet switch, which is connected to a cell-site control unit, at each cell site. The customer would then need a compatible terminal.

The obvious advantage of this data network over others is that it will have the same coverage as the cellular system and that it can be added to the service at a marginal cost of less than 10 percent of the cellular base equipment. Anticipated use includes checks on credit cards, security, monitoring, utility meter readings, and data logging via PC.

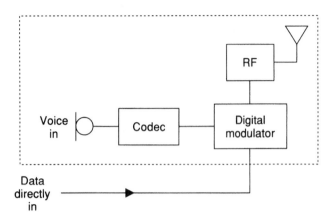

Figure 29.2   Digital modulator.

The Federal Communications Commission (FCC) permits cellular carriers to provide auxiliary services on the cellular network provided they can be shown not to interfere with the primary voice service.

DIGITAL

Digital systems can send data directly without a modem. The data are sent directly to the digital modulator (as shown in Figure 29.2), having bypassed the compression/decompression (CODEC).

Because errors can occur during transmission (due to noise pulses and non-linear transmission characteristics), a data transmission usually has some error-correction codes. These codes insert extra bits of information (redundant code) that have a definable mathematical relationship to the code bits containing the information. At the receiving end, the code bits are read, and then the mathematics done to check the integrity of the information received. The code is arranged so that not only can errors be detected but the original (correct) code can be reconstructed.

In a noisy environment like cellular radio, the error-correction codes need to be rather sophisticated, because there will be many errors. A re-send capacity is also needed for data blocks missed during handovers. For this reason, conventional land-line modems do not perform adequately on a cellular mobile.

Data losses occur during handoffs and local and multipath fades. These problems are not to be expected in land-line communications, and so purpose-built data protocols are the answer.

Data compression techniques are used to compress computer files and so increase the throughput. Compression ratios of 4:1 are available in some cases. It must be remembered, however, that any compression technique can only dispense with code redundancies.

This means that ASCII files, which are often filled with blanks, may be compressed by 4:1, but executable (EXE) and Component Object Model (COM) files, which are already relatively compressed, can only be further compressed to around 2:1.

The highest data rate that can reliably be expected from a cellular phone is 9600 baud in 1999, but much higher rates will potentially be available with 3G offerings.

PACKET SWITCHING

Packet-switched radio, as the name implies, sends packets of data in bursts, as opposed to a modem, which ties up a channel for the duration of the transaction. The channel is active only when there are data to be sent, and data are sent only if a full packet is ready or if a complete message, which may be smaller than a packet, is ready. Between packets the channel is free to be assigned to other users (including voice-channel users).

To see the need for packet switching it is worthwhile taking a brief look at the conventional (switched) alternative. If a conventional modem is used, for example by the police to query a registration plate, the channel will be held up for the time it takes the officer to key-in the details, send the data, and wait for and receive a reply. This whole operation may well take two minutes. The actual transaction might be:

Query: "Status registration CA 98723?"
Reply: "1999 Jaguar, registered to Mick S. Hunter, 1435 West Street, Medicino; expiry June 2000"

This message is 390 bytes long in Wordperfect format. If the data rate is 2400 bits/s then this message could have taken $8 \times 390/9600$, or 0.325 second, had an optimally efficient transmission mode been used. The query is 36 characters long, and the response is 87 characters.

Most messages written or sent by e-mail are less than 500 characters long. Because of this, a standard packet length is 512 characters. These packets can be sent in about 1 second. This means that the above message could be sent in two packets, taking a total on-air time of 2 seconds. A conventional modem would not have answered the ring tone in that time.

DATA LIMITATIONS

It appears that the interest in cellular data has not kept pace with the technology. There are limitations on mobile data that may not exist in the wireline environment. A computer connected to a large area network (LAN) may be able to communicate at data rates of 10^+ Mbits/s and transfer massive files quickly. A cellular-based connection that is able to achieve only 9600 bits/s is comparatively very slow. For large volumes of data, this can be *very* slow.

Cellular applications should be used for rapid transfers of modest amounts of data. Specific applications include law enforcement records, credit card checks, and security alarms.

SCADA SYSTEMS

Supervisory control and data acquisitions (SCADA) systems are networks of data-collection and telemetry devices that are integrated to link such things as water supply systems, sewerage, electrical transmission, and other wide-area networks to a common control center.

Usually SCADA points are remote and the links are by radio. Some SCADA devices have the ability to interface with a telephone, and these can be connected directly to a cellular phone. Polling will usually be arranged to occur on a regular basis or as requested by the control center.

CDPD

Cellular digital packet data (CDPD) is effectively an overlay network over the voice-switched circuit system. It uses idle channels to transmit packets of data in short bursts. The channel is returned to voice operations as required. All of this is accomplished by a physical control system at each base-station site, and a centrally located CDPD network controller. The data are essentially frequency hopped, as they move from channel to channel, in accordance with availability. CDPD is an efficient way of providing moderate speed packet data services, at a gross data rate of up to 19.2 kb/s. The net data rate throughput is approximately 10 kbps. Reed-Solomon codes are used for error correction and they result in a moderately robust transmission.

Because the standard is based on TCP/IP (an open operating standard) the technology easily lends itself to PC and internet services. It can be added quite readily to existing infrastructure.

As was said earlier, data are sent in short bursts (which may be up to 2048 bytes), and the protocol stipulates that the burst will cease within 40 ms if the channel being used is required for voice [most mobile database stations (MDBSs) are faster than this and will release the channel in about 10 ms]. The channel hopping that results means that efficient use is made of the channel's resources, but for simplicity, and to ensure that the network has at least some minimal data capacity, some channels can be dedicated to data.

The MDBS controls channel hopping, which may be planned (decided by the system time constraints on a channel) or unplanned (the channel is to be released for voice). There is a maximum time that CDPD can stay on an unused channel, because if it stays too long, the base station will treat the signal as interference and seal the channel. Of course, this period is system-dependent and depends on the network's system parameters.

CPDP uses the mobile data link protocol (MDLP), which is based on ISDN's LAPD, and it ensures error-free data, requesting retransmission of all bad frames. The network automatically buffers frames that are sent in the "dead" period caused by handovers.

With CDPD voice and data work on different platforms, but they share the spectrum and jointly enhance spectrum efficiency.

RF Hardware

The radio frequency (RF) channel management is done by the MDBS; there is one at each cell site. The MDBS monitors the site for channel activity. Access control for mobiles requesting CDPD and channel assignment and hopping, as required, are all functions of the MDBS.

Originally developed for AMPS, the application has been extended to D-AMPS which provides support for CDPD.

Applications include ATM machines, credit card validation, real-time courier tracking services and web access.

EMI

Electromagnetic interference (EMI) can be a significant problem in CDPD applications. The combination of a computer-based controller, bursty transmissions, and an RF environment can present all kinds of interference problems. To minimize these it is important to ensure that the CDPD equipment as far as possible has been well designed for the environment, and in particular that the hardware is well shielded and grounded.

DATA IN THE FUTURE

An early forecast on the popularity of data was made in the early 1990s by McCaw Cellular. They said that by 1997, 30 percent of revenues would be derived from data. That didn't happen, but in 1999 most of the pundits were predicting a rapid growth in data in the early years of the twenty-first century. Starting from the small base, these growth forecasts are being taken seriously by network operators as investment in data facility rises, and growth is being seen in the customer base. While data have yet to appeal to most users, being around 5 percent on the 2G networks, 3G, which has a lot to say about data, is targeting this traffic as a major revenue source.

The main difference between the current 2G networks and the 3G offerings will be its proported data capability. While all the major systems today offer some form of data interface (at least optionally, as not all networks are data equipped), they are slow. The GSM offering of 14.4 kbits (with minimal error correction) is the current (1999) pacesetter.

3G CDMA standards are made by the CDMA Development Group (CDG), and the advanced standards seem to be ahead of the International Telecommunications Union's (ITU) IMT 2000 timetables.

The future calls for the cdma2000 3G plan 3x to provide 1 and 2 Mbps data rates using three 1.25-MHz channels. In the interim there are a number of transitionary options that may or may not become widely accepted. For CDMA some networks can support 14.4 kbps, but currently (2000), it is not widely available, as network providers seem to be concentrating on the rollout of voice services.

The next step for cdmaOne is IS-95B, which will offer 64-kbps data rates. Some are calling the enhanced 2G data networks 2.5G, a reference to the fact that the main thrust of 3G is high-speed data. Enthusiasm for IS-95B is patchy, with most operators taking a wait-and-see approach. It is thought that many operators will wait for the 1xRTT part of the cdmaOne 3G proposal (known as IS-95C), which offers a more attractive 144-kbps data rate.

IS-95C will require new terminals and application-specific integrated circuits (ASICs), but there are other benefits that come with the new package, including increased channel capacity (doubling it), advanced packet data services, and better battery life; it is these additional enhancements that are most likely to ensure active manufacturer support for the new standard.

2.5G Offerings

Most GSM carriers are offering 9.6 kbps data, while some offer 14.4 kbps. However on most networks, due to noise, the net throughput is much slower than these figures would imply. Enhancements known as high-speed circuit-switched data (HSCSD) and general packet radio services (GPRS) allow the combining of time slots up to four at a time, with a net data rate of 57.6 kbps, initially with a future expansion to 115 kbps, and finally

to 172.6 kbps. GPRS, however, requires the building of virtually a parallel packet data network and the use of special handsets that are packet data capable.

GPRS can be seen as a wireless LAN, or a multiuser data network. It has been specified with various technologies in mind including GSM, Digital European Cordless Telecommunication (DECT), and Universal Mobile Telecommunications System (UMTS). Unlike the circuit-switched solutions, in the packet mode, customers will be charged only for packets of data sent, and, for applications where data activity is intermittent, this can be very cost-effective. The data packets are kept small and may vary from a few bytes to a few kilobytes. It is anticipated the GPRS will be cost-effective for applications ranging from two-way paging to video.

The operator can choose to implement GPRS in a number of ways. The most efficient way is to completely overlay an efficient packet network with a purpose-designed GPRS support node (GSN), or alternatively, and less efficiently, the GSN can be part of the mobile switching center (MSC), as seen in Figure 29.3. The base station system (BSS) radio path to the GSN can be tailored to the data density, with simpler and less costly interfaces being used in low traffic areas.

Most manufacturers have a GPRS product, and it remains to be seen how far the operators take to it.

For GSM operators a more attractive option may be to go directly to the Enhanced Data for GSM Evolution (EDGE) option. However, as EDGE can be evolved from GPRS, it may be necessary to have subscriber units that are multimode GSM/GPRS/EDGE. EDGE is essentially GPRS with an enhanced high-speed radio modulation technique.

EDGE will use a new modulation system, 8 PSK (phase shift keying), which will assist in the higher through-put, but will reduce the coverage. GMSK will

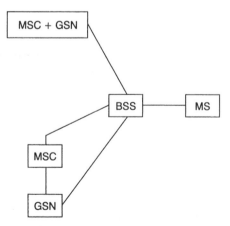

Figure 29.3 The GPRS implementation options.

be retained for its more robust properties in fringe areas.

EDGE, which is strongly supported by Ericsson and other major manufacturers, is also part of the IS-136 platform for the Universal Wireless Communications Consortium (UWCC), as accepted by European Telecommunications Standards Institute (ETSI) and the Telecommunications Industry Association (TIA) in the United States. This may give the standard an additional edge. The IS-136 plan, in accordance with the IMT-2000 guidelines, provides for a local-area 2-Mbps data rate as well as the 384-kbps wide-area rate.

As originally defined, EDGE required operators to clear 2.5 MHz of spectrum, a rather daunting task. As a result, EDGE Compact has been defined as operating using 1 MHz while still supporting 384 kbps data. The 2.5-MHz version of EDGE is now known as EDGE Classic.

Deployment of EDGE is being undertaken cautiously, due in no small part to the spectrum requirements. Some early GSM systems were the standard by the end of 1999, but time division multiple access (TDMA) operations are not expected until 2001.

EDGE and GPRS can support future 3G services.

Wireless Application Protocol

In 1997 the concept of wireless application protocol (WAP) began as a way of standardizing the interface between digital mobile phones and the Internet. The protocol is meant to cover all digital standards, including GSM, CDMA, TDMA, and 3G systems. The protocol is an open standard, and is basically compatible with existing hardware, although some software changes may have to be made; there also may be some intellectual property (IP) issues to be sorted out. However, the WAP standard is envisaged as an evolving standard that will revise its standards yearly.

WAP claims to offer reduced overheads in computational power, memory, and power consumption.

Networks that use WAP will have to have a WAP server that interfaces the mobile system with the Internet. It is anticipated that this will be used for advanced data services, including weather information, airline schedules, and e-mail.

WAP permits the manufacturers to maintain their own man–machine interfaces (MMI) and operating systems while standardizing the handling of services.

There has been considerable criticism that the WAP forum is proceeding too slowly and that many manufacturers have positioned themselves to "work around" the standard by developing their own protocols. However, the WAP forum has over 95 companies that endorse the technology. The first European user is SFR

in Switzerland, and a number of other major network operators have the technology under evaluation. Some handsets also offer WAP capabilities.

Wireless Knowledge

Wireless Knowledge is a system developed by Motorola and Qualcomm and launched late in 1998, which is similar to WAP and meant to provide wireless services to all cellular systems (GSM, CDMA, TDMA, etc.). The technology is being used for data on integrated dispatch enhanced network (iDEN).

BLUETOOTH

Bluetooth was envisioned originally as a replacement for cable connectivity between computers and their peripherals. As one of the many competing technologies for this link, Bluetooth, like its rivals, will need to find a wider application.

Bluetooth has a few things in its favor, not the least of which is that it is an open standard. The data rate of 1 M s/s (symbols per second) looks good in the year 2000, but may be a bit slow by the time it is widely implemented; a proposed extension to a gross data rate of 2 Mbits/second in a 3G version, may not save it (as gross data rates over such systems tend to be poor indicators of the net data rate). Operation in the 2.45 GHz ISM band has broad appeal, but it has its limitations too, including uncontrolled potential for interference. The noise immunity, quoted as -8 dB at 0.1% BER is indicative of a high error rate that is more applicable to voice communications than to data. What really counts against Bluetooth is the pathetically small nominal range of 10 meters, a potentially slow net data throughput rate, poor sensitivity (-70 dBm), and a lot of competition.

Although Bluetooth has wide support it is not without its competitors. Operations in the ISM (Industrial Medical and Scientific) band are highly attractive and have drawn many "solutions" to the interconnectivity problem. The ISM band is an unregulated band—anyone can use the band provided the maximum ERP of any device used does not exceed 100 mW (+20 dBm), with most devices operating at 1 mW (0 dBm).

Bluetooth uses frequency hopping with 79 hop frequencies 1 MHz apart, hopping at up to 1600 hops/s between 2.402 and 2.480 GHz (less in Japan, France and Spain). Typically the output power is 0 dBm (1 mW) with the option to increase to +20 dBm (100 mW) and decrease to -30 dBm (1 m w) using simple FM modulation. A Bluetooth transceiver can communicate with up to 7 other devices simultaneously.

There are some very impressive features here:

- Gross data rate 1 Mbit/s
- Low power consumption
- Lots of anti-interference features
- 2.45 GHZ "free band" (no license hassles)

So, yes, it looks good at first reading, but what about:

- Range 10 m (How many customers asked for that?!)
- RX sensitivity -70 dBm (Is that deaf or what?)

The Contenders

- The IEEE 802.11 wireless LAN standard, which evolved from the Ethernet as a wireless access protocol (and its recent iterations) has the capability to deliver up to 11 m/b/s (i.e., 10 Mb/s wired Ethernet substitution).
- Lucent's 'WaveLAN' is a potential "last meter" contender and has been implemented since 1998.

- DECT derivatives offering "virtual cable" solutions.
- PHS derived LAN systems.
- 3G itself is offering alternative solutions.

So regardless of the actual performance of Bluetooth, the vested interest is in the other technologies.

It seems we still haven't learned! The specification is still what is doable technologically rather that what the customer wants.

BUT DON'T BE FOOLED

The mobile data rates quoted for all of the foregoing systems are *gross* data rates. In a noisy mobile environment it is unlikely that a net data rate of 10 percent of these figures could be achieved.

Also much of the long battery life of digital systems is attributable to a very low duty cycle of the PA stage. These wideband systems require much higher duty cycles, much more battery power, and much better mobile PA stages.

30

PRIVACY

Privacy is not an inherent feature of analog cellular systems. Although some operators claim a degree of privacy due to the difficulty of locating or following a particular call, the fact remains that analog systems do not inherently provide the sort of privacy that is associated (sometimes unduly) with the plain old telephone system (POTS). Even so, it is widely rumored within the industry that many "shady deals" are transacted over the mobile telephone system rather than the POTS simply because of the difficulty of tracing a particular call. It is true that it is difficult to trace particular calls using a radio scanner, but full call accounting at the cellular switch ensures that details of the call are available to the proper authorities.

The security of a mobile is dependent to a large extent on the technology. Analog systems are rather easy to tune into. US-based digital systems that don't use the ANSI-41 encryption algorithm (or do use it but use a default A-Key) are somewhat harder to monitor, but nevertheless, they are no real challenge to today's technology. Digital equipment using ANSI-41 or Global System for Mobile Communications (GSM) is too hard to tap into off-air, but all the same, it can and is routinely monitored.

ANALOG SURVEILLANCE TECHNIQUES

Radio scanners are wide-band radio receivers that, under microprocessor control, can scan through a present group of channels and stop when a carrier (an active channel) is encountered. The receiver stops on that channel indefinitely (unless programmed otherwise) while there is a carrier present. Once the carrier has gone

(in the cellular instance, the call terminates), the scanner moves on through its list of programmed channels until a new carrier is detected, at which time it stops and proceeds as before.

For a few hundred dollars, anyone can purchase a scanner. There are a remarkable number of people who have not only done so, but who are content to spend hours every day listening to an interminable number of conversations on the off-chance of picking up some scrap of information, perhaps of an industrial, political, or even personal nature. The more organized "snoops" may have tape recorders attached to the scanners, and some may even have commercial outlets for various kinds of information. For this reason, it is wise to warn all analog-system users to avoid talking about secret business transactions, or at least to do so in a circumspect manner.

This may not be a problem for some people, as in the case of one rather well-known Australian politician who, when asked if the lack of privacy on his car phone was a problem, replied, "Heavens no, I never say anything that makes any sense anyhow."

Some cellular operators claim to provide reasonable privacy based on the fact that it is difficult to locate any particular mobile in the network and that handoffs produce a discontinuity in the conversation. It is true that it is difficult to lock onto a particular mobile, but it should be remembered that a snoop needs only to have a mobile at a high location to receive most base stations. Because some modern scanners are bus controllable and the signaling protocols are public knowledge, it would not be too difficult to build a scanner programmed to "track" a particular mobile. At least one instance of this, in the Netherlands, has been reported.

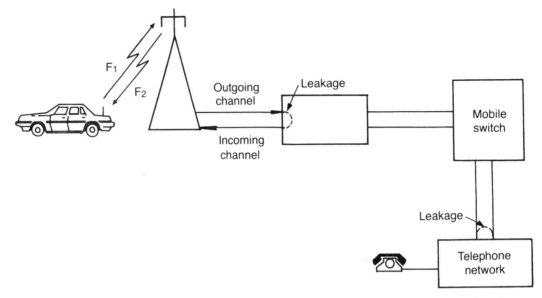

Figure 30.1 Leakage and internal automatic level controls enable scanners to hear both ends of a conversation even though it takes place over different duplex channels.

The government snoops, and cellular operators tracing fraudulent users, however, have more sophisticated equipment. Cellular call interceptors can monitor any control channel and listen for particular mobile identification numbers (MINs) or dialed numbers and then follow the call, including through handoffs. High tech interceptors like the Smith Myers CSM7706 can monitor 100 target MINs and 100 dialed numbers, using six independent channels to track multiple targets.

Another device, called a "swamp box," simulates a base station and forces mobiles in its location to give up their MINs and electronic serial numbers (ESNs). These devices are used by security forces, but also fraudulent operators to get the information they need to make a clone phone. A swamp box will often look very much like a regular transportable phone.

Handoffs provide some immunity to scanning, but they are not usually available in small systems, where there is nothing to hand off to, or when the mobile is a stationary handheld. In very dense networks, however, the claim is reasonably valid because handoffs do occur quite frequently. Still, a scanner programmed to follow a particular mobile could probably continue to do so.

There are basically two types of scanners on the market. Currently the cheapest type is usually limited to 512 MHz and therefore is unable to receive most cellular bands (and certainly not Advanced Mobile Phone Service (AMPS), Total Access Communications System (TACS), or Nordic Mobile Telephone NMT900).

Other, more expensive scanners cover the spectrum up to about 1.3 GHz (1300 MHz) or more and do cover the modern cellular bands. These receivers are usually not particularly sensitive and are therefore limited to relatively nearby base stations. However, there usually are still plenty of channels within range (10–15 km) to keep snoops, both amateur and professional, quite contented.

The way in which conversations are received warrants some attention. All conversations between a mobile and the base station are full duplex (that is, from the mobile, the transmission occurs on one frequency and the return signal, to the mobile, occurs on a different frequency, usually 45 MHz away). This fact has led many people to conclude that a scanner can therefore hear only half of a conversation, a conclusion that is, unfortunately, wrong. In fact, as shown in Figure 30.1, scanners hear both sides of the conversation.

As seen in Figure 30.1, although the two conversation paths appear to be separate, the path that transmits to the vehicle F_2 has a very wide-range automatic gain control (AGC) amplifier, which has the function of ensuring that constant audio levels are transmitted. This system works so well that side tone (or leakage from transmission to reception) is amplified to the level of normal audio.

Side tone is deliberately introduced into telephones (including cellular telephones) so that some response to the user's speech is heard in the receiver of the speaker. When this tone is not provided, the earpiece is silent during speech, and this silence is (usually) interpreted as a transmission failure (the telephone sounds "dead"). The leakage may or may not be related to the generation

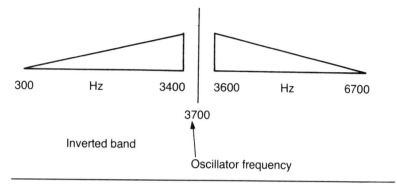

Figure 30.2 A simple frequency inverter.

of side tone, but it is invariably present; the mobile transmission as well as reception is transmitted by the mobile base at approximately equal levels. The clarity of the leaked path as received on a scanner varies from poor to very good, depending on the system, but is usually intelligible.

A number of encryption companies that can provide varying degrees of security have emerged. The market originally was aimed at the security sector, but this market is just too small. Besides, most secret services are so paranoid that they will not buy encryption from third parties for fear of leakage of the codes or keys from the vendors.

ANALOG ENCRYPTION

Criminals, "crazies," and underground organizations are in the market for encryption. This part of the market taints the whole industry and has perhaps led to suspicion on the part of potential corporate users.

The equipment available to date has been bulky and expensive as well as cumbersome to use. Commercially available devices that encrypt two-way radio communications have been available since the 1970s, but the market has been very thin.

There is, however, a need to protect corporate information, and it is this area that is currently being targeted.

ANALOG PRIVACY BY FREQUENCY INVERSION

Privacy has rarely been added as a standard feature to analog cellular systems for one simple reason—cost. To see how this cost arises, the simplest method of privacy or scrambling (frequency inversion) is examined here. Figure 30.2 shows how frequency inversion causes the speech spectrum to be inverted.

In Figure 30.2, the normal speech band of 300–3400 Hz has been inverted by a simple modulation technique so that the content of the high frequencies (3400 Hz) now appears as the low-frequency (300 Hz) part of the inverted band. Similarly, the 300-Hz component now appears as the high-frequency 3400-Hz component. This renders the speech virtually unintelligible.

The inversion can be achieved by a simple mixer and filter combination, as shown in Figure 30.3. The mixer produces the sum and difference of the input signal frequency and the 3700-Hz oscillator. If the mixer is followed by a low-pass filter that cuts off at around 3700 Hz, then all the sum components are filtered out and only the difference frequency is passed. Thus, if the input signal is 300 Hz and the oscillator is 3700 Hz, the output will be 3400 Hz (3700–300 Hz). Similarly, if the input is 3400 Hz, the output will be 300 Hz. These frequencies are now said to be inverted. Demodulation consists simply of "reinverting" and uses the same circuitry as shown in Figure 30.3.

This method has two disadvantages. First, it is simple and can therefore easily be decoded (unscrambled). This disadvantage can be overcome by splitting the audio band into a number of components and inverting them separately; this is commonly done. More sophisticated scramblers use this method, but go even further. They divide the speech band into many smaller segments, which are not only inverted separately but sometimes transposed according to a user settable pattern. Clip-on, acoustic-coupled devices with thousands of combinations are commercially available.

The second disadvantage—the inherent increase in signal-to-noise (S/N) performance—is more fundamental and has no easy solution. In its simplest form, noise necessarily introduced by the scrambling and unscrambling of this signal degrades the signal path by approximately 9 dB in S/N performance. This degradation reduces the useful range of a base station to about 60 percent of a base without scrambling. More elaborate

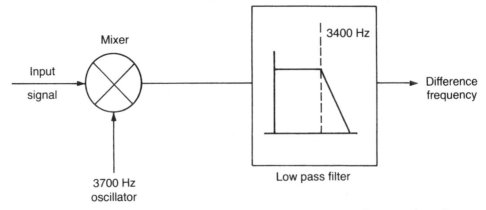

Figure 30.3 A double sideband AM modulator and filter produces frequency inversion.

decoders exist with filters that eliminate most of the noise caused by the inversion process. With such devices, the S/N degradation is reduced to about 1 dB, and so the range reduction is not so serious. When time-dependent inversion is used, then the filters must be switchable and follow the pattern of the encoder.

Unless scrambling capability is provided as a feature of the cellular switch, it should be noted that any "add-on" scrambler requires a decoder/encoder at each telephone (whether mobile or not) where scrambling and descrambling are to occur. Some switches, however, do incorporate a built-in scrambler. Most switch-based scramblers are proprietary devices and can therefore be used only with the supplier's mobiles. These devices usually use elaborate algorithms to scramble the inversion points in manners that are difficult to replicate. Typically, a random number sequence, which is "seeded" by the serial number of the mobile (which need never be transmitted), and a second number, supplied by the switch, determine the inversion sequence. Such systems are difficult for all but professional interceptors to decode.

Multiple splitting of the voice band is achieved as shown in Figure 30.4. The band is divided into two halves and then each half is separately inverted around its own center frequency. A more sophisticated version of this has the center frequencies and the split moving in real time, with up to 32 split points being commonly available. This encryption is known as variable split band (VSB) with rolling code.

Analog scrambling is effective and difficult to decode in real-time. However, it can be analyzed and the inversion points and frequencies determined from regularities in the waveform and so be decoded. More complex analog scrambling systems inherently introduce more noise and also degrade the speech quality.

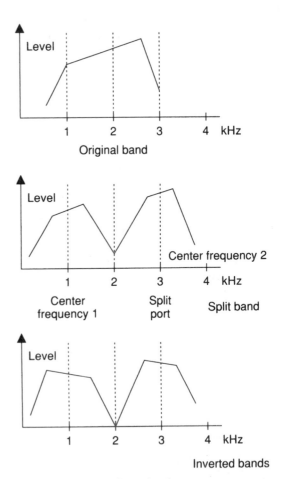

Figure 30.4 A two-inversion frequency encryptor.

DIGITAL ENCRYPTION (OF ANALOG SIGNALS)

The alternative is to use digital scrambling. Digital scrambling starts by sampling the analog signal and then processing it. Within some bounds the degree of "processing" will not be related to the voice quality, provided the reconstruction does not introduce errors. Transpositions in the time domain must be limited to a few milliseconds so that delays are not noticed by the receiver. In fact, nearly all digital systems use scrambling techniques to process the signals in a way that ensures that long strings of "1s" or "0s" do not occur, even when encryption is not the intention.

Other methods to discourage eavesdropping are in place. In most countries, it is either illegal to listen to cellular and other telecommunications, or it is illegal to use any information gathered by such eavesdropping. These laws are largely "paper tigers," however, because they are virtually unenforceable. There is little value in legislating that listening to the cellular bands is illegal unless that legislation can be enforced. Some administrations have gone a little further and prohibited import or manufacture of scanners that cover certain bands. This measure results in the marketing of scanners with discontinuous coverage. This method is also not particularly effective, because there are usually plenty of radio shops willing to "modify" these scanners for a relatively small fee so that full coverage is once again available.

One interesting way of discouraging eavesdroppers takes into account the way that the scanners actually work and turns scanner use into a disincentive in its own right. As noted before, scanners skip over unused channels and stop only on those that transmit a carrier. If all channels always transmitted, then the scanner could function only manually. By having all channels always transmitting, and by having them transmit loud and unpleasant sounds, the eavesdropper soon finds that the hours spent at the scanner are not particularly pleasant. This discourages a number of potential snoops.

On the negative side, however, there are quite a few costs. All channels must be turned on at all times. This significantly increases the power consumption of the base station and decreases the life of the power amplifier stage of the base-station channels. Unless some interlocking device stops this mode of operation during power failures, the time that the batteries can support a base station is significantly reduced.

For this method, base-station antennas must operate at all times at maximum power; this contributes to corrosion and antenna faults. Finally, because all channels are always transmitting, co-channel and adjacent-channel interference always operates at its maximum level, and contributes to a net decrease in the traffic capacity and availability of the system.

In general, one can see that, where attempts are made to add privacy to analog systems, the costs are high. This, of course, does not apply to digital systems, where very high levels of encryption can be applied at little or no cost. Digital systems can thus be expected to provide very high levels of privacy.

DIGITAL CELLULAR ENCRYPTION

By nature, digital cellular radio transmissions are sent to air encrypted. The degree of encryption is sufficient to eliminate the amateur with a scanner, but in general it will not be enough to ensure complete security of the voice path. The GSM code was broken by an off-air computer in mid-1993, and the US ANSI-41 security is somewhat lower than GSM.

The voice security offered by digital cellular is high compared to its analog counterpart, but it should not be assumed that the code is unbreakable.

It is safe to assume that digital signals that are properly encrypted will rarely be monitored successfully off-air. Even though the codes have been broken, this has not been done in real time, so the air interface of digital systems can rightly be considered rather secure.

There have, in fact, been a number of instances where the security forces have held up the implementation of GSM simply because they could not snoop (Australia was an example of this). However, the solution is surprisingly easy; simply put all calls to or from targeted mobiles onto an internal diversion via the switch. From the point of view of the security forces, this is far easier than off-air monitoring.

Digital systems, even when unencrypted, have a degree of security, in that the coding, which includes check bits and redundant data, means that at the very least it is necessary to understand and duplicate the encoding algorithm to make any sense of the message sent. While this will keep all but the most serious amateurs from eavesdropping, it is no real impediment to professionals. Some operators take too lightly the threats to the voice and system security, and do not even bother to activate built-in encryption security.

In the chapters covering particular technologies such as GSM and ANSI-41, the details of the specific encryption used by those protocols will be covered in some detail. In this section however, encryption has been covered in a more general sense.

31

RURAL AND OFFSHORE APPLICATIONS OF CELLULAR RADIO

ENVIRONMENTAL LIMITATIONS

In rural areas with little or no infrastructure, the environmental limitations can be daunting. The equipment must be able to perform while installed in areas with a wide range of temperatures, humidities, and altitudes. The reliability of power supplies (including brownouts) must also be considered. In some locations, vandalism (including deliberate attacks with high-powered rifles) may be encountered. Rural areas do not have good access to qualified maintenance staff or quality test equipment. Reliability is a vital factor and the installation should be a low-maintenance design.

SMALL SWITCHES AND REPEATERS IN RURAL AREAS

For rural areas that are remote from major switching centers, there are a number of possible ways to establish a cellular system. Trombone trunking (or backhauling) can be used to operate the remote bases from a distant switch, although doing so becomes less economical as the distance to the remote switch increases. Trombone trunking involves the use of a link from the centralized switch to the rural area for all calls (even local ones), as seen in Figures 31.1 and 31.2. In many rural areas, it may be economical to use repeaters to effectively extend the coverage area of a larger network to include the rural area.

CELLULAR PAY PHONES

Cellular pay phones are a practical way to provide a public service in remote areas such as islands, mountains, and freeways (emergency) where a nearby cellular system is already installed. Commercial units exist (some complete with attractive cabinets) that can meter calls from the remote site and report back when the coin box is becoming full or a power failure occurs. The units usually contain a "look-up table" that enables the meter rate (call charge rate) to be read from the digits dialed. These phones, being portable, can be moved into place quickly for emergencies or for catering to large temporary gatherings. This is especially true if the phones are solar powered.

The meter pulses are generated locally so that a conventional pay phone can be attached. This point is significant because pay phones are subjected to a great deal of vandalism, and if the handsets can be replaced by a standard telephone unit, maintenance becomes easier and cheaper. A standard mobile phone with a credit card facility is shown in Figure 31.3. Table 31.1 lists the specifications of a typical pay phone unit.

Wireless Technologies has a line of specialized pay phones for the cellular or mobile operator. Figure 31.4 shows a range of phones designed to be coin, token, debit, or credit card operated both indoors and outdoors.

Cellular phone booths will generally be subject to high calling rates, so if remotely powered will need a substantial solar cell. Figure 31.5 shows a vandal-resistant phone booth. The solar panel is claimed to be bullet resistant. It is backed with stainless steel and divided into

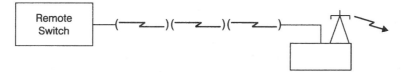

Figure 31.1 A remote base-station link.

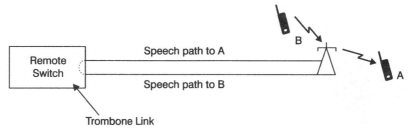

Figure 31.2 The trombone trunking.

Figure 31.3 A cellular pay phone with credit card facilities. (Photo courtesy of Wireless Technologies.)

TABLE 31.1 Specifications for a Typical Cellular Pay Phone

Voltage	115/220 VAC 50/60 Hz
Dialing	Tone or pulse (10 to 20 PPS)
Power	15 V DC @ 5.3 amps
Battery	12 V 10 amp-hours
Operational temperature	$-30°$ to $60°$C

modules, any one of which can fail without affecting the others.

RIGIDLY MOUNTED VERSUS MOBILE RURAL UNITS

There are thus two basic fixed cellular rural structures: mobile and Wireless Local Loop (WLL). In mobile systems, the customer units are conventional portables; in WLL, the customer's equipment is rigidly mounted, and external fixed antennas can be used.

Mobile systems, which involve no installation costs, generally appear to be less expensive than the fixed installations; however, the fixed units almost certainly require less maintenance, because they are not subject to the same environmental hazards. In mobile systems, the customer can deliver faulty mobiles to a specialist service center; WLL requires a visit by a repairer. It is difficult to determine which type has the lowest overall maintenance costs.

The external antenna mount available with a WLL system leads to a greater usable range. It is shown later

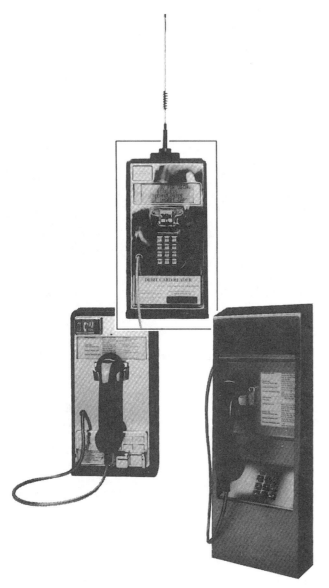

Figure 31.4   Public call boxes suitable for cellular radio connection. (Photo courtesy of Wireless Technologies.)

Figure 31.5   A vandal-resistant public phone booth. (Photo courtesy of Wireless Technologies.)

in this chapter that path-loss decreases of 24 dB can be achieved, but given the extra costs involved, this solution is not, on the face of it, economical. But other factors should also be considered. For example, the ability to use conventional handsets may, from a marketing perspective, be important to many companies. Rural services are generally provided at a loss or by way of some form of subsidy. Providing rural subscribers with state-of-the-art cellular phones can substantially spur demand, so the net increased loss (or subsidy) may exceed the initial gains made by lesser installation costs.

The ability of many fixed cellular phones to operate in

party-line mode could justify the extra cost (particularly if two party-line phones can replace two transportable units). Party-line facilities of eight or more phones are available as a standard option.

Where pay phones are considered, WLL is almost the only solution. Some administrations have a policy that all fixed phones must have the same handsets; such a policy dictates that WLL phones must be used.

When mobile system phones are used, there may be a marketing conflict between rural phones that are pro-

TABLE 31.2   Range and Area of a Base Station to a Conventional Mobile

Base Station Height (m)	AMPS/CDMA		GSM 900	
	Range	Area (km^2)	Range	Area (km^2)
30	9.3	271	4.8	72
50	12.6	498	6.4	129
70	15.6	764	7.8	191
100	20	1,256	9.7	295
150	26	2,123	12.7	506
200	31	3,019	15.5	775
300	35	3,848	20.5	1,320
450	43	5,808	28.6	2,569

TABLE 31.3   Network Cost per Area Covered by Bases with Towers Higher than 30 m*

Tower Height (m)	Additional Cost of Tower ($ 000)	Cost of Complete Base ($ 000)	Area Covered (km^2)	Area Covered ($/km^2)
30	0	150	72	2,083
50	20	170	129	1,318
70	40	190	191	995
100	70	250	295	847
150	120	300	506	593
200	170	350	775	452
300	270	450	1,320	341
450	420	600	2,569	234

*Note that the cost is for the base station only. Switch and transmission costs have not been included.

vided at a nominal cost and mobile phones (which are the same thing) provided at a premium cost. This may cause conflict even when the rural service is restricted to the local cell.

RURAL MOBILE NETWORKS

A network where the terminal equipment does not use an external antenna is one where, in most instances, a subscriber is normally equipped with a conventional transportable or handheld telephone. A few instances of more distant subscribers with external antennas may occur.

The usable range of base stations meant to operate this way is necessarily limited, but they are inherently capable of supporting a conventional "mobile" telephone population. The subscriber densities that economically can be supported depends on the base-station coverage. Because telephone density is normally quite low in rural areas, it is important to seek out base-station sites that are high. Table 31.2 gives the expected range and area covered for different base-station heights.

From Table 31.2, a few important deductions can be drawn about the economics of customer densities and the importance of tower height. Since the density is low, it is important to get maximum range out of the system. To determine if high towers are economical, let's examine a particular case. Assume the following:

- The cost of a base station with a 30-meter tower and 50 channels equals $150,000.
- The cost of tower extension beyond 30 meters equals $1000/m.

- The number of subscribers is directly proportional to the area covered.
- A Global System for Mobile Communications (GSM) System.
- The base capacity is 1500 customers.
- The cost of power supplies (assuming local power is available) and huts is included. The cost of links is not included.

It is clear from Table 31.3 that this very generalized case study implies that the cost per square kilometer covered decreases fairly rapidly with base-station height. Because a low uniform density is assumed, it follows that at low-density sites the cost per customer is less if greater tower heights are used instead of extra bases. Table 31.3 can be used to directly convert to cost per customer, once the customer density is known.

In practice, of course, this approach must be tempered with the reality of the actual terrain, which may not permit increased range with increased antenna height. But in general it can be said that tall towers or high sites need to be a feature of rural cellular systems (notice that the opposite applies to urban systems). As the customer density increases, the need for high towers decreases.

Mobile system installations use conventional transportables and handhelds, and so have lower terminal end-costs. Typically, the terminal end-cost is $100 each (2000)—the mobile unit is assumed to be a handheld. These installations are more economical in higher-density environments, where large towers are not required.

TABLE 31.4 Network Cost per Area Covered for Different Tower Heights, Assuming a 7-m Fixed Antenna and a GSM System with a 9-dB Yagi Antenna on the Mobile

Tower Height (m)	Cost of Base ($ 000)	Radius Covered	Area Covered (km^2)	Area Covered ($/km^2)
30	150	22.8	1,633	92
50	170	30	2,827	60
70	190	34	3,631	52
100	250	39*	4,778	52
150	300	43	5,808	52
200	350	47	6,939	50
300	450	60	11,309	39
450	600	74	17,203	35

* Note range >35 km by time slot stealing.

NETWORKS WITH EXTERNAL TERMINAL EQUIPMENT ANTENNAS (WLL)

An alternative to using conventional mobiles is to use specially configured "fixed" mobiles with high-gain (and elevation) antennas. This configuration is more economical in very low-density regions. It also precludes an effective mobile network from operating in the same district as a spin-off (except in a hybrid arrangement, as discussed later).

For relatively low antennas (<10 m), the height gain is approximately 6 dB when the height is doubled.

Because the price of an external antenna installation is not strongly dependent on height (a single pole is assumed for the structure), it is reasonable to assume that all external installations are mounted on a 7-m pole. The height gain for this height (over the conventional 1.5 m for vehicle mounting) is about 12 dB. A handheld is generally assumed to be about 6 dB down on a vehicular unit. Further, it is reasonable to use a high-gain Yagi antenna, (for example, 9-dB gain compared to the 3-dB conventional omni antenna). Provided low-loss feeders are used (less than 3 dB), the net gain now is $12 + 9 - 3 + 6 = 24$ dB. Table 31.4 shows the corresponding range and area and dollars per area, assuming the additional 18-dB height gain for an GSM900 system.

It is immediately obvious that this configuration is much more economical in coverage costs than the mobile solution; in fact, a 30-meter tower can produce a smaller cost per kilometer than a 450-meter tower in the mobile example. Table 31.4 shows only network costs and does not reflect net costs. A WLL network has more costs at the subscriber's end, however, and these must be added before the two systems can be compared. The

Figure 31.6 A Type II emergency call station in California.

systems must also be studied further to determine how they compare.

An example is seen in Figure 31.6, which shows an emergency cellular telephone on Highway 1 in California just north of San Francisco. Notice the small solar panel and antenna at the top. These stations are placed at intervals of about 1 km on both sides of the highway.

CALLING RATES AND CUSTOMER DENSITY

The calling habits of fixed subscribers differ markedly from those of conventional mobile subscribers and do tend to show higher peak-hour traffic rates, even when the average calling rate is about the same. The calling rates of rural subscribers are very hard to predict, but

TABLE 31.5 Maximum Customer Density/km for One Cell

Tower Height (m)	Mobile Tower max Density Customer/km^2	WLL Tower max Density Customer/km^2	Network $/Customer (500 Users)	Total $/Customer Mobile	Total $/Customer WLL
30	21	0.9	500	600	850
50	11.6	0.53	540	640	890
70	7.8	0.41	580	680	930
100	5.06	0.314	700	800	1,050
150	2.9	0.258	800	900	1,150
200	1.43	0.216	900	1,000	1,450
300	1.13	0.13	1,100	1,200	1,750
450	0.58		1,400	1,500	

in general they increase as the cost of the installation increases. For example, if a phone costs $2000 to have installed, the rural users will organize to buy fewer units than if they cost $300 each; but the fewer $2000 units are likely to be used much more heavily.

If it is assumed that the maximum capacity of a GSM900 cell is 1500 subscribers, then some limits can be placed on tower heights for a given customer density.

Assume that for any tower height, the maximum customer density is 1500/base.

Table 31.5 shows the maximum customer density per kilometer for one cell. Costs shown in the table include those for mobile and base stations only. Net costs would need to include switch and trunk costs and so would be a few hundred dollars higher than these figures. The additional costs would be the same for both mobile and WLL systems.

The cost of a WLL system compared to a mobile system is based on the following assumptions:

Cost of 7-meter pole and Yagi	100
Cost of WLL mobile	250
	350
Cost of mobile	100
Difference	250

Although the costs shown are rather high per customer, they can be decreased by a factor of 3 if the base station can be fully loaded.

Big WLL towers are economically justified only for very-low-density areas, and it can be seen that if the density is higher than 20/km^2, then even a 30-meter tower using a mobile configuration is not justified.

The cost of mobile base stations is decreasing rapidly, and these studies need to be seen in that light. As time

goes by the cost advantage of high towers even in low-density areas will decrease.

OFFSHORE COVERAGE

Many operators who are situated near the sea will have incidental coverage of the waterways near their main system. In many cases, foreseeing a demand from the boating fraternity, the operator may have deliberately enhanced the seaward coverage, and in extreme cases may have provided one or more base stations dedicated to improving the overwater coverage.

In determining sea coverage, the overwater propagation, which is quite different from land propagation, must be taken into account. Where land coverage requires a given signal level, a number of tests have shown that a figure of around 12 dB less is more appropriate in the low multipath environment over water. This bonus is attributable to two factors. First, in the water environment there are no buildings, hills, or other structures that contribute to much of the multipathing on land. Next, seagoing vessels do not usually move at anything like the speed of land vehicles, and recalling that a lot of the multipathing is generated by the movement of the receiver through the local signal nulls created by the multipathing, there will be a lot less noise generated by this mechanism.

These factors have a commonality that must be fully understood if the maximum benefit is to be derived from this benevolent environment. What has improved most is the effective mobile receive sensitivity.

To fully exploit the potentially improved propagation, it is necessary to do whatever is possible to improve the mobile-transmitter (TX)-to-base-station receive per-

Figure 31.7   The main earth station dishes at Petrocom's New Orleans switch site.

formance. Enhancing the mobile TX level is an attractive option. The cheapest and easiest way to get some real improvement in the mobile-TX-to-base-station receive link lies with the base-station receiver itself. A low-noise masthead amplifier will probably account for a 2–4 dB improvement in the receiver performance, which at practical heights for the receive antenna will translate to about a 10–15 percent improvement in range. The next possibility is to use higher gain receive antennas. Antenna gains of up to 12 dB in the omni configuration are commercially available, and have been proven to perform to expectations in practice.

An additional bonus can be had in the overwater environment because most vessels large enough to carry a cellular phone will probably have a convenient place to mount an antenna much higher from the ground plane than the tradition 1.5 meters provided by the average land vehicle.

PETROCOM

Although most operators will have only a small proportion of their business generated from offshore, one company in the United States derives its business from this sector. Petrocom, based in New Orleans, provides an offshore cellular service to gas and oil rigs based in the Gulf of Mexico. Its 23 base stations provide an over-the-sea coverage of an area that is 175,000 square kilometers.

In order to link its base stations to its switch in New Orleans, Petrocom uses a dedicated satellite transponder. The earth station, shown in Figure 31.7, is fed by a main dish and a smaller standby unit. When the standby has to be placed in service the satellite transmit power is turned up to compensate for the lower gain. The final stage of the earth station power amplifier, shown in Figure 31.8, consists of a pair of traveling-wave tube devices. Because of the relatively low power needed to energize the main dish, the tubes have an extended life expectancy, and in fact the original units are still in service.

Petrocom has a lot of problems that are unique to its operating environment. One is that the base stations, which are all situated on oil rigs, where space is at a premium, have to be very compact. Oil platforms are relatively hazardous places to operate, and platform owners are especially guarded about the potential liabilities that may be incurred by allowing other users

Figure 31.8 The final PA stage of the Petrocom earth station.

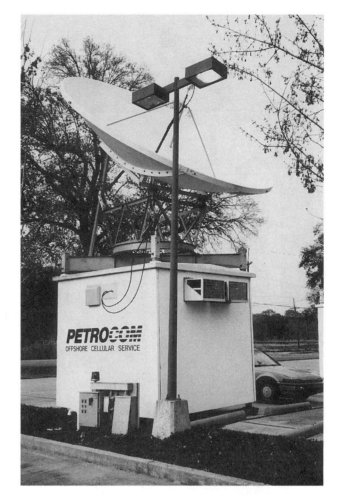

Figure 31.9 A complete Petrocom base station.

access to them. This makes the job of site acquisition, which is never easy, doubly difficult.

Since the traffic density is not high, the base stations do not need to have a large number of channels, and so they are usually designed to have only one rack of base-station channels. This means they can be quite compact, as can be seen in Figure 31.9, which depicts a base station that has been brought back to shore for an overhaul. Note that the satellite dish is mounted on top of the hut, and that two air conditioners (for redundancy) have been built into the sidewall. Positive air pressure is maintained at all times within the base station to keep the salt air out. This measure has reportedly been most successful.

The huts are all identical. This assists greatly in standardizing installation and maintenance, since familiarity with one means familiarity with all. Inside the hut, apart from the base-station rack (on the left) and the rack containing the earth station hardware (on the

right), there is one more surprise, namely the massive battery banks, which can be seen in Figure 31.10. These huge batteries with a 2000-ampere-hour capacity, can keep the base station and its air conditioner operational for up to three days. This long standby time is needed because during the hurricane season it is sometimes necessary to evacuate the oil platforms. Before this is done, the plant is turned off, which means that its power source, driven by surplus gas from the drilling operations, is also turned off.

Offshore maintenance is an especially costly and difficult task. Access is virtually only by helicopter, and because it may be needed at any time, the company has to keep the helicopter on continuous standby. A trip to a base station, counting the land leg, could take a whole day, so it is important that the maintenance crew take everything they might need with them. In this environment, the size of transportable equipment is limited by the helicopter cabin size, and this places a limitation on

Figure 31.10 Inside a Petrocom cellsite.

the gain of the antennas that can be used. High-gain antennas are necessarily longer than low-gain ones, and Petrocom has found that a 7.5-dB antenna is the optimum size.

The average platform will provide an antenna height of around 45 meters and this, with 7.5-dB omni antennas, will provide coverage to about a 40-kilometer radius.

32

INTERCONNECTION

Cellular radio was designed to be fully interconnected to the land-line facilities. These facilities have usually been in place for decades on a monopoly basis, and new competitors are not usually welcomed. This is a very difficult and often "emotional" part of cellular radio.

Generally there is no problem where the cellular operator is also the wireline carrier, but in most cases the cellular operator will be a new company that has to interconnect with the wireline facilities.

A major hurdle placed in front of a new operator is to get over the legal and bureaucratic barriers that will be put up by the wireline carrier.

It must be understood that the wireline operator traditionally believes that he has a "right" to the monopoly that has been enjoyed over the years and that at the very least some compensation for the loss of that status is in order. Because of this, interconnect agreements that strongly favor the landline operator are common.

The Federal Communications Commission (FCC) was involved in many disputes between cellular operators and wireline companies that were reluctant to interconnect with them. From these disputes some clear guidelines emerge from the FCC rulings on what "fair and reasonable" means in the context of interconnect.

The cellular operator is a co-carrier and not a subscriber. As such while traffic that originates from the cellular network may be carried by the public switched telephone network (PSTN), traffic from the PSTN can be terminated in the cellular network. Given that it is fair that the PSTN warrants compensation for terminated traffic, then equally the cellular operator should be compensated for PSTN traffic terminated in the cellular network.

Since the cellular operator is not a subscriber, the inter-operator charges that apply should be cost based. That is, the compensation should be based on the cost of the facilities used by the parties plus a fair return on the investment.

Determining the value of the cost of interconnection is difficult. First it is necessary to define the point of interconnection. Ideally, the two operators would meet at a halfway point. This is illustrated in Figure 32.1.

In most cases, the link between the cellular carrier and the PSTN will belong to the PSTN operator, and it will therefore attract a monthly rental charge at a rate that would have been established for other carriers or lessees.

The most difficult part is to determine what is a fair and reasonable rate for traffic flows between the carriers. Immediately we can rule out using the subscribers' call charges, as these are the retail cost of a call. However, this will set an upper limit for the fair cost of traffic.

Some carriers use a revenue-based settlement, but this is not accepted as a fair basis by the FCC. As an example, if two calls are terminated by a cellular carrier and one happens to be of international origin and the other is local, the cost to a cellular carrier of terminating those calls is independent of the origin and hence of their revenue value of the call to the PSTN.

In the reverse situation, where the cellular operator originates a local and an international call, then cost of delivering the call to the PSTN is again independent of the final destination. However, most interconnect agreements will require that each operator be responsible for all calls generated in their own systems and so the cost of revenue *collection*, which must include a provision for

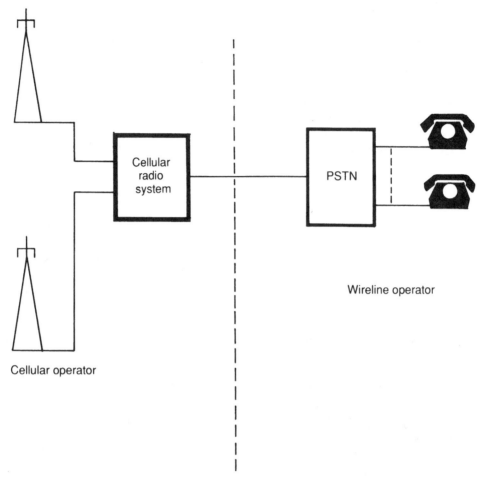

Figure 32.1 The ideal point of interconnection is at the half-way point between the carriers.

bad debts, is not the same for each call and cost recovery should allow for this collection cost.

KEEPING IT SIMPLE

There are a few basics of call charging and tariffing that need to be considered.

1. Tariffs should reflect costs; if not, distortions in user patterns will result so that the underpriced item will be overused and vice versa.
2. When multiple carriers are involved, consider the costs of each separately and make sure all costs are covered. Do not try to shortcut this by charging a surcharge on the "main areas of revenue" and applying a nominal charge to other services.
3. Do not guess user patterns; design a rate structure that does not need assumptions about the users' habits so that if they change the system still works.

The complexity of determining the actual true cost of traffic flows is very high. Apportioning the amortized costs of equipment, which may have an economic life of 25 years, to individual calls is not simple because the answer depends greatly on the initial assumptions made. In the United States this problem was circumvented by having the mobile users pay for incoming calls. This is not consistent with the user-pays principle and leads to distortions in user habits, but it is simple. Furthermore, the land-line carriers were just not conditioned to the idea that they should collect call charges and pass them on to their new competitors. However, elsewhere in the world the user-pays principle is mostly followed, although there are some exceptions.

In recent times even in the United States, there has been a shift to user pays.

COST PER MINUTE OF TRAFFIC FLOWS

Many schemes have been proposed to allow the determination of the fair charge for interconnect. Those based

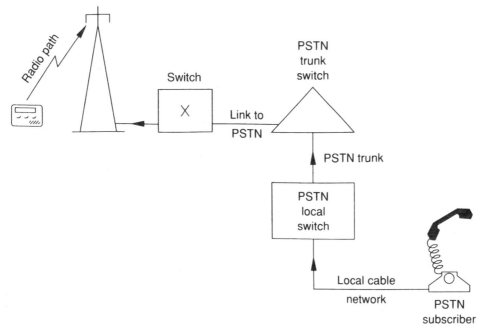

Figure 32.2   The physical path of a call to/from a cellular phone to a PSTN subscriber.

on the cost of a switch inlet or outlet ignore the fact that the first point of interconnect does not reflect the whole cost of carrying a call. The actual call path of a typical cellular phone to PSTN subscriber can be seen in Figure 32.2.

Because there is a very large number of different physical paths possible between the two systems, it is necessary to find some fair and reasonable "average path" and cost it. Next it will be necessary to determine the manner of measurement.

In general, the fairest way will be to base costs on minutes of actual usage, as this is very easy to measure and is easily verifiable by both parties. This differs from revenue, which will often be on a fixed cost per call, or for toll calls on a $1 + 1$ per-minute basis (that is, on answer a one-minute charge is already levied regardless of the actual talk time).

There are a number of ways that the cost of an average call could be determined. The simplest is on the the average-cost-per subscriber basis. For example, a 20,000-line cellular system may have cost $15,000,000 to set up, making the average cost per subscriber $750. To convert this to an annual cost we can allow for 10 percent depreciation, 10 percent maintenance and operations, and 15 percent return on capital investment to derive an annual return of 45 percent required. If the average subscriber uses 2000 minutes of airtime per year, then the cost of one minute of usage is $750 \times 0.45/2000$ or $0.17 per minute.

Here it must be appreciated that an originated call must involve the terminating party so that any call involves two subscribers. Assuming then that the average subscriber cost is set for calls purely within the cellular system, then the above cost should be halved to get the true per-minute cost of a single call. That is, the cost of a call minute in a cellular system is $0.085.

In a similar manner it would be possible to determine the cost of a call carried by the PSTN operator. This cost should not be much different, as the average cost of a new PSTN line is around $700 and the PSTN would have a larger percentage of depreciated assets, which would bring down the average cost per call to something below the cost of a cellular call.

A major difference between PSTN and cellular costs is the economic equipment life, which may be as high as 30 years for some PSTN equipment, but which ranges from 5 years to 10 years for cellular equipment. For this reason the average network book value of the PSTN per subscriber would be between $600 and $900, although very new and very old networks may lie outside these limits.

The cost of a call may seem high, but it must be remembered that the charges should be applied in both directions and so only the net settlement is important. Thus if the traffic were the same in both directions, there would be only a small net charge between the operators.

In general there will be a net imbalance in traffic between the PSTN and the cellular network such that

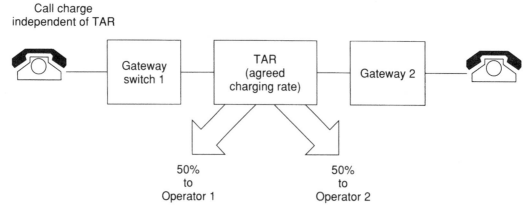

Call charge
independent of TAR

50%
to
Operator 1

50%
to
Operator 2

Figure 32.3   The TAR (assumed here to be $3.00) is usually split equally between the two operators.

the cellular operator will generate more traffic to the PSTN than it will terminate. This means that there will probably be a net settlement in favor of the PSTN operator; also the long-distance toll charges that are applied go largely to the PSTN operator.

DETERMINING FAIR COST FOR INTER-CARRIER TOLLS

International toll charges have long been established on the basis of an internationally agreed total accounting rate (TAR). This rate, which is usually somewhat arbitrary, is the agreed value between two international carriers of one call minute of traffic between them. The proceeds of the TAR are then split, usually 50/50, between the operators on the basis that the originator of the traffic pays. This settlement is illustrated in Figure 32.3.

TARs are often relics of an era when transmission costs were far higher than they are today. This means that some serious anomalies exist. In most cases, because of net settlement, the actual tariff is almost irrelevant; the net outflow is zero. Some countries do have large net outflows because of immigrant or migrant workers calling home. On the other hand, the immigrant sourcing countries often earn large net inflows of capital from this imbalance. In the United States in 1989 the net outflow settlement exceeded $US 2 billion.

Notice that the TAR need not bear any direct relationship to the actual toll charges paid by the PSTN subscriber.

For international calls outgoing from a cellular operator there is thus a clear basis for cost accounting. That is, the country portion of the TAR is the cost. For incom-

ing international calls the cellular operator need only treat them as local PSTN calls (at least for tariff purposes).

National calls may sometimes have an agreed cost value (as, for example, in the United States), but in most countries the only available basis for costing will be the published toll rates. In order to avoid too much debate over the true cost of a national toll call, the cellular operator will usually negotiate these calls at the published toll rate minus some discount (typically 20 percent to 40 percent).

Based on the above, a formula for settlement between a local cellular operator and a national PSTN carrier could be as follows:

Due to the Cellular Operator for PSTN-Originated, Cellular-Terminated Traffic

1. A fixed rate per minute of call determined on a cost basis regardless of the origin of the call.
2. Where there is an arrangement between the PSTN and the cellular operator that the PSTN operator will collect the air-time charge for PSTN originated calls, this will be added.

Note: There is an interesting philosophical difference between cases 1 and 2 above. In case 1, it is the cellular user who pays for airtime regardless of whether the call is originated or terminated. This has an advantage for some users who want to use the phone for business and do not want their customers deterred from ringing in by the high cost of making the call.

In case 2, the principle that is applied is "user pays," and this would be considered more fair by a lot of operators. Some complications that can arise if the user-pays principle is adopted are considered later in this chapter.

Due to the PSTN Operator for Cellular-Originated, PSTN-Terminated Traffic

1. For local traffic, fixed rate per minute as determined on a cost basis.
2. For national traffic, the normal published toll tariff minus a negotiated discount.
3. For international traffic, the national part of the TAR plus a reasonable return on the call cost. In no instance should this exceed the whole TAR.

Other Settlements

A charge is applicable for the use of the circuits that connect the cellular operator to the PSTN, and these should be accounted on a usage basis of cost-plus.

The carriers should add a collection fee to their subscriber bills to cover the cost of collections and bad debts. As far as possible, the PSTN operator and the cellular operator should agree on these charges so that the customers of each can be seen to be treated fairly and equally.

NATIONWIDE OR WIDE-AREA CELLULAR OPERATORS

The question of toll charging is made more complex when both the PSTN and the cellular operator operate over areas greater than the local charge zone. In countries where the PSTN is the mandated long-distance carrier or where all the trunks used by the cellular operator are provided by the PSTN carrier, the cellular system can be treated as a number of local-area systems.

In some cases the cellular operator will have an extensive trunk network designed to connect the base stations and switches comprising the cellular network. In this case, it may well be that the cellular operator is capable of hauling long-distance traffic over the cellular link networks. In fact, where the cellular system is substantial it may even be that in some areas the cellular operator has a more extensive trunk network than the PSTN operator does. This will be particularly the case when the cellular links are provided by medium-capacity digital radio or fiber-optic links. Figure 32.4 depicts a cellular system operating over three PSTN charge zones where both carriers have full trunk facilities. Notice that a call originated in zone 3 can be routed so that either carrier provides the toll route part or so that the toll portion is shared.

In addition to the local, national, and international agreed rates, it can be seen that it is also necessary that each inter-system trunk route will then have an agreed inter-operator access charge.

The respective operators would then charge the end-subscriber the same amount regardless of the routing or a route-dependent charge. Route-dependent charging is of course more complex. The inter-operator settlement would be separate from the individual call and would be a bulk settlement based on the trunk switch records.

USER-PAYS PRINCIPLE

When the user is the one to pay for the call, it is necessary for the PSTN to be able to bill the land-line caller to a cellular phone. While this presents no problems in the case where the cellular network belongs to the PSTN, there are some interesting legal and technical problems when the two networks are independently owned.

The PSTN and the cellular operator will both have a written agreement with their respective customers that obliges them to pay for the calls they are responsible for. However, there may be no legal way of making a PSTN customer directly responsible to the cellular operator for calls originated from the PSTN phone to a cellular phone. In fact, often it would not be possible for the cellular operator to determine the identity of the calling PSTN party. The PSTN customer will be obliged to pay for originated calls regardless of destination and so the PSTN operator can collect such revenue.

When roaming is available, an additional complication can occur. The PSTN will hand the call either to the nearest cellular switch in the roamable network or to the home switch of the cellular subscriber called.

If automatic roaming has occurred, the called cellular phone may be at neither the nearest switch nor the home switch but rather at some third switch. This information will be recorded in the billing tapes of the cellular operator, but it cannot easily be conveyed to the PSTN.

Most of these complications can be overcome if the two networks forward to each other all call details. In the case of the wireline network, the caller's number identifies the location. In the case of a cellular operator, the identity of the mobile gives no information about location because there is the possibility of roaming. To charge effectively, in general, the PSTN operator will need to know the actual base station being used in order to locate the mobile. This can be further complicated if the mobile is operating near a boundary of two switching areas and may, within the duration of a single call, band off to a new charge zone.

Hence in recent times most cellular operators have opted for a fixed-call charge nationwide.

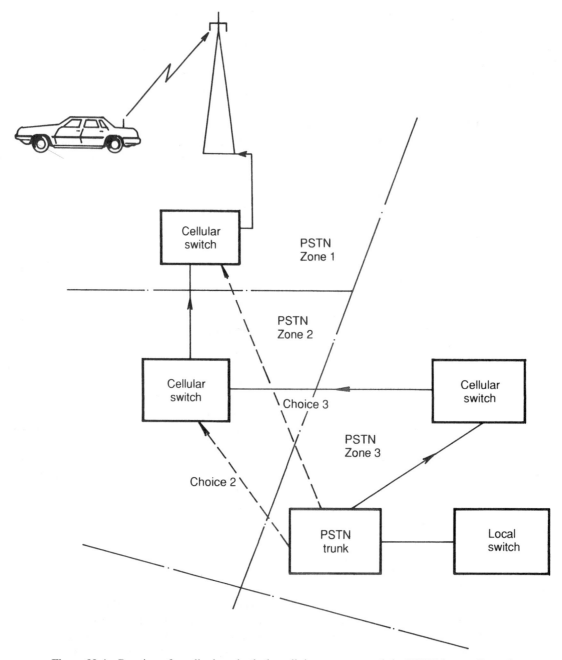

Figure 32.4 Routing of a call where both the cellular operator and the PSTN have toll trunk facilities can be complex as shown by the possible routes that the PSTN call in zone 3 can take to get to a cellular subscriber in zone 1.

THE GSM CHARGING PRINCIPLES AND THEIR APPLICATIONS TO OTHER NETWORKS

Where international roaming is envisaged this can become even more complex. One of the few Global System for Mobile Communications (GSM) procedures that is straightforward involves the call-charging principles.

They can be used as a starting point for call charging, with individual operators changing those parts that need to be separately dealt with to meet local constraints.

The general principle that the calling party pays is retained with a few exceptions that greatly simplify charging:

- Reverse charge calls are charged to the party called.
- The calling party is charged only for calls to the home network (HPLMN).

Note that for national networks involving non-wireline/non-GSM systems, it may be appropriate to replace the HPLMN with the "home switch." This means the PSTN does not need any information about the mobile's location, and thus greatly simplifies charging. The mobile will then pay for the transferred call if roaming is in use. Note that this is similar to call forwarding in that the caller will not be aware that the call is to be transferred outside the home area and so arguably should not have to pay for it.

- Forwarded calls are charged to the subscriber who initiated the forwarding. This still means that the caller pays the call path that would have been involved had call forwarding not been activated.

Because of the basic principle that each operator must be free to set its own tariffs, international roamers should for simplicity pay the same as local users when roaming. GSM provides for this, and the visited system (VPLMN) will automatically forward the charges accordingly.

Mobile Terminated Charges

Again in mobile terminated charges the user-pays principle is applied, but optionally (and simplest) for roam-ing mobiles the mobile subscriber can be charged for the cost of the call rerouting.

The Roaming Subscriber Bill

For originated calls a roamer will be charged the same per-call rates as a local subscriber. In GSM, within the first months, this principle led to some subscribers buying and registering mobiles on networks from nearby countries with the cheapest monthly access charges.

For incoming calls, based on the calling-party-pays principle, the international roaming subscriber will not get a bill for the *airtime* on incoming calls. This part of the call record is recorded as zero charge. However, the originating caller will be charged as though the call were terminated in the home network (HPLMN), and thus, in effect, the termination charge has been paid. The mobile roamer will pay the rerouting network costs for received calls.

Charging for Supplementary GSM Services

The GSM recommendations make it optional to charge for supplementary services. Additionally, activation of those services can be separately charged, although there should be no charge for deactivation. Call charges may apply to each use or duration of access of the services. No charges are to be applied for interrogation of the status of a service.

33

PREPARING INVITATIONS TO TENDER

Preparing an invitation to tender consists of writing a document that sets forth the commercial and technical specifications required from a supplier of equipment and/or services. An operator extends an invitation to tender in order to elicit several detailed bids so that the most suitable supplier can be selected. The bid selected will not always be the lowest one, a possibility that is usually stated in the tender offer. Price, of course, is for most operators a major consideration, but many other factors, such as the type and quality of the offer, as well as support and service, may affect the final choice. Invitations to tender may be issued on an open basis or only to select suppliers. Government agencies often require open bidding. Open invitations may be published in major newspapers.

TECHNICAL PREPARATION

It is important that the person preparing the invitation to tender has a clear idea of the type and quantity of equipment and services required. The invitation is not the place to "feel out" the market to find out what is available. The person(s) writing the invitation to tender should do the required homework and, if necessary, feel out the market before the invitation to tender is written.

Responses from suppliers are usually detailed, extensive, and reflect the original tender offer. Most suppliers prefer a clearly defined invitation to tender that specifies exactly what is to be supplied and leaves no loose ends. An operator cannot expect a supplier to guess what is really needed. An operator who is not confident of the system defined in the invitation to tender can invite suppliers to offer alternative configurations in addition to the one specified. This gives the tenderer a chance to input new or additional ideas without being disadvan-

taged with respect to other suppliers and without running the risk of having a nonconforming tender. It should be realized, however, that the preparation of a tender response is quite time-consuming and generally does not leave time for detailed analysis of alternatives.

It is to the operator's benefit to prepare a precise invitation to tender, as will become obvious when the bids are received. If the invitation to tender is vague and imprecise, then the responses will be likewise. Ultimately it is up to the operator, when assessing the bids, to draw conclusions about the best one. If an invitation to tender is not precise, then the responses are likely to be so different as to make comparing them difficult or impossible.

A clearly written invitation to tender invites a more serious response because it is obvious to the supplier that the operator is serious. Suppliers may decline to bid if they do not feel confident that they understand what the operator wants. Remember that preparing a cellular tender bid is an expensive, time-consuming process that is not undertaken lightly by a supplier.

An invitation to tender is usually divided into four parts, as seen in the sample, later in this chapter. This sample is for an imaginary company called *Supercell*, which intends to provide a system with a capacity of 6000 subscribers with 296 radio frequency (RF) voice channels in a town called *Mobile Town*. This invitation to tender assumes that *Supercell* will provide and select sites, and that it will also provide the microwave links, towers, shelters, and power (including rectifiers and batteries).

The four parts of the invitation to tender are as follows:

- *Terms of Tender* This part details the terms under which a bid will be accepted, the documentation required, and how the contract will be awarded. It also defines the terms used.

- *General Conditions* This part defines in more general terms deliveries, timing, acceptance of deliveries, schedules, payment, default, warranties, and other conditions that apply to the bid once it is accepted.
- *Technical Specification* This part gives technical specifications of the goods or services to be supplied. In the sample in this chapter, for example, this section clearly sets forth the number of subscribers (6000), bases (10), and voice channels (296).
- *Appendices and Specifications* Any additional information or attachments are included here.

All the features required are detailed so that the supplier has a clear idea of what is required. This specificity helps both supplier and operator.

The following sample tender is included only as an example, and its application to a particular case may be limited. The first two parts, "Terms of Tender" and "General Conditions," should be evaluated by legal experts in the country of application to ensure that local legal requirements and liabilities are covered. The technical part must agree with the actual technical requirements of the operator.

SAMPLE TENDER OFFER

I. Terms of Tender

1. **Invitation to Tender** Suppliers are invited to tender for the installation and provision of a Cellular Mobile Telephone System based on the Advanced Mobile Phone Service (AMPS) system for *Supercell* to be installed in *Mobile Town*.
2. **Alterations to Tender** No alternations to the tender documentation will be permitted but alternate proposals that offer enhancements may be considered.
3. **Closing Date**
   3.1. Tenders submitted up to *12 Noon October 10, 2000* will be considered valid.

      Tenders should be sealed and sent by verifiable means. No responsibility or allowance for delays will be made by *Supercell* for any tender documents sent by mail.

      The tender should be delivered to:

> *Supercell*
> *100 Hustler Street*
> *Mobile Town*
> *Phone* . . .
> *FAX* [facsimile] . . .
> *Email* . . .

Enquiries should be directed attention:

> *Mr. Fast Bucks*

The original plus one copy should be sent.

   3.2. Late and incomplete tenders may not be considered.
   3.3. The tender document must be signed by a duly authorized person acting on behalf of the tenderer.

4. **Documentation**
   4.1. The tenderer will answer each point listed in the tender and provide either the required information, or note each as either "fully complied with," "partially complied with," or "not complied with." In the last two instances, details should be given of the degree of noncompliance.
   4.2. All attachments and technical drawings will have drawing numbers that clearly identify them.

5. **Pricing** Separately itemized prices are required for the following. Equipment prices are free on board (FOB).
   a. Switch equipment:
      i. An initial 6000-line capacity
      ii. For 50,000 lines
      iii. For small (approx. 5000 line max.) switch(es)
      iv. For expansion on per subscriber basis and detailing any step functions
      v. Recommended peripherals
   b. Cell equipment for the initial sites:
      i. 10 voice channels, omni
         30 voice channels, omni
         3×12 voice channel, 120-degree sectored sites
      ii. Expansion on a per channel basis
      iii. Any step functions in expansion (for example, increases in switch capacity)
      iv. Site controller expansion
   c. Installation costs, including cabling
   d. Rectifiers, batteries, and power control panels (optional)
   e. Software packages and support
   f. Management information packages
   g. Training courses, to take place on *Supercell Premises*
   h. Prices for all recommended spare parts needed for an 18-month period
   i. Any specialized test equipment
   j. Cost and arrangements for repair
   k. Travel and living allowances

      Prices should be presented as an itemized price list for the system tendered and additional information (break points and expansions) presented separately.

6. **Lowest Tender not Necessarily Accepted** *Supercell* will not necessarily award the tender to the lowest bidder.

7. **Currency of Tender**   The prices should all be FOB prices in US dollars.
8. **Costs of Preparation of Tender**   The costs of preparation of tender will be borne totally by the tenderer and will in no part be the responsibility of *Supercell*.
9. **Subcontractors**   If the tenderer intends to use subcontractors for any part of the work, this should be clearly indicated.
10. **Validity of Offer**   The offer will be valid for a period of four months after the closing date, but may be extended by mutual agreement of both *Supercell* and the tenderer. During that period, prices offered are binding.
11. **Referee**   The tenderer should indicate any references who have purchased similar equipment and would be willing to confirm satisfactory operation.
12. **Inquiries**   All inquiries will be directed to the address indicated in clause 3.1.
13. **Responsibility for Information**   It is the tenderer's responsibility to ensure adequate information for submission of the tender not withstanding any information contained herein or afterwards supplied by *Supercell*.
14. **Award of Contract**   *Supercell* will evaluate the offers during the period *October 10, 2000 to December 10, 2000*.

    A short list of tenders will be produced on or about *December 15, 2000* and short-listed tenderers will be invited to final negotiations.

    On or about *February 15, 2001* the successful tenderer (the Contractor) will receive a letter of intent.

    Final negotiations will take place on or about *February 20, 2001* and the successful tenderer shall attend a final negotiation meeting. The representatives of the tenderer must have full negotiation powers over any part of the Contract.

    Tenderers who do not receive notification after this date should assume they have not been successful.
15. **Definitions**   *Supercell. Means Supercell Pty Ltd*, the purchaser.

    Contract Price. Means agreed price payable to the Contractor by *Supercell* under the Contract for full and proper performance by the Contractor in the execution of the works and provisioning as agreed.

    Contractor. Means the Contractor named in the Contract and includes any successors and permitted assignee(s).

    Contract Amendment. Means the document that changes prices, date of completion, and supplied items pursuant to clause 13.

    Tender. Means the offer submitted by the Contractor to *Supercell*.

Tenderer. Means one of the invited bidders. The successful tenderer becomes the Contractor.

Contract. Means the agreement concluded between *Supercell* and the Contractor named herein incorporating these conditions and includes:

a. Any Contract amendment
b. All agreed specifications, plans, drawings, and other documents that are prepared pursuant to the said agreement

II. General Conditions
1. **Acceptance**
   1.1. Acceptance will be deemed to have occurred 30 days from the issuance of a duly signed acceptance certificate.
   1.2. Each base station and the switch will be separately accepted.
2. **Rejections**
   2.1. *Supercell* reserves the right to reject goods that are damaged, spoiled, or do not meet specification.
   2.2. Upon written advice of rejection (FAX or other agreed method), the supplier will undertake to replace the rejected goods. The replacement is to be completed within 45 days. The Contractor will advise by return FAX the expected delivery dates.
   2.3. The Contractor will be liable for all shipping costs of rejected goods.
   2.4. In the instance where the replacements are not made good as required by 2.1, 2.2, or 2.3, then *Supercell* reserves the right to:
      a. Terminate the Contract and claim liquidated damages.
      b. Claim liquidated damages but replace the rejected goods with other suitable goods.
      c. Claim liquidated damages and terminate the Contract where the same item has been rejected and so has its subsequent replacement.
3. **Time of Essence**
   3.1. Upon receipt of a written order from *Supercell*, the Contractor shall supply and deliver in accordance with the delivery schedule.
   3.2. Where delivery dates are doubtful, the Contractor shall inform *Supercell* at the earliest time of any foreseen delays.
4. **Delivery Schedule**
   4.1. Switch—160 days from receipt of order.
   4.2. Base Stations—120 days from receipt of order.
   4.3. Other items—120 days from receipt of order.
5. **Packing and Documentation**   All items delivered shall be suitably packed for transportation and will be accompanied by a packing list detailing each part number and item delivered.

6. **Subcontractors** The Contractor will not, without written consent from *Supercell*, transfer any part of the Contract to subcontractors. All subcontractors will be wholly the responsibility of the Contractor.

7. **Liability for Personnel**

   7.1. The Contractor shall indemnify *Supercell* from any liability for injury to its staff or any staff employed or subcontracted by the Contractor while performing work on *Supercell*'s sites or on behalf of *Supercell*.

   7.2. Evidence shall be produced that the Contractor has suitable insurance to cover the liability mentioned in 7.1.

8. **Acceptance of Deliveries**

   8.1. *Supercell* will provide a note of acceptance on all deliveries made to it.

   8.2. Goods not rejected within 40 days of unpacking by *Supercell* will be deemed accepted.

9. **Terms of Payment of Equipment**

   9.1. *Supercell* will pay 80 percent of the costs of shipped goods within 10 days of receipt of all of the following:

   a. The goods

   b. Delivery documents detailing items delivered, order number, and price

   c. Shipping documents, including certification of contents by a suitable authority (to be agreed to between the Contractor and *Supercell*)

   9.2. A further 10 percent of the total price will be paid after acceptance of the goods by *Supercell* (as per clause 8).

   9.3. The final 10 percent, being the balance, will be paid within 30 days of the *system* acceptance.

10. **Payment for Services**

    10.1. Upon completion of each service, the Contractor will invoice *Supercell*.

    10.2. *Supercell* will, upon receipt of the invoice, confirm that the service has been satisfactorily performed and will pay 80 percent of the costs within 30 days.

    10.3. A further 10 percent will be paid within 30 days of the issuance of an acceptance certificate.

    10.4. The final 10 percent will be paid after acceptance as defined in clauses 1.1 and 1.2.

11. **Installation Schedule**

    11.1. All base stations will be installed and ready for acceptance within a period of 120 days from delivery.

    11.2. The switch and peripherals will be installed within 120 days from delivery.

12. **Personnel Qualifications** The tenderer will be required to submit the curriculum vitae of all supervisory personnel to *Supercell* for approval. Only approved supervisors will be employed.

13. **Default**

    13.1. If the Contractor fails to deliver any items within the specified periods (clause 4) or fails to obtain an acceptance note (clause 8), then *Supercell* may:

    a. Contact the supplier and require a satisfactory (to *Supercell*) explanation and assurance of delivery within 14 days, or

    b. Terminate the Contract forthwith, without penalty to *Supercell* by notification in writing to the Contractor. *Supercell* reserves all rights to claims of breach of contract and may obtain the outstanding items from alternate suppliers.

    c. Make the Contractor liable to any costs incurred in rectifying the shortcomings.

    13.2. Penalty *Supercell* may decide to take action on Contract termination due to late deliveries or failure to obtain an acceptance. In this instance, a penalty of 3 percent of the total Contract price per full 30 days of delay may be applied.

14. **Alteration of Contract** *Supercell* reserves the right to alter the Contract at any time. In the event of an alteration, a mutually agreeable amendment to price, and where relevant, delivery times, will be made by *Supercell* and the Contractor. Any such amendment will be agreed to and confirmed in writing by both *Supercell* and the Contractor before being considered valid.

15. **Bankruptcy** In the event of bankruptcy by the Contractor or such other events as detailed below, *Supercell* may summarily terminate the Contract.

    a. If the Contractor shall at any time be adjudged bankrupt, or shall have received an order for receivership or for administration of Contractor's properties or shall take any proceedings for liquidation or compensation under any Bankruptcy Ordinance for the time being in force or make any conveyance or assignment of Contractor's effects or compensation or arrangement for the benefit of Contractor's creditors or purports to do so, or

    b. If the Contractor, being a company shall pass a resolution or the Court shall make an order for the liquidation of its assets or a Receiver or Manager shall be appointed on behalf of each creditor or circumstances shall have arisen which entitle the Court or debenture holders to appoint a Receiver or Manager. Provided always that such determination shall not prejudice or affect any right or action or remedy which shall have accrued or shall accrue thereafter to *Supercell*.

16. Warranty

16.1. The Contractor will warrant *Supercell* for a period of 12 months, from the issuance of Acceptance as defined by clause 1.1, against all defects in workmanship and materials.

16.2. The Contractor shall make good without undue delay all defects in workmanship and materials as are brought to Contractor's attention by *Supercell* within the warranty period.

16.3. The warranty will cover all costs except freight and insurance charges for the shipping of items sent for repair.

16.4. Clause 16.3 does not apply to any items delivered but not accepted as per clause 8. Such items will be returned at the Contractor's expense.

16.5. The Contractor will undertake to make good all warranty claims with all possible speed, and in any case within 60 days of notice in writing from *Supercell*.

16.6. In the event of the failure to meet clause 16.5, *Supercell* may undertake to remedy any defects at the risk and cost to the Contractor.

16.7. The Contractor will be liable to *Supercell* under the warranty terms, regardless of whether or not the goods as supplied were in part or whole manufactured by Contractor.

The supplier shall satisfy *Supercell* that any goods or services supplied by subcontractors or other suppliers carry the same warranty as required by *Supercell*.

16.8. Software support for a period of 12 months will also be included in the warranty.

17. Hardware or Software Updates The tenderer will describe the procedure for the release of hardware and software upgrades and their implementation.

18. Copyright Indemnity The Contractor shall indemnify *Supercell* against any infringement of copyright, patents, trademarks, or registered design rights related to the equipment, system, or design supplied in relation to this Contract and shall affirm this in writing.

The Contractor will be liable for all costs of such infringement including negotiation and litigation costs.

19. Support The Contractor will guarantee support for any hardware and software provided for a period of eight years from acceptance.

20. Arbitration

20.1. In the event of any dispute or controversy in a matter of interpretation, implementation, and enforcement of this Contract, the same shall

be resolved through arbitration by the Chamber of Commerce of the City of *Mobile Town* and the decision will be final.

20.2. In the event that both parties shall, in writing, waive the right to avail themselves of the arbitration clause, then any controversy or dispute shall be referred to the Court of the City of *Mobile Town*.

21. Settlement of Disputes The laws of the land applicable to the City of *Mobile Town* shall apply in the interpretation, implementation, and enforcement of this Contract. In the event of any court action that any party may bring against the other under this Contract, the venue of such court action shall be in the Court of the City of *Mobile Town*.

22. Confidentiality During the term of this Contract and for a period of four years after its fulfillment, neither party shall intentionally disclose or permit to be disclosed to any third organization any information of a confidential or propriety nature concerning the other party.

23. Performance Bond The Contractor shall issue, within 60 days of acceptance of the Contract, an irrevocable Performance Bond (Letter of Credit) acceptable in the City of *Mobile Town* to the value of 10 percent of the total Contract value and to expire 12 months after the date of Acceptance.

24. Government Regulations

24.1. The Contractor will be responsible to ensure that all relevant government and local authority requirements and regulations are complied with in full.

24.2. The Contractor will indemnify *Supercell* against any penalty or loss arising from the Contractor's non-compliance with any such regulations, laws, or requirements.

25. Order of Precedence The order of precedence in the Contract is:

a. Part I, Terms of Tender
b. Part II, General Conditions
c. Part III, Technical Specification
d. Appendices and Specifications

III. Technical Specification

1. General *Supercell* is seeking an AMPS system to be installed in the City of *Mobile Town* which conforms to the following US specifications:

a. FCC—TITLE 47 CFR PARTS 2, 22
b. EIA—IS-3-D, IS-19-B, IS-20-A

The system will operate on the AMPS B band and be capable of operation on the extended B band.

Provision will be made for an initial installation of 6000 subscribers with 10 base stations and one switch. Digital compatibility will be required and the degree of compatibility should be clearly stated. A migration path to CDMA must be available.

Expansion to 50,000 lines is anticipated. Future rural expansion will require the provision of additional switches with capacities of about 5000 subscribers. Fully automatic roaming is to be provided between the switch areas.

A 7-cell frequency pattern will be employed (that is, $N = 7$).

Supercell will provide only the following:

a. Site selection, clearing, base station, switch shelters, towers, mains power, rectifiers, and batteries
b. Microwave 2 Mbits (or 2/8), 30-channel links, with channel equipment between bases and the switch
c. A 34-Mbit link from the switch to the public switched telephone network (PSTN)

2. **The Switch**
   2.1. **General**
      a. A stored program, full-availability (non-blocking) switch will be provided. Automatic roaming and interswitch handoff will be provided.
      b. The switch will be designed to interconnect at trunk level to the PSTN and to at least 10 other networks.
      c. Modularity of construction will be a switch feature.
      d. Digital voice announcement capability of at least ten 30-second announcements will be provided.
      e. Dual processors with complete redundancy will be provided. Automatic changeover in the event of processor error will occur and a detailed record of the changeover and its causes will be available.

      Provision will be made to take one processor off-line at any time for service or other purposes.
      f. A voice message facility with Dual Tone Multifrequency (DTMF) priority control will be provided. At least five categories of message will be available together with five retrieval categories.
      g. Both a positive and negative validation file will be provided.
      h. A printer will be provided which will:
         i. Print out all alarm details.
         ii. Print out all information input to the switch.
         iii. Print out details of any system reconfigurations whether automatic or manually instigated.

i. All information directed to the printer will also be recorded on a portable electronic medium. This tape will also record details of all calls and call attempts with full call accounting (that is, A & B subscriber number, call duration, and details of unsuccessful attempts). The call duration and airtime shall be separately recorded.

j. A management information system (MIS) will be required. Full details of compatible management information systems available will be submitted.

k. A means of loading programs and patches and adjusting any system parameters will be provided. A detailed description of the associated hardware and software is to be provided.

2.2. **Facilities** The following minimum facilities are required:

2.2.1. Real-time billing access will be provided for selected subscribers' billing details either at the switch site or at a remote location will be provided.

2.2.2. Call tracing will be provided, and details of any current call including base station, channel number, circuits used, and called or calling party, will be provided.

The tenderer should state if historical records are available of the above details and the period for which those details are available. It is envisioned that such details could be useful for fault tracing and malicious call traces.

2.2.3. A priority class subscriber category will be provided.

2.2.4. Call interception will be available.

2.2.5. Voice encryption will be available for some subscribers. The tenderer should give details of voice encryption operation and the proposed means of activation/deactivation.

This feature will be available to subscribers at extra cost. The tenderer should state the degradation in signal-to-noise (S/N) that will result from the use of the proposed encryption and the degree of security provided.

2.2.6. A facility to automatically access the cellular alarm and diagnostic system will be provided.

2.2.7. The following subscriber facilities will be offered as optional features to *Supercell's* customers.

a. Call Waiting

Some indication that another call is waiting will be provided.

b. Call Waiting Cancel

The subscriber can cancel the call waiting feature (a) from the mobile phone.

c. Follow-Me

The subscriber can redirect incoming calls to any other number. This will have two modes, one permanent and the other that automatically cancels at midnight on the day of activation.

d. Three-Way Conference Call

A subscriber can call a third party while holding an existing call. All parties can then interconnect in the conference mode.

e. Call Hold

An existing call can be placed on hold.

f. Voice Mail Transfer

The customer can transfer calls to the voice mail system.

g. Call Charge Information

A means of providing real-time billing information to the customer will be available. This will preferably be fully automatic.

h. No Answer Transfer

Calls are automatically diverted to another number if the mobile phone is not answered after a predetermined period.

i. Call-Received

A means of alerting the subscriber that a call was received while the mobile was unattended.

2.2.8. Both automatic periodic switching-in of redundant equipment and manual changeover will be available.

2.2.9. Alarms shall be categorized and a facility to output major alarms by audiovisual means will be provided.

2.2.10. A facility to monitor the activity of any trunk or link will be provided, which will include:

a. Audio trace

b. Call details

c. Loop test

d. Frequency-response test

2.2.11. Full details of diagnostic programs available will be supplied.

2.2.12. The system will handle local numbers of a minimum of 15 digits and international numbers of 24 digits.

2.3. Details to be Provided

2.3.1. A floor plan of a typical fully equipped switch room with all required peripheral equipment will be submitted. The plan will show the initial 6000-line installation and the subsequent expansion to 50,000 lines.

2.3.2. Power consumption, voltage, and heat dissipation for both the initial 6000- and ultimate 50,000-line switch will be stated.

2.3.3. Maximum floor loading and physical dimensions for each type of rack will be stated.

2.3.4. Operational temperature range for each type of equipment will be stated.

2.3.5. The number of alarms that can be handled from each base station and from other peripheral equipment is to be stated.

2.3.6. The degree to which the system can be reconfigured as a result of a fault will be detailed.

2.3.7. Details of the mean time between failures (MTBF) of the switch and of each of its component parts (such as boards and racks) will be provided.

2.4. Management Information Systems The MIS should be integral with the switch and will collect data for proper management of the system.

2.4.1. The minimum requirement for the MIS is that it will provide the following data:

a. Traffic information on all circuits incoming and outgoing to the switch

b. Circuit availability

c. Data error rates on links to base stations

d. Base-station transceiver performance (including availability)

e. Channel blocking both manual and automatic

f. Control channels usage and outages

g. Call attempts and call success rate

h. Call holding times

i. Total traffic and total traffic per cell as a function of time of day

Details of other MIS data available should be provided.

2.5. Hardware

2.5.1. The switch will be of modular design and be capable of upgrades without causing any system loss. In particular, upgrades of the following will be available without interruption unless the maximum capacity is provided at start-up:

a. Processor capacity

b. Number of input/output ports and peripherals

c. Memory capacity

d. Switch capacity

The tenderer shall state the start-up and ultimate capacity of the above items.

2.5.2. The tenderer will be responsible for the compatibility of any peripheral equipment offered and for the commercial availability of spare parts and consumables for a period of five years from delivery.

2.5.3. The central processor shall be capable of performing simultaneously all housekeeping and switching functions in both the installed and ultimate capacity (50,000) without overload resulting in lost calls.

2.6. Software

2.6.1. The software shall be fully documented and training provided for *Supercell* engineers to enable them to make any necessary patches. An agreement between the Contractor and *Supercell* to allow approved patches will be entered into.

2.6.2. The software should be of modular design and capable of partial updates without system downtime.

2.6.3. Support and updates of software for 12 months will be provided as part of the warranty.

2.6.4. All software updates shall be backward-compatible, tested, approved, and well documented.

3. Base Stations

3.1. General The base-station equipment to be provided consists of a site controller, transceiver racks (each of which contains at least eight channels), rectifiers, batteries, and associated equipment.

Supercell will provide all shelters, which will be transportable buildings with a floor plan of 3.1 × 5 meters and floor-to-ceiling height of 2.7 meters. The buildings will have a maximum floor loading of 1000 kg/m and be provided with single-phase 240, volts alternating current (VAC), 50-Hz power. *Supercell* will also provide towers, antennas, and feeders.

3.2. Facilities and Equipment

3.2.1. The base stations shall be provided with back-up batteries sufficient for three hours' operation when fully equipped (45 channels) including air conditioning load. Automatic and manual changeover to batteries will be provided. The tenderer will specify the required battery capacity for the equipment.

3.2.2. Complete equipment-monitoring facilities will be available at the site controller. The tenderer will state how manual access is gained to the controller and what additional hardware is needed.

3.2.3. The ability to quickly replace faulty channels will be required. The tenderer should state the method used to tune a replacement transceiver.

3.2.4. All critical components will be monitored by alarms that are accessible at the base station and remotely at the switch. At least four auxiliary alarms, such as entry and power, will be required.

3.2.5. The control channel will use a voice channel as a standby control channel in the event of control channel failure. The changeover will be able to be effected both automatically and manually.

3.2.6. Continuous field-strength monitoring will be provided. It is preferred that a standard channel unit (that is, transceiver or receiver) be used for field-strength monitoring so that special equipment is not required. It is preferred that a voice-channel receiver can be used as a standby signal strength receiver. The supplier will state details of signal strength measurement.

3.2.7. Diagnostics of local equipment faults will be carried out by the site controller and automatically sent to the switch.

3.3. Information Required The tenderer will provide the following information:

3.3.1.

a. Power consumption (for each base-station configuration), voltage, and air conditioning load of the base station

b. Floor loading, maximum weight per rack

c. Rack dimensions and empty weight

d. Environmental operational constraints

e. Degree of redundancy

f. Typical floor plan of a fully equipped base

g. MIBF of a base station and each of its modular parts and details of the method of calculation

3.3.2. The degree of remote control of base-station parameters will be clearly stated. Also, whether all base-station alarms and diagnostics are forwarded to the switch.

4. **Traffic and Planning Considerations**

   4.1. **General**

      4.1.1. The traffic capacity will be based on an average subscriber having a calling rate of 30 mE and a call holding time of 130 seconds. The grades of service will be 0.05 on the RF path and 0.002 on the switch to PSTN path.

      4.1.2. Based on 6000 subscribers at 30 mE, the total PSTN traffic of 180 E will be based on 212 circuits to the PSTN and 296 RF channels.

      4.1.3. The base stations will be provided with the following circuit capacities:

	Total Changes
2 × omni sites with 10 channels each	20
2 × omni sites with 30 channels each	60
6 × 120° sectored sites with 12 channels per sector	216
Total voice channels	296
Total control channels	22

      4.1.4. The links between the base stations and the switch will be via 2-Mbit digital radio links (30 voice and 2 control channels).

5. **Spare Parts**   The tenderer will provide a list of recommended spare parts for a period of 18 months after commissioning. Special attention should be given to consumable parts (fuses and lights), which will be provided at least at 100 percent of provisioning.

6. **Test Equipment**   The tenderer will list in detail any specialized test equipment needed to maintain the bases or switches. A separate list of other recommended test equipment will be provided.

7. **Repairs**   A repair contract will be sought for faulty boards and components. The tenderer will give details of repair facilities, including in particular:

   a. Minimum number of years for which repair by the supplier can be guaranteed

   b. Cost basis of repairs

   c. Expected turnaround time on a board swap basis

8. **Training**

   8.1. Training will be required for *Supercell* engineers in the following aspects:

      a. System overview (1 week)

      b. Switch operation, maintenance, commissioning, and software (approximately 8 weeks)

      c. Base-station installation, operation, commissioning, and maintenance (approximately 3 weeks)

      d. MIS (1 week)

   8.2. The tenderer will give details of proposed courses, costs, and maximum number of attendees per course.

   8.3. All course material and presentation will be in English.

9. **Commissioning**

   9.1. *Supercell* engineers will participate in the commissioning tests that will form part of the Acceptance test.

   9.2. Full commissioning test sheets will be provided to *Supercell* not less than two weeks before commissioning begins.

   9.3. *Supercell* may require additional or modified commissioning procedures.

10. **Installation**

   10.1. The Contractor will be fully responsible for installation, but will undertake to give on-site training to three *Supercell* staff in installation practices. The *Supercell* staff will be assigned to the tenderer for the duration of installation at no cost to the tenderer. All staff so assigned will have previously done the installation and commissioning course offered by the tenderer under clause 9.

   10.2. Program Evaluation Review Technique (PERT) and bar charts will be provided that detail:

      a. Activities from the award of contract

      b. Time interval for the various activities up to and including acceptance testing

      c. Delivery schedules

   10.3. The tenderer will state the number of staff hours for each activity phase and the installation team size for each activity.

   10.4. The tenderer shall submit monthly progress reports with details of any slip in timetable.

11. **Acceptance**

   11.1. Acceptance testing will be done by *Supercell* engineers. Conditional acceptance may be given where minor shortcomings are detected. When conditional acceptance is given, it will be subject to the rectification of the outstanding problems within 30 days (or as agreed).

11.2.[1] The tenderer will submit a copy of normal acceptance procedures, but should be advised that *Supercell*'s own test sheets (copies of which are available on request) will be used for Acceptance.

11.3. Acceptance will be deemed completed either after the issuance of an Acceptance Certificate or after the rectification of any outstanding, specified problems on a conditional test sheet.

IV. Appendices and Specifications

Attachments that *Supercell* should include with the offer to tender are:

1. Maps of the area to be covered with proposed coverage area

Details that should be available for more detailed discussions with the short-listed supplies:

2. Trunking plan showing how the base stations are connected to the switch and the switch to the PSTN

3. Signaling format (in detail) to PSTN

4. Coverage proposed from each site*

5. Coordinates of each base-station site*

6. Site plans*

7. Shelter drawings*

8. Power provisioning (in volt amps)*

9. Tower heights and above mean sea level (AMSL) height*

10. Tower plans and feeder size*

11. Switch room location and plans*

Note: The items marked (*) are not required for a turnkey system.

---

[1]See Installations for details.

34

MODULATION/DEMODULATION METHODS

Modulation is simply the process by which information is impressed on a carrier; demodulation is the opposite, involving extracting the information from the carrier. The earliest and simplest modulation was on-off keying (OOK). The transmitter either was sending or it was not. A detector could be very simple because all it had to do was detect the presence of a carrier.

To transmit voice a more sophisticated method was needed, and amplitude modulation (AM) fitted the bill. It simply involved making the radio frequency (RF) signal proportional in level to the voice level. Demodulation was simple (an antenna, tuned circuit, and crystal were all that was needed), and this form of modulation was dominant for decades. In the 1930s frequency modulation (FM) was developed. Although it was easy to modulate, it was not so easy to demodulate. In theory, FM uses an infinite bandwidth, but in practice most of the energy is contained in a relatively narrow bandwidth. FM had the advantage over AM that high-quality signals could be sent and received. This was due to the processing gain, obtained by spreading the modulation over a bandwidth wider than the base band (or original signal modulation bandwidth).

It is the processing gain that is at the heart of modern modulation methods whereby attempts are made to use a method that gives enhanced signal-to-noise (S/N) performance from a given carrier power level.

Most analog cellular systems use FM for speech and frequency shift keying (FSK) for data. Other modulation systems could be used, but this combination gives the best performance when S/N and simpler modulation methods are the main considerations.

Because of the threshold effect (discussed in the next chapter), the mathematics of noise performance above and below the threshold are very different. It is assumed that in cellular radio applications, all transmissions occur above the threshold level (approximately where the S/N of the off-air carrier is 17 dB). Therefore, the information in this chapter is valid only for values of S/N above 17 dB.

RECEIVER PROCESSING GAIN

The most usual way to get a processing gain is to use a wide-bandwidth (relative to the baseband) for the transmission. This can be done in both digital and analog systems, with FM being the best known of the analog techniques and CDMA being the equivalent digital approach.

Not all wideband systems are equally effective, and there are ways other than using a wider bandwidth to obtain a processing gain. Tape recorders have long used Dolby B and C to improve the ratio of the signal power to noise power; this process takes place within the baseband.

The baseband reference gain or processing gain (G_B) is defined as the S/N obtainable at the detector output compared to the S/N at the receiver input if the noise is considered to have the bandwidth of the baseband only. The signal-to-noise improvement factor is then

Equation 34.1

$$G_B = \frac{SNR_A}{SNR_R}$$

where

$$SNR_R = \text{received signal-to-noise ratio}$$

$$SNR_A = \text{actual signal-to-noise ratio}$$

This ratio becomes more meaningful when the results for various systems are tabulated, as shown in Table 34.1.

For SSB–SC and DSB–SC, $G_B = 1$, then, as a relative measure of noise performance, the S/N performance of any modulation mode compared to SSB–SC or DSB–SC can be used. The values of G_B for the major cellular systems are shown in Table 34.2, comparing results with other modulation systems.

Clearly, FM is superior in all cases, with the difference between the various systems being a function of their peak deviations. For this reason, FM was chosen for the speech channels on analog systems.

The data channels generally use FSK, which can yield excellent S/N performance particularly at low data rates. FSK is used frequently for signaling in noisy mobile environments. At the time most of the existing cellular systems were developed, techniques for good digital performance over reasonable bandwidths for speech channels had not emerged, and so only analog systems were considered for speech. Today, techniques are available that transmit good-quality speech over bandwidths less than the base bandwidth.

TABLE 34.1 Processing Gain of Different Modulation Systems

System	$G_B = \dfrac{SNR_A}{SNR_R}$
SSB–SC	1
DSB–SC	1
DSB	$\dfrac{m^2}{1+m^2}$
AM	$\dfrac{m^2}{2+m^2}$
PM	$(A\varnothing)^2$
FM	$\dfrac{3}{2}\beta^2$
Additional gain with FM pre-emphasis	$\dfrac{\pi}{6}\left[\dfrac{W}{f_1}\right]$
Compander	G_B is a function of level and the compression ratio being maximum at low levels of modulation

SSB = single sideband
DSB = double sideband
SC = suppressed carrier
AM = amplitude modulation
PM = phase modulation
FM = frequency modulation
and
m = modulation index $0 \leq m \leq 1$
$A\varnothing$ = maximum phase deviation for PM deviation
β = modulation index = $\dfrac{\text{deviation}}{\text{audio bandwidth}}$
W = baseband bandwidth of modulating signal
f_1 = 3-dB point for pre-emphasis and de-emphasis

TABLE 34.2 The Relative Noise Performance or Processing Gains of Different Modulation Systems as Signal-to-Noise Ratios

System G_B	Other Modes	NMT450/900 $D = 4.7$ kHz	TACS $D = 9.5$ kHz	AMPS $D = 12$ kHz
SSB	1			
DSB	1			
AM ($m=1$)	1/3			
PM*		2.4 (3.8 dB)	10 (10.0 dB)	16 (12.0 dB)
FM		3.6 (5.6 dB)	15 (11.8 dB)	24 (13.8 dB)

* PM using same bandwidth.

All cellular systems are assumed to have 3-kHz baseband.

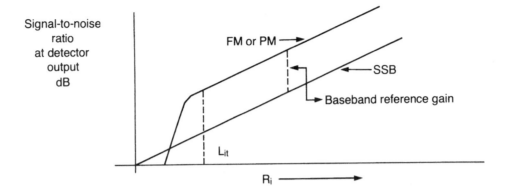

R_i = Receive input signal-to-noise ratio (dB)

Figure 34.1   Process gain of PM or FM over a linear system such as SSB.

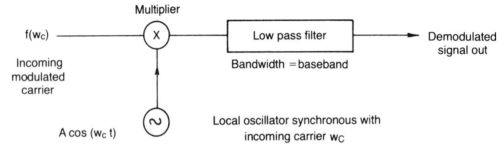

Figure 34.2   Synchronous product detector.

THRESHOLD EFFECT IN FM SYSTEMS

All angle-modulation techniques exhibit a threshold effect; the result of this effect is that a received signal moves from acceptable quality to unacceptable very rapidly as the signal level drops below a certain critical value. FM systems have a processing gain in S/N performance that is a function of their modulation index, which explains why FM is also chosen as the modulation method for high-quality commercial broadcasts. FM radio stations use 75-kHz deviation, which ensures a very high processing gain. The consequent lower threshold, however, requires a high input signal. A very noticeable improvement in commercial FM-received S/N can be noted with increased input signal level (e.g., by installing an external antenna), especially if the receiver is operating in a poor reception area.

At some level, L_{it}, the baseband reference gain, drops off sharply, and within a few dB of input level can drop to less than unity. It can be shown (by somewhat arduous theory or by practical measurement) that L_{it} is in the range of 10–20 dB for practical FM systems. This effect is not present in SSB or DSB systems, but it is present in

AM detectors that use envelope detectors. But because AM systems have processing gains less than unity, this effect is not so important. SSB systems that use synchronous product detectors do not exhibit this threshold phenomenon, so this mode can be used as a reference.

Figure 34.1 illustrates the threshold effect. Figure 34.2 shows a synchronous product detector.

All types of linear modulation (DSB, AM, SSB, VSB) can be detected by a synchronous product demodulator. For low-level data detection in AM systems, using a synchronous product detector can avoid the threshold effect. There is, however, little practical value in doing this for non-data circuits because the advantages are only realized at very low S/N.

The $f(w_c)$ for an AM signal can be represented as

$$K \cos w_m t \cdot \cos w_c t$$

where

$$w_m = \text{modulation frequency}$$

$$w_c = \text{carrier frequency}$$

TABLE 34.3 Relative Processing Gains at S/N Output of more than 35 dB for the Three Main Cellular Systems

System	Gain (dB)
NMT450/900	0
TACS	6
AMPS	8

TABLE 34.4 Bandwidth versus Channel Spacing for Cellular Systems

System	BT* (Bandwidth kHz)	Channel Spacing
NMT450/900	15.6	25
TACS	25.2	25
AMPS	30.2	30

*Assumes $f_m = 3.1$ kHz (audio speech bandwidth)

If the detector product,

$$P = K \cos w_m t \cdot \cos w_c t \times A \cos w_c t$$

$$= AK \cos^2(w_c t) \cdot \cos w_m t$$

is taken, we can expand

$$\left[\text{using} \cos^2(w_c t) = \frac{\cos(2w_c t) + 1}{2} \right]$$

$$P = \frac{AK}{2}(\cos 2w_c t + 1) \times \cos w_m t$$

Equation 34.2

$$P = \frac{AK}{2}(\cos 2w_c t \cdot \cos w_m t + \cos w_m t)$$

Because the term $\cos 2w_c t$ equals the second harmonic of the carrier frequency, a low-pass filter can easily remove this product, leaving only:

$$\frac{AK}{2} \cos w_m t$$

(the original modulation).

For a reasonably good-quality signal (acceptable S/N performance), the receiver must operate at an output S/N level of about 35 dB (30-dB S/N is considered the lowest level at which the noise is not obviously intrusive). This is well above threshold.

Because S/N performance determines range, the coverage from the same site of the three main analog cellular systems can vary significantly. Table 34.3 shows the relative gains of the systems at S/N output levels of more than 35 dB compared to the lowest, NMT450/900.

An advantage of 8 dB results in an improved coverage factor of about 1.7 times in range. There are, however, as always, trade-offs; in this instance, they are bandwidth and immunity to adjacent-channel interference.

BANDWIDTH

The bandwidth required for an FM system is given by Carson's Rule, which states that 98 percent of the power in the sidebands is transmitted if the bandwidth of the system is such that

Equation 34.3

$$B_T = 2(\Delta F + f_m)$$

where

B_T = bandwidth

ΔF = maximum deviation

f_m = maximum modulation frequency

Table 34.4 shows bandwidth for the three systems, and also lists the channel spacing. Notice that only Nordic Mobile Telephone (NMT) systems have any margin between the channel spacing and the bandwidth required. Because filters have less than ideal response, the channel filters need to be somewhat wider than 30 kHz or 25 kHz, respectively, for Advanced Mobile Phone Service (AMPS) and Total Access Communications System (TACS). As a result, adjacent-channel interference is a potential problem in AMPS and TACS systems.

PRE-EMPHASIS AND DE-EMPHASIS

The noise at the output of an FM detector has the following density function:

Equation 34.4

$$S_{(f)} = \frac{N_i \times f^2}{2P_r}$$

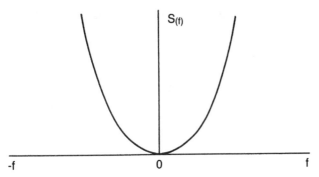

Figure 34.3 Noise power output of an FM detector as a function of instantaneous deviation.

where

$$S_{(f)} = \text{noise density function}$$

$$N_i = \text{noise power density at the receiver input}$$

$$f = \text{frequency}$$

$$P_r = \text{power received}$$

From this equation, it can be seen that the noise power is inversely proportional to the input power, and that this is responsible for the FM quieting effect (that is, the noise level decreases as the input carrier level increases). Figure 34.3 shows the noise power output of FM as a function of deviation. Also, the noise power has a parabolic spectrum, as it is proportional to f^2.

The detected S/N to input S/N can be shown as

Equation 34.5

$$SNR_{(\text{gain})} = \frac{3}{2} \times D^2 \times \left(\frac{B}{W}\right)$$

where

$$D = \text{peak deviation ratio}$$

$$B = \text{bandwidth}$$

$$W = \text{base bandwidth}$$

Thus, the noise power increases rapidly as the bandwidth increases. Fortunately, this can be partially compensated for by pre-emphasis. This involves boosting the transmitted signals in proportion to their frequency.

This is usually achieved using the transfer function

Equation 34.6

$$P_E = K\left(1 + j\frac{f}{f_1}\right)$$

where

$$P_E = \text{pre-emphasis}$$

$$f = \text{baseband frequency}$$

$$f_1 = \text{cutoff frequency}$$

$$j = \sqrt{-1}$$

The f_1 is the point at which the modulating signal is boosted by 3 dB. The slope of this curve approaches 6 dB per octave, which is the inverse of the noise spectral density function $S_{(f)}$. In practice, a 6-dB emphasis has been shown to be both readily realizable and capable of producing good results. Increasing the boost has not been found worthwhile. Naturally de-emphasis must be applied at the receive end, which has an inverse form:

Equation 34.7

$$D_E = \frac{S_o}{(1 + jf/f_1)}$$

where

$$D_E = \text{de-emphasis transfer function}$$

$$f = \text{frequency}$$

$$f_1 = \text{3-dB point}$$

$$S_o = \text{input signal de-emphasis circuit}$$

It can be shown that the improvement in S/N performance due to pre-emphasis and de-emphasis is given approximately by this equation

Equation 34.8

$$\text{Improvement} \approx \frac{\pi}{6} \times \frac{W}{f_1}$$

where

$$W = \text{base bandwidth}$$

$$f_1 = \text{3-dB point}$$

Nearly all FM systems employ some form of emphasis and de-emphasis, including commercial FM and cellular radio systems. A pre-emphasis network can be a simple *RC* network, as shown in Figure 34.4.

It is normal that $R_1 \gg R_2$, so that the time constant of this network is $R_1 \times C$ for the low-frequency cutoff point, and $R_2 \times C$ determines the high-frequency cutoff.

Figure 34.5 shows the de-emphasis circuit. This circuit is sometimes described by its time constant $R_1 \times C$ with values of 50, 75, and 100 μsec being common.

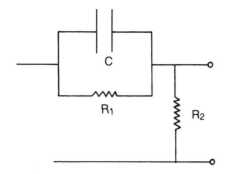

Figure 34.4   A simple *RC* pre-emphasis network for FM.

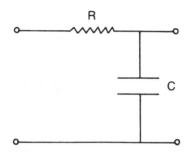

Figure 34.5   De-emphasis circuit.

SIGNAL-TO-NOISE IMPROVEMENTS WITH A PHASE-LOCKED LOOP

The ultimate performance of an FM system in high-noise conditions is determined by the threshold level, as can be seen in Figure 34.1. For some applications a threshold extension can improve overall performance

and range. Because of the high quality expected in cellular operations, this technique has limited applications in cellular, but could find application in fixed cellular installations where the received S/N is lower.

These techniques have particular application in satellite and deep space communications.

A phase-locked loop (PLL), which can be used in FM receivers, can be designed to improve S/N by 2.5 to 3 dB in the region below the threshold by incorporating a loop response that has spike suppression. Above the threshold, the processing gain of a PLL is the same as a conventional discriminator.

Figure 34.6 shows a simple PLL demodulator. Known as the FM feedback (FMFB) technique, it uses superheterodyne detection. Usually the detection will occur at the intermediate-frequency (IF) stage. The difference is that the local oscillator is replaced by a PLL, the instantaneous frequency of which is controlled by the frequency of the modulated signal.

The net effect is to reduce the effective bandwidth of the IF stage and so reduce the noise power at the output. For the equilibrium of this loop it is required that

$$\frac{d\phi}{dt} = \frac{d}{dt}\left(G \int_{-\infty}^{t} V(\lambda) \cdot dt \right)$$

and if

$$w = \frac{d\phi}{dt}$$

then

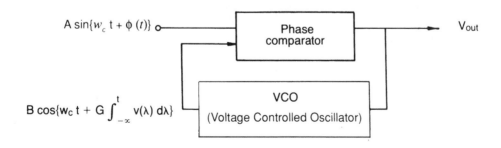

where w_c = carrier frequency
t = time
$\phi(t)$ = modulation frequency
G = constant
V = output voltage

Figure 34.6   Simple phase-locked-loop detector.

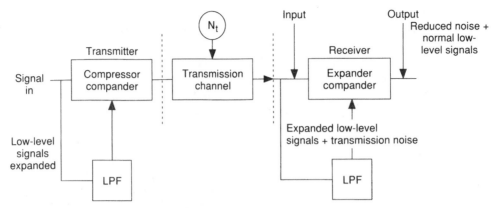

Figure 34.7 A compander system. The terms "compressor" and "expander" are somewhat misleading, as the compressor compander compresses signals above the mean level and expands those below it, whereas the expander compander does the opposite.

Equation 34.9

$$w = G \cdot V(t) \quad \text{or} \quad V(t) = \frac{W}{G}$$

so the output voltage is directly proportional to the modulation

where

$$w_c = \text{carrier frequency}$$

$$t = \text{time}$$

$$\phi(t) = \text{modulation frequency}$$

$$G = \text{constant}$$

$$V = \text{output voltage}$$

This demodulator is a first-order (or single-pole) device; a further improvement can be obtained by using a second-order transfer function.

Such a PLL can yield an additional 2.5- to 3-dB threshold improvement. More elaborate PLL detectors are available (with second- or third-order transfer functions).

COMPANDING

Companding is the process of compressing the signal before transmission and expanding it at the receiving end. Figure 34.7 illustrates this process.

A compander, as illustrated in Figure 34.7, reduces the distortion generated at the transmit end by reducing the voltage excursions of the modulating signal. It also improves the S/N of an analog linear system by boosting

the level of the low-level signals. The transmission noise (N_t) is added (logarithmically) to the signal on the transmission channel, but is reduced by the compander ratio at the receiver. In FM systems, a compander also improves the S/N performance by increasing the deviation of the low-level signal components. Figure 34.8 shows a typical transfer characteristic of a compander.

This same compression technique is used in some hi-fi tape recorders to reduce the relative S/N and is known commercially as HX noise reduction in some tape recorder systems.

SPREAD SPECTRUM

Spread-spectrum techniques have long been used by the military because of their high immunity to interference and their high security. It is the interference immunity that particularly appeals to the designers of future cellular systems.

In essence, the spread-spectrum principle is simple. As shown in Figure 34.9, the transmitted digital signal is multiplied by a pseudo-random sequence. The frequency of this sequence is such that it is significantly higher than the information signal, and so the resultant modulated waveform "smears" the modulation out over a wide spectral bandwidth. A conventional wide-band receiver would interpret the signal as noise.

In order to decode the signal, the receiver must use the same random sequence as the sender. By multiplying the received code with the decode key, the intelligence is removed.

However, if other signals, either wide-band or narrow-band, are received they will not correlate and so will be received as noise. With the use of robust error-correction techniques much of this noise can be filtered out.

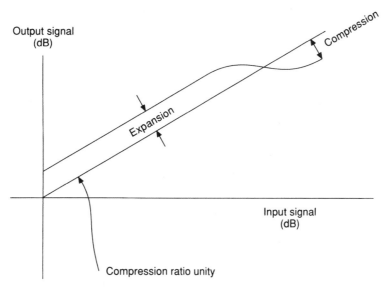

Figure 34.8 Typical transfer characteristic of a compander.

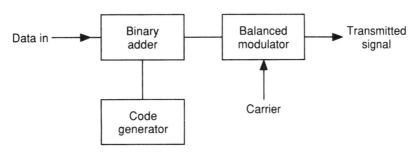

Figure 34.9 A simple spread-spectrum modulator.

Because the pseudo-random sequence is different for each call, many calls can be set up on the same bandwidth and transmitted simultaneously. Code division multiple access (CDMA) is an example of this form of spread spectrum.

In fact CDMA is an offshoot of military technology, and many of the originators of the technique were ex-military. Converting this technology to the high-volume, low-cost handsets, originally was a major technological challenge.

Cellular CDMA is mainly concerned with processing gain and self-interference immunity. The basic principles are the same, but the implementation is very different.

MODULATION

The usual modulation technique for spread spectrum is phase modulation using either 180 degrees phase-shift or binary phase-shift keying (BPSK). FSK is sometimes used, although it does have inferior S/N performance.

When the technique of encoding is a pseudo-random sequence multiplied by the data and then phase modulated, it is known as "direct-sequence" spread spectrum (DSSS). An important parameter of direct-sequence modulation is the "chirp" or the duration of the smallest bit in the code sequence. This determines not only the

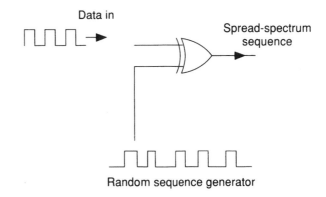

Figure 34.10 Exclusive OR gate.

spectral bandwidth of the modulated signal but also the processing gain of the receiver, which both increase as the chip duration decreases. The modulating pseudo-random sequence is called the *chip sequence*.

The bulk of the information energy is contained in a bandwidth of $F_0 - F_n$ to $F_0 + F_n$ where F_n = the center frequency of the transmission, and $F_n = 1/(t_c)$, where t_c = chip duration.

The modulator can be a simple exclusive OR gate, as depicted in Figure 34.10.

The chip sequence and data stream need not be synchronous, although if they are the clock recovery is simplified.

Other techniques include frequency-hopping (as employed in Global System for Mobile Communications [GSM]) and time-hopping where the time bursts have a pseudo-random duration.

Another technique known as *chirp modulation*, which consists of linearly sweeping over a wide frequency range, is also sometimes classified as spread spectrum, although the lack of a pseudo-random generator means that this is not true spread spectrum.

35

NOISE AND NOISE PERFORMANCE

All communications systems are noise-limited in the strict mathematical sense that noise determines the maximum data rate that can be transmitted over any fixed bandwidth and with any arbitrary error rate. The bit rate that can be communicated at an arbitrarily low error rate in a Gaussian noise environment was calculated by Claude Shannon in the early 1940s to be

Equation 35.1

$$C = W \log_2(1 + S/N)$$

where

$$C = \text{the baud (or bit) rate}$$

$$W = \text{the bandwidth in Hz}$$

$$S/N = \text{signal-to-noise ratio}$$

This equation is true for all signal-to-noise (S/N) ratios (including those <1), provided a suitable encoding (modulation) system is used.

All analog speech-encoding techniques are generally very inefficient because speech itself uses very high redundancy and makes poor use of the channel bandwidth.

From Equation 35.1, it can be seen that the information rate is improved with any increase in bandwidth and, for reasonably high S/N ratios, the improvement is almost directly proportional to the log of the S/N ratio. Frequency modulation (FM) was an early and successful attempt to exploit this relationship. In FM broadcasting, a wide bandwidth is used to gain an improved signal quality. Commercial FM broadcasting uses a bandwidth five times broader than the baseband (the modulation frequency).

For high frequency (HF, 3–30 MHz) the limiting noise is mainly man-made (for example, electrical and ignition systems), but also includes storms, solar flares, and, more importantly, other transmissions on the same frequency (if they are considered to be noise). Thus, with HF, the improvements in receiver technology cannot achieve much improvement in transmission over any single channel. Frequency-hopping techniques, however, can achieve a marked improvement.

Noise can have many origins, including galactic and extra-galactic, thermal, man-made, and even quantum mechanical. Modern radio receivers in the very high frequency/ultrahigh frequency (VHF/UHF) bands operate close to the theoretical limits of sensitivity imposed by these noise sources.

At VHF and UHF frequencies (30–3000 MHz), and particularly above about 300 MHz (where all cellular radio operates), the background noise is relatively low, and the limits of performance are set by the equipment. Galactic noise can be significant in the region of 40–250 MHz.

At these frequencies, the predominant source of noise is thermal. Modern receivers operate at very low-noise factors (levels of introduced noise) and, short of using very expensive technology (such as masers), few improvements in the basic sensitivity of these receivers are possible.

GALACTIC AND EXTRA-GALACTIC BACKGROUND NOISE

Galactic background noise (which originates within the galaxy) and extra-galactic background noise (which includes the microwave emissions left over from the Big Bang some 16 billion years ago) are significant radio noise sources.

Galactic noise has its origin in stars, supernovas, neutron stars, black holes, quasars, and other noise sources that are scattered throughout the galaxy. Particular noise sources include the sun, Cygnes A (thought to be a black hole that emits high levels of X-rays and radio, but only low-intensity light radiation), and Cassiopeia A (a particularly noisy star at radio frequencies). In addition, because the universe is composed of some 10^{12} galaxies, it is not surprising that some of the noise emanates from outside our local galaxy.

Antenna noise temperatures are used to define the background noise levels and have nothing to do with the actual temperature of the antenna itself. The noise power is directly proportional to the bandwidth of the receiver. It is important to note that the noise power is independent of antenna gain.

The radio noise is very much a function of frequency, as the Earth's atmosphere acts as an attenuator at high frequencies, while the ionosphere attenuates the lower frequencies. The quietest area is from 1 GHz to 10 GHz, where galactic noise is at a minimum. A low-noise region between 1 and 10 GHz is most amenable to application of special, low-noise antennas. Cellular radio systems are usually at 800–900 MHz and operate at antenna noise temperatures of about 2.5 K to 100 K. The large range of antenna temperatures is due to the fact that the antenna temperature depends on how the antenna is oriented at the time of measurement. At lower frequencies (below 1 GHz), the maximum temperature occurs when the antenna is pointed at the galactic poles. At higher frequencies, the maximum temperature occurs when the antenna is pointed at the horizon.

THERMAL NOISE

Because of random molecular movement caused by thermal energy, all passive and active components at any temperature above absolute zero generate a certain amount of wide-band energy. In electronic devices, this energy manifests itself as system noise and imposes fundamental limits on usable sensitivity of receiving and detecting systems.

Thermal noise (also known as resistor, Johnson, Nyquist, and circuit noise) is proportional to the absolute temperature of the conducting device. J. B. Johnson of Bell Laboratories first described this noise in 1927, and H. Nyquist first described it theoretically in 1928. It is best visualized as the random motion of electrons induced in a resistor by thermal energy. The energy is spread uniformly across the frequency band. This type of

noise is known as white noise (i.e., all frequencies are equally represented, as opposed to pink noise, in which there is a bias toward the lower frequencies).

In any conductor the available noise power can be determined by this relationship:

Equation 35.2

$$P = KTB \text{ watts}$$

where

P = available noise power

T = temperature in degrees absolute (Kelvin)

K = Boltzmann's constant = 1.38×10^{-23} joules/kelvin

B = the bandwidth in Hz

Notice that the noise power depends only on temperature and bandwidth and is not dependent on resistance.

Substituting into Equation 35.2, $B = 1$ Hz and $T = 290$ K (standard room temperature), the noise power is 4×10^{-21} watts/Hz, or -174 dBm/Hz.

Equation 35.2 seems to imply that infinite power is available in the noise source, as the bandwidth is unlimited. This, of course, is not the case, and as the bandwidth approaches optical frequencies at room temperature, quantum corrections come into play that band limit the independence on frequency so that the total power is constrained.

For a conductor with a resistance R (see Figure 35.1), it can easily be shown that

Equation 35.3

$$E^2 = 4 \, RKTB$$

where

R = the equivalent resistance

E = the equivalent noise voltage generator

Notice that the voltage is represented as E^2. The square of the actual value of voltage (that is, a value proportional to energy) is used in lieu of E, the actual voltage, because the average voltage is zero (that is, noise with negative-going pulses is just as likely to occur as noise with positive-going pulses).

The root-mean-square (RMS) value of voltage is

Equation 35.4

$$E_{\text{RMS}} = \sqrt{4 \, RKTB}$$

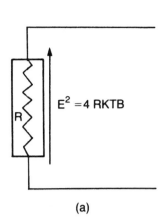
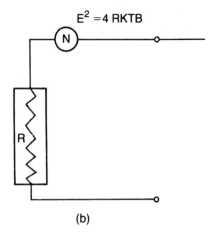

Figure 35.1 Equivalent voltage of a resistive noise source, where (*a*) is the voltage across a resistor, and (*b*) is the equivalent voltage generating circuit.

It is important to note that whereas the energy is directly related to the absolute temperature, the voltage is proportional to the square root of the resistance. In the case where a device has an input resistance of 1 mega-ohm at room temperature and a bandwidth of 1 MHz, it can be seen that the voltage across the input resistance is

$$E_{\text{RMS}} = \sqrt{(4 \times 10^6 \times 1.38 \times 10^{-23} \times 290 \times 10^6)}$$
$$= 126 \text{ microvolts!}$$

This means that all signals below 126 microvolts would be lost in noise.

For a 50 Ω input impedance, the noise would be

$$0.89\,\mu V$$

which for most modulation techniques sets the useable signal levels.

ATMOSPHERIC NOISE

Atmospheric noise is largely due to lightning discharges and is consequently very seasonal. It predominates in the frequency range up to about 20 MHz. Atmospheric noise is not generally a factor at cellular frequencies except in abnormal circumstances.

MANMADE NOISE

Due mainly to low-frequency devices such as motors, neon signs, power lines, and ignition, man-made noise sources tend to decrease rapidly in intensity with increasing frequency. Typically, suburban areas are about 15 dB quieter than city centers, and rural areas are about 15 dB quieter than suburban areas. This noise source is significant up to about 1 GHz, but it is generally not a serious problem above 500 MHz.

STATIC NOISE

Static noise is caused by ionospheric storms, which cause fluctuations in Earth's magnetic field. It is also caused by sunspot activity and can be a serious problem in long-distance communications.

SHOT NOISE

Shot noise is generated within active circuitry, and is caused by the movement of electrons under applied voltages, such as would be encountered within a transistor. Low-noise devices, which are used in critical parts of the circuitry such as the radio frequency (RF) stage, have been designed to minimize shot noise.

PARTITION NOISE

Partition noise is similar to shot noise and occurs when a current meets a junction within an active device, which causes the current to divide.

ABSOLUTE QUANTUM NOISE LIMITS

Without going into extensive theory, quantum mechanics demand that radio noise will exist in a vacuum even at 0 K, even though the vacuum is completely shielded from outside radio influences. In essence, the theory states that a condition of "absolute nothingness," free of any noise, is not achievable, even theoretically, in a perfect vacuum.

Some very sensitive measurements (like those used to measure gravity waves) are now approaching the limits of accuracy permitted by quantum mechanics. These noise effects are different from thermal effects (which are predictable by classical physics), and even though much smaller, they may one day limit the speed of future high-technology data communications by placing a fundamental limit on error rates. Such limits are now being approached in some experimental fiber-optic applications. New techniques to reduce quantum noise are being explored.

CROSS TALK

Cross talk is familiar to all cellular users and is caused by signal leakage from one circuit to another. It is the form of noise that ultimately limits cellular frequency reuse, and so is the dominant form of noise. Much of the design of cellular systems involves minimizing cross talk. The C/I ratio is a measure of how well the system can tolerate cross talk. It occurs when a signal from a distant user operating on the same channel interferes with the channel in use. Cross talk also refers to signal leakage between cable pairs in conventional wireline links.

SUBJECTIVE EVALUATION OF NOISE

An established way of measuring the effect of noise on a system is to measure its level relative to the wanted signal. The most common form of expressing this is as the logarithm of the ratio:

Signal-to-noise ratio

$$= 10 \times \text{Log} \{(\text{signal} + \text{noise power})/\text{noise power}\}$$

When the S/N is expressed as a ratio of voltage or current levels, the expression becomes

$$\text{S/N} = 20 \times \text{Log} \{(\text{signal} + \text{noise})/\text{noise}\}$$

The ultimate determination of S/N performance is the perception of the user. In the audio environment, it is possible to classify S/N in terms of quality. Table 35.1 shows some categories of everyday experience of S/N.

Another method that was developed by the mobile radio and amateur radio communities is to evaluate S/N

TABLE 35.1   Some Common S/N Levels in Everyday Systems

Type of Signal	S/N Ratio (dB)
Limit of operation of 5-tone sequential pager	0–3
Barely readable two-way radio	5–10
Telephone voice quality	25–40+
Hi-fi analog recording	55–65
Compact disc	80+

TABLE 35.2   Signal Quality as a Function of S/N Ratio

Signal Quality	Approximate S/N (dB)	Signal Number
Broken and unreadable	5	1
Broken and just readable	10	2
Readable with some difficulty	15	3
Readable with noise	20	4
Clearly readable	25+	5

on a scale of 1 to 5. Table 35.2 shows this method and its approximate S/N equivalents.

The fact that this table ends with "clearly readable" is indicative of mobile two-way standards, where a high-quality signal is not generally sought. An S/N of 20 dB would be a low limit of acceptability for cellular subscribers and would be acceptable in fringe areas only.

NOISE FACTOR

In order to look more closely at the noise performance of cellular receivers, it is necessary to introduce the concept of noise factor. It is important to understand that the noise factor is defined only for amplifiers or devices for which the input is a pure random noise source. Thus a 3-dB noise factor amplifier connected to a purely resistive load will increase the output noise level by the amplifier gain plus the 3 dB. Where the input contains additional noise (for example, from a noisy signal), the non-thermal noise will be increased only by the amplifier gain.

This manifests itself in such a way that the noise contribution of a series of cascaded amplifiers is largely determined by the noise factor of the first amplifier in the chain, as will be shown later.

Noise factor can be defined as

Equation 35.5

$$F = \frac{\text{available S/N power ratio at input}}{\text{available S/N power ratio at output}}$$

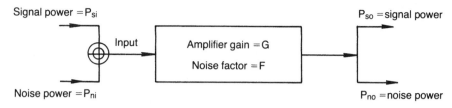

Figure 35.2 A single-stage amplifier, where the noise contribution of the noise factor F results in the amplifier adding to the output noise.

where

$$F = \text{noise factor}$$

Figure 35.2 shows the noise factor.

From Equation 35.5, you can see that

Equation 35.6

$$F = \frac{P_{si}}{P_{ni}} \times \frac{P_{no}}{P_{so}}$$

For linear amplifiers, F is always greater than 1 (that is, noise will be added):

$$G = \frac{P_{so}}{P_{si}}$$

where

$$G = \text{the gain of the amplifier}$$

Similarly, from Equation 35.6:

Equation 35.7

$$F = \frac{P_{no}}{GP_{ni}}$$

THE AMPLIFIER'S CONTRIBUTION TO NOISE (REFERRED TO THE INPUT LEVEL)

It is often useful to determine the contribution of the amplifier to the overall noise of a system. The output noise referred to the input is P_{no}/G. So, the noise contributed by the amplifier (referred to input level) is

Equation 35.8

$$\text{AMP Noise} = \frac{P_{no}}{G} - P_{ni}$$
$$= FP_{ni} - P_{ni}$$
$$= P_{ni}(F - 1)$$

CASCADED AMPLIFIERS

Amplifiers connected in cascade (series) result in an overall noise factor that includes the contributions of each stage. Figure 35.3 shows cascaded amplifiers.

From Figure 35.3, it can be seen that:

Equivalent noise at the input of amplifier 2
= noise input to amplifier by amplifier 1
plus the contribution to the total noise
by amplifier 2 referred to the input

$$= G_1 \times P_{ni1} \times F_1$$

$$= P_{ni2} \times (F_2 - 1)$$

(from Equation 35.7 and Equation 35.8)

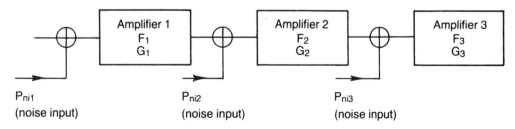

Figure 35.3 Amplifiers connected in cascade can be regarded as having inputs that are the sum of the previous stage outputs, regarded as NF = 1 stage and the net noise contribution.

Hence,

Noise output power of amplifier 2
$= G_2 \times$ the total noise input power

Therefore, the noise output power is

Equation 35.9

Noise output of stage 2, P_{no2}

$= G_2 G_1 \times P_{ni1} F_1 + P_{ni2}(F_2 - 1)G_2$

If the amplifier input and output impedances are matched, and amplifier 2 is at the same temperature as amplifier 1, then

$$P_{ni1} = P_{ni2} = KTB \text{ watts}$$

But, from Equation 35.7,

$$F = \frac{P_{no}}{GP_{ni}}$$

which results in the overall gain for series amplifiers 1 and 2 being

$$G = G_1 \times G_2$$

From Equation 35.9:

$$P_{no2} = G_2 G_1 \times P_{ni1} F_1 + P_{ni2}(F_2 - 1)G_2$$

Then, from Equation 35.7:

$$F_o = \frac{G_2 G_1 \times P_{ni1} F_1 + P_{ni2}(F_2 - 1)G_2}{G_1 G_2 P_{ni1}}$$

But $P_{ni1} = P_{ni2}$ for impedance-matched amplifier stages, so the cascaded noise factor F_o is

Equation 35.10

$$F_o = F_1 + \frac{F_2 - 1}{G_1}$$

It can similarly be shown that for additional cascaded amplifiers

Equation 35.11

$$F_o = F_1 + \frac{F_2 - 1}{G_1} + \frac{F_3 - 1}{G_2 G_1} + \cdots$$

Two important conclusions are now drawn from these equations:

- In the cellular mobile environment, the first RF stage virtually determines the noise performance of the receiver. A typical mobile RF stage has the following specification:

Gain $= 15$ dB (ratio $= 31.62$)

$F = 1.5$ dB (ratio $= 1.41$)

The next stage may have a gain of 30 dB (1000) and a noise factor of 10 dB (ratio 10). So, the overall performance, denoted F_o is

$$F_o = 1.41 + \frac{10 - 1}{31.62}$$

$$= 1.41 + 0.284$$

$$= 1.694$$

The overall noise factor in dB is 2.28. In other words, the noise factor has not been significantly increased by the addition of a noisy (10 dB) second stage.

It can be further shown that additional amplifiers can be of progressively lower quality (with a higher noise factor and therefore cheaper) without significant degradation in the overall system performance. This fact is most fortunate because the mixer stage in a superheterodyne is very noisy indeed and so, in all high-performance receivers, the mixer is preceded by a low-noise amplifier.

- The noise factor of a typical UHF receiver can be calculated from S/N ratios using Equation 35.4, once measurement details are known. Assume the following:

- S/N ratio $= 12$ dB
- Channel bandwidth $= 30$ kHz
- Modulation is 1 kHz at 1 kHz deviation
- Measured sensitivity $= 0.2$ microvolts
- Processing gain at these conditions $= 5$ dB

Then:

Actual input noise

$$= \frac{1}{2}\sqrt{4 \, RKTB} \text{ volts } RMS$$

$$= \frac{1}{2}\sqrt{4 \times 50 \times 1.38 \times 10^{-23} \times 290 \times 30,000} \text{ volts}$$

$$= 0.0774 \, \mu V$$

where

$T = 290$ K or 17°C (an accepted standard room
temperature for noise calculations;
in practice this temperature may be
somewhat high or low, depending
on location)

Hence,

$$S/N \text{ at the input} = 20 \log\left(\frac{0.2}{0.0774}\right)$$

$$+ \text{ processing gain}$$

$$= 8.24 + 5.53 = 13.77$$

Some measurements of S/N ratio may yield
negative noise factors (or less than unity if ratios
are used). Negative noise factors can occur when
using deviation ratios higher than unity and noise-
reduction techniques such as emphasis and companding. The lesson is that S/N ratios mean very
little unless the conditions of measurement are
clearly stated.

Further, when the measurements are made at
very low S/N levels (less than 15 dB at the receiver input), threshold effects can mask the true
nature of the S/N performance.

NOISE FACTOR OF AN ATTENUATOR

The mathematics of the noise contribution of an attenuator, such as a transmission line or coaxial cable, are a
little complex and involve thermodynamic considerations. Only the result is listed here. Figure 35.4 shows the
noise factor contributed by cable loss.

The noise factor of an attenuator F is

Equation 35.12

$$F = 1 + (L - 1)\frac{T_c}{T_o}$$

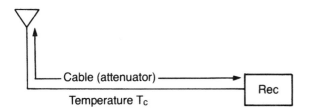

Figure 35.4   Noise factor contributed by cable loss. The feeder
cable to an amplifier is equivalent to a series attenuator.

where

$L = $ the attenuation factor

so that

$$L = \frac{\text{power out of attenuator}}{\text{power into attenuator}}$$

$$T_o = 290 \text{ K}$$

$$T_c = \text{cable temperature}$$

When $T_c \approx T_o$, then $F \approx L$ (the attenuator). Hence, the
noise factor introduced by the feeder cable will be similar
to the attenuation of the cable itself.

As an example, these calculations can be used to determine whether a low-noise masthead amplifier would
be of value in the mobile environment, given the need to
operate in a fringe area. Assume the following:

- The mobile receiver has a noise factor of 6 dB.
- The cable loss is 3 dB.
- A masthead amplifier of 15-dB gain and 4-dB noise
  factor is available. (This is a low-grade wide-band
  amplifier.)
- Receiver gain is 70 dB.

In the case of no masthead amplifier, the overall noise
factor is shown in Figure 35.5
For $T_c = T_o$:

$$F_{overall} = F_c + \frac{F_R - 1}{G_c} \text{ (from Equation 35.11)}$$

$$= 1.99 + \frac{3.98 - 1}{0.502}$$

$$= 7.92, \text{ or 9 dB}$$

Now, consider the use of a masthead amplifier, as shown
in Figure 35.6:

$$F = 2.51 + \frac{1.99 - 1}{31.6} + \frac{3.98 - 1}{31.6 \times 0.502}$$

$$= 2.51 + 0.031 + 0.187$$

$$= 2.72, \text{ or 4.35 dB}$$

So, an improvement of 4.65 dB in S/N performance
would result in this example and, in marginal areas,
could well be worth the effort.

It should be noticed here that the masthead amplifier
used is of better quality than the receiver, resulting in an
improvement that exceeds the cable loss. When the am-

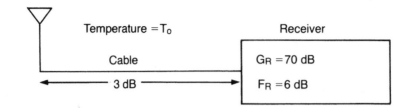

Figure 35.5 Noise factor of a receiver and feeder cable. The values shown are for a practical receiver with a 3-dB cable and a 70-dB receiver gain with a noise factor of 6 dB.

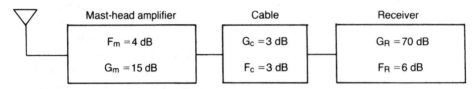

Figure 35.6 A typical masthead amplifier is placed as close as possible to the receiving antenna.

plifiers are of similar quality (that is, about a 4-dB noise figure on the first stage of the receiver), the improvement will be similar to the cable loss, because the contributions to the overall noise factor (F), after the first term, tend to be small. This system is not necessarily what would result in practice because masthead amplifiers with noise factors in the range 0.7–2 dB are available.

This suggests that masthead amplifiers are useful for survey purposes, particularly when low-powered transmitters are used as a source and measurements are done near the limits of the noise performance of the receiver. Masthead amplifiers can be difficult to install and may be prone to lightning strikes. In a mobile environment, because they are usually wide-band devices, they are somewhat prone to intermodulation, which can limit their utility. They generally (but not always) cannot be used on antennas that are connected to a duplexer.

PROCESSING GAIN AND NOISE

The processing gain, shown in Table 35.1 for an FM system, is

Equation 35.13

$$G_B = \frac{3}{2}\beta^2$$

where

$$\beta = \Delta f_d / \Delta f_m = \frac{\text{deviation frequency}}{\text{modulation frequency}}$$

However, this formula is valid only at high input S/N levels, where the noise levels are not high enough to cause the noise spikes familiar in FM systems operating in the threshold region. At very high noise levels (at the receiver input port), noise spikes are generated by the noise components, which are of sufficient level to cause a phase reversal in the incident waveform. Figure 35.7 illustrates the locus of the vector sum of the carrier and noise signals.

The signal out of the receiver, S_o, is such that

$$S_o \propto d\theta / dt$$

where

θ = the phase of the resultant of the incident signal and noise

t = time

From Figure 35.7, it can be seen that a phase reversal produces a pulse. If a plot is made of $\beta(t)$, $\theta(t)$, and $d\theta/d(t)$, the form of S_o can be determined. Here, we assume a steady carrier and a noise signal of approximately constant amplitude (but larger than C) that is rotating uniformly with respect to C.

Figure 35.8 shows a typical FM discriminator output filter.

Figure 35.9 shows that this phase reversal produces a pulse. Similarly, it can be shown that when the locus of the resultant does not encircle the point P (as shown in Figure 35.7), a different pulse form arises (as seen in Figure 35.10).

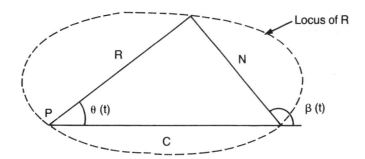

Figure 35.7 The locus of an FM noise pulse. The received signal is the vector sum of the carrier level (C) and noise level (N). This figure shows the locus of the resultant (received) vector (R), which causes the phase reversal associated with FM "picket fencing."

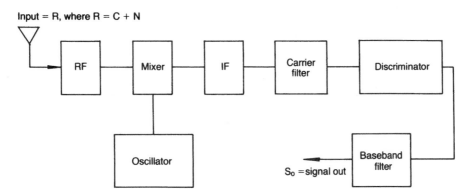

Figure 35.8 FM receiver block diagram.

The noise pulse in Figure 35.10 can be shown to be less energetic than the noise pulse in Figure 35.9. Thus, the effect of noise increases rapidly with deterioration of the incident signal. The actual S/N under these conditions can be shown to be

Equation 35.14

$$(S/N)_o = \frac{(3/2)\beta^2 (S/N)_i}{1 + (12\beta/\pi)(S/N)_i \times \exp\left[-\frac{1}{2}\left\{\frac{(S/N)_i}{\beta+1}\right\}\right]}$$

Note that for large S/N (as encountered in service) the exponential tends to zero and $(S/N)_o \to 3/2\beta^2(S/N)_i$, as before.

The term in the denominator of Equation 35.14 can be seen to tend to 1 when $(S/N)_i$ is large. To determine the levels at which this term becomes effective, it is traditional to define the threshold as the point where the effect of these terms is to reduce the $(S/N)_o$ by 1 dB, as shown in the following equation:

$$(S/N)_o = \frac{1}{1 + (12\beta/\pi)(S/N)_i \times \exp\left[-\frac{1}{2}\left\{\frac{(S/N)_i}{\beta+1}\right\}\right]}$$

$$= 10^{-0.1}$$

$$= 0.7943$$

This function can now be plotted for the three major systems [Advanced Mobile Phone Service (AMPS), Total Access Communications Service (TACS), and Nordic Mobile Telephone (NMT900)]. The values for commercial FM are also tabulated.

Using Equation 35.14, the S/N, as measured at the discriminator output of systems with various modulation indexes, can be determined. This function was plotted for AMPS and NMT systems and for commercial FM in Figure 35.11.

Note: the S/N used in the equation is a power ratio *not* the dB level. So for 20-dB S/N, for example,

$$S/N = 10^{\frac{20}{10}} = 100$$

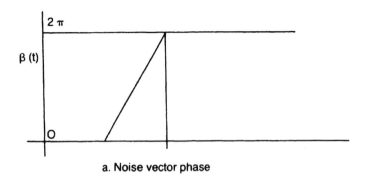

a. Noise vector phase

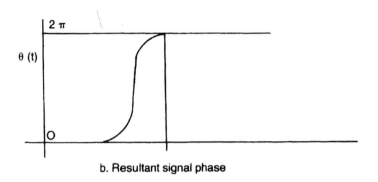

b. Resultant signal phase

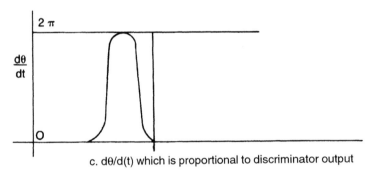

c. dθ/d(t) which is proportional to discriminator output

Figure 35.9   Noise output pulse generated by a noise pulse that causes the resultant locus to pass through 2π radians.

A plot for TACS would be very similar to the AMPS plot, but 2 dB down. Notice that the AMPS system has a significant relative processing gain up to the threshold (18 dB $(S/N)_i$), but this advantage decreases rapidly as the $(S/N)_i$ decreases below the threshold. Above the threshold, all the systems have significant gains over a zero processing-gain system such as single sideband (SSB). However, at very low levels of $(S/N)_i$ (for example, below 12 dB), the SSB system will out-perform the FM systems.

Notice that the threshold for the NMT system is 14 dB or 4 dB below that of the AMPS system.

The various systems have a processing gain that tends to unity in the range 12–13-dB S/N output. For this reason, it is usual to measure S/N at 12-dB signal to noise and distortion (SINAD), because this measure reflects the quality of the hardware (noise figure) and will be the same for equal-quality systems regardless of their processing gains.

In all real receivers (as opposed to the theoretical receivers considered previously), the processing gain does not improve the S/N indefinitely, and a threshold is reached where increases in S/N input do not result in increased S/N output.

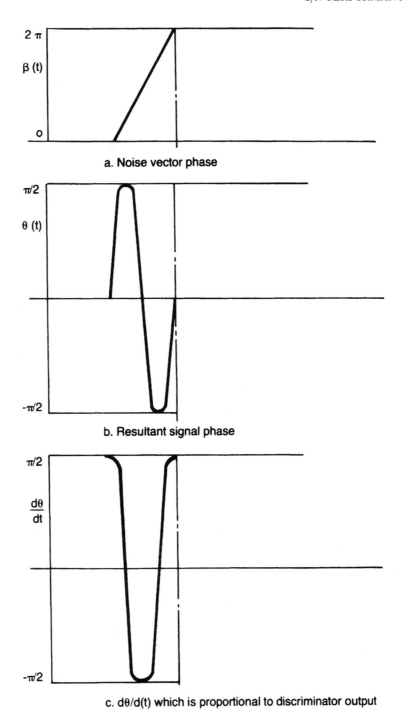

a. Noise vector phase

b. Resultant signal phase

c. dθ/d(t) which is proportional to discriminator output

Figure 35.10 Noise output pulse from a noise vector N, which is smaller than the carrier and rotates through 2π radians.

S/N PERFORMANCE IN PRACTICE

Because of the lower operating frequency, an AMPS antenna has a slightly larger aperture than a TACS/NMT900 antenna of the same gain, and this contributes an additional 0.6-dB system gain to the AMPS system.

The effect of this gain in practice is, however, minor. For modern cellular systems, the requirements of good handheld coverage and high S/N levels mean that those systems will always be operating above the threshold. Handhelds are the weak link in the cellular chain and are usually assumed to be operated from stationary posi-

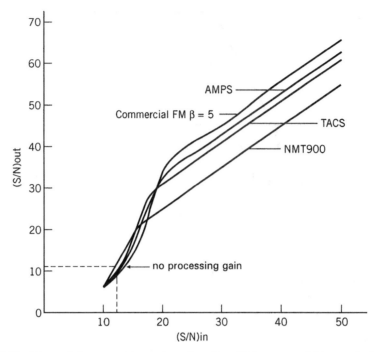

Figure 35.11 Plots of the processing gains of various FM systems. This graph shows S/N performance versus carrier signal to noise of the AMPS and NMT900 systems from Equation 35.14.

tions inside buildings. Thus, handhelds do not ordinarily operate in high multipath environments. In general, the full benefits of the processing gain of the FM systems is usable in modern cellular systems.

The situation is similar, but a little different, if mobile units are considered. With mobile units, there is some significant multipath, and because of occasional excursions below the threshold (contributing relatively high levels of noise), there is some reduction in the relative processing gain. The other requirement of good S/N performance, however, minimizes this effect.

A very different situation occurs in mobile communications systems where voice quality is secondary to range (such as citizens band (CB) and amateur transactions). In this instance, the systems are often operated below the FM threshold, and the performance depends only marginally on deviation. Such systems are usually designed to have low deviations and narrow channel bandwidths.

36

DIGITAL CELLULAR

SOME HISTORY

By late 1993 dozens of public digital systems had been installed. It seemed the whole world just couldn't wait to go digital, and countries rushed to be first with the latest. Almost universally the systems installed were somewhat disappointing. None achieved the minimum target of "being at least as good as analog." The possible exceptions to this were in Germany and Portugal, where the analog "C" system previously in place didn't offer much of a challenge. Some systems were so bad they were quickly withdrawn from service; others, such as the Australian and UK Global System for Mobile Communications (GSM) systems were left to linger with extensive and expensive nationwide coverage with little or no promotion.

In Australia advertising for GSM was stopped almost on the day of the launch, and most dealers have stayed away from the technology. With 1,400,000 analog phones in late 1994 and a connection rate for analog at 50,000 a month, the GSM network had a sorry 800 users after six months of service. The story was the same in the UK: With around 2,000,000 analog users, the GSM system attracted only a few thousand users.

Hong Kong, the "gadget capital" of the world, opened with both a GSM and a digital AMPS (DAMPS) system, and the result was the same—virtually no customers. Hong Kong was a particularly telling case, because as the analog capacity was actually exhausted, the only real alternative to digital was to go without a service. In the United States dozens of operators installed DAMPS but the customers stayed away. At least one US operator took out full-page advertisements to announce that it had tried digital, didn't like it, and would not be offering it.

It seems that in the whole cellular world the only ones happy with digital cellular were the Germans and French, so conditioned as they were by almost a decade of substandard analog service that they didn't expect too much.

Where did digital go wrong? For a start, the implementation of most of the systems was politically driven. Governments wanted to be perceived as "up to date and digitally oriented." What they really proved in the end was something everyone already knew: namely, that the governments have no idea what they are talking about when they launch into speeches on digital technology. More worrisome, the regulatory bodies who advise the governments are seen to be somewhat inadequate.

There were many reasons for the initial poor acceptance of digital technology. Probably the biggest problem has been the haste with which the technology was introduced, which resulted in equipment being delivered with a lot of software bugs that took years to eliminate. These problems caused the system to perform well below specification.

Overly enthusiastic salespeople built expectations that could not be achieved by the technology, resulting in disappointment by both the operator and the end user. Foolish comparisons between digital sound and compact discs did not help. Digital cellular is anything but hi-fi, and by its use of compressions/decompressions (CODECs), it in fact precludes clear reproduction of anything but speech.

The digital technologies that went public at great expense in the early 1990s were developed with too much haste and too little caution. Telecommunications companies today are largely run by managements who have

499

scant regard for technology, and the mistakes were inevitable.

Time division multiple access (TDMA) systems (which include GSM and DAMPS) have been proven to be prone to be sources of interference (caused by its intrinsic pulsed-mode of operation), and some people have questioned whether such a hostile modulation system should be permitted to operate at all. It is certain that if not for the political nature of cellular, this issue would be much more prominent.

A major initial political objective of GSM was to permit Europe-wide roaming. Desirable though this objective is, time should have been allocated for a mature technology to evolve. The primary objective for GSM in the early 1990s was clearly to sell German and French hardware to the world. The marketing was superb; unfortunately, the product was not.

By 1997 quite a different picture had emerged. Most of the software bugs had been fixed and GSM was widely deployed across the world. Customer acceptance of the digital technology was still not up to expectations, largely because the "new" technology was still more expensive than the analog it replaced. GSM had the added disadvantage of having a low-path budget, so that in countries where it had to compete with the Advanced Mobile Phone Service (AMPS) or the Total Access Communications System (TACS), it offered reduced coverage.

The debate in the United States, which effectively held up the adoption of any one standard, was seen by the Europeans as a marketing godsend, and they took full advantage of it to disperse GSM worldwide. In hindsight, the doubts and questions raised in the United States were justified and the wait-and-see attitude was the right way to go, technically.

By 2000 the dominance of GSM was well established. The relative role of analog was diminishing rapidly.

The United States

In the United States TDMA development, together with that of Narrow-band Analog Mobile Phone Service (NAMPS) (arguably the best technology compromise between voice quality and capacity to come out of the first half decade of the 1990s) was stalled, on the promise of code division multiple access (CDMA). The implementation of CDMA was subject to many delays, with the first major commercial operation being in Hong Kong in October 1995. This commercial launch, initially with 10,000 subscribers, was to prove only partially successful. Like all digital systems at that time, the voice quality left something to be desired, and while not being as good as analog, it was an improvement on GSM.

Hong Kong, with its cellular-hostile terrian, is a good proving ground for the capacity claims of the various systems. GSM was able to show only a 40% increase in capacity/MHz over AMPS.

PCS/PCN

The original concept that personal communication service (PCS) (personal communications network [PCN]) was to be complementary to cellular (and in some ways different), has all but faded away. In an effort to get hardware into the field as soon as possible, all the players had to accept that PCS would be nothing more than high-frequency, low-powered cellular.

By 1997 an unsustainable number of licenses for PCS had been issued by most of the world's major countries. Free enterprise was left to sort out the victors, the amalgamators, and the loosers. It would appear, to most reasonable observers, that licenses have been issued as revenue-raising excercises without any real regard for need or viability.

In 2000, "silly" bids for 3G Spectrum have replaced the PCS/PCN frenzy.

Rather than intimidate the cellular operators, PCS seems to have motivated them to innovate and dominate. In the United States, this has resulted in a serious effort (for the first time) to activate wide-area roaming. In the UK and elsewhere it has brought on price wars, and operators vie with one another to give away phones at the lowest prices (prices well below cost).

This leaves the PCS operator with a serious uphill battle. First, the PCS operator has to contend with an established and determined cellular operator in a market with a "give-away" mentality. Then it is necessary to work within the expectation that the lesser coverage of PCS makes the service a "poor man's cellular" system, and that the charges should reflect this. When the bottom line is drawn, it will be found that to establish a PCS operation in any region the size of an average city *costs more* than to do the same thing in cellular, because many more low-range PCS sites are needed to equal the service area of one cellular site.

An interesting part of the whole PCS equation is that the large number of sites needed means that on day one, the basic capacity will be high (often 100,000 or more), so that, in order to generate *some* revenue flow, it is often tempting to offer even more incentives (subsidies) to get customers signed up.

Despite the problems of PCS, two operators—Orange in the UK and Bouygues Telecom in France—have not only provided effective nationwide PCN coverage but have done so admirably well.

COMPLEXITY

In complexity (which can be crudely measured by the necessary memory capacity of the mobiles), the first-

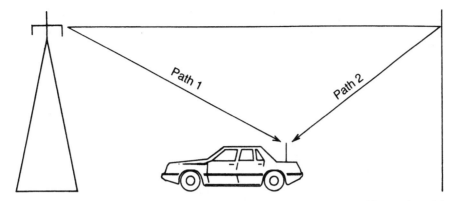

Figure 36.1 The two radio paths are different in length and so will have different time delays.

generation digital systems are about four times more complex than analog mobiles and require about 250 K of memory. As memory becomes cheaper, however, it can be used more freely, so the cost probably depends more heavily on the complexity and hence software than any other factor. A better measure of relative complexity may be the number of instructions per second, which is measured in millions of instructions per second (MIPS). Analog AMPS systems operate at about 0.25 MIPS, whereas GSM phones require 40 to 85 MIPS. G3 offering may be 10 to 100× more complex, yet again.

PATH DELAY

Before 1980, the problems of operating a mobile digital radio were beyond the scope of affordable technology. The main problem was synchronizing the data stream as the mobile moved through it in a multipath environment. Digital systems read the incoming data in blocks called *timeslots*. The decoder must recognize the synchronization bits and then extract the information contained within the timeslots. Figure 36.1 shows that in the multipath environment, bursts of data from different paths travel different distances and therefore have different propagation delays. Accommodating this requirement was easily accomplished on fixed services by introducing the appropriate delays. The multipath environment, however, causes big differences in the propagation delays experienced by signals received over the different paths. These time delays must be equalized in real time for each data frame. In fact even today, this frame sychronization puts upper limits on digital mobile path lengths. How to do so was not immediately obvious in the mobile environment.

The first thing that must be done in the mobile environment is to make the timeslots short because large differences in propagation delays can occur over very small distances (and so can occur quickly). Next, the

propagation delay must be accounted for in real time. This problem has now been solved with the introduction of adaptive equalizers that effectively continuously adjust the equalization as required.

The solution to this problem brought with it an unexpected bonus. Not only was the adaptive equalizer able to set the equalization for each burst of code, it also could do so in a multipath environment. In fact, it performed better in a multipath environment than in a single-path environment, because for each burst of code, it synchronized with the best of the multipath alternatives. This produced bit error rates of an order of magnitude better than could be achieved without use of the multipath as a diversity mode. The rake receiver in a CDMA system very effectively takes advantage of multipath.

NAMING SYSTEMS

The naming of the various systems has not been well coordinated at all. GSM originally meant Groupe Speciale Mobile (a French name despite the fact that in its original form it was simply the digital version of the notorious German C system, and in fact the Germans called it the D system). Commercial pressures were applied and the name was changed to mean Global System for Mobile Communications.

The US industry settled on "TDMA" as a name for its first digital system. I will use quotation marks in this section when referring to the cellular system as distinct from TDMA as a form of modulation. This is necessary, as "TDMA" is an unfortunate name choice and, as will be seen, is misleading. Almost all digital radio systems built in the 1980s were TDMA, including most microwave links. GSM is a TDMA system, as is Japanese digital cellular (JDC). However, "TDMA" is also a frequency division multiple access (FDMA) system (as is GSM and JDC). Because of this conflict, I have throughout this book referred to the US "TDMA" system as DAMPS, or digital AMPS, one of the earlier—but at least less ambiguous—names for it.

JDC is at least a clear and unambiguous name, if not terribly imaginative.

Qualcomm's CDMA was in danger of having the same fate as TDMA as far as its name is concerned, until the recent change to cdmaOne, which clearly identifies the system and differentiates it from the techhnology.

DIGITAL CAVEAT EMPTOR

Digital is a "buzz word" today, and those who use it as such seem to have little or no sense of history. Digital is not new (it's been around now for almost a century in the guise of things like Morse code and decadic switching). Despite this fact, there are always legions who cite digital as the only way to the future. They are probably correct in this, but it is the rate of progress that needs to be viewed with some caution.

Integrated Switched Digital Network (ISDN) is much touted as the ultimate solution for the telecommunications industry (if only someone can find the problem).

I remember seeing my first ISDN switch in Sydney in 1975. It was a most impressive sight, with rows of flashing lights making it look like the control deck of the Starship *Enterprise*. The people running this machine spoke in hushed reverence about the capabilities of this "ultimate universal switch" and how it would revolutionize data communications. This miracle switch had only two problems: first, it didn't really work; second, no one seemed to be able to identify any real customers who could use it even if it did work!

Digital cellular, more than 20 years later, features ISDN compatibility. But ISDN is also still largely a buzz word, and many years will pass before there is wide consumer acceptance of these switched data networks. The concept of ISDN is illustrated in Figure 36.2 (full compatibility for GSM and partial for the US and Japanese systems).

It is rather ironic that the industry is still touting the 1970s ISDN concept of two voice and one data channel from 64 kbits as a desirable objective. Wide-band services are slowly becoming available and they are set to make ISDN obsolete. Despite this, I am aware of a least one major telco who is conducting limited "consumer acceptance" testing in the year 2000!

NO DUPLEX FILTER?

Digital systems can remove some of the bulk from cellular mobiles because a radio frequency (RF) duplex filter is no longer required. In analog systems, to have duplex speech paths, it is necessary to insert a duplex filter to enable the simultaneous operation of the transmitter and receiver. Digital systems can operate sequentially (that is, transmission and reception in different timeslots), so the filter is no longer needed. Recent advances in duplexer design (and size), however, have minimized this advantage, particularly as smaller, simpler, and cheaper duplexers are becoming available. But even small, low-cost filters take a heavy toll on battery talk time because of their high power loss. It is interesting to note, however, that until about 1998, most GSM mobile phones had duplexers installed to perform a filtering function.

BATTERY TALK TIME

One of the greatest shortcomings of analog systems is that, despite improvements in battery technology, the handheld talk-times are limited; most, even moderately heavy users, find it is always necessary to carry a spare battery. Digital systems can significantly improve the battery talk and standby times.

The removal of the duplexer is a major factor, as the insertion of a duplex filter typically results in a 3-dB loss. The removal of this component alone cuts the power requirements in half during transmission without reducing the effective radiated power (ERP).

Because digital paging transmissions can be made synchronous, it is not necessary for the mobile to "listen" continuously; the receiver can "sleep" during periods when data transmission is not relevant. Using this technique, it is possible for a mobile on standby to "awaken" for only a small amount of the standby time, thereby reducing the standby current drain to a small fraction of the consumption of its analog counterpart, which must "listen" continuously.

Digital systems are designed for small cell operations, so low mobile ERPs are not only possible but are necessary to reduce interference problems. Taken together, removal of the duplexer and use of low mobile ERPs improve the effective battery discharge cycle by a factor of around 10×.

VOICE QUALITY

The voice quality of the digital systems, initially promoted as being a strong point of these systems has been a disappointment. Numerous field trials have shown that the best digital systems are not equal to analog. Instead of the noise that is familiar to analog users in poor signal areas, digital systems deliver "metallic voice," splats, and shrieks when the bit error gets to high. The voice

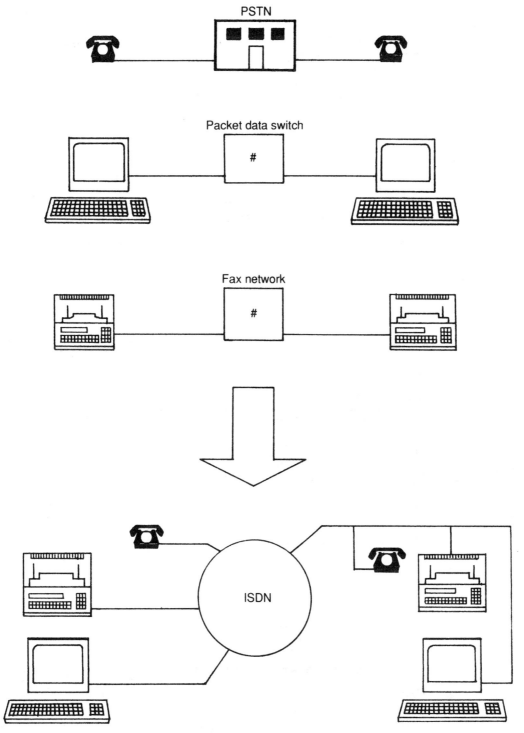

Figure 36.2   The ISDN concept.

quality of any radio system is a function of its field strength; a subject comparison is shown in Figure 36.3.

While the digital systems do seem to be more immune to extraneous noises than analog system, it is the "artificial" sound, even in high field strengths, that limits their quality. It is the very immunity to noise that can cause subscriber irritation in low signal areas. While analog systems will degrade gradually in a way that warns the user that operations are marginal, digital systems, and particularly GSM, give no such warning and

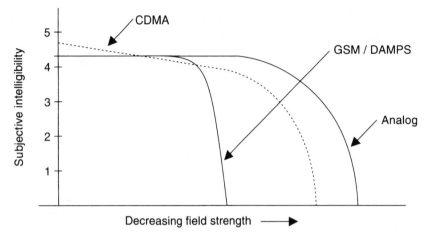

Figure 36.3 The relative intelligibility of a digital versus an analog system.

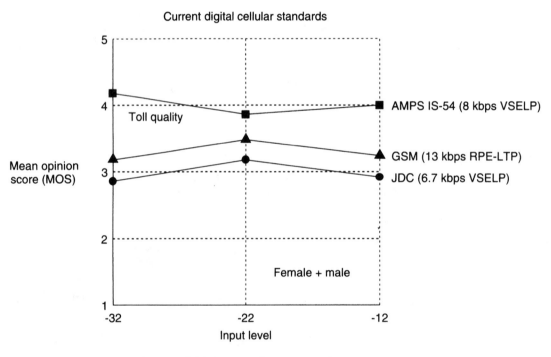

Source: DBP Telekom Research Institute

Figure 36.4 Subjective relative quality of JDC, GSM, and DAMPS.

will abruptly drop the call with little notice. This is particularly annoying when the phone is used in a high-multipath environment (such as the interior of a high-rise building). With analog in such an environment it is possible to move to an area of better reception; with digital it seems that there are simply places that work and others that don't.

Figure 36.4 shows a subjective study of JDC, GSM, and DAMPS. Other studies comparing DAMPS to AMPS generally find the subjective quality of the two about equal. However to date I have not heard a digital system with a voice quality that is the equal of analog. Some of the later CDMA systems seem to come closest.

THE FUTURE OF ANALOG CELLULAR IN A DIGITAL WORLD

The future for cellular is exciting and dynamic. Even as some existing urban analog systems are today slowly reaching maximum capacity, it is evident, by extrapolation, that the first-generation digital systems, soon after their introduction, will face the same fate.

The ever-decreasing cost of mobile units is set to continue as more and more customized chips are manufactured for cellular purposes. The earliest handheld units were made from semi-customized chips that were really developed for other industries, notably the computer and control industries. As speed was more important than volume for most of these applications, the chips generally were parallel-bus devices. As a result, interconnection required the use of high-level multilayer printed circuit boards, and 10-layer boards were not uncommon.

Because a significant demand for cellular phones is now evident, a number of chip manufacturers have begun to produce specialized cellular customized chips that generally use serial data buses to reduce the total number of connections. Lower chip counts mean lower fault rates and lower costs.

Developments in passive duplexer technology are also leading to cheaper and smaller RF hardware. The volume has halved in two years.

DIGITAL ADVANTAGES

The driving force behind digital cellular was spectral efficiency, which most systems have failed to deliver to any significant extent. Proven advantages include greatly enhanced privacy; and enhanced services, such as voice recognition and voice messaging, together with better battery life.

INTERFERENCE FROM TDMA SYSTEMS

For the purposes of TDMA interference, GSM, DAMPS, and E-DAMPS can all be regarded as TDMA. Specifically excluded from this group is CDMA, which works quite differently. The term TDMA refers to the way that multiple channels share one frequency by having timeslots assigned to them in turn. These timeslots in turn are grouped into frames, which contain a number of timeslots. In the case of GSM the frame length is 8 timeslots; it has a repeat rate of 217 frames/s, whereas the DAMPS frame rate is 150 frames/s. The frames are sent in bursts of transmission, followed by breaks during which the mobile is expected to respond.

After the time for the mobile time frame has elapsed, the next frame will be sent. Effectively, the system emits bursts of radiation at the frame rate.

It is this frame rate that produces the annoying buzzes associated with TDMA. GSM, for example, with a frame rate of 217 frames/s will produce noise at 217 Hz and its harmonics, 434 Hz, 651 Hz, etc. This low-frequency component of the RF carrier ends up as a buzz in hearing aids, conventional telephones, analog mobile phones, and other devices. Because it is carried by the same carrier as the wanted signal and demodulated by the device experiencing the interference, it is very hard to eliminate. This buzz will get into analog radios (both cellular and broadcast).

Interference with hearing aids is a problem. Tests carried out by the EMC laboratories in Denmark show interference to hearing aids up to 30 meters away from an 8-watt GSM transmitter. At 1.5 meters, levels were measured at 130 dBA—the level of a jet engine on takeoff! That level is sufficient to cause damage to hearing.

Hearing-aid users are therefore excluded from the potential users of TDMA phones. Car drivers are also at risk. Examples of modern microprocessor-controlled cars being unable to start while the GSM phone is turned on and cars misbehaving in the vicinity of GSM base stations have been reported. Fears have been expressed by some auto manufacturers that TDMA could cause serious accidents by causing brakes and airbags to malfunction.

Computer users also should take care. Digital phones have been blamed for actually destroying equipment used in close proximity to a working telephone.

While medical research into the effect of RF radiation has not turned up much in the way of definitive results, the area of most concern remains the extra-low-frequency (ELF) fields, which are suspected to have a direct chemical interaction by causing resonance in some organic molecules. The TDMA frame buzz qualifies as a relatively high-powered ELF, and this could be a cause for some concern in the future.

THIRD-GENERATION SYSTEMS

Todays GSM and IS95 CDMA are regarded as second-generation (2G) standards. Their successors are intended to be Internet ready, and with wide-band digital capabilities such as Web browsing and video links. The lack of uniformity of standards in 2G systems has been an impediment to successful roaming, and has increased the cost of handsets (indirectly by thwarting the opportunities for economy of scale).

In essence, the 3G will provide more bandwidth and greater flexibility. By some estimates its aim is to provide the same facilities as wireline, including Internet and video capability. Of course, all of this will require much more bandwidth than is available for mobile services today.

It will also require some nearly miraculous feats of network engineering to reduce noise levels to the point where the BERs will be low enough for a respectable through-put. Data rates in 1999 are typically 14.4 kbps (gross) for circuit-switched data, 19.2 kbps for cellular digital packet data (CDPD), and up to 64 kbps for some high-end digital systems (but this in not widely in service). These rates are fine for moderate data speeds, but far too slow for real-time video. The perceived situation today is that the world is moving from a spectrum shortage to a surplus, and this is reflected strongly in the bidding prices in the United States, which have fallen from a high of $208 M/Mhz, to the 1999 low of $1 M/Mhz.

It can be expected that 3G handsets will be from 20 to 30 times more complex than their 2G counterparts and will need about 1 GB of memory for onboard storage. The first phones are expect to be from 90 grams to 150 grams (by then the 2G phones could be closer to 30 grams).

It is generally recognized that there will be no abrupt transition to 3G services, but rather that there will be an evolution beginning with 2G+ and that services and facilities will be added as the demand requires. In the meantime the 3G standard itself will be evolving.

Although the 3G concepts evolved from cellular considerations, the expectation today is that the standards will be adopted for wireline, satellite, and other services as well.

3G Band Allocations

Once again there is a discrepancy in the frequency allocation of the United States and Europe (although in case of 3G, it is more like the United States and the rest of the world). While a 120 MHz spectrum allocation has been designated for 3G, the United States has already auctioned some of this spectrum for PCS1900 applications.

The International Telecommunications Union's (ITU) initial stance on 3G (covered by the IMT-2000 program) was that it should be capable of delivering at least 144 kbps to a fast-moving vehicle, 384 kbps to slow-moving ones, and 2 mbps to stationary users. The ITU has abandoned their initial stance that there would be only one 3G standard and have now accepted a family of standards. This basically occurred because there was no sign of agreement between the CDMA and TDMA camps.

The European standard for 3G wireless services is known as the Universal Mobile Telecommunications Systems (UMTS). The standard will have three modes, namely, direct spread (DS), multicarrier (MC), and time division duplex (TDD).

A somewhat unexpected agreement between the Universal Wireless Communications Consortium (UWCC) and the GSM North American Alliance to converge the two TDMA standards (GSM and D-AMPS) would seem to have strengthened the position of the GSM 3G standard, leaving cdmaOne out on a limb. The ultimate aim is GSM to TDMA interoperability and the standardization of 3G development. What seems to be assured is that wide-band CDMA (W-CDMA) is the way of the future.

It is anticipated that the emerging 3G standards will have applications not only in cellular, but also in satellite mobile and wireless local-loop applications. This will encourage the production of chip sets with a universal application that will drive prices down with economies of scale.

Smart antenna technology (see Chapter 12) will be an integral part of 3G technology.

Who Needs 3G?

This often asked question does actually have an answer. While everyone agrees it would be nice to have more bandwidth and higher data rates, most users can live without them. A practical driving force behind 3G is in fact just capacity, and in Japan's megacities capacity problems are being experienced. In the less densely populated US and European cities this problem is not as urgent. It is probably fair to say that Japan needs 3G sometime soon, Europe by 2003 (approx.), and the United States some years later; for most of the rest of the world there is no real need to hurry.

Japan's Telecommunications Laboratory (NTT) DoCoMo has announced that it plans to begin W-CDMA operations in 2001.

But does anyone really want to browse the Net on their mobile phone? With 3G, the user pays only for the data packets used, but sending video and graphics will require a lot of data, and will be costly. In the meantime, the wireline networks are getting better at providing high-speed data (particularly over optical fibers) at ever decreasing prices.

There is no real agreement among the marketing gurus on the demand for wide-band mobile services. Forecasts range from "virtually no demand" to a $100 billion market by the year 2005.

W-CDMA

It is virtually certain that 3G mobiles will be based an a W-CDMA platform. The bandwidth will be significantly wider than the 1.25-MHz IS95 standard, and is expected to be in the range from 5 MHz to 20 MHz. Smart antennas are expected to emerge as the antenna of choice, for improved capacity, range, and immunity from interference. Smart antennas can be fitted to 2G equipment, but the ad hoc nature of their interfacing makes them both expensive and power inefficient. By specifying the smart antenna interfacing in the 3G equipment, these limitations will be overcome. Once the smart antennas achieve economies of scale, they have the inherent potential of being very inexpensive.

3G RF

Some people feel that the real challenge of 3G will be in the RF part rather than in the signaling and control. This conception is brought about by the need for multiband operation, which by one estimate could require up to 30 RF filters in the RF and intermediate frequency (IF) sections (together with other attendant hardware).

However, as the processing demands increase, the speeds rise to levels where designs resembling RF techniques will be required for the logic circuitry. This will become evident as soon as clock speeds approach the 1-GHz rate.

Additionally wide bandwidths (5 to 20 MHz) present real challenges in power amplifier (PA) design, with linearity being difficult to maintain. This can lead to spurious responses and interference. As recently as 1999, suppliers such as Spectrian were offering W-CDMA, 3G linear amplifiers capable of 25 watts and operational over a 5-MHz bandwidth, so the problems are being addressed.

Generally speaking, linear power amplifier offerings only marginally meet the specifications required of them and they are very prone to failure. Costs of these units are high, as they require extensive alignment and testing in the manufacturing stage due to wide variability in the production parameters. Improvements in linear power amplifier (LPA) technology must be forthcoming for 3G to succeed.

Spectrum

The spectrum needed to operate the 3G systems has yet to be decided upon. A meeting of the World Administrative Radio Conference (WARC) in late 2000 is expected to firm up the 3G allocations. The ITU is trying to define a universal frequency range, but this is proving difficult. In most countries it is not clear where the spectrum will come from, except that some of it will come from the bands currently being used for PCS/PCN. It is estimated by the ITU that an additional 160 MHz will

be needed by the 3G services in 2010. Some industry estimates are considerably higher than this. Frequencies in the 1800–2200-MHz range are candidates.

There is still some considerable conflict between the spectrum preferred by the United States and the preferences in Europe. The rest of the world has an easier option of accommodating, without too much difficulty, whatever is decided upon.

Defining 3G Traffic

Much of the talk of 3G revolves around gross data rates, but that is only part of the story. The nature of the traffic varies considerably and it can broadly be grouped this way:

Error Sensitive	Error Tolerant
Video	Voice
Image Transfer	
File Transfer	
Transaction Processing	

Delay Sensitive	Delay Tolerant
Voice	Image Transfer
Transaction Processing	File Transfer
Video	Voice Messaging

Here, we notice that only voice is error tolerant and delay sensitive.

Within the context of broadband services there will also be another distinction. Between multi-rate services, different users operate at different data rates and variable rate services, for such things as ATM, where the user's rate during a single connection varies in real time with demand.

An additional constraint is that 3G services are intending to compete head-on with wireline, where bit-error-rates are typically 10^{-11}, compared to mobile data where 10^{-3} is considered acceptable, and often is about all that can be expected. Most of the wideband applications are error sensitive while the mobile medium's intrinsic error rate represents a serious challenge. Achieving better BER rates will come at the expense of either bandwidth or throughput, both of which are in limited supply in 3G applications.

3G Challenges

By 2000, with 3G services still far from completion, it was apparent that a lot of development was still required on the screen technologies (the demands of portability and screen size are in direct conflict) and the battery

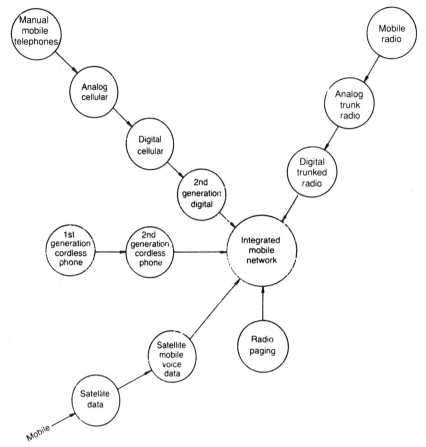

Figure 36.5 Ideally, future mobile technology will be integrated. The convergence of the mobile technologies is a worthwhile goal—but is it just a dream?

capacity (nothing new about this one … it has long been the bane of cellular); however 3G mobiles will require significantly more power than their 2G counterparts.

3G systems will be expected to support GSM 900, 1800, and the IMT-2000 bands. This will present a serious design challenge for the antenna engineers. The original dual band antenna problem was not as difficult as the PCS/PCN frequencies, which are nearly twice that of GSM 900.

The wide range of frequencies also presents some serious challenges for coping with self interference. It has long been recognized that triple conversion receivers, while great for selectivity, have serious limitations brought about by the use of the three mixers, each generating their own set of internal interferences. This can be directly observed by connecting a dummy load to the antenna, and scanning the band for "signals"; any found this way will probably be self generated. In a cellular transceiver, both the transmitter itself and the mixers will produce intermodulation products which may fall in

the RX band. With only one frequency of operation, this can be controlled by tweaking IF frequencies and oscillator frequencies. But with three diverse receivers in the one box, there is a need to engineer these products out more directly.

The I/O capabilities of the 3G handsets will need to be fully voice activated; this will replace the clumsy "press *104 and SEND", then on response "press 6 followed by 1" to get your voice mail. Future systems will need to be more oriented to the Internet. The incorporation of IP (Internet Protocol) and, probably ATM, is a must.

UNIVERSAL MOBILE RADIO

There is much talk in the industry about the convergence of the mobile technologies (such as paging, packet radio, data, and cellular) into a single personalized communicator. This vision acknowledges that, given the current state of the art, universal communications are not economically practical for the general population, but it fails to accept that the demand for improvements in the existing technologies is so high that standardization is being neglected. With so many variations and basic incompatibilities within the same technology subgroups, it is possibly a little optimistic to expect that other mobile technologies will halt development long enough to achieve integration. Figure 36.5 illustrates this concept.

The dream of the cellular equipment designer is emerging in the 3G proposals, from the basic concept seen in Figure 36.6.

One area that is rapidly being integrated into cellular systems is paging. As cellular phones become ubiquitous, short messaging is leading to the demise of paging. Paging systems world-wide are shedding customers at an alarming rate. However trunked radio remains separately viable; but the cost of terminal units limits its appeal.

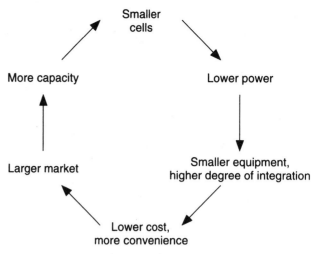

Figure 36.6 Cellular circle concept.

37

GSM PAN-EUROPEAN CELLULAR

BACKGROUND

The Global System for Mobile Communications (GSM) specifications are comprehensive and voluminous. Manufacturer compatibility has been assured by precisely specifying the system at a number of interface levels. Perhaps this preciseness has something to do with GSM's having originated with a 1981 French-German study and later evolving into a Pan-European standard.

In the 1980s a Conference of European Posts and Telegraphs (CEPT) formed a study group called Groupe Speciale Mobile (the group's original name; GSM) to study the development of a pan European mobile telephone system.

In 1990 the European Telecommunications Standards Institue (ETSI) published the specification for GSM phase 1.

The complexity of GSM is quite astounding, and it is arguably the most complex non-military development in radio that has ever been undertaken. There are many in the industry who insisted that GSM is short for "Giant Software Monster." Even though there was nothing essentially new in GSM, all previous like developments were for low-volume, high-cost military applications. Bringing this technology to the domestic marketplace in the early 1990s, in a way that would compete effectively with a mature analog market, was a massive undertaking.

The traditional method of shrinking costs and price has been to develop application-specific integrated circuits (ASICs) for the application. This is the region where most of the development occurred.

When GSM was actually turned on in late 1991, the development of mobiles was lagging behind infrastructure to the point that some in the industry were wearing badges that read "GSM" in large letters with "God Send Mobiles" in small letters underneath. By mid-1992 there were plenty of mobiles but, with the sole exception of Germany, very few buyers. The UK, with 1.6 million analog mobiles, had a thousand GSM users; Australia, with 700,000 analog lines, had 800 GSM lines. Hong Kong, which had reached saturation in the analog system, was selling the remaining analog lines and phones at a premium of 35 percent above GSM digital Advanced Mobile Phone Service (DAMPS) (both systems are used in Hong Kong), and was being swamped with buyers.

A large number of countries had adopted GSM before it was proven and then had to make it competitive, or in some cases even just work. Most took comfort in the numbers. With so many systems around "surely someone will get them working" soon was the refrain of the faithful. Many operators had invested hundreds of millions of dollars and still had few customers, the word was that GSM stood for "God Send Money."

By about 1996 most of the bugs were ironed out and GSM began to flourish. By late 1998 it became the dominant system. Everyone forgot its checkered start, and connection rates soared. Meanwhile cdmaOne and Iridium, ignoring the lessons of GSM, were both launched, without an adequate supply of mobiles.

The original version of GSM in the 900-MHz band is now know as GSM 900, and the high frequency versions are known as GSM 1800 (PCN; personal communication network) and GSM 1900 (PCS; personal communication service) in their 1800- and 1900-MHz incarnations, respectively.

The first "live" call in the GSM system was demonstrated by Motorola at the Hanover Fair in March 1991.

The demonstration comprised a production-type base station, two mobiles and a switch simulator. This is an indication of how fast GSM was implemented.

WHAT GSM OFFERS

The voice quality of GSM, with the enhanced compression/decompression (CODEC), is comparable to that of analog but even by 2000 was not actually as good. Like all digital systems GSM offers good voice security and some fraud immunity.

GSM base stations are significantly less expensive than eight channels of their analog counterparts. However, this may not mean that GSM is cheaper to implement, because more base stations are needed to cover the same area to the same standard. As most bases will necessarily be sectored, the minimum size of a base station will be 24 channels (pooling of carrier timeslots across a sector is not an option).

It is the very complexity of GSM that is both its strength and its weakness. If all of the GSM features were implemented and working satisfactorily, the system would probably be rather attractive, but most of the features are in fact optional extras; few systems in fact even implement most of them.

IMPLEMENTATION

GSM was implemented in two phases. Phase 1 had a limited feature set and did not include frequency hopping for the network infrastructure. Although the system can function without this feature, frequency hopping produces a processing gain similar to space diversity, and it is really essential for good handheld and rural coverage. Switches must be interconnected with signaling system No. 7 (also known as SS7) if automatic roaming is to be achieved.

ENCRYPTION

A significant advantage of digital systems is the ability to provide a secure speech path. GSM uses the A5 algorithm, which has been proven to be a very high-security encryption code. It should be realized, however, that national security organizations generally will not permit the use of any public mobile encryption that they cannot decipher. An example is the Australian GSM system, which was to have been switched on April 1, 1993, but had its implementation delayed indefinitely because

ASIO, the Australian security organization, complained that they could not decipher it. Eventually they were supplied with "full access" and the system was turned on. This meant that ASIO had direct access via the switched network to any phone they wanted to snoop on.

The Europeans initially proved to be very protective of their own security and exhibited a great reluctance to issue the A5 algorithm outside the original memorandum of understanding (MOU) countries. An alternative algorithm, the A5X, is available for use in countries that cannot obtain an export license for the A5. However this is not really an issue in 2000.

THE RADIO FREQUENCY INTERFACE

The radio frequency (RF) interface, also known as the Um interface, has been designed to use the same spectrum originally allocated in Europe for the analog systems. The RF interface has been specified allowing for the need for high capacities and fast, accurate handoffs. For this reason, both the mobile and the base station are involved in handoffs.

Some RF parameters are as follows:

- GSM 900 (900 MHz); GSM 1800 (1800 MHz); GSM 1900 (1900 MHz).
- Channel spacing: 200 kHz for 8 channels or 16 half-channels.
- RF power: 32-watts base station/carrier; 13-watts mobile transmit.
- Sensitivity for 1 percent bit error rate (BER): -104 (approximately), but -107 is widely achieved in practice.
- Power steps: 15 of 2 dB each.
- Carrier to interference target: 10–13 dB.
- Operates at C/N of 10–12 dB.
- 1000 full-rate (16 kbit/second) or 2000 half-rate channels.
- A 3/9 cell or 4/12 (125 channels) cell structure is possible but most systems use $N = 4$ or $N = 7$.

The RF interface uses slow frequency hopping (217 hops/second). The resulting frequency diversity results in a reduced bit-error rate at a given incident signal-to-noise ratio (SNR) and improved immunity to interference. The interface also provides for full-rate channels, designated Bm, and half-rate channels, designated Lm. The half-rate CODEC was tried, but the quality was unacceptable.

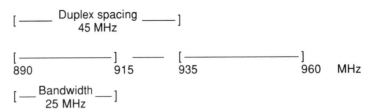

Figure 37.1 The frequencies for GSM.

GSM ENHANCED ENCODER

The propagation delay equalization can allow for up to a 233 microsecond absolute time delay. This is necessary to ensure data synchronization. The system can also cope with a 16-microsecond dispersion.

A carrier-to-interference ratio (C/I) of 12 to 14 dB is more realistic than the 10 to 13 dB originally specified. In addition to this, it is prudent to allow an additional 8 dB fade margin. The degradation of C/I immunity is very marked, as the threshold is approached, and GSM is quite sensitive to C/I.

Co-channel interference is a problem with GSM, and it will be necessary to ensure that channels at any one site are separated in frequency by one channel width (200 kHz).

GSM 900 FREQUENCY USAGE

The frequency band of GSM 900, which spans 890 MHz to 960 MHz, is shown in Figure 37.1. Base-station transmitter (TX) is 890 to 915 MHz, and the base receiver (RX) is 935 to 960 MHz.

The GSM 900 spectrum has 25 MHz (×2) for the duplex channels and 45 MHz duplex spacing between the send and recieve channels. The 25 MHz is further subdivided into 124 channels, each 200 kHz wide. These 200-kHz-wide channels, in turn, use time division multiple access (TDMA) techniques to derive 8 voice channels.

In most countries there are three operators, and because of the low tolerance of GSM to adjacent channel interference, problems arise when adjacent channels from different operators are allocated. In general, a guard band between carriers will be essential. Blocking can occur on the uplink, which can be countered with suitable filtering, but downlink blocking to the mobile is nearly impossible to control. The problem is worst when a mobile is using a distant cell or is being blocked by the nearby adjacent carrier of another operator. Perhaps the only solution to this problem is coordinated frequency planning.

MODULATION

The modulation method use by GSM is Gaussian minimum shift keying (GMSK).

Like all digital systems, GSM uses a CODEC to sample the speech and encode it in a way that takes advantage of the redundancies in speech to compress the code, which is transmitted. (See discussion in Chapter 48, "Digital Modulation," on CODECs for a full description of their operation). It should be noted that because the CODECs tailor the encoding to speech, they are not transparent to analog data signals (such as that from a modem). If data are to be sent, the CODEC must be bypassed. Provision has been made for data rates of 12 kbits/s (9600 kbits/s of data), 6 kbits (4800 kbits/s of data), and 3.6 kbits/s (2400 kbits/s of data or less).

The CODEC input is at the rate of 8000 samples per second of a 13-bit pulse duration modulation (PCM). This is converted to 13 kbits/s basic voice data rate at the CODEC output in 20 ms bursts of 260 bits of data each. The rate is then increased to 22.8 kbits/s once the error correction code is added. The CODEC mode is regular pulse excitation with long-term prediction (RPE-LTP). To this is added synchronizing and signaling information and guard bands, as shown in the time slot structure (Figure 37.2). These additional bits increase the net bit rate to 33.9 kbits/s.

A battery-saving feature of GSM is that it uses discontinuous transmission (DT), in which a voice-detection device is used so that the transmission can be turned off when speech is not present. The CODEC output is then processed to add parity bits and time slots

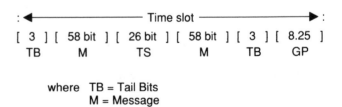

Figure 37.2 Timeslot structure.

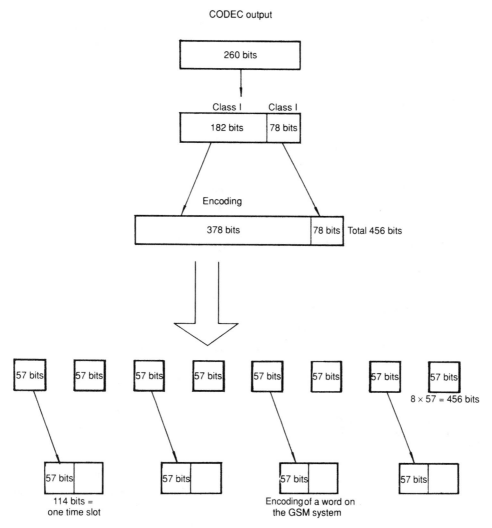

Figure 37.3 The GSM CODEC structure.

with 2×57 bit words are produced as shown in Figure 37.3.

The CODEC output is transmitted in a timeslot along with tail bits and training bits, which are used to extract timing information. This timing information is used to model the transmission channel so that the appropriate equalization can be added. Apart from channel information, other data may appear in a timeslot, as seen in Figure 37.4.

The access burst is a shortened burst used by the mobile for its first, unsynchronized attempt to access the system. The frequency correction burst transmits the instructions to frequency hop. The periods of quiet that occur during the time that the transmitter is turned off were found to be disturbing to the mobile user, and so it was necessary to inject "comfort noise" in the quiet

periods. Figure 37.5 illustrates this injection, as well as the main speech processing functions.

GSM FRAME STRUCTURE

The GSM frame structure has as its basic component the timeslot. A frame consists of a group of eight timeslots plus the guard bands. Each channel consists of the 148 bits of information plus the guard time (8.25 bits long), making the total channel length equal to 156.25 bits, or 0.577 msec. The eight timeslots in the frame are 4.62 msec long, producing a frame rate of 217 frames/s.

The frame structure, as seen in Figure 37.6, consists of multiframes (26 traffic or 51 control channels), superframes (1326 frames), and finally hyperframes (2048

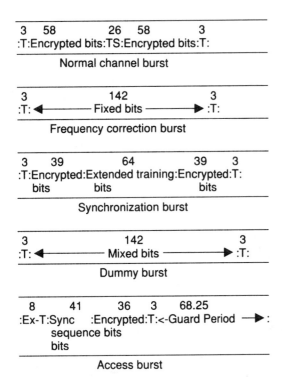

T = Tail bits
Ex-T=Extended tail

Note that to each of the frames above, (except the access burst) the guard period of 8.25 bits should be added.

Figure 37.4 The forms of a timeslot.

superframes). The frame number is used in the encryption process.

DATA TRANSMISSION

The data rates originally supported by GSM are 3.6, 6, and 12 kbits/s. There are two transmission modes, known as transparent and non-transparent.

In the transparent mode, data are sent at a constant rate and the error correction is limited to the capabilities of the forward error correction (FEC). The data rate does not compensate for radio noise or environment, and the data flow through as they would along a pair of wires (hence the term transparent—i.e., it does not "see" the radio environment).

In non-transparent transmission, in addition to the FEC, when the error is such that it cannot be corrected by the FEC, an automatic retransmission request (ARQ) is sent. In this mode the rate of transmission will vary with the quality of the link, which in the radio environment can be highly variable. Bad data bursts are resent until they are received properly.

GSM data are supported through the interworking function (IWF), which is typically located at the mobile switching center (MSC), and is essentially a modem pool that interfaces with the public switched telephone network (PSTN), Integrated Service Digital Network (ISDN), Internet, and the X.25 public switched packet data network (PSPDN), with the most common data rate being 9.6 kbps.

The short message service (SMS) and the associated data services, such as cell broadcasting (SMS-CB), weather, and traffic services, use control channels and are sent at low data rates.

Phase 2+ offers mostly improved data services, including the high-speed circuit-switched data, which uses multiple bearers to achieve gross bit rates up to 96 kbps.

LINK INTEGRITY

Lost and dropped calls have become an accepted part of analog cellular. Advanced processes in GSM will address this problem in an effort to improve link integrity with a reduction in lost calls and interference. Procedures exist that will allow dropped calls to be re-established. Call quality is monitored by both the mobile and the base station. Monitoring includes the levels of co-channel and adjacent channel interference, which are measured as bit-error rate degradation and level. Hand-off to adjacent cells can even be initiated by the switch on the basis of call congestion. However by the year 2000 GSM link integrity has proven to be "about the same" as analog.

The encoding is done to identify and correct errors that may have occurred in transmission. There are three coding processes. Block coding is used to allow easy identification of errors. This is then followed by a convolution process, which adds 227 bits to the code and enables error correction. The data then are interleaved to spread them over a number of bursts so that individual burst errors can be corrected. Next the processed code is subjected to ciphering to ensure voice privacy, and finally it is coded in the channel format, as shown in Figure 37.7.

PHYSICAL AND LOGICAL CHANNELS

GSM distinguishes between physical channels, which are the timeslots (there are eight channels in each frame),

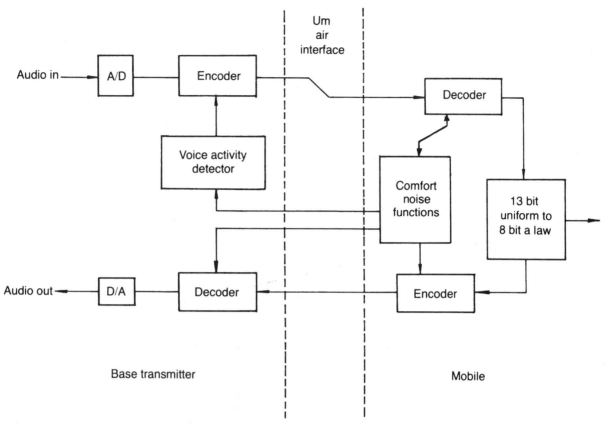

Figure 37.5 Speech processing includes the injection of "comfort noise." The speech path is first passed through an analog-to-digital (A/D) converter and then encoded. At the encoder, "comfort noise" is added to give an impression of path continuity during the "blanking" that occurs when no voice activity is present.

and logical channels, which are contained in the time-slot. As has been seen, the timeslots usually contain data for two traffic channels. These traffic channels are known as *logical channels*.

Other logical channels include broadcast channels (BCH), synchronization channels (SCH), broadcast control channels (BCCH), paging channels (PCH), and access grant channels (AGCH).

SYNCHRONIZATION

Because of the exacting timing requirements of GSM, it is essential that the parts of the system be stably locked to a precision reference oscillator. A reference with a stability of 0.05 part-per-million is provided, and all components of the network are synchronized to this reference. The mobile takes its reference from the base station.

HANDLING MULTIPATH

Multipath problems represent a considerable challenge for digital cellular. This is particularly true when applications such as use on a high-speed train (moving at up to 300 km/hour) means that synchronization needs to be very systematic. GSM uses 20 synchronizing bits in each time-slot, which enables the receiver to calculate the transfer function that should be used to correct for the multipath. The complexity of this process is such that the channel equalizer uses 50,000 Complementary Metal Oxide Semiconductor (CMOS) gates. The very short symbol duration, of less than 3.7 microseconds, is somewhat less than the typical delay between multipath signals. To allow for this, a training burst is included at the beginning of each burst to allow synchronization. Multipath delays of up to 16 microseconds can be accommodated.

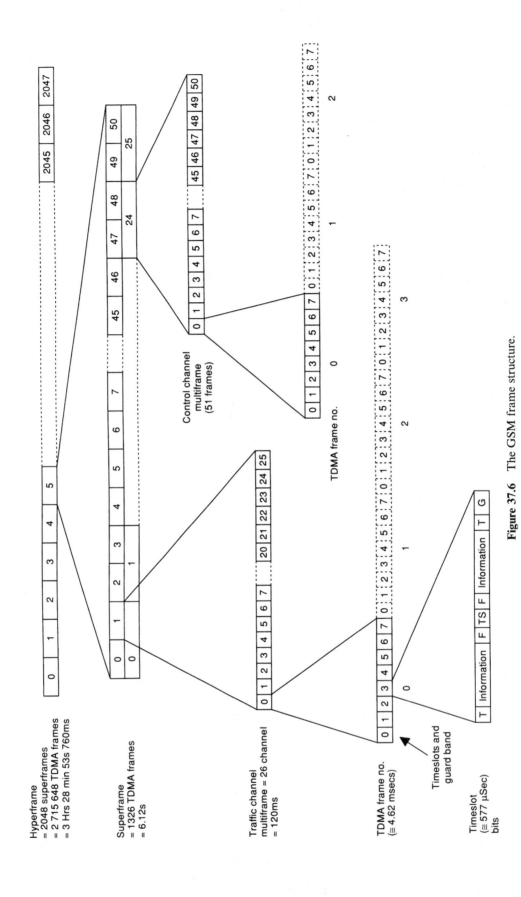

Figure 37.6 The GSM frame structure.

516

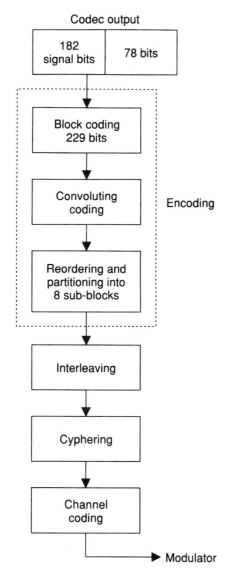

Codec output

| 182 signal bits | 78 bits |

Block coding 229 bits

Convoluting coding

Reordering and partitioning into 8 sub-blocks

Encoding

Interleaving

Cyphering

Channel coding

→ Modulator

Figure 37.7 The coding and ciphering process.

SYSTEM CONSIDERATIONS

The introduction of GSM has brought with it the need to develop a number of enhancements over its analog counterparts. The complexity means that more sophisticated management and maintenance software needs to be developed. Hardware fault diagnosis needs to be much more detailed and reliable.

New wide-range, fast-acting AGC systems, which can cope with the necessarily rapid response demanded by the short burst times, with dynamic ranges of the order of 100 dB have been developed.

Simple preamp (PA) design cannot achieve the linearity needed for the rapid ramping of GSM, and so a number of innovative quasi-linear designs have emerged.

The actual GSM network can be considered to consist of three separate hardware platforms. It shares with analog systems the first two, namely the switch and base-station parts, while the third platform consists of the computers that hold the databases. Computer-based parts include the home location register (HLR), Authentication Center (AuC), and equipment identity register (EIR), which are known collectively as the GSM Network Data Bases. These parts are often produced by systems information specialist groups that make generic products that can be integrated with the switches of any manufacturer.

In some cases the main switch may also (optionally) provide one or more of the HLR, AuC, visitors' location register (VLR), or EIR functions. As an example, the AT&T 5ESS switch can perform some or all of these functions. In fact, the 5ESS can be configured as a dedicated HLR, with a capacity of one million entries. For smaller systems it may perform all four functions and still have an MSC capability.

The GSM message functions are also computer based. They include the Short Message Service Center (SMSC) and Cell Broadcast Service Center (CBSC). The Operations and Maintenance Center (OMC) and operations support system (OSS) functions are also essentially computer-based systems, which again may be produced independently of the system supplier. It should be noted, however, that because standard interfaces for these parts do not exist, there needs to be some customized integration.

GSM TERMINOLOGY

An essential part of the GSM specification is the standardization of the terminology used to describe the component parts of the system. The multinational nature of GSM has made the use of an extensive uniform vocabulary a practical necessity. But mastering this terminology then becomes essential in order to follow discussions about the system. Figure 37.8 illustrates the logical parts of the GSM system with the appropriate terms.

GSM is divided into three basic "intelligent networks" (INs). These are the

- *SMS* (service management system), which controls network software downloading, network performance records, traffic management, customer registration, and authorizations.
- *SSP* (service switching point), which determines the services that are needed and controls the switching of those services.

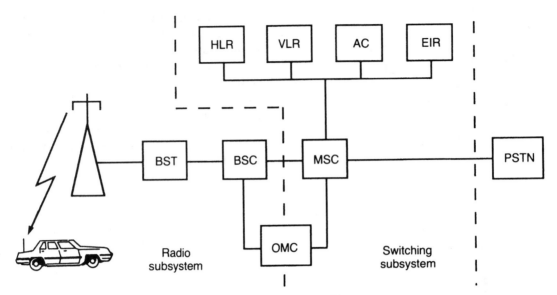

Figure 37.8   GSM system architecture.

- *SCP* (service control point), a database management system that validates requests that may be made in the call progress.

These are the functional parts of the GSM network subsystem (NSS) system:

- *MSC (Mobile Switching Center)* The mobile switching center controls switching and much of the system processing. It may jointly be a local or trunk switch. The other switching subsystem components are designed to be operated either as part of the switch or as physically separate entities. The MSC will probably have been designed originally as a PSTN switch; the basic functions are the same as those required in a trunk switch. The MSC controls a number of base-station controllers.

- *HLR (Home Location Register)* The HLR contains the permanent records of the subscriber and is the main user database. Full access is available via the Operations and Maintenance Center (OMC). Typically, this register has a capacity of 100,000 to 1,000,000+ subscribers. The HLR also contains a completely up-to-date record of the subscriber's current location. The HLR is connected to the MSC/ULR by a high-speed mobile application part (MAP) interface.

- *AuC (Authentication Center)* Here the "authentication" parameters are generated for subscribers' access, and a cipher key is obtained for encryption

of data communications between the system and the mobile.

- *EIR (Equipment Identity Register)* This register is a database of mobile stations and performs identification functions. It will identify unauthorized attempts to access the network and trace and control the use of unauthorized equipment. It can be fully accessed by the OMC.

- *VLR (Visitor Location Register)* The VLR is a temporary database for subscribers. It is used for both local (HPLMN) and roaming subscribers. It has an extensive data exchange with the subscriber's HLR. The VLR is accessed by the MSC for each call setup, and therefore the MSC will always have an associated VLR. Extensive signaling occurs between the MSC and VLR, and so they will usually be co-located. Often the VLR will be integral with the switch.

Equipment Identity Register

The EIR is a register of all mobile stations (as opposed to the subscriber identity module (SIM) card). It contains the international mobile station equipment identities (IMEIs), which is composed of the equipment serial number plus the type approval code. All EIRs have access to a central EIR, which can distribute this information throughout the network. Mobiles can then be grouped into three groups:

- Blacklist, containing all stolen mobiles.
- Gray list, containing mobiles that are causing interference or have been marked for special attention.
- White list, containing all approved mobiles, and used to ensure that only properly type-approved units get onto the system.

The EIR is fully accessible to any OMC. The EIR is connected to the MSC via a Consultative Committee on International Telegraphy and Telephony (CCITT) No. 7 MAP interface. The EIR is usually a standard computer-based platform, which may be integrated with the AuC.

SIGNALING AND INTERFACING

Most of the signaling in GSM is done using a subset of the Common Channel Signaling System No. 7. This subset is used by the mobile application part (MAP). MAP procedures are used for signaling between the MSC, VLR, and HLR. The systems accessed via MAP are essentially databases, and access is usually on an inquiry and response basis. System designers need to ensure that a suitable 64-kbits/s link is available for this purpose. MAP procedures can in turn be broken down into functional subsets.

BASIC SUPPORT SERVICES

The basic support services involved in setting up a call, including paging, security procedures, mobile categories, and MSC handover, are supported by MAP. Some of these are described below.

Call Setup

MAP supports the retrieval of subscriber data. For a call to be set up, the authentication for service request is sent to the VLR for category (e.g., are outgoing calls permitted?). For incoming calls, the HLR must indicate the current VLR and MSC with which the mobile is currently registered. The VLR will also provide the international mobile subscriber identity (IMSI), which may not necessarily be the same as the subscriber's telephone number.

Access Management

After an incoming call has been paged and has responded, the VLR will request the subscriber's data so that authentication can be done and a temporary mobile station identifier (TMSI) allocated before connection is allowed.

Handover Inter-BSC

Inter-base station controller (BSC) handover is handled by the MSC, which uses the pre-processed information of the handover criteria that is sent from the BSS. (Note that handovers intra-BSC are controlled by the BSC.)

Handover Inter-MSC

Sometimes a handoff is required to take place between two MSCs. When this happens, the first MSC maintains control of the call but the subscriber information must be sent to the new VLR. Should the subscriber continue to roam to yet a third MSC, the first one relinquishes control of the call to the second, while the new VLR must receive the subscriber's data.

Subscriber-Activated Services

Subscriber-activated services, such as call forwarding, conference facilities, password handling (for special services), and other subscriber communications between the HLR and MS are supported by MAP.

Short Message Service

The short message is controlled from the short message service center (SMSC), which stores, delivers, and acts as a transit center to other public land mobile network (PLMNs) for the short messages. The short message service center is connected to the Gateway MSC by an X25 protocol link. However, the message routing to or from the visited MSC, VLR, and Gateway MSC are handled by MAP.

MSCs must be connected to each other by a common channel signaling (CCS) No. 7 link if the roaming capabilities are to be invoked. Signaling to the PSTN need not be CCS No. 7, but it should be, if possible, so that the full potential of the ISDN capabilities of GSM can be used.

THE OPERATIONS SUPPORT SYSTEM

The OSS is a centralized controller, usually based on a standard computer platform that centralized the network operations and control from more than one OMC.

The operations support system in GSM is an integral part of the system and comprises the following parts: data postprocessing system (DPPS), personalized center for SIMS (PCS), security management center (SMC), network management center (NMC), and operations and maintenance centers (OMCs).

Data Post Processing System

Until the advent of GSM, the billing system was very much an add-on to the cellular network, which was handled by a billing specialist company. Specialist billing companies are still involved in most systems, and they have integrated their products into the GSM infrastructure. The GSM concept embodies the billing structure into the system together with the subscriber and marketing support modules, which are a part of the more advanced billing systems.

The DPPS includes the following functions:

- Billing and invoicing
- Customer administration
- Accounting, credit control, payments, and bad debts management
- SIM card management
- Sales and marketing information
- Tariff maintenance

The interface between the MSC and the billing system uses the X.25 protocol. The call accounting records (CARs) are written onto a disk or tape, usually with call record backup of at least a few days in the event of a link failure.

Personalized Center for SIMS

The SIMS card is central to the security of the GSM network. Very elaborate safeguards have been put in place to prevent the manufacture of illicit cards. The PCS includes the robotic system to personalize the cards, including the laser en-graving, quality assurance, and issuance of PIN/PUK (personal unblocking key) letters.

Security Management Center

The SMC controls data encryption, generation, and distribution, as well as updating of the master keys. Secure communications with other parts of the network, including authentication, data integrity, and confidentiality, are controlled here.

Network Management Centers

The NMC holds information on the PLMN configuration, and network-wide data. The production of network statistical reports and management tasks are done here. The NMC also handles traffic control and network reconfiguration.

The Operations and Maintenance Center

Each GSM system or subsystem will have an operations and maintenance center. These systems provide network supervision functions, including alarms, traffic and overload controls, fault reporting, statistics collection and analysis, system inventory and control, and network administration.

Most major operators originally, from day one, decided to start with two systems suppliers, perhaps for security of supply or maybe to gain a wider experience of different products. This approach is not without its problems, and many have since returned to a single supplier.

The degree of open specification of GSM is very limited and effectively is only on the A, MAP, and Um interface. This means that any switch can control the base-station controller of another manufacturer. However, the BSC can only drive base-station equipment from the same manufacturer.

The OMC is itself not part of the open GSM specification, which leads to the operational problem that in a system using more than one supplier, multiple OMCs are needed. The Bundespost (the PSTN operator in Germany), which has two suppliers, has solved this dilemma by introducing an OSS, which interfaces with the supplier-specific OMCs, and allows access to most OMC functions from one center, as shown in Figure 37.9. The cost of an OSS, or alteratively the difficulty of operating two OMCs, means that small systems would be well advised to stay with a single manufacturer.

The Bundespost OSS is much more than a simple device to tie together two OMCs. It has a DPPS, and also provides customer administration, billing and invoicing, accounting, payment and credit management, SIM management, maintenance of tariffs and GSM data tables, management of sales organization, and sales and marking information functions. The personalized center for SIMS fully automates the SIMS production, controlling the ordering, physical control of the SIM cards, robotic production, and quality assurance. The customer parameters stored in the HLR are entered by the OSS. Although the Bundespost OSS is an elaborate example of OSS centralization, conceptually other OSSs will be similar.

BASE-STATION SUBSYSTEM

The base-station subsystem (BSS) defines a complete local base-station network, which operates into one of the few "open specification" interfaces in GSM, namely the A interface. At this level, it should be possible to interconnect a BSS from any supplier to the MSC of any

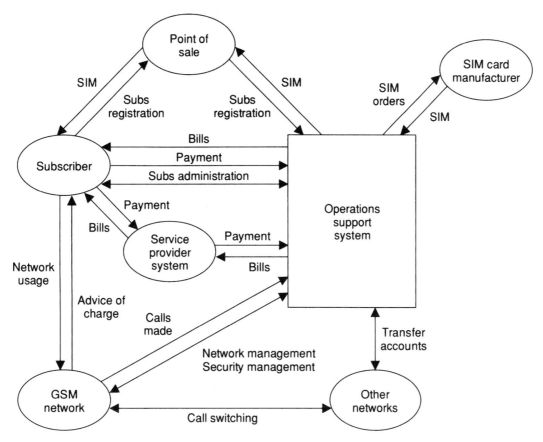

Figure 37.9 The functional structure of the Bundespost OSS.

other. It is well to be aware that the OMC that monitors and controls the BSS functions will be manufacturer-specific and will need to be from the same stable as the BSS. The BSS structure can be seen in Figure 37.10.

Where the BSS has suitably configured parallel controllers, as is the case with the AT&T system, BSS software can be updated in the standby controller while current calls are handled by the active one.

TRANSCODER

The transcoder provides rate adaption to the 64 kbits/s required by the A interface. It takes the air interface data, which may be 13-kbit/s speech or 3.6/6/12-kbits/s data, and converts it to 64 kbits.

The transcoder can be placed either at the BTS or remote from it. When co-located it has no real significance, as it is simply a wire link. When the transcoder is located at the BTS the Abis interface will operate at 64 kbits/s. When remote the Abis interface link includes control bits for operations between the transcoder and channel coder. This link can operate at 16 kbits/s so

that 4 channels can be multiplexed onto one 64-kbits/s channel. This can be used to considerable cost advantage when operating remote base stations over a long trunk route.

BASE-STATION CONTROLLER

The primary function of the BSC is call maintenance. Each BSC can control a number of BTSs, which may or may not be co-located. The signal level from the BTS being used by the mobile equipment (ME) is monitored, and the mobile scans the nearby BTS's BCCH for signal levels during the time slots not allocated to it for transmission. These levels are constantly reported back to the BSC [about every half second on the associated control channel (ACCH)] from the MS so that a decision can be made on handoff. While the alternatives are within the BTSs controlled by the BSC, the decision can be made and acted on without reference to the switch. The BSC also makes decisions about and controls Um power levels.

The BSC size will depend on the manufacturer. The

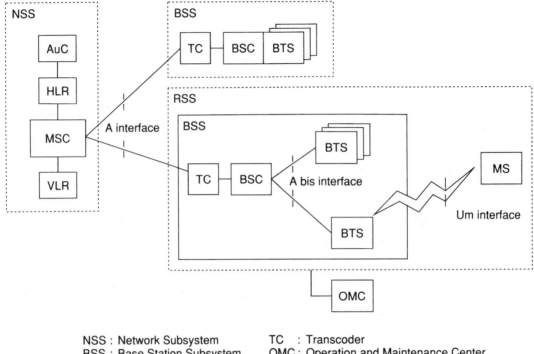

NSS : Network Subsystem
BSS : Base Station Subsystem
MSC : Mobile Switching Center
BSC : Base Station Controller
BTS : Base Transceiver Station
RSS : Radio Subsystem

TC : Transcoder
OMC : Operation and Maintenance Center
AuC : Authentication Center
HLR : Home Location Register
VLR : Visitor Location Register

Figure 37.10 The base-station controller can control a number of base stations and the air interface.

Ericsson CME 201 BSC, for example, can be used to control up to 256 sites with up to 512 cells. Because the BSC can control the handover functions, large BSCs have the advantage of minimizing BSC-to-MSC signaling.

For small systems the BSC will probably be co-located with the MSC. Generally the hardware of the BSC and MSC will have much in common, so co-location will utilize spare parts and expertise most efficiently. Transmission between BTS and BSC can be at 16 kbits/s if the transcoders are located at the BSC.

Intra-BSC handovers are organized by the BSC acting alone, whereas inter-BSC handovers will involve the MSC, which will receive the target list of handover cells and instruct the original base to issue the MS with a handover instruction.

The BSC will perform the following measurements for the OMC:

· Average number of busy TCHs
· Traffic channel congestion time
· Number of incoming inter-cell handovers within the BSS
· Number of outgoing inter-cell handoffs within the BSS

· Number of BSS handovers
· Total number of access attempts
· Number of attempted standalone dedicated control channel (SDCCH) seizures in a period
· SDCCH congestion time
· Average number of busy SDCCHs

BASE TRANSCEIVER STATION (BTS)

The BTS consists of the radio transceivers, coupling equipment and antennas, the baseband section (which among other things controls the frequency hopping, where installed), and the Abis interface to a remote BSC (where applicable). It should be noted that the Abis interface is not an open specification and will be manufacturer-specific.

Other functions controlled from the BTS are rate adaption, channel coding, interleaving, encryption/decryption, TDMA framing, antenna diversity (when used), signal strength monitoring, time alignment to compensate for propagation delays, and frequency hopping.

TABLE 37.1 GSM 900 BTS Specifications

Parameter	Specification
Frequency	900 MHz
Channel spacing	200 kHz
Frequency stability	±0.05 ppm
RF ramp time	28 msecs
Receive sensitivity	−104 dBm

The BTS can be equipped with a drop-and-insert switch (Transmission Radio Interface), which allows the BTS to be used as a transit point for other BTSs.

The BTS will conform to the specifications shown in Table 37.1. Base-station TXs are typically rated at 20 watts per carrier. There is a six-step power control function for the TX, each nominally 2 dB, which enables the OMC to control the base-station effective radiated power (ERP).

The exacting frequency standards required of the base station will be provided by a reference generator module (RGM), consisting of a rubidium oscillator (often with a redundant standby).

MOTOROLA'S FIRST IMPLEMENTATION

Figure 37.11 shows one of Motorola's first base stations. Each cabinet can hold five transceivers (giving a total channel capacity of 40). The BTS consists of the radio channel units (Figure 37.12), a base-station control unit, power supplies and cooling fans (a lot of heat is generated), and combiners.

The base station will be controlled by a BSC, which may be a separate rack of equipment remotely located, or may, as is the case in Figure 37.13, be configured in the same equipment rack as the BTS.

OVERLAY/UNDERLAY CELLS

The BTS will be able to support overlay/underlay cells, which permit co-sited cells to operate at different cell radii, as controlled by the ERP of each cell. Mobiles operating close to the cell are directed to the underlay cell, whereas those more distant may be directed to the overlay.

NETWORK CONFIGURATION

A network consists of a number of systems, with a hierarchy as seen in Figure 37.14. Broadly, the system can be broken up into base-station systems, switching systems, and PSTN. Although these are shown in a sample of the possible configurations, it could be, for example, that an operator chooses to co-locate all three systems in the same building.

Figures 37.11, 37.12, and 37.13 Fig. 37.11: An early Motorola base station in Chicago, 1994; Fig. 37.12: A radio channel unit (transceiver); Fig. 37.13: A base-station control cabinet (BSCC) consists of a BTS and a BSC combined. (Photos courtesy of Motorola.)

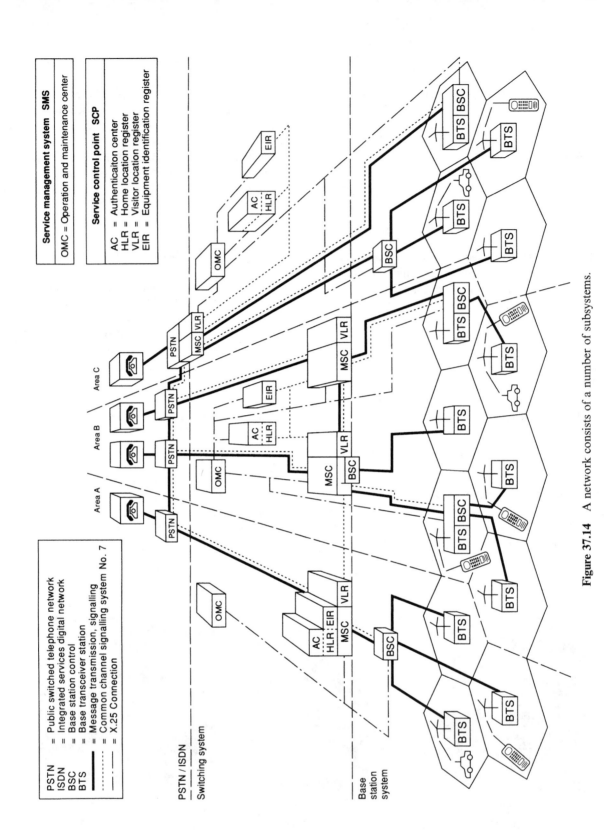

Figure 37.14 A network consists of a number of subsystems.

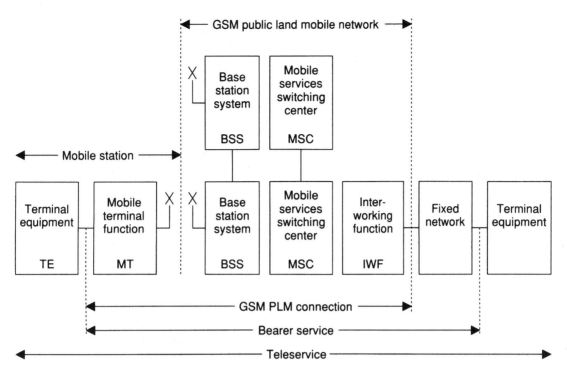

Figure 37.15   The component parts of the GSM network.

The GSM network is divided into functional parts, as can be seen in Figure 37.15. The GSM PLMN is defined to be the MSCs, BSSs, and their interface to the PSTN. This definition is similar to the analog network concept. Additionally, GSM defines the GSM PLMN connection, which is essentially the network plus the air interface.

The bearer service, which can be to or from a mobile, may also involve a conventional PSTN subscriber, and so is defined to be from the air interface to the PSTN-to-subscriber interface.

The teleservice is from subscriber to subscriber, whether PSTN or landline based.

FREQUENCY HOPPING

GSM base stations are state-of-the-art devices with military-like specifications on a commercial budget. One of the earliest problems encountered was that of how to provide frequency-hopping, which must be done simultaneously at the base station by each BTS transceiver for eight channels. The net rate of change of frequency is 1700 times a second.

This means that within a cell the frequency hopping must be correlated so that clashes do not occur. However, in order to improve immunity to interference the sequence from adjacent cells will be deliberately uncorrelated.

Frequency hopping provides a system gain that is similar to diversity gain. This is because different frequencies have different multipaths and so have different fade patterns. By hopping in frequency, it is possible to ensure that only small portions of the signal are subject to fade at any one time. Redundant code and error correction can in fact completely restore bits of data that are lost.

In the phase 1 equipment, frequency hopping at the base stations was optional whereas it was always mandatory for the mobiles. The reason is that while the mobiles can take advantage of the seven unused time slots to accomplish the hop, the bases must do it much more quickly. Hopping occurs at a rate of 217 hops per second per channel, with a hop occurring every 4.615 msec. The net result is better bit error rates and increased immunity to interference.

Frequency hopping is optional on all phase 2 GSM base stations, but it is available on most equipment supplied. Activating frequency hopping will improve the signal quality and range by reducing the BER.

The effect of frequency hopping is a function of the mobile's velocity, being more effective for slowly moving mobiles. At 50 kmph, frequency hopping, with antenna diversity, will reduce the C/I requirement for a 0.02 BER by about 4.5 dB. For a mobile at walking pace, the corresponding improvement is about 8 dB. From this it can be seen that frequency hopping offers big improvements

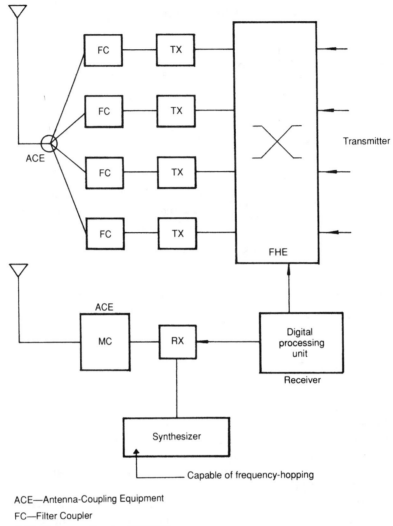

ACE—Antenna-Coupling Equipment
FC—Filter Coupler
FHE—Frequency-Hopping Equipment
MC—Multicoupler

Figure 37.16 The GSM base-station structure.

in areas of heavy frequency reuse. In rural areas the immunity to interference is less important than the higher effective sensitivity, but the gains are of the same order.

In high-density areas this can improve capacity by around 40%. At RF frequencies, high rates of frequency hopping can be achieved but not cheaply. The most favored solution today is to fix the BTS channels while switching the data streams. The base-station structure is shown in Figure 37.16.

DISCONTINUOUS TRANSMISSION

Discontinuous transmission (DTX), available in both the up- and downlink, causes a break in transmissions corresponding to speech pauses. It improves the quality of the network, as the total transmitted energy is reduced by a factor of 40–50%. This will also increase the battery life of the mobile.

The downside of DTX is that when it is used with power control, it can make the measurement of signal strength difficult, and some operators will not use both together.

THE LOCATION REGISTRATION

GSM networks are organized into a number of geographical locations zones. Within these zones, the mobiles will listen out on the BCCH for the identity of its location. When changes are noted, the mobile will issue a location update request (LUR) that will register the mobile in the new location. Consequently, the customer database in each location, although needing continuous updates, can be held at a controllable size. If the mobile roams into an area supported by a different VLR, then the VLR will issue a new mobile subscriber roaming number (MSRN). The HLR is then given the information about the mobile's new MSRN. Alternatively, the VLR can inform the HLR of the current MSC location and the MSRN can then be transacted between the MSC and the HLR. At certain time intervals, the mobile will be requested to provide periodic registration (PR) to update the file on its location and status.

A feature of GSM, and probably of all future digital systems, is that it is the mobile and not the base stations that reports on the field strength. Digital systems can use the idle time between active timeslots for other purposes. The mobile scans the adjacent channels during the idle time.

DIGITAL BEARER SERVICES

GSM can support digital services in a manner that is similar to ISDN. The ISDN concept uses the standard twisted-pair cable to derive two 64-kbits/s "voice-channel" lines called B channels plus one 16-kbits/s D channel for signaling. This means that a subscriber upgrading to ISDN from a conventional analog phone gets two lines plus a signaling channel in its place.

Of course, the bandwidth of a GSM channel is very much restricted relative to an unloaded cable, and so the equivalent structure offers only one B channel (known as a Bm channel), which is 13 kbits/s for speech or 12 kbits/s for data. One D channel (known as Dm) for slow-speed data, which is capable of 382 bits/s, is available. In its most familiar operation, the Dm channel is used for the short message, where it can be used in parallel with the speech channel.

Although called bearer services, it must be realized that the slow data rate of a maximum 9600 kbits limits the applications to relatively slow-speed data transfer. This rate is reduced significantly if the communications occurs in a noisy environment.

The services that can be supported are described below.

Interworking with the PSTN

- Data rates asynchronous
  300, 1200, 1200/75, 2400, 4800, and 9600 bits/s
- Data rates synchronous
  1200, 2400, 4800, and 9600 kbits/s

Interworking with the Packet Switched Public Data Network

- Data rates asynchronous
  300, 1200, 1200/75, 2400, 4800, and 9600 bits/s
- Data rates synchronous
  2400, 4800, and 9600 bits/s
- A 12-kbits/s unrestricted digital stream that can be used alternately with speech.

These rates assume that transparent transmission mode is used. Where non-transparent communications is required, the maximum rate of transmission is limited to 4800 bits/s.

Using these standards it is possible to support alphanumeric message services, group 3 fax (with a long-term possibility to support group 4), and electronic mail.

TELESERVICES

Teleservices include voice, short messages, cell broadcasts, and facsimile.

Short Messages

Short messages can be mobile originated or mobile terminated and have a maximum length of 160 characters. This service can be viewed as an advanced paging service. In the event that the mobile is switch off, the message will be stored and forwarded to the subscriber on switch on.

It is envisaged that this service will have an associated paging bureau, which can dispatch messages in the same manner as a conventional paging service. This service is proving popular in the UK but is little-used elsewhere.

Short Message Cell Broadcast

A message can be sent to all mobiles operating with a certain cell group. This can be traffic information, cellular service updates, or, more worrisome, commercial information. There is no mobile acknowledgment and the maximum message length is 93 characters. As a general broadcast service, this does not have any special subscription.

SHORT-MESSAGE FEATURE

GSM incorporates the ability to send short messages of up to 140 alphanumeric characters from the PSTN. This service uses a store-and-forward capability in the switch, which automatically cancels messages not accessed within 24 hours.

GSM ROAMING

Roaming in GSM networks is both its strength and weakness. A significant part of GSM fraud is enabled mainly by the lack of direct routes between the electronic data exchange (EDX) and the networks around the work. Those that are not directly linked exchange data tapes or discs, and so introduce a delay into the verification process for roamers.

GSM roaming, even when it is automated is a trombone-trunked arrangement that can tie up two voice links from the home HLR to the roamer network as seen in Figure 37.17. Additionally, should the roamer activate conditional call forwarding, the same kind of configuration will arise as the home system routes the call to the foreign network, and finding the mobile not available, the foreign network then routes the call back to the voice message system (VMS). Again two international trunks are used in the call. A call in a roaming area from a roamer mobile to another in the local area causes a link to be established back to the home HLR. The further the roamer roams, the more that link will cost. The high price of GSM roaming is because of this configuration.

There is, of course, no inherent reason that the roaming needs to be so routed. It should be possible for the roamer to make a call, and for the foreign network to then confirm the validity over a data link. Having done this, if the call is to a local subscriber, then the call could be connected locally without the international links. This is what, in fact, is being proposed for future GSM roaming.

Charging for the existing routing is simple. The calling party pays for the call from the originating network to the home PSTN of the called mobile and the roamer (the called party) pays for the leg back to the roamed network. In fact, the difficulty in the charging mechanism for optimal routing (one that avoids tromboning) is comparable to the technical challenges that it poses.

GSM MOBILE UNITS

Mobiles are classified according to their power levels, as shown in Table 37.2. Although these powers seem high at first, especially when compared to the analog mobiles, it has to be remembered that the TDMA structure is such that the *average* power is one-eighth of these figures. Even so, few manufacturers market class 1 mobiles. The mobile station (MS), under the control of the radio

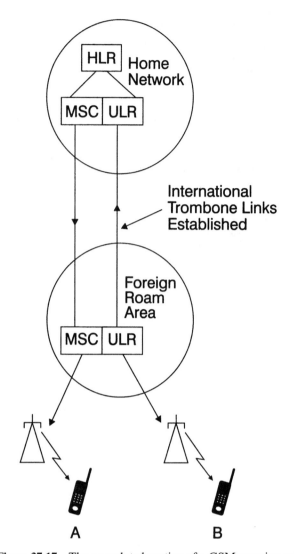

Figure 37.17   The convoluted routing of a GSM roaming call.

TABLE 37.2   Mobile Classifications

Mobile		Power Level (W)
Class 1	Vehicle or portable	20
Class 2	Vehicle or portable	8
Class 3	Handheld	5
Class 4	Handheld	2
Class 5	Handheld	0.8

base station (RBS), can have its power output controlled in nominally 2-dB steps down to 20 mW.

Mobiles are designed to have internal phase and frequency synchronization that will allow for Doppler shifting at up to 250 km/hr and propagation delays that will allow for the maximum cell radius of 35 km. This limit is not expected to be a limitation in itself, as the sensitivity of the mobiles makes the prospect of transmissions beyond 35 km very slight except for fixed units with high-gain directional antennas.

There are mandatory features for mobiles, which include the use of a SIM card and a means to confirm the authority to use the SIM by means of a PIN number.

Upon switch-on, the mobile attempts to access its HPLMN. The mobile constantly identifies which network it is locked onto and which other PLMNs (and their countries of origin) are available. The user is able to select the network provider from the mobile station.

The called number is displayed and a minimum key set comprising 0–9, *, #, and +. The "+" is used to signify access to the international network. This overcomes the need for the customer to keep track of the various access codes used in different countries.

Where automatic redial is included, the number of retries is limited.

Optional features are up to the manufacturer. The following is a list of some of them.

- Data interface
- Abbreviated dialing
- Barring of outgoing calls based on class (e.g., all calls, all long-distance calls, or all international calls)
- Handsfree operation
- Signal level and bit error rate indicator
- Short message display screen

Figures 37.18(*a*) and 37.18(*b*) show Motorola's first GSM handheld and "transportable", circa 1992. In 1999 Motorola had handhelds well under 100cc.

The DTX transmit ensures that the handportable only transmits when speech or data are to be sent. The discontinuous receive (DRX) utilizes a group paging feature that allows the receiver to be in a low drain cycle standby mode for up to 98 percent of the time. An active power control feature, which controls the output power over 30 dB in 2 dB steps, not only saves power, but minimizes interference.

ACCESS CONTROL

Like the analog counterparts, GSM has a number of subscriber classes (see Table 37.3), which are not meant to be used in normal service, but are available to allow the network operator to exercise control over access in special cases. The access classes can be activated on a cell-by-cell basis. This allows dynamic reconfiguration of congested cells as the occasion demands.

Figure 37.18 (*a*) Motorola's first GSM handheld (1992). (Photo courtesy of Motorola.)

SMART CARDS

A novel feature of GSM mobiles is the smart card or SIM, which enables the user to plug into any GSM mobile and to use it as a personal phone. Physically the SIM card comes in two forms. One card is the size of a standard credit card and meant to be used in much the same way (this size is rare today); the other is about one-quarter that size (25 mm × 15 mm) and is meant to reside, semi-permanently in a mobile. The smaller card would be likely to get easily misplaced if used in the same way as a credit card.

The advantage of the smart card is that the user buys a service rather than a phone. Any smart card can be used on any phone, so the user has unrestricted portability. The information contained on the card is sufficient to identify the user in any GSM or PLMN.

Figure 37.18   (*b*) Motorola's first GSM transportable (1992). Note the slot at the front for the smart card. Photo courtesy of Motorola.

TABLE 37.3   GSM Subscriber Classes

Class	Allocation
0–9	These classes are allocated randomly to the normal subscribers based on the IMSI assigned.
11–15	These classes are allocated to high-priority users such as emergency services, security, and PLMN staff.
Emergency calls	Allows/disallows all users, even those without a valid IMSI, to access a restricted range of emergency numbers.

A PIN number gives added security, and the network security algorithm, which is interactive between the card and the network, minimizes the extent of fraudulent use. The PIN number is requested by the ME upon insertion of the SIM into the ME or at each switch-on of the mobile.

The PIN number is a 4- to 8-digit number. This number can easily be changed by the user at any time. When requested, the user has a total of three attempts to give the correct PIN. After three tries the card is "blocked" to prevent future use. Unblocking can be done by the legitimate user, who inserts the 8-digit PUK. The PUK may be given to the user with the SIM card or it may be required that the SIM be represented to be unlocked, dependent on operator preference. Only 10 incorrect attempts to input the PUK will be permitted before the SIM is permanently blocked. For those who desire it and are prepared to accept the risk, the PIN can be disabled during the SIM personalization.

The SIM card contains a single integrated circuit (IC) that supports not only all the functions of the GSM and DCS1800 (a PCN version of GSM) networks, but has a considerable amount of reserve capability. This will allow future use of the smart card as a credit card, phone card, and identification (ID). The fact that the data on the card can be securely modified on-line means that updates of the features, cancellation of validity, and addition of new capabilities can be done easily and cheaply.

Because the SIM is interactive, it can store some of the features of the GSM network, as listed below.

- Short messages can be stored in the card.
- Abbreviated dialing (with alphanumeric capability).
- A list of up to eight preferred PLMNS.

The exact number of short messages and abbreviated dialing numbers will depend on the actual SIM card and is not specified.

The SIM card contains two encryption algorithms that are operator defined rather than being European Telecommunications Standards Institute (ETSI) standards. They are the A3 authentication algorithm and the A8 cipher key. It also contains the authentication key "Ki."

Other information stored includes:

- BCCH information
- Access control and the one of 15 access classes assigned to the SIM
- Forbidden PLMNS (up to four)
- IMSI and TMSI
- Approved services

The Cards

There are two kinds of smart cards. A simple one, which basically only holds monetary units, and a microprocessor-based smart card, which is designed for GSM (where it is known as a SIM card) and other applications. Other digital cellular systems like code division multiple access (CDMA) and TDMA have not used smart cards (although in principle they could).

The GSM smart card is available in full sized (about the size of a credit card) and as a plug-in version, which is about the size of a (small) postage stamp.

It is possible to update a smart card over the air, using GSM. However, to do this requires significant extra add-on infrastructure.

A controversial possible use of smart cards is to utilize their ever-increasing surplus capacity to store medical and other personal information.

SECURITY

GSM offers a high level of subscriber security, which ensures subscriber authenticity and speech/data path confidentiality Using a "Challenge Response Scheme," data are passed over the air path, which confirms the identity, even though the identity itself is never sent to air.

When a call is first being set up the mobile needs to identify itself by means of its TMSI or IMSI.

The TMSI is assigned by the NSS and written into the SIM card. It is periodically updated by the VLR. Consisting of up to 15 digits, the IMSI is stored in the SIM card and is structured as:

MCC: The country code
MNC: HPLMN code
MSIN: Identification of the subscriber in the HPLMN

The IMSI is sent unencrypted and is sent only once. The network will preferentially use the TMSI, as this helps protect the identity of the user. However, in the case when the mobile is first turned on, or when it roams to a foreign network, it will need to use the IMSI on first contact. It will immediately be allocated a TMSI, which will be used for all future transactions. The assigned TMSI is valid only in the VLR in which it was assigned and may be changed by that VLR from time to time.

Additionally, a local mobile subscriber identity (LMSI) is issued by the VLR. This is mainly used locally to speed up access to subscriber data.

The IMSI/TMSI identity does not carry any level of proof and will later be challenged. The network will respond with a randomly generated number (RAND). The mobile now takes this random number and processes it through the A3 algorithm to produce a response called Signed Response, as seen in Figure 37.19. In the meantime the network does the same calculation and then compares the expected response to the actual. The comparison takes place in the VLR in the PLMN being accessed.

It can be seen that the operator must be keenly aware of the need for the A3 algorithm to be one that is difficult to decipher and to keep it secure, as the whole network security is based on it. The A3 has some ETSI restrictions, which includes maximum calculation time and data stream length (RAND is 128 bits and SRES is 32 bits).

Transactions between the mobile and the network do not use the IMSI but rather a TMSI. The mobile can be allocated a TMSI after the initial authentication, and this TMSI can be changed from time to time at the discretion of the operator. When changed, it will be done as shown in Figure 37.20.

The first stage of the voice encryption process is to generate the cypher key, known as "Kc." This is done using the A8 algorithm, which is also operator defined. Often the A3 and A8 algorithms are the same. The mobile identifies itself with its TMSI and as a response is sent a random number, as shown in Figure 37.21.

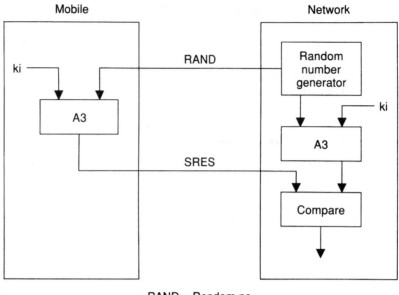

RAND = Random no
SRES = Signed response

Figure 37.19 The A3 algorithm generates the response to the initial challenge for the mobile to authenticate its ID.

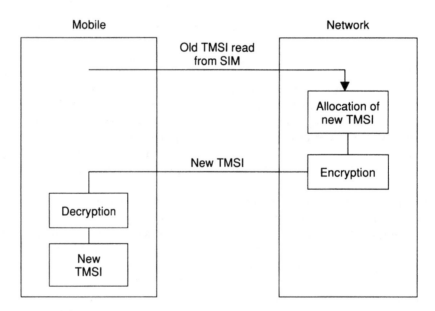

TMSI = Temporary Subscribers Numbers

Figure 37.20 The subscriber's temporary number can be changed from time to time.

Finally, this cipher key, "Kc," is used together with the TDMA frame number (which has values from 0 to 2715647, and repeats approximately every 209 minutes) to ensure that the encryption code does not frequently repeat. Figure 37.22 shows the implementation of the A5.

INCOMING PSTN CALLS

Each mobile will have a telephone number that corresponds to the ISDN number, known as a Mobile Subscriber ISDN number (MS-ISDN). This number comprises a country code, network code, and subscriber's

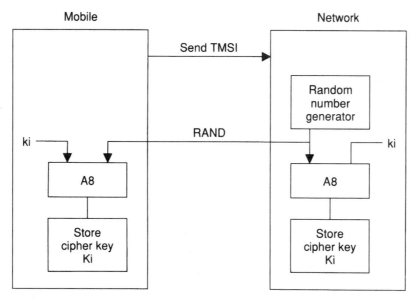

Figure 37.21 The voice encryption begins with the generation of the cipher key.

number. The VLR will allocate the mobile a MSRN, which is similar to the MS-ISDN.

A call from the PSTN will be routed to the nearest GSM gateway switching center (GMSC). The GMSC then interrogates the associated HLR for the subscriber and determines its current MSC. The information on the subscriber's location is contained in the location area identification (LAI), which has the following structure:

MCC: Country Code
MNC: PLMN code
LAC: Location area code, a 2-byte code

A more precise location is contained in the cell global identification (CGI), which consists of:

The LAI

The CI (cell number in the LAI (2 bytes))

The call is then forwarded to the current MSC for termination.

COMPLEXITY

GSM mobiles are far more complex than their analog counterparts. A crude measure of complexity is the 500,000 bytes of software code required compared to the 60,000 bytes for analog radio. In hardware the story is much the same with a total of around 300,000 gates re-

quired for the GSM very large scale integration (VLSI) compared to 15,000 for an analog terminal.

PCS

For practical purposes, PCS first became a factor in the US market in 1996. By the end of 1998, new PCS connections had equaled new cellular ones. With a lack of roaming and limited coverage, but massive capacity (in order to have reasonable system a large number of sites are needed, so the minimum-build capacity is very high), the way ahead for the PCS market was aggressive pricing.

Typical of the deals available is Sprint's 1999 plan, offering free long distance (within the United States) and a monthly plan starting at $29.99 including 120 minutes (with additional minutes at 35 cents). That amounts to 25 cents per minute for the included airtime. For heavy users at the top of the plan, 1500 minutes are available at 10 cents per minute. At these prices the limited mobility of the PCS ceases to become a serious issue, and the service challenges the fixed wireline service on both price and convenience.

AT&T, however, set the pace with its "Digital One Rate Plan," which offered nationwide calling in three pricing options; 600 minutes for $90, 1000 minutes for $120, and 1400 minutes for $150. These rates saw the PCS operators pick up many new customers, while creating churn from the analog services.

Most other carriers were unable to match these prices (particularly with the elimination of roaming fees), but nevertheless responded with aggressive pricing.

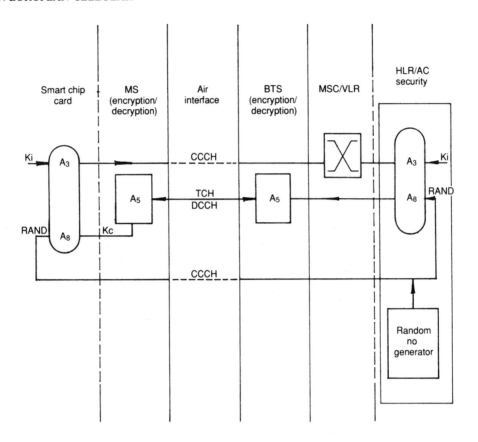

A₃, A₅, A₈—GSM Algorithms
CCCH—Common Control Channel
DCCH—Dedicated Control Channel
Kc—Key Cipher
Ki—Individual Subscriber Key
SRES—Signed Response
RAND—Random Number
TCH—Traffic Control Channel

Figure 37.22   The smart card concept for GSM.

With aggressive rollout plans the PCS carriers have in many cases matched the coverage of the analog service providers, but they continue to focus on being the "cheaper alternative," rather than on building a market identity separate from cellular.

IN-BUILDING COVERAGE

While many large public buildings present a challenge for 800/900-MHz mobile coverage, it is much harder for PCS. The exception is that large unglazed windows can actually present little attenuation to PCS frequencies, as the apertures are large compared to the wavelength.

However, internal walls will soon attenuate any incoming RF to unusable levels.

The solutions to this problem are threefold. For low to moderate traffic areas there is a fiber-optic distribution system that is a kind of cell extender. For higher traffic areas, it will be necessary to place an array of microcells in the building (which is simply an extension of the outdoors network). Finally, for some limited circumstances involving short runs, leaky cables may prove effective.

A fiber-optic distribution system will consist of a base site (or repeater) where the RF is directly modulated into an optic fibre for transmission to a remote site, where it is retransmitted as RF from a local antenna, which may

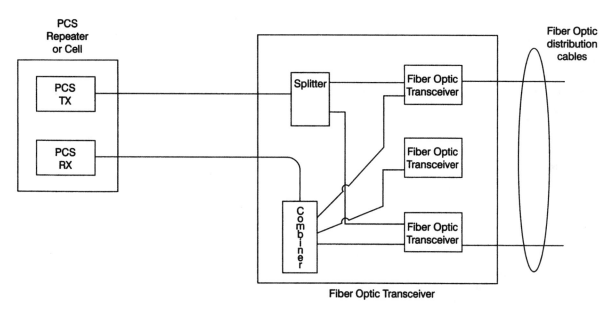

Figure 37.23   A PCS cell can be a donor to multiple repeaters.

be only about 10–15 cm in diameter, such as the system depicted in Figure 37.23. Optical fiber is cheap, and has a small diameter, so it is usual that a pair of fibers will be used, one for transmit and the other for receive.

Since the fiber is effectively lossless (less than 1 dB per km) for the distances concerned, the cost of actually running the cable will be the only limitation on how far away the cell extenders can be placed.

Because the RF energy is directly modulated as an optical signal, the signal is an amplitude modulated baseband signal. A simpler (and cheaper) modulation technique is hard to imagine.

Because the fiber-optic transmission is insensitive to the actual RF format, a number of signals, even of different carrier types, can be sent along one cable.

The RF transmitter depicted in Figure 37.24 is simply a diode encoder/decoder for the fiber-optic signal and a local amplifier.

The limitation of the optic-fiber cell extender is that, while it gives wide-area coverage, it still loads the donor

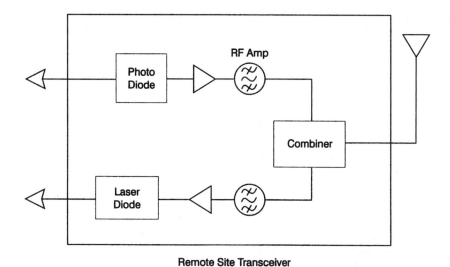

Figure 37.24   The PCS repeater is relatively simple as the RF signal is transmitted directly via a fibre optic cable, and is amplified without the need for processing.

cell with any traffic that it collects. When this limit becomes a problem, it will be necessary to add a new donor cell, or as is often the case, locate one at a new site to feed still more cell extenders.

Frequency reuse is rarely a problem, as the low powers permit reuse on alternate floors or after a few walls horizontally.

PCS Antennas

PCS antennas, which are small compared to regular 800/900-MHz antennas, can be incredibly small for special applications. For indoor applications like shopping malls and stadiums small antennas can be the size of a regular power outlet plate (in fact some are mounted behind regular switch plates.)

One omni-directional antenna with a 3-dBi gain (about 1 dBd), from Xertex, is $120 \times 67 \times 12$ mm and covers a bandwidth from 1930 MHz to 1990 MHz.

Site Sharing

Increasingly, sites are becoming harder to get, particularly as the number of operators increase, and visual pollution is becoming a social issue. In many countries site sharing is mandated, but this is rarely a problem because it is often convenient to share. Spectra Share has an innovative combiner, which allows up to six PCS systems to share a common antenna (see Figure 37.25).

The specifications for the combiner are:

Isolation TX-TX >40 dB

Isolation TX-RX >75 dB

Isolation RX-RX >20 dB

TX insertion loss <1 dB (excluding lightning protection feeder and jumper cables)

Total power handling 200 watts/BTS.

Roaming in the United States

While roaming from PCS into other networks is a worldwide problem, it is particularly so in the United States, where there are islands of PCS coverage within an extensive AMPS network. Dual-mode mobiles are a part of the answer, but unless something is done at the network level, the user is left to make the arrangements and effectively have two services. This presents problems with incoming calls and messages, which may be sent to either network. For ANSI-41-based systems (CDMA, TDMA, AMPS, NAMPS), the solution is simply to interconnect the networks. For GSM 1900 the problem is more complex. Although the principles of ANSI-41 and

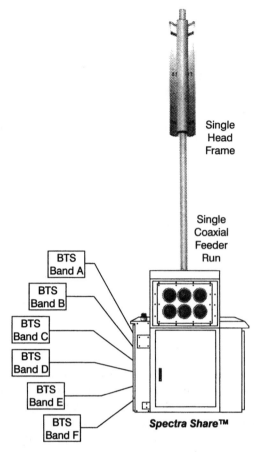

Figure 37.25 A Spectra Share 6 BTS combiner.

the GSM MAP protocols are similar, they are totally incompatible and direct interconnection is not possible.

A company called Synacom has an answer, which is called RoamFree. This is effectively a gateway between the ANSI-41 network and the GSM MAP, as seen in Figure 37.26. The GSM 1900 roamer still needs a second mobile (or a dual-mode one), but as a roamer the following enhancements are available;

- Registration of the GSM subscriber on the ANSI-41 network
- Registration of an GSM subscriber arriving onto a GSM network from a ANSI-41 network
- GSM call delivery to the ANSI-41 network
- Call forwarding from the GSM network to ANSI-41
- Call waiting
- Voice messaging

With this offering, the ANSI-41 network appears to the roaming GSM subscriber to be the same as the GSM HLR.

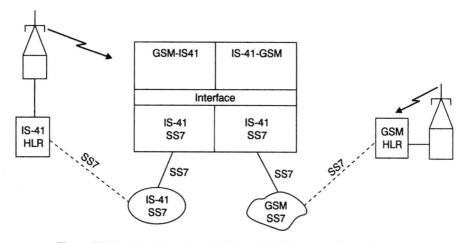

Figure 37.26   The intersystem GSM to ANSI-41 system from Synacom.

U.S. PCS: The Bidding Rules

PCS licenses were auctioned in 1996 and 1997 under the following rules. The Federal Communications Commission (FCC) allocated a total of 160 MHz at 1850–1970, 2130–2150, and 2180–2200 MHz for PCS services. This is four times the spectrum originally allocated for the cellular telephone service. The major elements of the Commission's decision are:

- 120 MHz was allocated for licensed PCS services (1850–1890/1930–1970 MHz and 2130–2150/2180–2200 MHz)
- 40 MHz was allocated for unlicensed PCS devices (1890–1930 MHz)
- The licensed allocation was channeled into two 30-MHz channel blocks, one 20-MHz channel block and four 10-MHz channel blocks, as follows:

Channel Block	Frequency (MHz)	Service Area
(30 MHz)	1850–1865/ 1930–1945	Major Trading Area A
(30 MHz)	1865–1880/ 1945–1960	Major Trading Area B
(20 MHz)	1880–1890/ 1960–1970	Basic Trading Area C
(10 MHz)	2130–2135/ 2180–2185	Basic Trading Area D
(10 MHz)	2135–2140/ 2185–2190	Basic Trading Area E
(10 MHz)	2140–2145/ 2190–2195	Basic Trading Area F
(10 MHz)	2145–2150/ 2195–2200	Basic Trading Area G

The unlicensed allocation is in two 20-MHz blocks, one for devices that will provide voicelike services and one for devices that will provide datalike services, as follows: voice (synchronous) 1890–1900 and 1920–1930 MHz, data (asynchronous) 1900–1920 MHz.

The PCS service areas adopted were major trading areas (MTAs) and basic trading areas (BTAs), generally as defined by the Rand McNally Atlas. There are 51 MTA- and 492 BTA-based service areas.

The licensing term is set at 10 years, with provisions for renewal similar to those that currently apply to the cellular service. Cellular licensees were permitted to participate in PCS outside of their existing service areas or in any area where the cellular licensee serves less than 10 percent of the population of the PCS service area.

Cellular licensees were defined as entities that had an ownership interest of 20 percent or more in a cellular system and were permitted to compete for one of the 10-MHz PCS channels in their existing service area. Local exchange carriers were permitted to apply for PCS licenses on the same basis as other applicants, except insofar as they held interests in cellular operations.

Eligibility for channel blocks C and D was addressed in the companion "Notice of Proposed Rule Making" award of PCS licenses through competitive bidding processes. That notice proposed licensing preferences for small businesses, rural telephone companies, and businesses owned by minorities and women.

Licensees are authorized to aggregate up to 40 MHz in any one service area and can aggregate markets (service areas). Cellular licensees are restricted to only one 10-MHz channel block in their cellular service area.

Under these rules the three largest bidders GWI, Nextwave, and Pocket all filed for bankruptcy soon after the auction closed.

In 1999, 356 of those licences were re-auctioned and yielded prices from a few cents to a few dollars per pop, compared to the average $40.00 per pop bids in the original auction.

Protocols

Since the FCC put no restrictions on the system to be used, all the major systems have appeared in the United States and Canada. GSM in 1999 had 3.6 million users in North America and was the system of choice of 17 PCS carriers. CDMA and TDMA are also used.

Existing Microwave

At least 5000 in band microwave paths were in existence at the time of the award of the PCS licenses. The FCC required a "coexistence" that would see the incumbent operators continue to use their systems until they could economically be phased out. The time-lines and terms were to be determined on a case-by-case basis. Although most PCS operators want to clear the spectrum quickly, others took a more conservative approach that involved medium-term coexistence.

In the United States it is largely the utility companies like gas and water, and railroads and local government that are the existing occupiers of these bands.

In many cases the problem was compounded by the fact that not only in-band link allocations might clash, but it was found that adjacent channel assignments up to 10 MHz away could cause problems.

PCN

PCN is much the same as PCS, but the frequencies are at 1800 MHz (GSM 1800), and these systems are mostly located outside North America. Generally Europe has extensive, good-quality PCN coverage.

Orange in the UK was the first to cover a significant area of the country and take on a large customer base of 100,000 virtually overnight. Orange has gone on to expand its operations throughout the UK, to become a major competitor to the GSM 900 networks, and, at the same time, to earn the respect of the cellular community for quality service.

The initial studies for a nationwide GSM 1800 system for Germany revealed that 14,000 base sites would be needed to cover the country completely. At approximately $60,000 per PCN cell, the cost of this network will be around $840 million, not a large sum for a nationwide telecommunications system, but a large sum for a startup minimal installation.

Assuming each base station had three sectors, each with one carrier (the minimum sectored installation),

that would amount to 18 × 14,000 or 252,000 voice channels, for a net capacity of around 5,000,000 users.

It is thus obvious that nationwide PCS carriers are faced with two immediate problems. First, the massive infrastructure must be financed, and second, once the system is built, the huge inherent capacity means that marketing must be geared up to match the challenge.

GPRS

General packet radio service (GPRS), while offering more bandwidth, compromises the performance of the GSM link. For a start, the regular GSM configuration lets the mobile use 6 of the adjacent time slots to measure the field strength of adjacent base stations to permit mobile assisted handoff (MAHO). If these time slots are used for traffic, this function is not available. Also, GPRS gets 14.4 kbps from a basic 9.6 kbps channel by sacrificing some of the error-correction code, so the link will be less robust.

In the regular single channel mode, a mobile has a 1/8 duty cycle which is significant for TX/RX isolation, TX power considerations, and battery consumption. Once a full 8-channel GPRS link has been established, the mobile effectively has to perform the functions of a base station channel. The power consumption therefore can be as high 2 watts continuous for GSM 900 and 1 watt for GSM 1800/1900. Given a PA efficiency of not more than 60%, this can be a real challenge for the handset designer.

In a single channel operation, the GSM channels are time duplex (by three time slots) as well as frequency duplex. This significantly simplifies the front-end design of a mobile and reduces losses and power, when the RF duplexer is eliminated. When multiple channels are used, time duplexing is no longer available and an RF duplexer (with its associated losses) will be required. Dual band phones with duplexers require three duplex filters so the problem in this case is more accute. The synthesizer settling time in single channel implementation is assisted by having the time between channel slots for settling. When that time is no longer available, the demands on the synthesizer are more exacting.

TWO-SITE FREQUENCY REUSE

Because GSM can operate with base-station C/I of around 9 to 10 dB, an $n = 3$ cell site configuration can be achieved. Motorola has gone further than this and introduced a (patented) plan for $n = 2$. Figure 37.27 shows the proposed plan, which at first sight appears to be a four-cell structure (there are a, b, c, and d cells). On closer examination, it will be seen that the spectrum is in

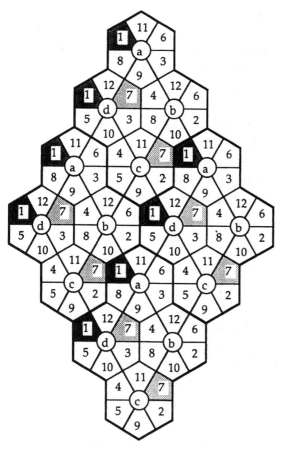

Figure 37.27 Two-site frequency reuse. (Courtesy of Motorola.)

TABLE 37.4   Basic Services Codes

All Services Including Bearer Services	Number Code Required
Teleservices	
All teleservices	14
Telephony (speech)	11
All data services	12
Fax	13
Telex	15
Short message service	16
Message handling service	17
All data services except SMS	19
Bearer services	
All bearer services	20
All async services	21
All sync services	22
All data circuit async	25
All data packet sync	26
All packet assembly disassembly (PAD) access	27
12 kbits/s unrestricted	29
Supplementary services	
Unconditional call forwarding	21
Call forward on busy	67
Call forward on no reply	61
Call forward on no answer	62
Block all outgoing calls	33
Block all international calls	331
Block all international calls except home country	332
Block all incoming calls	35
Block all incoming calls when roaming	351

fact divided into 12 groups and that each site uses six groups, hence $n = 2$. Thus the four cells share the same group of 12 channels, but there are four configurations of those channels.

INTELLECTUAL PROPERTY RIGHTS

The intellectual property rights (IPR) issue has become a major stumbling block that still is fraught with difficulty. The GSM specification calls for the rights to intellectual property (essentially, but not limited to patents) to be readily available on a reasonable and non-discriminatory basis. Although the intent of this specification is clear, the path to implementation is not. The bulk of the property rights are held by Motorola and Philips.

SUPPLEMENTARY SERVICES

Although IP rights have been successfully negotiated in "bulk" for Europe, in other countries they are still handled in an ad-hoc case-by-case basis. The standard-

ization of supplementary services is not mandatory, but if the full potential of the SIM card for international roaming is to be made usable, it is highly desirable that it is. This allows a user from one network to activate a service using the same procedure as would be used in the HPLMN.

One the strengths of GSM is the large number of supplementary services that can be offered. The complexity that accompanies the choices, however, is one of the weaknesses. Without standardization across networks, access to these services by the average subscriber will be minimal. The services offered and their respective number codes are listed in Table 37.4. Calling procedures are shown in Table 37.5, using the number codes from Table 37.4.

The complexity of the list shown in Table 37.5 is such that there is clearly room for the development of an alphanumeric scrolling, either of the commands or the complete function, as an optional feature of the mobile phone. There should be little doubt, however, that standarization is essential.

TABLE 37.5 Calling Procedures for Emergency Calls and Supplementary Services

Type of Call	Calling Procedure
Emergency calls (may be done even without SIM card)	122SEND
Set up call	DN SEND
End call	END (or on hook)
Call forwarding	
All calls forwarded	**NN*DN#SEND
Forward on basic service	**NN*DN*BS#SEND
Cancel "all calls forward"	##NN#SEND
Cancel one service	##NN*BS#SEND
Cancel all forwarding	#002#SEND
Interrogate call forwards	*#NN#SEND
Interrogate one service	*#NN*BS#SEND
Calling facilities with password control	
Register with password	**NN*PW#SEND
Register one service with password	**NN*PW#SEND
Activate only with password	*NN*PW#SEND
Activate with password on service	#NN*PW*BS#SEND
Deactivate only with password	#NN*PW#SEND
Deactivate only with password one service	#NN*PW*BS#SEND
Erase barring with password	##NN**PW#SEND
Erase with password one service	##NN*PW*BS#SEND
Deactive all bars	#330#SEND
Interrogate for call bars	*#NN*#SEND
Interrogate for one basic service	*#NN*BS#SEND
To change password	**03*NN*OldPW*NewPW *NewPW#SEND
Password Control	
Call barring	**03*330* Old PW*New PW*New PW*#SEND
All services	**03* Old PW*New PW*NewPW#SEND
PIN control	
Enter PIN	Pin No#
Change PIN	**04*Old PIN*New PIN*New PIN#
Unblock SIM	**05*PUK*New PIN*New PIN#

Note: DN = directory number; NN = supplementary service code; BS = basic service code; PW = password; PIN = PIN number.

GSM TERMS

A interface   Interface between MSC and BSS

A3   Authentication algorithm

A5   Stream cipher algorithm

A/D   Analog to Digital (converter)

Abis   Interface between BSC and BTS

ABR   Answer Bid Ratio

ABS   Administration and Billing System

AC   Authentication Center

ACCH   Associated Control CHannel

ADC   Administration Center

ADCCP   Advanced Data Communications Control Protocol

ADM   ADMinistration processor

ADMIN   ADMINistration

AFC   Automatic Frequency Control

AFN   Absolute Frame Number

AGC   Automatic Gain Control

AGCH   Access Grant CHannel

AI   Artificial Intelligence

AM   Amplitude Modulation

AM/MP   Cell broadcast Mobile terminated messages

AMA   Automatic Message Accounting (processor)

AOC   Automatic Output Control

ARQ   Automatic ReQuest for retransmission

ARP   Address Resolution Protocol

ASE   Application-Specific Entry (TCAP)

ASE   Application Service Element

ASIC   Application-Specific Integrated Circuit

ASP   Alarm and Status Panel

ASR   Answer Seizure Ratio

ATB   All Trunks Busy

ATTS   Automatic Trunk Testing Subsystem

AuC   Authentication Center

AUTO   Automatic mode

BCCH   Broadcast Control CHannel

BCF   Base-station Control Function

BER   Bit Error Rate

BES   Business Exchange Services

BHCA   Busy Hour Call Attempt

BLLNG   BiLLiNG

Bm   Traffic channel (full rate)

bps   bits per second

BCU   Base-station Control Unit

BSC   Base-Station Controller

BSC   Base-Station Code

BSIC   Base-Station Identity Code

BSS   Base-Station System

BSSAP   BSS Application Part (DTAP and BSSMAP)

BSSMAP   Base-Station System Management Application Part

BSSOMAP   BSS Operation and Maintenance Application Part

BSU   Base-Station Unit

BST   Base-Station Transmitter

BT   Bus Terminator

BTC   Bus Terminator Card

BTS   Base Transceiver Station

C7   Signaling System No. 7 (CCITT#7)

C/I   Carrier to Interference Ratio

CAMP   Control, Administration, and Maintenance Position(s)

CBI   Control Bus Interface (MSC/LR)

cc   call control

CCCH   Common Control CHannel

CCD   Common Channel Distributor

CCH   Council for Communications Harmonization (referred to in GSM Recommendations)

CCITT   International Telegraph and Telephone Consultative Committee

CCLK   Common Channel Link (MSC/LR)

CCM   Common Channel Manager (MSC/LR)

CCP   Cell Coordination Processor

CCP   Connection Control Processor

CCS   Hundred call-seconds. The unit in which amounts of telephone traffic are measured. One call that lasts for one hundred seconds constitutes one CCS.

CDR   Call Detail Records

CEPT   Conference of European Postal and Telecommunications operators

CGM   Cell Group Manager

CKSN   Ciphering Key Sequence Number

CLR   CLeaR

CM   Connection Management

CMP   Central Maintenance Processor

CMR   Cellular Manual Revision

COM   COMmunications processor

COMM   COMMunications

CONF   CONFerence circuit

CONFIG   CONFIGuration Control Program

CPF   Call Processing Frame (MSC)

CPM   Call Processing Manager

CPTD   Call Processing Tone Detector

CPU   Central Processing Unit (processing complex)

CRC   Cyclic Redundancy Check

CRT   Cathode Ray Tube (video display terminal)

CSC   Cell Site Controller

CSPDN   Circuit Switched Public Data Network

CTD   Call processing Tone Detector

D/A   Digital to Analog (converter)

DAN   Digital ANnouncer (for recorded announcements on MSC)

DAS   Data-Acquisition System

DB   DataBase

DBMS   DataBase Management System

Db/No   Energy per bit/Noise floor

DBPROC   subscriber DataBase PROCessor

DC   Direct Current

DCC   Digital Channel Controller

DCF   Data Communication Function

DCCH   Dedicated Control CHannel

DCCH   Digital Control CHannel

DCN   Data Communication Network

DDS   Direct Digital Synthesis

DFE   Decision Feedback Equalizer

DIA   Drum Intercept Announcer

DIA   Disk Interface Adaptor (MSC/LR)

DLCI   Data Link Connection Identifier

Dm   Signaling channel

DMA   Deferred Maintenance Alarm; an alarm report level; also see PMA, MI; action is required but can be deferred until normal working hours

DMF   Digital Maintenance Frame (MSC)

DMX   Distributed Electronic Mobile Exchange (Motorola's networked EMX family)

DN   Directory Number

DPNSS   Digital Private Network Signaling System (BT Standard for PABX interface)

DRX   Discontinuous Reception

DS-2   Term for PCM Interface in Germany for Digital Span of 2.048 Mbit/s; alias Megastream in UK; alias T1 Span in the United States

DSC   Digital Switch Corporation

DSP   Digital Signal Processor

D-TACS   Extended TACS (analog cellular system, extended)

DTAP   Direct Transfer Application Part

DTF   Digital Trunk Frame

DTI   Digital Trunk Interface

DTMF   Dual Tone MultiFrequency (tone signaling type)

DTX   Discontinuous Transmission

E   Erlang

EC   Echo Canceler

EIR   Equipment Identity Register

EIRP   Equipment Identity Register Procedure

EIRP   Effective Isotropic Radiated Power

EMC   ElectroMagnetic Control

EMF   ElectroMotive Force

EMX   Electronic Mobile Exchange (Motorola's MSC family)

en bloc   Fr.- all at once (a CCITT#7 Digital Transmission scheme); en bloc sending means that digits are sent from one system to another en bloc (i.e., all the digits for a given call are sent at the same time as a group). En bloc sending is the opposite of overlap sending. A system using en block sending will wait until it has collected all of the digits for a given call before it attempts to send digits to the next system. All the digits can then be sent as a group.

EOT   End Of Tape

Erlang   A unit of telephone traffic that is numerically equal to percentage occupancy. It is obtained by multiplying the number of calls by the length of the average call in fractions of an hour. One Erlang is equal to 36 CCS. This unit was named for Agner K. Erlang of the Copenhagen Telephone Company. In the United States, this unit is also known as a "traffic unit" (TU).

EPROM   Erasable Programmable Read-Only Memory

ETSI   European Telecommunications Standards Institute

ETX   End of Transmission

FAC   Final Assembly Mode

FACCH   Fast Associated Control CHannel

FCCH   Frequency Correction CHannel

FFS   For Further Study

FIR   Finite Impulse Response (filter type)

FISO   Fault ISOlation subsystem

FM   Frequency Modulation

FN   Frame Number

FOA   First Office Application

FS or FFS   For Further Study (used throughout GSM documents)

FS   Frequency Synchronization

FTP   File Transfer Program

GDS   GSM DSP board (part of BSC)

GHz   GigaHertz (10^9 Hertz)

GMB   GSM Multiplexer Board (part of BSC)

GMSC   Gateway Mobile services Switching Center

GMSK   Gaussian Minimum Shift Keying

GND   GrouND

GPC   General Protocol Converter

GPRS   General Packet Radio Service

GSM   Groupe Special Mobile (former name; now Global System for Mobile Communications)

GWY   GateWaY (MSC/LR) interface to MTN

H-M   "Human-Machine" Terminals

HAD   HLR Authentication Distributor

HAP   HLR Authentication Processor

HLR   Home Location Register

HSN   Hopping Sequence Number

HSM   HLR Subscriber Management

HW   HardWare

IO   Input/Output

IA5   International Alphanumeric 5

IAM   Initial Address Message

IBT   Interrupt and Bus Terminator

IC   Integrated Circuit

ICMP   Internet Control Message Protocol

ICT   Interrupt Control Terminator (MSC/LR)

ID   IDentification

IEEE   Institute of Electrical and Electronics Engineers

IF   Intermediate Frequency

IMACS   Intelligent Monitor And Control System

IMEI   International Mobile station Equipment Identity

IMM   IMMediate assignment message

IMSI   International Mobile Subscriber Identity

IN   Intelligent Network

INS   IN Service

IP   Internet Protocol

IP   Intermodulation Products

IPR   Information Problem Report

ISC   International Switching Center

ISDN   Integrated Services Digital Network

ISUP   ISDN User Part

IWF   InterWorking Function

IWMSC   InterWorking MSC

K   Kilo (1000)

kb   kilobit

kbit/s   kilobits per second (×1000)

kbps   kilobits per second (×1000)

Kc   ciphering Key

kHz   kiloHertz

Ki   individual subscriber authentication Key

kW   kiloWatt

L2ML   Layer 2 Management Link

LAI   Location Area Identification (identity)

LAN   Local Area Network

LAPB   Link Access Procedure "B" (balanced) channel

LAPDm   Link Access Procedure "Dm" (mobile "D") channel

LC   Inductor Capacitor (type of filter)

LCM   Land Call Manager (MSC)

LCN   Local Communication Network

LCS   Land Call Sequencer (MSC)

LED   Light-Emitting Diode

LF   Line Feed

LF   Low Frequency

LIFO   Last In First Out

LLC   Logical Link Control

Lm   Traffic channel (half rate)

LMS   Least Mean Square

LPC   Linear Predictive Code

LPROC   Location control PROCessor

LR   Location Register

LTM   Line Trunk Manager

LTP   Line Trunk Processor

L2R   Layer 2 Relay function

M   Mega (1,000,000)

MA   Mobile Allocation

MAC   Medium Access Control

MAF   Mobile Application Function

MAIDT   Mean Accumulated Intrinsic Down Time

MAINT   MAINTenance

MAIO   Mobile Allocation Index Offset

MAP   Mobile Application Part

MAPP   Mobile Application Part Processor

MAT   MATrix (MSC)

MBM   Mobile Busy Manager

MCC   Mobile Country Code

MCS   Mobile Call Sequencer

MCSS   Mobile Call Sequencer Selector

MCM   Mobile Control Manager

ME   Maintenance Entity (GSM 12.00)

MEF   Maintenance Entity Function (GSM 12.00)

Megastream   UK Term for PCM Interface; alias T1 Span in USA; DS-2 in Germany; for Digital Span of 2.048 mbit/s

MF   Mediation Function block

MF   MultiFrequency (tone signaling type)

MFTX   MultiFrequency Transmitter

MHS   Mobile Handling Service

MGMT   ManaGeMenT

MGR   ManaGeR

MHz   MegaHertz (1,000,000 Hertz)

MI   Maintenance Information; an alarm report level; also see PMA, DMA; no immediate action required

MLP   Mobile Location Processor

MM   Mobility Management

MMD   Mobility Management Distributor

MMI   Man Machine Interface

MML   Man Machine Language

MMP   Matrix Maintenance Processor

MNC   Mobile Network Code

MNT   MaiNTenance

MO/PP   Mobile Originated Point-to-Point messages

MOMAP   MOtorola OMAP

MoU   Memorandum of Understanding

MPC   Matrix Port Controller

MPD   Mobile Page Distributor

MPI   Matrix Port Interface

MPT   Mobile Page distributor and Trunk idle list

MPX   MultiPleXed

MRN   Mobile Roaming Number

MS   Mobile Station

MSC   Mobile-services Switching Center

msec   millisecond (0.001 second)

MSF   Mass Storage Frame (MSC/LR)

MSIN   Mobile Station Identification Number

MSISDN   Mobile Station international ISDN number

MSRN   Mobile Station Roaming Number

MT   Mobile Termination

MTC   Magnetic Tape Controller (MSC/LR)

MTD   Matrix Timing Distribution

MT/PP   Mobile Terminated Point-to-Point messages

MTBF   Mean Time Between Failures

MTL   Mobile Trunk idle List processor

MTN   Message Transport Network

MTNC   Message Transport Network Controller

MTP   Message Transfer Part

MTTR   Mean Time To Repair

MUX   MUltipleXer

MV   MSC + VLR

MVHE   MSC + VLR + HLR + EIR

μS   microsecond ($\times 0.000001$ second)

NAP   Network Agent Processor (68010-based communications board)

NE   Network Elements

NEF   Network Element Function block

NLK   Network LinK processor(s)

NM   Network Management (Manager)

NMC   Network Management Center

NMT   Nordic Mobile Telephone

NSP   Network Service Provider

nW   nanoWatt (Watts $\times 10^{-9}$)

O&M   Operations and Maintenance

OACSU   Off Air Call SetUp

OFL   % OverFLow

OMAP   Operations and Maintenance Application Part (previously was OAMP)

OMC   Operations and Maintenance Center

OMCR   Operations and Maintenance Center—Radio part

OMCS   Operations and Maintenance Center—Switch part

OML   Operations and Maintenance Link

OMSS   Operations and Maintenance SubSystem

OSF   Operations Systems Function block

OOS   Out Of Service

OSS   Operator Services System

overlap   Overlap sending indicates that digits are sent from one system to another as soon as they are received by the sending system. A system using overlap sending will not wait until it has received all of the digits for a particular call before it starts to send the digits to the next system. This is the opposite of en bloc sending, in which all the digits for a given call are sent at one time.

PA   Power Amplifier

PABX   Private Automatic Branch eXchange

PAD   Packet Assembly Disassembly facility

PBX   Private Branch eXchange

PCH   Paging CHannel

PCM   Pulse Code Modulation; also see T1 span, DS-2, and megastream, which are the physical bearer of PCM.

PDN   Packet Data Network

PDF   Power Distribution Frame (MSC/LR)

PDU   Protected Data Unit

PEDC   Pan European Digital Cellular

PID   Process IDentifier

PIM   PCM Interface Module (MSC)

PIN   Personal Identification Number

PLMN   Public Land Mobile Network

PMA   Prompt Maintenance Alarm; an alarm report level; also see DMA, MI; immediate action is necessary

PMUX   PCM MUltipleXer

PN   Permanent Nucleus (of GSM)

POTS   Plain Old Telephone Service (basic telephone services)

pp   peak to peak

ppm   part per million ($\times 0.000001$)

PROM   Programmable Read-Only Memory

PSPDN   Public Switched Packet Data Network

PSTN   Public Switched Telephone Network (alias Telco)

PSW   Pure Sine Wave

PTO   Public Telecommunications Operator

PUP   Peripheral Utility Processor

PVP   Path Verification Processor (MSC)

PWR   PoWeR

Q-Adaptor   Used to connect MEs and SEs to TMN (GSM 12.00); also alias QAF

QAF   Q-Adaptor Function block

QOS   Quality Of Service

RACCH   Random Access Control CHannel

RACH   Random Access CHannel

RAM   Random-Access Memory

RAND   RANDom number

RAx   Rate Adaption

RBDS   Remote BSS Diagnostic Subsystem

RCVR   ReCeiVeR

RCU   Radio Control Unit

RELP   Regular Exited Linear Predictive

RELP-LTP   RELP-Long Term Prediction

RF   Radio Frequency

RLP   Radio Link Protocol

RMS   Root Mean Square

RMSU Remote Mobile Switching Unit

ROM Read Only Memory

RM4-16 Random Memory Card (MSC/LR) Provides 16-Mbyte additional memory to "MP"processors.

ROSE Remote Operations Service Element (a CCITT specification for O&M)

RR Radio Resource management

RSL Radio Signaling Link

RSS Radio SubSystem (replaced by BSS)

RSSI Received Signal Strength Indication

RU Rack Unit

Rx Receive(r)

RXLEV_D Received signal level downlink

RXLEV_U Received signal level uplink

RXQUAL_D Received signal quality downlink

RXQUAL_U Received signal quality uplink

SACCH Slow Associated Control CHannel

SAGE a brand of trunk test equipment

SAPI Service Access Point Indicator (identifier)

SAW Surface Acoustic Wave (filter type)

SC Service Center

SCCP Signaling Connection Control Part

SCEG Speed Coding Exports Group (of GSM)

SCF Service Circuit Frame

SCH Abis Signaling CHannel

SCH Synchronization CHannel

SCI Status Control Interface

SCIP Serial Communication Interface Processor

SCM Status Control Manager

SCP Service Control Point—an intelligent network entity

SDC Abis traffic channel

SDCCH Standalone Dedicated Control CHannel

SDR Special Drawing Rights—an international "basket" currency for billing

SE Support Entity (GSM 12.00)

SEF Support Entity Function (GSM 12.00)

SEL SELector

SFCC Serial Four-Channel Communications card (two types, -2 & -3)

SIM Subscriber Identity Module

SIM Subscriber Information Management

SLNC Serial Link

SMAE System Management Application Entity (CCITT Q795, ISO 9596)

SME Switch Matrix Extension processor

SMGR Statistical analysis cluster ManaGeR

SMM System Matrix Manager

SMP Switch Matrix Processor (MSC/LR)

SMS Short Message Service

SMT Switch Matrix Tester (MSC/LR)

SND SeND

SNDR SeNDeR

SNR Serial NumbeR

SP Special Product

SPROC Security related information PROCessor

SRD Service Request Distributor

SRES Signed RESponse

SS Supplementary Services support

SSP Service Switching Point—an intelligent network entity

SS7 CCITT Signaling System No. 7

SSS Switching SubSystem (MSC + LRs)

STAN STatistical ANalysis (processor)

STAT STATistics

STG System Timing Generator (MSC/LR)

STP Signaling Transfer Point

SVM SerVice Manager

SW SoftWare

SYS SYStem

SYSGEN SYStem GENeration

T1 span The basic 24-channel 1.544-Mb/s pulse-code modulation system used in the United States (e.g., one T1-span carries 24 conversations and/or data links); alias megastream in the UK

TA Terminal Adaptor

TA Timing Advance

TAC Type Approval Code

TACS Total Access Communications System

TAS Tone Announcement Sequencer (MSC); used to combine tones and voice announcements)

TBD To Be Determined

TCAP Transaction Capabilities Application Plan

TCH Traffic CHannel

TCP Transmission Control Protocol

TDC Time of Day Clock (MSC); alias TODC

TDM Time Division Multiplexed

TDMA Time Division Multiple Access

TE Terminal Equipment

TEMP TEMPorary

TEST TEST control processor

TGC Transport Group Controller

TIC   Transport Interchange Control (part of MTN in MSC)

TIN   Transport INterchange (part of MTN in MSC)

TIS   Transport Interchange Supervisor

TKP   TrunK Processor (MSC/LR)

TM   Traffic Manager

TMM   Traffic Metering and Measurement

TMN   Telecommunications Management Network

TMSI   Temporary Mobile Subscriber Identity

TNC   Transport Node Controller (MSC/LR); provides interface between cluster and MTN's GWY.

TODC   Time Of Day Clock (MSC); alias TDC or TOD

TODS   Transaction-Oriented Database management System (MSC/LR software)

traffic unit   This unit is equivalent to an Erlang

TRANS   TRANSlation processor, alias TRP

TRP   TRanslation Processor, alias TRANS

TRX   Transceivers

TSA   TimeSlot Acquisition

TSI   TimeSlot Interchange

TTL   Transistor Transistor Logic

TTY   TeleTYpe (refers to any terminal)

TU   Traffic Unit

TUP   Telephone Users Part

UBT   Universal Bus Terminator (MSC/LR)

UDP   User Datagram Protocol

UHF   UltraHigh Frequency

UI   Unnumbered Information frame

Um   air interface

VA   Viterbi Algorithm

VAD   Voice Activity Detection

VCO   Voltage Controller Oscillator

VDBM   VLR DataBase Manager

VLR   Visited Location Register

VLSI   Very Large Scale Integration (IC)

VOX   Voice-Operated transmission

VSM   VLR Subscriber Management processor

VSP   Vehicular Speaker Phone

WAP   Wireless Application Port

WS   Work Station

WSF   Work Station Function block

XC   Transcoder

XCB   Transceiver Control Board (part of transceiver)

38

DAMPS

Digital Advanced Mobile Phone Service (DAMPS), defined by the standard IS-135, is a derivative of Global System for Mobile Communications (GSM), with much of the complexity removed. DAMPS is sometimes known as time division multiple access (TDMA) in the United States. It was purpose designed to be a evolutionary path from analog AMPS to a more spectrally efficient digital system. DAMPS can be added to an AMPS network on a per-channel basis (with one analog channel being replaced by three TDMA).

Although DAMPS originally came into being to meet a demand in the United States, it is now widely deployed around the world. Dual-mode analog/TDMA phones are widely available. A version of the cellular digital packet data system is supported by TDMA.

DAMPS is a contender for the IMT 2000 specification and has demonstrated most of the capabilities required by that standard.

The initial installation enables three channels to be used in the same bandwidth as a single channel analog. Compression/decompression (CODECs) able to encode with a spectral efficiency of 3 bits/s/Hertz have been tried. Such encoders take intelligent advantage of the idle time and redundancy in normal speech, and are thus able to compress the spectrum needed to transmit voice intelligence.

The digital system uses the same bands as the AMPS network, with no additional spectrum allocated. In some areas, where only 666 channels of the expanded 832-channel band have been used, there is a relatively easy path to digital by first using the spare channels. Where there are no spare frequencies, the initial change to digital means that some frequencies will need to be recovered.

NETWORK STRUCTURE

The hub of the network is the mobile switching center (MSC). This needs to have SS7 signaling into the public switched telephone network (PSTN) in order to gain full advantage of the network capabilities. The MSC controls all interfacing with the PSTN and it controls the subscriber functions of registration, authentication, and records through the associated home location register (HLR). The visitor location register (VLR) is a fast-access register that contains details of recently registered subscribers. It is used for all subscribers and not—as the name might imply—for only visiting users.

The essential configuration of DAMPS is shown in Figure 38.1.

BASE-STATION CONTROLLER

The base-station controller (BSC) controls handoffs as well as the audio compression and decompression. It interfaces with the switch through an open interface known as the A interface.

BASE TRANSCEIVER STATION

The base transceiver station (BTS) generates the air interface to the mobile. In most instances it will be dual mode (AMPS plus DAMPS). It includes the transceivers and the multiplexer equipment for the digital link connecting the BTS to the BSC.

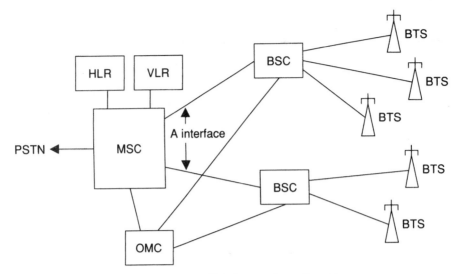

Figure 38.1   U.S. digital network configuration.

The Operations and Maintenance Centers

The operations and maintenance centers provide interactive access to the rest of the network, the collection of statistics and faults, as well as a remote network management center.

STRUCTURE

Figure 38.2 illustrates the concept of coexistence of digital and analog systems.

THE DAMPS TDMA FRAME STRUCTURE

The basis of DAMPS is to get three digital channels on the analog channel that is replaced. The mobile samples the audio and sends it out in a 20-millisecond timeslot with interleaving, as shown in Figure 38.3.

The frames are divided into six timeslots, each of which contains speech, synchronization, and error correction code, as shown in Figure 38.4. A full rate frame will contain one transmitter (TX), one receiver (RX), and one idle timeslot.

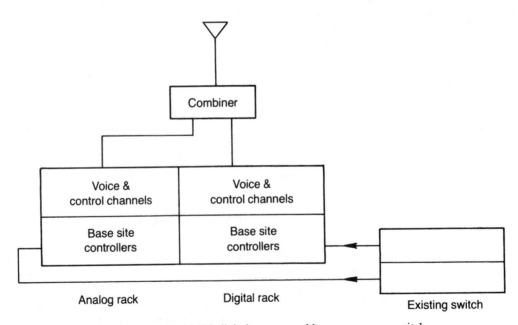

Figure 38.2   An AMPS digital system could use a common switch.

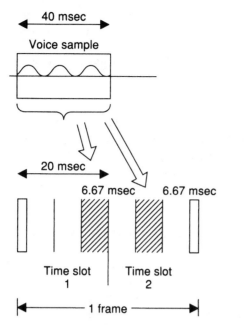

Figure 38.3 The analog signal is sampled and sent out in a compressed time frame over two interleaved timeslots.

The timeslots are structured according to their functions of forward, reverse and shortened burst. The signaling direction is defined with reference to the base station.

The forward traffic channel, depicted in Figure 38.5, shows the slot beginning with some synchronization bits.

The reverse channel differs from the forward channel in that it allows the mobile a guard and ramp time before data are transmitted, as seen in Figure 38.6.

The shortened burst structure (Figure 38.7) is used when communication initiates. The shortened burst is more robust in time synchronization. The link can stay in this mode until the base station has established the necessary time offset to be used.

The guard time is needed to prevent burst overlaps. The ramp time allows the power amp (PA) of the mobile to rise to full power gradually, and so reduce the potential harmonic distortion. This is needed only for the reverse channel because the mobile PA is turned on only as needed. The base PA is always switched on and thus does not need a ramp.

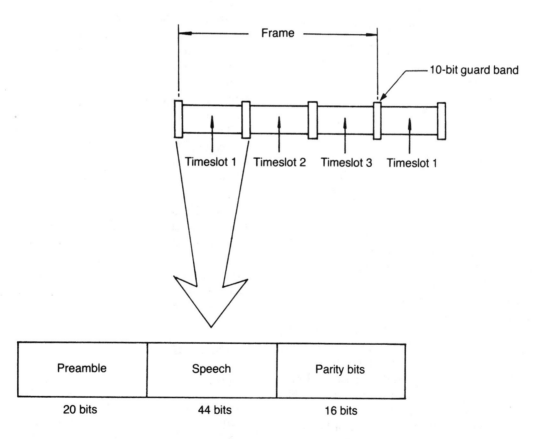

Figure 38.4 TDMA timeslot structure.

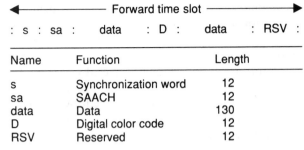

Figure 38.5 The forward channel slot structure.

Name	Function	Length
s	Synchronization word	12
sa	SAACH	12
data	Data	130
D	Digital color code	12
RSV	Reserved	12

◄──────── Reverse channel time slot ────────►

:G1:ra: data1 : s : data2 :sa:D: data2 :

Name	Function	Length
G1	Guard bits	6
ra	Ramp bits	6
data1	Data	12
data2	Data	122
sa	SACCH	28

Figure 38.6 The reverse channel time slot structure.

◄──────── Shortened burst slot ────────►

:g1:ra: s : D :V: s : D:W: s :D:X: s :D: Y : s : G2 :

Name	Function	Length
g1	Guard band	6
ra	Ramp	6
s	Synchronization word	28
V	OH	4
W	OOH	8
X	OOOH	12
Y	OOOOH	16
G2	Guard time	22

Figure 38.7 The short-burst frame structure.

A sychronization word is used to allow time alignment of the base and mobile. The slow associated control channel (SACCH) is used for short messaging. DAMPS uses a digital voice color code in much the same way that AMPS uses Supervisory Audio Tones (SAT) tones to identify foreign carriers.

Time synchronization is achieved by measuring the delay from the synchronization bits and then instructing the mobile to advance or retard its transmit rate. The default delay between the mobile and the base is 88 bits. This delay can be advanced or retarded by up to 30 bits in 1-bit increments. One bit is 20.55 microseconds long. This sets an upper limit on the usable range of DAMPS, which can be calculated as:

Total synchronization delay $= 30 \times 20.55$ microseconds

The guard time of 6 bits can be used additionally for long-range communications, so the total delay is 36 bits.

Total round path $= 300,000 \times 36 \times 20.55 \times 10^{-6}$ km

$$= 221.9 \text{ km}$$

Maximum path length is half of this, or 111 km. Care should be taken when repeaters are used to extend the range of a DAMPS system so that the total propagation distance is not exceeded (even though this would be a rare problem, given the DAMPS path budget).

VOICE-CHANNEL PROCESSING

The voice signal is first passed through an analog/digital (A/D) convertor to produce a 64-kbits/s analog signal. This signal is then passed through a vector sum excited linear prediction (VSELP) speech coder, which compresses the signal to 7950 bits/s. Error correction is then added, which brings the net bit rate up to 13 kbits/s. Generous though it may seem to add almost 100 percent redundancy in error correction, it seems that in real life trials it is the lack of robustness in the error correction, coupled with the low bit rate encoder (which means that it has little redundancy), that initially accounted for the poor sound quality of DAMPS in heavy multipath. In fact, in very low signal-strength areas with some significant multipath, it developed a low-frequency rumble, which increased both in frequency and volume as the signal degraded. This problem has since been fixed.

The error protected signal is later combined with the associated channel data and then sent for channel encoding at a net bit rate of 16.2 kbits/s, as can be seen in Figure 38.8. It is at this stage that the SACCH channel is added.

The channel audio signal is next sent to a storage register, which records (digitally) the whole sample. This sample is then held until its timeslot is active and the correct delay (which has been determined by path delay measurements) has been added. At this time the stored signal is pulsed out *at three times* the original data rate, as seen in Figure 38.9.

SIGNALING

Control Channel Signaling

Both the analog and digital parts of the base station use a common control channel with frequency shift keying (FSK) signaling, although the secondary control channels (channels 696 to 716 in the A band and 717 to 737 in the B band) have been reserved for digital-only sig-

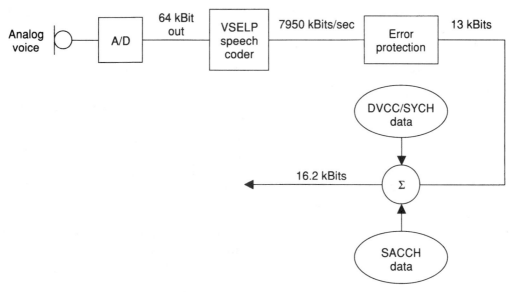

Figure 38.8 The audio level processing of a speech channel.

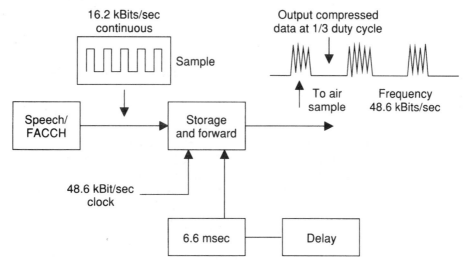

Figure 38.9 The store and forward procedure use in DAMPS results in a 3:1 time compression.

naling. The control channel signaling is very similar to AMPS, except that a dedicated bit indicates if dual mode is available and a new format is used for digital channel assignment.

Analog Voice-Channel Signaling

Voice-channel signaling is virtually the same as AMPS, except that an additional message to cause handoff to a digital channel has been added.

Digital Traffic Channel Signaling

The actual commands on the digital channels are very similar to AMPS, although they are often sent in an entirely different way. Dual Tone Multifrequency (DTMF) signaling, for example, which in AMPS is done in-band on the voice channel, is sent by a combination of in-band and out-of-band signaling. The in-band DTMF signaling that passes through the CODEC results in twist, which can render the tone unable to be decoded at the far end. Twist occurs when the relative levels of the two tones that make up a DTMF signal are amplified differentially and thus have incorrect relative levels. The SAT tone is replaced by a functionally equivalent voice digital color code (CDVCC). The CDVCC is looped back to the base station in the same way as the SAT tone in AMPS.

The SACCH Message

A short message is 132 bits long, and when sent over the SACCH channels it will have a maximum equivalent data rate of 300 bits/s, with an effective message-carrying rate (after redundancy coding) of 218 bits/s full-rate and 109 bits/s half-rate. The SACCH is sent so that each 12 bits of data are sent over two interleaved timeslots.

The FACCH Message

Longer messages can be sent by replacing the speech data with the fast associated control channel (FACCH) message. This mode gives an effective data rate of 2.4 kbits/s for full-rate channels and 1.2 kbits/s for half-rate.

Call Procedures for a Dual-Mode Mobile

Initialization   Like its analog counterpart, a mobile when first switched on will seek out the control channels, read the overhead messages, and then log on. A dual-mode mobile will first scan the control channels for digital capability. If this is not found, it will then scan the secondary control channels for digital capability. If this is still not found, it will revert to the primary control channels in the analog mode.

Idle   In the idle state both analog- and digital-mode mobiles monitor the system messages, awaiting a call or the need to log on should the mobile roam to another system.

System Access   System access is also done much the same way as AMPS, except that the electronic serial number (ESN) is not sent directly, but rather is encoded with secret shared data and sent encrypted.

Conversation   The conversation mode in the digital system is a little more complex than is the case for analog. There are three kinds of messages that can be expected. The short message can be sent on the SACCH simultaneously with a voice message. The voice channel itself can be carrying a voice message *or* a data (FACCH) message. The formats of these two messages are different. The mobile will first attempt to decode a voice

message. If it detects that the frame structure is not matched, it will then decode as a FACCH message.

Control Channel Capacity

Mixing digital and analog systems on the one control channel adds to the capacity demands on that channel. Given that most systems that will be adding digital will do so because of the need for additional capacity, it can reasonably be assumed that the control channel load is already high.

The control channel can be split into an access and paging channels to increase the overall capacity. (Note, however, that the second channel will reduce the number of voice channels available.) In an analog-only configuration, although it is possible to split the control channel, there is little to gain because the paging function does not use the reverse channel and the access channel makes relatively little use of the forward channel.

When separate paging and access are available in the dual-mode system, the digital functions will be assigned in reverse to the analog ones (that is, the analog paging channel will be the digital access channel and vice versa) so that the overall usage will be smoothed out.

Linking the Base Station to the Switch

For digital channels there are a number of options for linking the base-station channels to the switch. In the simplest implementation, the base-station channel can be multiplexed back to a 64-kbit/s stream and connected directly into the base-station multiplexer.

Additional efficiency can be obtained by transmitting the base-station channel information in a compressed form. For very small bases (with only a few channels of digital), it is effective to take the three-channel bit rate of 48.6 kbit/s and pass it through a rate adapter so that it occupies one 64-kbit/s channel, as seen in Figure 38.10. The channel de-multiplexer separates out the three channels and the channel coder, and de-interleaves the channels.

For larger bases it will be more efficient to take the voice channel at the 8 Mbits/s level via an 8 kbits/s to 64 kbits/s multiplexer (MUX). These MUX devices are available commercially from a number of suppliers. This information still needs to pass through a CODEC and finally an A/D convertor for switching, as seen in Figure 38.11.

THE RF ENVIRONMENT

Although the main vehicle for additional traffic density is the splitting of the channel bandwidth, another objec-

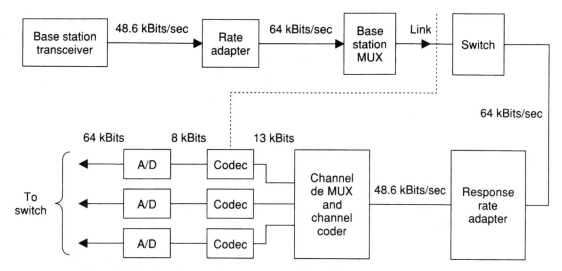

Figure 38.10   A scheme to increase transmission efficiency for small digital bases.

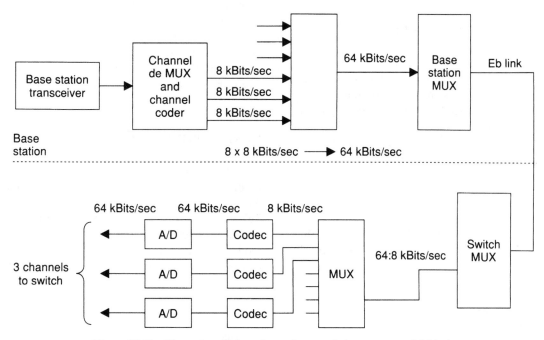

Figure 38.11   The most efficient channel transmission occurs at 8 kbits/s.

tive is to improve the C/I ratio (the ratio between the level of interference that can be tolerated to the desired signal level). If the cells can be made more interference tolerant, then they can be operated closer together; and more efficient frequency reuse follows.

It does appear that the target C/I level of 13 dB is not achievable (compared to 18 dB for AMPS). Any additional interference immunity will make the 4-cell pattern much more attractive. One consequence for operators

of analog cellular systems using the 7-cell pattern is that 4-cell base sites have more channels per site, and analog systems that have been shoehorned into a small site may not be able to fully exploit this advantage.

TDMA has been designed to be compatible with the existing 4- and 7-cell patterns. This is a vital point because the digital overlay is only viable if the digital channels can be added without adversely effecting the service area or interference immunity.

TABLE 38.1   Change of State

Symbol (degrees)	Bits	
+45	0	0
+135	0	1
−135	1	1
−45	1	0

MODULATION

The modulation technique used is $\pi/4$ differential quadrature phase-shift keying ($\pi/4$ DQPSK). The prefix $\pi/4$ refers to the fact that the changes of state of the phase in any one transition are limited to the four states, which are $\pm\pi/4$ and $\pm 3 \times \pi/4$ from the current state. The serial bit stream that is fed into the modulator is combined into pairs of bits that cause the change of state shown in Table 38.1. A state change is called a *symbol*.

RADIO FREQUENCY AMPLIFIER

The radio frequency (RF) amplifier is a linear amplifier, and as such will have an efficiency of about 30 percent. This is compensated for to some extent by the fact that the duty cycle is 1/3, and so the amplifier is mostly not in service. The transmitter is isolated from the receiver by a simple switch that allows either the RF amplifier or the receiver to access the antenna, but not both together.

THE RECEIVER

The receiver is similar to an AMPS receiver, except that instead of a frequency modulated (FM) detector it will have a $\pi/4$ DPQSK demodulator. One additional complication is the need to equalize the incoming signals for path delays. The delays could be caused by any or all of the following: path length, multipath, and doppler shift due to a moving vehicle. To accomplish delay equalization it will have an adaptive equalizer, which will use the synchronization word (which is of a known format) to determine the delays being experienced by the timeslot and to adjust the delay accordingly.

The receiver audio processing is virtually the reverse of the transmission, as can be seen in Figure 38.12.

AUTHENTICATION

As has been widely recognized, fraudulent use of mobile access has been the greatest problem with analog cellular. The DAMPS digital mobile has an authentication scheme that scrambles the electronic serial number (ESN) and dialed digits on each call by using what is known as the cellular authentication and voice encryption (CAVE) algorithm. The algorithm combines a personal identification number (PIN) number, a shared secret data number, and a random number (RAND), which is sent out on a continuous basis by the network, to ensure that the mobile identification (ID) is secure.

DUAL-MODE MOBILES

The compatibility of DAMPS with analog extends to the mobile. The Dual-mode Digital Cellular Standard specification EIA-IS54 was released in December 1989 by the Telecommunications Industry Association (TIA). This specification calls for the initial introduction of dual-mode (analog and digital) mobiles to allow subscribers to the digital network to preferentially access a digital

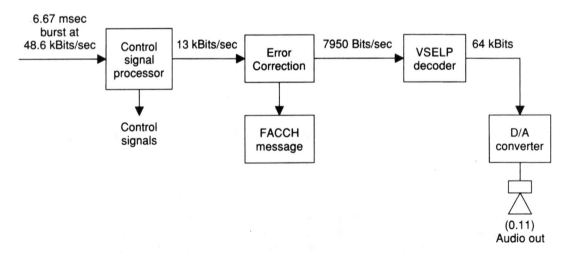

Figure 38.12   The receiver audio processing.

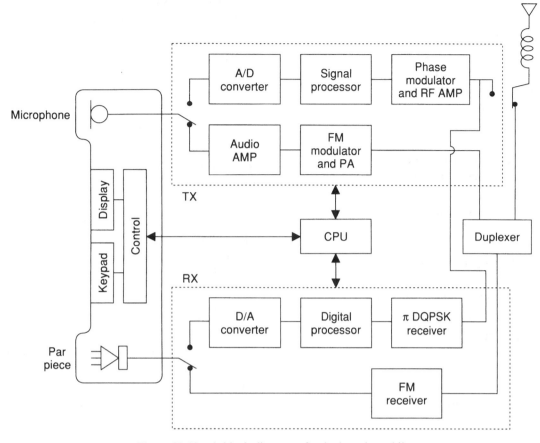

Figure 38.13 A block diagram of a dual-mode mobile.

channel, but to use the extensive existing analog network where digital is not available.

The minimum cell size that is possible in AMPS is to a large extent limited by the minimum RF power to which the mobile can be switched. The EIA-553 specification calls for the mobile to have a minimum power level of 7 mW, which in practice limits cell sizes to around a 0.5-kilometer radius.

A dual-mode mobile, as seen in Figure 38.13 will contain virtually two transceivers, one analog and one digital. It is possible to design the RF stage so that it can be used in common for both modes. A limitation is that the linear amplifier needed for digital transmission is much less efficient than the class C analog FM amplifier, so to conserve battery talk time the PAs are usually separate.

THE CODEC

The full rate CODEC operates at 7950 bits/s and has as input a 64-kbits/s channel derived from an A/D convertor. The CODEC uses the VSELP method, which is a type of code-excited linear predictor (CELP). In each

20-msec sample the CODEC outputs 159 bits. Error protection adds 101 extra bits per sample. The voice quality of the CODEC is quite good without multipath, but in the real-world multipath environment it can sound anything but natural. The problem appears to be that even though a good deal of error correction is provided, it is simply not enough.

The complexity of the CODEC is such that it is a significant contributor to the total power consumption of the first-generation dual-mode mobiles. The earliest CODEC consumed 600 mW, but this has dropped significantly in second-generation devices; and this problem was soon addressed.

MOBILE-ASSISTED HANDOFF

Digital mobiles have the added feature of mobile-assisted handoff (MAHO), which speeds up handoff procedures and reduces processing overhead at the base station. The mobile can monitor both the field strength and the received bit error rate for up to 12 channels during the timeslots when it is not receiving data. It reports the results of this scan to the base station on the SACCH or the FACCH channels.

DAMPS TERMS

CCLIST   Control Channel LIST

CDVCC   Coded Digital Voice Color Code

CELP   Code-Excited Linear Predictive

CMAC   Control Mobile Attenuation Code

CPA   Control Paging and Access

CSS   Cellular Subscriber Station

DQPSK   Differential Quadrature Phase Shift Keying (the modulation technique used in DAMPS and Japanese digital cellular)

DTC   Digital Traffic Channel

DTX   Discontinuous Transmission

FDTC   Forward Digital Traffic Channel

FIRSTCHA   FIRST CHannel for Access

FIRSTCHP   FIRST CHannel for Paging

FOCC   FOrward Control Channel (analog)

FVC   Forward Voice Channel (analog)

IEC   InterExchange Carrier

IVDC   Initial Voice Designation Channel

LASTCHA   LAST CHannel for Access

LASTCHP   LAST CHannel for Paging

LEC   Local Exchange Carrier

MAHO   Mobile-Assisted HandOff

NAWC   Number of Additional Words Coming

RDTC   Reverse Digital Traffic Channel

RECC   REverse Control Channel (analog)

RTC   Reverse Traffic Channel (digital)

RVC   Reverse Voice Channel (analog)

SACCH   Slow Associated Control CHannel

SBI   Short-Burst Indicator

SDCC (1, 2)   Supplementary Digital Color Codes

SID   System IDentification

TA   Time Alignment

TDD   Time Division Duplex

TIA   Telecommunications Industry Association

VMAC   Voice Mobile Attenuation Code

WFOM   Wait For Overhead Message

39

NAMPS

Although narrow-band advanced mobile phone service (NAMPS) is not a digital technology it was used as an interim solution to the capacity problem for the United States and other countries. It is based on a concept from Motorola to use frequency division multiple access (FDMA) to expand the capacity of the existing AMPS spectrum by using frequency division to get three channels in the standard 30-kHz single-channel spectrum. NAMPS will triple the capacity without significant degradation in the signal quality. The whole concept is similar to the narrow-band Japanese Total Access Communications System (JTACS), or Narrow-band Total Access Communications System (NTACS), which uses two 12.5-kHz channels in the place of the standard 25-kHz JTACS channels, but is backwards-compatible with the standard bandwidth systems.

The JTACS narrow-band equivalent of NAMPS, NTACS has been operational since mid1990.

STANDARDS

NAMPS is described in the Telecommunications Industry Association (TIA) standards for narrow-band analog cellular phones, IS-88, IS-89, and IS-90.

NAMPS Enhancements

NAMPS not only offers the operator more capacity, but for the consumer it means a higher grade of service (NAMPS phones first attempt to access NAMPS channels and, if not available, they have a second chance to access AMPS channels), fewer dropped calls, and a messaging system. All this comes in the same-sized, same-priced package.

NAMPS is more than simply a narrow-band AMPS system, as it addresses some of the shortcomings of AMPS. AMPS signaling on the voice channels uses "blank and burst techniques," which literally means that the voice channel is blanked out while the signaling takes place. Not only is this disconcerting to the user, but the switching causes splutter onto adjacent channels, thereby degrading the overall system performance. This will occur every time there is a handoff (about once a minute in built-up areas). NAMPS uses subaudible digital signaling tones (below 300 Hz) to allow the in-band signaling to occur without disruption to the call. This signaling is done on a 200-bits/s dedicated signaling channel at the rate of 100 Hz continuous. The only time a NAMPS call is muted is during handover.

Messages in AMPS are transmitted five times, and the best three of the five are used. In NAMPS the message is transmitted once and an automatic repeat request (ARQ) is activated if the message is received with errors.

One of the problems identified with AMPS was the total occupied bandwidth, which has a tendency to spill over to adjacent channels during signaling. The bandwidth needed for a frequency modulated (FM) channel can be found from

$$\text{Bandwidth} = 2 \times M + 2 \times D$$

where

$$M = \text{the highest modulation frequency}$$

$$D = \text{the maximum total deviation}$$

For AMPS, this is $2 \times 6 + 2 \times 14 = 40$ kHz. Note that the 6 kHz Supervisory Audio Tones (SAT) tone is the highest modulating frequency. For NAMPS, it is $2 \times 3 + 2 \times 5.7 = 17.4$ kHz.

The three supervisory SAT tones provided in AMPS to identify individual cells are replaced by seven Digital Supervisory Audio Tones (DSAT) in NAMPS. With only three tones it is possible that a mobile in a high-frequency reuse environment can lock onto the wrong channel or drop a call because of interference in low-signal areas. The greater number of DSATs will reduce the chance of locking onto the wrong number, and the more robust nature of the digital signaling will avoid mistaken cells.

Although AMPS has ways of detecting interference at the base station, it is well known that interference is far from reciprocal. Often the land-line will be interference-free, while the mobile is experiencing talkover that makes the call unusable. NAMPS gives the mobile a chance to report interference and request another channel if needed. At the mobile both signal strength and bit error rate (BER) are monitored. This feature is called mobile-reported interference (MRI). By monitoring the error rate in the DSAT signal, the mobile can infer interference. Similarly, the base station can monitor the base-station carrier-to-interface ratio (C/I) by the same method, and in either case allocate a new channel when required. This means that both inbound and outbound signals are continuously monitored for quality, and a handoff can be initiated if required.

Measurement of both signal strength and BER enables the operator to determine separately problems caused by interference and weak signals. Interference will show as BER problems in a good field strength, and frequent incidences of this may indicate co-channel interference, which may warrant a system retune or redesign.

ADDITIONAL SERVICES

Additional services can be provided on NAMPS, including a short message that allows up to 14 characters of text to be sent. By requiring the mobile to respond to paging messages, a record that the page was sent *and received* can be obtained.

Voice mail can be used, even on AMPS, but NAMPS enables the voice mail status to be automatically displayed (up to 14 characters), so that it is necessary to interrogate the system just to find "no messages."

Calling-line identification, a feature of TACS but not AMPS, is supported in NAMPS. This is a 32-character field.

Enhanced paging allows up to 32 digits to be sent.

Some operators are marketing the above feature package as "a cellular phone with a built-in pager and answering machine."

NETWORK CONFIGURATION

Motorola provides NAMPS hardware equipped with channels that have dual-mode capability. A linear power amp (PA) with class A amplification is used to reduce the harmonic component of the output to the extent that a tuned filter on the output stage is not needed. Although the use of a linear PA is not an essential feature of AMPS, the added flexibility by eliminating combiners and the need for tuning make the linear PA approach most attractive.

Although the system was developed by Motorola, other manufacturers have been licensed to produce mobiles.

HOW DOES IT PERFORM?

NAMPS has a performance very similar to AMPS. The most noticeable difference is the voice quality at high signal levels (around −50 dBm). Compared to AMPS the sound is somewhat fuzzy, but nevertheless it is entirely intelligible. At lower signal levels the performance is much the same as AMPS, and this applies right down to the threshold level of approximately −115 dBm.

The NAMPS channels can directly replace AMPS and provide an equivalent service with better spectrum efficiency. Because it can operate at the same C/I as AMPS, NAMPS can be expected to provide useful frequency efficiency gains where they are most needed.

The costs of NAMPS channels are similar to those of AMPS, although it is reasonable to assume that it costs around 80 percent of an AMPS channel.

Like all dual-mode systems, the dual-mode mobiles around have much better system access, measured as grade-of-service (GOS), than an AMPS phone because the actual GOS will be the product of the GOS offered to the NAMPS phones and the GOS offered to the AMPS phones.

NAMPS is ideal for "hot-spot" traffic relief, as it can be introduced on a needs basis to areas where high density is a problem. There is no need to consider upgrading old base stations that do not have a traffic capacity problem, although it may in some cases prove cost effective to expand some base stations in NAMPS. Consideration has to be given to the cost of new antennas and combiners (or RF PAs), which will mean that for some base stations, small incremental NAMPS expansions may not be cost effective relative to AMPS.

Also, because only some mobiles have NAMPS capability, there will be a problem of serving only a portion of the total market with NAMPS. This, of course, means that generally it can be expected that all base stations will also be dual mode.

MOBILES

Mobiles will be divided into three classes:

Class I   6 dBW effective radiated power (ERP) (4 watts)
Class II   2 dBW ERP (1.6 watts)
Class III   −2 dBW ERP (0.6 watts)

The mobiles have eight power levels, which nominally reduce the power by 4 dB per level.

THE HARDWARE

The hardware configuration for the channels was initially for 20 channels per rack. The existing combiners cannot be used and will either have to be replaced or more probably eliminated with the use of a wide-band amplifier. Motorola claims that NAMPS is about 25 percent cheaper per channel than standard AMPS. For a comparison of NAMPS with AMPS, see Table 39.1.

A compelling point in favor of NAMPS was that it was readily AMPS compatible and can be configured so that the three NAMPS channels can be allocated dynamically if the calling phone was AMPS, so that a normal AMPS connection could be made. However phones with NAMPS capability will automatically request a 10-kHz channel, if it is available, but can use a 30-kHz channel if the base station is only capable of AMPS. NAMPS mobiles are able to handoff to and from 30-kHz channels. This backward compatibility is a proven feature of the Japanese system.

Unlike the digital alternatives, NAMPS had a readily available handheld of convenient size and cost, where again the development was tested in the Japanese network.

TABLE 39.1 How NAMPS Compares to AMPS

Parameter	AMPS	NAMPS
Supervisory tone	6 kHz	200 B NRZ
Supervisory tone deviation	2 kHz	700 Hz
Peak voice deviation	12 kHz	5 kHz
Average voice deviation	2.9 kHz	1.5 kHz
Channels	832	2412
RF sensitivity	−116 dBm	−118 dBm
Channel spacing	30 kHz	10 kHz
Signaling	blank and burst 10 kbit	sub-audible 100 Hz
Supervisory tones	3	7

Note: NRZ = non-return to zero.

40

E-TDMA

In early 1991, Hughes announced its entry into the cellular terminal market with a new mobile that would not only be analog and digital compatible, but would also be equipped with a new standard called *extended* time division multiple access (E-TDMA). Hughes is better known as a satellite manufacturer, and it was in this field that the company gained its experience in TDMA, which is the technology used in its very small aperture terminals (VSATS). The new standard claimed to have a capacity 15 times that of analog systems, which ranked it with the (then) claimed capacity of CDMA.

All E-TDMA mobiles are triple-mode Advanced Mobile Phone Service (AMPS)/TDMA/E-TDMA. Because the E-TDMA standard is built around the TDMA standard, there is not a lot of extra complexity in building triple-mode instead of dual.

The method of achieving this higher capacity is to use half-rate encoding [that is, to use a state of the art, 4.8-kHz compression/decompression (CODEC) utilizing the code-excited linear predictor (CELP) algorithm, see Chapter 49] and to combine this with digital speech interpolation (DSI). In the DSI concept (which is used in satellites), advantage is taken of the quiet passages in a normal conversation. The method takes advantage of the fact that normal speakers are talking for only about 40 percent of the time. Six speech channels can thus be compressed into three speech paths. The user may maintain a channel whenever there is speech activity, but between pauses is retuned to an assignment channel; upon new activity from either party a new assignment is made. It can be seen that a by-product of this technique is a form of frequency hopping, which will reduce susceptability to interference and so lower the acceptable carrier-to-interface ratio (C/I) as well as improve the multipath capacity.

Normal speech is divided into active periods called *spurts* and *pauses*. The pauses can be due to the natural spacing between words or to periods when the talker is listening. In normal speech the spurt duration is around 1 to 2 seconds, and the spurt duration is exponentially distributed. The proportion of the total talk time occupied by spurts is called the *voice activity factor* (VAF). For mobiles the VAF from the mobile is typically 0.35 to 0.50, whereas in the land-line direction it is 0.35 to 0.45. It is usually slightly higher from the mobile than from the land-line end.

A fast switch can truncate a pool of voice channels into a smaller number of traffic channels and so save on circuits. The basic principle of this compression is shown in Figure 40.1, in which six voice circuits are carried by three traffic channels.

Additional efficiencies can be had by increasing the size of the pool of channels in the physical stream. Studies by Hughes suggest that the DSI gains shown in Table 40.1 are achievable.

In the E-TDMA implementation, each original analog channel has six half-rate TDMA channels. There are some slots for pool control and some for voice so that if there are N analog channels available, the distribution is

$$Vs + Nc = 6N$$

where

$$Vs = \text{speech timeslots}$$

$$Nc = \text{control timeslots}$$

Apart from the gains directly attributable to pooling, other gains accrue due to the ability to treat the send

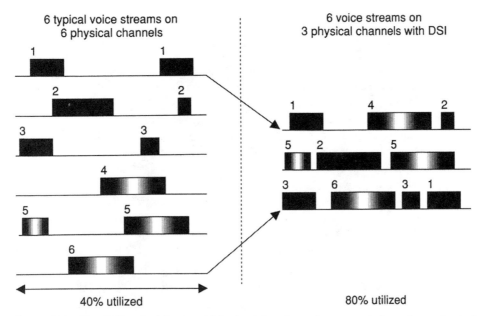

Figure 40.1 The DSI principle, in which six voice channels are carried on three physical streams.

TABLE 40.1   DSI Gains

DSI Gain	Pool Size
2	3
2.85	8
3.1	12
3.3	19

and receive paths separately so that, for example, fax machines that mainly transmit and so leave the return channel open will allow that channel to be used by another service.

To produce a mobile with DSI, it was necessary to undertake some development and testing of a voice-activation detector (VAD) that would operate in the noisy mobile environment. The device had to be capable of working on a handsfree phone in a car with a lot of wind noise and background chatter and noise that might be encountered in the environment of a handheld. The VAD has to have a response time to voice of a few milliseconds.

As with all digital systems, the most cost-effective way to gain capacity is through the use of an efficient vocoder. E-TDMA uses a half-rate vocoder, which was originally targeted to be at least as good as the first DAMPS full-rate vocoder. This raises the obvious question of how a half-rate vocoder could be as good as a full-rate device. The answer is in the technology. The original VSELP algorithm, although it was considered sophisticated when it first appeared, has been shown to have many short-

comings in practice, and it is not difficult to conceive of one that is both better and more efficient.

As a consequence of pooling, a DSI channel will intrinsically have slow frequency hopping, a technique known to improve multipath and noise immunity. This will improve both range and voice quality.

The base stations connect to the switch (which is an Alcatel S12) at the encoded level, so that the increased circuit efficiency is also available to the links connecting the base station to the switch. Compared to an analog service the decreased number of links (base to switch) will be directly proportional to the capacity gain. For example, if it is assumed that the E-TDMA system is 10 times more efficient than an analog system, then the link requirement will also be one-tenth that required for a like-sized analog service.

The E-TDMA transceiver is about the same size as an AMPS transceiver and includes the slot processor so that it can function in trimode. A transceiver unit is shown in Figure 40.2.

SOFT CAPACITY

The capacity of a cellular system per MHz can easily be determined in a single-site system, as it is merely the traffic capacity of the channels provided. In the case of E-TDMA, even this is not so easy to determine. By allowing a controlled level of clipping to occur, the capacity of a base station can be increased at the expense of speech quality.

Figure 40.2   A Hughes trimode E-TDMA transceiver unit. (Photo courtesy Hughes.)

In a large pool of users the total number of spurts per pool required at any one time will be normally distributed with a mean of VAF times the number of users assigned to that pool. The probability (or percentage of time) that any given number of users will generate more spurts than the pool capacity can easily be calculated using standard probability theory. When the pool size is exceeded, the system can do a number of things:

- It can drop all spurts in excess of the pool capacity; thus, some conversations will be missing a word or a syllable. This is called *clipping*. Given that spurt occurs on average once in $(1/\text{VAF}) \times \{\text{average spurt duration}\}$, for a typical spurt duration of 1.5 seconds and an average VAF of 0.4, a spurt will be generated once every 3.75 seconds. For a 180-second call there will be 180/3.75 or 48 spurts per call. This means that 2 percent clipping is equivalent to dropping one spurt per call, which may be hardly noticeable to the average user.
- The spurt overload can be queued for the next available timeslot. This will introduce timing and delay problems, but will avoid losing information.
- A combination of the above can be used.

CAPACITY GAINS

Hughes claims system capacity gains over analog ranging from 7× for minimal pools of 3 channels to 11× for pools of 19 radio frequency (RF) channels. This is made up of three factors.

$$\text{Capacity gain} = D \times DSI \times 2$$

where

$$D = \text{digital gain (same as DAMPS)}$$

$$DSI = \text{DSI gain}$$

and the factor 2 is due to the half-rate vocoder.

E-TDMA was estimated to have infrastructure costs that were 40 percent less per subscriber than DAMPS, with most of the savings being in base-station costs. A capacity of 10 calls per analog channel has been demonstrated, as has the "trimode" (AMPS/DAMPS/E-TDMA) handoff.

Figure 40.3 An MSU with 96 subscriber capacity and 3 E-TDMA channels. (Photo courtesy Hughes.)

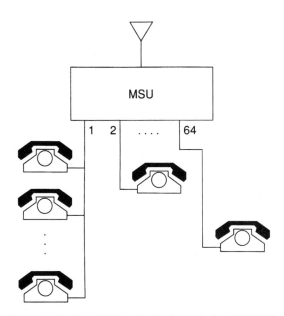

Figure 40.4 The MSU with 64 channels for E-TDMA.

CHANNELS TO SPARE

The utilization of all these techniques leads to the possibility of a base station where 20 analog channels are converted to give $6 \times 20 = 120$ half-rate TDMA channels. By utilizing DSI, in turn, each 6 voice channels can be truncated into 3 service channels. The result is that 240 new E-TDMA channels are made available for the recovery of only 20 analog channels.

Calls will not be lost as the result of a channel failure, because the signal loss will simply cause the DSI to send the call to a new channel.

FIXED WIRELESS APPLICATIONS

Most E-TDMA applications have been in WLL. A multisubscriber unit (MSU), catering for up to 96 local phones (each with its own number) is available, as seen

in Figure 40.3. It enables parallel party-line services on any of the numbers as required, as shown in Figure 40.4. Proposed applications are offices, apartment complexes, remote sites such as mining camps, rural towns, and third world villages. A single-subscriber version with party-line capacity also is available, as seen in Figure 40.5.

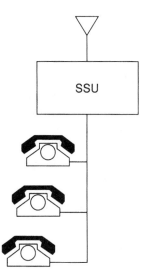

Figure 40.5 The single-subscriber unit (SSU) for E-TDMA.

41

CODE DIVISION MULTIPLE ACCESS (CDMA)

Code division multiple access (CDMA) is the second standard adopted for digital systems within the United States and designated IS-95. IS-95 defines the CDMA air interface. The standard is derived from the work of Qualcomm which today is known as cdmaOne.

CDMA has a demonstrated capacity of about 6 times that of Advanced Mobile Phone Service (AMPS), which gives it a significant capacity advantage over all other digital systems except extended time division multiple access (E-TDMA). Some subjective tests show that the voice quality is superior to analog (although not all tests concur on this point). As a long-term user of both AMPS and CDMA, I would rate CDMA as "nearly as good as AMPS."

TRANSMISSION

CDMA signals are sent over the same carrier channels at the same time, unlike frequency division multiple access (FDMA), which separates the signals in frequency, and time division multiple access (TDMA), which does this in time. In both the time domain and frequency domain the CDMA signals are transmitted virtually on top of each other. The spectrum is decoded in a code domain using a correlator that "sees" only correlated signals. The other superimposed signals will still be present, but only as low-level background noise.

CDMA COVERAGE

Demonstrations in Australia in 1999 attempted to show that cdmaOne can achieve coverage that matches or exceeds that of AMPS. By improving the processing

power at the base station, and so extending the search window, cdmaOne has been shown to achieve ranges in excess of 120 km. These early tests did indeed show CDMA out-ranging AMPS.

However, it was later found that the demonstration was to some extent rigged! The base station was otherwise totally unloaded and the terrain chosen was such that it would generate significant multipath.

Later, more extensive testing revealed that CDMA *on average* can almost match AMPS coverage but falls short in low multipath areas (e.g., over water), and of course has less coverage when it is heavily loaded.

CHANNEL BANDWIDTH

The channel bandwidth is 1.25 MHz, which came about because it was exactly the amount allocated to Qualcomm by Pactel for the initial field trials of the technology. The signals are spread over 1.2288 MHz. This 1.25-MHz bandwidth was chosen to be one-tenth of the extended AMPS (E-AMPS) spectrum (A and B bands both = 12.5 kHz), as well as to be wider than the bandwidth normally associated with selective frequency fading at 800 MHz, which is about 200–300 kHz. Guard bands of 270 kHz are needed between adjacent CDMA or adjacent AMPS channels, which reduces the number of CDMA channels that can be provided. There must be guard bands between the A and B bands, and it is suggested that the separation be 44 analog channels, or 1.32 MHz. The AMPS extended bands cannot be used within these constraints. As can be seen in Table 41.1, the useful resultant bandwidths are 322 channels (9.66 MHz) in the A band and 289 channels (8.67 MHz) in the B band. Given that a channel width is 1.25 MHz

TABLE 41.1 The Capacity of a Reverse CDMA Channel for a Propagation Characteristic of 4.0, as Derived by the TIA Subcommittee TR 45.5

Eb/No IN dB	CHN Activity = 0.411		CHN Activity = 0.5	
	OMNI	3 Sector	OMNI	3 Sector
6	35	83	30	70
7	27	65	23	54
8	21	50	18	43
9	17	39	14	33
10	13	30	11	26

Note: TIA = Telecommunications Industry Association.

plus 0.270 MHz guard band, a channel will occupy 1.52 MHz. This means that the A band can be six channels wide and the B band five. This is significantly less than the 10 carriers originally anticipated from each band.

SOFT HANDOFF

Because CDMA has the capability to transmit simultaneously from two cells, handoffs can be made with no interruption to an existing call. CDMA systems are effectively $N = 1$ designs, so adjacent channels will be using the same frequencies. A handoff can be done so that the call is handled by both the current and target cell together until the process is completed. Rake receivers are used to process both signals simultaneously.

The advantages of this are that the handoff will be virtually undetectable (a boon for data communications) and dropped calls will be minimized. Because handoffs occur about once a minute, this is a major advantage of this technique.

A mobile with a call in progress will continuously monitor the adjacent cell, and when a handoff becomes desirable the mobile will request it from the switch. The switch will then establish a path to the target cell while still maintaining the original connection. The parallel connection is released only when the mobile is firmly established on the new cell.

At call initiation the mobile is given a set of tailored handoff criteria and a list of candidate handoff cells. During a call all pilots are monitored (with emphasis on the candidate cells) and details are kept of all pilots above the threshold levels. This information is available to the switch on request.

The switch will issue the command to lock onto a second cell, which is then monitored in conjunction with the original cell. A decision to stay on the original cell or to swap to the second one is made on the basis of the carrier energy-to-noise level of the pilot. The decision is made more robust by requiring that a handover be initiated only if the second cell exceeds the signal performance of the first by a definable amount and that it has done so for a certain time.

It should be noted that the soft handoff process effectively parallels channels and will reduce overall capacity, by how much is uncertain.

The soft handoff is distinct from the hard handoff, which occurs when the mobile is being directed to a different frequency of different frame offset. The hard handoff is much like an analog handoff.

A third type of handoff is CDMA to analog, which occurs when the mobile is directed to change systems.

THE SOFTER HANDOFF

A version of the soft handoff, known as the softer handoff, is used for inter-sector handoff. In this case the cell site takes control of the handoff and provides a parallel path between the sectors. The two signals from the different sectors are diversity-combined for maximum signal quality. The switch is informed of the activity, but is not an active participant.

CDMA MULTIPATH ENHANCEMENT

Multipath is the word used to describe the many different signal paths that may present themselves at the antenna. In analog systems these paths will sometimes add constructively and sometimes destructively, causing clicks or even complete loss of signal. In CDMA systems the signals resulting from multipath can be identified by their differing times of arrival. By using a bank of correlators, each of which locks onto one path (as identified by its delay), the various signals can be separately processed and then added constructively. It should be noted, however, that not all multipath signals can be independently correlated, and thus fading is not *completely* constructive even with CDMA techniques. In particular, signals derived from different paths that have arrival times of less than 1 microsecond (μsec) apart cannot always be constructively recombined.

In fact, CDMA exploits multipath to the extent that it can be deliberately induced at the base station by creating time delays in a distributed antenna system, as seen in Figure 41.1. Delays of approximately 2.5 μsec can be used between the antennas, with a system gain in low multipath environments such as buildings of around 10 dB.

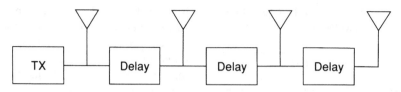

Figure 41.1 The distributed antenna concept.

RAKE RECEIVERS

Rake receivers or parallel correlators are used to take full advantage of the multipath environment. Instead of having just one correlator to lock onto the best signal, three are used in the mobile and four in the base station. When multiple signals are present they can be detected by the spread in arrival times, and individual arrivals can be processed parallel. During early testing it was found that one or more rake receivers were used for 70 percent of the time. The actual statistics for three rake receivers were:

1 finger	30% of the time
2 fingers	50% of the time
3 fingers	20% of the time

The multipath separation was typically 1–2 μsec.

ENHANCED SERVICES

CDMA provides the following services.

Wireless Data

- Short message service
- Store and forward mail
- Wide-band services (greater than 19.2 kbits/s)

Wireless Fax

- Group 3 and (in the future) group 4

Wireless in Building

- Interoffice and mobile cellular service
- Microcells

Integrated Services Digital Network (ISDN) Services

- Simultaneous voice and data

Video

- Wide bandwidth permits unparalleled video services

DIVERSITY

Virtually every form of diversity can be exploited. Time diversity can be achieved by interleaving, frequency diversity is inherent because of the wide bandwidth used on each channel, and space diversity is utilized in three ways:

- Multiple signal paths from two or more cells during handoff
- Multiple antennas
- Use of rake receivers in the signal processing

CDMA OVERLAY

CDMA can be expected to provide an increase in capacity of the order of 5 to 8 times that of analog for a given amount of spectrum. Each CDMA channel requires 1.23 MHz, together with a guard band on either side of 0.27 MHz, making a total of 1.77 MHz for one carrier. Where the full spectrum available in the A or B band is used, the analog system would have a total of 11.85 MHz, so the overlay of one CDMA channel represents the recovery of 15 percent of the analog spectrum.

CDMA is usually introduced initially in capacity hot spots. To do so the 1.77 MHz needs to be recovered not only from the new CDMA service area, but also from a geographic guard zone around the CDMA coverage area. This is because the CDMA uplink in particular is very susceptible to interference from analog mobiles. The greater this guard zone the better, but it must be at least one cell wide, as seen in Figure 41.2. This guard zone can contribute significantly to the total analog spectrum that is to be recovered; it also means that as a hot-spot relief mechanism CDMA is limited and some consideration must always be given to blanket coverage.

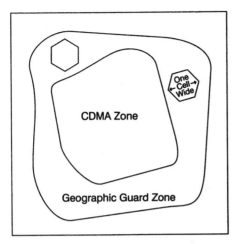

Figure 41.2 The concept of a geographic guard zone.

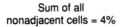
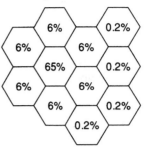

Figure 41.3 The interference contributions of distant cells in a CDMA system.

CAPACITY CONSIDERATIONS

The capacity of an analog system can, in theory, be increased virtually indefinitely, provided the interference levels can be contained to tolerable levels. In practice, limits are set by the power-level control range of the mobile and the cost of infrastructure. Interference in analog or TDMA systems is essentially a result of interference from a small number of immediately adjacent channels on the same frequency/timeslot.

CDMA systems, being wide band with a large number of users on the interfering carrier, are governed by the average behavior of a large number of other co-frequency users, and thus will be affected by the *average* usage of the population. Mathematically, it can be said that the statistical law of large numbers applies to the interference. This means that as long as the wanted signal exceeds the threshold value, an acceptable signal can be decoded. TDMA/analog systems, on the other hand, need a healthy margin above the threshold to perform satisfactorily.

The factors that determine CDMA system capacity include processing gain, bit energy-to-noise ratio, voice duty cycle, frequency reuse plan, and the number of sectors used.

Probably the dominant result of system overload in CDMA systems is the breathing effect. As a site loads up, the range of that site contracts. Having a system that has no fixed boundary can be a source of much frustration for system designers. Customers might find that they have coverage in some areas in off-peak periods that disappears during traffic peaks; they can interpret this as unreliable coverage.

Voice has been long known to be a process with lots of pauses, and it is exploited for extra capacity in the E-TDMA system. CDMA can take advantage of the pauses by reducing the transmission rates when there is no speech, thereby reducing the total interference to other users. Because the total noise from other users limits capacity, this technique will increase the total number of users per cell. The duty cycle of a typical speaker is 35 percent, so a reduction in net interference of more than 2 can be achieved while at the same time gaining a similar increase in transmitter (TX) talk time.

Interference from adjacent cells in a CDMA network accounts for around 36 percent of the total, as shown in Figure 41.3, while the cell itself produces 65 percent of the total interference. Only 4 percent of all interference is contributed by non-adjacent cells.

Sectoring has the net effect of reducing the total interference in direct proportion to the area covered by the sector. Thus a 120-degree sectorization would reduce the total interference to 1/3. In practice, the reduction is 1/2.55 due to coverage overlap.

The very wide bandwidth used in CDMA systems permits very powerful high redundancy coding to be used. This is not possible in TDMA systems because of the relatively narrow bandwidth used and the requirement for a high bit energy-to-noise (E_b/N_o) ratio restricts the total number of error correction bits. CDMA requires an E_b/N_o of 7.5 dB for acceptable voice quality.

SOFT CAPACITY

Other cellular systems have a well-defined capacity at each site (defined by the number of channels fitted). In the CDMA case there is an upper limit of 62 voice channels per CDMA channel (1.25 MHz), but the actual number of channels that can be used is a function of the total noise and is about one-half that number. Should it be desired during the busy hour to increase the number of users on a particular base station, this can be done at the cost of a small degradation in bit error rate (BER) per channel.

The soft capacity feature will lead to a big reduction in lost calls, which occurs due to the unavailability of channels in the peak busy hour. Because the extra channel allocation can be made dynamically assignable, the degraded BER will persist only while the demand for the extra channels exists. This peak demand will be sporadic and so can be accommodated with little effect on the average call quality.

In fact, there is no simple relationship that determines the "ideal" channel capacity of a CDMA cell, as the capacity is a function of interference rather than the number of channels. Thus, a rural base station with no neighboring cells and only its self-generated interference to contend with will be able to carry around 40 percent more traffic than one operated in a multicell area (the 40 percent being the interference generated by neighboring cells). In the same way, cells that cover highways that attract high traffic during drive time to and from work will generally be surrounded by cells that will be lightly loaded during these same periods. This means that more traffic can be carried where it is needed *without degradation of the BER*. Typically, gains of 20 percent capacity can be made available in this way.

The converse of this is that high-quality circuits, virtually error-free, can be made available to users who want (and are prepared to pay for) them. Radio broadcasters and data services may find this feature a boon.

POWER CONTROL

Because the capacity of the network is directly related to the average power of the interferers, it is important to keep the power of the mobile just high enough to ensure the desired signal quality. Higher signal levels will give a moderate increase in voice quality at the cost of lower system capacity.

The mobile monitors the signal from the base station and estimate the path loss, setting its initial power level accordingly. The actual rather than a relative path loss can be calculated because the synchronization channel transmits the power level being used by the pilot channel. This permits cells of different powers and coverage to be used in the network concurrently. Should there be a sudden change in the signal level, an analog power adjustment can permit a power change in the mobile TX power with dynamic range of 85 dB. The response time for this change is a few microseconds, this rate being limited by the closed-loop response of the base station. This analog level control, which results from measurement and the calculated response, is known as open-loop control.

OPEN- AND CLOSED-LOOP POWER CONTROL

In analog and TDMA systems the path loss in one direction is considered to be a good measure of the reciprocal path loss. Although this will hold true for the *average* path loss, Rayleigh or multipath fading will be independent of the forward and receive channels. This is because the 45-MHz separation between send and receive channels is sufficient to ensure that local multipathing will instantaneously be different for the two channels. It should be noted that in measuring the pilot-channel field strength the mobile will be seeing the multipath faded level of the forward channel and so is not getting a direct measurement of its own performance. Thus, if rapid signal strength adjustments are to be made, they must take account of this difference, and this is done by closed-loop control.

Closed-loop control is used to fine-tune the power levels, while at the same time accommodating the different multipath fading levels. At the cell site the BER is measured and compared to a reference rate. If the actual rate is lower than the reference, the base station will instruct the mobile on the outbound channel to reduce its power level by 0.5 dB. This process is repeated at a rate of once every 1.25 msec, until the desired BER is achieved. The target BER can be assigned from the switch.

Because the mobile is able to read the effective radiated power (ERP) of the base station pilot channel, its own measurement in the forward direction will be a good indicator of the path loss. By transmitting this reading to the base station, an appropriate TX level can be determined. Some fine-tuning is still required, because for a given path loss the ideal base-station TX level will not be fixed but will depend on the average noise level in the area. This fine-tuning is done by having the base station periodically reduce its power level until the BER at the mobile becomes high enough that the mobile requests additional power. This is done at a much slower rate than is the case for the reverse channel, with a change of only 0.5 dB per 15–20 msecs. The dynamic range of this adjustment is also limited to only 6 dB.

PILOT CARRIER

Each cell has a unique pilot carrier signal, which is used to provide initial sychronization as well as time and phase tracking of the signals from the cell site. The cell coverage can be controlled by varying the power of the pilot. Each pilot uses the same code, but has different phase offsets by which they are distinguished. Since the same code is used by all cells, sychronization can

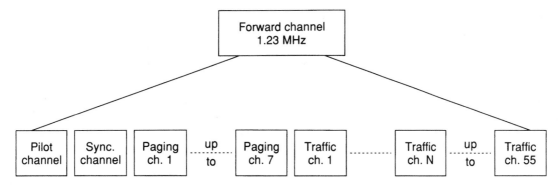

Figure 41.4 The structure of a typical CDMA forward channel.

be done with a single search through all possible code phases.

CHANNEL STRUCTURE

The nominal bandwidth of a CDMA channel is 1.25 MHz. However, the actual frequency assignment is 30 kHz × 41 or 1.23 MHz. The pseudorandom (PN) chip rate is an exact multiple of 9600 kbits/s, being 128 × 9600, or 1.2288 MHz. The PN codes are generated by a shift register, which gives a code with a period of 32,786 chips. The transmitted signal is digitally filtered with a bandpass filter having 3 dB points at 1.23 MHz.

The forward channel (base to mobile) consists of a pilot channel, a sychronization channel, and 63 traffic channels (up to 7 of which can be paging channels), as seen in Figure 41.4.

The reverse link (mobile to cell) uses the same PN chip rate as the forward link but additionally has a fixed code phase offset, which is used to distinguish a particular mobile. With $2^{42} - 1$ possible sequences, all of which

can be valid addresses, the address range is extremely large and the degree of privacy high. The reverse channel structure is seen in Figure 41.5. This channel can be composed of up to 32 access channels and 64 traffic channels.

Synchronization Channel

The synchronization channel has the same phase offset as the pilot channel and can be decoded together with the pilot. It contains cell site identification (ID), the pilot power level, and the system time sychronization. It is used during the initialization phase, immediately after power-up. It will not ordinarily be accessed after that until the next power-up. The synchronization channel carries only one message, called the *Sync Channel Message*, which contains, among other things, the station PN sequence.

Paging Channel

Once the mobile has the Sync Channel Message, it will switch to the paging channel. The paging channel conveys four messages:

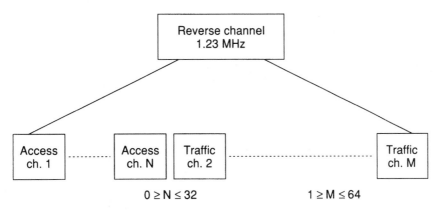

Figure 41.5 The reverse channel structure.

- *System Parameter Message* Contains registration parameters and pilot-channel details.
- *Access Parameter Message* Contains information on the access channels.
- *Neighbor List Message* Contains information on the offset of neighboring channels.
- *CDMA Channel List* Contains the frequency assignments of the paging channels.

The essential function of the paging channel is to call a mobile when an incoming call is waiting. It also acknowledges registrations and can lock out rogue mobiles. To preserve battery life, a paging mode, called *slotted mode*, can be used. It assigns mobiles to certain timeslots at registration so that the mobile will be called only in a predetermined timeslot and so can power down at other times. These slots can be from 2 to 128 seconds apart.

The paging data rate is variable and can be 2400, 4800, or 9600 bits/s. At 9600 bits/s, up to 180 pages a second can be handled. There are a maximum of 7 paging channels, so a total maximum paging rate of 1260 per second is possible.

Access Channel

Each paging channel has an associated access channel on which the mobile replies to a page or an instruction issued on the paging channel and on which the mobile requests network access. All access transmissions occur at a bit rate of 4800 bits/s.

Traffic Channel

The traffic channel can dynamically send data at 9600, 4800, 2400, or 1200 bits/s, depending on the demand. During pauses the frame data rate is reduced, and it is increased when required. A traffic frame is 20 msec long. Like the analog systems, CDMA signals on the traffic channel, but it does so in two different ways. The first is the familiar blank-and-burst, in which a traffic frame is replaced by a signaling one. The other way is called dim-and-burst, in which a frame, transmitted at 9600 kHz, is arranged so that half of the frame is used for signaling and half for voice traffic. This technique produces a negligible degradation in voice quality. The messages that are sent on the traffic channel are controls for handoff, forward link power, call handling, and authentication.

REGISTRATION

Registration of the mobile occurs on power-up and it can occur at other times. There is a trade-off in system overhead between the resources needed to support frequent registration (which permits the network to be well-informed of the mobile's location, and so permits paging to be restricted to just a few base stations, but which ties up access channels) and infrequent registration (which requires the system to page all calls on all bases because it is unaware of the actual mobile location).

CDMA Supports Eight Types of Registration:

- Power-up registration.
- Power-down registration. This is regarded as the least reliable of the registration forms, as local fades and driving out of range may cause the system to miss the call. However, a positive power-down will indicate that there is no need to waste overhead capacity paging the mobile, so it may still be valuable.
- Time-based registration. The mobile automatically registers at certain time intervals.
- Distance-based registration. The mobile notes all new base-station coordinates (in longitude and latitude) and calculates the distance moved since the last registration. When it has moved a predefined distance it will re-register.
- Zone-based registration. The bases will have a zone identity and the mobile can register whenever a new zone is encountered.
- Parameter-based registration. The mobile re-registers whenever a parameter, such as timeslot, is changed.
- Ordered registration. Registration on a command from the base.
- Implied registration. Each attempt to use the access channel can signify a registration.

The base station can control and disable any of the first six forms of registration (which are known as autonomous, as they are automatically initiated by the mobile). The actual registration regime used in any particular network can be software optimized.

SERVICE OPTIONS

The CDMA system is structured to provide layers of service options, which are requested at the mobile end. Service options can be primary services, known as Service Option 1, such as voice, data, or fax. Secondary traffic may also be carried on the same option, and this may include packet data, which is transmitted in parallel with the voice but at a rate that varies with the rate of the voice data.

AUTHENTICATION

CDMA uses the Electronic Industries Alliance (EIA)/ TIA/IS-54-B, cellular authentication and voice encryption (CAVE) algorithm. The mobile station has its own A-key and secret shared data.

VARIABLE RATE VOCODER

The initial voice vocoder is an 8-kbits/s code-excited linear predictor (CELP) device, which subjectively gives a good voice quality. However, the CDMA structure is not defined around any particular baseband data rate and it will be possible to have a wide variety of data rates corresponding to the services offered, such as ISDN, fax, voice services of various qualities, and data.

The encoding is done at the switch level. A pool of vocoders with different data rates can be assigned by the switch on a demand basis. Lower data rate vocoders on transmitting from the mobile to the base station achieve efficiency by essentially punching holes in the transmitted waveform. Thus, a vocoder that usually operates at 9.6 kbits/s when transmitting at 4.8 kbits/s continues to generate data at 9.6 kbits/s, but gates the transmitter on and off in a pseudorandom manner so that it is off for half the time.

Mobiles

By early 1999 the CDMA Development Group (CGD) claimed that there were more than 70 different CDMA phones available from 14 different manufacturers including Audiovox, Hyundai, LGIC, Motorola, NEC, Nokia, OKI Telecom, Philips, Qualcomm, Samsung, and Sony.

The mobile receiver is essentially a duplex transceiver, complete with duplexer, which allows the mobile to tune to any of the twenty 1.25-MHz channels. Working as a standard superheterodyne, the intermediate frequency (IF) stage is passed through a 1.25-MHz bandpass filter and then directly to an analog-to-digital (A/D) converter. Mobile phones became available in experimental quantities in early 1993 and, as can be seen in Figures 41.6 and 41.7, the mobiles were comparable in size to analog phones of the same vintage. By 2000 CDMA phones (as shown in Figure 41.8) were available in sizes similar to other systems.

Apart from the radio frequency (RF) hardware, all that is needed additionally is contained in two application-specific integrated circuit (ASICs) (which can be purchased from Qualcomm). The heart of the system is the Mobile Station Modem (MSM). The MSM is configured in the mobile, as shown in Figure 41.9. It

Figure 41.6 A handheld dual-mode (CDMA/AMPS) phone in 1993. (Photo courtesy Qualcomm.)

consists of three main parts: the demodulators, the subscriber modulator, and the Viterbi decoder.

Demodulators

The subscriber module contains three single-path receivers (known as fingers), which are implemented as a rake receiver. Each of the three fingers is an independent demodulator, capable of frequency tracking, time tracking, and data demodulation. These provide the decoding for up to three multipaths and a significant processing gain, which enhances the signal-to-noise (S/N) ratio. The signals from the fingers are combined in a maximal ratio manner, which combines the signals proportionally to their S/N levels. The pilot signal from the base station can be used to determine phase relationship so that the combining can be done coherently.

A fourth demodulator, known as a *searcher*, is used to continuously scan for multipath signals and to assign the strongest signals to the fingers. The searcher also is used to direct handoff, choosing the best path between two cells.

Subscriber Modulator

The subscriber modulator modulates the transmitter at the baseband level. It contains the convolutional en-

Figure 41.7   A dual-mode (CDMA/AMPS) mobile phone from Qualcomm in 1993. (Photo courtesy Qualcomm.)

Figure 41.8   A Hyundai dual-mode CDMA weighing 145 g, circa 2000.

coder, the block interleaver, and the sequence spreader. The modulator output is then power controlled by both the control processor and the analog receiver, and then it is modulated to 850 MHz. The modulator also contains the deinterleaver for the receiver.

Viterbi Decoder Module

The Viterbi decoder module decodes using the Viterbi algorithm. The outputs of the maximal ratio diversity combiner are first deinterleaved and then decoded using the Viterbi decoder.

The whole chip is a 144 pin thin quad flat pack (TQFP) $20 \times 20 \times 1.4$ mm with a power consumption of 50 mW on standby and 300 mW when processing full RX/TX functions.

CELL SITE EQUIPMENT

The cell site can have two or more receive antennas for space diversity. The degree of space diversity can readily be increased by the use of additional antennas. The signal processing is much the same as in the mobile, except that because no pilot signal is available from the mobile, the multipath combining will be incoherent. The base station uses two searcher receivers.

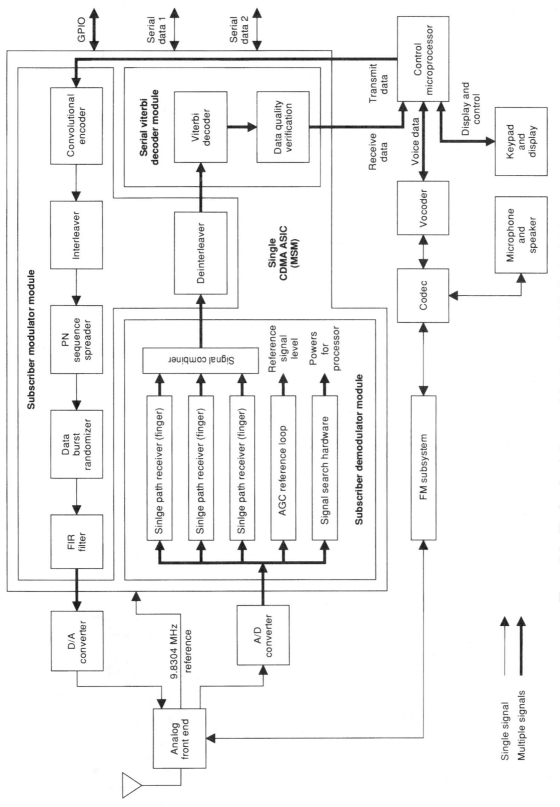

Figure 41.9 The functional diagram of Qualcomm's Mobile Station Modem.

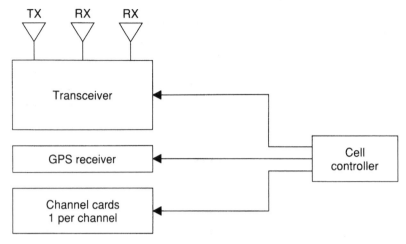

Figure 41.10   The structure of a single-cell CDMA base.

This simplicity even carries over to the network, where a base station can be built around three similar ASICs. These chips, which are functionally similar to the MSM, are the base-station demodulator (BSD), the base-station modulator (BSM), and the serial viterbi decoder (SVD).

A base-station channel uses four BSDs, which are configured to form a rake receiver, with four demodulation paths.

A basic cell site is shown in Figure 41.10. Each base has a digital shelf, which does the processing for the channels. These can be either voice or control channels. The channel cards process the signal to the IF level before directing the signal to the transceivers. The cell controller (CC) allocates resources for traffic, collects statistics, and distributes the timing information. A Global Positioning System (GPS) receiver (or a redundant pair) is used to derive the precision timing needed by the system. Commercial GPS timing receivers have an accuracy of $\pm 3 \times 10^{-12}$, and will mostly have a 10-MHz and a 1-second clock output. They cost about $2000 each.

The full set of chips to implement a base station and mobile phone circa 1995 are shown in Figure 41.11. Two chips are required for a mobile and three for a base station.

FREQUENCY PLANNING

With an $N = 1$ system, meaning that all frequencies are reused at all sites, it may be easy to conclude that frequency planning will become a thing of the past. It is not, however, so simple. In the case where a system is being installed in a new area (where there is no analog on the same band) there may be advantages in adopting, for a start, an $N = 4$ frequency plan. This would eliminate interference from co-channels and so would increase the capacity of each cell site by around 40 percent.

In the more probable environment in which CDMA is inserted alongside an existing AMPS system, an $N = 1$ plan is all that is available. Here, there will be—at least initially—no frequency planning, but every effort should be made to minimize the effect of co-channel interference, as it will reduce network capacity. Bases, whether analog or digital, should be examined to see if co-channel interference is likely, as the CDMA carrier is not at all tolerant to interference from other systems.

Figure 41.11   The complete CDMA chip set for base-station and mobile implementation. (Photo courtesy Qualcomm.)

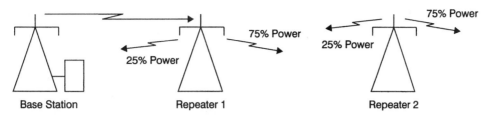

Figure 41.12 A cascaded repeater.

The biggest challenge for the frequency planners is to eliminate adjacent channel interference. Guard bands must be provided (which means eliminating even more AMPS channels from adjacent bands). It is likely that the adjacent channels will cause interference, or if they are near enough, they may even be able to block the CDMA channel.

As the AMPS band recovery process continues, for small and medium-sized cities, base-station capacity gains can be achieved by adopting an $N = 2$ frequency plan.

SWITCHING

The switch performs all the usual public switched telephone network (PSTN) type functions as well as providing a vocoder for each speech path. The voice signals from the base station are digital and will need to be passed through a vocoder, which converts the voice to either standard 64-kbits digital voice or to analog voice as required. If the base-station signal is involved in a handoff, the switch will use a selecter (one associated with each vocoder) to dynamically choose the best base station on a frame-by-frame basis.

In the case of voice from the PSTN, the switch also assigns a vocoder to digitize the signal. The switch determines and controls the appropriate allocation cells to mobiles and the PN code to be used for the call.

Repeaters

Because any CDMA base station will have a lot of capacity (around 30–50 users simultaneously), there is a wide scope for the use of repeaters. Effective use of repeaters in low-to-medium traffic densities can lead to network savings from 20–40 percent, so it is well worth considering.

Repeaters are typically limited to two per base station. Two configurations recommended by Repeater

Technologies are cascaded and parallel repeating. In cascaded repeating the two repeaters repeat in a chain with the power distribution, as shown in Figure 41.12. In flatter, more open terrain, where the repeaters are typically further away and have higher gain receiving antennas, the repeaters are co-located in a parallel configuration, as seen in Figure 41.13 is recommended.

Along major highways in rural areas repeaters can be used to replace as many as two-thirds of the base stations, saving the cost of the E1/T1 links as well as the cost of the base station. Such hybrid networks are called *repeater hybrid networks* (RHN).

RHNs have been built in the United States in rural areas, with as few as 6 base stations and 24 repeaters making up the network. A network that extensively uses repeaters will cost in the vicinity of 30–50 percent less than one based entirely on base stations. The costs are tempered by the need to have more sites when using repeaters than would have been the case had the network been all base stations. Nevertheless you will still find CDMA repeaters in the traditional places—outer suburban areas, rural areas, along highways and, in building applications.

Repeater Technologies of California have a 6.3-watt unit (the OA850C), which offers a gain in the range 65–95 dB and a noise figure of 5 dB. The unit has diversity built in, which has been shown to improve the link budget by around 4 dB. These repeaters are all centrally

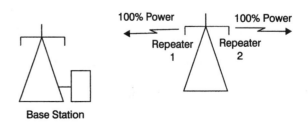

Figure 41.13 A parallel repeater.

monitored on a regular basis. The Repeater Technology hardware comes with a design package that includes spreadsheets that calculate link budgets, repeater antenna isolation, and the other critical parameters that are needed for a successful design.

Generally a CDMA system will initially be installed with a single carrier, and that soon the time will come when, for traffic reasons, the second carrier must be installed. Where the second carrier can be installed over the entire service area, this is simple. But difficulties sometimes arose when the second carrier was needed only in a few local hot spots, because in some early versions of CDMA, inter-carrier handoff was not well supported.

One solution is to install pilot beacons at neighboring sites. The mobile assisted handoff (MAHO) measurements then see the beacons on the second carrier and force a handoff to the first carrier in the area that has an acceptable Carrier to Interference level. The handoff is "blind" and automatic, that is, the handoff is not to the best cell in the region. This can result in high failed handoff or dropped calls, sometimes running as high as 10 percent.

The second method is to install the second carrier over a wide area (known as the "wedding cake") and hand off at the edges using a transition sector. In this case, the handoffs are also blind, and the dropped call rate can be high.

New Developments

All cdmaOne networks are synchronized, and almost universally this is done with GPS timing receivers. Base stations are required to be synchronized to within ± 10 μsec, and this is well within the capabilities of GPS. The advantages of a synchronized network are many, including faster initial mobile access, reduced mobile "awake" time (the mobile knows when to come out of the idle mode, which in turn means longer standby time), and faster handoffs. The disadvantage is that the whole network depends critically on the GPS timing.

GPS time is traceable to universal time coordinated (UTC), differing only in not having the leap seconds that are occassionally introduced to UTC.

Base transmitter stations (BTSs) cannot be directly coordinated as they cannot see each other's transmissions due to self-jamming.

One way around this, which has been proposed by Qualcomm, is for the network to use information that is already being gathered for synchronization. When using the soft handoff the mobile senses the relative pilot strengths and their arrival times in a pilot strength measurement message (PSMM). These relative arrival times can be used to calculate any error in the clocks between

the base stations measured. Field testing has shown that an accuracy of around 1 μsec can be obtained in this way.

Networks designed to be self-synchronizing would still largely use GPS timing, but the criticality of the clocking (and hence need for redundancy) could be greatly reduced. Others have suggested that a way to avoid dependence on GPS is to make 3 G systems asynchronous, which solves the dependency problem, but disallows the advantages of synchronisation.

CDMA BASE-STATION SIMULATOR

A base-station simulator, designed to be portable and readily erected, has been designed by Berkeley Varitronics. The base station has an built-in GPS clock, and is locked to a Rubidium clock, so that accurate PN offsets can be preset.

The unit is designed to be used in the field to test and optimize site locations, for new base stations.

CDMA TERMS

Access Channel The reverse channel used by the mobile for requests for system access and response to paging and registration requests.

Access Channel Slot The assigned timeslot for an access probe on the access channel.

Access Probe One access channel transmission consisting of preamble and message.

A-Key A secret 64-bit pattern stored in the mobile and used to generate the mobile stations shared secret data, for authentication purposes.

Authentication The procedure used to validate a mobile.

Autonomous Registration A method of registration whereby the mobile registers without a command from the mobile.

Candidate Set The set of pilots that have been received and noted to have a usable field strength, but which have not yet been activated.

CDMA Code Division Multiple Access.

CDMA Channel Number An 11-bit number corresponding to the center of the CDMA frequency assignment.

Convolution Code An error correcting code in which the input data are convoluted with an impulse generator function.

Cyclic Redundancy Code (CRC) An error correcting code that generates check bits by division with a polynomial generator. The remainder gives the check bits.

De-interleaving The transmitted signals are interleaved in time to achieve time diversity. Interleaving is the reverse process.

Dim-and-Burst Signaling on a voice channel in which the signaling is multiplexed with the voice or data.

E_b The energy of a bit of information.

Frame A block of data. For the access, paging, and traffic channels this is 20 msec, and for the sync channel it is 26.666 msec long.

Hard Handoff A handoff that causes temporary disconnection of the channel in a manner similar to analog handoffs.

Padding A sequence of bits at the end of a message used to fill a frame. Usually "0s" are used.

Pilot Channel An unmodulated direct-sequence spread-spectrum channel emitted by every CDMA base station, which provides voting, timing, and phase information for the mobile.

Shared Secret Data A 128-bit pattern stored in the mobile in a semi-permanent register. It is made of two parts, each of 64 bits. One is used for authentication and the other for generating the encryption.

Slotted Mode A battery-saving option that allows the mobile to monitor only a designated paging time-slot.

Sync Channel A forward channel that provides the sychronization information.

42

JAPANESE DIGITAL

Japan has recently had a significant surge in cellular connections and has now caught up with the rest of the developed world in market penetration. The analog networks are still functional (in 1999) and are profitable. The analog networks are mostly the Japanese Total Access Communications System (JTACS), although a local analog service known as (Hi-CAPs) is still providing some service. The bulk of the market (around 90 percent) is on the time division multiple access (TDMA) digital system known as personal digital cellular (PDC).

CdmaOne (code division multiple access) has been introduced by the DDI Cellular Group, although it is in the JTACS band. The operation in a non-standard band has two undesirable outcomes. Regional roaming is made difficult, as all other operators use the US 800-MHz-band plan, and handset availability is highly limited. In order to achieve nationwide roaming with the initial introduction of cdmaOne, dual-mode CDMA/JTACS phones have been used. With the dual-mode capability comes some undesirable bulk, which has hampered the initial acceptance of the CDMA system.

Nippon Ido Tsushin (IDO) are the second CDMA licensee and have roaming with DDI.

The solution to the roaming problem that is actively being pursued is to introduce dual-band CDMA phones. An agreement has been reached with Korea to have activated roaming by 2002, in time for the World Cup. Under this agreement between DDI, IDO, and Shinseiki Telecom of Korea roaming is to be implemented at the network level.

ROAMING

Roaming from Japan's custom mobile networks is difficult, but NTT DoCoMo has introduced a call-forwarding system for those willing to pay the premium. This system forwards calls to known numbers on other networks.

PERSONAL HANDY PHONE SYSTEM

After a successful implementation in 1995, the growth of the personal handy phone system (PHS) in Japan was brought to a halt, after peaking in 1998 at 7 million subscribers, as the cellular networks became too competitive.

In an attempt to increase the attractiveness of PHS, NTT DoCoMo has released a dual-mode PHS-cellular phone called *doccimo*. NTT also has a printed circuit (PC) card for PHS that allows the simultaneous use of voice (8 kbps) and data (24 kbps).

The philosophy of the Japanese digital development (known as Japanese digital cellular, or JDC) is that, because of the high research costs the new digital system must have higher capacity, lower equipment costs, more features, and smaller handsets. This contrasts sharply with the Global System for Mobile Communications (GSM) objective to be "at least as good as analog" and even with the US objective to achieve increased capacity with backward compatibility. Because of the more lofty objectives it might have been expected that Japanese digital would take somewhat longer to develop. However, politics again drove development, and compromise was the order of the day.

In 1989 Japan's Ministry of Posts and Telecommunications (MPT) began the process of evaluation of the digital options. The Japanese Digital Cellular Radio System Committee was formed the same year, organized by the MPT. Another body, the Japanese Research and Development Center for Radio Systems (RCR) is responsible for standardization. The development of cordless phones is also a responsibility of these bodies.

One major concern for the Japanese was to identify an efficient and robust compression/decompression (CODEC), which is the key to spectrum efficiency. The vector sum excited linear prediction (VSELP) algorithm was chosen by the RCR.

DECLINE OF PHS

The marketing targeted the price conscious, and it was relatively too easy to make a business case for a service that was more reasonably priced than the inflated cellular prices at the time.

For some years PHS prospered, but falling cellular prices put an end to its success, and by 1998 the net subscriber rate on PHS was falling.

For a more detailed look at PHS, see Chapter 44, "Cordless Telephone Technologies."

PERSONAL DIGITAL CELLULAR

The Japanese digital system, which is an overlay to the analog service, is marketed as personal digital cellular (PDC). It provides a capacity gain of around 3:1 compared to analog cellular. The sound quality is very similar to that of DAMPS.

TECHNICAL

In many ways the Japanese digital systems resembles US digital. JDC is TDMA based on three timeslots per 25-kHz carrier. Frequencies allocated to this system are in the 810–830-MHz (base transmit) and 940–960-MHz (base receive) bands. The duplex channel spacing then becomes 130 MHz.

The modulation method is quadrature phase-shift keying (QPSK), $\pi/4$ shifted. The similarity to the US digital system can be seen in Table 42.1.

The network architecture of the Japanese digital is somewhat simpler than GSM and consists of three main parts (see Figure 42.1):

- the BS or base station

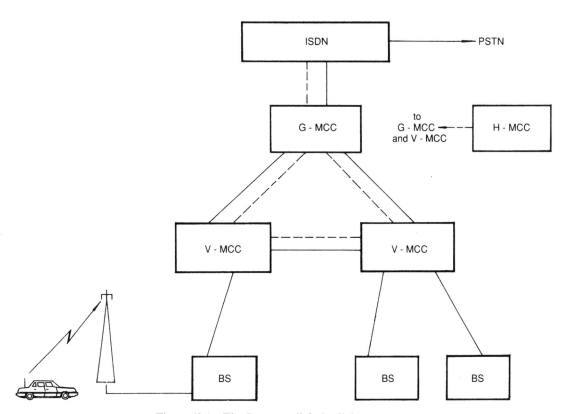

Figure 42.1 The Japanese digital cellular structure.

TABLE 42.1 Comparison of United States and Japanese Digital Systems

Feature	United States	Japan
RF carrier width	30 kHz	25 kHz
Cell radius	0.5–20 km	0.5–20 km
Multiplex	TDMA	TDMA
Channels/carrier	3	3
CODEC	VSELP	VSELP
Transmission bit rate in kb/s	48.6	42
Modulation	$\pi/4$ QPSK	$\pi/4$ QPSK

- the H-MCC or V-MCC—the home- or visited-mobile control center
- the G-MCC or gateway mobile control center

The G-MCC performs the interface function to the public switched telephone network (PSTN). The H-MCC stores location and subscriber files and communicates this data to the V-MCC, which is in control of setting up and terminating calls.

TIMING AND OPERATIONS

The CODEC operates at both 6.7 kbits/s and 4.5 kbits/s.

One of the major problems with digital systems is inter-symbol interference caused by the non-linearity of the power amplifier (PA) on the final stage of the radio frequency (RF). Usually this is overcome by backing off the power levels and thereby reducing the PA efficiency and to use a modulation scheme that minimizes abrupt transitions. The $\pi/4$ QSPK modulation scheme is the same as the US solution to containing the extent of the phase transitions. An additional improvement over the US digital proposal is to incorporate a non-linear compensation circuit in the PA that will extract the full power efficiency of that stage without significant spreading of the signal, which causes inter-symbol interference.

43

SATELLITE MOBILE SYSTEMS

The earliest satellite repeaters were simply passive reflectors from which Earth stations bounced off microwave signals to distant receivers. By the 1960s satellite-borne microwave repeaters were in space and global links were made possible.

Most communications satellites are in the Clark Belt, or geostationary orbit, at which level the speed of rotation just matches that of Earth. This makes them appear to be stationary to an Earth-bound observer, which has the advantage that a large ground station with a fixed dish can be used.

EARLY SYSTEMS

By 1990 there were a number of satellite-based mobile telephone systems in the testing phase. Practical applications were limited mainly to aircraft systems such as Toshiba Corporation's which provides three cordless phones and a fax. The cost of this system was around $700,000 per installation, a price that would surely limit the market potential.

After extensive testing by Japan and British Airways (among others), Imarsat announced its first commercial aircraft phones in early 1991.

The American Mobile Satellite Corporation (AMSC) was granted a Mobile Satellite Service license in 1989. AMSC is a conglomerate of eight companies and the license is exclusive. It was operational with the launch of its first satellite in 1994. This system uses geosynchronous satellites, and so is not a LEO. The initial service area is the United States and its coastal areas, including Alaska, Puerto Rico, and Hawaii. The service is premium priced at around $2.50 per minute.

The system uses the L-band for satellite-to-mobile communications using FDMA and the Ku-band for satellite to gateway. The voice channels are 6 kHz wide with a 4.8-kbit CODEC.

Only vehicle-mounted mobiles are available because of the 5–25-watts power requirement and the need for a 600-mm antenna, which rules out handhelds.

To see how far satellite receiver technology has come, we do not have to look back too far.

AUSSAT, an Australian concept, was implemented in 1994 and is marketed as Mobilesat. This system uses ($\pi/4$ QPSK) modulation and is essentially the same as that proposed for the United States. The AUSSAT test bed was the "mobile" mounted in the four-wheel drive seen in Figure 43.1. This mobile had been used in trials throughout Australia and elsewhere in the world. The AUSSAT mobile, was afflicted with the same leviathan malady that was then part of the digital scene. Figure 43.2 shows the original prototype mobile hardware, and it is clear that anything smaller than this trial vehicle would not be able to carry it. Notice the surprisingly small rooftop antenna (Figure 43.1), which consists simply of two droopy dipoles (dipoles inclined at about 45 degrees from the horizontal to improve the skyward radiation pattern). When this photograph was taken, the mobile was being demonstrated and the small ampere meter on the power supply was indicating that 13 amps was being drawn. The engine had to be kept running at 1000 rpm to keep the battery voltage up. Of course, the final commercial product is a lot smaller, being about briefcase size.

In 2000, the equivalent mobiles are still large sized, and cost $3000. Monthly rates are $30, and call rates are $1.30 p.m peak or half that off-peak.

Figure 43.1   The AUSSAT Mobilesat phone being demonstrated at Sydney's Darling Harbour Convention Centre.

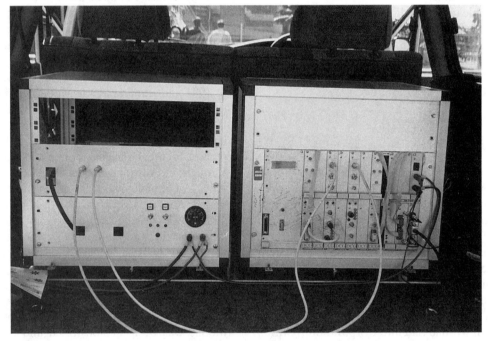

Figure 43.2   The AUSSAT test bed hardware in 1991.

SATELLITE ORBITS

The satellite orbits need to avoid the Van Allen belts, the first of which at 1500 to 5000 km, is outside the low earth orbits (LEOs) while the medium earth orbits (MEOs) are placed between the inner belt and the outer one, which is from 13,000 km to 20,000 km. Geostationary earth orbits (GEOs) at 35,780 km (exactly) are outside both belts.

At an altitude of 35,000 kilometers, the satellites are a

long way from Earth and therefore need high effective radiated power (ERPs) and large antennas for effective communications. Ground stations are relatively large high-powered devices requiring complex, bulky antennas. They also will introduce significant delays in the speech path, as the signal can travel only at the speed of light (300,000 km/sec). Geostationary satellites are expensive (costing hundreds of millions of dollars) and heavy (weighing several tons), and expensive to launch as well.

Geostationary satellites used well-proven technology, and very little handoff is needed between satellites. Three such satellites can provide worldwide coverage.

Mobile telephone systems based on these satellites have been available for some time for use on ships and aircraft. The Intelsat stationary satellite telephone service became available in 1965. The large equipment size has meant that mobility was extremely limited.

LEOs reduce the losses between the satellite and the mobile by the expedient of bringing the satellite closer to Earth. The orbits range from a few hundred to a few thousand kilometers. Because of the well-known inverse square law (the signal strength varies inversely with the square of the distance), it can be seen that reducing the satellite orbit by a factor of 10 will decrease the path loss by a factor of 100. These systems are also known as global mobile personal communications by Satellite (GMPCS).

LEOs have been in service for military and scientific purposes for a considerable time, becoming commercially viable with the advent of miniaturization and low-priced launch facilities. LEOs orbit within the drag of the outer edge of Earth's atmosphere and thus will require constant repositioning as the drag slows the satellite and changes its orbit. The life of the satellite is limited by the fuel reserves of the repositioning rockets to about five years.

The large number of LEOs that are needed to provide effective service means that there is a significant time lag between the launch of the first satellite and the time when the full system is in service.

LEOs are robust systems, and a reasonably good service can be provided even if a few of the satellites fail. However, the handoff regime between them is very complex, and the Doppler effects in particular complicate equalization.

Two types of LEOs are proposed. First, there are the satellites meant for conventional two-way or mobile phone applications, including voice, data, positioning, and fleet management. These are proposed to operate in the L-band (1.6 GHz). The primary allocations are in bands 1525–1530 MHz in regions 2 and 3, plus 1492–1525 MHz in region 2 (the Americas), and additionally the bands 1610–1626.5 MHz (uplink) and 2483.5–2520 MHz (downlink).

The second type of LEO is meant for two-way data operation below 1 Ghz. These are known as "little LEOs." They operate in primary frequencies of 137–137.025 MHz, 137.175–137.825 MHz, 148–149.9 MHz, and 400.15–401 MHz. Secondary allocations are 137.025–137.175 and 137.835–138 MHz. These frequencies were allocated at the WARC-1992.

Satellites offer virtually universal coverage, but the price is paid in low spectral efficiency and high equipment costs.

It is instructive to consider who the users of satellite cellular are expected to be. The first thing that is certain is the technology can only be complementary to the conventional terrestrial service, if for no other reason than capacity. The cost of the satellite mobiles is around $1000, which is expensive compared to a conventional cellular phone. The power requirements for most systems are considerably higher than for conventional cellular, which means the units are more bulky.

All of the satellite proposals for voice services include dual-mode mobiles. The second mode will be variable and includes a choice of any of the major land based systems currently in use. The wide variance in technology (and in particular frequency) used by satellite versus conventional mobiles will mean considerable duplication is required, which in turn means bulk and, to a lesser extent, cost.

Although throughout the world a high proportion of the population is now covered adequately with conventional cellular services, there is a small but significant portion that is not. In developed countries this underserved population is probably less than 10 percent; in developing countries it may range from 25 percent to almost 100 percent. In arid regions, mountain areas, and on the seas it is generally not practical or economical to provide conventional coverage. Of course, in most of these areas there is not enough activity to support a cellular service, and for this reason they may never have one unless it is by satellite.

Satellites can provide a valuable service to mining, mining exploration, remote communities, travelers to places such as outback Australia, the Rockies, and almost anywhere in India, where there is simply no conventional cellular alternative and mostly no other means of communications except perhaps high-frequency (HF) radio, which at best is noisy and unreliable. With HF, while it is usually possible to contact somebody in an emergency, it is often impossible to contact a particular party due to the vagaries of HF propagation. This applies particularly to low-powered portable units. When the choice is between a relatively high-priced, somewhat bulky mobile and no communications at all, satellite cellular looks attractive. What is clear is that satellite cellular is not for everybody—in fact, it's not even for most people. But it will find a market that is complementary to the terrestrial system.

Although satellite propagation is generally thought of as line of sight, it is actually somewhat more complex than that. In an open field there may be a direct line of sight to the satellite, but in the forest there will be blanketing by the trees and in the built-up areas, the buildings and other obstructions will interfere with the path. This is helped to some extent by having multiple satellites that offer a wider range of service paths.

The other considerations are covered in the following sections.

Specular Reflection

Specular reflection is reflection from Earth's surface. The amount of reflection will depend on frequency and the smoothness of the ground. In some circumstances it can be significant and poses a serious problem for handheld communications (directional and parabolic antennas have a natural immunity to this reflection). The phase of the reflection will vary randomly, and can either add to the signal or interfere destructively with it.

Diffuse Reflection

Because the satellite beam is very wide there are numerous opportunities for terrain scattering, and this scattering at any point will be mostly highly phase-incoherent multipath. This will result in fast fading (again the mobile antenna is highly vulnerable).

Building Penetration

The problem here is similar (but a bit more complex, because the frequencies are higher and the up- and downlinks are not correlated) to conventional cellular. There needs to be about a 30-dB link budget surplus to ensure coverage within buildings. In general this is not available on any of the current mobile satellite systems. The best that can be hoped for is that the systems will work near windows and in wooden buildings.

Body Losses

Blockages caused by the body (for example, wearing the mobile on a belt or carrying it in a pocket) will cause link budget losses of around 15 dB.

Interference

The current mobile satellite link budgets do not have significant margins, so many radio frequency (RF) sources may interfere. These include microwave ovens, medical equipment, and other transmitting devices.

In general, from all of these propagation risks, a fixed terminal will fare better than a mobile one.

Traffic Density Problems

As LEOs and MEOs move across the surface of the earth, so the traffic that they can carry will vary widely. Those over the oceans probably carry very little traffic, while those over certain land areas carry heavy traffic.

This traffic non-uniformity will not only have an effect on the total traffic capacity of the satellite network, but it may also degrade the quality of service in the dense traffic areas, directly by adding to the total noise in CDMA systems and indirectly through interference in others.

A way around this problem, takes advantage of the fact that there is a considerable area of overlap in the coverage of adjacent cells, (i.e., "double coverage areas".... even in the case where the overlap is due to more than two satellites). Schemes can be devised to force traffic from the densely trafficed cells to the less dense ones. The simplest of these schemes is to widen the coverage area of the lightly trafficed cells, while contracting the coverage of the denser ones.

LEO Velocity

A LEO operating at 1500 km, has a velocity of 7.1 km/s or 25,500 km/hour, and so is much faster than even a military jet. Globalstar at 1414 km, is close to this orbit, while Iridium at 780 km is somewhat lower. As such, relative to a LEO satellite any earth bound user can be regarded as essentially stationary.

The Doppler effect will be very pronounced at this speed, and low orbit satellites cause abrupt changes in the frequency as they pass over, which must be compensated for. One easy way to compensate is for the ground station to loop back its own TX signal via the satellite and measure the frequency shift.

Inter-Satellite Links

Inter-satellite links (ISL) require steerable antennas, however, the complexity can be reduced considerably if the links are all between satellites in the same orbital planes; in which case very little antenna steering is necessary. Conversely linking adjacent orbits requires quite sophisticated antenna steering. Iridium uses both link modes.

By October 2000, only Globalstar was operational and Iridium had ceased to function. A plan was in place to bring down the satellites. Improved mobile unit capabilities have increased the desirability of some of the other proposals, including those involving medium Earth orbit (MEOs) and geosynchronous Earth orbiting satellites (GEOs).

IRIDIUM

Iridium was a mobile satellite system run by the Washington-based company Iridium LLC. The concept of Iridium originated from Motorola, whose holding of almost 19 percent of Iridium LLC stock made it the largest single shareholder.

The concept of Iridium was first proposed by Motorola engineers in 1987, and the project was officially announced in 1990.

The satellites were launched on a number of rocket platforms, and Figure 43.3 shows the Delta 2 rocket launch vehicle deployment of one of the satellites. Figure 43.4 shows a close-up of one of the satellites, with the solar panels unfurled.

It was originally planned to have 77 satellites in LEO. The name Iridium was derived from the fact that the element iridium (Ir) has position 77 in the Periodic Table of Chemical Elements. Iridium is also associated with the extraterrestrial theory of the dinosaur extinction, which proposes that the extinction was brought about by an unlucky encounter between Earth and a massive meteor, which brought about years of "nuclear winter" conditions. The unusually large concentrations of iridium found in soil deposits dated from the extinction period are thought to have originated from the debris of that meteor. Most of the iridium on Earth's surface is thought to be of extraterrestrial origin, as it is commonly found in meteorites. All in all, a very satisfactory and interesting name.

Unfortunately, in 1993 Motorola revised the number of LEOs to 66 (which will reduce the project cost almost in direct proportion to the satellite count). Element number 66 on the chemical table is the rather boring and little known dysprosium (Dy). It does not seem likely that a name change is forthcoming.

Dryprosium is a metal at room temperature that is highly reactive with water and oxygen. It has a high melting point and a strong ability to absorb neutrons, and so finds use in control rods for nuclear reactors; it also has some limited applications in electronic hardware manufacturing. Iridium likewise is a metal that has few applications in its pure state, being mainly used as an alloy with platinum to make jewelry and electrical contacts; the iridium alloy (typically 5–10 percent iridium) is harder and more chemically resistant than pure platinum. The international standard kilogram (the only standard that uses a physical piece of material) is 90 percent platinum and 10 percent iridium.

Iridium featured both a mobile phone system and a pager with 200 characters per message display; the pager can run for about one month on its disposable batteries.

By November 1998, and with an expenditure of $6 billion, full commercial service was activated with all 12 ground stations and 66 satellites with 6 in-orbit backups.

THE LAUNCHING

The Iridium satellites were launched by three different countries, the United States, Russia, and China. In the United States, McDonnell Douglas used Delta 2 class rockets, with five satellites being launched on a single launch vehicle. Twenty-one were sent into orbit on Proton rockets by the Russian Federation's Khrunichev Enterprise. China launched satellites with the Chinese Long March 2C rockets (two satellites per rocket).

MOBILES

The first working phones were developed by Motorola in 1992, and its componentry covered a large table. The phone was shrunk, and the handheld models worked virtually anywhere that there was a direct line of sight to the satellites. Even small obstructions could be significant, and so it was important that the antenna be above the head when in use (the phone is in fact disabled if the antenna is not properly deployed). This also meant that indoor use was in general not possible.

The phones were available as satellite only or as dual-mode phones that were connected to the roaming partner's cellular network.

There were a limited number of phone suppliers, with Motorola and Kyocera being the first to make product available. Motorola's dual-mode phone, which was capable of GSM 900, code division multiple access/Advanced Mobile Phone Service/Narrow-band Analog Mobile Phone Service (CDMA/AMPS/NAMPS), weighs about 450 grams and has a talk time of 2 hours or a standby time of 24 hours. It initially cost about $4000.

Mobile phones for Iridium were not dissimilar in size to the cellular phones of the 1994 period, and the Tecom Iridium phone for example weighed 0.5 kg. Much

Figure 43.3   The Delta 2 rocket deployment of an Iridium satellite. (Photo courtesy Burson-Marsteller.)

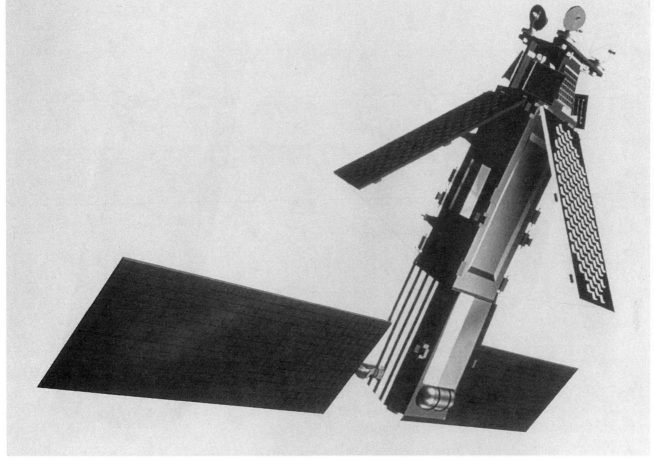

Figure 43.4 The Iridium satellite as it would look in service. (Photo courtesy Burson-Marsteller.)

heavier than the lightweight 100-gram cellular phones that have become the norm, but still very manageable.

Motorola phones offered dual-mode capability with GSM 900, AMPS, NAMPS, and CDMA. The Japanese supplier Kyocera had compatibility with GSM 900 and personal digital cellular (PDC).

All Iridium phones were distinguishable by the thick antenna structure. The two initial offerings from Kyocera, a multimode handset (Figure 43.5) and an Iridium-only set (Figure 43.6), both had the identifiable antenna.

Charges

The charges varied from place to place, but indicative charges from Australia were a U.S. $120 activation fee and a monthly charge of U.S. $50. Calls within the country were charged at U.S. $2.20 per minute (compared to about 12 cents a minute for land line), and international calls were about double that. Calls from the public switched telephone network (PSTN) to an Iridium phone were least 50 percent higher than the above figures.

Numbering

All phones were given an international number with an 8817 prefix (for satellite only) or 8816 for dual mode.

Frequencies and Modulation

The mobile service links were in the L-band (1616–1626.5 MHz), while the inter-satellite links were in the Ka-band (23.18–23.38 GHz). The links to the ground stations were at 19 GHz for the downlinks and 29 GHz for the uplinks.

The modulation system was frequency division multiple access (FDMA). The voice channels had an 8-kHz bandwidth and used 4.8-kbit compression/decompression (CODEC). The data bearers were 2400 bits/s.

Figure 43.5 The multimode handset from Kyocera. (Photo courtesy Burson-Marsteller.)

Figure 43.6 The Iridium-only phone from Kyocera. (Photo courtesy Burson-Marsteller.)

The satellites were arranged in six polar orbits at 421.5 nautical miles (780 km) above Earth's surface. The concept was that there would be sufficient intelligent satellites so that it was always be possible to handoff to one of two adjacent counterparts, which will be 2173 nautical miles away. Each satellite, weighing 700 kg, had 37 cells, and each cell had a service area (footprint) around 360 nautical miles in diameter.

The low-level orbits meant that low-power mobiles could be used. Equally important, the propagation delay times were much smaller than is usual on satellite links, so the delays were not noticeable. The more conventional geostationary satellites are positioned at 36,000 km, or at 54 times the distance of the Iridium system, and thus experience much greater path losses and delays (typically 0.5 second per link).

The original plans called for a fade margin of 6 to 12 dB, but field tests revealed that 16 dB was needed in cluttered environments. This meant higher gain antennas. The satellites were equipped with 48 beams each. The beams are arranged so that they present a constant pattern at Earth's surface. Because the satellites themselves were moving, the beams were constantly in motion.

The handoff was from beam to beam. In a way it can be considered to be the reverse of conventional cellular, in which the mobile moves in relation to a fixed base station and thus causes handoffs. Here, for practical purposes, the mobile is fixed and the handoff occurs because the base station moves out of range. Handoffs occurred at a rate of about one per minute.

It is the complexity of the linking that has drawn the most criticism from Iridium's detractors. The critical timing and the need to use point-to-point lasers and masers made the system very complex. As an offset to the complexity, Iridium unlike the other systems is completely self-contained and does not rely on intelligent input from ground stations.

The system capacity was determined by the maximum number of users that can be accommodated in the busiest area. Advantage could have been taken of the different world time zones to optimize traffic efficiency. Unused beams or those positioned over low traffic areas could be turned off. This improved frequency reuse and meant that at any one time about 68 percent of the beams were turned on. The reuse pattern was based on a 12-beam reuse, which meant 180 reuses worldwide. The modulation was quadratic phase-shift keying (QPSK), and a combination of TDMA and FDMA multiplexing was used.

The satellite links include four cross-links per satellite (for handoffs) and up to four to the gateways. The up- and downlinks were in the L-band and cross-links between satellites as well as between satellites and gateways were in the Ka-band. The gateways interfaced the 4.8-kbits/s Iridium voice rate to the 64-kbit PSTN. All links were circularly polarized.

The gateways use dishes consisting of a minimum of two 3.3-meter tracking units. In areas of heavy rainfall (such as the tropics), three or even four units separated by at least 30 km were needed.

The Iridium project was envisioned as having a maximum capacity of 2,300,000 subscribers, with 16,800 subscribers per cell.

Iridium Markets?

An initial push by Iridium, aimed at the international roaming market, was abandoned very early. Accumstomed to mobile phones, the international business traveler soon became disillusioned with the handset size, talk time, and difficulties using the handsets indoors and in some urban environments.

Because of a lack of customers, Iridium filed for Chapter 11 relief (temporary protection against debtors) on August 13, 1999. However, service continued.

The marketing in late 1999 was aimed at the traditional satellite phone user, such as maritime workers, forestry workers, mining workers, governments, and military personnel. It was hoped that these segments would be more familiar with the limitations of the technology, and would be more prepared to adopt it.

In the meantime, prices dropped to $3.00 per minute for international calls, and to $1.50 to $2.50 for regionwide calls. Iridium-to-Iridium calls dropped to a flat $1.50. At the same time, the pricing of the handsets fell below $1000, and pagers to below $350.

DEMISE OF IRIDIUM

At its peak Iridium only even had 55,000 customers. This apparently was not enough to justify running costs as the system was being prepared for a shut-down by April 2000.

The plan is to drop the satellites four at a time, in order to clear the space and effect a controlled disposal.

Iridium is an example of technology for the sake of technology, with little consideration for the actual needs of the user. It is sure not to be the last example of this kind of folly.

GLOBALSTAR

An alternative proposition to Iridium is Globalstar, from Loral Qualcomm Satellite Sources (LQSS). Globalstar is operated by Space Systems/Loral (SSL) and LQSS, Globalstar is a 48 satellite network commenced operations in late 1999.

The system satellites are LEOs in orbit at 1410 km and the satellites are expected to have a lifetime of 7.5 years in this orbit and weigh 250 kg. The satellites are in 8 orbital planes and have 6 satellites in each plane.

The orbital period is 114 minutes. This means that a user terminal can be served by one satellite for a period of 10 to 15 minutes of the orbital time.

Globalstar has kept costs low by just using simple so-called bent-pipe transponders that repeat the radio path to the mobile back to the nearest gateway in a single hop.

Globalstar uses a code-excited linear prediction (CELP) variable rate vocoder. The voice processing will incorporate a background noise canceling procedure. The overall objective is to equal or exceed the quality of the IS-96 CDMA standard.

A strategy for low signal strength areas is to lower the peak data rate from 4.8 kbits/s to 2.4 kbits/s; this will provide intelligible, but degraded communications, where it otherwise might not have been possible. At this altitude satellite life is expected to be 7.5 years. The satellites will be in eight circular orbits, each with six satellites per plane. The orbital period is two hours. The coverage will be almost global, extending from 72 degrees North to 72 degrees South. The total system cost is estimated at $2 billion. The initial installation was based on 24 LEOs to provide North American coverage.

The modulation method is CDMA, and voice, data, fax, messaging, and position location are supported. Globalstar differs substantially in concept from Iridium in that the satellites will have relatively little intelligence and will function simply as repeaters. Also, there will be no direct satellite-to-satellite links. The uplink will be L-band, and the downlink S-band. The structure of Globalstar is shown in Figure 43.7. Mobile units will be dual-mode with either GSM or a U.S. digital option available.

The capacity of the initial 24-satellite service was expected to be 5000 simultaneous users (which translates to around 250,000 subscribers). The complete sytem of 48 satellites is expected to support 100,000 simultaneous users, or a subscriber base of around 5 million.

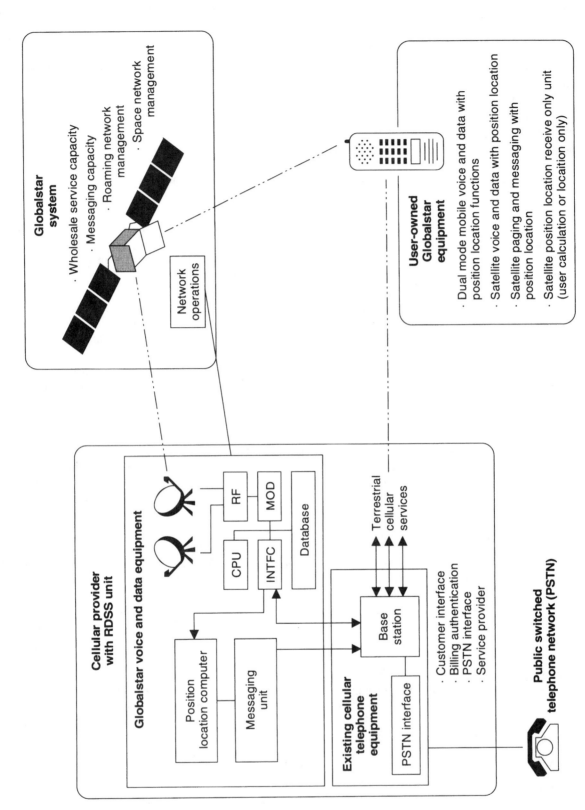

Figure 43.7 Structure of Globalstar.

Globalstar system
- Wholesale service capacity
- Messaging capacity
- Roaming network management
- Space network management

Network operations

User-owned Globalstar equipment
- Dual mode mobile voice and data with position location functions
- Satellite voice and data with position location
- Satellite paging and messaging with position location
- Satellite position location receive only unit (user calculation or location only)

Cellular provider with RDSS unit

Globalstar voice and data equipment

RF
MOD
CPU
INTFC
Database

Position location computer
Messaging unit

Terrestrial cellular services

Existing cellular telephone equipment

Base station
PSTN interface
- Customer interface
- Billing authentication
- PSTN interface
- Service provider

Public switched telephone network (PSTN)

Data services are provided at 7.2 kbits/s, with a higher bit rate at 9.6 kbits/s to some terminals.

The registration procedure locates the user to an accuracy of 10 km. This function is used mainly to ensure assignment to the relevant ground station. As an extension of this service, the user terminal position can be made available to an accuracy of 300 meters, provided at least two satellites, which are separated by at least 22 degrees, can be seen.

Coverage

The coverage is worldwide from 72 degrees North to 72 degrees South, and so excludes the polar regions. This is not "whole of earth" coverage, but it certainly includes most populated areas.

Frequencies

Globalstar uses the L-band to the mobile terminal and receives S-band on the uplink.

From the gateway to the satellite the frequency band used is 5090–5250 MHz, while the downlink (from the satellite to the gateway) uses 6875.95 to 7052.8 MHz. By using both right- and lefthand circular polarization, the eight frequency blocks can deliver 16 beams from the satellite.

Transmission

The air interface uses a modified version of IS-95, CDMA. Diversity will be provided by allowing the gateway to use multiple satellites and the user terminals (UT) to use all satellites within sight.

The link budget is such that operations indoors are expected to be marginal, and the unit may work in a timber-frame house, but signal level variations of as much as 30 dB are expected as a result of even minimal movement.

Handoff

Handoff is from one satellite to another, and handoff of the uplink does not necessarily imply handoff of the downlink (this is because of the large frequency difference between the send and receive frequencies, making the two paths uncorrelated.

Handoff may be soft, in which case two or more links are simultaneously demodulated, and the new link is activated using a new pilot before the old link is terminated. Hard handoffs are characterized by a temporary disconnection of the traffic channel before the new link is established.

The Globalstar PSTN interconnection scheme envisages that direct interconnection is made to the PSTN operators in various countries, with those operators also having exclusive rights to the Globalstar service in their regions. The PSTN interface is via the gateway, which is designed to be unmanned. The main requirement of the gateway is that it can provide an unobstructed path for the antennas.

The gateway has up to four parabolic antennas 5.5 at least meters in diameter, and with positioning mechanisms.

Capacity

The capacity of the system is such that it can support 100,000 simultaneous conversations, which translates to about 3–7 million users (depending on the usage rate).

Mobiles

Globalstar features dual-mode mobiles, and in built-up areas the subscribers will first access the land-based cellular system. Initially phones were available from Qualcomm, Ericsson and Telital. Prices in late 1999 began at $880.00, and call charges were around $1.50 per minute.

Mobiles are available as Globalstar-only, Globalstar/ GSM, and Globalstar/CDMA/AMPS units. The dual-multimode units do not feature handoff, and so a user transitioning the service boundary will have to reinitiate the call.

Mobiles have a cold start-up acquisition time of between 10 and 20 seconds.

OPTICAL SATELLITE LINKS

LEO systems have a serious problem communicating with the ground station because their low altitudes and high transit speeds means that they are constantly moving away from any ground reference point (and that includes any ground stations that they may be in contact with). One solution to the problem is to have the LEOs communicate with the ground via a GEO system.

Optical systems have the advantage of very efficient frequency reuse, low cost, the small power of the transmitters, and a very wide bandwidth that is protocol independent.

This system is being studied by the European Space Agency (ESA), using 2.488 Mbits/s infrared 60-mW links. The problems in alignment that this poses are not minor. The 40,000 km (a typical distance, given that its orbit is 36,000 km) that separates a LEO from a GEO is enough to cause the laser beam to diverge by around 300 meters; in that same time the satellite moves around 2000 metres.

The really hard problem is the initial alignment. The ESA proposal is to initiate contact with a 700-mW "broad-band" laser array from the GEO to establish initial contact. From there the positioning for data communications should take place with a 95 percent probability of success in 2 minutes. Once contact is established, calculations and corrections can keep the satellites in communications.

MEDIUM EARTH ORBIT SATELLITES

MEO satellites operate between LEOs and geostationary levels and are midway in cost and size between these two systems. They operate in a region of high radiation and need to be specially designed to operate in that environmment.

A proposal that is intermediate between the LEO concept and the older geostationary system is the Odyssey system. The concept calls for 12 satellites in medium orbit to give global coverage. The system will use CDMA and is planned to be in service by 1998. These satellites cost millions of dollars and require an Atlas-class rocket for launch. The orbit is within the Van Allen radiation belt, and so the satellites have to be specially hardened. The system is able to support 900,000 users.

An interesting newcomer to this field is Hughes (space and communications group), traditionally a company concentrating on the manufacture of conventional satellite technology. In 1990 Hughes announced its intention to enter both the cellular radio and satellite-based mobile areas.

Hughes' North American Mobile Satellite System (MSAT) is an early system. The two-satellite system provides coverage of the United States, including Alaska and Hawaii, and Canada (also spillover coverage of Mexico is possible).

The satellites each have 3200 radio channels, although the mobiles are not necessarily able to access all of these.

The spacecraft is three-axis body-stabilized and uses two 5-by-6-meter mesh-reflector antennas. The service life is expected to be 10–12 years.

Inmarsat

Inmarsat was formed in 1979 and is an international organization owned by 72 countries. It operates five satellite services known as Inmarsat-A, Inmarsat-B, Inmarsat-C, Inmarsat-Aero, and Inmarsat-M. They offer global coverage from geostationary satellite networks, except for the polar regions above 70° latitude (North and South).

Recently, Imarsat was privitized and is now known as Imarsat Holdings. Imarsat Holdings is to initially operate as a private company with the original shareholders.

Each orbital location has at least one hot-standby satellite to provide a high degree of security. The land networks are provided by a competitive network of mobile earth stations (MESs), which provide technical and commercial diversity.

All communications are on the L-band (1.5–1.6 GHz).

Inmarsat-A Inmarsat-A was the first service introduced; it provides telephone, fax, and data. Data services using Integrated Services Digital Network (ISDN) at 64 kbits are available. The portable terminals weigh about 20 kg, come in a suitcase, and have a fold-out antenna that is manually pointed at the satellite. The units come with telephone, fax, and telex ports, and are ordinarily mains operated (at 240 or 110 volts). The power requirement is about 100 watts.

Maritime systems are bigger and have parabolic dishes that are from 0.9 meter to 1.2 meters in diameter and power requirements of from 500 watts to 1000 watts. The dishes are gyroscopically stabilized.

Costs for the terminals range from $25,000 to $50,000, and call cost varies depending on the ground-station charges, but are in the vicinity of $5.00 per minute peak, and around 50% less for off-peak. There are 21,000 terminals in use, with 20 percent being land-based.

Inmarsat-B Inmarsat-B was put into service in 1994, and is the successor to Inmarsat-A. It is fully digital, and although the services offered are similar to those for Inmarsat-A, compression techniques and narrow-band operations make it more economical.

The terminals are similar in size and functionality to those of Inmarsat-A.

Inmarsat-C Inmarsat-C is a store-and-forward system for data. It can handle American Standard Code for Information Interchange (ASCII) and binary files, and is designed primarily for low-volume, low-budget data services. It operates in the L-band and has a bit rate of 600 bits/s. It is often bundled with an Internet service.

Messages, which are limited to 23 kbits in length, can be addressed to individual users or sent to a group of users.

The terminal is relatively small, with the electronic package weighing 3–4 kg, and operating with an omni-directional antenna.

Data costs are around 20 cents per 256 bits.

Inmarsat-M Inmarsat-M is Imarsat's answer to the LEOs. The phone is smaller, though at the time of writing it was still briefcase size, and but not cheaper (at around $6000). It supports data at 2400 bits/s, voice and fax. The power consumption of the unit in 1999 was 20–40 watts, and various models are available with weights ranging from 2.2 to 13.6 kg. In 1999, there were around 50,000 users worldwide.

It is planned that in the year 2000 a "mobile-phone"-size unit will be introduced. Connection rates are close to half that of the other Inmarsat products.

ICO Global Communications

The ICO Global system will consist of 12 satellites in an interconnected mode and a network of land–earth stations (LESs), and was due to be in service in 2000. The network is budgeted at $2.6 billion. In August of 1999, however, ICO, like Iridium, filed for bankruptcy. Ongoing planning was that the network rollout should continue. By July 2000, ICO was still operational but none of the new satellites had been successfully launched.

ICO is a MEO with two orbits at 10,355 km, which are inclined at 45 degrees to the equator. Each orbit has five active satellites and one spare. The satellite transceiver array has 164 antennas and can accommodate 4500 radio channels.

The downlink is at 2.2 GHz, while the uplink is at 2 GHz. The ground-station-to-satellite link is at 5 GHz and 7 Ghz.

There are 12 ground stations, all of which are networked.

PAGING SERVICES USING LEOs

The one thing that all the little LEOs seem to have in common is that they propose to use 24 satellites.

Starsys is proposing a system called Starnet, which will provide worldwide paging and positioning at 1 GHz. Using 24 satellites, messages will be restricted to 32 characters, but the mobiles are to cost a mere $100 each.

Orbcomm is offering an upmarket paging system with messages of any length, a global positioning system, and tracking service. Terminals will cost about $500 each. A conventional automobile AM/FM antenna will be used.

Leosat proposes a messaging, positioning, and fleet-management system. Short messages of up to 32 characters displayed on a light-emitting-diode (LED) screen will be available. The 24 satellites are proposed to be at 1000 km. The launch will cost $250,000 and the satellites cost $750,000 each. This system will have ground-based gateways.

Little LEOs will not in any way be providing services that could be considered competitive to cellular voice transmissions, as they are limited to short messages. However, because they are cheap, the chances are that they will account for much of the mobile data market that is only just starting to be addressed by digital cellular systems as a value-added service.

SATELLITE LIFETIMES

All satellites carry fuel for re-orientation. The amount of fuel that they carry and the total number of manoeuvers that they need to do to stay pointed in the right direction determines how long they will remain in a useful orbit.

Geosynchronous satellites nominally stay fixed relative to the ground, but in fact gravitational pulls by the sun, moon, and Earth combined distort the orbit. Even the rotation of the sun adds a new force for the satellite to experience. The process of keeping the satellite pointed the right way is called *stationkeeping*.

For geosynchronous satellites the fuel is used approximately in the following proportions: 1 percent for latitude excursions, 3 percent for east–west correction, and 96 percent for north–south corrections.

TECHNICAL ADVANCES

Phased Arrays

Antennas   Phased arrays have become the standard for LEOs such as Iridium, Globalstar, and MEOs like ICO–Global, as well as for general-purpose wide-band data services. These applications require multiple steerable beams, and the economics of an array become apparent as the number of beams required increases. The wide beamwidths that the antennas are required to steer through, also tend to favor arrays over complex mechanical arrangements to steer dishes. It is not unusual for a LEO antenna to need to scan angles of ±60 degrees from nadir (the direction of the line connecting the satellite to the earth's center).

Costs   The cost of a phased array for space applications is directly proportional to the number of elements in the array and varies from a few hundred to a few thousand dollars per element. A typical phased array may have about 1000 elements, and so the cost will be in the $100,000+ to $1,000,000+ range.

Multiple arrays can produce improvements of a few dB, but in 1999, the costs of this were generally regarded as prohibitive.

Path Loss   The array will have a service area that describes a cone around the nadir direction. The gain will decrease with increasing angle from the nadir and falls off at a rate of $\cos(\theta)^k$, where k is in the 1.2 to 1.5 range and theta is the angle from the nadir. For a LEO this factor typically will amount to about 4 dB. It is at the edge of the coverage that the rain effects are also greatest, and so the path gain is determined at this point.

Three factors contribute to path loss, namely altitude (and therefore distance), scan angle, and rain density. It is interesting to note that to some extent two of these factors operate to compensate each other. Increasing the altitude will increase the path loss, but it will also decrease the scan angle to any given point, except the point that is coincident with the nadir, and so will decrease the scan loss. This factor means that for any given constellation size (i.e., number of satellites) there will be an optimum height for minimum path loss.

Constellation Size From the preceding discussion it naturally follows that a larger constellation will have smaller scan angles, and thus have a path loss advantage. The gains, however, are small. For a 50-satellite LEO, doubling the number of satellites to 100 will only give a path loss advantage of about 3 dB in clear skys, and as little as 2 dB in rain. As might be expected for large constellations, very large increases in numbers are required to produce modest returns.

Hardness The technology of choice for space-based phased arrays is gallium arsenide (GaAs), which is used for the amplifiers and phase shifters. GaAs devices are inherently quite radiation tolerant, but the arrays' controllers are still silicon-based integrated circuits (ICs), which are not.

44

CORDLESS TELEPHONE TECHNOLOGIES

Cordless telephones evolved in the 1970s, and they were designed to work with a home telephone. The principle of operation was to use a small, low-power duplex radio link to connect to a base station, which in turn connected to the telephone. As these became popular, the shortage of frequencies became a problem and so did misuse. The signaling was Dual Tone Multifrequency (DTMF), and there was usually no security check that would prevent another user on the same frequency from making calls.

Later more frequencies were allocated to this service and phones were manufactured with built-in security codes, which prevented false (or illegal) line seizures. However, a shortage of frequencies still limited the use of these devices, particularly in apartment blocks and offices, often to as few as four users per area.

The multiuser cordless telephone concept, with frequency or timeslot sharing, was first mooted in 1985 when the Conference of European Posts and Telegraphs (CEPT) called for proposals for the implementation of wide-area cordless phones.

The proposals that prospered were a British proposal, which later became cordless telephony generation 2 (CT2), and one from Ericsson and Televerket [Swedish Post Telephone and Telegraph administration (PTT)], which became Digital European Cordless Telephone (DECT).

CORDLESS IS NOT CELLULAR

Systems like PHS and DECT have blurred the distinction between cordless technology and cellular but there is a subtle difference. Cordless technology is essentially simplex and relies on time division to separate the send and receive paths. Cellular systems on the other hand even if they use time division, also have frequency duplex spacing of 45 MHz or more between the send and receive.

In essence, it is generally true that simplex technology is simpler and cheaper to implement. The down side is that there are much tighter constraints on the range, as with simplex operations, intersymbol interference is the major constraint to achieving range. So simply increasing the TX power will not necessarily do much to increase the range.

A common factor in the modern cordless phone technologies is the concept of microcells, which have a small range (around 100–500 meters) and the capability of high-density frequency reuse. To achieve this, very low-powered base stations are used, and it is expected that most cells will be operating in the range of 0.01 to 0.1 watt. In order to reduce cost and size, the mobile units will also be relatively low-powered units.

Microcells are designed to serve slow-moving traffic and so have a slow handoff, which makes them simpler to build and less demanding on processor time.

Picocells are meant for the office and other high-density environments (with a range of less than 200 meters). They are designed to use adaptive channel allocation. This is done by scanning the channels in use in the area and allocating a free one. This means that the microcells are not part of the frequency planning exercise and that they can potentially use any frequency.

CT1

Cordless phone technology began with cordless telephony generation 1 (CT1), which was an 8-channel, radio-based phone for home use. The base stations use frequencies of 1.642–1.782 MHz to transmit with 20-kHz channel spacing, and 47.45625–47.54375 to receive with 12.5-kHz channel spacing. Under some conditions the base transmitter (TX) can be heard on broadcast receivers as an image response. The antenna is quite inefficient (being very small compared to a quarter wavelength). The voice quality from a CT1 is at best only fair.

The original CT1s had no security coding, and there were instances of people driving around high-density suburbs until they found a dial tone and then making unlimited free calls. These phones were also likely to charge almost anyone in a high-rise building rather than the actual originator of the call. Current CT1s all have an identification (ID) built in that prevents unauthorized use.

CT2 Public Systems

The CT2 system (basically a cordless phone with outgoing access only) was introduced with some haste into the UK in the 1980s, to mixed reception. The name CT2 signifies "second-generation cordless phone." Obviously, the lack of an incoming call facility led to the general conception that the CT2 system was second rate. Coupled with the relatively low cost of cellular terminal equipment, this concept led to low market penetration for CT2. The CT2 system as a network was based on small (about 200 meters) service areas with relatively low terminal equipment costs (around $250 each), as well as a low call charge rate (for the period) of around 18 cents per minute.

CT2 was the first operational digital cordless telephone system. It can hardly be regarded as a success story for the new breed of mobile communications. Originally introduced with much fanfare, CT2 has failed to stimulate the market. In the UK where the largest investments were made in this technology, today all of the four original operators have gone out of business.

By 2000 virtually all public CT2 systems had closed, but the story of CT2 is nonetheless interesting.

Security

A feature of CT2 is the elegant security procedures that virtually eliminate fraud. Only encrypted IDs are sent to air, so that off-air cloning is practically eliminated. A penalty is paid, however, in the complex registration procedures. CT2 has a "ZAP" facility that enables a stolen mobile to be temporarily or permanently disabled on registration.

Complaints

Users complained of difficulties finding base stations, poor coverage, and operational complexities of the handsets. In general, the operators located their terminals in busy areas, which also tended to have a lot of background noise. This, combined with the lightweight handsets, which had poor audio coupling and no volume control, made conversations very trying.

Too Many Operators the Problem?

By 1989 there were four licensed operators in the UK, each with their own communications protocol. This meant that the systems were incompatible and that users on one system could not use, or transfer to, another. Because of this the operators developed "common air interface" (CAI) so that the mobiles on the different systems would at least be compatible with each other.

The CT2 operators in the UK were all consortiums. This was because the technology was seen as a high-risk and high-cost venture, and while all the major players wanted to participate in the development, few were prepared to commit too heavily to this concept.

Operators were Mercury with Call Point, Ferranti with Zone Phone, and BYPS Rabbit, which waited for the CAI. Another operator was the UK's Phonepoint, which was a consortium of British Telecom, STC, Bell Atlantic (of the United States), French Telecom, and Germany's Bundespost. By the end of 1990 Phonepoint had 1000 bases and was operational in 12 cities. Many of the stations were mounted on top of British Telecom's public payphone boxes. At its peak in 1991, the total subscriber base was 7,400. Two years later, having spent $150 million, Phonepoint announced its withdrawal from the market. By the end of 1993 the market was left to Rabbit (which had a real knack for hiding their service points, see Figure 44.1), a Hutchison-led venture, which ceased operations in 1994. All of the original players had left the marketplace. The Rabbit system, which ironically was the only one to use the CAI standard, was the last to go. But what's the point of a CAI when there is only one system?

Other countries that tried and abandoned CT2 include Canada, Germany, France, Netherlands, Spain, Belgium, Finland, Spain, Singapore, Malaysia, Hong Kong, Australia, and New Zealand. Most of these trials were half-hearted, as the operators had seen the monumental failures elsewhere, but still insisted on repeating the exercise for themselves. Australia's trial was based in Brisbane, the country's third biggest city (approximately 1.5 million) but the one with the highest cellular penetration per capita. The cellular system in Brisbane, by most measures, is the most successful in Australia.

Figure 44.1 The Rabbit symbol marks the station location and antenna is usually not far away.

Telstra launched its service with 600 base stations. To successfully cover greater Brisbane, the sprawling built-up area of around 1000 square kilometers would need to be serviced. The 600 base stations, with an average range of 200 meters, would cover 75 square kilometers. With 7.5 percent coverage, it is not difficult to predict that the system will not offer any threat to cellular.

The marketing philosophy in Brisbane was not to attempt to cover the whole city, but rather to provide coverage at all of the commercial outlets of some major retailing organizations, including post offices, McDonalds, a few of the major retail chain stores, and a major petrol seller. Coverage was based on covering all of these easily recognized outlets, marking them with a sign like the one in Figure 44.2. This system also failed.

Singapore and Hong Kong were once the success stories. By mid-1993 both cities had around 20,000 customers, and the customer base was growing. These same cities had the highest per capita use of pagers in the world and high cellular telephone densities. Both cities are compact and densely populated, and the people seem to be captivated by electronic gadgets. However, these systems have also since closed.

Despite the demise of CT2 elsewhere in the world, Korea Telecom's CT2 network remained operational in

1999, and was being enhanced to enable limited two-way calling. However, it has 440,000 subscribers and the number is falling.

In Bangkok a new CT2 network consisting of 30,000 base stations has been installed by TelecomAsia, but the system, which was due to be turned on in late 1998, has not been turned on.

CT2 Networks

In Europe CT2 operated in the ultrahigh frequency (UHF) spectrum in the 861–865-MHz range. The band is divided into 40 channels, each of which is 100 kHz wide. The channels are frequency division multiplexed (FDM). However, since these frequencies are not universally available, not all CT2 systems operated on the same band. In the UK the frequency used is 864–868 MHz.

Figure 44.2 Sign in Brisbane indicating CT2 service.

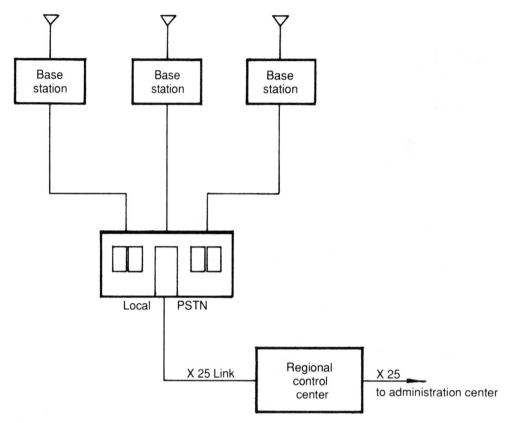

Figure 44.3 The structure of the telepoint network.

Although CT2 is digital, it is an frequency division multiple access (FDMA) system.

Methods are used on the voice level to improve the spectrum efficiency, and use is made of the high level of redundancy in normal speech. The speech is "packaged" and compressed into 1-ms timeslots in such a way that the send and receive timeslots are alternated. This eliminates the need for a duplexer and thus leads to a smaller and cheaper transceiver.

The data rate is a total of 75 kbits/s, which is subdivided into 2 by 32 kbits/s (one each for send and receive) plus overhead bits.

A base station consists of a maximum of six channels, but in high-density areas it is possible to co-site one or more bases, up to the full 40-channel capacity.

The base stations record user information, including the user ID, personal identification number (PIN), called party, date, time, and duration of the call. This information is later transferred to a centralized processor for billing. The network is shown in Figure 44.3.

The user was required to "log on" by entering a PIN. In the UK the users were prevented from "roaming" by the requirement to log on at each new base station. This restriction was regulatory and not a technological limitation, and is a prime example of just how silly regulators can be.

Modulation

The modulation uses binary frequency shift keying (BFSK) with a Gaussian shaping. The deviation is 14.4 and 25.2 kHz. The bit transitions are constrained to be phase-continuous.

The speech is digitized at 32 kbit/s, using adaptive differential pulse code modulation (ADPCM), which conforms to the Consultative Committee on International Telegraphy and Telephony (CCITT) Blue Book standard G721, but is transmitted at 64 kbits.

Mobiles

The power output of a mobile is only 10 milliwatts, which contributes to a lightweight mobile unit (a typical mobile weighs around 130 grams) and to very extensive frequency reuse possibilities.

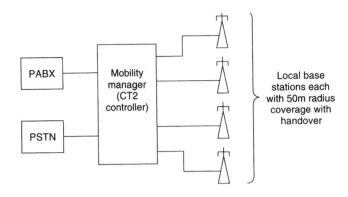

Figure 44.4 The simple structure of a CT2 network.

Cordless PABX

A mainstay of the market post-2000 for CT2 is in cordless private automatic branch exchange (PABX) applications. In an office environment, a cordless PABX can be implemented as shown in Figure 44.4. A local controller can handle from 10 to 80 base stations, each with a range of around 50 meters and with full handover and incoming-call capability. This is the essence of a cordless PABX. If desired, the CT2 phones can also be registered on the public network and/or used at home with a home base unit.

As an extension of the this concept, it is possible for a business with a number of offices to network their cordless phone systems so that a user can access (or be accessed by) the system from any of the offices. In this case a manual log-on procedure would be required.

A large proportion of CT2 phones sold are used in standalone cordless PABX applications.

Home Base Stations

The most popular application of CT2 is the home base station. The signal is much cleaner than the old CT1 and it is less prone to interference. Unlike other digital systems it seems to have a clean record with regard to electromagnetic interference (EMI).

I have used a CT2 phone (see Figure 44.5) as a cordless phone for about 8 years. The voice quality is good, but the range of around 20 meters leaves a lot to be desired (my house is 33 meters long). The talk time is at least 5 hours & the standby time several days.

Limitations

The real limitation of the CT2 concept for network applications lies in the expense and impracticality of pro-

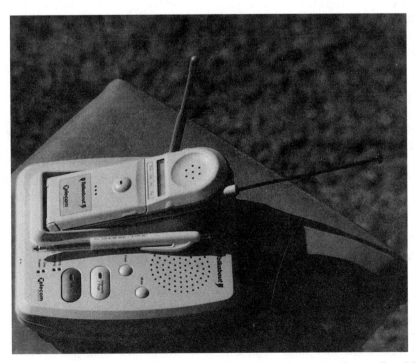

Figure 44.5 A Motorola CT2 phone circa 1993, sitting in a cradle that makes it a cordless PSTN phone.

viding a wide area of coverage. For example, a typical modern city could have an area of 600 square kilometers; to effectively cover this area (completely) with a CT2 system using 200-meter cells would require half a million bases.

A limited form of handover was available by 1992, but as CT2 was conceived as a device for pedestrians, the handover speed was limited to 20 km per hour. Therefore, it could be used in vehicles, even if the coverage were to be improved so that it was otherwise practical.

Marketing Philosophy

Most of the early attempts to market CT2 as a product competitive to cellular proved to be dismal failures. By 1993 the view of the market was that it was essentially a cordless **PABX** market with a home-based cordless phone facility, *supplemented* by a public switched telephone network (PSTN) interface capability.

CT2 Plus

CT2 Plus is an enhanced version of CT2 as proposed by Northern Telecom, Canada. The standard, known as protocol control information (PCI), is a nonproprietary evolution of the CT2 standard. It provides compatibility with the CT2 CAI interface, but with enhanced facilities through the introduction of common-channel signaling. The signaling format of the system is able to identify a standard CT2 phone and connect it accordingly, so that both standards will be supported. By using one or more of the designated channels for signaling it can provide handoff, frequency agility (which means that the equipment can hop in frequency between a number of disjoint bands of allocated spectrum), greater battery life, and location recording through registration. It can thus provide some level of incoming call capability, with the provision that registration has occurred. It also features low-speed data capability.

It initially operated the bands 864–868 MHz, 930–931 MHz, 940–941 MHz, and later on, 1.8 GHz. These bands have been selected to be compatible with the available spectrum in the United States.

DECT

DECT is a standard that is similar to the personal handy phone system (PHS), and that was meant to rival the CT2 standard. It was originally seen as the European cordless mobility product that was set to dominate the market in the way that Global System for Mobile Communications (GSM) has. As a cordless technology, it is meant mainly for cordless home and office (cordless PABX) use, but its role as wireless local loop (WLL) and a "marginal" alternative to cellular was overlooked.

The mobile in the DECT specification is responsible for channel selection and handoff. This greatly simplifies the base-station hardware and eliminates the need for frequency planning.

In 1997 Ericsson released a dual-mode GSM-DECT handset (GH337), which was essentially two radios (with two external antennas).

The dual-mode handset was used in a trial in China for a hybrid DECT/GSM network. While umbrella coverage was provided by GSM, local office, home, and some public areas were covered by DECT.

The study showed that under optimum conditions built-up areas needed 20–25 base stations, and in rural areas about 5 per square kilometer. The DECT mobiles could work with vehicle speeds of up to 50 kmph.

DECT therefore is a candidate for WLL applications.

DECT SPECIFICATIONS

Frequency	1880–1900
Traffic channels	12 channels/carrier
Radio frequency (RF) carriers	10
Access method	TDMA/time division duplex (TDD)
Voice channels	32 kbits/s
RF efficiency	144 kHz/voice channel
Transmission power	10 mW

Dynamic Channel Allocation

There are 10 carrier frequencies, and each carrier is further divided into 24 repeating timeslots using TDMA; the timeslots are duplexed so that 12 timeslots are used for reception and 12 for transmission. This form of duplexing is called time division duplexing (TDD), and so DECT systems are often referred to as TDMA/TDD systems.

Thus, DECT has 12 speech channels on each of the 10 carriers, for a total of 120 speech channels. Each mobile can access all of these channels and constantly monitors the air interface, selecting a better channel when appropriate. Because the handover is completed before the mobile leaves the service channel, the handover is seamless.

DECT transmits in 400-μs bursts, once every 10 ms, that is, at a frame rate of 100 Hz. The system does not attempt to cope with multiple reflections or the Doppler effect caused by moving vehicles, (which any truly mobile technology must).

Voice coding is ADPCM with encryption. It is specifically designed to interface with GSM and Integrated Services Digital Network (ISDN) networks. However, it would now seem that DECT applications have been limited by the success of GSM1800 and GSM1900, which perform much better in the public mobile arena.

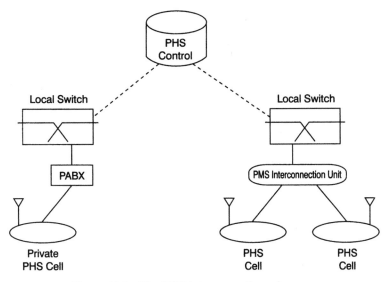

Figure 44.6 The PHS interconnection scheme.

PHS

PHS was initially turned on in Japan in early 1994, and was expected to be a big success in the country. One of the driving forces behind the PHS networks was the *artificially* high cost of cellular phones in Japan at the time. Originally, the enthusiasm for PHS was very high and three licenses were issued for each region in Japan, under which the licensees were obligated to achieve 50 percent of the total residential population covered by the service within 5 years.

The initial optimism was short-lived, however, and the capacity of the system peaked in 1998, with a falling share of the market.

PHS is essentially based on analog cordless phone technology, made digital by the addition of compression techniques and digital modulation at 32 kbits/s; in fact, it is much like CT2. As such, its power is limited to 10 mW, with a frequency of 1.9 GHz, which limits the range to about 100–150 meters. The low power means that a phone can provide over 200 hours of standby time and 5 hours of talk time.

As a consequence of its cordless phone heritage, PHS phones were always small in size compared to cellular phones, and in 1997 there was even a wrist phone.

A base station can be from 100 to 500 mW, and is typically mounted on a rooftop or utility pole. This gives a range of about 100 to 500 m. Essentially the system is designed for pedestrian use rather than for moving vehicles, although coverage to "slow-moving cars downtown" is part of the PHS specification.

Conceptually, PHS differed in some very significant ways from the cellular networks. In place of the fully self-contained infrastructure of cellular, PHS uses the existing wireline and even PABX networks, connected by a modified ISDN protocol (see Figure 44.6). Because of this reuse of the existing infrastructure, PHS was estimated to cost from one-half to one-third as much as cellular. Initial prices of PHS services were at about one-fifth the Japanese cellular prices. While this may seem to be a huge discount, nevertheless these PHS charges were compatible with cellular charges outside of Japan.

There are two network interfaces defined, the first being the interface between the public cell station (CS), called X1, and the second being the interface between the first point of interconnection to the point where the network becomes digital, called X2, as seen in Figure 44.7.

The PHS phone is intended to be networked while mobile and act as a stand-alone cordless phone in the home or office.

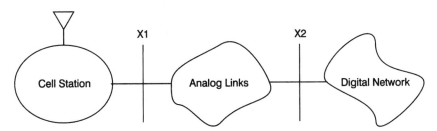

Figure 44.7 The defined interface points between the CS and the PSTN.

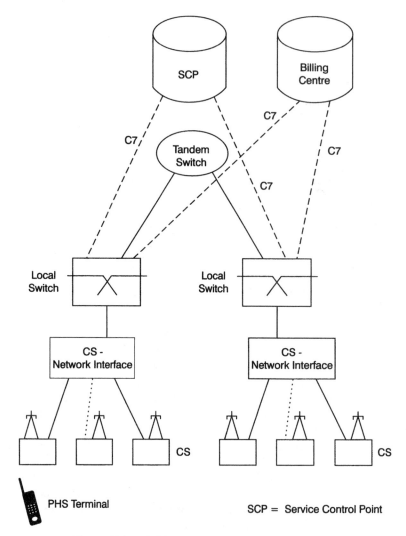

Figure 44.8   PHS integrated into the PSTN network.

There are two different network configurations for PHS. It can be either part of the wireline network, or it can be operated in a stand-alone configuration. In the PSTN configuration it is linked by the SS7 network, as shown in Figure 44.8.

At the local switch level, paging, call information gathering, authentication, and operatives and maintenance (O&M) takes place. The service control point (SCP) centralizes the location registration, authentication, and O&M.

The stand-alone system works similarly, but is a little simpler. A PHS adapter takes over from the local exchange switch and there is only a gateway interconnection to the PSTN, as seen in Figure 44.9.

The configuration of the local system is seen in a bit more detail in Figure 44.10. The PHS server provides paging, authentication, and handover, and the connection to the PSTN using SS7 or some other signaling format such as MFC R2.

PHS supports the following;

Authentication
Location registration
Handover
Roaming

In Japan a total of 9.6 MHz was initially allocated for PHS services. Frequency allocations include a spectrum reserved for both private and public use. Channel 1 starts at 1895.150 MHz, and then channels are allocated every 300 kHz up to channel 77 at 1917.950 MHz.

Modulation is MC/TDMA/TDD (multicarrier, time division multiplexing, time division duplexing quadrature phase-shift keying) (QPSK). There are 77 RF carriers with a frame rate of 200 per second and eight time slots per frame (four each way).

The PHS systems uses SS7 signaling and ISDN for interconnection. Like the CT2, the handover capa-

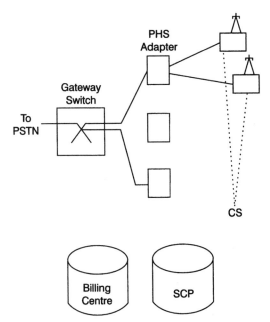

Figure 44.9 The stand-alone PHS network.

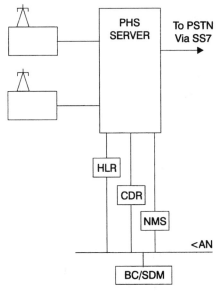

HLR = Home Location Register
CDR = Call Detail Recording
NMS = Network Management System
BC/SDM = Billing Centre / Subscriber Data Management

Figure 44.10 The detail of the PHS system configuration.

bility is limited and it cannot be used in fast-moving vehicles.

Two of the eight timeslots are used for control channels, leaving three duplex channels per carrier. The system is distributed with each base station managing its available channel resources.

Currently in Japan, PHS is seen as the system of choice for data users, as it can offer a 28.8-kbits/s packet data rate, and to counter this Japan's looks likely to be the first country to implement IS-95B, with its 64 kbits/s data rate.

The PHS system, which was an early success story in Japan, began to flounder in 1998, with the subscriber base shrinking for the first time. From a high of over 7 million subscribers, the base in June 1999 had fallen to 5.7 million. PHS has the (perhaps temporary) advantage of higher data rates (32 kbits/s) than cellular, and as the subscriber base falls, PHS operators in Japan are trying hard to offer ever better deals to their existing subscriber base and to push the data capability.

PHS SPECIFCATIONS

Frequency	1895–1918 MHz
RF carriers	77
Channels	4 per carrier
Carrier spacing	300 kHz
Access mode	TDMA/TDD
Voice mode	32 kbits/s ADPCM
RF efficiency	75 kHz/voice channel
Transmission power of CS	80–800 mW (peak) or 10–100 mW (average)
Handset power	80 mW (peak) or 10 mW (average)

The CS is very small, weighing less than 2 kg with a volume of less than 2 liters and a power consumption of about 10 watts. Although the base stations are small, however, a large number of them are needed and according to one Japanese Telecommunications Laboratory (NTT) estimate the number of CS sites would be as shown in Figure 44.11. The original PHS rollout plans for Japan called for 250,000 cells nationwide by the year 2000.

The CS comes in modules of 4 channels. Thus, the first CS has a 3-voice channel capacity, and each additional CS adds 4 voice channels.

Of the 23 MHz assigned for PHS by the Ministry of Posts and Telecommunications (MPT) of Japan, 12 MHz was assigned for public use, while 11 MHz was assigned for shared private and public use.

Because each base station dynamically assigns channels from the entire available spectrum (choosing one that is not currently in use), there is no need for frequency planning.

While PHS is clearly losing the mobile market battle in Japan, it is now seen perhaps somewhat as a case of wishful thinking, as a frontrunner for the WLL market. WLL was originally seen as a potential for PHS, but mainly outside of Japan. That is still the way it is marketed.

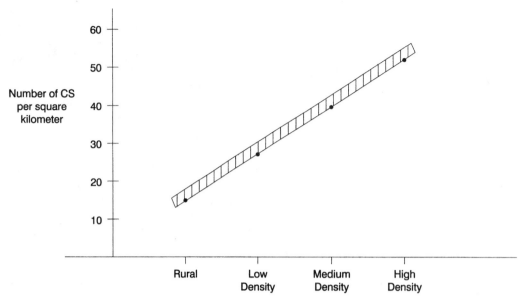

Figure 44.11 The CS cell density.

Dual-mode PHS-GSM handsets have been made by a number of manufacturers.

PACS

Personal access communications systems (PACS) is an enhanced cordless phone technology capable of 32 kbits/s data, and is expandable to 64 kbits/s. This is essentially the United States's answer to PHS and DECT/CT2. It is designed to work in the (PCS) bands.

In most instances the primary market for PACS is seen to be in the enhanced WLL environment, working alongside rather than directly competing with cellular and wireline.

The PACS structure is shown in Figure 44.12.

PACS SPECIFICATIONS

Frequency	1850–1910 MHz (up)
	1930–1990 MHz (down)
Channels	200 carriers in multiples of 8 channels
Modulation	time division multiple access/ frequency division duplex (TDMA/FDD)
Compression/decompression (CODEC)	32 kbits/s ADPCM
Channel bandwidth	75 kHz
Base-station power	100 mW
Portable power	25–200 mW
Typical coverage zone	300–500 meters
Mobility	up to 65 mph

HOME BASE STATIONS

There are a number of products on the market that allow the user to use a regular mobile phone from the household wireline connection. Ordinarily, number portability applies, so that the user can have the same number while roaming as used at home. The home base station is simply a low-powered single-channel base site that calls the mobile for an incoming call, and accepts outgoing calls.

Another variant of this is the PABX, which allows the mobile phone to be used in the office environment in the same way.

CORDLESS GSM

Cordless Telephone System

A specification for a cordless telephone system based on GSM is actively being pursued by the European Telecommunications Standards Institute (ETSI), which is scheduled to release the standard 1999. The concept has met with mixed reactions from the major vendors, some of which are enthusiastic, while many are cautiously awaiting the final specification details.

ATM via GSM

A consortium comprising Alcatel, Gemplus, and Mondex International has produced a GSM interface to the Mondex automatic teller machine (ATM) device that also allows voice calls.

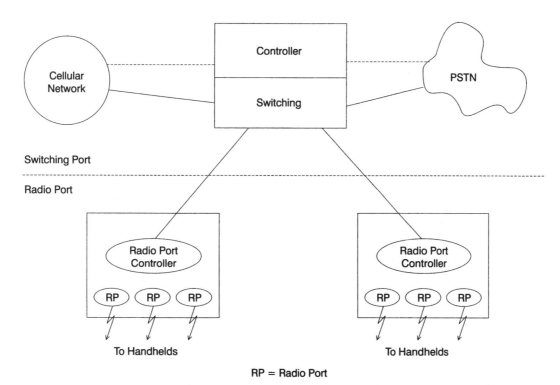

Figure 44.12 A PACS network.

HOMEBASE

In 1998 a cordless telephone system (CTS) GSM phone known as *HomeBase* was introduced. HomeBase is an open system and allows the user to attach a HomeBase to the regular house PSTN line and then receive calls from it in much the same manner as a cordless phone. The range of the system is about 200 meters.

Although the subscriber can maintain the two separate telephone numbers, calls to the subscriber from the GSM network will be routed via the HomeBase (and hence the PSTN) whenever the subscriber is within range, or via the regular GSM air interface otherwise.

HomeBase can support multiple users each with their own number.

HomeBase does not have a special frequency allocation, but uses an adaptive frequency allocation (AFA) and dynamic channel selection (DCS) algorithms to find a "free" frequency within the band of the operator. Therefore HomeBase does not require any frequency planning.

THE FUTURE FOR CORDLESS

What I have classified here as cordless technology, would seem to have proved itself consistently unsuitable for use as either a cellular alternative or a cellular supplement. Despite this, cordless technology flourishes and it still has a place as a mobility device for landline and PABX extensions. But does it have a future, in either this role or as a cellular supplement?

Much of the original arguments for cordless "cellular" were based on the high subscriber densities that cordless could deliver and the relatively low cost of the technology. Today cellular volumes are such that the price issue is no longer of any consequence, and it has been proven worldwide that while the market is content with voice-only services, there is no spectrum shortage; in fact it may be fair to say that there is an over supply.

G3 technology is all about high gross data rates over otherwise conventional cellular technology. It can be expected that no matter how high the gross rate can be pushed the throughput of cellular data using conventional technology will always be sluggish. Bluetooth and other related technologies are attempts to address this limitation for data.

It may well be that in the future there will be a serious market for data over mobile, but if so, it will need a new approach. It is possible (although not probable) in the next few decades that high GHz technologies will solve the mobile data problem. It seems almost certain that there will be attempts to use it (as indeed in the year 2000 a lot of WLL loops systems operating at around 30

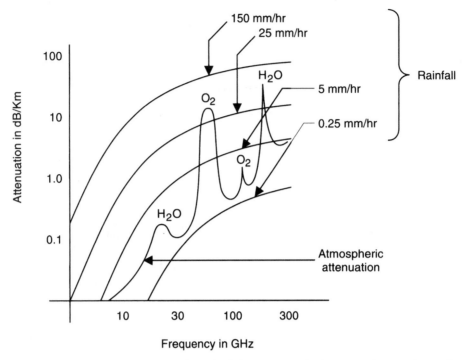

Figure 44.13 The attenuation in dB/km of millimeter waves in the atmosphere and at various rainfall levels.

GHz are being used as high-speed data links). If this spectrum is to be exploited for mobile applications, it can be expected to be used, first for cordless technology and later grafted into a supplementary role in cellular services. The following text provides an overview of that technology.

THE POTENTIAL OF MILLIMETER-WAVE CORDLESS

Millimeter-Wave Potentials

Millimeter waves are frequencies above 30 GHz, so-called because their wavelength is less than 1 centimeter and so will be measured in millimeters. These wavelengths are largely unexploited, mainly because hardware at those frequencies is not readily commercially available. A good deal of research is being undertaken in this band, particularly with regard to future point-to-point and point-to-multipoint applications.

The attenuation at these frequencies is very significant and is mainly due to interaction with molecules of oxygen and water vapor. As can be seen in Figure 44.13 there are absorption lines for water at 22 and 183 GHz, while the oxygen lines are at 60 and 119 GHz. It has

been suggested that these high attenuation points, far from limiting the potential of millimeter waves, in fact enhance the possibilities for frequency reuse.

Rain attenuation has been fairly extensively studied and from Figure 44.13 it can be seen that, as would be expected, the attenuation is strongly dependent on the rainfall rate. Even the light drizzle (0.25 mm/hour) produces an attenuation comparable with that of the atmosphere. In heavy rain of 150 mm/hr even at 30 GHz the attenuation is in excess of 20 dB/km. This means that occasional outages in heavy rain would have to be expected for any millimetric hops of more than a few kilometers.

As well as attenuation, the atmospheric absorption produces a frequency-dependent dispersion, which will result in pulse distortion for data systems operating with pulse durations shorter than 2 nanoseconds. This is particularly a problem for frequencies near the 60-GHz oxygen absorption band.

THE TECHNOLOGY

At present the technology is best described as discrete-component based. Hybrid silicon gallium arsenide (GaAs) devices are the norm. Vacuum tubes still offer the highest powers, but they are very expensive.

The plumbing at these frequencies is high-precision microengineering, and, for example, a 100-GHz waveguide needs to be accurately tooled to dimensions of around 1 mm × 2 mm. High-performance devices operating at a quarter wavelength can have dimensions around 0.25 mm.

Despite these problems RF amplifiers and mixers are available commercially with noise figures as low as 4 dB at 60 GHz. Radio telescopes can use liquid-nitrogen-cooled devices with noise figures around 2 dB, and where money is no object, superconducting devices with noise figures of around 1 dB are beginning to appear.

Practical applications research to date is centered on the bands below 70 GHz. Some work is being done on the development of integrated circuits (ICs) in these bands. For frequencies much above 70 GHz, an almost total lack of hardware rules out early commercial exploitation.

By going to very high frequencies, where true ray-like propagation occurs, small cells can be contained. Of course, the opposite problem (unpredictable "dead spots") could well mean that a mix of the two techniques may be necessary. High-density coverage would be provided by very high-frequency cells with second-choice lower-frequency umbrella cells provided to overcome the dead spots.

It has been suggested that the optimal frequency range for high-density systems is the 59–66-GHz band, where atmospheric attention peaks at around 15 dB/km and refraction and diffraction effects are minimal. Such frequencies would be ideal for large office buildings and other high-density areas.

At least 2 GHz of spectrum could be made available. With existing channel frequency densities, this amount of spectrum would accommodate 80,000 channels. Of course, digital systems can achieve satisfactory voice performance over channels much narrower than this. These channels could be reused 0.5 km away without any real prospect of interference.

Additional interference protection is provided by the inability of these frequencies to pass through obstacles, which at lower frequencies are virtually transparent. As little as 3 cm of wood, concrete, or brick would form an impenetrable barrier. One of the most difficult problems at this frequency is that the human body is similarly opaque. Therefore, diversity transmission is needed to avoid the situation where the mobile transceiver is carried on a person in such a way that the person's body is between the base and the transceiver. Because of these difficulties, the use of umbrella cells at lower frequencies would be essential.

Transceivers operating at 60-GHz frequencies present technological problems which have yet to be solved. The antenna could be a phased array and the array would only be a few centimeters across (1 wavelength = 0.5 cm). Diversity could be provided by placing a second array, 10 wavelengths away (5 cms). Steerable arrays are used today, but the cost is prohibitively high. A specialized monolithic integrated circuit would need to be developed to perform this function.

HIGH GHZ BASE-STATION ANTENNAS

Omni-directional antennas are difficult to make and not really practical at these frequencies for reasons previously discussed. At frequencies of 60 GHz, normal leaky feeders are very lossy; and some new types of transmission medium will be required. A type of transmission known as "image line," with characteristics similar to an optical fiber, appears most promising.

INTEGRATED PCN SYSTEM—THE FUTURE

Figure 44.14 shows a typical GHz+ installation. Naturally, the usable range of a system operating at 60 GHz would be very limited, so any practical application would have to work in conjunction with some type of overlaying digital cellular system. The natural extension of this concept is to have an additional satellite overlay to provide almost universal coverage.

The network would be always programmed to select the highest-density option, so that the basic function of high-penetration components is mainly to fill in areas that otherwise would be difficult to cover. Handoffs between systems would greatly enhance the applications and traffic capacity available. Handoffs would require the integration of three technologies into a single handset.

Windows may be able to "radiate" in the future, using a technique that is opposite to the technique for preventing radiation from microwave-oven windows. Future buildings will need to be "wired" for the new technologies, and leaky cable will become a standard part of construction.

A certain block of spectrum could be reserved for satellite operation, so that the mobile has three choices:

- 60 GHz: local (1st choice)
- 1.5–10 GHz: umbrella (2nd choice)
- 2–10 GHz: satellite (3rd choice)

This segregation of the spectrum could ensure worldwide coverage and is probably indicative of cellular radio around the year 2010.

Satellites will always have their limits and brute power in space is only a partial answer. Although high-power

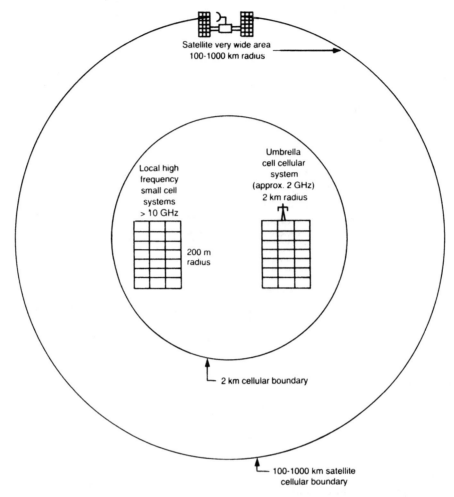

Figure 44.14 The ultimate PCN. Local coverage using GHz+ frequencies can be complemented by digital cellular radio that, in turn, is complemented by satellite coverage. This blend will be the ultimate PCN (once handoffs between systems become possible).

satellites with enormous gain antennas are practical, the high power will limit the ability to reuse frequencies. Conversely, high-gain spot beams will enhance frequency reuse and dynamic assignment of spot beams is promising.

Even at lower frequencies like 18 GHz it will be necessary to develop the ability to use arrays of directional antennas for both the base stations and the mobile transceivers. By exploiting digital techniques and the potentially wide bandwidths available at high GHz frequencies, a wide range of applications for data as well as voice transmission can be made available in these future networks.

The most difficult problems to be resolved in this area relate to the development of reasonably sized and priced

semiconductors that can operate in the 10-GHz+ frequency range. The new techniques, such as bipolar heterojunction transistors (which exploit the quantum mechanical behavior of electrons) and the older GaAs technologies, which can now offer noise figures below 1 dB, are promising.

MOBILE UNITS

The mobiles would operate on a low band for cellular and satellite access (about 2–15 GHz) and on a high band (60 GHz) for the immediate vicinity and with automatic switching as in a conventional handoff.

45

iDEN

Integrated dispatch enhanced network (iDEN) is a system that is both cellular and trunked, and was developed by Motorola from largely Global System for Mobile Communications (GSM) concepts. It is marketed in competition with both cellular and trunked radio.

Formerly known as Motorola Integrated Radio System (MIRS), iDEN is a digital narrow-band time division multiple access (TDMA) (squeezing 6 channels into 25 kHz) system, using a multilevel linear modulation scheme called *M-16QAM*. This particular modulation method has proven quite robust, and has the ability to tolerate severe differential path delays, without needing an adaptive equalizer for timing corrections. Adaptive equalizers are used in GSM and IS-54 to reduce intersymbol interference. However, when iDEN was under development, these equalizers added significant cost complexity and power consumption to the mobile, and for these reasons it was avoided.

To avoid using an adaptive equalizer, the iDEN technology sends the signal out over several 25-kHz channels at a relatively slow symbol rate, which means that the spread caused by multipath does not cause serious intersymbol interference spreading.

The spectral efficiency of iDEN comes from its modulation method. It should be noted that since most modern trunk systems operate at 12.5 kHz per channel, the actual capacity gain over analog is only three times. However, this is still significant, given that frequency reuse is very limited in trunked systems.

TRANSITIONING FROM ANALOG

The original concept was that analog operators in the 800-mHz band would migrate gracefully from analog to iDEN, recovering a few channels at a time (and gaining six digital channels per analog channel). It is claimed that no guard band between the digital and analog channels is required.

Its Motorola proprietary system and mobiles cost about $600.

HYBRID CELLULAR/TRUNK?

iDEN has all the features of a trunked radio system plus most of the features of cellular. Overall it is more cellular-like than trunklike. Facilities provided include:

- Short alphanumeric messaging
- Paging capabilities
- Automatic redial
- On-hook dialing
- Call forwarding
- Voice mail
- Busy transfer
- Call restriction

While all of these features are available on other systems (such as MPT 1327), it does seem that the iDEN system is effectively trunked radio that was envisaged as being directly competitive with cellular. This is particularly evident in the size and capabilities of the switch.

The implementation of iDEN leaves no doubt as to its origins (which were GSM). The system was initially offered based around Northern Telecom's DMS-100 central office switch. The basic structure of iDEN also looks a lot like GSM, as can be seen in Figure 45.1

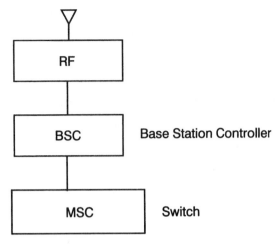

Figure 45.1   The structure of iDEN.

The essential switching structure of GSM has been retained, while the achilles heel of GSM, the TDMA modulator, has been replaced by what seems to be a more mobile friendly 16 quadrature amplitude modulation (QAM) system.

However, a new complexity has been added. Dispatch calls are not routed in the same way as public switched telephone network (PSTN) calls. A separate packet-switched network is overlayed onto the system to provide a more efficient way of handling short-duration calls. While they use the same radio frequency (RF) resources (channels, antennas, etc.), they are switched by an independent packet switch.

DESIGN

The design concept of iDEN is somewhat like that of cellular. The thrust seems to be to use a large number of smaller cells to enhance frequency reuse, and so increase

capacity. While this can be achieved, it is at the cost of economical wide-area group calling. The more sites that are used to cover an area, the more channels there are that will be tied up in a group call.

The small cell approach also increases the switching complexity (and hence the switch cost), because site-to-site handoff of calls is allowed in iDEN. The handoff facility (which operates like a cellular handoff) is very demanding of switch processing time.

PERFORMANCE

In the United States the major user of iDEN is Nextel, which has been rolled out nationwide. Service began in California, with mobiles costing around $850. Monthly fees are $27 for a basic trunked service that includes 75 minutes free airtime. Airtime is calculated in seconds of talk time, and measured only when the Press to Talk (PTT) is pressed. Telephone access is available at $40 per month. The other main user is Onecomm, which offers service in Colorado.

A number of systems are in operation in the United States and there are quite a number around the world.

Japan Shared Mobile Radio (JSMR), which as one of the main analog trunk radio operators has over 200,000 customers, operates an iDEN system in Japan in Joint Venture with Motorola. This system was initially delayed as permission is needed from the PTT before the paging and "cellular alternative" services can be offered. Service is charged at 3700 yen (approx. $45) per month.

Another major system outside the United States is in Beijing, where it is operated as CATCH.

The performance of iDEN, in terms of coverage, is claimed to be better than that of 800-mHz analog by about 30 percent. Voice quality subjectively is claimed to be consistently better than IS-54 TDMA. Such claims of

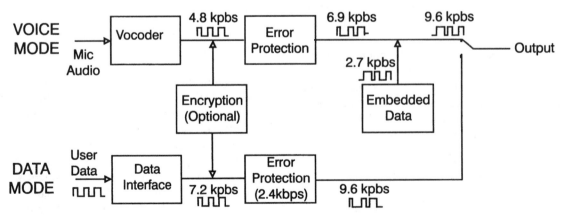

Figure 45.2   The digital channel structure.

Data:
7.2 KB/s Data by replacing voice segments
2.4 KB/s Error Correcting Code
    EXAMPLE:
        File Transfer or FAX

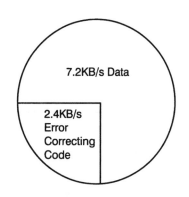

Voice & Data:
4.8 KB/s Voices
2.1 KB/s Voices ECC
1.6 KB/s System Signalling
0.8 KB/s Encrypt Sync
0.3 Kb/s Low Speed Data
    EXAMPLE:
        Alpha Messages "under" voice

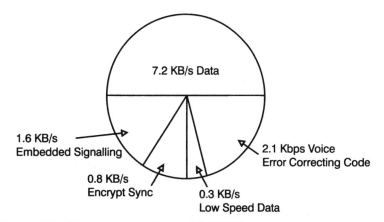

Figure 45.3 The two data modes used in iDEN.

more range and better voice quality for digital systems are often made, but they are rarely justified; users report less than agreeable voice quality.

The modulation method is quite different from TDMA, although the vector sum excited linear predictive coding (VSELP) compression/decompression (CODEC) operating at 7.2 kbits is similar to that used in the IS-54 standard. Error correction accounts for 2.4 kbits/s, while the digitized voice data stream is 4.8 kbits, as shown in Figure 45.2.

The voice quality is fairly consistent within the coverage area, and like most digital systems the quality degrades rapidly at the boundaries of coverage, and the link fails soon after.

DATA

iDEN provides for two data modes, the first is effectively equivalent to using a voice channel for data, while the second permits slow-speed data to be carried on top of a voice signal. These two modes are seen in Figure 45.3.

MOBILES

The first handhelds weighed 450 grams. They feature paging and data communications through a built-in RS232 port.

46

WIRELESS LOCAL LOOP

WIRELESS LOCAL LOOP, INCLUDING LMDS AND PTMP

Wireless local loop (WLL) is generically defined as a radio path for the last mile. It is the last link in the phone distribution network, connecting the subscriber to the public switched telephone network (PSTN) (a function that has traditionally been accomplished by copper cables). Local loop, however, is a term used in the wireline world to mean the connection between the wireline switch (or a node off it) and the subscriber.

WLL

There are many names that this technology referred to here as WLL, is known as. In the United Kingdom, it is called radio fixed access (RFA), it is also known as fixed radio access (FRA), radio in the local loop (RLL), digital radio concentrator (DRCS), wireless local loop (WiLL), local multipoint distribution systems (LMDS), point-to-multipoint (PMP), local multipoint communications services (LMCS), and many other names.

While there is no agreement on names, there appears to be even less on the technologies used, including the frequencies, bandwidth, modulation type, and even on what the real purpose is. While everyone agrees that WLL is meant to carry POTS traffic (voice), there is much more uncertainty about data (and its rate), cable TV, Internet, and other services.

While this uncertainty exists, there can be no standardization and, and this means that the costs will re-

main high, and probably that the acceptance will stay low. Although cellular offerings fall well short of the PSTN in monetary value from a customer's perspective, currently (early 2000), there seems little doubt that the lack of standardization in the WLL industry will lead to the dominance of cellular technology for WLL in the first decade of the new century.

Currently WLL offerings can be divided into four groups

- Cordless phone derivatives
- Cellular derivatives
- Proprietary systems
- High capacity systems (often video based).

The last category includes the Nortel Proximity, Tadiran Multigain, Lucent Airloop, DCS, Airspan, and Phoenix. The first three systems named here operate in the 3.4–3.6 GHz bands.

In the near future it can be expected that the cellular WLL systems will survive because of their cost effectiveness. High capacity WLL systems will have a place where data rate is important (assuming that 3G cellular will be slow in offering cost-effective significant data rate improvements).

Cordless phone systems fail dismally when coverage is an issue, and proprietary systems are unlikely to be cost effective. While cordless phones often quote ranges up to 5 km, these are rather optimistic and assume ideal line-of-sight conditions. The same phones have a range of only around 200 meters when operated in the mobile mode, and the range can be compromised in the WLL mode by foliage, poor maintenance of antennas and cables. New obstructions (like new buildings) can limit the range in practice to much less than the "optimum."

WLL is seen as a way for the new players in telecommunications to get direct access to the consumer, quickly and efficiently. In most developing countries the dominant carrier (often government owned) has a long established wireline network, which potentially results in a stranglehold on the end-user service.

WLL is playing a role in two ways. In developed countries the WLL solution addresses both the voice market (often with narrow-band products) and the wide-band data market with video bandwidths. In response to this the wireline carriers are upgrading their networks with fiber optics, offering potentially unlimited bandwidth.

Narrow-band WLL has worldwide applications, but is primarily seen as providing service in developing countries, with Asia being the dominant market in the next decade. Countries that currently have low telephone densities are experiencing a massive growth in demand and WLL is seen as an important part of the answer.

In poor countries, there is still a significant problem with the theft of copper wires (which is exacerbated by the fact that most of the copper is above ground, on poles, and highly visible), making the maintenance of traditional wireline services even more difficult. It is doubtful if this WLL application is realistic as conventional cellular is establishing a firm hold even in the poorest of countries.

DEFINING LOCAL LOOP

The term *local loop* is derived from the way that early telephone connections where established. A plain old telephone (POT) was traditionally hardwired to the telephone switch (or exchange), and contact was established in one of two ways. An incoming call generated a low-frequency tone that activated the bells via a capacitor. When the subscriber answered, the act of lifting the receiver, looped the circuit to the exchange and caused a direct current (DC) path to exist.

Similarly, to initiate a call the subscriber picked up the receiver, which activated a switch, causing a DC path to exist (hence, the term looping the line), and this was used to indicate to the exchange that the dial tone should be sent. Pulse dialing is also accomplished by making and breaking that loop.

In essence what a WLL does is to simulate the copper connection to a telephone. Most WLL connections will not literally "loop" the line, but if provision is made to connect a regular telephone, or some other device that "looks" like a regular phone (such as a fax), then a wired connection must be simulated.

REVENUES

While the costs of providing a wireless line today may be much the same as providing a wireline, there is still a discrepancy in revenues. For example in the United States, cellular providers can expect to earn about 15 cents per minute, while wireline returns only 2–5 cents per minute.

The advent of the personal communications service (PCS) and its attendant surplus of capacity has driven the prices down both in the United States and Europe. Sprint PCS, for example, in early 1999 was offering a subscription rate of $29.99 for 120 minutes of airtime per month, which includes free long-distance calls within the United States. This breaks down to 25 cents per minute, while for high-volume users who take the 1500 minute plan, the cost is just 10 cents per minute. Similar plans are available in other countries with extensive PCS. For example, Bouygues Telecom, a DCS1800 operator in France, offers calls (including nationwide at no additional cost) for around 1 French franc (approximately 13 cents) per minute on its 240-minute/month plan.

Some Asian countries offer calls for as little as 5 cents per minute (notably in India and Bangladesh), but these prices have been widely regarded as costing aberrations, a situation that is largely confirmed by the red ink on the balance sheets of those companies. In 1999 some upward correction in this pricing was underway.

Thus, while there is evidence that the revenue discrepancies are getting smaller, there is still no doubt that even the cellular "bargains," will generally return higher revenues per minute than wireline. WLL has to face up to the fact that not only will subscribers demand wireline quality and reliability, but that they will also be reluctant to pay too high a premium for the service.

WLL THE CHALLENGE

When confronting wireline directly in the central business district (CBD), WLL has many disadvantages. In 1999, WLL cost between $500 and $850 per subscriber, while wireline service was priced at $600 to $780. Also, while copper loops have virtually no associated maintenance costs, a wireless terminal can be expected to have a service life of less than five years. However, if WLL cannot easily compete on cost, bandwidth, data integrity, or capacity, there are niches for it outside the CBD.

On the other hand, when WLL vendors quote a cost

per line they are usually reluctant to give specific values for customer traffic/subscriber and design grade of service (GOS). Without these qualifiers, the cost per line of a wireless service means little. This is particularly true today when the Internet is pushing up the average call holding time.

Making it still harder to assess the true cost of the WLL solution is the cost of licenses, spectrum, and subscriber equipment, which because of individual preferences and requirements, is not necessarily a fixed cost.

At the other extreme, there are low-density rural areas like outback Australia where the automation program of the 1970s and 1980s saw extensive use of WLL, mostly in the form of the digital radio concentrator systems (DRCS). When, toward the end of the program (when the really difficult subscribers were being connected), the cost rose to around $10,000 per subscriber, wireline in most cases was not a viable option.

The real opportunities for WLL today are between these extremes, and are still mostly outside the CBD, in the low- and medium-density areas. This is especially true in developing countries, where wireline has yet to make an impact and WLL can be a way to establish a service quickly and economically.

BANDWIDTH

Because wireless costs have fallen dramatically over the last decade, it might be reasonable to assume that the competitiveness of WLL is now relatively high. It would be higher still if the objective was mainly to replace the POT. It is perhaps the advent of the Internet that has been the biggest impediment to WLL. The Internet is simply too demanding on data rates and hence bandwidth.

As the Net moves from a message service to a multimedia entertainment circus, the graphic images increasingly demand higher data rates. For WLL high data rates translate directly into bandwidth, and bandwidth today is expensive.

The problem of bandwidth is being addressed by all players. The most permanent solution is fiber optics, but wide-band services are at the heart of all modern telecommunications development, including 2G+ and 3G cellular, broad-band microwave, and enhanced techniques to get more data onto the existing copper-wired networks.

The Cost of Bandwidth

The bandwidth (B) that is required to transmit any giving bit rate (C) is mathematically determined by Claude Shannon's relationship

$$C = B \times \log_2(1 + S/N)$$

where

$$\log_2 = \text{logarithm to the base 2}$$
$$S/N = \text{signal to noise ratio}$$

Thus, for the WLL case, there will be a one-to-one relationship between data rate and bandwidth. Making matters worse in any frequency reuse environment, other users will appear as noise, thus decreasing the value of S/N, and so further limiting the bit rate per Hz of bandwidth.

However, this equation also implies that in areas of low-frequency reuse (and hence high values of S/N) it will be possible to either increase the data rate for a given bandwidth or decrease the bandwidth for any given data rate. WLL systems should be designed to exploit this advantage in an asynchronous way, that is, if the interference in either the send or receive direction is low, then increase the data rate, or alternatively, if an increased data rate is not required (for example, in a voice transmission), then use less bandwidth. This should be done dynamically for maximum effect, and code division multiple access (CDMA) exploits this well.

Today it is widely becoming accepted that spectrum is a resource, and a valuable one at that. As such, governments are seeing spectrum as a source of revenue and are auctioning it off. It is not cheap today and it is likely to get even more expensive in the future (although some of the ridiculous prices paid in the United States for PCS spectrum that ultimately bankrupted many bidders may have a sobering effect).

WIRED LOCAL LOOP

WLL must compete head-on with wireline services, and advances in wireline mean that the task is not only a difficult one, but is made even more difficult as wireline advances.

At the time of writing, most data (Internet) interconnect connections are 33 kbit/s with some 56 kbits/s available; 56 kbits/s is the benchmark that WLL technologies need to achieve to be able to transparently take the place of wireline for applications like fax and data.

Over the analog wireline network, a modem is needed to interface digital data to the transmission line. A modem works by converting the digital data stream into audible signals, which are sent to line as audio. The bit rate of the conventional modem is limited by the audio bandwidth of the copper pair, which has traditionally been considered to be 4 kHz. The 4-kHz bandwidth is necessitated mainly by the wide use of loading coils, which are placed in the lines to improve the frequency characteristics at voice frequencies.

Wireline connections typically have a bit error rate (BER) of around one part in 10 billion, which is far better than can be expected over WLL.

The term xDSL is used to describe the various technologies that offer digital subscriber lines over copper cable. One of the most promising of these technologies is asynchronous digital subscribers line (ADSL), which offers in the range of 1.9 to 9 M bits/s downstream and 16 to 640 K bits/s upstream, plus a POT service. The high downstream rate takes account of the fact that the main users will be accessing the Internet, and will spend more time downloading than they do sending.

xDSL

xDSL technology will only work over twisted pairs for short distances. As a result, for general purpose applications, it is will be necessary to use optical fiber for the local distribution from the PSTN switch to a pillar that is located not more than 500 meters from the twisted pair termination point. This significantly increases the roll-out cost of the technology and begs the question of full implementation of fiber optics, to the subscriber end.

Other technologies include high-bit-rate digital subscriber line (HDSL), symmetric digital subscriber line (SDSL), and very-high-bit-rate digital subscribers line (VDSL). Table 46.1 shows the relative capacities of these technologies.

HDSL evolved from an extension of the Integrated Services Digital Network (ISDN) to deliver a T1 bandwidth at the customers premises. HDSL operates at 784 kbits/s, full duplex on each cable pair; two cable pairs give a full 1.544 Mbits/s.

Note that in some cases multiple twisted pairs are needed to get the rates quoted.

In 1998, a consortium consisting of Compaq, Intel, and Microsoft, in conjunction with large regional telephone companies, announced the introduction of a new high-speed service using asymmetric DSL technology. The term asymmetric here refers to the fact the these

lines will permit higher data rates in the direction incoming to the subscriber than it will outbound. With some limitations, this will allow data rates of up to 400 kbits/s. The requirements are that the local subscriber loop line (the line from the phone to the switch) be copper, be from 3 to 8 kilometers in length (the actual length being determined by the cable characteristics), and that the loading coils be removed (these can be dispensed with for short cable runs). These restrictions mean that the service will only be available to about 40 percent of US subscribers.

The ADSL systems will operate at some hundreds of kilohertz and will require that a special modem be installed at both the subscriber and PSTN sides. As a bonus, because the ADSL operates at frequencies well above the audio spectrum, it will be possible to simultaneously use the line for voice and data.

Trials of this technology began at about one-half the final rate in 1997 with general availability in 1999.

Fiber optics is adding to the available bandwidth, first with wave-division multiplex (WDM), and more recently with dense wave-division multiplex (DWDM), which offers up to 5000 wavelengths on a single fiber and the possibility of private wavelengths.

ALTERNATIVE TECHNOLOGIES

There are numerous technologies that have to be used for WLL. Mostly WLL is derived from other technologies, such as cellular and trunk radio. The reason for this is largely economic: WLL can then benefit from the enormous resources that have been put into the development of chips and software for the mobile market. Also the healthy state of the mobile market ensures cost reductions associated with high production volumes that WLL on its own probably could not generate.

There have been a few instances where WLL has arisen from purpose-designed technologies like cordless telephony generation 2 (CT2), CT3, and Digital European Cordless Telecommunication (DECT). These purpose-built technologies had two things in common: first, they were low-power, short-range solutions, and

TABLE 46.1

	ADSL	HDSL	SDSL	VDSL
Transmission rate	1.5–1.9 Mbits/s downstream	1.544 Mbits/s	1.544 Mbits/s	13–52 Mbits/s
	16–640 kbits/s upstream	2.048 Mbits/s		1.5–2.3 Mbits/s
Range	6000 m	4000 m	3000 m	300 m

TABLE 46.2

	97	98	99	00	01	02	03	04	05
Cellular/PCS	0.8	1.4	2.2	3.8	5.8	8.8	12.7	17.2	22.3
DECT/PHS	0.3	0.7	1.4	2.3	3.8	5.6	7.8	10.3	12.9
Specialized	0.2	0.4	0.8	1.5	2.5	3.9	5.5	7.5	10.0
Satellite	0.0	0.0	0.1	0.2	0.3	0.6	0.9	1.3	1.7

WLL customers in 1,000,000s.

second, they were mostly outstanding failures in the market place. More recently PHS is being seen as a serious contender for WLL.

The Strategis Group has forecast the technology mix for WLL as in Table 46.2.

More recently Pyramid has revised the forecasts for 1999 down to a total of 2.7 million. Over the last decade, WLL forecasts seem to be constantly being revised downwards.

CELLULAR-BASED WLL

When using cellular hardware for WLL there are a few adjustments in thinking that need to be made. First, the connection is usually line of sight, or very nearly so. This means that the propagation loss will be characterized by something close to 20 dB per decade rather than the 40-dB characteristic of the mobile environment. As such, significantly greater range can be expected although some digital systems cannot exploit this well.

Additionally, since there is minimal multipath to contend with, the carrier-to-interface ratio (C/I) that produces an acceptable signal will be significantly lower than in the mobile environment, which means that even more range can be expected and that frequency reuse will be easier than in the mobile environment.

Once the fact that the user-end antennas are often high gain and direction is taken into account, there is yet another lowering of net interference, which will lead to both greater range and greater ease of frequency reuse.

If CDMA is to be used, a significant traffic increase of about 100 percent per carrier can be anticipated. However, this figure will be reduced since CDMA has a soft capacity that depends mostly on Signal (bit) Energy/Interference Energy (Eb/Io) and for the reasons stated above.

Taken together, the propagation is a very different thing from the regular mobile situation, and it is important that the link budgets be calculated accordingly.

There have been many attempts to derive WLL from cellular technologies, including Advanced Mobile Phone Service (AMPS), Total Access Communications System (TACS), digital AMPS (DAMPS), Global System for Mobile Communications (GSM), cdmaOne, E-TDMA, and personal handyphone system (PHS). One way to do this is to add a "black box" that converts the cellular signaling to POT signaling and interfaces with a phone or fax via an RJ11 socket. While this is technically easy to do, there are some problems. The first problem is that because of the additional hardware that is produced in relatively small volumes (compared to cellular phone production), the cost of the subscriber unit—which now consists of a cellular phone plus an interface plus a power supply plus a regular phone—will be higher than a regular cellular phone by at least 50 percent. It would be reasonable to ask why the customer would want to pay 50 percent more for a cellular phone that lacks portability.

There have been a number of successful small-scale operations using this approach, however, and prices as low as $600 per customer have been achieved, which is certainly competitive with wireline costs.

Regulators and customers alike tend to expect to pay a premium for cellular mobility, but expect the fixed services to come cheap. Where the operator derives a fixed service from a cellular infrastructure, the same costs will be incurred whether the customer is fixed or mobile, and yet the revenue may well be down. Indeed the costs of providing service to a fixed customer may well be higher than for a mobile customer, as the operator may have to take on some of the wireline responsibilities for the maintenance of the customer equipment.

Cellular networks have not to date been able to provide anywhere near the reliability of the wireline network, nor have they been able to match the voice quality.

It is probably fair to say that to date, WLL has been driven more by what is possible technically rather than any serious concern for what it is that the customer wants.

CDMA has been shown to be able to achieve approximately twice the traffic density in fixed WLL applications as compared to the mobile environment, mainly because of the lower fade rate and the much reduced handover overheads.

CDMA, and in particular IS-95B (with its 64 kbit/s capability) and later IS-95C, offers considerable promise for future WLL applications, with 1-MHz+ bandwidth.

Data are already being marketed over cellular WLL; for example, Sprint PCS is offering a phone called Neopoint 1000, which is a product of Phone.com.Inc that offers Web browsing. However, like most 2G products, the browser is slow and the screen small, which limits the attraction.

NON-CELLULAR SOLUTIONS

Probably the most prominent of the non-cellular solutions to WLL is DECT. DECT has the advantage of using an unlicensed spectrum, so that deployment can be rapid. In addition, it has a very low cost base station and high voice quality. Potentially it also provides a high data rate, although to date only 4.8 kbits/s are available.

The down side for DECT is that, due to the low power of the systems, the range is very limited and so a large number of base stations are needed with consequent high cost for links and backhauling. The licensing situation can be a two-edged sword, as the operator has no control over other users in the band, and this may result in interference or limitations on spectrum usage options. Finally, the short range that the technology allows often dictates that many of the users have to use conspicuous external antennas.

It turns out that DECT has a range to fixed access units (FAU) of around 500 meters in built-up areas and as much as 5 km in open rural areas. However, DECT was originally designed for indoor private automatic branch exchange (PABX) applications, and it does exhibit some problems with delay spread when used outdoors, which tends to limit the coverage.

DECT WLL is said to be able to provide fax and data at 28.8 kbits/s, and FAUs with up to four lines.

PHS

PHS experienced a big decline in Japan, beginning in 1998. However, at the same time the WLL growth experienced a surge. In October 1998, NEC had 350,000 lines of PHS-based WLL in 11 countries.

In Japan the PHS Memorandum of Understanding (MOU) group has standardized the WLL systems on the Radio Industries and Businesses' (ARIB's) RCR STD-28 for the PHS air interface.

For a fixed base station with an external antenna, a range of up to 10 km is claimed. One base site can service about 90 users at a per line cost of around \$600–\$700.

Because much of the WLL infrastructure is dependent on aid funds, it is significant that the Japanese government is supporting the push for a WLL role for PHS.

The PHS WLL structure is shown in Figures 46.1 and 46.2.

Point to Multipoint

Point-to-multipoint systems rely on directional receive antennas for the terminal units, and typically have 64-kbits/s data rates. A number of vendors including SR Telecom (Canada), Alcatel, AT&T, and NEC have such systems, which are purpose-designed for rural telephony. They mostly use time division multiple access (TDMA) technology, and use 2- or 4-Mbits/s microwave trunks at 1.7 or 2.4 GHz.

The subscriber link is 64 kbits/s both for speech and data, and encryption is rarely used. Because the systems are targeted for rural areas, spectrum efficiency is not a major consideration, while long-range (50 km or more) is essential.

AT&T's Airloop is a CDMA WLL product.

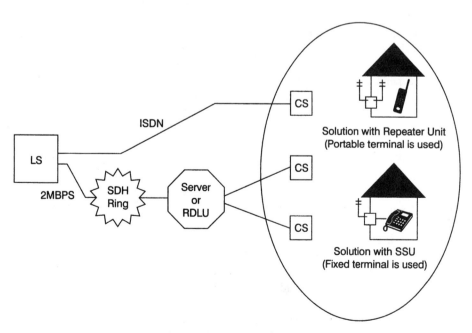

Figure 46.1 WLL using PHS.

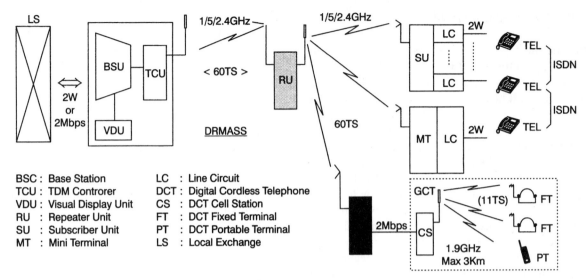

Figure 46.2 Rural WLL with PHS.

For smaller rural services there are other point-to-multipoint offerings, such as OptaPhone's STAR, which operates in the ultrahigh frequency (UHF) band and is designed to serve 96 subscribers.

Point-to-Point Loop

Point-to-point loop is a "last kilometer" replacement for the subscribers' cable. Because this link can be dedicated, it can have a relatively high capacity. One offering is from Telemobile, which offers a digital link that occupies only 25 kHz, and yet delivers a 19.2-kbits/s digital carrier. The link can provide up to two standard voice channels and one for data. The telephone line can be 2 wire loop start or 4 wire ear and mouth (E&M). Encryption and compression are used.

The range of the device depends on the power output, which varies between 1 and 35 watts, giving a maximum range of around 50 km.

Ionica

Ionica was a WLL system that used custom hardware. After being launched with much fanfare in 1993 to compete for wireline services in the UK, it has now receded into history. The system closed in 1998.

Ionica was based on the Nortel Proximity I, which is a TDMA-based system with 300-kHz radio frequency (RF) channels operating at 2.5 and 3.5 GHz.

Ionica used 64 kbits/s for voice and data and was wireline compatible, connecting to fax machines and modems at rates up to 28.8 kbits/s. Encryption was used to ensure privacy, and the system was ISDN compatible

(2B + D capable). Each subscriber unit had two RJ11 sockets.

The radio system had a nominal range of 5 km in urban cell sites and 15 km in rural areas, so it could be regarded as medium range. Base stations had 60 traffic channels and could be sectored for greater capacity.

WLL in Southeast Asia

Southeast Asia is widely regarded as the area where most of the WLL activity will be concentrated.

About 10 years ago many countries in Southeast Asia had telephone penetrations of around the 1 percent mark. This compared poorly with developed countries, which typically had penetration rates of around 50 percent. At the same time, most of these countries were aware that the lack of communications was a significant factor in impeding economic growth. Most of those same countries now have telephone penetration rates of around 10+ percent so much progress has been made.

The Philippines was a particularly interesting case. They moved from one dominant carrier (PLDT) to many carriers, with the new carriers being required to provide a service in a designated rural area. Since many of these carriers were also cellular carriers, it was natural that they look to the possibility of using the cellular network to discharge their obligations. After some considerable thought, however, most decided to go down the conventional wireline route. The problems with cellular WLL were these:

· How do you prevent your WLL customer from going mobile?

- How do you service the WLL power requirement (many of the places to be covered lacked alternating current (AC) mains, or if they had them, they were not reliable).

- If you provide a cellular WLL customer with a service at the regulated (low) cost, as required, how do you later convince the regulator that the higher charges for cellular (mobile) are justified?

- Since it will be necessary to provide the WLL phone below cost, how do you prevent a market based on re-selling the WLL phone as a cellular phone?

- There were also serious doubts about the ability of the network to sustain the traffic, particularly that generated by long-duration local calls.

- There were no suppliers of cheap cellular-based WLL loop phones.

Singapore and Hong Kong both had an early fling with CT2, and on a per population basis, they were initially very successful. Because both countries are physically very small, it was potentially possible to provide extensive coverage. In both countries the marketing thrust was as a cellular alternative rather than a wireline substitute. The problem was one of image as much as anything else. CT2 became thought of as a "poor man's cellular phone," and in these very image-conscious countries this was to spell doom.

Indonesia tried an extended TDMA (E-TDMA) WLL system with some success, but like everything else in Indonesia at that time, politics dampened most ventures (including all the cellular ones), and conclusions about how it might have fared in a free market are hard to draw.

Local Multipoint Distribution Service

The local multipoint distribution system (LMDS) is a broad-band service, offered in the United States in the 28-GHz band and with 1.3 GHz of spectrum. Worldwide it operates from 2 GHz to 42 GHz. It is basically a "last mile" solution, and competes directly with wireline, cable, and other systems that are wired into the home.

Because of the high frequencies used, there will be considerable fade during heavy rain. Also, in 1999, there was only a limited knowledge of fading at these high frequencies. Most of the detailed studies of the past were done by the International Telecommunications Union (ITU) and others have been confined to frequencies below 10 GHz.

LMDS is also sometimes referred to as a local multipoint communications service (LMCS) or simply as point to multipoint (PMP).

Typically, LMDS offers connection rates of 64 kbits/s,

but it can be as high as 512 Mbits/s. The systems commonly use FDMA combined with TDMA to give a service radius of around 5 km. The asynchronous transfer mode (ATM) and Internet protocol (IP) are available.

The hardware is available in most bands from 2 Ghz to 42 Ghz. Frequency reuse is done in a "cellular-like" manner. Services include E1/T1 and multiples of these levels.

Unlike most WLL solutions, LMDS uses directional antennas at the customer terminals (CTs).

Design of LMDS systems is done mostly on a planning tool like EDX, which uses a terrain database and propagation models to predict coverage. At 28 GHz, propagation in zero rain conditions is very predictable, and provided the terrain models are good, the planning tools should produce satisfactory results.

Since LMDS is mostly packet switched, there can be problems with voice quality caused by the timing constraints of packet switching. Although most systems give priority to voice, there can still be breaks in transmission. In 1999, Bell Labs released a product that removes this timing constraint.

LMDS and microwave multipoint distribution system (MMDS) are widely considered to be broad-band WLL technologies, and they are being sold as alternatives to the broad-band market. In 1999, the Stategis Group predicted that these systems will account for 20 percent of the broad-band market by 2004.

NEW TECHNOLOGIES

It seems that what WLL needs is new technology and new thinking. The advent of the Internet means that bandwidth (and data throughput) will be an issue that won't go away. The system needs to have significant capacity and range. Since these two things are almost mutually exclusive, the system probably needs to be dual mode.

Most of all, though, any successful WLL system needs to be flexible. With technology rendering hardware obsolete after a few years, the system must be implemented as much as possible in software. Recent developments in digital signal processor (DSP) technology might just make this practical. The simplest and earliest DSPs were just analog-to-digital converters. Today it is possible to get an off-the-shelf DSP that can interface with an RF device at the intermediate frequency (IF) level. From then on all signal processing can be under software control.

DSPs will allow "over-the-air" software updates that can keep the subscriber equipment performing at the state-of-the-art level. As new compression and processing techniques are developed, advantage can be taken of the improvements by updating the mobile performance "over the air."

Because a DSP can also do all the RF filtering tasks, the bandwidth can be allocated dynamically. The available bandwidth can be under the control of a central processor, which can restrict bandwidth at peak times, but then allow high-speed data when the system is not so busy.

Voice-processing algorithms are getting better all the time, and these can be updated as they are released.

ONGOING COSTS

Failure to adequately account for the ongoing costs of WLL has been the downfall of human operators. For example, in his 1998 book *Introduction to Wireless Local Loop*, William Webb quoted the cost of maintaining underground copper at 5 percent of investment, and the cost of WLL as 1.25 percent, based on "comparisons with cellular systems." The figure is derived from a "1 percent maintenance" cost for cellular systems.

The life span of a cellular mobile is about 18 months. So if it is reasonable to extrapolate that the cellular network costs to WLL (and it probably is not), then why not also use the terminal maintenance costs? If we do that, noting that the terminal cost is the most significant cost in WLL, then the maintenance rises from 1.5 percent to around 50 percent of costs.

Because the WLL subscriber's unit is fixed, it will undoubtedly outlast a mobile phone, but it can hardly be expected to last more than 5 years (if old age does not get it, obsolescence will), so the maintenance cost will be about 20–50 percent (this figure takes account of the relatively much higher cost of a technician attending a subscriber terminal, to the relative cost of attending a base-station site). In the case of a subscriber's unit, the cost of simply visiting the site of a faulty unit would probably exceed the cost of the repair.

Picking a Winner

All systems are limited by Shannon–Hartley theorem, that capacity of a channel is related to the bandwidth and noise by the following relationship:

$$C = B \times \log_2(S/N + 1)$$

where

B = bandwidth

C = Channel capacity in bits/s

S/N = signal to noise ratio

Surprisingly this limit is being approached by current computer modems for land lines. If we take the bandwidth of a PSTN line to be 3 kHz, and the S/N ratio to be 40 dB (10,000:1), and substitute in the equation, we will find.

$$C = 3000 \times \log_2(10,001)$$

so,

$$C = 39,863.6$$

From this it would seem that a 33.6-kBps modem is close to the limit and a 56-kBps has passed it! In fact, a 56-kBps modem relies on the fact that most of the network today is digital and a better S/N can be expected (sometimes). Often a 56-kBps modem will actually be running at 33.6 kBps. Another minor enhancement is to push the bandwidth of the line to about 3.2 kHz, which also helps the throughput.

For short wireline connections (1.5 kilometers or less), the loading coils (that are put in to ensure linearity across the used frequency band) can be left out, and so a much wider bandwidth can be achieved. Pushing the wireline to its limits, HDSL has been demonstrated at 800 kBps over a distance of 4 kilometers (in the 1990s by PairGain Technologies) and a massive 8 MBps over 1.6 kilometers by John Cioffi of Stanford University. This last technique operates at around 1 MHz, and derives 256 virtual 4-KHz channels.

The downside for wireline technologies is that the longer the distance, the slower the throughput. To make matters worse, the wireline network was not designed to operate above a few kilohertz. So it does not necessarily have good noise immunity at high bandwidths (this is particularly noticeable during storms). Any impressed noise will slow the throughput.

Cable TV services have a number of users using the same cable effectively in parallel. As the usage rises, so the data rate available to an individual falls. Cable TV service installations standards are generally not high by telecommunications standards and service interruptions are relatively frequent. However where the cables are already installed, the marginal cost of a WLL service will be small. High frequency point-to-multipoint systems like LMDS have limitations in rain, and have difficulties penetrating foliage. However they can quickly provide high capacity services. Satellite systems have come under serious questioning since the difficulties of Iridium and other service providers have come to light. Mobile telephone solutions will be relatively expensive (per megabit), but widely available and ubiquitous. From personal experience, I see a major driving force for mobile access being the wide variety of dial and access modes used by hotel PABXs, making the collection of email from a hotel room via the room telephone a major chore.

From a purely technical point of view, fiber-optic cable directly to the household must be picked as the ultimate solution. It offers the widest bandwidth (a mere few fibers have the capacity to carry *all* Internet traffic), best security, and lowest noise. The downside is that, at least in the short and medium terms, it is the most expensive and has the longest lead times. In 1999, Sprint estimated the cost of fiber at $48,000 per kilometer.

Wave division multiplex enables one cable to be used for many optical links. Systems of around 100 wavelengths were available in 1999, while experimental systems in 2000 were at 1000 wavelengths. The challenge for the first decade of 2000 is to develop flexible optical switches to avoid the costly down- and up-conversions that are necessary when conventional switching and cross-connect functions are needed.

All of this leads to the conclusion that there is no one solution for now. Probably most or all of these technologies will obtain some degree of success in most countries over the next few decades. There will be a wide variety of niche markets. Because the relative advantages are not clear cut, in the short term the winners may simply be determined by the amount of money thrown at the technology and the enthusiasm of the marketers.

47

THE TECHNOLOGY

RADIO FREQUENCY AMPLIFIER

The technology of choice for radio frequency (RF) stages right through the 1990s was gallium arsenide (GaAs). The standard operating voltage today is 3.5 volts, and the operating voltage is expected to fall. This is because there is an inherent power savings for the digital circuitry if it operates at lower voltages. For digital circuits the power relationship with voltage is

$$P = k \cdot C \cdot V^2$$

where

P = power

C = device capacitance

V = operating voltage

k = a factor depending on the frequency and switching rate

In order to get these efficiencies, semiconductor companies are targeting a 1-volt operating system. While no corresponding benefits apply to the RF part, in the overall interests of device efficiency it is best if all componentry operate at the same voltage (otherwise inverters or regulators will need to be used).

Current output stages contain two or three stages of amplification in the microwave monolithic integrated circuits (MMICs). Next generation devices will include small direct current–direct current (DC–DC) converters that will provide an active bias for the cutoff needed by the depletion-mode devices.

RF MMICs operating at 2 volts, and later at 1.2 volts, have been demonstrated and it is felt that lower voltages are achievable.

Today, basically because of their overall flexibility the heart of the MMIC is GaAs technology. Ga (gallium) and As (arsenic) are groups III and V, respectively, in the periodic table. Other elements in the III–V semiconductor groups, as well as the IV group (which includes the semiconductors of choice for low-frequency applications: silicon and germanium) show promise as replacements for the relatively expensive GaAs. As production volumes rise, the relative cost of GaAs becomes a liability, but the vast wealth of knowledge about this material, combined with its high frequency and high-temperature capabilities, has so far kept GaAs in the forefront; some changes may be expected early in this century. In particular, some alternative materials offer lower operating voltages and are looking attractive.

On the other hand, the ever-increasing demands of such things as wide-band code division multiple access (W-CDMA) requiring high linearity and low-voltage operation, and other applications in the 6–20+ GHz range, invariably turn to GaAs as the technology of choice.

Silicon technology has recently become competitive at frequencies up to 2.4 GHz (and beyond), and for low-voltage operation in the 2.4- to 3-volts region.

Silicon germanium (SiGe) is one alternative that is being explored, and devices with cutoff frequencies up to 160 GHz have been demonstated. SiGe devices can be manufactured with the familiar complementary metal oxide semiconductor (CMOS) techniques, while it readily yields higher cutoff frequencies and lower noise figures. The only real difference between Si-based devices and SiGe is that some Ge is added to the base region.

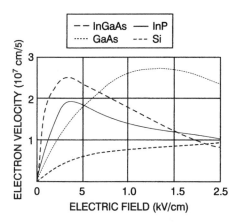

Figure 47.1 The electron velocities for various applied electric fields for some semiconductor materials.

For high-frequency applications (millimeter-wave applications) indium phosphide (InP) shows a good deal of promise. The high electron velocities as a function of the applied electric field (see Figure 47.1) are the reason for this. InP technology also results in low DC power consumption. For any given semiconductor device there will be a trade-off between frequency of operation and the breakdown voltage (increasing frequency can be traded for lower breakdown voltage), but the inherently high breakdown voltage offered by InP gives an advantage here.

An early application for InP technology is expected to be the linear amplifiers for 3G cellular phones. This area is currently dominated by GaAs technology, but InP offers low operating voltages, an advantage of 5–10 dB in the level of third- and fifth-order intermodution products, and potentially lower costs (due to a relatively high power density capability). InP technology is expected to likewise play a big role in local multipoint distribution system (LMDS) power amplifiers.

Differential Amplifiers

Differential amplifiers are widely used in cellular technology because the symmetry offers some real advantages. Consider the differential amplifier shown in Figure 47.2. Generally in order to minimize distortion, an amplifier is designed with as much headroom as possible between the supply rail and the out signal level. A single-ended output would have only 1.5 volts of headroom from the 5-volt supply rail to the 2-volt peak to peak signal, while the balanced output has both output sections operating on 5-volts peak. Thus, with each supplying a 1-volt output signal, there are 2 volts of headroom, which leads to lower distortion.

From an RF perspective the differential amplifier has a 6-dB improvement in the 1-dB compression point, which significantly improves the third-order intercept (IP3) and greatly reduces distortion.

Balanced circuits tend to be more immune to external noises, because any induced signal will be impressed on each input approximately equally and so will tend to cancel out.

SAW Filter Technology

Surface-acoustic wave (SAW) device production in 1999 was more than 800 million components, and a good deal of the production was directed at the cellular market. Leaky SAW (LSAW) devices are widely used in duplexer applications for TDMA and FDMA systems, where they measure around $1.4 \times 0.6 \times 0.2$ cm. Typically fractional bandwidths from 2 percent to 5 percent are required for this application.

SAW devices basically convert the electrical energy of the signal into an acoustic (mechanical) signal. This is readily done using piezoelectric crystals, which convert the incident signal to a surface wave that propagates at a speed from 3 to 4 km per second.

The most common application for SAW filters is as high performance IF filters, which are used for both wideband and narrow band applications, and exhibit very good temperature stability.

Until recently the lack of differential RF amplifiers has meant that SAW filters have mostly operated in the unbalanced mode. When this is done the stray capaci-

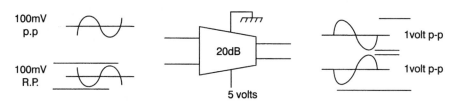

Figure 47.2 A differential power amplifier has move headroom with respect to the supply rails.

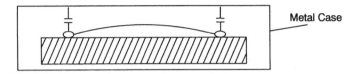

Figure 47.3 Stray capacitance limits the unbalanced mode upper frequency rejection of a SAW filter.

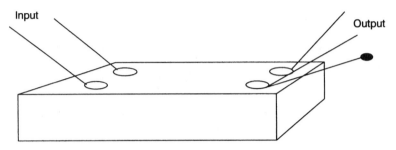

Figure 47.4   A SAW filter is inherently differential.

tance puts an upper bound on the effectiveness of the filter (see Figure 47.3). If it is operated in the differential mode, the effect of these capacitances tends to cancel, thus making the filter more effective.

Surface acoustic wave (SAW) filters have mostly been used in the unbalanced mode, but they are inherently differential devices, and in fact work better that way. A SAW filter is essentially a quartz resonator with two inputs and two outputs as shown in Figure 47.4.

The whole concept can be put together to form a differential receiver, as shown in Figure 47.5. The SAW filters provide the intermediate frequency (IF) bandpass characteristics and harmonic filtering.

Linear Amplifiers

Linear amplifiers should not be confused with wide-band amplifiers. A wide-band amplifier may have a very broad operational bandwidth (for example, it may be rated from 800 MHz to 1.5 GHz), but it is intended for general-purpose use with a single carrier. Linear amplifiers usually will be wide-band amplifiers, but they are designed for multiple carriers.

Until recently it has been very difficult to make a linear amplifier that is economical (at least for applications outside the military where cost does not seem to be a issue). These amplifiers are also known as multicarrier

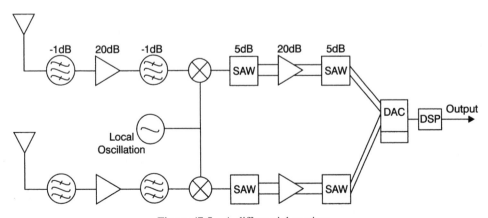

Figure 47.5   A differential receiver.

power amplifiers (MCPA). The advantage of the linear amplifier is that the output of a number of transmitters can be combined and amplified in a single amplifying stage. The difficulty in the past has been to get the intermodulation products between the carriers down to a manageable level. In recent times a number of companies, including Power-Wave (United States) and Wireless Systems (UK), have produced linear amplifiers that have intermodulation products below −65 dBc and that can meet both Federal Communications Commission (FCC) and European Telecommunications Standards Institute (ETSI) standards.

These amplifiers are generally available for all mobile protocols and represent a new challenge for the operator. All linear amplifiers will produce some intermodulation products, and to be sure that these products fall within specifications, it will be necessary to measure the intermodulation. This is best done using a multitone generator that generates in-band signals with randomly varying phase and controlled variation in signal levels. This will give a good simulation of the real world that the amplifier is subject to.

With linear amplifiers the combining is done at low power levels, where losses are of little importance.

The power rating of the linear amplifier needs to be adequate for the load to be carried, but peak power capability must be considered. The average power of the carriers will be significantly less than the peak power, because the signals will mostly be in a random-phase relationship with each other. Occassionally they will be in phase, which is when the peak power is encountered. For this reason the peak power rating should be at least six times greater than the average power.

3G Amplifiers

What is demanded for 3G RF applications is linearity, power efficiency, and cost. For signal processing it is much the same: more speed, less power, and lower cost. To some extent all three of these factors are mutually exclusive. The easiest way to improve linearity is to increase the headroom, which means more power. So the real objective to make the hardware better *and* cheaper at the same time is only feasible if the volume is there to pay for the development costs. So far, the industry is providing that growth, and this trend seems bound to continue, at least into this decade.

THIRD-ORDER INTERCEPT

The third-order intercept (often written IP3) is the level of the signal out of a non-linear device (like an amplifier) that is high enough to cause the third-order intermodu-

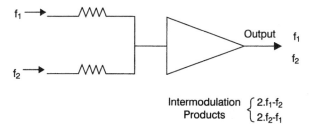

Figure 47.6 A simple amplifier is to some extent non-linear & will produce distortion products as shown.

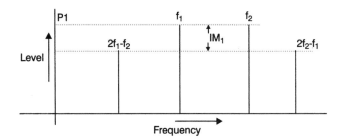

Figure 47.7 The spectrum of the intermodulation products.

lation products to appear at the output at the same level as the input.

To see how this works let's look at an amplifier that has two input signals F_1 and F_2, as seen in Figure 47.6. These signals will intermodulate in the amplifier and produce intermodulation distortion (IMD), in particular the third-order products $2 \cdot F_1 - F_2$ and the products $2 \cdot F_2 - F_1$. Of course, other higher intermodulation products are also produced, but for the purposes of this explanation they are best ignored.

If you were to view these signals on a spectrum analyzer, the result might look something like Figure 47.7. If we were to plot the levels of the fundamentals and the intermodulation products, then they would look something like Figure 47.8. The scale in Figure 47.8 is logarithmic, and from that it can be seen that the intermodulation product increases its relative level with increasing input levels of the fundamentals.

The IP3 point can be found directly from the measurement in Figure 46.9 by the relationship

$$IP3 = P_1 + IM_1/2$$

where P_1 is the power level at the output of the device, and IM_1 is the difference in level between P_1 and the level of the intermodulation products.

To see how this relationship is derived, consider Figure 47.9. Here we select any arbitrary point for the P_1. The nth harmonic product will have a slope of n, while the fundamental will have a slope of unity. From Figure 46.10 it can be seen that at the nth-order intercept

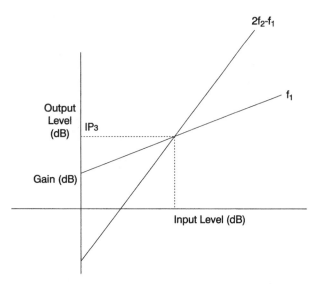

Figure 47.8 The relative power levels of the intermodulation varies with the input level.

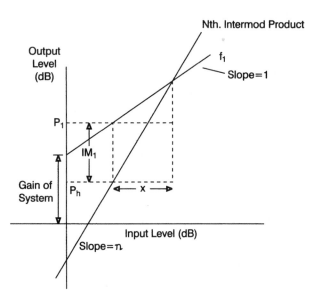

Figure 47.9 A generalised nth order intermodulation level diagram.

$$P_1 + x = P_h + n \cdot x$$

so

$$P_1 - P_h = x \cdot (n - 1)$$

or

$$x = (P_1 - P_h)/(n - 1)$$
$$x = IM_n/(n - 1)$$

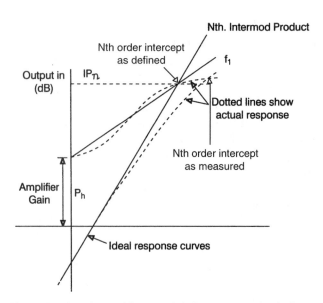

Figure 47.10 The real intermodulation response is similar to but different from the idealized representation.

where IM_n = the difference between the values of the point P_1 and the corresponding point P_h, the power level of the nth harmonic, but

$$IP_n = P_1 + x$$

so

$$IP_n = P_1 + IM_n/(n - 1)$$

Hence we now have a general expression for the nth-order intercept.

When the IMD products of a broad-band amplifier "spill over" into the spectrum of other transmitted channels the result is often termed *spectral regrowth*.

It might have occurred to some readers that a simpler way to find the nth-order intercept would be just to turn up the signal levels of F_1 and F_2 on the signal generator until the levels of the distortion product in question and the fundamental were equal, as seen on the spectrum analyzer. In fact, this cannot be done, because as the amplifier approaches saturation, it does so in a way that distorts the output more at the higher frequencies of the intermodulation products than it does at the fundamental frequencies. If we were to load up the amplifier to saturation, we would see something more like Figure 47.10. From this it can be seen that the IP_n figure is a theoretical construct and not something that can be directly measured on a device.

The spectrum analyzer, particularly if it is presented with high signal levels, may itself overload and produce intermodulation products. Unless this is taken into account erroneous results might occur. To check the analyzer, measure the IMD as before, and then redo the measurement with a 10-dB pad between the device under test and the analyzer. If the two results are the same, then the analyser is okay. If the unpadded measurement gives a higher level of IMD, then the analyzer itself is producing IMD, and the measurement level must be decreased until this effect disappears.

For high-quality linear amplifiers and most passive devices, it may well be that the signal generator produces significant levels of IMD. This needs to be checked for and eliminated by putting band-pass filters in series with the generators if the problem is found. A circulator between the two signal generators to improve isolation is also recommended.

Intermodulator

Analog systems (and co-sited analog transmitters of all kinds) will mostly use class C output stages (see Appendix C). These stages are highly non-linear and generate a lot of harmonics. They are also potential sites for intermodulation should any stray signal get into the circuit at the PA stage.

The LNAs operate in class A (see Appendix C) on all systems, and this ensures low intermodulation, provided the signal levels are low. The IP3 is a measure of how well the LNA will handle overload, higher intercepts indicating better immunity. A good figure for an LNA is 40 dB. High-level signals, which may come from co-sited transmitters, will introduce intermodulation as the stage becomes overloaded. The preselector filter is the line of defence against this.

Power Amplifier Linearity

Analog systems relied mostly on the highly efficient, but very non-linear class C amplifier, followed by a lot of filtering. Digital systems require high linearity in all the amplifying stages and especially so in the power amplifier. The two requirements of linearity and power efficiency are somewhat in conflict. Class A amplifiers are best for linearity, but poor in efficiency; however, they are widely used, especially in lower powered applications. Class AB amplification is a reasonable compromise in most cases.

An amplifier can be typified by a number of characteristics. One of these is known as the 1-dB compression point (known as P 1 dB) and is the power level where the amplifier saturation has caused the output of the amplifier to drop to one dB below its "ideal" ouput, as seen in Figure 47.11.

The IP3 is another figure used to characterize the

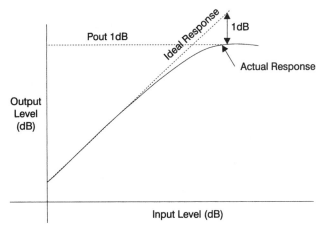

Figure 47.11 The actual response of an amplifier compared to its ideal response defines the 1-dB compression point.

linearity of an amplifier. To meet Global System for Mobile Communications (GSM) standards, for example, an amplifier should have a IP2 of less that −35 dBc (dB below the carrier level) and an IP3 of less than −45 dBc.

The IP3 is perhaps the best indicator of the linearity of an amplifier, as different amplifiers with the same P 1 dB point may induce very different amounts of intermodulation, due to different characteristics away from the P 1 dB point.

In practice, it will also be found that the IMD is not the same for all fundamentals, and will change, particularly as the spacing between the fundamentals is varied. The reason for this is that the distortion products that take the form of a polynomial will have different constants at different frequencies. When the two frequencies are close together the constants will be nearly the same, but they will diverge as the frequencies move apart.

Additionally, low-frequency signals (produced by the difference between the fundamentals) can cause non-linearities in the DC supply circuitry that impair the amplifier operation.

The linearity of the amplifier is usually rather high at low signal levels, but it is not necessarily so. When there is any doubt, instead of using single-point measurements for IMD and IP_n, it is advisable to take a number of measurements and determine the best-fit slope for the fundamental and third-order lines.

Adaptive Digital Predistortion

A relatively simple way to linearize an amplifier is to pre-distort the input, so that regardless of the level or frequency of the input, the output is maintained at the correct level. This can be done using a lookup table that has been precalibrated for each level and each frequency being sent. It can be seen in Figure 47.12 that the

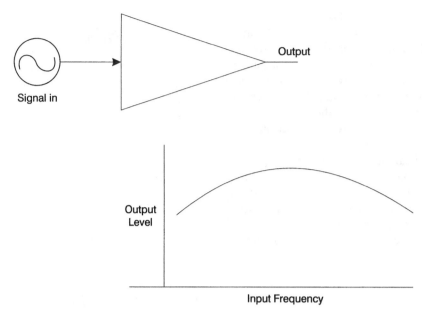

Figure 47.12 The typical non-linear frequency response of a broad-band RF power amplifier.

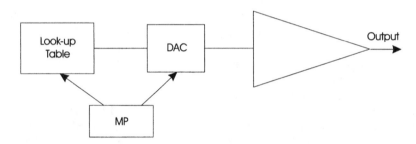

Figure 47.13 The input levels can be tailored to compensate for the nonlinear output.

output of the power amplifier (PA) will be frequency dependent. By measuring this dependence, and varying the driver level accordingly, the output can be linearized. This is usually done by having a database of the PA response that is consulted before the transmitter driver level (derived from a digital-to-analog converter) is set, as seen in Figure 47.13. The database will be unique to each amplifier, as the characteristics can be expected to vary widely across the samples. For cellular applications, a lookup table of several hundred settings will usually suffice.

For many applications it may be necessary to compensate for both frequency and phase nonlinearity in the PA, as well as for temperature.

Feedback Linearization

Linearization of the amplifier is mostly done using a pair of feedback amplifiers. The first feedback amplifier samples the input and delays it, and then subtracts it from a sample of the output. The result of this process is essentially the net IMD products. This input is then applied to another feedback amplifier, which delays the spectrum appropriately and them applies the IMD in the correct phase to the output to ensure nearly complete cancellation of the unwanted products.

The IMD of linear amplifiers is controlled by active feedback loops, and because of this there will be a well-defined range over which the IMD is well controlled. It is important to ensure that the IMD rejection band includes the whole receive band as well as the transmit band.

Delay Lines for Feedforward Power Amplifiers

The simplest of all delay lines is a coaxial cable. It can be used for both low and high power applications and has good phase linearity and stability. The downside is that to achieve the delays required (around 15 nanoseconds), a length of about 3 meters is required. For reasonably low loss cables, this can amount to a volume of about 80cc for a loss of around 2 dB. Smaller lines, using smaller diameter cables, result in higher losses and more fragile cables, which are difficult to manufacture.

Delay line filters are another alternative, and these are basically wideband filters with optimized group delay and linear phase, for the frequency band of interest.

Cavity filters are the solution of choice for many high power applications. For a given loss, these will be smaller than a cable solution, and the lower loss can enable the manufacturer to back off the PA even more to improve IMD (which typically is improved by 6 dB for every 1 dB reduction in power).

Where high power is not an issue, and reasonably large losses (>3 dB) can be tolerated, surface-mounted delay line filters using ceramics and combined lumped and distributed elements are available and cost effective.

Cool Amplifiers

Amplifiers running at low temperatures will in general have low noise figures (NFs) and higher gains. Until recently cooling amplifiers to cryogenic temperatures has been costly, but new heat pumps are beginning to make cooling to liquid nitrogen temperatures (77 K or −196°C) an economic reality. The life cycle of the coolers was in the past a serious limitation, but new techniques allow life expectancies of 3 to 5 years. A typical GaAs field-effect transistor (FET), with a room temperature noise factor of 0.5 dB, can be expected to improve to around 0.2-dB NF in liquid nitrogen

DSP (digital signal processor)

In late 1998 Lucent announced a DSP-based design package for GSM mobile phone manufacturers. The package includes the hardware platform, software tools, and the protocol stack. Customizing can be done in software and the man-machine interface (MMI) allows individual designs to develop.

In order for a DSP to be effective a high-speed wideband A/D converter is needed. A/D converters are becoming faster, and in 1998 National Semiconductor announced its CLC5956, 12-bit 65 mega samples per second model. This device, which was meant for cellular signal processing, operates from DC to 250 MHz, and so can be introduced after the first IF stage. This means fewer mixers and amplifiers and less intermodulation.

Intermodulation can never be totally eliminated, but it can be controlled. In the transmitter path, isolators control modulation by acting as a one-way valve (much as a diode does). The degree of isolation (measured in dB) that obtained is very nearly a measure of the reduction in of the intermodulation level.

Transceiver Architecture

Increasingly, transceivers are digital signal processor (DSP) based, and the chipsets are being purpose designed to meet this requirement. Figure 47.14 shows the basic configuration of a modern transceiver.

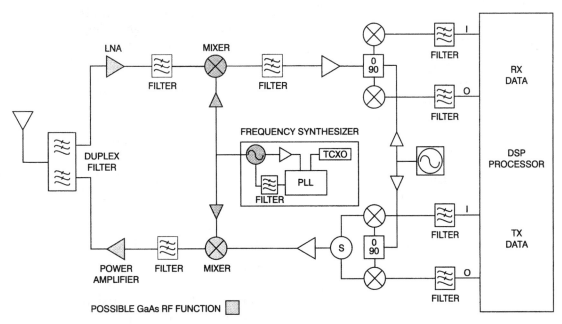

Figure 47.14 A modern DSP transceiver based on custom ICs.

Mixed-Signal Technology

Until very recently all integrated circuits (ICs) were either analog or digital, but in recent times there has been a trend toward mixed-signal hardware, and this is particularly the case with the DSP. There are two main types of mixed-signal technology: analysis and synthesis.

Analysis is a process that starts with an analog signal, then digitizes and processes it in the digital domain. Such things as digital cameras and fax machines are examples.

Synthesis is where an analog output is generated internally by the equipment: musical instruments and computer sound cards are examples of this.

At the heart of a DSP is an analog-to-digital (A/D) converter, which samples the analog signal at regular intervals and converts the measured value at the particular time interval into a numeric value. It is most important to understand that the sampling must be done at the Nyquist rate, which is twice the rate of the highest frequency to be sampled. This means that if the DSP was sampling a 25-kHz-wide channel, the DSP would need to sample at a rate not less than 50,000 samples per second (sps).

The reverse process is called digital-to-analog conversion (DAC), where the binary representation is converted to an analog voltage. To recover the sampled signal perfectly as an analog one, all that is needed additionally is a perfect low-pass filter, with a cutoff frequency equal to half the sampled frequency. A perfect filter will have no loss for frequencies from zero to the cutoff frequency, and then an infinite loss for all higher frequencies. Of course, perfect filters do not exist, so in practice, to allow for the rolloff, a band-pass filter with a cutoff of 40 percent of the sampled frequency is used. This in turn means that a perfect replica of an analog signal that has been digitized is not possible.

The importance of the Nyquist rate can be seen if an attempt is made to recover a signal that has been sampled at less than the Nyquist rate. In this case, the spectral images of the samples will overlap (in a sense the samples interfere with each other ... a process called *aliasing*) and the mixing is irreversible. Information has been lost and the analog signal can never be recovered perfectly.

There are a number of fundamental limitations in the digitizing process that affect performance:

- Quantization noise
- Harmonic distortion
- Clock jitter
- Switching transients
- Zero-order hold effects

Quantization noise is easily understood by looking at a basic A/D conversion process. If the A/D converter is an 8-bit one, then it can represent 256 (2^8) levels. Since most analog signals are continuous, whatever level is actually sampled, if it is to be digitized, then it can only be approximated by one of the 256 levels that the A/D can distinguish. This "error" is known as quantization noise and results in a background noise floor that always accompanies digitization. For a sine wave, this noise is directly related to the sample rate and the result is known as the signal-to-quantization noise ratio (SQNR), which is

$$SQNR = 6.02B + 1.76 \text{ dB}$$

where B = the number of bits of the A/D.

For an 8-bit A/D this figure is about 50 dB. It should be noted that this expression assumes that the sinusoidal signal spans the full voltage range of the A/D.

Harmonic distortion is produced by non-linearities; there is no one-to-one relationship between an analog voltage level and its binary representation (that is, the digitization process is imperfect). There will be an inherent distortion and this will produce harmonics.

The sampling process requires a clock, and imperfections in that clock will cause a modulation in the signal sampled. Clock jitter can be periodic or random.

Switching transients occur because a true ideally sharp switching process is not possible. There will always be some transition time caused by capacitance, impedance mismatch, and line inductance, which makes the process less than ideal.

Finally, the digitization process represents a continuous analog signal as a series of transitions from quantized states. Mathematically this is equivalent to passing the desired signal through an impulse filter, which produces errors known as the *zero-order hold effects*.

Because, as has just been seen, there are many sources of noise in an A/D process, a convenient way of accounting for these is to rate the chip with an effective number of bits (ENOB) which when used in the formula

$$SNR = 6.02(B\text{-}ENOB) + 1.76$$

where SNR = signal-to-noise ratio, this equation gives the net noise floor from all the sources (except zero-order hold effects, which can be mathematically cancelled).

Software Dependent Radio

The concept of a DSP based radio has got so popular that it has its own acronym—SDR—software-defined radio. While the future is certainly all about SDR, limitations today with both processing speed and memory significantly restrict its application.

IMPLEMENTATION OF IF A-TO-D CONVERSION

The dynamic range of a mobile phone link is in excess of 100 dB. Digitizing this order of dynamic range would require a converter in excess of 18 bits. While this may be acceptable for base station implementation, given the power and cost restrictions of handset implementation 18 bits is not an option for mobile phones. To address this problem, the dynamic range is reduced prior to conversion by means of a variable gain amplifier (VGA) employed in an automatic gain control system (AGC). When the AGC has been implemented the conversion process can be achieved using an 8 or 10 bit converter. The sampling ADC may be characterized by parameters more relevant to the digital receiver application.

Some Parameters to Consider

The signal-to-noise ratio (SNR) of the sample and hold (S/H) and ADC together is the ratio of the full scale analog input (rms) to the wideband noise (rms). Wideband noise includes quantization error and any other nonharmonic spurious or background noise. It is generally specified at specific input frequencies and sampling rates. Signal-to-noise + distortion (SINAD) is defined as the previous SNR but includes generated harmonic components. For both SNR and SINAD, noise is the rms sum of the spectral components below the Nyquist frequency and excludes DC.

In the case of a perfect or "ideal" ADC, the only noise would be due to quantization. Quantization is the difference between the digitally representative output and the analog input. It has the classic staircase construction where each 'tread' represents one least significant bit (LSB). Thus any analog input level can have uncertainty of ± 0.5 LSB. Given the rms summation of the error, the SNR of the "ideal" converter has a value of $6.02n + 1.76$ dB where n equals the number of bits.

It follows that the ideal SNR can be tabulated against n:

n	"Ideal" SNR dB
4	25.84
6	37.88
8	49.92
10	61.96
12	74.00
14	86.04

An often-applied rule of thumb for a realizable SNR is to multiply the resolution by 6 (e.g., 6-bit converter SNR = 36 dB). This rule should not be applied for converters above 12 bits as different noise mechanisms dominate (e.g., flicker noise).

Although not telling the whole story, the SNR does indicate how closely the practical noise performance reaches to the "ideal" performance.

The SINAD performance tells more of the story in a typical signal sampling application. As it includes both noise and discrete spurii, it quantifies all signal products that appear in the converted output. The harmonic part of the SINAD expression can also be dimensioned separately as total harmonic distortion (THD). This is the rms sum of all the harmonics in the output signal bandwidth of interest. In practice, only the first 3 or 5 are considered, as these represent most of the harmonic distortion present. By careful selection of the sampling rate and IFs, the impact of harmonic distortion in a digital receiver can be minimized.

An overall assessment of the sampling converter can be expressed in terms of the effective number of bits (ENOB). It is derived from the SINAD expression and includes not only (quantization + wideband) noise and harmonics but also AC/DC non-linearities, and phase or aperture jitters ENOB is usually shown as:

ENOB is also seen with SNR substituted for SINAD but is a less, all-encompassing definition. More correctly, a second term should be added to this formula

$$[20 \log(\text{full scale amplitude}/ \text{actual input amplitude})]/6.02.$$

This term shows ENOB increasing with decreasing input amplitude although it is a less-significant factor, unless the input is considerably below full scale.

Of considerable importance in IF sampling applications is spurious free dynamic range (SFDR). This parameter characterizes the converters in-band harmonic performance. It is the ratio of the rms value of the fundamental signal to the rms value of the largest distortion component. Although this is usually assumed to be a harmonic of the fundamental, other large spurii may be observed (e.g., the converter clock may couple into circuit due to poor layout, ineffective isolation, or cross talk). SFDR is another qualification of the non-linearity of the ADC processes and is important, as it may be the limiting factor of the receiver's dynamic range. In this way, signals of interest may not be recovered or spurs may appear to be adjacent channel signals.

As SFDR is a prime system performance determining parameter, the designer may make the decision to trade a lower SNR for an improved SFDR. This dynamic range improvement can be obtained by restricting the full scale input range. The designer must ensure that the SFDR figure obtained before the trade-off adjustment is due to conversion non-linearities and not "pick up" spurii. The latter would be unaffected by a reduction in input range.

A second method to trade SNR against SFDR is by the inclusion of dither which is achieved by adding an out-of-band noise signal. Although the amount of dither required is ADC specific, the technique applies to all converters. The inclusion of dither randomizes the coherent spur energy that occurs in the ADC. Although the level of spur energy is unaltered, it is transformed (spread) into a small increase of floor noise-level over the entire ADC bandwidth. Indeed, the addition of dither helps to distinguish those spurs that are caused by ADC non-linearities and those that are externally generated.

Static noise is another parameter that is frequently seen as a measure of ADC performance. Static noise relates to the bit variation that occurs with a fixed DC input. This would normally be expected to change by no more than 1 LSB. High-speed converters can have static noise figures of 5 or even 7 dB. This parameter is rarely of concern in wireless sampling applications as DC signals are infrequently processed. However, to prevent this parameter reducing dynamic range, it is recommended that the ADC input is AC coupled and not externally biased up to mid-rail in the usual way.

SAMPLE AND HOLD FUNCTION

Although the SINAD, ENOB, and SFDR characterizations include noise and distortion mechanisms that occur in both the S/H and conversion section, it is valuable to consider the requirements of the S/H section independently. Aperture jitter (a_j) results from noise that is superimposed on the 'Hold' command (i.e., it affects its timing). It is specified as an rms value. A 'rule of thumb' may be applied to this parameter (i.e., the signal must not change by more than $\pm 1/2$ LSB during the aperture jitter time).

Table 47.1 shows the frequency (f MHz) capability for a range of aperture jitters by ADC resolution.

MAKING A MOBILE

A mobile made today is made to sell in the mass market. Price is all-important and the aim for a digital phone is to be below $200; for analog phones, the target is below $100. Small is in. However, one must be careful that the keyboard has not gotten so small that it is difficult to use. Really small phones must be voice controlled.

Weight should be kept down, but talk time must be up. Clever use of power-consuming resources is a must for talk and standby time. The three-cell (3- $\times$ 1.5-volt

TABLE 47.1

Aperture (Jitter a_j (ps))	ADC Resolution			
	6	8	10	12
1	2487	621	155	38
2	1243	310	77	19
5	497	124	31	7.6
10	248	62	15	3.8
20	124	31	7.7	1.9
50	49	12	3.1	0.76
100	24	6	1.5	0.38

Note: The Nyquist criterion requires that the ADC sampling rate be at least twice the frequency shown here.

batteries) design is beginning to replace the four-cell design as chips become available that have been designed to operate at the lower voltage (some early three-cell devices had to use up-convertors to power some of the chips).

Next generation devices will drift towards 1.0 volt operation and 0.7 volts is being targeted.

AN EXAMPLE OF A GSM PHONE CONSTRUCTION

The following is a few notes on the way a GSM phone was designed by Voxson (an Australian company) in 1998–1999. The phone reached type-approval but never got into production. The phone uses a single six-layer board, and is virtually a discrete component design, making it ideal to study as an example of the technology. The phone has a peak power consumption of 2.6 amps, and can operate from −20°C to +60°C (which is 5 degrees better than the GSM specifications). It is interesting to note that the lower temperature limit is set by the clock on the processor, which when very cold, starts to run slower, and finally stops at −25°C. The chip count is 457, which seems high, but it must be remembered that this is a discrete design. The average industry chip count for a GSM phone is around 300 (in 1999).

Clock

The clock on GSM phones must sychronize with the frequency correction channel (FCCH) of the base station once every frame, so the clock accuracy need not itself be very high. The phone clock runs at 13 MHz, and has an internal accuracy of 2 ppm. In the Voxson phone, this produces a frequency error that is mostly better than 30 Hz.

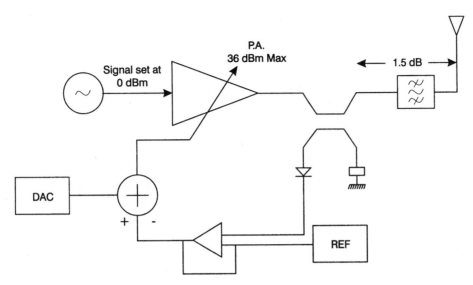

Figure 47.15   The PA and power ramp control.

CODEC

GSM now has effectively three official voice coder/ decoders (CODECs). The original full rate, the half-rate (which is almost never implemented on the networks), and the enhanced full rate (EFR). It is the half-rate CODEC that requires the most processing power, requiring 70 million instructions per second (MIPS), compared to 36 MIPS for the full rate. Some phones are capable of running all three CODECs, but the processing rate requirement means that the Voxson phone, like many others, does not do the half-rate.

Talk Time

The talk time depends on a number of factors, one of which is the network. A GSM phone has a capability of registering up to 28 neighboring cells. In the standby mode, the mobile re-registers when the signal drops below a preset level and a scan is initiated. Depending on the base-station layout, the number of re-registrations can vary significantly enough to affect the standby time. During a conversation the network controls the hand-offs, and again depending largely on the network parameters and base-station layout, the number of hand-offs can vary significantly, to the extent that different networks in the *same* service area may have very different talk times.

Level changes occur in discrete 2-dB steps at the rate of 1 per frame. Where power levels are changing often, the time spent switching up and down at appropriate levels can be both performance and talk-time affecting.

Transactions on the control channels take place without power control (in fact, in general, voice conversations on the same carrier as the control channel take place without power control).

The Power Ramp

In the Voxson phone the power ramp is controlled by the gain control of the PA. As seen in Figure 47.15. The PA is a variable-gain device that is fed a signal at a constant dBm. The PA, which is 49 percent efficient at full power, has a maximum power level of 36 dBm (4 watts) and has its output adjusted under the control of the DAC, which ensures that the ramp profile is followed. In turn, the output level is sensed and compared to the DAC level via a feedback loop.

Duplexer

It was a bit of a surprise to find a duplexer in the front end of the mobile, and a bit more of a surprise to learn that most GSM mobiles have a duplexer. Theoretically any digital mobile can get by without a duplexer by having the send and receive functions separated in time. There are, however, three reasons that a duplexer is used. First, it is used as a low-pass filter to keep the second and third harmonics suppressed to −60 dBc. Next, and most importantly, filtering is necessary to keep the transmitter (TX) from generating noise that may interfere with other mobiles. To meet the GSM specification, the noise floor at the receiver (RX) frequencies that is directly generated by the TX must be

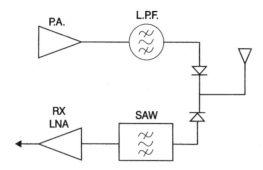

Figure 47.16   A GSM front end without a duplexer.

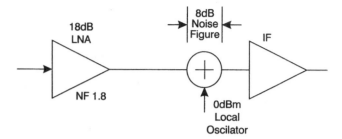

Figure 47.17   The receiver front end.

79 dB down. Finally, to prevent the GSM mobile low-noise amplifier (LNA) from blocking (due to compression), the receiver must be able to reject a signal of 0 dBm from 980 to 1275 MHz. The duplexer provides the necessary filtering for all three of these requirements.

The duplexer contributes 1.5-dB loss on the TX side and 2.5-dB loss on the RX side.

It is not mandatory, however, that a duplexer be used, and it can be avoided if, instead of up-converting the TX frequency, an offset phase-locked loop (PLL) is used to generate the 900-MHz signal. This method, which allows the direct generation of the 900-MHz carrier, produces a much cleaner signal that requires less filtering. However, filtering is still needed and the isolation that the duplexer provides is now provided by positive intrinsic negative (PIN) diodes, as seen in Figure 47.16.

The Receiver Front End

The receiver front end consists of an LNA and mixer, as seen in Figure 47.17. The high-gain LNA is needed to overcome the high noise figure (8 dB) of the mixer. The LNA noise figure of 1.8 dB is not exactly state of the art, but it does represent a sensible compromise between cost and performance.

Because of the tight specifications for GSM (or any digital mobile phones) a lot of expensive test equipment is used. Figure 47.18 shows the network analyzer and R & S, CMD test sets that were used in the development.

Voxson Charger

There are many battery types on the market and most of these have very different chemistries. The aftermarket today is very active and a manufacturer cannot be sure of the actual chemistry of future batteries that may be used with the handset. This presents a problem for the designers of battery chargers as the user is not likely to be aware that the aftermarket battery is of a different type.

The Voxson GSM phone has been developed with the battery charger within the phone itself, and uses a resistor packed with the battery for the phone to be able to determine the battery type. To detect the type of battery attached to the phone, a 10-bit ADC is used onboard with a simple resistive divider network. This allows detection of the battery and so its charging chemistry. The battery voltage is monitored via a second 10-bit ADC (giving around 1.1 mV resolution) and temperature with another 10-bit ADC (through a thermistor configuration) giving 0.5° resolution.

The different battery chemistry that the charger supports are:

NiMh
LiIon
LiPoly (software under development)

The charge algorithms are outlined in following text for the different battery chemistries.

NiMh   With NiMh, each cell has a 1.5 V at maximum charge therefore the maximum voltage is 4.5 V for a three-cell design. The phone can operate down to 3.0 V before switching off. To charge the battery again, the charger will output a constant current of around 350 mA. When NiMh approaches fully charged, either a small negative voltage gradient is evident (i.e., 4.53 V back to 4.5 V) and/or there is a sudden increase in battery temperature. Therefore when charging NiMh, it is necessary to monitor both voltage and temperature, and control the shut-off of the charger accordingly.

LiIon and LiPoly   There are two types of LiIon cells currently available: a 4.2 V cell and a 4.1 V cell; (it appears that most manufacturers of LiIon are moving towards 4.2 V by the end of 2000). The cell must be charged to within 1% of this maximum voltage—any over-charge can cause explosion of the cell. To charge the cell, a constant current of around 350 mA is used until it is detected that the cell is at its maximum voltage.

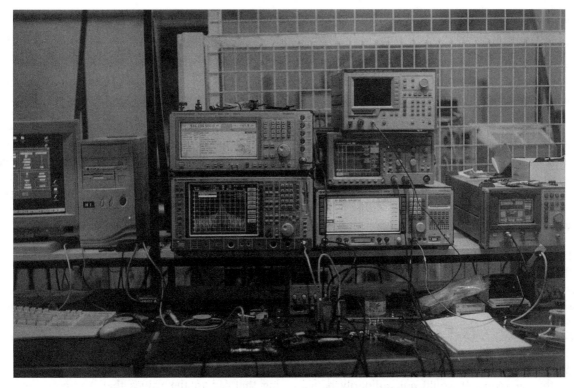

Figure 47.18   The test sets used to develop a GSM phone.

Once this is detected, the charger is switched to a constant voltage charge and the current adjusted accordingly to maintain the maximum charge voltage.

LiIon cells do not change temperature during charge, unless there is a problem, so the temperature is still monitored. Another characteristic of LiIon is that if the cell has been over discharged (to say 2.0 V) for some reason, then the charger must slow charge (around 70 mA) the cell until 3.0 V is reached before the fast charge is started.

CONNECTORS

Connectors are chosen for their physical and electrical characteristics. Among the important electrical characteristics are the power-handling capability, the voltage standing-wave ratio (VSWR), the passive intermodulation distortion (PIMD), and RF leakage. For most mobile applications the power carried is relatively small, and most connectors can handle 10 watts without stress, while higher power applications may require specialized connectors. In this respect it is worth noting that connectors that are mounted directly onto a metal surface will have better heat dissipation than those that are not, and so can carry higher power levels. For some applications peak voltage (leading to breakdown) is a more serious consideration than power levels.

In most RF applications the connector will be part of a coaxial transmission line, which it will match well. Where the connector is interfacing between circuit hardware (which is mostly strip lines or microstrips) and the outside world, the matching to the connector will often be challenging, and will have more effect on the net VSWR than the connector itself.

Where matching is critical, right-angle connectors, which have inherently high VSWRs, should be avoided.

RF radiation from a connector will range from 70 dB to 100 dB down on the carried signal, and while this is not often a problem, it can in some cases lead to cross talk.

Mechanical stress is often applied excessively to connectors, most commonly occurring in unsupported cable runs. This can result in premature failure of the connector. Stress is often induced by applying incorrect torque to the connector; in this regard, it is most important to follow the manufacturer's guidelines, although all too often this is not done.

TNC Connectors   TNC connectors abound in the industry and have been a part of the RF scene for decades. The TNC connector evolved from the bayonet BNC, which is still commonly used for frequencies below 1 GHz. The screw-coupled TNC is more stable and predictable than the earlier BNC, and in its original form was rated to 11 GHz. Its small size comes at a cost of a complex dielectric interface and critical pin location, and there are six interface dimensions (three on the male and a corresponding three on the female connector) that must be tightly met.

The popularity of the TNC has meant that variants have been spawned to meet demands of tighter impedance control and higher frequencies. Most of these variants look alike, but they are not compatible.

An early upgrade saw the TNC frequency extended to 16 GHz, this upgrade being done in a way that permitted backwards compatibility. However, a further extension to 18 GHz saw a total redesign of the dielectric interface, in essence creating a new connector. At the time of writing, the latest designs permit operation at 19 GHz. A considerable number of TNC variants have appeared recently, there being more than a dozen types. Within these variants there are American standards and European ones, with the International Electrical Commission (IEC) connectors conforming to European standards.

The performance of TNC connectors of different types when plugged together is unknown. The result may vary from lowered cutoff frequency to actual physical damage to the connector. It would not ordinarily be a problem that connectors from one system supplier would be incompatible, but today the industry has diverse sourcing, and test cables of doubtful origin abound. The potential for mixing types is significant.

Faced with this situation it is tempting to avoid TNC connectors altogether. However, their widespread use makes this difficult to accomplish. When new connectors are brought into the organization, the safest thing to do is to obtain a detailed copy of the specifications and confirm compatibility (remember that differences between incompatible connectors are still small). Additionally a pin gauge should be obtained that can be used to confirm that the connectors as supplied conform to the specifications (small pin protrusions can cause serious damage and connector mismatch if they are mated).

A set of connector gauges for TNC connectors are available from Maury Microwave Corporation. These gauges measure the three critical dimensions of a TNC connector, for all of the main variations of the standard. It is recommended that the gauges always be used after assembly of a connector, as well as periodically on in-service equipment.

DIRECT FAULT MONITORING

At the RF level there are two traditional ways to detect faults. The first is to monitor the RF path directly. This is relatively inexpensive and will detect all RF faults, but has the disadvantage of introducing losses and disturb-

ing the VSWR. The other method is to monitor the internal biasing currents, which will change mostly in response to an RF fault.

Recently, a new method, which monitors the temperature of all terminations, has been developed by Florida RF Labs. The terminal sensor provides a resistance to ground that varies proportionally to its temperature (in this way, the temperature, and so the resistance, is directly proportional to the power carried by the termination). By biasing the sensor with a DC voltage, a current that is proportional to temperature can be obtained. For systems where changes in ambient operating temperature can be expected, an optional ambient bias resistor can be used.

WHY 50 OHMS?

The true reasons for the "standardization" (actually the more you look into it, the more "standards" you find) of impedances are probably lost in history. Some very plausible explanations have surfaced over the years, however, and they probably contain some of the thinking that led to the values that we accept as standard today.

The original coaxial cables were made from commercially available copper water pipe and had spacers that made them basically rigid, air-gap cables. The impedance of a cable is found from:

$$Z_0 = \sqrt{(L/C)}$$

where

Z_0 = the characteristic impedance

L = the inductance per unit length

C = the capacitance per unit length

However, this relationship can also be expressed in terms of the diameters of the inner and outer conductors as

$$Z_0 = 138.16/(\sqrt{\varepsilon}) \times \log D/d$$

where

ε = the dielectric constant of the medium between the conductors

D = the diameter of the outer conductor

d = the diameter of the inner conductor

For air "ε" is 1.

The most likely early use for coaxial cables would have been for high-powered transmitters, where using low-loss coaxial cable would minimize the power loss in delivering the signal to the antenna. Calculating some values for impedance, assuming that the inner conductor was a one centimeter water pipe, we find the following:

Outer Conductor (cm)	Impedance (Ω)
1.65	30
2.3	50
3.5	75
5.3	100
7.4	120
12.2	150

From these figures (which are a little contrived to get the more frequently encountered "standard" impedances), it is fairly clear that the whole range of standard impedances encountered today are easily matched with some easily constructed cables.

Within these impedance options there are a number of different optimums. The maximum power that can be transmitted is derived by considering the value of $V^2 \times Z_0$ for various impedances, where V is the breakdown voltage of the cable. From this consideration, the optimum value occurs at $Z_0 = 30$ ohms.

However, maximum power carrying capacity is not necessarily the most desirable characteristic, as at the low impedance of 30 ohms the power losses to ohmic heating of the cable are high, and better efficiency can be achieved at higher impedances.

A common figure for impedance in the early days was 77 ohms. When polyethylene was put into cables instead of air, the standard-size cables had their impedances reduced by the dielectric constant of polyethylene (2.3), which made new impedance

$$Z_0 = 77/(\sqrt{\varepsilon}) = 51 \text{ ohms}$$

Fifty-one-ohm cable is available today, as are 51.5, 52, and 53 ohms. Other "standard" impedances include 70, 75, 93, 120, and 125 ohms.

48

CODING, FORMATS, AND ERROR CORRECTION

DIGITAL SIGNALING

Digital systems can be constructed so that each repeater, or link, not only amplifies but can regenerate the whole code. Provided sufficient error correction is used, the reconstructed code can have an arbitrarily small error rate.

For digital data, fast-frequency shift keying (FFSK) commonly is used. This consists simply of transmitting a 1200-Hz tone to represent a logical "1" and a 1800-Hz tone for "0." Detection can be completed in one and a half cycles.

The mobile environment is a particularly harsh one for the transmission of high-quality signals, and thus, particularly at high bit rates, it is inevitable that some bits of data will be lost or distorted to the extent that erroneous messages will be received. Because of this, any workable mobile digital code will either have to be very slow or incorporate a robust error correction technique and in most cases both.

In two-way communications automatic repeat requests (ARQs) are commonly used. Error correction is complete if only error detection is used because a detected error can prompt a resend. Although the basic techniques used for error correction and error detection are similar, error correction is much more complex.

WORDS

Words are groups of code containing the data to be transmitted. A word can be constructed of any number of bits of data. The distance between two code words is defined to be the number of position-bits within the words that differ. Of particular importance is the minimum distance between any two words, which defines how closely the words can match. Where the number of errors exceeds the minimum distance, an erroneous word can be misread as a different but yet valid word.

Digital data are generally structured into words, which are in turn grouped into blocks. Where error correction code is included in the blocks, it is known as block code.

PARITY

Parity is the simplest form of error detection and correction. In its most basic form, a digital word of, for example, 8 bits might contain one parity bit. This bit is inserted to ensure that the total number of 1s in the word is either odd or even (depending on the parity selected). In this example the information part of the word is 8 bits. To illustrate how this works, assume that the digital information

1011001

is to be sent.

The total number of 1s is 4. To construct the word for even parity the last bit would be a 0, that is, the 8-bit word is

10110010

while for odd parity the last bit would be 1, and so the word is

10110011

638

In a simple ARQ system this is all that is needed, as once an error is detected then a request to repeat the word can be sent. Notice that this code will identify a single error only and can incorrectly accept a word with two errors if they cause the parity to remain correct. Despite these limitations, this method is sufficient for many applications.

Extending this concept, it is possible, by using two parity bits per word, to correct actual errors even in the forward error correction (FEC) mode, which corrects errors received rather than relying on a resend.

To simplify, consider a word of 4 bits total. If each word contains 1 parity bit, and each block has one parity word, then the actual bit in error can be detected. For even parity the word sequence may be

	Information	Parity
Word 1	101	0
Word 2	111	1
Word 3	001	1
Word 4	010	1
Block parity	001	1

Note that even parity has been used for both the word and block check bits. It would be equally valid to use odd parity for the word check bits and even for the block (or vice versa).

Assume now that there was an error in the second bit of word two (that is, the X is a zero instead of one). We now have

	Information	Parity
Word 1	101	0
Word 2	1X1	1
Word 3	001	1
Word 4	010	1
Block parity	001	1

error in parity

and it can be seen that the error can be detected at the intersection of the error check column and check row.

The block code may consist of any number of message bits, k, together with $n - k$ error detection bits (where n is the total number of bits in the block). Such code is known as (n, k) *code*. The dimensionless ratio $r = k/n$ $(0 < r < 1)$ is known as the *code rate*.

CONVOLUTIONAL CODES

Convolutional codes differ from block codes by viewing the data as a continuous message sequence, and will generate parity bits continuously with the data flow. This type of encoding is distinguished from block codes by the use of memory, so employed that the encoding is dependent on code previously transmitted.

Convolution codes are relatively low-efficiency codes (typically $r = 0.5$), which, because they can correct a continuous string of errors, are ideal for correction when error bursts, due to impulse noise, are likely to be encountered.

HAMMING CODES

Because mobile receivers work in a high noise environment, where an error rate of more than 1 in 16 is common, more sophisticated techniques are needed. One commonly used technique is the Hamming code.

The Hamming code, devised by Richard Hamming in 1950, is capable of correcting multiple errors. Like most early error correction techniques, the Hamming code is largely a product of trial and error rather than a systematic and rigorous mathematical approach. The number of redundant bits is determined by the formula

$$2^n \geq k + n + 1$$

where

k = the number of bits in the data word

n = the number of redundant bits

The code rate for this technique is $r = k/n = 1 - 1/n \times \log_2(n + 1)$, and it is a reasonably efficient code for long code words.

The redundant bits can be placed anywhere in the word, but depending on the transmission mode there may be positions that will improve the noise immunity.

The Hamming bits are determined by the position of each bit and then determine the resulting exclusive-or (XOR) value with a value for each information bit with a value of 1. At the receiver the Hamming bits are extracted and XORed with all 1s. The result gives the position of single bit errors.

MODULO-2 ARITHMETIC

At the heart of most modern error correction codes is modulo-2 arithmetic. In essence it is simply described by the following rules for binary numbers:

a. All digits must be 1 or 0.

b. $1 + 1 = 0$

c. $1 + 0 = 0 + 1 = 1$

d. $0 + 0 = 0$

e. Addition of binary numbers is accomplished with the above rules plus the rule that there is no carry function.

Addition

For example, add 111001 to 100011, as below

$$
\begin{array}{r}
111001 \\
+ \ 100011 \\
\hline
011010
\end{array}
$$

The rules a to d above are just those performed by an XOR gate.

Subtraction

Subtraction is identical to addition.

Division

Division is carried out in the same manner as ordinary division. For example, divide 111 into 11001

$$
\begin{array}{r}
101 \\
111 \ \overline{)\ 11001} \\
111 \\
\hline
101 \\
111 \\
\hline
010
\end{array}
$$

Multiplication

Multiplication is performed the same as conventional multiplication except that the addition is modulo-2. For example, multiply 111 by 110

$$
\begin{array}{r}
111 \\
\times \ 110 \\
\hline
000 \\
111 \\
111 \\
\hline
10010
\end{array}
$$

The operation of modulo-2 arithmetic will not produce results that agree with conventional arithmetic but, importantly, it will produce consistent results that can be reproduced in easily constructed hardware. The most commonly used codes, which are subsets of the cyclic block codes (to be discussed later), were derived in the mid-1950s, when the first mathematically derived codes began to appear. Using modulo-2 arithmetic, the ease of realization is not so much a coincidence, but at the time may have been a necessity. Of course, today with microprocessor technology, there are few restrictions on the mathematical complexity of the calculations.

In many texts you will find that modulo-2 operations are described using encircled conventional arithmetic signs, as shown in Figure 48.1.

Modulo 2 operation		Symbol
$+$	$=$	$\oplus$
$-$	$=$	$\ominus$
$\times$	$=$	$\otimes$
$\div$	$=$	$\oslash$

Figure 48.1 Modulo-2 arithmetic operators.

CYCLIC BLOCK CODES

The most commonly used codes today are derivatives of the cyclic block codes. A cyclic code is one in which any code word can be end-about shifted to form another code word. Consider the code formed by the message bits m1, m2, m3, m4 and the code check bits c1, c2, c3 to form the code word

m_1	m_2	m_3	m_4	c_1	c_2	c_3
1	0	0	1	0	1	1

Shifting one place to the left results in the word

$$0 \ 0 \ 1 \ 0 \ 1 \ 1 \ 1$$

where 0010 is a valid message word and 111 the corresponding check sequence. The prominence of cyclic codes is due mainly to the fact that they can be easily implemented by straightforward modulo-2 hardware circuits. The word and its corresponding check bits are generated by dividing the message bits (padded with zeros to the total word length, n) by a binary number of length $(n - m + 1)$; here m = message bits per word. This binary number is often referred to as the *generator polynomial*. The important part of this division is the remainder, which forms the check bits. A characteristic of the word so generated is that it is a perfect multiple of the divisor. So a simple check for error by the receiver is to divide the received word by the divisor and any remainder other than zero indicates an error. With a careful choice of divisor, error correction is also possible.

TABLE 48.1 BCH Codes up to Word Length 31

n	m	t						Divisor			
7	4	1								1	011
15	11	1								10	011
15	7	2							111	010	001
15	5	3						10	100	110	111
31	26	1						11	101	101	001
31	21	2				1	000	111	110	101	111
31	16	3			101	100	010	011	011	010	101
31	6	7	11	001	011	011	110	101	000	100	111

Note: n = total word length; m = message bits per word; t = maximum number of detectable errors.

To see how this works, it is instructive to consider the analogous situation in conventional arithmetic. If the data bits were equal to 13 and the divisor were 5, then by performing the division 13/5 we find a remainder of 3. To make the code word perfectly divisible by the divisor (5) it would be necessary to *subtract* the remainder, making the code word 10. However, recalling that in modulo-2 arithmetic addition and subtraction are identical, then *adding* the remainder will produce a number perfectly divisible by the divisor.

The next task is to select a suitable divisor. There are no simple rules to select a divisor, and the codes have high redundancy. The codes are divided into classes based on the choice of divisors.

BCH CODES

BCH codes (BCH stands for the inventors, Bose, Chaudhuri, and Hocquenghem, who derived the code in the 1950s) are the most commonly used codes today. The code is specified by its total word and message length as (n, m).

BCH codes are such that they can be devised to correct any given number of random errors per code word. For block lengths of up to a few hundred bits they are among the most efficient in terms of total block length and code rate. Table 48.1 gives the BCH divisors for word lengths up to 31.

POLYNOMIAL CODES

The most general class of cyclic codes is known as polynomial codes. A general code word can be described by a polynomial as

$$f(x) = 1 + x + x^2 + x^3 + \cdots + x^{n-1}$$

where x can only take a value of 1 and the power of x signifies its position in the word. It is important to realize that this polynomial is not a conventional one, and for the purposes of this text it can be regarded as a shorthand descriptor of the code bits. The factor x^k, where $x = 0$, is implied for all values of $k < n - 1$, where the kth power is not in the equation.

For example, the Consultative Committee on International Telegraphy and Telephony (CCITT) V41 $(256, 240)$ code uses the 17-digit divisor

$$10001000000100001$$

which could be described by the polynomial;

$$g(x) = 1 + x^4 + x^{11} + x^{16}$$

The generator polynomial can get much more complex as, for example, the generator used for the code division multiple access (CDMA) paging, sync, and access channels, which is

$$g(x) = x^{30} + x^{29} + x^{21} + x^{20} + x^{15} + x^{13} + x^{12}$$
$$+ x^{11} + x^8 + x^7 + x^6 + x^2 + x + 1$$

For the forward CDMA traffic channel, the generators are

a. 9600 bps

$$g(x) = x^{12} + x^{11} + x^{10} + x^9 + x^8 + x^4 + x + 1$$

with a frame length of 192 bits in 20 msecs

b. 4800 bps

$$g(x) = x^8 + x^7 + x^4 + x^3 + x + 1$$

with a frame length of 96 bits in 20 msecs
The reverse channel generator is

$$g(x) = x^{16} + x^{12} + x^5 + 1$$

GOLAY CODE

The Golay code is generated by the divisor, or generating polynomials:

$$g(x) = 1 + x^2 + x^4 + x^5 + x^6 + x^{10} + x^{11}$$

or

$$g(x) = 1 + x + x^5 + x^6 + x^7 + x^9 + x^{11}$$

The Golay codes can detect any combination of three errors in a 23-bit block.

POCSAG

POCSAG, the international code format for paging, is a $(31, 21)$ BCH code, with the generating polynomial:

$$g(x) = x^{10} + x^9 + x^8 + x^6 + x^5 + x^3 + 1$$

To this 31-bit code word is added one extra even parity check bit for the whole code word.

INTERLEAVING

Depending on the nature of the transmission path, an improved performance measured in bit error rate can sometimes be obtained by interleaving bits of words so that in small-duration fades the corruption of any one word can be kept to within the capabilities of the error correction code. Interleaving can thus improve performance without adding redundancy. Because paging generally is used for only relatively short messages, with little opportunity to interleave, this technique is more commonly applied in longer data streams.

49

DIGITAL MODULATION

Digital modulation can be divided into three basic types: amplitude shift keying (ASK), in which the carrier is either turned on or off; frequency shift keying (FSK), where the carrier-center frequency is changed from one value to another; and phase-shift keying (PSK), where the phase of the carrier is switched. All of these forms of modulation are used in telecommunications, but PSK is becoming the most extensive.

Although in a random-noise environment ASK and FSK have the same performance, ASK does suffer from additional errors due to the amplitude variation that is characteristic of radio signal fading. PSK can be shown to have a 3-dB advantage over ASK in a random-noise environment, and the information rate can be doubled, in the same bandwidth, by using 4-level [or quadrature phase-shift keying (QPSK)] phase shifting.

Digital modulation came into vogue in telecommunications in the 1960s. At that time the use of the technique was severely limited by the bandwidth needed to transmit intelligible voice. The standard voice channel of 64 kbit was derived from the Nyquist consideration that a signal needs to be sampled at twice the highest frequency in the sample (nominally 4 kHz) and that an 8-bit word containing 256 levels was required to give a subjectively acceptable dynamic range. Hence the 64-kbit is equal to $4000 \times 2 \times 8 = 64,000$ bits/second or 64 kbits/s. This form of encoding is covered by Consultative Committee on International Telegraphy and Telephony (CCITT) standard G711.

The general method of encoding as an N-digit binary number is known a pulse-code modulation (PCM). The basis of PCM is to represent the analog level of the signal by a pulse-coded number. For voice using basic PCM the 64-kHz bandwidth required can be seen to make rather inefficient use of the spectrum compared to the 3.4-kHz needed to transmit the same data as an analog signal.

The main advantage of digital transmission over analog is its noise performance. With analog systems each repeater, each amplifier, and in fact every component of the network will add some degradation to the signal-to-noise ratio (S/N). Digital systems offer the possibility of noiseless regeneration, and with the use of an appropriate error correction code a repeater can "clean up" an incoming signal, making it possible to communicate at any arbitrarily small bit error rate (BER) desired (provided sufficient bandwidth is available).

Digital systems also offer the following advantages over analog:

- Integration of voice and data services
- Store-and-forward capability
- Bandwidth efficiency if compression/decompressions (CODECs) are used
- Security
- Immunity to interference

In order to make digital transmission more spectrally efficient it is necessary to incorporate a considerable amount of memory and processing power. With the advent of mass-produced digital integrated circuits (ICs) and amplification-specific integrated circuits (ASICs) in the 1970s commercial digital transmission became economically practical for point-to-point applications, such as microwave and fiber-optic applications.

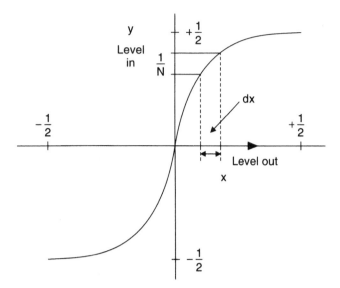

Figure 49.1 A digital compander transfer characteristic.

DIGITAL ENCODING

Digital voice PCM signals are encoded as an N-binary level and decoded by passing the reconstructed signal through a low-pass filter. The output power of the low-pass filter is directly related to the sample duration, and it is in order to make the filter effective that a sample rate of 8 kbits/s (rather than 2×3.4 or 6.8 kbits was originally chosen).

As each sample must be rounded off to a discrete binary level, there will be quantizing errors that will produce noise and distortion. Where the quantizing results in errors, distortion arises. This distortion is particularly severe for low signal levels. In practice, to minimize this source of noise, companding is used to increase the level of low-level signals and to decrease the level of the highs, as seen in Figure 49.1.

It can be shown that if the slope of the companding curve is proportional to the input level, then the quantizing distortion will be independent of the input level. That is, $dy/dx = kx$ where k is a constant. Such a relationship is satisfied by $y = k \cdot \log x$. It should be noted, however, that as x approaches 0, so $\log x$ tends to minus infinity, and so the relationship needs to be modified for very small input levels.

The two widely used log compression laws in digital transmission are the A Law and the Mu Law.

The A Law

The A Law is defined as

$$\text{For } 0 < x < 1/A \cdots y = Ax/\{1 + \ln A\}$$
$$\text{For } 1/A < x < 1 \cdots y = \{1 + \ln(Ax)\}/\{1 + \ln A\}$$

TABLE 49.1 The Spectral Efficiency of a Number of N-ary Systems

		Spectral Efficiency bits/sec/Hz (max)
BPSK	.......64 kbits/s	1
QPSK	.......32 kbits/s	2
16QPSK	.......16 kbits/s	4
64QPSK	.......64 kbits/s	6
256QPSK	......256 kbits/s	8
1024QPSK	.....1024 kbits/s	10

The Mu Law

The Mu Law is defined as

$$y = \{\ln(1 + \mu x)\}/\{\ln(1 + \mu)\}, \qquad \text{for } x \text{ positive.}$$

Practical values for μ and A can be obtained by using the rule of thumb that each quantum step should be at least 16 dB below the root-mean-square (RMS) level of the speech and that the quantum step should not exceed 1.4 dB. Both these laws can only approximate the uniform distortion criteria because of the need to avoid a singularity at $x = 0$.

A half-rate encoding known as adaptive PCM (ADPCM) using 32 kbits/s sampling rate is widely used.

Simple binary encoding, which generally takes the form of a binary phase-modulated signal (BPSK), is not spectrally efficient. Soon higher levels of phase modulation were used and the hierarchy shown in Table 49.1 established.

Note that these modulation types are also known as 4-ary, 16-ary, etc., with a general system with m phases being known as m-ary. Higher-ary systems have lower costs per channel, but also have higher BERs for the same signal levels.

This modulation is illustrated in the Figures 49.2 and 49.3.

MODULATION SYSTEMS

The main modulation techniques used in digital links are

- QAM (quadrature amplitude modulation)
- PSK (phase-shift keying)
- FSK (frequency shift keying)
- CP-FSK (continuous phase frequency shift keying)
- FFSK (fast-frequency shift keying, which is really a special case of CP-FSK)

These techniques are best illustrated by signal constellation diagrams that illustrate the phase and level rela-

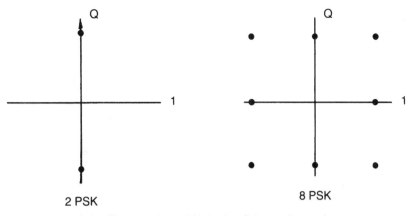

Figure 49.2 Phase-shift keying (Phasor diagram).

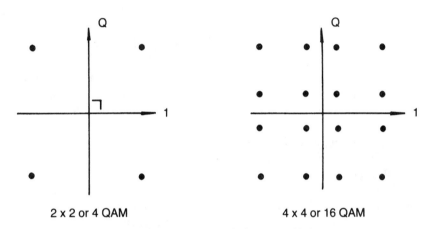

Figure 49.3 QAM modulation constellations.

tionships between the system states. Figure 49.2 illustrates PSK; Figure 49.3 illustrates QAM.

Digital systems are inherently reasonably secure because a scrambler is needed to prevent adjacent channel interference due to spectral peaks that may occur during quiet passages, synchronization bursts, or other times when a repetitive pattern may be sent. A typical digital encoder is shown in Figure 49.4.

In the mobile radio environment it is very desirable to have an encoding system that is both spectrally and power efficient. It is also desirable, as well as a regulatory requirement, to avoid interference to the adjacent channels. Economically priced high-power, high-efficiency power amplifiers, of the type that can be expected to be found in a mobile phone, are non-linear. This means that they will have a gain that rolls off with increasing output

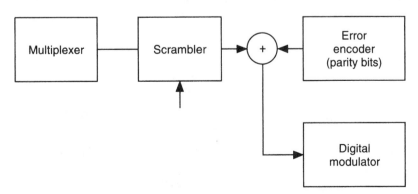

Figure 49.4 A digital encoder.

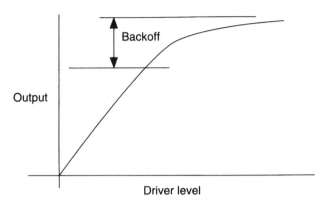

Figure 49.5 The output stage (PA) of a power-efficient amplifier will be non-linear.

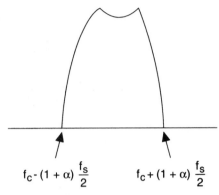

Figure 49.7 A raised-cosine filter characteristic.

level. The gain of a typical power amplifier (PA) is shown in Figure 49.5. The non-linearity of the PA stage can be reduced by "backing off" the peak output power by 3 or 4 dB, as shown in Figure 49.5.

For many applications, particularly in the field of satellites, QPSK is the most common choice. Because of this QPSK is often the standard by which the relative merits of other modulation systems are judged.

The unfiltered spectrum of a QPSK signal is shown in Figure 49.6. Notice that the spectrum is not uniform within the base bandwidth, and that the side-lobes are only about 16 dB down from the center frequency level. The non-linearity of the PA can boost the sidebands to almost the level of the center frequency, and so a band-pass filter must be inserted to attenuate these frequencies.

Usually a raised-cosine filter will be used as a band-pass filter, and it will have an idealized response, as shown in Figure 49.7. To a considerable extent the shape of this filter will determine the spectral efficiency (defined as the bit rate to the bandwidth) of the system. In practice efficiencies of around 1.7 bits/s/Hz can be achieved with $\alpha = 0.2$.

In turn, the spectrum of a QPSK signal filtered by a raised-cosine function will have an idealized waveform, as depicted in Figure 49.8.

The filtered spectrum does result in an attenuation of the side-lobes to 40 dB below the center frequency level, but it needs to be kept in mind that they are still there and can appear at the output at much higher levels due to PA non-linearities if precautions are not taken.

SPECTRALLY EFFICIENT ENCODING

It has been seen that the envelope variation in QSPK becomes a spectral problem at the PA stage. Spectrally efficient modulation can be achieved by employing techniques to smooth the envelope. QPSK allows phase transitions of ± 90 and ± 180 degrees, which when band-limited will cause very large envelope variations. By limiting the phase transitions (Orthogonal Keyed QPSK) or smoothing them (MSK), it is possible to obtain an envelope, as shown in Figure 49.9. The modulation used by Global System for Mobile Communications (GSM) is derived from minimum shift keying (MSK) modulation.

An MSK modulator differs from a QPSK modulator in that the rectangular-shaped pulses are converted into half-sinusoidal pulses by a pulse-shaping filter. MSK is a special case of continuous-phase frequency shift keying.

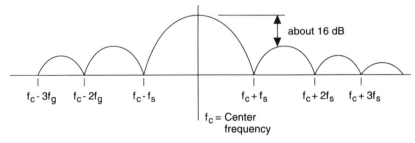

Figure 49.6 The spectrum of a QPSK signal; includes significant levels of sidebands.

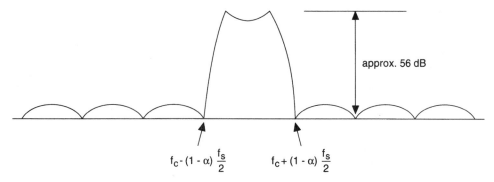

Figure 49.8 The spectrum of a QPSK signal, which has been passed through an idealized raised-cosine filter.

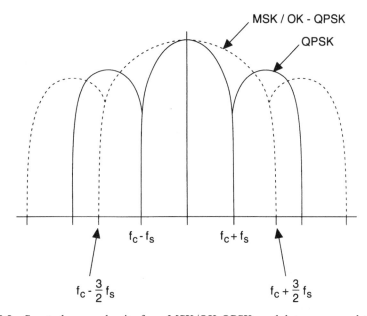

Figure 49.9 Spectral power density for a MSK/OK-QPSK modulator compared to QPSK.

DEMODULATION

Demodulation can be *coherent*, meaning that the receiver uses the transmitted frequency and phase information, or *incoherent*, meaning that it does not. A coherent system has superior noise performance, but is more complex and costly to implement.

GSM

GSM uses a form of modulation known as Gaussian minimum shift keying (GMSK). This is much the same as MSK, except that a filter with a Gaussian transfer function is used.

$\pi/4$ QPSK

Digital Advanced Mobile Phone Service (DAMPS) and Japanese digital uses a form of modulation called $\pi/4$ QPSK. This modulation method has spectral advantages over 8 QPSK and yet is simple to implement, it is not patented, and in particular it is very simple to demodulate. In fact, a standard frequency modulated (FM) discriminator can be used as a demodulator. As can be seen in Figure 49.10, the $\pi/4$ QPSK constellation looks like a pair of 4 QPSK diagrams superimposed. The phase excursions are limited by requiring that changes of phase must alternate from one constellation to the other. Thus instead of up to 180-degree changes being permitted in 8 QPSK, changes are limited to 135 degrees. This can be

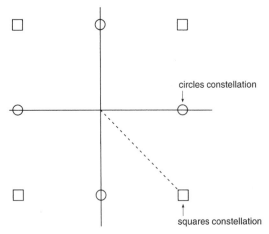

Figure 49.10 $\pi/4$ QPSK can be visualized as two superimposed 4 QPSK constellations. Transitions must alternate between the phases.

verified by considering Figure 49.10 and starting from either constellation; for example moving from the "circles constellation" from any starting point it can be seen that a transition to the "square constellation" can only involve a 45- or 135-degree phase shift. The penalty that is paid is that the information content is reduced from 3 bits/phase in 8 QPSK to 2 bits/phase.

SPREAD SPECTRUM OR CDMA

The definition of a spread spectrum signal is one that

- Has a transmitted signal, which has a much wider bandwidth than the baseband (message bandwidth).
- Has an occupied bandwidth, which is independent of the baseband signal.
- Has a baseband spreading generated by a code sequence that is independent of the baseband.

Under this definition GSM, with its wide-band frequency-hopping, conforms, but wide-band FM does not (because of the bandwidth dependence on the baseband signal).

The three basic methods of generating a spread spectrum are

- Encoding based on encryption with a reproducible pseudorandom digital sequence (which is the essence of code division multiple access [CDMA]).
- Frequency-hopping where a large number of frequencies can be selected in a pseudorandom manner.
- Chirp, where the carrier frequency is swept with each pulse.

Spread-spectrum techniques are widely used in the military and to some extent in civilian satellites.

The interest of the military was initially based on the relative security that the techniques provided, because it is practical to transmit useful spread-spectrum intelligence at levels that would be below the noise threshold of conventional receivers. Equipped with the right receivers, however, it is not as difficult as might at first be suspected to intercept spread-spectrum messages. This is particularly true of frequency hopping where the sweeping can be used to detect it.

Another military application of spread spectrum is in radar ranging, which can be done with precision without the use of high-powered, short-duration pulses.

Spread-spectrum signals can co-exist to some extent with narrow-band emissions, and these emissions will be seen as just background noise. Other spread-spectrum signals that co-exist with it will be seen as noise also, provided they are not correlated to the wanted signal. In practice there will always be some correlation, and as the number of interferers increases, so the chance of correlations increases.

It is the interference immunity that particularly appeals to the designers of cellular systems.

One of the things most difficult to grasp about spread-spectrum techniques is the ability of the system to work at very poor signal-to-noise (S/N) levels. Shannon's theorem can be used to illustrate how this can be done. Consider the relationship

Equation 49.1

$$C = W \log_2[1 + SNR]$$

where

$$C = \text{the maximum bit rate}$$
$$W = \text{bandwidth in Hz}$$
$$SNR = \text{the signal-to-noise ratio}$$

This tells us that any bit rate can be achieved provided the correct mix of S/N and bandwidth are adhered to. If it is desired to operate at a very poor SNR, for example, sending 2400-baud data over a channel with $SNR = 0.01$, Equation 49.1 gives, after transformation

Equation 49.2

$$W = C/\{\log_2[1 + SNR]\}$$
$$= 2400/\{\log_2(1.01)\}$$
$$= 167,185 \text{ Hz}$$

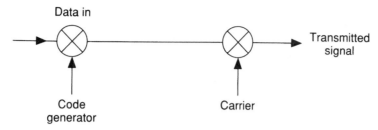

Figure 49.11   A spread-spectrum modulator.

So it appears that it is possible to communicate over such a noisy channel if 167 kHz of bandwidth is used. This expression can be simplified by noting that

$$C/W = \log_2\{1 + (SNR)\}$$

and changing bases of the logarithmic expression using the identity

$$\log_a F = \log_b F / \log_b a$$

then

$$C/W = \log_2 e \times \log_e\{1 + (SNR)\}$$

or

$$C/W = 1.44 \times \log_e\{1 + (SNR)\}$$

Note that $\log_e\{1 + (SNR)\}$ expands to

$$SNR - (1/2)[SNR]^2 + (1/3)[SNR]^3$$
$$-(1/4)[SNR]^4 + (1/5)[SNR]^5 \cdots \text{etc.}$$

For very noisy signals, *SNR* will be small (less than 0.1), and thus to a good approximation the higher powers of *SNR* can be ignored. Thus the relationship reduces to

Equation 49.3

$$C/W = 1.44 \times (SNR)$$

or

$$W = C/(1.44 \times SNR)$$

Using Equation 49.3 for the conditions in the previous example,

$$W = 2400/(1.44 \times 0.01) = 166,666 \text{ kHz}$$

which is a useful approximation to the required bandwidth.

CDMA

In essence, the CDMA principle is simple. As shown in Figure 49.11, the transmitted digital signal is multiplied by a pseudorandom sequence. The frequency of this sequence is such that it is significantly higher than the information signal, and so the resultant modulated waveform "smears" the modulation out over a wide spectral bandwidth. A conventional wide-band receiver would interpret the signal as noise.

In order to decode the signal, the receiver must use the same random sequence as the sender. By multiplying the received code with the decode key the intelligence is removed.

However, if other signals, either wide-band or narrow-band, are received, they will not correlate and so will be received as noise. With the use of robust error correction techniques much of this noise can be filtered out.

Because the pseudorandom sequence is different for each call, many calls can be set up on the same bandwidth and transmitted simultaneously. In this form spread spectrum is an example of CDMA.

MODULATION

The usual modulation technique for spread spectrum is phase modulation using either 180-degrees phase shift or BPSK; FSK is sometimes used, although it does have inferior S/N performance.

When the technique of encoding is a pseudorandom sequence multiplied by the data and then phase modulated, it is known as "direct-sequence" spread spectrum (DSSS). An important parameter of direct-sequence modulation is the "chip" or the duration of the smallest bit in the code sequence. This determines not only the spectral bandwidth of the modulated signal but also the processing gain of the receiver, both of which increase as the chip duration decreases. The modulating pseudo-random sequence is called the *chip sequence*.

The bulk of the information energy is contained in a bandwidth of $F_0 - F_c$ to $F_0 + F_c$, where F_0 = the center frequency of the transmission, and $F_c = 1/(t_c)$, where t_c = chip duration.

The chip sequence and data stream need not be synchronous although if they are, the clock recovery is simplified.

Other techniques include frequency hopping (as employed in GSM) and time hopping, where the time bursts have a pseudorandom duration.

Another technique known as *chirp modulation*, which consists of linearly sweeping over a wide frequency range, is also sometimes classified as spread spectrum, although the lack of a pseudorandom generator means that this is not true spread spectrum.

DEMODULATION

Demodulation proceeds in the reverse order of modulation (i.e., just multiply the received signal by a synchronized copy of the chip sequence). The additional complications that the frames must be synchronized by using an autocorrelation pseudorandom code (one that will still correlate even if a phase shift has occurred) and by allowing the received and the regenerated pseudorandom code to slip in phase (by adjusting their clock frequencies) until the best correlation is found.

Because direct-sequence demodulation is mainly done in the digital domain, the decoders are simple and cheap compared to frequency-hopping demodulators.

The demodulation can be done at the RF, IF of baseband levels. The demodulator is sometimes known as a "despreader," and indeed the baseband demodulators consist simply of a chip-sequence multiplier followed by a low-pass filter the purpose of which is to filter out the noise that is still spread.

FREQUENCY HOPPING

Frequency hopping involves sending bursts of code called *chips* on a number of different frequencies, determined by a pseudorandom sequence generator that controls a frequency synthesizer. The processing gain (improvement in S/N) of a receiver is equal to the number of available channels or to the total radio frequency (RF) bandwidth divided by the intermediate frequency (IF) bandwidth.

GSM uses this form of spread spectrum.

CHIRP SPREAD SPECTRUM

Chirp spread-spectrum modulation is largely confined to radar techniques that improve the range and resolution for a given power.

TESTING

The Federal Communications Commission (FCC) has permitted spread-spectrum systems under Part 15 of the Federal Communications rules governing the operation of radio systems without an individual license. The frequencies permitted are 902–928 MHz, 2.4–2.4835 GHz, and 5.725–5.850 GHz using direct-sequence or frequency-hopping techniques. Transmit power is limited to 1 watt, although no limit is placed on the antenna gain.

MULTIPATH IMMUNITY

Systems operating at frequencies around 900 MHz and greater have distinct nulls within the local interference patterns, which have wavelengths around 15 cm. Because DSSS operates over a wide bandwidth, it becomes unlikely that all of the signal will be in a null. The wider the operational bandwidth, the greater the immunity.

PROCESSING GAINS

The processing gain of a DSSS system is simply B_{rf}/B_s, where B_{rf} is the RF bandwidth and B_s is the signal or base-band width.

CODECs

CODECs, which are digital compression devices, are the key component leading to the spectral efficiency of digital systems. It has long been known that speech contains a high degree of redundancy. This can easily be demonstrated in a noisy room full of people talking all at once (such as at restaurants). Despite the background of babble, most people will not have too much trouble following the conversation of the person next to them. Even though a lot of the actual sound will be missed, the listener can reconstruct the original sound from the bits received. It was therefore recognized quite early in the development of electronics that clever encoding could reduce the bandwidth needed for speech.

Bandwidth compression of speech in the analog domain is possible, and the first practical example of this was in 1936 when "Dudley's Vocoder" was demonstrated. It consisted of ten 300-Hz band-pass filters, which were used to extract the speech envelopes and then to pass these envelopes through a 25-Hz band-pass filter. The original 3000 Hz was thus reduced to 250 Hz (plus guard bands).

These ten "representative" levels were then transmitted to a receiver, which reconstructed the original signal. The speech was said to be intelligible.

Truly controlled and efficient bandwidth compression of speech of a commercial standard had to await the development of economical digital processing. At the heart of current efforts in improvements in spectral efficiency is the CODEC, which compresses the bandwidth of digitally encoded speech. The redundancy of speech is such that the theoretical maximum compression is about 1.5 kbits/s. This limit is rapidly being approached.

Modern CODECs fall into three main families: waveform, source, and hybrid coding. The first of these, waveform coding, is not usually considered to be a compression technique, but on deeper analysis it is.

Waveform Coding

The early work on waveform coding was done by Bell Systems in the 1930s and 1940s. Commonly known today as PCM, it compresses voice into 64 kbits. As is well known, this encoding samples at the 8-kbit rate and uses 8-bit bytes. The 8 bits allow for only 64 levels, and it was the recognition that voice levels are perceived by the ear not linearly but logarithmically that permitted effective use of 8-bit coding. Use was made of companding to expand low signal levels and compress higher levels, as shown in Figure 49.12, so that more bits are used to represent the lower-level signals.

Use was also made of the by then well-known fact that telephone systems transmitted good intelligible

speech over bandwidths of around 3400 Hz. So the input audio signal was first passed through a band-pass analog filter. The Nyquist theorem says that the sample rate should be at least twice the bandwidth of the signal being sampled, and therefore 8-kbits/s sampling would be adequate.

To illustrate how this represents compression, consider the example of a compact disk that samples at 44,000 Hz and uses a 16-bit digital-to-analog conversion (DAC) process for an overall sampling rate of 704 kbits/s. Compared to this, PCM represents an 11 : 1 compression.

Because of this early work, PCM became a network standard, and most modern CODEC systems today are designed to multiplex channels into 64-kbits/s streams (or into higher multiples of 64-kbits/s).

The next step was a 2 : 1 compression technique to derive two 32-kbits/s channels from one 64-kbits/s channel. The method used is called adaptive differential PCM (ADPCM). Instead of sending the actual signal, ADPCM sends the difference between successive samples. The subjective quality for voice is equal to PCM, and the technique has been given international recognition by the CCITT under recommendation G.721.

Another form of waveform encoding is continuously variable slope delta (CVSD) modulation, which has been used for voice at rates from 12 to 16 kbits/s with reasonable success, but with high background noise levels.

Source Coding

Source coding involves modeling speech in a way that imitates the way the sounds were formed. The sounds of the human voice lie in the 100- to 10,000-Hz band and have an average power of a mere 10 to 20 microwatts. Compare the voice power level to that of a drum, which can produce about 25 watts peak power, or an orchestra, which can generate 100 watts. The vocal cords are responsible for producing the raw sound energy. They resonate at frequencies of 200–1000 Hz for women and 100–500 Hz for men. The sounds produced are rich in harmonics, and it is the filtering of the generated sounds by the cavities in the mouth, throat, and nose that produces the characteristic sounds of vowels. The lips, tongue, and teeth are used to produce consonant sounds.

The ear is a very selective receiver, with the perceived sound level being both frequency and level dependent. All frequencies have a threshold level below which they will not be heard; if the level is high enough, sounds will not only be heard but they will be felt. The ear has a response to sounds varying from around 25 Hz to 18 kHz, but the range is very age-dependent. From about 20 years of age the frequency range of the average person begins to deteriorate in frequency, linearity, and dynamic range until about 60 years of age, when the average hearing

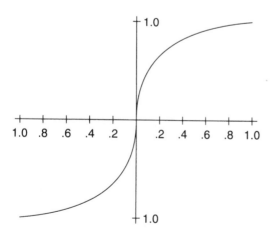

Figure 49.12 A companding curve.

bandwidth is only about 5 kHz (which accounts in a large part for the frequent complaint of older people that the youngsters "don't speak clearly anymore"). People subjected to high noise levels over long periods of time, such as in heavy industry, will have premature hearing impairment.

It is well established that for people with average hearing, good-quality intelligible speech can be transmitted within a bandwidth limited from 300 to 3400 Hz. This bandwidth is now the internationally accepted "toll quality" range.

The earliest of the source coding techniques is known as linear predictive coding (LPC). The system models voice as two types of sounds, vowel-like (called voiced sounds) and consonant-like (called unvoiced sounds) plus a time-varying digital filter.

An intelligent source coding CODEC breaks down the components of speech into its original parts by modeling the vocal tract. The vocal chords are seen as the source and the vocal tract a filter. The speech components of English can further be broken down into

Vowel sounds: A E I O U
Nasal sounds: M & N
Fricatives: S Z T H
Plosives: K P T

The compressed voice is sent in frames (timeslots) that transmit the following:

• The digital filter model
• The sound type (voiced or unvoiced)
• Amplitude
• Pitch information

Commercial CODECs that can achieve around 4 kbits/s and use LPC are available but they still leave something to be desired, particularly with female voices. LPC relies on the fact that speech is highly predictable in that no rapid changes in level or frequency are expected and the frequency range is quite limited. It is thus possible to sample speech, predict the next sample, and transmit only the difference between the prediction and the expectation. LPC CODECs, although spectrally efficient, are only useful for modeling voice sounds and are not suitable for transmission of other sounds. Even multiple speakers talking simultaneously can confuse the algorithm. Subjectively, the voice quality is only fair.

As the degree of LPC encoding increases, the CODECs will become more specific to voice and maybe even to particular characteristics of voice, so that a CODEC that sounds fine when used by a native speaker of,

for example, English may not sound so good when the speaker uses Chinese. The GSM CODEC, for example, was initially tested only with the five main European languages (which have commonality—English and German or French and Spanish). It is interesting to speculate on how other sound sources such as music or background noises like traffic noise will sound after they have been processed by a CODEC built for speech!

Also with increasing efficiency the CODEC algorithms become more complex, and so require more complex processing and more complex coding. This all translates to a higher chip cost.

Hybrid Coding

Hybrid CODECs use both waveform coding and source coding. By the mid-1980s there were three main hybrid techniques in common use: residual excited linear prediction (RELP), adaptive predictive coding (APC), and multipulse coding. Each of these are time-domain coders, which perform processing on the signal in real time. Like source coders, the hybrids use source models, but they use it differently. The model is used to identify and remove the redundancies.

RELP uses linear prediction to model the vocal tract, which is then used to inverse-filter the speech. It is the remaining (residual—from the first letter of the name RELP) signal that is sent. This signal can then be put through a low-pass filter (of around 800 Hz) and transmitted. It can be reconstructed by a reverse process at the receiver end. This method does have problems with the high frequencies found in female voices.

APC also uses a linear predictive residual, performing a number of time-domain processing operations to minimize the information that needs to be sent to transmit this information.

Multipulse compression is similar to APC, but additionally the residual is sent in pulses that are spaced in time and length depending on the information content. A multipulse code known as regular pulse excitation (RPE) is used by INMARSAT for its Aeronautical Mobile Satellite Terminals. GSM also uses a version of RPE.

A more advanced coding, adaptive transform coding (ATC), makes use of the different spectral content of speech (as shown in Figure 49.13). Voiced sounds are at lower frequencies than unvoiced sounds. The ATC algorithm allocates most transmission bandwidth to the spectral components that are most strongly represented after the LPC process and reverse-filtering. This produces good fidelity for both high- and low-pitched tones, and unlike most of the other techniques, can transmit fax and

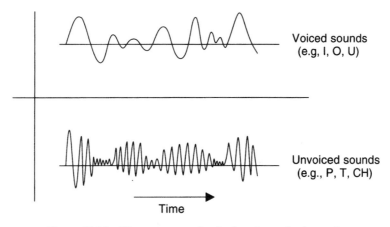

Figure 49.13 The spectrum of voiced and unvoiced sounds.

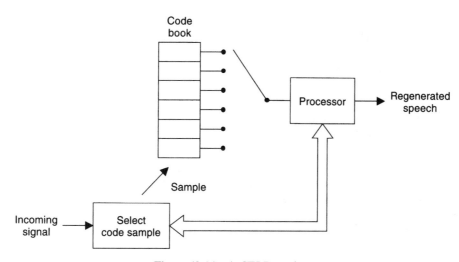

Figure 49.14 A CELP receiver.

modem tones. Because it processes a frame at a time, it does not need significant amounts of memory.

Code-Excited Linear Prediction

Although ATC produced good quality voice at rates as low as 9.6 kbits/s, its limitations below this rate limit its application. Code-excited linear prediction (CELP) has become the technique showing the most promise for lower rates. It is the method chosen by the Telecommunications Industry Association (TIA) for the US digital cellular specification. This technique is the most complex of the hybrid methods.

CELP does not use a vocal tract model, but rather it compares the sample with a library of individual generators and then transmits the one that is nearest to the

sample. At the receive end the same sample (called a *code*) is accessed and then processed to regenerate synthetic speech, as can be seen in Figure 49.14.

Time-Domain Harmonic Scaling

Time-domain harmonic scaling (TDHS) is a hybrid procedure that does not fall clearly into any of the coding schemes mentioned so far. It relies on the periodic nature of voiced sounds and the fact that adjacent samples have similar waveforms. It compresses by sending the average pitch for a number of surrounding samples. It has problems with short sounds and at transition points between voiced and unvoiced sounds.

A new encoding technique known as Multi-band excitation (MBE) has been developed, which offers enhanced performance over the LPC technique. Commercial units

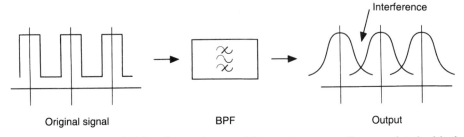

Figure 49.15 Inter-symbol interference is caused by spectrum spreading associated with the band-pass filters.

operating at 4.2 kbits are available and laboratory units are operating at 3 kbits.

DIGITAL SYSTEM PERFORMANCE

While analog system performance is easily quantified by S/N measurements, the situation is a little more complex for digital systems. Because of the increased complexity, there are more ways of performance impairment of a digital radio system. The contributing factors are

- Additive white noise
- Inter-symbol interference
- The transmission-medium band-pass characteristic
- Co-channel interference
- Adjacent-channel interference

- Timing errors
- Non-linearities of phase and amplitude
- Fading

Inter-symbol interference is caused by the inevitable spreading of the signal spectrum when it is passed through the filters that are part of the system. The spreading effect is illustrated in Figure 49.15.

Fading can be either flat (wide-band), as is the case in rain attenuation, or selective (due to multipath), where only some frequencies are affected.

The effect of these factors can be measured by the following parameters:

- BER
- Error-free seconds
- Transmitted power spectrum

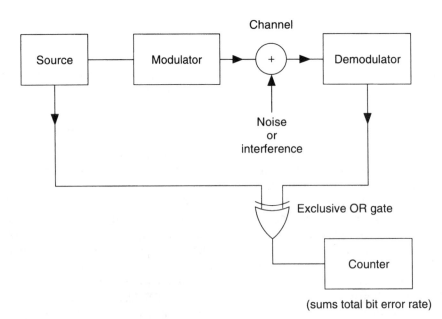

Figure 49.16 A simple BER counter.

The easiest thing to measure is BER, and in its simplest form the measurement can be done using only an XOR gate and a counter, as shown in Figure 49.16.

Acceptable BERs in real systems are

System	BER
Mobile radio	10^{-2}
Good voice	10^{-4}
Microwave	10^{-6}
Quality data	10^{-8}

For a given transmitted power the BER will increase as the data rate or total bandwidth increases. In order to compare the performance of digital systems, a measure that is independent of bandwidth and pulse rate is needed. Such a measure can be defined by dividing the signal energy at the receiver by the noise power spectral density.

The bit energy at the receiver can be defined as

$$E_b = CT_b$$

where

E_b = the bit energy at the receiver input

C = the received carrier power

T_b = the bit duration

Further, if the noise power spectral density, N, is defined as

$$N = \{\text{noise power}\}/\{\text{bandwidth}\}$$

then E_b/N is a performance measure that is independent of both bandwidth and pulse rate.

50

OTHER MOBILE PRODUCTS

A cellular operator will eventually begin to look at the potential of other mobile products. The similarity between public mobile radio (PMR), paging, and cellular radio cannot easily be overlooked, and cellular operators would do well to evaluate the opportunities presented by each.

Voice-message/mail systems which are telephone network value-added services, are often added to cellular operations. Short messaging is essentially paging over cellular but with verification of receipt.

Although many operators are enthusiastic about the "one-stop shop" concept for mobile products, others are not. From the customer's perspective, the one-stop shop has advantages. So why are some operators so cool on the idea? The answer has much to do with the degree of similarity. Admittedly, these other services have a lot in common with cellular operations, but they also have a lot that is not similar. Operators tend to regard cellular as the "senior service" and the other activities as add-ons. As a result, operators tend to overlook the opportunities presented by these other activities. Without properly evaluating the resources to operate these new services, it is easy to underestimate and consequently underfund them.

A cellular operator usually has a fairly substantial infrastructure network of shelters, communications links, emergency power supplies, and so on, all of which are necessary for the successful operation of other mobile services. By using these already established features, the cost of adding "other services" can be about half what it would be if the other service was provided completely independently.

Since the staff skills and mix are also closely related, the ongoing operating costs of "other services" can also be integrated in a very cost-effective way. This chapter explores these options and examines their implications for the cellular operator. The advantages of expanding into new services include better use of resources (both staff power and tangible resources), and lower start-up and maintenance costs than for non-cellular operators.

PUBLIC MOBILE RADIO

Because many cellular operators might consider PMR to be a competitive system, they may be reluctant to enter the field. In practice, many large cellular operators not only also operate PMR but are themselves extensive users of the system. To understand why, it is necessary to look at the differences between cellular radio and PMR from the customer's viewpoint.

PMR includes simple two-way, as well as trunked and wide-area, radio systems. Among the advantages of PMR are the following:

- The annual operating cost is substantially lower.
- Group calls can be made. All vehicles, or some subgroup thereof, can be called simultaneously. This capability is essential for many operations, such as taxi and parcel-pick-up-and-delivery services.
- Although telephone access is usually not as convenient as with cellular, it is much cheaper on PMR.
- PMR is more suitable for fleet operation.
- An incoming local call may be identified and charged as a local call only.

PMR can consist of a simple, conventional two-way radio repeater, or it can be a much more sophisticated (and expensive) trunked network. In all cases, automatic telephone access and selective calling can be provided.

Selective calling is the ability to call either a particular vehicle (by alerting it with a ring tone) or a predetermined group of vehicles. Normally, the system is configured so that only the called group can hear (or take part in) the conversation. Cellular radio systems can be thought of as having selective call capability to only one mobile unit (car telephone) at a time.

Group calling can be used to call a predetermined group. For example, a cellular-telephone company can call, as a group, all technicians or all salespeople or all staff. A large number of groups and subgroups is possible. For example, a group of all technicians and all salespeople could also be called. Of course, individual calls are also possible within this scheme.

Automatic telephone access is also available as a feature of PMR systems. There are three modes of operation: full-duplex, half-duplex, and simplex. In all cases, fully automatic in-dialing (to the mobile) and out-dialing (from the mobile) are possible.

Full-Duplex Mode

Full-duplex mode is the same as that used in a cellular phone. It enables simultaneous talk-and-receive signals in the same way as a conventional telephone. Although this mode is the one preferred by most users, it is not the usual mode for PMR, because full-duplex requires the simultaneous operation of the transmit-and-receive functions of the mobile radio. This capability results in a higher-cost mobile (because continuous transmit duty requires more elaborate heat sinks) and requires the addition of a duplexer to enable transmission and reception simultaneously from one antenna, which is not only costly and bulky but significantly reduces the useful range of the mobile.

Half-Duplex (Two-Channel Simplex) Mode

Half-duplex is the usual compromise adopted by PMR operators. A press-to-talk (PTT) switch is needed in this mode because simultaneous transmit-and-receive on the radio path has been eliminated. In half-duplex, separate frequencies are used for transmission and reception (this allows the repeater function to be used, as described earlier).

When telephone conversations are conducted on a half-duplex system, only one party can talk at a time. This is because the mobile can only receive or transmit, but it cannot do both at one time. In some systems, control is given to the mobile, so that if the PTT switch is pressed, land subscribers cannot transmit. In most systems, however, land subscribers can still transmit on top of the mobile, though the transmission cannot be heard. Half-duplex telephone interconnections, although useful, cannot provide the same quality of service as full-duplex or as a cellular system.

Simplex Mode

In simplex mode, only one radio frequency is used for transmission and reception and so only one path of conversation is possible at any one time. The direction for telephone interconnect is usually chosen by a voice voting system so that once a signal is heard (for example, voice), then that direction has priority until the conversation ceases. In less sophisticated systems, this can be somewhat embarrassing when, for example, a recorded announcement with very few pauses is dialed, and the system locks-up on that announcement. The system, however, can have land party or mobile party priority.

Such systems are often referred to as VOX (Voice Operated Switch, X = transmit as in Tx). Although they can work well if they are carefully set up, they can be the source of some frustration if the VOX operates erratically. VOX operation can clip the first syllables or even short words of a user's sentence if it is incorrectly adjusted (too long attack time prevails).

Simplex mode is the mode least-preferred by both users and most operators, but it is most-preferred by frequency-management authorities because it uses only one frequency.

Trunking

Trunking offers more sophisticated features than a simple mobile repeater system and is used for circuit efficiency and wide-area coverage. As discussed earlier, an advanced trunking system has most of the features of cellular radio but can be quite expensive to implement.

The mobiles used for trunking are usually more sophisticated but priced at about the same level as those used on a simple repeater. Trunked systems become economical only when used on a fairly large-scale basis and usually only in reasonably high-density areas. Because a trunked network is very similar to a cellular network, it is the type of system most suited to be included as part of a network package. Trunked systems may become economical when five or more mobile channels are justified on the basis of traffic.

iDEN

The integrated dispatch enhanced network (iDEN) is, in fact, a hybrid cellular/trunked radio system, and is

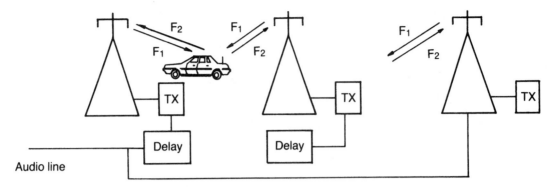

Figure 50.1 In simulcast transmission, all bases use the same frequencies. Destructive interference presents a challenge when coverages overlap at similar levels occurs. The delays ensure that signals that overlap signals of similar levels are in phase.

currently competing rather successfully with both technologies. (See Chapter 45.)

Other Wide-Area Systems

When traffic density is low, other methods of obtaining wide-area coverage (more than the area covered by one base) can prove just as effective as trunking and are somewhat cheaper.

Simulcast (also known as quasi-synchronous) can be used to parallel a number of transmitters/receivers on one frequency. This technique fell out of favor some years ago when it was found to be difficult to synchronize the transmitters in a way that would avoid "dead spots" caused by interference, particularly due to audio phase drifts. Current technology can handle these problems so that simulcast is now the cheapest wide-area system. It is effective, but of all the techniques it still requires the most maintenance. Figure 50.1 illustrates a simulcast transmission.

Race voting (the term used to signify the way in which the base receivers are selected) is also popular and offers the wide-area coverage of simulcast without the attendant difficulties in synchronization. Race voting, however, consumes more frequencies and requires more complex multichannel, microprocessor-controlled mobiles. This technique is illustrated in Figure 50.2.

Race voting uses base stations with different transmit but identical receive frequencies. In this way, the need to phase synchronize the base transmitters is removed. When a mobile transmits, it can be received by one or more base receivers. Each receiver measures the field strength, and then attempts to seize control at a speed that increases with increased received field strength. In this way, the receiver that has the highest receive level seizes the line and locks out all others. This "race" occurs

each time the mobile presses the PTT switch, or it can be forced to occur at fixed time intervals for duplex systems.

The receiver that wins the "race" inhibits all other receivers on the seizure line and then takes control of the audio both to land lines and to the other transmitters. Ordinarily only three base transmit frequencies are required.

A more sophisticated voting system using a centralized comparator to compare the signal-to-noise (S/N) levels at each base station can be used. This system requires individual links back to the voting system from each base station. Although these extra links make the centralized voting more costly, in practice centralized voting has proven more effective and reliable than the race-voting technique.

Marketing

As has already been noted, PMR appeals to a different market segment than cellular customers, and so no strong conflicts are expected. A plus for marketing strategy is that corporate target groups identified for cellular radio, if not in the market for cellular, will probably be in the market for PMR, and vice versa. The potential PMR customer is usually (today) the owner of a small- to medium-sized business who undoubtedly will benefit from some form of mobile communications. Thus, whether the potential customer is targeted by advertising, direct marketing, displays, or any other medium, the same potential customer has a good chance of wanting one of the two systems. Hence, marketing costs can be contained.

Indeed, a good sales force should be able to adapt very readily to the additional technology and, although a specialized salesperson would probably still be the norm, each salesperson should be able to sell the other product

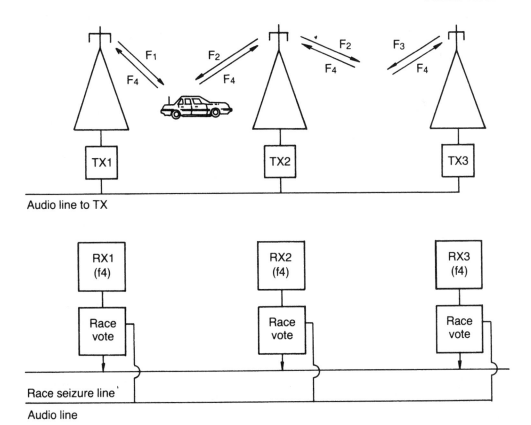

Figure 50.2 Race voting typically uses three base-station transmit frequencies and one TX frequency. The race-vote device seizes the line at a rate that increases with receive level. All receivers are on the mobile TX frequency, so the one with the best receive level seizes the line and controls the audio for the whole system.

when the opportunity arises. The main difference in marketing emphasis is that the PMR salesperson typically seeks the fleet customer, and the cellular salesperson usually targets the single-unit buyer. This distinction alone justifies a separate sales structure, even though cross-sales can be anticipated. PMR also has a greater diversity of options that tends to require specialist sales attention.

Any serious PMR operator should have an experienced, technically qualified sales engineer who can discuss in detail the system requirements of the customers. Most customers will be relatively unclear about what they really want and will have only the vaguest idea of the available options. The sales engineer should therefore back up sales personnel and directly handle only the bigger, more technical customer accounts.

Financing

The financial structure of a PMR system is very similar to that of a cellular operation. The operation is rather capital-intensive, with the PMR mobiles costing sig-

nificantly more than a cellular phone, but base stations are much cheaper. The marginal cost of expansion is similar (on a per-channel basis).

Billing is sometimes on a straight monthly fee per mobile unit, but it is often on an air-time basis. Telephone calls are usually charged separately. The cellular billing system can usually be adapted to PMR operators.

The income per mobile is about the same as that expected from a cellular system.

Engineering

Not all cell sites make ideal PMR sites because the frequency reuse requirement of cellular radio leads to less prominent sites. On the other hand, PMR usually uses large cells, and since it usually uses half-duplex mode, very large cells can be achieved.

With some care, a good number of cellular sites can be used for PMR, but some additional, special sites will probably be required. To some extent this requirement reduces the attractiveness of operating both systems.

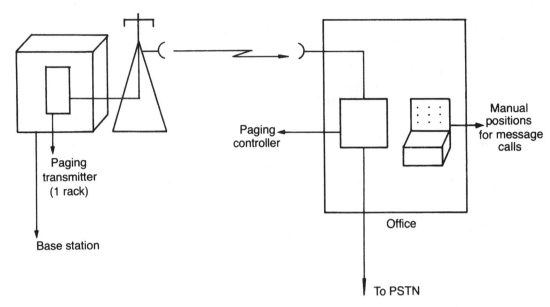

Figure 50.3   A basic paging configuration.

Notice, though, that there is no conflict in rural areas, where cellular radio also uses high sites.

The technology used to operate and maintain a PMR network is very closely related to the cellular radio requirements and, with a minimum of additional training, engineering staff can handle both systems. The test equipment employed for cellular radio installation and maintenance should be readily applicable to PMR. Radio survey and installation techniques are similar in both systems and could use the same staff and facilities. PMR repair, however, probably requires a specialist technician, but this work can be contracted.

PAGING

Paging is a relatively mature industry and widely seen as one in decline. The extensive availability of cellular phones, and to a lesser extent SMS, has led to its demise. Because many of the resources needed to run a successful paging business are closely related to those needed for cellular radio, most operators of either business at some time will probably consider the possibility of entering the other domain. Figure 50.3 shows a typical paging configuration.

Paging can come in alert (tone) only, numeric, or alphanumeric forms. Recent advances in paging have concentrated on reducing the size and increasing the capacity of the paging receiver. The price has decreased only slowly over time and ranges from about $50 for a simple alert pager to about $300 for a sophisticated message pager (at retail prices). Paging seems to be re-

treating into mini systems for local use in areas with hospitals and factories.

Operating

A paging system ordinarily requires a manual operator, which can be a major expense. If the cellular system already has an operator, these functions can be combined. Approximately one operator per 150 manual subscribers is required.

Paging Formats

Four paging types may need to be supported:

- Alert
- Alert and voice
- Numeric
- Alphanumeric

The alert paging type usually gives an audible alarm that indicates a page. There may be more than one tone so that different conditions can be indicated. Most alert pagers have either one, two, or four separate tones. Some also have a vibrating mode, so that calls can be received in meetings or other situations without disturbing others. Many users agree that the most valuable function of the alert pager is the test switch that simulates a paging call and can be used as an excuse to escape from especially trying meetings to attend to an "urgent call." Early versions of these pagers were analog (tones such as two of five, or two of seven), but more recent systems are digital.

The alert-and-voice paging type gives the audible alert followed by a voice message. This type usually, but not necessarily, involves a manual operator who takes a message and relays it. At times, a most inappropriate message can be broadcast over these systems. For example, a quote may be relayed while the customer is within hearing range. Alert-and-voice pagers must be used with care. Some systems have a message-storage facility that works much like a telephone answering machine.

A numeric pager operates much as the alert pager, but it can also register a number. This number can be sent either by a manual operator or from a conventional touchtone telephone. The most common use is to transmit the telephone number to which the return call is expected.

Alphanumeric (*alpha* refers to letters and *numeric* to numbers) systems initially call in the same way as an alert pager, but the caller can leave an alphanumeric message. These pagers usually have memories, and some can store more than 20 messages or 400 characters. The message can be entered by a manual operator or, in some cases, from a personal computer via an appropriate modem. The personal computer usually needs specialized software to access the paging controller.

VOICE MAIL

Voice mail enables messages to be stored and retrieved at the user's convenience. Because more than half of all business calls are known to be unsuccessful, it makes sense for cellular users to have access to a voice-mail service. This service can be likened to a sophisticated answering machine.

Additional revenue is earned from the charges for additional airtime, so it may not be necessary to also charge for storage or access. This decision depends on air-time charges, but usually they are adequate to cover voice-mail costs with only a nominal rental charge of, for example, $5–$10 per month.

The concept and value of voice mail (even like the concept of cellular itself) sometimes can be difficult to explain to the general public. Advertising may be of limited value unless a "free trial" is offered so that the users can become familiar with the service. Although the acceptance rate is high, the most usual complaint is that the system is "too difficult" to use. Customer education programs, which can be either a voice-mail tutorial or a follow-up on those who are paying for the service, but not using it, are generally successful.

The voice-mail system should be co-located with the switch, because, like the switch, it requires 24-hour attention. Cellular switches can work directly with voice-messaging systems; operators who have added this feature report cellular customer demand at around 30 percent.

The voice-message system is basically a voice mail box; it differs from an answering machine only in its flexibility. Incoming messages can be categorized as urgent, recent, or even filed according to a particular caller. Once a message has been received, it can simply be stored to await an access request or to initiate a pager or telephone call.

Automatic purging of the files after a specified period means the system holds only up-to-date information. Recorded announcements have also been exploited; systems can carry information about the weather, news, sports, the stock market, traffic, restaurants, theaters, and other items of interest, all available on request from a menu. Selections are made using Touch-Tone dialing to indicate the menu item sought.

PACKET RADIO

Packet radio systems are used to send "packets" of data over mobile radio equipment. From a control console, an operator can send information specifically addressed to one or more receivers. Because dedicated land links are quite expensive, shared alternative data links can be attractive. In its present form, packet radio systems are suitable for point-to-point-to-multipoint thin-line (low-capacity, usually single-channel) systems. For this reason, they are especially attractive to smaller companies and to operations in rural areas, where high capacity is not a requirement.

Reliability in these systems is built into the error correction code and is quite high. Often, however, the system is used to replace dedicated land links, and conventional mobile two-way equipment is used for the radio frequency (RF) link. This equipment has a mean time between failure (MTBF) that is low compared to leased lines. As a result, for high-security applications, it is necessary to build in a good deal of redundancy (which usually means 100 percent duplication and hence doubling of costs).

The data sent is synchronous and requires a handshake reply (that is, a confirmation that the data have been correctly received and decoded). This precaution ensures very low data error rates together with positive assurance that the message was received.

Packet switching can be accomplished in a single hardware unit that forms the interface between the computer terminal (usually an RS 232 bus) and the radio transceiver. A packet of information contains the

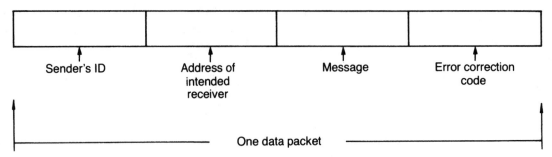

Figure 50.4   Typical data packet. Notice that the actual message may be broken up into a number of packets.

sender's identification, the receiver's address, the message, and a block of redundant error correction code. The process of assembling and disassembling data for transmission is called *padding*. Data bursts typically last 250 seconds; this short period offers some measure of security. Figure 50.4 shows a typical data packet.

The cost of adding this technology to an existing mobile system is quite low (less than $1000 without the personal computer). The cost of a complete dedicated network, however, is such that payback periods of five years (when compared to leased lines) should be anticipated.

Cellular mobile operators often have their own mobile radio networks (to provide an independent operational link between the base stations and switch) and may find incidental internal uses for data-packet technology.

51

SAFETY ISSUES

INTRODUCTION

Numerous tests have been done to establish the safety of people working in the radio frequency (RF) environment. Many of these tests are epidemiological case studies of personnel exposed over long periods to high RF fields in the course of their normal working environment.

Anecdotal evidence seems to be that those (usually men) who work in intense RF fields (such as near TV and frequency-modulated (FM) broadcast antennas, which only a little more than a decade ago were routinely worked on "live" produced high proportions of female offspring. I have never seen this confirmed scientifically, but the belief is widely held and appears to be held in many different countries.

A comprehensive study was completed in 1991 by a panel of experts from the Institute of Electrical and Electronics Engineers (IEEE) in the United States. The study involved a review of 321 previous studies.

Although some of the studies identified that high levels of prolonged exposure to 1-GHz-plus frequencies could accelerate some chemically induced cancers in mice, it surprisingly also found that the same exposures to healthy mice actually produced longer life spans. Of course, the increase in life span was small.

Caution must be expressed when considering mice as subjects because the effect of a particular frequency on an organism is dependent on its size. The closer the organism is to the resonant frequency, the more energy it will absorb. A mouse might be resonant at 1-GHz-plus frequencies, but humans resonate at around 80 MHz. Also, since all animals are made up of collections of individual cells, it is possible that at certain frequencies certain organs could be selectively affected.

Even at absorption rates as high as 10 watts/kg, little effect was found on health at frequencies used by cellular radio (800–900 MHz). This level is well above potential radiation levels in the cellular RF environment.

RF radiation is known as non-ionizing, to distinguish it from high-energy radiation, such as X-rays and atomic decay particles. These latter have such high energy that they literally knock electrons off the atoms, leaving behind highly reactive ions, which will rapidly recombine with other atoms and thereby change the chemical composition of the surrounding tissue. The main effect of RF power at cellular frequencies is to energize electrons still in their orbitals, which has the net effect of raising the body temperature. This in itself will cause some measurable metabolic changes, and some of the internal chemical processes are quite temperature-dependent. However, body temperature is also raised by physical activity or sitting in the sun or in front of a heater.

Mobile phones operate below the exposure levels set by bodies such as the IEEE (US) and the National Council on Radiation Protection (NCRP), which have set levels below which they are satisfied that there is no scientific evidence that the exposure poses any health risk. The human brain operates at about 20 watts, or one-fifth of the total body energy at rest. Most of that power is dissipated as heat. A 0.6-watt phone operating near the head can be shown to be virtually as efficient a radiator as one operating in free space. This means that only a small proportion of the total energy is transmitted into the head. It has been established that at least 50 percent of the incident RF energy at cellular frequencies is reflected by the head. Combining these effects means that at most 0.2 watt would be absorbed, which is only 1/100th of the brain's normal operating power.

Evidence from the UK suggests that the maximum absorption from a cellular phone of 0.03 watt per square centimeter is less than one-tenth of the level that is considered to have any effect on humans by the UK National Radiological Protection Board.

The level of exposure to RF has risen by a factor of at least 10 times over the last 40 years, but the incidence of cancer on a per population basis for any given age group has remained constant. In particular, the incidence of brain cancer has remained constant at around 5 to 6 cases per 100,000 population, and the location distribution remains constant. There is no evidence of a shift toward locations nearer to the antenna location. One simple test would be to compare the brain cancer rates of cellular users with that of non-cellular users. This study has yet to be done.

In 1993 the Cellular Telecommunications Industry Association (CTIA), convinced that there is no health hazard but nevertheless concerned about some of the claims and worried about having the issue inexpertly decided in the courts, instigated a four-point study of the possible health issues. The study involves:

- A literature search. Review of the already extensive body of research done on the effects of RF on the human body.
- Phantom Head Study. A model head, filled with a gelatin-like substance to simulate the electrical properties of a head, can be studied in detail for evidence of localized effects on brain tissue. In particular, the dosage levels can be quantified.
- Involvement of the cancer institute in an epidemiological study of brain cancers, which would include but not be restricted to a survey of patients on their usage of cellular phones.
- Experiments on tissue and living subjects.

At a later time the study was extended to include solutions to mobile interference to medical devices.

The study was conducted by the Wireless Technology Research (WTR) program, and took six years to complete. The results were disappointing to most and were basically inconclusive, with the WTR chairman George Carlo claiming that the studies did not suggest any link with cancer and then later claiming that they did.

Ongoing studies by the CTIA have been budgeted for, but as of 1999 no firm projects existed; it is being suggested that the Food and Drug Administration (FDA) will take a lead in future studies.

UK Study

A study in Britain at the Bristol Royal Infirmary, using 36 volunteers with transmitters attached to their heads to simulate 30 minutes of mobile phone use, has shown a link between short-term memory loss and the microwave radiation. The cells involved in short-term memory are near the right ear. These tests are being reproduced in other trials.

Natural Nonionizing Radiation

Sunshine exposes the body to about 80 mW per square centimeter, and nearby lightning to megawatts of radio frequency interference (RFI). The sun is a strong source of radio energy all the way up to the ultrahigh frequency (UHF) band, as is indeed the galaxy. The big bang itself left behind a microwave background (admittedly of low level). So the claim that an RF environment is not "natural" cannot be supported. What has actually changed is the level; and particularly the level from handheld phones, due mostly to their close proximity.

In fact a mobile phone produces a field about 1 billion times stronger than the background RF levels.

Media Balance?

Perhaps the biggest problem with mobile-phone safety is getting a balanced view from the print and broadcast media. In 1993 when a woman in Florida sued for damages, claiming that a mobile phone had exacerbated her tumor, the worldwide press covered the story with enthusiasm. Later in 1995, however, when the judge dismissed the case citing "insufficient evidence" the media showed no interest. The public therefore is left with the impression that the claims were valid.

Most people in the industry dismiss claims of a health risk from mobile phones as being either non-existent or absolutely minimal. The scare campaign sponsored by the media, however, has left many with lingering doubts, and some with an outright fear of the consequences of exposure to radiation. Perhaps as much as anything, the lack of an aggressive response from the industry has caused the levels of public fear and suspicion to rise to the unacceptable level that it has today. Spurred by pseudoscience and half-baked "facts," a small army of people today are mounting a "safety" campaign against the industry, which the industry seems determined to ignore.

The protesters support the argument that, proven or not, the only way to be sure is to stop or miminize all exposure to RF. They rarely suggest that the high-power (100,000s of watts) city-based TV and frequency-modulated (FM) stations be closed down; instead, they form protest lobbies over the siting of mobile-phone base stations. In this regard it is worth noting that the exposure from a base station placed 100 meters away is miniscule compared to the exposure one would get from making a few calls a day with a handheld mobile phone.

In fact, this issue is what really separates the protesters from the "serious" researchers. While the former protest over base-station sites, the serious researchers are concerned for the health of the long-term high-call-rate mobile-phone users.

It is extremely difficult to find an objective opinion anywhere. Researches in both camps are playing for large stakes (multimillion dollar research grants), and they invariably release reports recommending *further* research (i.e., more funding). The reports are rarely conclusive and many are contradictory.

Radiation Scare Fraud

In mid 1999, when the Department of Health and Human Services' Office of Research Integrity released the draft analysis concluding that Robert P. Liburdy, a biochemist of Lawence Berkely National Laboratory "intentionally falsified data" to show that low-level electromagnetic fields affected cultured cells, the popular press ignored the announcement.

In the report, investigation revealed that Liburdy had spent most of a \$6 million federal grant between 1990 and 1999 and concluded that Liburdy had presented only 7.1% of the data points to support his theory that the low-level fields alter the movement of calcium ions in cell membranes. It also contended that he had fabricated readings to make the apparent errors look smaller than they in fact were. The report notes that Liburdy manipulated calculations to show an apparent effect when none existed and made up data to conceal the manipulation. This information was totally ignored by the popular press. But what a noise it made when Liburdy first released "evidence" of potential cell damage from low-level electromagnetic radiation.

The Relativities of Exposure

The exposure levels of various RF sources can be referenced to the most common source (a GSM handheld). If this is given a unit value of exposure, we find the approximate relative power levels are

- Base-station antenna at 1 meter = 10
- GSM handheld (average) = 1
- Base-station antenna at 6 meters = 1/10
- CDMA handheld (average) = 1/100
- Base stations (at close proximity on the ground i.e., within 100 meters) = 1/10,000
- Natural background radiation = 1/100,000,000

WANT TO KNOW MORE ABOUT EMR?

The literature on the subject is extensive and perhaps the best way to get a good overview is to look at the World Health Organization (WHO) Web site at www.who.int/emf/. Below are a few extracts from that page.

> The International EMF [electromagnetic force] Project was established by WHO in 1996. For the Project, EMFs are defined as electromagnetic fields with frequencies from 0 to 300 GHz.

> Static fields (0 Hz): Magnetic levitation trains for public transportation, magnetic resonance imaging devices used in medicine, and electrolytic devices using direct electric currents for materials processing in industry.

> Extremely low frequency (ELF) fields (>0 to 300 Hz): Trains for public transport ($16\frac{2}{3}$ to 50 or 60 Hz, plus harmonics), any device involved in the generation, distribution or use of electric power (normally 50 or 60 Hz).

> Intermediate frequency (IF) fields (>300 Hz to 10 MHz): Anti-theft and security devices, induction heaters and video display units.

> Radiofrequency (RF) fields (>10 MHz to 300 GHz): Mobile telephones or telecommunications transmitters, radars, video display units and diathermy units.

Two reviews of the genotoxic (damage to genes, and hence to DNA) potential of RF were published in 1998. Verschaeve and Maes concluded that:

> According to a great majority of papers, RF fields, and mobile telephone frequencies in particular, are not genotoxic: they do not induce genetic effects *in vitro* [in cell culture] and *in vivo* [in animals], at least under non-thermal conditions [conditions that do not cause heating], and do not seem to be teratogenic [cause birth defects] or to induce cancer.

Standards

> Specifically, the ICNIRP standard is 0.40 mW/cm-sq for cellular phone frequencies and 0.90 mW/cm-sq for PCS phone frequencies, while the NCRP guideline is 0.57 mW/cm-sq for cellular phone frequencies and 1.00 mW/cm-sq for PCS phone frequencies.

> Guidelines for Evaluating the Environmental Effects of Radiofrequency Radiation (FCC 96-326), Federal Communications Commission, Washington, D.C., 1996. Available from the FCC web page.

> Specifically, the new FCC standard is 0.57 mW/cm-sq for cellular phone frequencies and 1.0 mW/cm-sq for PCS phone frequencies.

> **UK standard**: The UK standard is 0.57 mW/cm-sq at 900 MHz and 1.00 mW/cm-sq at 1800 MHz. There are multiple transmitting antennas at different frequencies, the method for assuring adherence to the ANSI or FCC standards is complex.

These radio frequency standards are expressed in "plane wave power density," which is measured in milli-

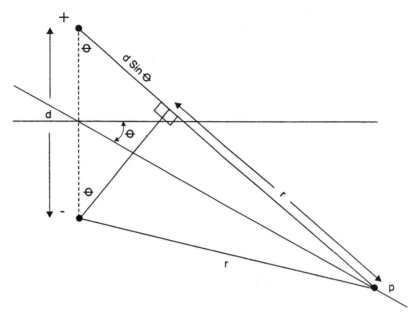

Figure 51.1 Dipole fields like electric and magnetic fields vary as an inverse cube due to the geometry as shown.

watts per square centimeter (mW/cm^2). For personal communication service (PCS) antennas, the 1992 American National Standards Institute (ANSI)/IEEE exposure standard for the general public is 1.2 mW/cm^2. For cellular phones, the ANSI/IEEE exposure standard for the general public is 0.57 mW/cm^2. The ICNIRP standards are slightly lower and the NCRP standards are essentially identical.

Safety standards for uncontrolled (public) exposure could be violated if antennas were mounted in such a way that the public could gain access to areas within 20 feet (6 meters) of the antennas themselves at the same level (height) as the antenna. This could arise for antennas mounted on, or near, the roofs of buildings.

If there are areas accessible to workers that exceed the 1992 ANSI or Federal Communications Commission (FCC) standards for uncontrolled (public) exposure, make sure workers know where the areas are, and what precautions need to be taken when entering these areas. In general, this would be areas less than 20 feet (6 meters) from the antennas (but in the plane of the antenna beam).

Electric and Magnetic Fields

Electric and magnetic fields are rather different from the EMR discussed earlier. If left to themselves, EMR fields will propagate out into space, with the field strength decreasing as an inverse square law function. Electric and magnetic fields, however, are likely to be found around power lines and transformers, and the magnetic fields

collapse as soon as their power source is removed. Furthermore, they decrease in strength by an inverse cube law (because their sources are dipoles).

Consider a field as seen in Figure 51.1 that is generated by the charges flowing through a power line feed. Each of the charges exerts a field at a point P that is inversely proportional to the square of their distance to it. If we let the distance from one of the points be r, as in Figure 51.1, then

$$F = K/r^2 - K/(r + d\sin\{\phi\})^2$$
$$= K((r + d\sin\{\phi\})^2 - r^2)/(r^2(r + d\sin\{\phi\})^2)$$
$$= K(r^2 + 2\,dr\sin\{\phi\} + (d\sin\{\phi\})^2 - r^2)/$$
$$(r^2 \cdot (r + d\sin\{\phi\})^2)$$

In the limit as r tends to infinity

$$F = (2 \cdot K \cdot d \cdot r \cdot \sin\{\phi\})/r^4$$

or

$$F = k/r^3$$

Thus dipole fields that are caused by power lines, transformers, and electrical appliances fall as an inverse cube of the distance, meaning that they are significant only at very short range.

Like radio-wave radiation, there is no conclusive evidence of health risks to humans from these fields.

PMR

Compared to two-way radios, which may be 25 watts for vehicle-mounted units and 5 watts for handhelds, the 0.6 watt of an analog mobile phone is low power. However, the net exposure by the users of the two systems are equalized by the fact that cellular users tend to have longer usage times in terms of both call holding time and total talk time per day. Analog cellular phones, being duplex, transmit even when the user is not talking. Digital phones, some of which do turn off during speech pauses (and so, like two-way radios with "press-to-talk" (PTT) operation, do not transmit except while the user is talking), have power levels similar to but less than two-way radios.

However, it is still wise to avoid working for more than 10 minutes within 1 meter of a working PMR base-station antenna and to take sensible precautions to minimize RF exposure.

TDMA

Time division multiple access (TDMA) systems, like Global System for Mobile Communications (GSM) and Digital Advanced Mobile Phone Service (DAMPS), which have a low frequency frame rate (around a few hundred hertz), could be more of a problem than their analog counterparts. Because they are pulsed on and off at the frame rate, they will emit an ELF radiation as well as the carrier frequency. There is a body of evidence (although it is not definitive) that suggests that low-frequency fields can be dangerous, even at low levels. The theory is that the low frequencies (meaning frequencies around power-main frequencies of 50 or 60 Hz and up to a few hundred hertz) cause resonances in a number of biological molecules and so speed up (or slow down) their chemical interactions. More than a hundred lawsuits have been filed in the United States alleging health problems resulting from exposure to electricity company ELF. There is however no definitive evidence of this "problem".

TDMA systems can interfere with the normal operations of modern motor vehicles, causing engine malfunction. Although there has been no documented evidence of safety problems, some European car manufacturers are warning against using GSM in moving vehicles, saying that *potential* problems are possible with anti-skid brakes and airbag activation systems. There is plenty of evidence of interference with engine control systems and car radios.

GSM is banned from use in Australian hospitals because of the interference with sensitive medical equipment.

NOT INTRINSICALLY SAFE

There has been some recent concern that mobile phones are not intrinsically safe units. Intrinsic safety refers to the construction of the unit itself and not to its effect on the human body. This means that there *may* be minor sparking when the unit is switched on, batteries are changed, or when some fault conditions arise, which *could* cause ignition of combustible gases. This possibility has led at least one Australian petroleum distributor to ban the use of mobile phones [and citizen bands (CBs)] at filling stations where flammable vapors may be present.

USE OF MOBILE PHONES WHILE DRIVING

Definitive tests have been done that show that levels of stress that accompany the use of mobile phones in moving vehicles are minimal if a handsfree unit is used. Dialing while driving remains hazardous and has in fact caused accidents. Therefore, in most countries it is illegal to use a handheld or a handset while driving.

A study in Australia has found that people who use mobile phones (without handsfree devices) frequently while driving have an elevated risk of accident, which has been likened to driving on the 0.05 blood alcohol limit (a common legal limit). Another study in Kuwait found that users who make more than three calls per trip are three times more likely to have an accident than nonusers.

A 1991 study by the AAA Foundation in the United States found that drivers using a car phone were 20 percent less likely to notice a dangerous traffic situation. In 1995 a study conducted by the U.S. Rochester Institute of Technology found that drivers with cellular phones are 34 percent more likely to be involved in traffic accidents than nonusers.

Although some of these results are a little inconsistent, they do point to an overall increased driving hazard with the use of mobile phones.

MOBILE USE IN AIRCRAFT

In most countries the use of mobile phones in aircraft is prohibited during flight. There has been no direct evidence of interference being caused to navigational equipment by analog mobile phones, and the restriction is widely seen as a conservative reaction by the aircraft industry. There has been some suggestion that cellular operators, fearing widespread blocking of control channels by airborne cellular users, were quick to point out the "potential dangers of interference to aircraft equipment."

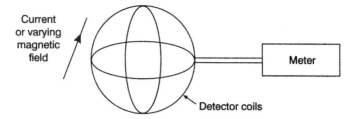

Figure 51.2 A magnetic field detector consists of three orthogonal detector cells.

The FCC has approved the "on ground" use of cellular phones and their installation in private aircraft. Even in private aircraft, notices must be attached to the phone indicating the conditions of use, which include any FAA restrictions. Use in the air is prohibited. Violators are subject to suspension of services and fines.

In Australia, the Civil Aviation Authority (CAA) has issued a rather bland statement, which, in keeping with the scant evidence of any problem, merely states that the pilot will ensure that the use of mobile phones does not inhibit aircraft operations.

Although any serious interference is unlikely to be encountered with analog equipment, digital sets, in particular GSM, are potentially hazardous.

There have been hundreds of reports of electromagnetic compatibility (EMC) attributed by the airline authorities to such bizarrely diverse pieces of electronic equipment as CD players and PCs. Reportedly, the aircraft experiences strange interference to navigational aids and cockpit display screens. The "interference" causes everything from windspeed indicators to fuel gauges and autopilot systems to malfunction.

Pilots insist that they have experienced such interference, quoting instances of the culprit device being located by cabin staff, and when taken to the cockpit and turned on and off by the pilots revealed the correlation between the device and the interference. It is interesting that this "interference" has yet to be replicated in controlled experiments using similar devices and aircraft.

The theories that "explain" the phenomenon are as unlikely as the problem itself. The most plausible is that the devices in question have been butchered during repairs and now leak large levels of EMC. No real devices with such repair defects have been produced. Other theories include resonance between the device and aircraft equipment, shielding faults in the devices, and even localized RF ducting between the device and the aircraft electronics. More probably what is happening is that the glitches in the aircraft system are caused by much more powerful external sources of interference, such as atmospheric or even galactic noise bursts. When these occur the cabin staff immediately run around and accuse the first spotted PC or CD user of causing the problem.

They instruct the user to turn the device off and the interference goes away (as it would have anyhow because these noise bursts are usually short-lived). In other words, a false correlation has been established.

EMR MONITORING

It is good practice to monitor the electromagnetic radiation (EMR) fields at base stations on a regular basis, and to keep permanent records of these levels, if for no other reason than to have some protection against future litigation.

By the standards in 1999, the EMR that is likely to be experienced in a base station is not dangerous to human health, except perhaps at distances of 1 meter or less from a transmitting antenna. Recently more conservative standards refer to 6 meters.

Site field-strength surveys can be carried out with purpose-designed equipment that measures the electric and magnetic fields at various points on the site.

The electric-field-strength measures a field, which with a device like the Wandel & Golterman EMR-20, will cover the frequency range of 100 kHz to 3 Ghz, with a resolution from 0.1 V/m to 800 V/m. The magnetic field strength, on the other hand, will be measured using a set of three coils at right angles to each other as in Figure 51.2. The magnetic detector will *not* be measuring the magnetic component of the RF, because it has a limited frequency response (typically in the range 5 Hz to 30 kHz).

RF radiation levels are measured indirectly as induced electric or magnetic fields. In the far field of any electromagnetic radiator, the electric (E) field and the magnetic (H) field are directly related by the expression $E = c \times H$, where c is the speed of light. In the near field (that is, within about three wavelengths of the radiator), some account must be taken of the radiator itself. If it is an antenna, then the currents flowing in the antenna will create both electric and magnetic fields. These fields will then be added vectorially to any induced fields from the radiation.

This means that if a measurement is being taken of a site, any measurements taken closer than about three wavelengths are not very meaningful. Some manuals on radiation measurement will suggest that in the near field the magnetic and electric components should be measured separately (because they no longer seem to follow the relationship above). In fact, in the far field what is being measured is, indirectly, the power density of the RF field. In the near field, on the other hand, there will be electric and magnetic fields generated by the antenna itself *together with* an RF field. A gross measurement of the magnetic or electric field at any point will reveal the vector sum of these two different sources of energy and will not necessarily give up much information on either (that is, you cannot directly deduce the RF field or the magnetic or electric field at that point).

Test equipment is available for the measurement of electric and magnetic fields for all frequencies from ELF up to at least 20 GHz. It is a good idea for operators to have one of these test sets and probes for the bands in which they operate in order to be able to confirm that the sites and equipment are within the applicable limits set by the local standards association. This may particularly be necessary on shared sites where there is limited knowledge of the emissions of the other users.

AC VOLTAGES

Nearly all celluar radio networks run from a main power source. Serious injury or death can result from quite small currents passing through the human body. As little as 10 mA can cause muscular paralysis, which in itself may be distressing, but more importantly, can cause the victim to be unable to release their grip. No live mains equipment should be worked on, and even when the equipment has been turned off and checked, it is wise to test it first by touching it with the *back* of the hand; as currents induced this way will cause the hand to move away from the source, as an automatic reflex.

At 30 mA, respiratory paralysis can occur and breathing may stop. Usually breathing will start again once the current has been removed, but resuscitation may be necessary. The most dangerous currents are between 75 and 250 mA. At this level, once the exposure exceeds a few seconds, ventricle fibrillation can be induced, causing the heart to beat in an uncoordinated manner, which renders it unable to circulate the blood. The fibrillation is likely to continue even after the source of the current has been removed, and the victim may well die unless first aid is rendered immediately.

Curiously, at currents higher than about 300 mA, the heart is clamped into a state that probably will set it beating normally once the current is removed. Serious injury or death can still result from prolonged exposure to these currents or, in the case of very high currents, from burns. This level of current is unlikely to be encountered in distribution voltages from the main power source.

When working on high voltage equipment there should always be at least two people present, and they should have been trained in cardiopulmonary resuscitation (CPR). This training needs to be repeated regularly to ensure currency. Because irreversible brain damage will occur within 4–6 minutes of heart or lung failure, time is of the essence. Brain death occurs in about 10 minutes.

52

BUYING USED HARDWARE

In 1993 about 5 percent of the cellular equipment sold was secondhand. This percentage is increasing as operators upgrade to more capable, smaller, and more power-efficient hardware. Digital upgrades may also result in some surplus analog equipment, although at the time of writing this was not a significant factor.

Used equipment may not even actually have really been used. Once the manufacturer has shipped equipment it will be considered used. Even if it goes directly from the manufacturer to a store and is then sold, the warranty that applied to the first buyer will usually not be transferable. Rarely will used equipment fetch more than 75 percent of the new price.

While bargains can be found, there are problems. Buyers of used equipment cannot often be choosy about brands and models. Since spare parts have to be kept, the more mixed the network is, the bigger the spare parts inventory. Training and test equipment costs will go up, and because of the added maintenance demands, the overall network survivability will go down. By how much is largely up to the operator.

The main considerations for the used equipment buyers has to be:

- Check the availability of spares
- Verify original supplier support
- See the equipment working
- Don't pay too much
- Ensure that repairs and maintenance can be done
- Obtain full documentation
- Compare the cost of modifying to your actual requirement

BASE-STATION HARDWARE

There is a thriving market for base-station hardware. A lot of it has become available because of a system buy-out, in which a supplier has purchased the complete network of a competitor to establish a foothold. The supplier, who now holds some of the competitor's stock, will usually elect to pass it on through a broker rather than risk getting directly involved in the sales and support of the product. Other stock becomes available as operators decide to replace early generations of equipment with the latest model. Finally, digital upgrades are slowly leading to the redundance of some analog hardware.

Older base stations should cost from 20 percent to 50 percent of the new price, depending largely on the age of the equipment. Early generations of base stations were about twice the size and consumed twice the power of current equipment. The mean time between failures (MTBF) of current base stations may be as high as 10 years per channel, compared to 12 months for first-generation equipment. Remote tuning and sophisticated diagnostics are recent innovations not likely to be available on used equipment.

Taking account of the negatives, used base equipment is most suitable for low-density applications in areas that are reasonably accessible for maintenance.

USED SWITCHES

Most of the early switches were relatively small, and those on the market will probably have less than 50,000

670

lines capacity. In general, except for the very early switches, interswitch working will be a feature readily available in used switches. Interswitch working takes up a good deal of processing time, and four 10,000-line switches connected in a parallel network cannot be expected to render a capacity of 40,000 lines. Limitations on the processor will mean that between 25,000 and 30,000 lines is all that can be expected. In fact, many of the switches that come onto the market will be from operators who have discovered that simply adding more small switches is not a good way to build up a large network.

Mostly, the used switches will be capable of running the current versions of software, and because they are available at about half the price of a new item, they can represent very good value. Unlike base stations, switches are invariably kept in climate-controlled environments and can be expected to be in very good condition. Because they are centralized and will have on-site maintenance staff, the somewhat greater maintenance load of a used switch will not necessarily be a serious problem. The power consumption of older switches can be expected to be twice that of a new one.

The main reservation with a used switch would be the degree and cost of supplier support to the second owner. In particular, switches are often sold with leased software, which may not be transferable. This should be determined before any commitment is made. Software updates (a lot of which are to fix bugs) must be available and their price must be reasonable.

MICROWAVE EQUIPMENT

Digital microwave, used equipment can be purchased for about half the new price. It can be a bargain, but there are traps. Microwave equipment is made to order, and new applications must fit in with the existing frequency plans and usage. In general, older equipment is less frequency agile than its newer counterpart, and so it will be more constrained.

Microwave frequencies and dish sizes are link-specific. For example, low frequencies are usually used for the longer hops; higher frequencies for short hauls. High rainfall can mean that 10+ GHz systems may not be effective. Big and expensive dishes used for long hops in the original system may become a tower-loading problem for a purchaser who only wants to operate over a moderate distance.

Interfacing older equipment by drop and insert may be a problem, as the equipment to do it may no longer be available.

RECTIFIERS

Rectifiers sometimes become available as operators select a few big units to replace banks of smaller ones or when they change from the bulky conventional rectifiers to the smaller switched-mode supplies. Conventional rectifiers have a very long life, and provided they are sold as working units, they can be expected to give good service. However, they often have "tricky" circuitry, and unless servicing is assured, it would be wise only to buy units with complete service manuals. Be wary of switch-mode rectifiers that come onto the market, as those made before 1990 were notoriously unreliable and very difficult to maintain.

BATTERIES

Used batteries are the most questionable item to buy on the recycled market. Batteries that could have a useful life of 15 years can have that life reduced to 18 months if they are continuously deeply discharged. Because the life of the battery is usually declared to be exhausted when the capacity reaches 80 percent, it can be difficult to pick a battery on its last, few discharge cycles.

TOWERS

Towers are very poor resale items because in most cases the greater part of the original cost is buried in the massive foundations. Masts are not much better because the guy wires are usually cut for the specific site and cannot easily be reused.

Both towers and masts are bulky and will require special equipment to dismantle. A used tower may not bring the seller what it costs to dismantle it, and so the prospective seller would be well advised to consider selling the tower as a going concern or leasing space on it.

If you buy a tower, it is best to buy it on-site, where it can readily be inspected. Ensure that the original drawings and calculations are available, as they may well be required for building approvals at the new site. The price should be around 10 percent of the new price.

The cost of dismantling a guyed mast is around $200 per meter, and for a tower around $350 per meter.

New bolts and foundations must be included in the costing.

CABLES AND ANTENNAS

Some administrations have a policy of not reusing cable or antennas even when those items become available on in-house equipment. This policy is a little extreme, but should serve as a warning that plenty can be wrong with used equipment. Used antennas and cables can have corrosion and/or minor perforations, which will lead to long-term problems.

Again, it is best to inspect the equipment on-site and to at least do a voltage standing-wave ratio (VSWR) check on the cables and antennas. In fact, it would not be wise to purchase these items unless they can be purchased on-site and the buyer has some control over their dismantling and shipping. Flexible waveguide and super-flex cables are the most likely to suffer from transportation, and thus are not good propositions.

Parabolic (microwave) antennas have very critical surface tolerances, which can be exceeded with a little rough handling. For moving, these dishes must be crated in the same manner as new ones.

EQUIPMENT UPDATING

Used equipment will probably no longer be supported by the original supplier to the extent that it can readily be upgraded. The life of a new model of equipment tends to be three to four years. During the model life cycle upgrades are constantly being added, but are often not backwards compatible. Even when compatibility applies, it may not be economical to upgrade the old equipment. If old equipment is purchased, the buyer should be ready to use it long-term in the configuration supplied.

APPENDIX A

RF PROPAGATION ROUTINE

A non-copyright BASIC program to calculate range and path loss is included below. This program has limited validity in regard to range, antenna heights, and frequency. However, it is usable over most of the ranges normally encountered in cellular radio. From the layout and comments you should be able to follow the structure and adapt it, if need be, to special-purpose applications.

You will notice that a function has been used to calculate logs to the base 10. This is because the original program was written in GWBASIC, which does not have a log base 10 function built in. If your BASIC has it, then do not bother with lines 200 to 230, and replace the statement FN BLOG(x) with the expression your BASIC uses for log to base 10 of x, wherever it occurs. For example in Power BASIC the expression is LOG10(x).

A much more comprehensive range of engineering software, featuring radio frequency (RF), switching, traffic, and power engineering routines for cellular system design called *Mobile Engineer* has been written by the author. Featuring more than 100 routines and running on Windows PCs it is available from John Wiley & Sons, Inc.

MOBILE PATH LOSS AND RANGE CALCULATION PROGRAM

```
10 BLK$= STRING$(30, " ")
20 CLS
30 Locate 7,5:PRINT"THIS PROGRAM DETERMINES RANGE OR PATH LOSS"
40 LOCATE 8,5:PRINT "OF A MOBILE SYSTEM GIVEN THE RESTRICTIONS"
50 LOCATE 9,5:PRINT"1. THE RANGE IS 1-20 KM."
60 LOCATE 10,5:PRINT"2.THE BASE HEIGHT IS 1-20 Km.
70 LOCATE 11,5:PRINT"3.THE VEHICLE ANTENNA HEIGHT IS 1-10 METRES"
80 LOCATE 12,5:PRINT"4.THE FREQUENCY IS 450-2000 MHz"
90 LOCATE 13,5:PRINT"5.THE % BUILDINGS TO LAND IS 3-50%"
95 LOCATE 14,5:PRINT"6.TERRAIN FACTOR IS ZERO"
100 REM frequency
110 f=850
120 REM base station height(meters).
130 H=30
140 REM vehicle height
150 V=1.5
160 REM % of buildings to land
170 BL=15
180 REM terrain factor
190 T=0
200 REM this is to take account of BASIC without base 10 logs
210 DEF FN BLOG(D)=LOG(D)/LOG(10)
220 REM assumes LOG(x) is the natural log of x
230 REM FN BLOG(x)=LOG to base of x

240 DELAY 2
250 CLS

300 LOCATE 3,50:PRINT "TO MAKE CHANGES ENTER THIS"
310 LOCATE 4,50:PRINT "NUMBER AND PRESS ENTER"
320 LOCATE 3,12: PRINT "THE DEFAULT VALUES ARE"
330 LOCATE 5:PRINT"_____"
```

```
370 LOCATE 7,10:PRINT"1.frequency in MHz";F
380 LOCATE 9,10:PRINT"2.base station height";H
390 LOCATE 11,10:PRINT:"3. % buildings to
land";BL
400 LOCATE 13,10:PRINT"4.vehicle antenna
height";V
410 LOCATE 15,10:PRINT"5.terrain factor";T
420 LOCATE 17,10:PRINT"6.accept above"
430 LOCATE 19,10:PRINT"7.change all of the above"

450 LOCATE 22,20:INPUT "SELECTION";ops

490 ON ops GOTO 500,520,540,560,580,600,500
500 LOCATE 7,10:PRINT BLK$
505 LOCATE 7,10:INPUT"1.frequency in MHz";F
510 IF ops GOTO 250
520 LOCATE 9,10:PRINT BLK$
525 LOCATE 9,10:INPUT"2.base station height",H
530 IF ops GOTO 250
540 LOCATE 11,10:PRINT BLK$
545 LOCATE 11,10:INPUT"3. % buildings to land";BL
550 IF ops GOTO 250
560 LOCATE 13,10:PRINT BLK$

565 LOCATE 13,10:INPUT "4.vehicle antenna
height";V
570 IF ops GOTO 250
580 LOCATE 15,10: PRINT BLK$
585 LOCATE 15,10:INPUT"5.terrain factor";T
590 IF ops GOTO 250
600 CLS

650 LOCATE 10,10:PRINT"SELECT A RANGE OR PATH
LOSS CALCULATION"
660 LOCATE 12,10:PRINT "R for RANGE"
670 LOCATE 14,10:PRINT"L for loss"
```

```
680 LOCATE 18,10:INPUT"your selection";T$

690 IF T$="L" OR T$="1" THEN GOTO 900
700 LOCATE 20,10:INPUT"allowable path loss";L
710 GOSUB 1000
BEEP
720 LOCATE 22,5:INPUT "another calculation Y/
N";P$
730 IF P$="Y" OR P$="y" THEN GOTO 250 ELSE END

900 REM THIS CALCULATES LOSS
910 LOCATE 20,10:INPUT"range is";R
920 GOSUB 2000
BEEP
930 LOCATE 22,5:INPUT"another calculation Y/N";P$
IF P$="Y" OR P$="y" THEN GOTO 250 ELSE END


END


1000 REM THIS SUBROUTINE CALCULATES RANGE
IF F>1000 THEN L=L-3*F/1000
S=30-25*(FN BLOG(BL))
A=(1.1*(FN BLOG(F))-0.7)*V-1.56*(FN BLOG(F))+0.8
RA=10^((L+S+T-69.55-26.16*(FN BLOG(F))+13.82*(FN
BLOG(H))+A)/(44.9-6.55*(FN BLOG(H))))
LOCATE 21,45:PRINT"RANGE IS ";RA
RETURN


2000 REM THIS SUBROUTINE CALCULATES PATH LOSS
S=30-25*(FN BLOG(BL))
A=(1.1*(FN BLOG(F))-0.7)*V-1.56*(FN BLOG(F))+0.8
L=69.55+26.16*(FN BLOG(F))-13.82*(FN BLOG(H))-A+
(44.9-6.55*(FN BLOG(H)))*(FN BLOG(R))-S-T
IF F>1000 THEN L=L+3*F/1000
LOCATE 21,45:PRINT"PATH LOSS IS  ";L
RETURN
```

APPENDIX B

ISO MODEL

A standard structure for the implementation of data/communications networks has been put forward by the International Standards Organization (ISO), which allows a structured definition of any system. This structure has been used to develop the enormously complex Global System for Mobile Communications (GSM) recommendations, which are more than 5000 pages long.

The ISO model divides the functions of a system into layers, which can be independently logically structured. The ISO system was meant to apply to computer and data networks, but can be applied to digital communications, which increasingly are becoming the same thing.

The ISO framework is shown in Table B.1, with the functionality for cellular radio shown.

Layer 1 comprises the actual physical and mechanical components of the circuits; although layers 2 and 3 are mainly implemented in hardware, they can involve considerable software. The higher layers are essentially software, except for the user part.

TABLE B.1 The ISO Structure Applied to Cellular Radio

Layer Number	Definition	Function
7	Application	End user
6	Presentation	End user
5	Session	Call setup/clear down
4	Transport	The user-to-user link
3	Network	Radio resource management
2	Data Link	Structuring the layer 1 information into meaningful messages
1	Physical	Coding
		CODECs
		Modulation types
		Frequencies

APPENDIX C

AMPLIFIER CLASSES

A lot of discussion in cellular radio centers around the final power amplifier (PA) stage and the class of amplifier. This is particularly true of digital systems, which will work only if the final amplifier is of a very high quality and low distortion. There are six amplifier classes: A, AB, B, C, D, and E. These classes originated in the early days of radio and were used to differentiate between the various valve amplifier configurations. However, these distinctions are just as meaningful when applied to solid state devices.

CLASS A

Class A is the simplest of all the amplifiers and consists of an active device, such as a transistor or field-effect transistor (FET), which is always conducting. The active device must be biased to ensure that it operates in its linear region. By the nature of its operation, it will draw a significant current even when there is no input to it. This means that it is not power efficient. For many applications in which good linearity is essential (as in most digital applications), class A is still the choice (see Figure C.1).

Most active devices will have a transfer function that looks something like the curve shown in Figure C.2. A high-quality class A amplifier will be biased to operate in the linear region. In this region the *average* power into the output load will be a constant regardless of the input signal level (provided the input signal can be regarded as random). In fact, early textbooks on class A amplifiers recommend checking class A amplifiers for distortion by putting a direct current (DC) ammeter in series with the output load and then observing the meter reading as a

signal is applied. No fluctuations mean "no distortion," but variations in the ammeter with applied signal level indicate non-linearity.

When driven hard the device will tend to saturate, and when driven too lightly the device will again go non-linear. It is the lightly driven non-linearity that causes problems with other, more power-efficient amplifier types.

The significance of all of this for cellular is that class A amplifiers are widely used in the PA stages of digital transmitters (because of the need for low distortion). As a consequence of the inefficiency (in regard to the power) of this type of amplifier, in dual-mode mobiles it will be usual to have a second, more power-efficient amplifier for the analog part.

CLASS B

Class B amplifiers overcome the efficiency problem by using two active devices, each switched for one half of

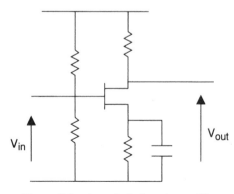

Figure C.1 A typical class A amplifier.

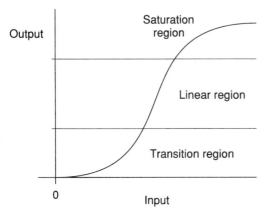

Figure C.2 The transfer function (output versus input) of a typical class A active device.

when they have no input. This keeps them operating more linearly than a straight class B amplifier, but they are still a compromise compared to class A for linearity. Typically, the quiescent current is 2–10 percent of the maximum working current.

CLASS C

Class C amplifiers are similar to class B, but they are self-biased. This means that the active device does not begin to conduct until the input reaches a high enough level to bias the device on. These amplifiers are very non-linear, but they are highly efficient and tend to be used in applications where linearity doesn't matter much, such as in frequency-modulated (FM) and phase-modulated (PM) transmitters.

CLASS D

Class D is a switching-mode amplifier, which by its nature is very efficient (typically around 80 percent and wide-band). It is used widely in amplitude-modulated (AM) broadcasting transmitters. A typical AM class D amplifier is shown in Figure C.4.

CLASS E

Class E amplifiers are a subset of class D with tuned circuit outputs, which improve the efficiency. The trade-off is that this amplifier will be a narrow-band device.

the cycle (180 degrees). When a class B amplifier has no input, both devices are off, and so it draws no current. It is therefore efficient, but because it requires the active devices to operate from full on to full off, it necessarily causes them to operate in regions where their linearity is poor. The most common example of a class B amplifier is the push-pull amplifier, which was widely used in audio circuits for decades from the 1930s when it first appeared. An example of a transistor push-pull amplifier is shown in Figure C.3.

CLASS AB

This is essentially a class B amplifier with some forward bias so that the active devices draw some current even

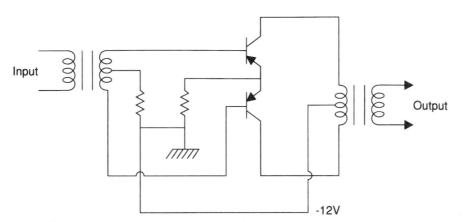

Figure C.3 A typical class B push-pull amplifier.

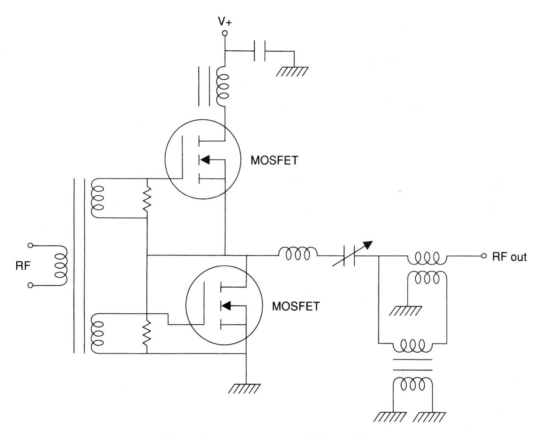

Figure C.4 A typical class D RF amplifier.

APPENDIX D

911 LOCATION REQUIREMENTS

EMERGENCY CALL LOCATION

In the United States, the FCC has mandated that the carriers incorporate the ability to locate accurately the position of emergency calls (calls to 911 in the United States). In 1996 the FCC issued a mandate (revised in 1997), docket number FCC 97-402, which includes the following:

> Not later than five years after the effective date of the rules (i.e., by October 1, 2001) carriers are required to have the capability to identify the latitude and longitude of the mobile units making 911 calls to within a radius of no more than 125 meters, using root mean square calculations (which roughly equate to success rates of approximately 67 percent).

Although most digital systems have some ability to determine the location of a caller, it is never to the degree of precision required for 911. For most carriers to comply, it will be necessary to install an overlay system for the tracking. In principle, the process of position location is simple. From a number of sites the mobile to be tracked is monitored and the time difference in arrival of the signal is measured, and from that the position is located.

In practice it is a bit more difficult. Multipath will cause signals to arrive from the mobile station (MS) at different times, and it will be necessary to distinguish the direct signal, which arrives first, from those following reflected paths. In a cellular environment, the frequency reuse may mean that more than one signal is present, and the detector will need to be able to discern the wanted signal. The timing must be precise, as any errors in this will translate directly to positional errors.

CELLTRAX SOLUTION

Celltrax is a company that began in the tracking business by developing methods of tracking lightning strikes. Anyone who has used an amplitude-modulated (AM) radio during a storm, will have heard the crackling that accompanies a lightning strike. These radio pulses can be used to locate the position of the strike.

The Celltrax system uses a network of remote sensors that have time difference of arrival (TDOA) capabilities; these sensors then send the information via a low speed data line to the position locator, and then onto displays for the customer. The all-important timing is derived from the Global Positioning System (GPS) network, with some algorithms to compensate for the errors in the GPS, the result being a timing stability in the range of 10 to 20 nanoseconds.

Celltrax has been designed to work with Advanced Mobile Phone Service (AMPS), time division multiple access (TDMA), and GSM protocols, and code division multiple access (CDMA) is being developed. In theory, because of the architecture, more than one system could be monitored by the one Celltrax system, so the possibility of a co-operative tracking system for any region exists.

Another solution by Corsair uses both TDOA and angle of arrival (AOA) to determine location and is currently available for AMPS and TDMA, with other protocols being developed.

Position Location

While the 911 location compliance may be expensive, position location can also be a business. Some cars are already fitted with GPS location alarms as factory options. These alarms ensure that stolen vehicles are quickly recovered. Additionally, best-route information, routing assistance, and personal security are target services.

Position location can be a vital planning tool. When inner-city cell sites need relief, it is often only educated guesswork that is used to determine the best location for the microcells that are deployed to carry the traffic. When precise position location is available, then the true extent and location of the hot spots can be accurately determined and the microcells positioned accordingly.

One US company, Protection One, has already marketed a mobile phone, called NavTalk, that includes a GPS, with a regular GPS screen, and an emergency call button that immediately alerts the control center of the caller's location.

Position Location for GSM

For GSM, accurate position location is an add-on extra, but facilities are already available. In 1999, Ericsson released its mobile position system (MPS), which works with unmodified GSM phones and on any GSM network. The system uses a mobile location center (MLC), which is server based. Interfacing to the MLC is an application programming interface (API) that allows the development of custom applications.

GPS ONBOARD

It is emerging that GPS onboard the mobile may be the cheapest and easiest way to implement this requirement. Today an accuracy of about 10–15 meters is possible. Providing the cost of the GPS unit can be brought down sufficiently, this seems to be an effective solution.

APPENDIX E

DISTORTION AND NOISE

NOISE AND DISTORTION

In a perfect world, where all devices are linear, there would be *no* distortion. A perfectly linear device, like an ideal amplifier, might have a response characteristic like that shown in Figure E.1. It is a perfect straight line. Most real-world devices are less than perfect, and so they may have a response more like that shown on the dotted line. To see what effect that may have, it is necessary to recall that any curve can be approximated to any arbitrary degree of accuracy by a polynomial. A linear response curve takes the form

$$\text{Output} = \text{Constant} \times \text{input}$$

or

$$Y = K \cdot X$$

where

$$Y = \text{the output}$$
$$K = \text{a constant}$$
$$X = \text{the input}$$

A non-linear response can be characterized as the polynomial

$$Y = K \cdot (X + e \cdot X^2 + f \cdot X^3 + g \cdot X^4 + \cdots)$$

where, because the distortion is small, the terms e, f, g, etc., will also be small.

Now consider the effect of the first distortion term only and let the input be sinusoidal, that is

$$x = K \cdot \cos(2 \cdot \pi \cdot f \cdot t)$$

where

$$t = \text{time}$$
$$f = \text{frequency}$$

Then the equation

$$Y = K \cdot (X + e \cdot X^2)$$

becomes

$$Y = K(\cos(2 \cdot \pi \cdot f \cdot t) + e \cdot \cos^2(2 \cdot \pi \cdot f \cdot t))$$

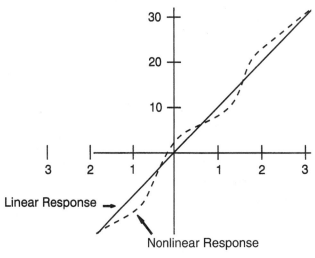

Figure E.1 A linear and nonlinear response curve.

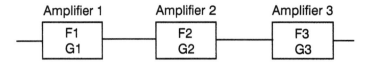

Figure E.2   Cascaded amplifiers.

recalling that

$$\cos^2(A) = 1/2(1 + \cos(2A))$$

Then

$$Y = K(\cos(2 \cdot \pi \cdot f \cdot t) + e/2 + e/2(\cos(4 \cdot \pi \cdot f \cdot t))$$

So now we have a term with an amplitude equal to $K \cdot e/2$, which is *twice* the frequency of the fundamental; that is, the second harmonic. It is significant here that the *magnitude* of the intermodulation product is proportional to *one half* the magnitude of the level of the signal that produced it. This can be used to determine the origin of intermodulation; if the signal causing the intermodulation is external to the receiver, then a 3-dB reduction in the incoming level will cause a 6-dB reduction in the intermodulation.

Proceeding in the same way, the third and higher harmonics can be derived from the higher powers of X.

It is worthwhile noting that the *magnitude* of the second and higher harmonics will be a function of the actual non-linearity of the device, and in general it will be very hard to quantify, except when caused by very predictable devices. However, one thing is certain: there will be *some* higher harmonics in any real-world device.

There is one other term in the equation that is worth a second look at. The term $K \cdot e/2$, in the expression for Y represents a direct current (DC) component, so in addition to producing harmonics, a non-linear device will also rectify (that was how the old-fashioned crystal set worked). When the carrier signal has some amplitude modulation (AM), such as an AM radio station or the frame buzz in Global System for Mobile Communications (GSM), K itself becomes a low-frequency variable that the non-linear device will detect.

For this reason, an unshielded audio amplifier (being a very non-linear device, especially at broadcast frequencies) in the vicinity of a powerful AM station may demodulate that signal and render it audible. GSM frame buzz (at around 217 Hz) is likewise demodulated by a wireline telephone handset, audio circuit, hearing aid, and, according to some reports, even by some dental fillings.

CASCADED AMPLIFIERS

The overall noise performance of an amplifier (or a receiver) can be measured by its noise factor, and the lower it is, the better. Amplifiers can be characterized by their gain (usually expressed in dB) and their noise figure (again usually expressed in dB). When a number of amplifiers are connected in series, as in Figure E.2, the overall noise figure is given by

$$F_0 = F_1 + (F_2 - 1)/G_1 + (F_3 - 1)/(G_1 \times G_2)$$
$$+ (F_4 - 1)/(G_1 \times G_2 \times G_3) + \cdots$$

where values of F (the noise factor) and G (the amplifier gain or loss) are expressed as ratios (not the more usual logarithmic form), and F_0 is the overall noise figure.

To see why the LNA (low noise amplifier) is important, we will look at the case where an attenuator is connected in series with a good-quality LNA. The noise figure will be dependent on whether the attenuator is connected before or after the LNA.

Consider the arrangement in Figure E.3. Let's assume that the attenuator has a loss of 3 dB. This translates to a gain of 0.5 $(10^{3/10})$ and a noise factor of 2. We assume that the LNA has a gain of 10 dB $(10\times)$ and a noise figure of 0.6 dB $(1.148\times)$. Substituting these values in the above equation

$$F_0 = 2 + (1.148 - 1)/0.5 = 2.295$$

so the noise figure in dB is $10 \times \log 2.295 = 3.6$ dB.

Now reverse the connections and put the LNA first, as in Figure E.4, and substitute

$$F_0 = 1.148 + (2 - 1)/10 = 1.248$$

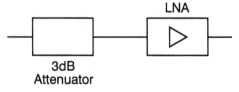

Figure E.3   An LNA preceded by an attenuator.

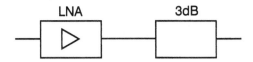

Figure E.4 An LNA followed by an attenuator.

or, in logarithmic form, the overall noise figure is

$$10 \times \log 1.248 = 0.962 \text{ dB}$$

The important thing to notice is that the contribution of the loss and high noise figure of the attenuator *after* the LNA was reduced in significance by a factor proportional to the gain of the LNA.

Because of this the closer the LNA is to the antenna, the lower will be the overall noise figure of the system. Mounting the LNA on top of the tower before the feeder losses will improve the noise figure by the amount of the feeder losses.

APPENDIX F

RECOMMENDED FURTHER READING AND SOURCES OF INFORMATION

- Agonsta, John and Russell, Travis, "CDPD", McGraw-Hill Series on Communications, 1996, ISBN 0-07-000600-8.
- Gallagher, Michael D. and Snyder, Randall A., *Mobile Telecommunications Networking with IS-41*, McGraw-Hill, 1997, ISBN 0-07-063314-2.
- Aaron Kershenbaum, *Telecommunications Network Design Algorithms*, McGraw-Hill, Computer Science Series, 1993, ISBN0-07-034228-8.
- *Microwave Journal*, Horizon House Publications, www.journal.com. A free magazine that covers the technology behind microwaves (which includes all mobile technologies) and has some brilliant articles.
- *RF Design*, Intertec Publishing, www.rfdesign.com. This is a free magazine with good up-to-date information on the hardware that drives the technology and includes an excellent tutorial section.
- *RTT Systems*. Great web page on 3G technology, see www.rttsys.com. The web page has a monthly "Hot Topic" on 3G that is well-worth reading, current, and well thought-out. Also from RTT, some great courses are held on 3G technology in general, and in particular its application to mobile handset technology. I highly recommend the Oxford series of courses. Contact them through their web site.

APPENDIX G

INTERNET PROTOCOLS

The International Engineering Task Force (IETF) is defining internet protocols (IP) at a pace rarely seen before. Proposals posted on their web site (www.IEFT.org) are either adopted as defacto standards after six months or they are dropped. The body, first formed in 1986, and operating in an open and accessible manner, is proving far more efficient and effective than the traditional, formal bodies that have in the past set international standards.

Versions of IP are increasingly being seen as the way of the future for both wireline and wireless applications. The mobile IPV6 is being developed to address the issue of IP address mobility. It's development is looking attractive. The new IP protocol will be used right across the mobile network. The IP is structured to tie together the GSM MAP and IS41 networks and to interface the HLR/VLR registers into a common database. More interestingly, the protocol enables mobiles themselves to act as routers for other mobiles, thus forming a "network" that is nearly independent of the MSC/BSC network. The mobiles do this by local negotiation, and so there is less need for centralized intelligence.

AD-HOC NETWORKS

The following description of an ad-hoc network was largely sourced (with permission) from RTT Systems web site (see Appendix F). Devices that use the IPs do not necessarily have to be connected to the Internet; and they can indeed form local stand-alone (or virtual) networks. This is not a new idea, and can be seen as a concept in TETRA, iDEN, and PHS, where designated mobiles can form "relays" for other mobiles in a local network. Of course the IP proposal is somewhat more sophisticated in that it will allow large numbers of mobiles to participate in a call between two distant users.

Ad-hoc networks do present a problem for the operators, who may find it difficult to charge for stand-alone usage. In fact in the extreme case, the networks themselves may become almost redundant, providing only verification and spectrum availability. Prepayed usage is one option for the network operator, that could be used to allow revenue collections even for usage in ad-hoc networks, but it does not seem reasonable that such collections would be tolerated for too long.

Peripheral nodes are nodes whose minimum distance to the node addressed is exactly equal to the zone radius. This provides an efficient basis for providing inter-zone routing using a process known as *border casting*.

In this example shown in Figure G.1, Node A has a datagram to send to node L. We assume a uniform routing zone radius of 2 hops. Since L is not in A's routing zone (which includes B, C, D, E, F, and G), 'A' bordercasts a routing query to its peripheral nodes, D, F, E, and G. Each one of these peripheral nodes checks whether L exists in their routing zones. Since L is not found in any routing zones of these nodes, the nodes bordercast the request to their peripheral nodes. G bordercasts to K which realises that L is in its routing zone and returns the requested route (L, K, G, A) to the query source, A.

Using IPV6, a zone routing protocol can then be built both for inter-zone and intra-zone routing using the destination address (the 32-bit IP address of the destination host), the next hop address, the next but one hop address and so on, the number of hops is described using a 4-bit hop count field—the complex routing description

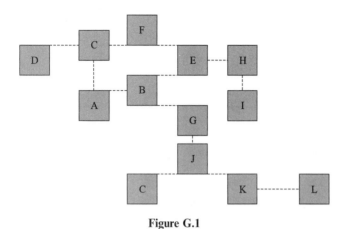

Figure G.1

becomes the basis for the intra-zone and inter-zone routing tables.

These routing protocols can then be refined into proactive or reactive protocols. Proactive protocols continuously evaluate the routes within the network. When a packet needs to be forwarded, the optimal route is already known. This has the advantage of minimal routing delay. It would be used, for example, in a user group in which "on to channel" access times were critical. The disadvantage is that the protocol absorbs network bandwidth. A reactive protocol invokes a route determination procedure on demand; implicitly this will involve a channel set-up delay which will be variable and ill suited to real-time or time-critical applications. However, we can use local memory bandwidth in the IP appliance to give us the benefits of proactive protocols (fast and deterministic channel access) without incurring high signaling bandwidth overhead. To do this, we use route discovery. In route discovery, the source node broadcasts a route request packet. Each node hearing the request passes on the request and adds its own address to the header. The forwarding of the route request propagates out until the target of the request is found. The target then replies. All source routes learned by a node are kept in a route cache. When sending a packet, a note will only do route discovery if no suitable source route is found in the local cache memory (i.e., memory bandwidth has been used to reduce the signalling bandwidth overhead). As a node overhears routes being used by other nodes for route discovery or route maintenance, the node may insert these routes into its route cache—this is called "promiscuous snooping."

Many refinements are possible over and above these basic intra-zone and inter-zone routing protocols; any or all of the protocols can be used either to transparently fill in the gaps between base stations and/or extend the

coverage of an instant self-contained ad-hoc network. An example is a cluster-based routing protocol—a cluster is a group of mobiles which is "clustered" round a base station or in a self-sustaining cluster (sometimes also described as a "cloud," although clouds tend to be more mobile than clusters). A cluster head is elected for each cluster to maintain cluster membership information (a kind of club membership directory which includes access and membership rights); this sort of optimization can improve the scalability of the protocol thus avoiding flooding and address storms.

The scalability and stability of routing protocols can be measured and described in terms of convergence. *Convergence* describes the condition in which a state of equilibrium is approached in which all nodes of the network agree on a consistent collection of states about the topology of the network and in which no further control messages are needed to establish the consistency of the network topology (i.e., a steady state condition). *Convergence time* is the time required for a network to reach convergence after an event (typically the movement of a mobile node) has changed the network topology. Other key performance measurements in ad-hoc networks include the *distance vector* (how many hops to store in the routing header), *goodput* (the total bandwidth used less the protocol overhead), *laydown* (the relative physical location of the nodes within the network), the *mobility factor* (the relative frequency of node movement compared to the convergence time of the routing protocols, and the *security parameter index* (the security context between defined router pairs). *Payload* is the description of the actual data within the packet.

This brings us to one of the intriguing opportunities when we come to qualify the use of IP routing protocols in mobile radio networks—an IP packet can be (more or less) as long as you want. As the packet gets larger, the IP address overhead reduces as a percentage of the payload. A typical PTT (press to talk) transaction on a mobile radio is several seconds—the IP header overhead on such a transaction is (more or less) trivial. Longer transactions can include *IP jumbograms* (a jumbo payload is defined as a payload between 65, 536 and 4, 294, 967, 295 octets!). IP datagrams can therefore deliver exceedingly impressive goodput. IP address and routing protocols therefore provide a very elegant and flexible mechanism for managing mobile networks including ad-hoc self-configurable networks.

Those familiar with George Orwell's book 1984, could be forgiven for thinking that the terminology above came from the tenth edition of the *Newspeak* dictionary (a dictionary devised to eliminate all redundant words, and to form new words by combining two or more old ones), so you get words such as, "goodthink" (correct thinking), "speakwrite" (a machine for taking dictation) (ungood—meaning bad, but at the same time eliminating the unnecessary word "bad") and "double-plusungood" (really, really, bad).

IEFT: THE FUTURE

The IEFT protocols are open, and unlike many of the alternatives are readily available to all developers. This alone may be enough to ensure survival of the standard over its protected rivals. Because it is likely that the IETF standards will become increasingly important in the mobile world, it may be worthwhile taking time to become familiar with a few of the sometimes irreverent acronyms that are used by the IETF. The following list was taken from their web-site www.IEFT.org.

IEFT Area Abbreviations

- APP: Applications
- INT: Internet Services
- IPNG: IP: Next Generation
- MGT: Network Management
- OPS: Operational Requirements
- RTG: Routing
- SEC: Security
- TSV: Transport
- USV: User Services

Acronyms

- ANSI: American National Standards Institute
- ARPA: Advanced Research Projects Agency
- ARPANET: Advanced Research Projects Agency Network
- AS: Autonomous System
- ATM: Asynchronous Transfer Mode
- BGP: Border Gateway Protocol
- BOF: Birds Of a Feather
- BSD: Berkeley Software Distribution
- BTW: By The Way
- CCIRN: Coordinating Committee for Intercontinental Research Networks
- CCITT: International Telegraph and Telephone Consultative Committee
- CIDR: Classless Inter-Domain Routing
- CIX: Commercial Information Exchange
- CNI: Coalition for Networked Information
- CREN: The Corporation for Research and Educational Networking

- DARPA: U.S. Defense Advanced Research Projects Agency (now ARPA)
- DDN: U.S. Defense Data Network
- DISA: U.S. Defense Information Systems Agency
- EGP: Exterior Gateway Protocol
- FAQ: Frequently Asked Question
- FARNET: Federation of American Research NETworks
- FIX: U.S. Federal Information Exchange
- FNC: U.S. Federal Networking Council
- FQDN: Fully Qualified Domain Name
- FYI: For Your Information (RFC)
- GOSIP: U.S. Government OSI Profile
- IAB: Internet Architecture Board
- IANA: Internet Assigned Numbers Authority
- I-D: Internet-Draft
- IEN: Internet Experiment Note
- IESG: Internet Engineering Steering Group
- IETF: Internet Engineering Task Force
- IGP: Interior Gateway Protocol
- IMHO: In My Humble Opinion
- IMR: Internet Monthly Report
- InterNIC: Internet Network Information Center
- IPng: IP: Next Generation
- IR: Internet Registry
- IRSG: Internet Research Steering Group
- IRTF: Internet Research Task Force
- ISO: International Organization for Standardization
- ISOC: Internet Society
- ISODE: ISO Development Environment
- ITU: International Telecommunication Union
- MIB: Management Information Base
- MIME: Multipurpose Internet Mail Extensions
- NIC: Network Information Center
- NIS: Network Information Services
- NIST: National Institute of Standards and Technology
- NOC: Network Operations Center
- NREN: National Research and Education Network
- NSF: National Science Foundation
- OSI: Open Systems Interconnection
- PEM: Privacy Enhanced Mail
- PTT: Postal, Telegraph and Telephone
- RARE: Reseaux Associes pour la Recherche Europeenne (no longer exists)
- RFC: Request For Comments
- RIPE: Reseaux IP Europeens

- SIG: Special Interest Group
- STD: Standard (RFC)
- TERENA: Trans European Research & Education Networking Association
- TLA: Three Letter Acronym
- TTFN: Ta-Ta For Now
- UTC: Universal Time Coordinated
- WG: Working Group
- WRT: With Respect To
- WYSIWYG: What You See is What You Get

IP DELAYS

For all its advantages, IP transmission has a disadvantage; there will be delays. It is expected that the 3G applications will experience a delay from 100 ms to 500 ms. Decreasing the delay becomes increasingly expensive, and in the limit (no transmission delays due to processing) we are back to switched circuits.

COMPRESSION

At the application layer, we are seeing an order of magnitude increase in compression ratio every 5 years. In 2000, it is not unusual to see image and video streaming compression ratios of 70 to 1 in proprietary implementations. In 5 years, this is likely to be closer to 700 to 1. This content will be undeliverable over a 1 in 10^6 error rate radio channel; because the compression is achieved at the expense of redundancy, the compressed signal is increasingly intolerant of errors. The error rate could be reduced by error coding (send-again protocols) but this would produce unacceptable, variable delay. The wireline delivery systems with their BERs of 10^{10} or greater can support this degree of compression.

INTERNET TRAFFIC

By 1999, the total number of messages passed on the Internet equalled the total of the conventional services (telephone and fax) and was set to surpass it. By its nature (with lots of images) the Internet is inherently a wideband service. At the time of writing (2000) most users have to contend with a maximum through-put of 56 kBits. Internet traffic is quadrupling every year, while the regular voice traffic grows at around 10% p.a. It is therefore evident that the dominance of the Internet, will be the only real consideration in the next few decades. There can be no real doubt that the future is in wideband services, and the questions are more when and how rather than if.

APPENDIX H

ERLANG B AND C TABLES

Although many cellular operators use Erlang B tables for base-station dimensioning, the underlying assumption of the Erlang B relationship—that all congested calls upon failing are removed from the system (i.e., there is no retry)—clearly is inappropriate. There may, however, be some switch-to-switch links, where for practical purposes this assumption holds.

Base stations carry randomly generated traffic that almost invariably will experience a redial if congestion is encountered. Traditionally, the Erlang B table has been used to dimension these channels, but as that table assumes that calls that strike congestion are removed from the system (i.e., if the call does not get through the first time, the caller gives up), it does not agree with reality. The Erlang C table on the following pages is a modified form of the Erlang B that assumes that all calls that strike congestion will retry. For this reason it is a conservative approach and will nearly always require more circuits for a given traffic and grade of service (GOS) than the Erlang B table.

The GOS in the cellular situation is not a measure of lost calls because if all (or most) callers immediately retry, they will eventually get through on all but the most congested circuits. Rather, GOS measures the probability that the caller will need to redial.

The table includes some very poor grades of service (up to 0.2), as in practice these are often encountered.

ERLANG B TABLE

Number of Traffic Circuits	Offered Traffic in Erlangs for the GOS shown					
	0.001	0.002	0.01	0.05	0.1	0.2
1	0.001	0.002	0.01	0.05	0.11	0.25
2	0.046	0.07	0.15	0.38	0.60	1
3	0.19	0.25	0.45	0.90	1.3	1.9
4	0.44	0.53	0.87	1.5	2.0	2.9
5	0.76	0.9	1.4	2.2	2.9	4.0
6	1.1	1.3	1.9	3.0	3.8	5.1
7	1.6	1.8	2.5	3.7	4.7	6.2
8	2.1	2.3	3.1	4.5	5.6	7.4
9	2.6	2.9	3.8	5.4	6.5	8.5
10	3.1	3.4	4.5	6.2	7.5	9.7
11	3.7	4.0	5.2	7.1	8.5	10.9
12	4.2	4.6	5.9	8.0	9.5	12.0
13	4.8	5.3	6.6	8.8	10.5	13.2
14	5.4	5.9	7.4	9.7	11.5	14.4
15	6.1	6.6	8.1	10.6	12.5	15.6
16	6.7	7.3	8.9	11.5	13.5	16.8
17	7.4	7.9	9.7	12.5	14.5	18.0
18	8.0	8.6	10.4	13.4	15.5	19.2
19	8.7	9.4	11.2	14.3	16.6	20.4
20	9.4	10.1	12.0	15.2	17.6	21.6
21	10.1	10.8	12.8	16.2	18.6	22.9
22	10.8	11.5	13.7	17.1	19.7	24.1
23	11.5	12.3	14.5	18.1	20.7	25.3
24	12.2	13.0	15.3	19.0	21.8	26.5
25	13.0	13.8	16.1	20.0	22.8	27.7
26	13.7	14.5	17.0	20.9	23.9	28.9
27	14.4	15.3	17.8	21.9	24.9	30.2
28	15.2	16.1	18.6	22.9	26.0	31.4
29	15.9	16.8	19.5	23.8	27.1	32.6
30	16.7	17.6	20.3	24.8	28.1	33.9

Erlang B Table (*continued*)

Number of Traffic Circuits	Offered Traffic in Erlangs for the GOS shown						Number of Traffic Circuits	Offered Traffic in Erlangs for the GOS shown					
	0.001	0.002	0.01	0.05	0.1	0.2		0.001	0.002	0.01	0.05	0.1	0.2
31	17.4	18.4	21.2	25.8	29.2	35.1	84	61.3	63.2	69.1	78.9	86.6	100.7
32	18.2	19.2	22.0	26.7	30.2	36.3	85	62.1	64.1	70.0	79.9	87.7	102.0
33	19.0	20.0	22.9	27.7	31.3	37.5	86	63.0	65.0	70.9	80.9	88.8	103.2
34	19.7	20.8	23.8	28.7	32.4	38.7	87	63.9	65.9	71.9	82.0	89.9	104.5
35	20.5	21.6	24.6	29.7	33.4	40.0	88	64.7	66.8	72.8	83.0	91.0	105.7
36	21.3	22.4	25.5	30.7	34.5	41.2	89	65.6	67.7	73.7	84.0	92.1	106.9
37	22.1	23.2	26.4	31.6	35.6	42.4	90	66.5	68.6	74.7	85.0	93.2	108.2
38	22.9	24.0	27.3	32.6	36.6	43.7	91	67.4	69.4	75.6	86.0	94.2	109.4
39	23.7	24.8	28.1	33.6	37.7	44.9	92	68.2	70.3	76.6	87.1	95.3	110.7
40	24.4	25.6	29.0	34.6	38.8	46.1	93	69.1	71.2	77.5	88.1	96.4	111.9
41	25.2	26.4	29.9	35.6	39.9	47.4	94	70.0	72.1	78.4	89.1	97.5	113.2
42	26.0	27.2	30.8	36.6	40.9	48.6	95	70.9	73.0	79.4	90.1	98.6	114.4
43	26.8	28.1	31.7	37.6	42.0	49.8	96	71.7	73.9	80.3	91.1	99.7	115.6
44	27.6	28.9	32.5	38.6	43.1	51.1	97	72.6	74.8	81.2	92.2	100.8	116.9
45	28.4	29.7	33.4	39.6	44.2	52.3	98	73.5	75.7	82.2	93.2	101.9	118.2
46	29.3	30.5	34.3	40.5	45.2	53.5	99	74.4	76.6	83.1	94.2	103.0	119.4
47	30.1	31.4	35.2	41.5	46.3	54.8	100	75.2	77.5	84.1	95.2	104.1	120.6
48	30.9	32.2	36.1	42.5	47.4	56.0	101	76.1	78.4	85.0	96.3	105.2	121.9
49	31.7	33.0	37.0	43.5	48.5	57.3	102	77.0	79.3	85.9	97.3	106.3	123.1
50	32.5	33.9	37.9	44.5	49.6	58.5	103	77.9	80.2	86.9	98.3	107.4	124.4
51	33.3	34.7	38.8	45.5	50.6	59.7	104	78.8	81.1	87.8	99.3	108.5	125.6
52	34.2	35.6	39.7	46.5	51.7	61.0	105	79.6	82.0	88.8	100.4	109.6	126.9
53	35.0	36.4	40.6	47.5	52.8	62.2	106	80.5	82.8	89.7	101.4	110.7	128.1
54	35.8	37.2	41.5	48.5	53.9	63.5	107	81.4	83.7	90.7	102.4	111.8	129.4
55	36.6	38.1	42.4	49.5	55.0	64.7	108	82.3	84.6	91.6	103.4	112.9	130.6
56	37.5	38.9	43.3	50.5	56.1	65.9	109	83.2	85.5	92.5	104.5	114.0	131.8
57	38.3	39.8	44.2	51.6	57.1	67.2	110	84.1	86.4	93.5	105.5	115.1	133.1
58	39.1	40.6	45.1	52.5	58.2	68.4	111	85.0	87.3	94.4	106.5	116.2	134.4
59	40.0	41.5	46.0	53.6	59.3	69.7	112	85.8	88.3	95.4	107.5	117.3	135.6
60	40.8	42.4	46.9	54.6	60.4	70.9	113	86.7	89.2	96.3	108.6	118.4	136.8
61	41.6	43.2	47.9	55.6	61.5	72.1	114	87.6	90.1	97.3	109.6	119.5	138.1
62	42.5	44.1	48.8	56.6	62.6	73.4	115	88.5	91.0	98.2	110.6	120.6	139.3
63	43.3	44.9	49.7	57.6	63.7	74.6	116	89.4	91.9	99.2	111.6	121.7	140.6
64	44.2	45.8	50.6	58.6	64.8	75.9	117	90.3	92.8	100.1	112.7	122.8	141.8
65	45.0	46.7	51.5	59.6	65.8	77.1	118	91.2	93.7	101.1	113.7	123.9	143.1
66	45.8	47.5	52.4	60.6	66.9	78.4	119	92.1	94.6	102.0	114.7	125.0	144.3
67	46.7	48.4	53.4	61.6	68.0	79.6	120	93.0	95.5	103.0	115.8	126.1	145.6
68	47.5	49.2	54.3	62.6	69.1	80.8	121	93.9	96.4	103.9	116.8	127.2	146.8
69	48.4	50.1	55.2	63.7	70.2	82.1	122	94.7	97.3	104.9	117.8	128.3	148.0
70	49.2	51.0	56.1	64.7	71.3	83.3	123	95.6	98.2	105.8	118.9	129.4	149.3
71	50.1	51.8	57.0	65.7	72.4	84.6	124	96.5	99.1	106.8	119.9	130.5	150.6
72	50.9	52.7	58.0	66.7	73.5	85.8	125	97.4	100.0	107.7	120.9	131.6	151.8
73	51.8	53.6	58.9	67.7	74.6	87.0	126	98.3	100.9	108.7	121.9	132.7	153.0
74	52.7	54.5	59.8	68.7	75.7	88.3	127	99.2	101.8	109.6	123.0	133.8	154.3
75	53.5	55.3	60.7	69.7	76.8	89.6	128	100.1	102.8	110.6	124.0	134.9	155.6
76	54.4	56.2	61.7	70.8	77.8	90.8	129	101.0	103.7	111.5	125.0	136.0	156.8
77	55.2	57.1	62.6	71.8	78.9	92.0	130	101.9	104.6	112.5	126.1	137.1	158.0
78	56.1	58.0	63.5	72.8	80.0	93.3	131	102.8	105.5	113.4	127.1	138.2	159.3
79	56.9	58.8	64.4	73.8	81.1	94.5	132	103.7	106.4	114.4	128.1	139.3	160.5
80	57.8	59.7	65.4	74.8	82.2	95.7	133	104.6	107.3	115.3	129.2	140.4	161.8
81	58.7	60.6	66.3	75.8	83.3	97.0	134	105.5	108.2	116.3	130.2	141.5	163.0
82	59.5	61.5	67.2	76.9	84.4	98.3	135	106.4	109.1	117.2	131.2	142.6	164.2
83	60.4	62.4	68.2	77.9	85.5	99.5	136	107.3	110.0	118.2	132.3	143.7	165.5

Erlang B Table (*continued*)

Number of Traffic Circuits	Offered Traffic in Erlangs for the GOS shown						Number of Traffic Circuits	Offered Traffic in Erlangs for the GOS shown					
	0.001	0.002	0.01	0.05	0.1	0.2		0.001	0.002	0.01	0.05	0.1	0.2
137	108.2	111.0	119.1	133.3	144.8	166.8	190	156.4	159.8	170.1	188.1	203.3	232.9
138	109.1	111.9	120.1	134.3	145.9	168.0	191	157.3	160.8	171.0	189.2	204.4	234.1
139	110.0	112.8	121.0	135.4	147.0	169.3	192	158.3	161.7	172.0	190.2	205.5	235.3
140	110.9	113.7	122.0	136.4	148.1	170.5	193	159.2	162.6	173.0	191.3	206.6	236.6
141	111.8	114.6	123.0	137.4	149.2	171.7	194	160.1	163.6	173.9	192.3	207.7	237.9
142	112.7	115.5	123.9	138.4	150.3	173.0	195	161.0	164.5	174.9	193.3	208.8	239.1
143	113.6	116.4	124.9	139.5	151.4	174.3	196	161.9	165.4	175.9	194.3	209.9	240.4
144	114.5	117.4	125.8	140.5	152.5	175.5	197	162.9	166.4	176.8	195.4	211.0	241.6
145	115.4	118.3	126.8	141.5	153.6	176.8	198	163.8	167.3	177.8	196.4	212.1	242.9
146	116.3	119.2	127.7	142.6	154.7	178.0	199	164.7	168.2	178.8	197.5	213.2	244.2
147	117.2	120.1	128.7	143.6	155.8	179.3	200	165.6	169.2	179.7	198.5	214.4	245.4
148	118.1	121.0	129.7	144.6	156.9	180.5	201	166.5	170.1	180.7	199.5	215.5	246.7
149	119.0	121.9	130.6	145.7	158.0	181.7	202	167.5	171.0	181.7	200.6	216.6	247.9
150	119.9	122.9	131.6	146.7	159.1	183.0	203	168.4	172.0	182.6	201.6	217.7	249.2
151	120.8	123.8	132.5	147.7	160.2	184.2	204	169.3	172.9	183.6	202.6	218.8	250.4
152	121.8	124.7	133.5	148.8	161.3	185.5	205	170.2	173.8	184.6	203.7	219.9	251.6
153	122.7	125.6	134.5	149.8	162.4	186.7	206	171.2	174.8	185.5	204.7	221.0	252.9
154	123.6	126.5	135.4	150.8	163.5	188.0	207	172.1	175.7	186.5	205.8	222.1	254.1
155	124.5	127.5	136.4	151.9	164.6	189.2	208	173.0	176.6	187.5	206.8	223.2	255.4
156	125.4	128.4	137.3	152.9	165.7	190.5	209	173.9	177.6	188.5	207.9	224.3	256.6
157	126.3	129.3	138.3	153.9	166.9	191.7	210	174.8	178.5	189.4	208.9	225.4	257.8
158	127.2	130.2	139.2	155.0	167.9	193.0	211	175.8	179.4	190.4	209.9	226.5	259.1
159	128.1	131.1	140.2	156.0	169.0	194.2	212	176.7	180.4	191.4	211.0	227.6	260.3
160	129.0	132.1	141.2	157.1	170.2	195.4	213	177.6	181.3	192.3	212.0	228.7	261.6
161	129.9	133.0	142.1	158.1	171.3	196.7	214	178.5	182.2	193.3	213.1	229.8	262.9
162	130.8	133.9	143.1	159.1	172.4	197.9	215	179.5	183.2	194.3	214.1	230.9	264.1
163	131.7	134.8	144.1	160.1	173.4	199.2	216	180.4	184.1	195.2	215.1	232.0	265.4
164	132.7	135.8	145.0	161.2	174.6	200.4	217	181.3	185.0	196.2	216.2	233.1	266.7
165	133.6	136.7	146.0	162.2	175.7	201.7	218	182.2	186.0	197.2	217.2	234.2	267.9
166	134.5	137.6	146.9	163.3	176.8	202.9	219	183.2	186.9	198.2	218.2	235.3	269.1
167	135.4	138.5	147.9	164.3	177.9	204.2	220	184.1	187.8	199.1	219.3	236.4	270.3
168	136.3	139.4	148.9	165.3	179.0	205.4	221	185.0	188.8	200.1	220.3	237.5	271.6
169	137.2	140.4	149.8	166.4	180.1	206.7	222	185.9	189.7	201.1	221.4	238.6	272.9
170	138.1	141.3	150.8	167.4	181.2	207.9	223	186.9	190.7	202.0	222.4	239.7	274.0
171	139.0	142.2	151.7	168.4	182.3	209.2	224	187.8	191.6	203.0	223.4	240.8	275.3
172	139.9	143.2	152.7	169.5	183.4	210.4	225	188.7	192.5	204.0	224.5	241.9	276.6
173	140.9	144.1	153.7	170.5	184.5	211.7	226	189.6	193.5	205.0	225.5	243.1	277.8
174	141.8	145.0	154.6	171.5	185.6	212.9	227	190.6	194.4	205.9	226.6	244.2	279.1
175	142.7	145.9	155.6	172.6	186.7	214.2	228	191.5	195.3	206.9	227.6	245.3	280.4
176	143.6	146.8	156.6	173.6	187.8	215.4	229	192.4	196.3	207.9	228.6	246.4	281.6
177	144.5	147.8	157.5	174.7	188.9	216.7	230	193.3	197.2	208.8	229.7	247.5	282.9
178	145.4	148.7	158.5	175.7	190.0	218.0	231	194.3	198.1	209.8	230.7	248.6	284.2
179	146.3	149.6	159.5	176.7	191.1	219.2	232	195.2	199.1	210.8	231.8	249.7	285.3
180	147.3	150.6	160.4	177.8	192.2	220.5	233	196.1	200.0	211.8	232.8	250.8	286.6
181	148.2	151.5	161.4	178.8	193.3	221.7	234	197.1	201.0	212.7	233.9	251.9	287.8
182	149.1	152.4	162.4	179.8	194.4	223.0	235	198.0	201.9	213.7	234.9	253.0	289.1
183	150.0	153.3	163.3	180.9	195.5	224.1	236	198.9	202.8	214.7	235.9	254.1	290.3
184	150.9	154.3	164.3	181.9	196.6	225.4	237	199.8	203.8	215.6	237.0	255.3	291.5
185	151.8	155.2	165.2	182.9	197.8	226.6	238	200.8	204.7	216.6	238.0	256.4	292.8
186	152.8	156.1	166.2	184.0	198.9	227.9	239	201.7	205.7	217.6	239.1	257.5	294.1
187	153.7	157.1	167.2	185.0	199.9	229.2	240	202.6	206.6	218.6	240.1	258.6	295.3
188	154.6	158.0	168.1	186.1	201.1	230.4	241	203.6	207.5	219.5	241.1	259.7	296.6
189	155.5	158.9	169.1	187.1	202.2	231.7	242	204.5	208.5	220.5	242.2	260.8	297.9

Erlang B Table (*continued*)

Number of Traffic Circuits	Offered Traffic in Erlangs for the GOS shown					
	0.001	0.002	0.01	0.05	0.1	0.2
243	205.4	209.4	221.5	243.2	261.9	299.2
244	206.3	210.4	222.4	244.3	263.0	300.3
245	207.3	211.3	223.4	245.3	264.1	301.6
246	208.2	212.2	224.4	246.3	265.2	302.8
247	209.1	213.2	225.4	247.4	266.3	304.0
248	210.1	214.1	226.4	248.4	267.4	305.3
249	211.0	215.1	227.3	249.5	268.5	306.5
250	211.9	216.0	228.3	250.5	269.6	307.8
251	212.9	216.9	229.3	251.6	270.7	309.1
252	213.8	217.9	230.2	252.6	271.8	310.4
253	214.7	218.8	231.2	253.6	272.9	311.6
254	215.6	219.8	232.2	254.7	274.1	312.9
255	216.6	220.7	233.2	255.7	275.2	314.0
256	217.5	221.7	234.1	256.7	276.3	315.3

ERLANG C TABLE FOR DIMENSIONING BASE-STATION CHANNELS

Circuits	GOS					
	0.001	0.002	0.01	0.05	0.1	0.2
1	0.001	0.002	0.01	0.05	0.1	0.2
2	0.045	0.06	0.15	0.34	0.5	0.74
3	0.19	0.24	0.43	0.79	1	1.4
4	0.43	0.51	0.81	1.3	1.7	2.1
5	0.73	0.86	1.3	1.9	2.3	2.8
6	1.1	1.3	1.8	2.5	3.0	3.6
7	1.5	1.7	2.3	3.2	3.7	4.4
8	2.0	2.2	2.9	3.9	4.5	5.2
9	2.4	2.7	3.5	4.6	5.2	6.0
10	2.9	3.2	4.1	5.3	6.0	6.9
11	3.5	3.8	4.7	6.0	6.8	7.7
12	4.0	4.4	5.4	6.8	7.6	8.5
13	4.6	5.0	6.0	7.5	8.4	9.4
14	5.2	5.6	6.7	8.3	9.2	10.2
15	5.8	6.2	7.4	9.0	10.0	11.1
16	6.4	6.8	8.1	9.8	10.8	12.0
17	7.0	7.5	8.8	10.6	11.6	12.8
18	7.6	8.1	9.5	11.4	12.4	13.7
19	8.3	8.8	10.2	12.2	13.3	14.6
20	8.9	9.5	11.0	13.0	14.1	15.5
21	9.6	10.1	11.7	13.8	15.0	16.3
22	10.2	10.8	12.5	14.6	15.8	17.2
23	10.9	11.5	13.2	15.4	16.7	18.1
24	11.6	12.2	14.0	16.3	17.5	19.0
25	12.3	12.9	14.7	17.1	18.4	19.9
26	13.0	13.6	15.5	17.9	19.2	20.8
27	13.7	14.4	16.3	18.7	20.1	21.7
28	14.4	15.1	17.0	19.6	20.9	22.6
29	15.1	15.8	17.8	20.4	21.8	23.5
30	15.8	16.5	18.6	21.2	22.7	24.4
31	16.5	17.3	19.4	22.1	23.6	25.3
32	17.2	18.0	20.2	22.9	24.4	26.2
33	18.0	18.8	21.0	23.8	25.3	27.1
34	18.7	19.5	21.7	24.6	26.2	28.0
35	19.4	20.3	22.5	25.5	27.1	28.9
36	20.2	21.0	23.3	26.3	27.9	29.8
37	20.9	21.8	24.2	27.2	28.8	30.7
38	21.7	22.6	25.0	28.0	29.7	31.7
39	22.4	23.3	25.8	28.9	30.6	32.6
40	23.2	24.1	26.6	29.8	31.5	33.5
41	23.9	24.9	27.4	30.6	32.4	34.4
42	24.7	25.6	28.2	31.5	33.3	35.3
43	25.4	26.4	29.0	32.4	34.1	36.2
44	26.2	27.2	29.8	33.2	35.0	37.2
45	27.0	28.0	30.7	34.1	35.9	38.1
46	27.8	28.8	31.5	35.0	36.8	39.0
47	28.5	29.6	32.3	35.8	37.7	39.9
48	29.3	30.3	33.1	36.7	38.6	40.8
49	30.1	31.1	34.0	37.6	39.5	41.8
50	30.9	31.9	34.8	38.5	40.4	42.7
51	31.6	32.7	35.6	39.3	41.3	43.6

Erlang C Table (*continued*)

Circuits	GOS						Circuits	GOS					
	0.001	0.002	0.01	0.05	0.1	0.2		0.001	0.002	0.01	0.05	0.1	0.2
52	32.4	33.5	36.5	40.2	42.2	44.5	106	76.8	78.6	83.1	88.9	91.8	95.3
53	33.2	34.3	37.3	41.1	43.1	45.5	107	77.7	79.4	84.0	89.8	92.8	96.2
54	34.0	35.1	38.1	42.0	44.0	46.4	108	78.5	80.3	84.9	90.7	93.7	97.2
55	34.8	35.9	39.0	42.9	44.9	47.3	109	79.4	81.1	85.8	91.6	94.6	98.1
56	35.6	36.7	39.8	43.8	45.8	48.2	110	80.2	82.0	86.7	92.5	95.6	99.1
57	36.4	37.5	40.7	44.6	46.7	49.2	111	81.1	82.9	87.6	93.4	96.5	100.0
58	37.2	38.3	41.5	45.5	47.6	50.1	112	81.9	83.7	88.5	94.4	97.4	101.0
59	38.0	39.2	42.4	46.4	48.5	51.0	113	82.8	84.6	89.4	95.3	98.4	101.9
60	38.8	40.0	43.2	47.3	49.5	52.0	114	83.6	85.5	90.2	96.2	99.3	102.9
61	39.6	40.8	44.0	48.2	50.4	52.9	115	84.5	86.3	91.1	97.1	100.2	103.8
62	40.4	41.6	44.9	49.1	51.3	53.8	116	85.3	87.2	92.0	98.0	101.2	104.8
63	41.2	42.4	45.7	50.0	52.2	54.8	117	86.2	88.0	92.9	98.9	102.1	105.7
64	42.0	43.2	46.6	50.9	53.1	55.7	118	87.1	88.9	93.8	99.9	103.0	106.7
65	42.8	44.0	47.5	51.7	54.0	56.6	119	87.9	89.8	94.7	100.8	104.0	107.6
66	43.6	44.9	48.3	52.6	54.9	57.6	120	88.8	90.6	95.6	101.7	104.9	108.6
67	44.4	45.7	49.2	53.5	55.8	58.5	121	89.6	91.5	96.5	102.6	105.8	109.5
68	45.2	46.5	50.0	54.4	56.7	59.4	122	90.5	92.4	97.4	103.5	106.8	110.5
69	46.0	47.3	50.9	55.3	57.7	60.4	123	91.4	93.3	98.3	104.5	107.7	111.4
70	46.8	48.2	51.7	56.2	58.6	61.3	124	92.2	94.1	99.2	105.4	108.6	112.4
71	47.6	49.0	52.6	57.1	59.5	62.2	125	93.1	95.0	100.0	106.3	109.6	113.3
72	48.5	49.8	53.4	58.0	60.4	63.2	126	93.9	95.9	100.9	107.2	110.5	114.3
73	49.3	50.6	54.3	58.9	61.3	64.1	127	94.8	96.7	101.8	108.2	111.4	115.2
74	50.1	51.5	55.2	59.8	62.2	65.1	128	95.7	97.6	102.7	109.1	112.4	116.2
75	50.9	52.3	56.0	60.7	63.2	66.0	129	96.5	98.5	103.6	110.0	113.3	117.1
76	51.7	53.1	56.9	61.6	64.1	66.9	130	97.4	99.4	104.5	110.9	114.3	118.1
77	52.6	54.0	57.8	62.5	65.0	67.9	131	98.2	100.2	105.4	111.9	115.2	119.0
78	53.4	54.8	58.6	63.4	65.9	68.8	132	99.1	101.1	106.3	112.8	116.1	120.0
79	54.2	55.7	59.5	64.3	66.8	69.8	133	100.0	102.0	107.2	113.7	117.1	120.9
80	55.0	56.5	60.4	65.2	67.7	70.7	134	100.8	102.8	108.1	114.6	118.0	121.9
81	55.9	57.3	61.2	66.1	68.7	71.6	135	101.7	103.7	109.0	115.5	118.9	122.9
82	56.7	58.2	62.1	67.0	69.6	72.6	136	102.6	104.6	109.9	116.5	119.9	123.8
83	57.5	59.0	63.0	67.9	70.5	73.5	137	103.4	105.5	110.8	117.4	120.8	124.8
84	58.3	59.8	63.8	68.8	71.4	74.5	138	104.3	106.3	111.7	118.3	121.8	125.7
85	59.2	60.7	64.7	69.7	72.4	75.4	139	105.2	107.2	112.6	119.3	122.7	126.7
86	60.0	61.5	65.6	70.6	73.3	76.3	140	106.0	108.1	113.5	120.2	123.6	127.6
87	60.8	62.4	66.4	71.5	74.2	77.3	141	106.9	109.0	114.4	121.1	124.6	128.6
88	61.7	63.2	67.3	72.4	75.1	78.2	142	107.8	109.9	115.3	122.0	125.5	129.5
89	62.5	64.1	68.2	73.4	76.1	79.2	143	108.7	110.7	116.2	123.0	126.5	130.5
90	63.3	64.9	69.1	74.3	77.0	80.1	144	109.5	111.6	117.1	123.9	127.4	131.5
91	64.2	65.8	69.9	75.2	77.9	81.1	145	110.4	112.5	118.0	124.8	128.3	132.4
92	65.0	66.6	70.8	76.1	78.8	82.0	146	111.3	113.4	118.9	125.7	129.3	133.4
93	65.9	67.5	71.7	77.0	79.8	83.0	147	112.1	114.2	119.8	126.7	130.2	134.3
94	66.7	68.3	72.6	77.9	80.7	83.9	148	113.0	115.1	120.7	127.6	131.2	135.3
95	67.5	69.2	73.5	78.8	81.6	84.8	149	113.9	116.0	121.6	128.5	132.1	136.2
96	68.4	70.0	74.3	79.7	82.5	85.8	150	114.8	116.9	122.5	129.5	133.1	137.2
97	69.2	70.9	75.2	80.6	83.5	86.7	151	115.6	117.8	123.4	130.4	134.0	138.1
98	70.1	71.7	76.1	81.5	84.4	87.7	152	116.5	118.7	124.3	131.3	134.9	139.1
99	70.9	72.6	77.0	82.5	85.3	88.6	153	117.4	119.5	125.2	132.2	135.9	140.1
100	71.7	73.4	77.8	83.4	86.3	89.6	154	118.2	120.4	126.1	133.2	136.8	141.0
101	72.6	74.3	78.7	84.3	87.2	90.5	155	119.1	121.3	127.0	134.1	137.8	142.0
102	73.4	75.1	79.6	85.2	88.1	91.5	156	120.0	122.2	127.9	135.0	138.7	142.9
103	74.3	76.0	80.5	86.1	89.0	92.4	157	120.9	123.1	128.8	136.0	139.7	143.9
104	75.1	76.8	81.4	87.0	90.0	93.4	158	121.7	124.0	129.7	136.9	140.6	144.8
105	76.0	77.7	82.3	87.9	90.9	94.3	159	122.6	124.8	130.6	137.8	141.5	145.8

Erlang C Table (*continued*)

Circuits	GOS						Circuits	GOS					
	0.001	0.002	0.01	0.05	0.1	0.2		0.001	0.002	0.01	0.05	0.1	0.2
160	123.5	125.7	131.6	138.8	142.5	146.8	214	171.3	174.0	180.9	189.3	193.7	198.7
161	124.4	126.6	132.5	139.7	143.4	147.7	215	172.2	174.9	181.8	190.2	194.6	199.6
162	125.3	127.5	133.4	140.6	144.4	148.7	216	173.1	175.8	182.7	191.2	195.6	200.6
163	126.1	128.4	134.3	141.5	145.3	149.6	217	174.0	176.7	183.6	192.1	196.5	201.6
164	127.0	129.3	135.2	142.5	146.3	150.6	218	174.9	177.6	184.5	193.1	197.5	202.5
165	127.9	130.2	136.1	143.4	147.2	151.6	219	175.8	178.5	185.5	194.0	198.4	203.5
166	128.8	131.0	137.0	144.3	148.2	152.5	220	176.7	179.4	186.4	195.0	199.4	204.4
167	129.6	131.9	137.9	145.3	149.1	153.5	221	177.6	180.3	187.3	195.9	200.3	205.4
168	130.5	132.8	138.8	146.2	150.0	154.4	222	178.5	181.2	188.2	196.8	201.3	206.4
169	131.4	133.7	139.7	147.1	151.0	155.4	223	179.4	182.1	189.1	197.8	202.2	207.3
170	132.3	134.6	140.6	148.1	151.9	156.3	224	180.3	183.0	190.1	198.7	203.2	208.3
171	133.2	135.5	141.5	149.0	152.9	157.3	225	181.2	183.9	191.0	199.7	204.1	209.3
172	134.1	136.4	142.4	149.9	153.8	158.3	226	182.1	184.8	191.9	200.6	205.1	210.2
173	134.9	137.3	143.4	150.9	154.8	159.2	227	183.0	185.7	192.8	201.5	206.0	211.2
174	135.8	138.1	144.3	151.8	155.7	160.2	228	183.9	186.6	193.7	202.5	207.0	212.2
175	136.7	139.0	145.2	152.7	156.7	161.1	229	184.8	187.5	194.7	203.4	208.0	213.1
176	137.6	139.9	146.1	153.7	157.6	162.1	230	185.7	188.4	195.6	204.4	208.9	214.1
177	138.5	140.8	147.0	154.6	158.6	163.1	231	186.6	189.3	196.5	205.3	209.9	215.1
178	139.3	141.7	147.9	155.6	159.5	164.0	232	187.5	190.2	197.4	206.3	210.8	216.0
179	140.2	142.6	148.8	156.5	160.4	165.0	233	188.4	191.1	198.3	207.2	211.8	217.0
180	141.1	143.5	149.7	157.4	161.4	165.9	234	189.3	192.0	199.3	208.1	212.7	218.0
181	142.0	144.4	150.7	158.4	162.3	166.9	235	190.2	192.9	200.2	209.1	213.7	218.9
182	142.9	145.3	151.6	159.3	163.3	167.9	236	191.1	193.8	201.1	210.0	214.6	219.9
183	143.8	146.2	152.5	160.2	164.2	168.8	237	192.0	194.7	202.0	211.0	215.6	220.8
184	144.7	147.1	153.4	161.2	165.2	169.8	238	192.9	195.6	203.0	211.9	216.5	221.8
185	145.5	148.0	154.3	162.1	166.1	170.8	239	193.8	196.6	203.9	212.9	217.5	222.8
186	146.4	148.8	155.2	163.0	167.1	171.7	240	194.7	197.5	204.8	213.8	218.4	223.7
187	147.3	149.7	156.1	164.0	168.0	172.7	241	195.5	198.4	205.7	214.8	219.4	224.7
188	148.2	150.6	157.0	164.9	169.0	173.6	242	196.4	199.3	206.7	215.7	220.4	225.7
189	149.1	151.5	158.0	165.8	169.9	174.6	243	197.3	200.2	207.6	216.6	221.3	226.6
190	150.0	152.4	158.9	166.8	170.9	175.6	244	198.3	201.1	208.5	217.6	222.3	227.6
191	150.9	153.3	159.8	167.7	171.8	176.5	245	199.2	202.0	209.4	218.5	223.2	228.6
192	151.7	154.2	160.7	168.7	172.8	177.5	246	200.1	202.9	210.4	219.5	224.2	229.5
193	152.6	155.1	161.6	169.6	173.7	178.4	247	201.0	203.8	211.3	220.4	225.1	230.5
194	153.5	156.0	162.5	170.5	174.7	179.4	248	201.9	204.7	212.2	221.4	226.1	231.5
195	154.4	156.9	163.4	171.5	175.6	180.4	249	202.8	205.6	213.1	222.3	227.0	232.4
196	155.3	157.8	164.3	172.4	176.6	181.3	250	203.7	206.5	214.0	223.3	228.0	233.4
197	156.2	158.7	165.3	173.3	177.5	182.3	251	204.6	207.4	215.0	224.2	229.0	234.4
198	157.1	159.6	166.2	174.3	178.5	183.3	252	205.5	208.3	215.9	225.1	229.9	235.3
199	158.0	160.5	167.1	175.2	179.4	184.2	253	206.4	209.3	216.8	226.1	230.9	236.3
200	158.9	161.4	168.0	176.2	180.4	185.2	254	207.3	210.2	217.8	227.0	231.8	237.3
201	159.7	162.3	168.9	177.1	181.3	186.1	255	208.2	211.1	218.7	228.0	232.8	238.2
202	160.6	163.2	169.8	178.0	182.3	187.1	256	209.1	212.0	219.6	228.9	233.7	239.2
203	161.5	164.1	170.8	179.0	183.2	188.1							
204	162.4	165.0	171.7	179.9	184.2	189.0							
205	163.3	165.9	172.6	180.9	185.1	190.0							
206	164.2	166.8	173.5	181.8	186.1	191.0							
207	165.1	167.7	174.4	182.7	187.0	191.9							
208	166.0	168.6	175.3	183.7	188.0	192.9							
209	166.9	169.5	176.3	184.6	188.9	193.8							
210	167.8	170.4	177.2	185.5	189.9	194.8							
211	168.7	171.3	178.1	186.5	190.8	195.8							
212	169.6	172.2	179.0	187.4	191.8	196.7							
213	170.4	173.1	179.9	188.4	192.7	197.7							

APPENDIX I

CONVERSION OF UNITS USED FOR CELLULAR RF

A lot of errors are made by simple conversion mistakes between units. The following table has the most common units used in cellular listed for a large range of values that are commonly encountered.

TABLE I.1   Conversion Table

dBm	μV	dBμV	Freq.MHz (dBμV/m) 150	450	900
10	707107	117	129	138	144
1	250891	108	120	129	135
0	223607	107	119	128	134
−1	199290	106	118	127	133
−2	177617	105	117	126	132
−3	158301	104	116	125	131
−4	141080	103	115	124	130
−5	125743	102	114	123	129
−6	112069	101	113	122	128
−7	99881	100	112	121	127
−8	89019	99	111	120	126
−9	79338	98	110	119	125
−10	70710	97	109	118	124
−11	63021	96	108	117	123
−12	56167	95	107	116	122
−13	50059	94	106	115	121
−14	44615	93	105	114	120
−15	39763	92	104	113	119
−16	35439	91	103	112	118
−17	31585	90	102	111	117
−18	28150	89	101	110	116
−19	25089	88	100	109	115
−20	22360	87	99	108	114
−21	19929	86	98	107	113
−22	17761	85	97	106	112
−23	15830	84	96	105	111

dBm	μV	dBμV	Freq.MHz (dBμV/m) 150	450	900
−24	14108	83	95	104	110
−25	12574	82	94	103	109
−26	11206	81	93	102	108
−27	9988	80	92	101	107
−28	8902	79	91	100	106
−29	7933	78	90	99	105
−30	7071	77	89	98	104
−31	6302	76	88	97	103
−32	5616	75	87	96	102
−33	5005	74	86	95	101
−34	4461	73	85	94	100
−35	3976	72	84	93	99
−36	3543	71	83	92	98
−37	3158	70	82	91	97
−38	2815	69	81	90	96
−39	2508	68	80	89	95
−40	2236	67	79	88	94
−41	1993	66	78	87	93
−42	1776	65	77	86	92
−43	1583	64	76	85	91
−44	1410	63	75	84	90
−45	1257	62	74	83	89
−46	1120	61	73	82	88
−47	998	60	72	81	87
−48	890	59	71	80	86
−49	793	58	70	79	85
−50	707	57	69	78	84
−51	630	56	68	77	83
−52	562	55	67	76	82
−53	501	54	66	75	81
−54	446	53	65	74	80
−55	398	52	64	73	79
−56	354	51	63	72	78
−57	316	50	62	71	77
−58	281	49	61	70	76

TABLE I. (*continued*)

dBm	μV	dBμV	Freq.MHz (dBμV/m)			dBm	μV	dBμV	Freq.MHz (dBμV/m)		
			150	450	900				150	450	900
−59	251	48	60	69	75	−95	4	12	24	33	39
−60	223	47	59	68	74	−96	3.5	11	23	32	38
−61	199	46	58	67	73	−97	3.2	10	22	31	37
−62	177	45	57	66	72	−98	2.8	9	21	30	36
−63	156	44	56	65	71	−99	2.5	8	20	29	35
−64	141	43	55	64	70	−100	2.2	7	19	28	34
−65	125	42	54	63	69	−101	2	6	18	27	33
−66	112	41	53	62	68	−102	1.8	5	17	26	32
−67	100	40	52	61	67	−103	1.6	4	16	25	31
−68	89	39	51	60	66	−104	1.4	3	15	24	30
−69	79	38	50	59	65	−105	1.3	2	14	23	29
−70	70	37	49	58	64	−106	1.1	1	13	22	28
−71	63	36	48	57	63	−107	1	0	12	21	27
−72	56	35	47	56	62	−108	0.89	−1	11	20	26
−73	50	34	46	55	61	−109	0.79	−2	10	19	25
−74	45	33	45	54	60	−110	0.71	−3	9	18	24
−75	39	32	44	53	59	−111	0.63	−4	8	17	23
−76	36	31	43	52	58	−112	0.56	−5	7	16	22
−77	32	30	42	51	57	−113	0.5	−6	6	15	21
−78	28	29	41	50	56	−114	0.47	−7	5	14	20
−79	25	28	40	49	55	−115	0.4	−8	4	13	19
−80	22	27	39	48	54	−116	0.35	−9	3	12	18
−81	20	26	38	47	53	−117	0.32	−10	2	11	17
−82	18	25	37	46	52	−118	0.28	−11	1	10	16
−83	16	24	36	45	51	−119	0.25	−12	0	9	15
−84	14	23	35	44	50	−120	0.22	−13	−1	8	14
−85	13	22	34	43	49	−121	0.2	−14	−2	7	13
−86	11	21	33	42	48	−122	0.17	−15	−3	6	12
−87	10	20	32	41	47	−123	0.16	−16	−4	5	11
−88	9	19	31	40	46	−124	0.14	−17	−5	4	10
−89	8	18	30	39	45	−125	0.126	−18	−6	3	9
−90	7	17	29	38	44	−126	0.122	−19	−7	2	8
−91	6.3	16	28	37	43	−127	0.1	−20	−8	1	7
−92	5.6	15	27	36	42	−128	0.089	−21	−9	0	6
−93	5	14	26	35	41	−129	0.079	−22	−10	−1	5
−94	4.5	13	25	34	40	−130	0.071	−23	−11	−2	4

APPENDIX J

COUNTRY CODES

The following country codes have been assigned by the International Telecommunications Union—Radio (CCIR) to help in the recognition of roaming mobiles. Since most countries now have multiple networks, these codes have been extended to include the local network number. So, for example, the first three Australian (505) networks have the following codes

Telstra	505-01
Optus	505-02
Vodafone	505-03

TABLE J.1 International Mobile Network Prefixes Recommended by CCIR

Zone 2	
Code	Country or Geographical Area
202	Greece
204	Netherlands
206	Belgium
208	France
212	Monaco
214	Spain
216	Hungary
218	German Democratic Republic
220	Yugoslavia
222	Italy
226	Romania
228	Switzerland
230	Czechoslovakia
232	Austria
234	United Kingdom
238	Denmark
240	Sweden
242	Norway
244	Finland
250	Soviet Union
260	Poland
262	Federal Republic of Germany
266	Gibraltar
268	Portugal
270	Luxembourg
272	Ireland
278	Malta
280	Cyprus
284	Bulgaria
286	Turkey

Zone 3	
Code	Country or Geographical Area
302	Canada
308	St. Pierre and Miquelon
310	United States of America
330	Puerto Rico
332	Virgin Islands (USA)
334	Mexico
338	Jamaica
340	French Antilles
342	Barbados
344	Antigua
346	Cayman Islands
348	British Virgin Islands
350	Bermuda
352	Grenada
354	Montserrat
356	St. Kitts
358	St. Lucia
360	St. Vincent and The Grenadines
362	Netherlands Antilles
364	Bahamas
366	Dominica

Zone 3 (*continued*)

Code	Country or Geographical Area
368	Cuba
370	Dominican Republic
372	Haiti
374	Trinidad and Tobago
376	Turks and Caicos Islands

Zone 4

Code	Country or Geographical Area
404	India
410	Pakistan
412	Afghanistan
413	Sri Lanka
414	Myanmar
415	Lebanon
416	Jordan
417	Syria
418	Iraq
419	Kuwait
420	Saudi Arabia
421	Yemen Arab Republic
422	Oman
423	Yemen Democratic Republic
424	United Arab Emirates
425	Israel
426	Bahrain
427	Qatar
428	Mongolia
429	Nepal
430	United Arab Emirates (Abu Dhabi)
431	United Arab Emirates (Dubai)
432	Iran
440	Japan
450	Korea
452	Vietnam
454	Hong Kong
455	Macao
456	Kampuchea
457	Laos
467	Democratic People's Republic of Korea
470	Bangladesh
472	Maldives

Zone 5

Code	Country or Geographical Area
502	Malaysia
505	Australia
510	Indonesia
515	Philippines
520	Thailand
525	Singapore
528	Brunei Darussalam
530	New Zealand
535	Guam
536	Nauru
537	Papua New Guinea
539	Tonga
540	Solomon Islands
541	Vanuatu
542	Fiji
543	Wallis and Futuna Islands
544	American Samoa
545	Gilbert and Ellice Islands
546	New Caledonia and Dependencies
547	French Polynesia
548	Cook Islands
549	Western Samoa

Zone 6

Code	Country or Geographical Area
602	Egypt
603	Algeria
604	Morocco
605	Tunisia
606	Libya
607	Gambia
608	Senegal
609	Mauritania
610	Mali
611	Guinea
612	Ivory Coast
613	Upper Volta
614	Niger
615	Togolese Republic
616	Benin
617	Mauritius
618	Liberia
619	Sierra Leone
620	Ghana
621	Nigeria
622	Chad
623	Central African Republic
624	Cameroon
625	Cape Verde
626	Sao Tome and Principe
627	Equatorial Guinea
628	Gabon Republic
629	Congo
630	Zaire
631	Angola
632	Guinea-Bissau
633	Seychelles
634	Sudan
635	Rwanda
636	Ethiopia
637	Somali
638	Djibouti
639	Kenya
640	Tanzania
641	Uganda
642	Burundi
643	Mozambique
645	Zambia

Zone 6 (*continued*)	
Code	Country or Geographical Area
646	Madagascar
647	Reunion
648	Zimbabwe
649	Namibia
650	Malawi
651	Lesotho
652	Botswana
653	Swaziland
654	Comoros
655	South Africa

Zone 7	
Code	Country or Geographical Area
702	Belize
704	Guatemala
706	El Salvador
708	Honduras
710	Nicaragua
712	Costa Rica
714	Panama
716	Peru
722	Argentina
724	Brazil
730	Chile
732	Colombia
734	Venezuela
736	Bolivia
738	Guyana
740	Ecuador
742	Guiana
744	Paraguay
746	Suriname
748	Uruguay

GLOSSARY

2G (Second Generation) The mobile phone services characterized by the digital systems of the 1990s (GSM, TDMA, and CDMA).

2G+ (Second Generation Plus) This mostly refers to GSM evolution to 3G from 2G and effectively means enhanced 2G services. However, this is also applied to advanced data services on TDMA and CDMA.

3G The next generation of mobile services that offers advanced high-speed data services over the mobile phone.

Access Overload Class An identifier used to separate the customers into priority groups to be used in the event of system overload.

ACD Automatic Call Distributor. A device that routes incoming calls automatically to a number of specialized answering positions.

ACELP Algebraic Code (book) Excited Linear Predictive. A CODEC algorithm.

ADC Analog to Digital Converter. Takes an analog signal as an input and digitizes it. Increasingly used in cellular phones to take the IF signal and digitize the baseband signal, which in turn is processed by a DSP.

Adjacent Channel A channel with a frequency assignment one channel away (the next channel in the group).

AI Artificial Intelligence. A class of computing that does not rely on a set of programming instructions, but which is self-programming. An AI machine will have inputs and outputs, and some guidelines describing the desired output. It will then organize itself internally to achieve those objectives. Common modes are genetic algorithms and neural networks. Generally based on neural networks or genetic algorithms, these systems are task oriented and can be "trained" to perform specific tasks, without purpose-specific programming.

Airtime The total time that a channel is occupied, including call time, call setup, and cleardown time.

Algorithm A mathematical model or process.

AMPS Advanced Mobile Phone System.

Analog A system that processes electrical signals by representing the original signal as a continuous function of the original signal. Thus the sound pressure at the input of a microphone may be represented as a voltage level at the terminals of an amplifier, which then amplifies the voltage. Analog differs from digital in that the digital representation of a signal is discrete; that is, all signals are represented as discrete numbers.

ANSI American National Standards Institute. The main US national public standards group.

Antenna Gain The gain of an antenna compared to a dipole or quarter-wave antenna. Sometimes the gain is compared to an isotropic antenna, and this is referred to as dBi. dBi = dBd + 2.1.

ANSI-41 American National Standard. This is an evolution of IS-41.

APCO Association of Public safety Communications Officials.

Area Code Usually a two- or three-digit number that identifies the area of a telephone outside the home area of a caller.

ARQ In data transmission (including digital cellular) this term refers to an automatic request for a retransmission of a frame, that is so badly errored that the original data cannot be recovered.

ASIC Application-Specific Integrated Circuit. Much like a VLSI except that it has been designed for a particular application only. Increased use of ASICs lead to smaller size and lower volume cost; however, an individual ASIC can be very expensive to develop. The ultimate cellular ASIC is one where the whole cellular radio can be put on one chip.

Asynchronous Transfer Mode (ATM) Data transfer in the asynchronous mode (as opposed to most digital systems which are synchronous).

Authentication The process by which the identity and validity of the mobile are determined.

Authentication Center (AC) The function part of the system that transacts authentication procedures with the mobile.

Authentication Key (A-Key) A secret (private) 64-bit key used in the ANSI, CAVE algorithm to encrypt and de-encrypt information. The A-key is used to generate the temporary private key, known as the shared secret data (SSD). The A-key is known only to the mobile and the authentication center.

Backhauling Refers to remotely run base stations that are linked back to a switch that primarily serves a different area. An example would be an RSA served from an MSA switch located 100 kilometers away. In general, it will be necessary to switch the calls in the MSA, but as most will terminate in the RSA, the calls will need to be "backhauled" for completion.

Base or Base Station A site that contains the cellular radio equipment. It can have one or more cells.

Baseband The signal that is presented to the modulator for transmission.

Bent-Pipe A term in satellite technology that refers to a simple repeater, that is, the incoming signal is simply repeated back (usually at a different frequency). This requires virtually no processing power at the satellite.

Bit Error Rate (BER) The number of errors, expressed as a fraction of the total number of bits sent, of a digital signal.

Blocking Calls lost because of the lack of switching or carrying capacity. Also means base-station channels taken out of service by interference.

BOT Build-Operate-Transfer. A system whereby a sponsor is granted a license to operate a system for a fixed period of time (typically 10 years) and at the expiration of the license, the whole system is transferred to the government, usually at no charge.

Boundary (Of coverage) The defined limits of a particular cell. Usually defined to be 39 dBμV/m (-96 dBm) for AMPS system, and variously at all levels, around -84 dBm for GSM. However, some coverage is usually available well outside the boundary.

Called Party The phone that receives the call.

Called Party The phone that initiates the call.

Carrier-to-Interference Ratio (C/I) The ratio of the power in the carrier signal to the total power of the interfering signals. Usually expressed as a log of the ratio in dBs.

Cavity A resonant device, usually drum (or cylinder) shaped, that acts as a filter in cellular systems.

CBD Central Business District (city center).

CCIR International Radio Consultative Committee (ITU).

CCITT International Telegraph and Telephone Consultative Committee (ITU).

CDMA Code Division Multiple Access. A wide-band spread-spectrum system whereby many RF users can share the same spectrum simultaneously, discriminating the signals by the code that is sent.

Cell A group of co-located channels that cover the same area. A base can have one or more cells by using directional antennas.

Cell Extender A cellular repeater that repeats base-station channels. It is essentially a linear amplifier.

Cell Site The location of the cell.

Cellular Authentication and Voice Encryption (CAVE) The general term for the encryption algorithm that is used in the ANSI authentication process. It makes use of the mobiles MON, ESN, A-Key, and SSD to verify the authenticity of a mobile or base station. Also refers to a collection of algorithms that are used in a number of systems to encrypt user information and speech.

Cellular Digital Packet Data (CDPD) A system used by AMPS that sends data packets over channels that would otherwise be idle.

Cellular Operator The owner and/or operator of a cellular network.

Cellular Telecommunications Industry Association (CTIA) An international organization of wireless telecommunications carriers and manufacturers.
Also the reference could be more loosely to "the CTIA," meaning the cellular industry trade fair that the CTIA organizes annually (highly recommended).

Centrex A virtual PBX or PABX provided by a central PSTN switch.

CEPT European Conference of Posts and Telecommunications Administrations.

Channel A path, either unidirectional or bidirectional, for communications. In analog systems the channel is a pair of frequencies; in digital it might be a timeslot or a code sequence.

C/I Carrier-to-Interference Ratio. The power ratio, usually expressed in dB, of the wanted carrier to the net interference.

Coaxial Cable A pair of conductors consisting of a central conductor surrounded by an outer conductor. These cables are used because of their immunity to interference and relatively low power losses at high frequencies.

Co-channel A channel that is on the same frequency.

Collinear Antenna A gain antenna with dipoles stacked vertically.

Combiner A device for combining several (usually four or five) transmit channels.

Common Channel Signaling A method of signaling that uses a dedicated channel to transmit signaling information to a number of channels.

Coupler A device for connecting two or more sources of RF energy to a single cable or port.

Coverage The area over which the service is of an acceptable standard.

CT1 The original cordless phone. Meant for household use only.

CT2 An attempt to produce a commercial public version of CT1.

DAMPS Also know as TDMA. A digital system that uses dual-mode AMPS/TDMA transmission for cellular. The TDMA standard is to some extent a simplified version of GSM.

dB (DECIBELS) A unit for expressing the relative intensity of sound. This is equal to

$$10 \times \log \frac{\text{Power referred to}}{\text{Power of a reference level}}$$

or

$$20 \times \log \frac{\text{Voltage referred to}}{\text{Voltage of a reference level}}$$

dBc Decibels below the carrier. Often used to specify things like noise levels relative to the carrier level.

dBd Gain relative to a dipole antenna.

dBi Gain relative to a hypothetical isotropic antenna. (It is 2.1 times higher than dBd for the same antenna.)

DECT Digital European Cordless Telecommunications. The European cordless phone standard.

Demand Assigned Multiple Access (DAMA) A system where a pool of channels is shared by a number of users. Cellular systems are an example, but the term DAMA is mostly used for WLL and satellite applications.

Deviation The amount of frequency change from the center frequency in a modulated FM system (expressed in kHz or 000 Hz).

D-GPS Differential GPS (see GPS). A method for locally increasing GPS accuracy from about 15 meters to around 1 meter, or in some instances even centimeters (it depends on the methods used).

Diffraction Propagation around an obstructing object.

Digital A processing system whereby the processing is done using a number of discrete levels to represent the signal. The simplest digital representation is a binary one; that is, one that has two levels or states—for example, on and off. Any number can be translated into a binary number consisting only of a string of 1s and 0s.

Distributed System A system where the resources are shared over a device or system. For example, a distributed switch will be modular, with the switching and processing parts self-contained with a number of modules, so the a partial failure will not affect the rest of the system.

DTMF Dual Tone MultiFrequency. The signaling used on modern push-button telephones. Combinations of two tones represent various numbers.

DS0 (Digital Signal Level 0) A 56 kbits/s channel.

DS1 (Digital Signal Level 1) 1,544,000 bits/s data rate, also known as T1.

DS3 (Digital Signal Level 3) A level that can carry up to 28 DS1 signals.

DSP Digital Signal Processor. A device that takes the output of an ADC and processes the resultant signal digitally. Typically, it will do filtering, channel separation, and detection tasks.

DTX Discontinuous Transmission. Mostly applies to digital systems that can sense periods of no audio (or data) and can then cease transmission (shut down the RF part) in order to save power and conserve spectrum.

Dual-Mode Mobile A mobile that can operate in more than one mode. Mostly refers to AMPS, and DAMPS, or AMPS/CDMA capability. Could also refer to satellite/land-based mobile systems.

DWDM Dense Wave Division Multiplex. The provision of large numbers of wavelengths on an optical fiber. 5000 or more wavelengths can be multiplexed together.

E1 A carrier that has 32, 64-kbits/s channels, of which 30 are voice channels.

EDGE Enhanced Data for GSM Evolution. Effectively GPRS with a faster modulation method for higher packet data speeds.

EMI ElectroMagnetic Interference.

Encryption A method of securing data by "scrambling" the signal in a way that is intended to make it difficult for anybody other than the intended recipient(s) to decode it.

Erlang A unit of telephone traffic such that 1 Erlang is one occupied circuit per hour. Also sometimes called 36ccs = 36 × 100 call seconds, and so translates to 1 call for 1 hour. Erlangs may also be defined instantaneously to be the number of circuits activated at a particular time.

ERP Effective Radiated Power. The power, expressed in watts that is radiated in the direction of maximum antenna gain calculated by multiplying the power at the antenna terminals by that gain. It is often a theoretical power, as antenna gains often neglect antenna losses.

ESN Electronic Serial Number that is contained in the NAM (Number Assignment Module).

E-TACS Extended TACS, essentially the same as TACS but using an extended band of frequencies from the original allocation.

E-TDMA Extended TDMA. A Hughes system to extend the capacity of the US digital TDMA system by a more elaborate processing of the audio signal.

ETSI The European Telecommunications Standards Institute. The body behind GSM.

Extended Cell A cellular repeater that uses frequency translation.

FCC Federal Communications Commission. The regulatory body in the United States.

FDMA Frequency Division Multiple Access. Cellular systems are FDMA systems. In these systems, the users are assigned a particular pair of frequencies (channels) on request for the duration of a particular call.

Feeder A coaxial cable or waveguide connecting a transmitter/receiver to an antenna.

Feedline Same as Feeder.

FLMTS Future Land Mobile Telecommunications Systems. The ITU program that became its input to 3G and was later renamed IMT-2000.

FM Frequency Modulation. A very common analog modulation technique noted for its excellent signal-to-noise (S/N) characteristics. The frequency of the carrier varies in proportion to the amplitude of the modulating signal.

Free Space An idealized concept of an antenna far from any objects that could interfere with its radiation pattern.

Frequency The rate at which the electric and magnetic fields of a radio wave vibrate per second. Frequency is usually expressed in MHz (1,000,000 Hz); 1 Hz (hertz) = 1 cycle per second.

Frequency Translation The process of converting a signal to a different frequency. Often used in repeaters.

FSK Frequency Shift Keying. A modulation method using frequency changes (in steps) to transmit data. Usually only two frequencies are used.

Full-Duplex Any system that allows simultaneous talk and listen functions. The opposite is simplex, as in push-to-talk mobiles.

Full-Rate Most digital systems have been introduced with CODECs that are still under development. These systems hope to double their throughput or capacity in the future by using CODECs that are twice as fast. These channels are known as half-rate.

Fuzzy Logic A set of software rules that makes decisions on a set of fuzzy rules. Fuzzy logic uses relative terms like "about the right size," "too big" and "far too big," hence the term "fuzzy."

Gain The factor, usually expressed in decibels (dB), by which the signal received is amplified or improved (in the case of an antenna).

GEO Geostationary Earth Orbit. A satellite orbit, the rate of which makes it appear stationary in the sky.

GOS Grade Of Service. The probability that a call will fail due to the unavailability of links or circuits. In cellular systems typical figures are a GOS of 0.05 for base-station channel availability and 0.002 for switch-to-switch circuits. In cellular systems GOS really means the probability of a "resend" in most instances, rather than the probability of call failure.

GPRS General Packet Radio Services. A packet-switched data service that, in its GSM version, can use up to four time slots.

GRAN Generic interface to GSN Networks. A interface defined as part of the 3G specification to allow interfacing between GSM and other air-interface systems.

Ground Plane The area directly below a quarter-wave or other unbalanced antenna. It should be of low resistance and at least a quarter wavelength in radius from the antenna.

GSM Originally an abbreviation for *Groupe Speciale Mobile*, renamed Global System for Mobile Communications. The Pan-European digital system.

GSN (GPRS Support Node) See GPRS.

Half-Rate Channels that are compressed so that they can be sent in half the time of a full-rate channel.

Handoff The ability of a cellular mobile to be able to move through the coverage area, handing off from cell to cell in order to maintain a good signal quality. The handoff is ideally not perceptible to the user.

Handsfree   A voice-operated system that allows hands-free (no handset) telephone operation (most loud-speaker telephones are of this type).

Hard Handoff   A break-before-make handoff that results is some discontinuity of the channel.

Hexadecimal   A number system based on 16.

High-Speed Circuit Switched Data (HSCSD)   The name given to the 14.4-kbits/s (and potentially to the 64-kbits/s) data circuits for GSM.

Home Base   A GSM open specification system for cordless telephone operations from GSM phones.

Home Location Register (HLR)   The register that contains the main database for the subscriber's records.

HTS   High Temperature Superconductor. Mostly for cellular applications, this refers to superconductors that can operate in liquid nitrogen.

IC   Integrated Circuits. The building blocks of modern electronic devices.

IETF   International Engineering Task Force. The body that sets current and future Internet protocols and standards. Expect to hear a lot from this body in the near future.

Inband Signaling   A method whereby control signals are sent over the voice channel, often without interference to the voice path, as things like sub-audible or super-audible tones are used.

IMD   Intermodulation Distortion. The distortion that occurs when two or more signals of different frequency are passed through a non-linear device, producing harmonic products of the fundamental frequencies.

IMT-2000   International Mobile Communications-2000. The ITU's 3G program.

Intelligent Network (IN)   A distributed network with a number of nodes, where each of the nodes has processing as well as switching capabilities.

Interactive Voice Response (IVR)   Effectively an interactive voice-messaging machine, used for voice mail, credit card authorization for payments, e-mail, and other applications.

Interference   The reception of unwanted signals that are impressed on the desired signal. In cellular most interference comes from other parts of the same network.

Intermodulation   The mixing of two or more signals in a non-linear device (and for practical purposes, all devices are to some extent non-linear), to form new signals that are linear multiples of the sums and differences of those signals. For example, two signals of

frequency A and B will form products with frequencies of $A + B$, $A - B$, $2A - B$, $2B - A$, $3A - B$, $3A - 2B$, etc. The order of intermodulation is the sum of the multipliers; thus, the first two products are second order, the next two are third order $(2 + 1)$, etc. Fortunately it is the general case that the magnitude of these signals decreases as the level rises.

Integrated Access Device (IAD)   A device that concentrates and supports a variety of different user services, such as POTS, video, X.25, and LANs. IADs typically combine the functions of a digital cross-connect and a multiplexer.

IP   Internet Protocol. The protocol used by the Internet, but widely tipped to be used extensively in future cellular networks both on the air and other interfaces; these cannot efficiently survive as 64-kbps circuits when the air interface is IP.

IP$_3$ Third-Order Intercept. The power level at the output of a non-linear device where the third-order intermodulation product, measured at the output, has the same level as the original signals.

IS-41   The interim standard for inter-system handoff.

IS-54   The interim standard for the dual-mode AMPS/TDMA system.

ISDN   Integrated Digital Services Network. Essentially the digital standard for the PSTN, which has been extended to include other services like cellular. At the subscriber level this often refers to the provisioning of a 64-kbits/s line.

ISM Band   Industrial Scientific and Medical Band. The FCC has defined three bands in which unlicensed operation is permitted, provided the long-term average power output is limited to 1 watt and spread-spectrum technology is used. The ISM band 902 to 928 MHz is widely used for spread-spectrum links and cordless technologies. Other bands include 2400–2583.5 MHz and 5725–5850 MHz.

Isolator   A unidirectional RF device, which allows the signal to pass in one direction only.

Isotropic   The same in all directions; in antennas, equal radiation in all directions.

JTACS   Japanese TACS. A version of TACS on different frequencies to the conventional TACS and used only in Japan.

kHz   KiloHertz. One-thousand (kilo) cycles per second; 1000 kHz = 1 MHz.

Leaky Cable   A cable that is designed to deliberately leak RF energy. This is often used to provide coverage in tunnels and basements.

LEO   Low Earth Orbit. A satellite at a height that just clears most of the drag of the Earth's atmosphere. Used for Iridium and Globalstar.

LNA Low Noise Amplifier. A pre amplifier, specifically designed for low noise that is used as the first RF amplifier in the RX stage with the objective of improving the overall sensitivity.

LMCS Local Multipoint Communications Service. A point to multipoint service.

LMDS Local Multipoint Distribution Service. A WLL point-to-point service.

LPA Linear Power Amplifier. A wide-band amplifier that can amplify a number of signals at different frequencies concurrently.

Maintenance Restoring a unit to working order by replacing it or replacing an integral module (for example, a panel).

Mast A guyed structure meant to support antenna(s).

MDLP Mobile Data Link Protocol. The protocol used for CDPD, which is based on ISDN's LAPD protocol.

Memory Integrated circuits that store information such as telephone numbers.

MEO Medium Earth Orbit. A satellite operating between the GEO orbit and the LEO band.

Microcell A small cell with a coverage of 200–500 meters radius that is meant to cover a very localized area (like a city block) only. Cells smaller still, designed perhaps to cover an office, are called picocells.

Milliwatt One-thousandth of a watt.

MIS Management Information System. A software package containing billing and management information. Usually sold as a package to the cellular operator, but sometimes sold as a service by third parties.

MMIC Microwave Monolithic Integrated Circuit. A microwave RF circuit, typical of the PA of an RF system containing one or more amplifying stages.

Mobile Positioning System An interface to a mobile network that allows accurate location of the mobiles in the network. Generally this will work with unmodified radios for digital systems.

Modem Modulator/demodulator that converts binary-to-analog signals and analog-to-binary signals. Used to connect digital devices like computers over analog telephone lines.

Modulation The method by which the transmitted signal is impressed on the carrier.

MSA Metropolitan Statistical Area (US). Basically, the major city cellular service areas.

MSC Mobile Switching Center. The mobile radio switch.

MSS Mobile Satellite Services. The generic term for all satellite-based mobile communications systems.

MTBF Mean Time Between Failure. Usually measured in years.

MTSC Mobile Telephone Switching Center. A term used to describe the mobile switch in most places outside the United States.

MTSO Mobile Telephone Switching Office. A term used to describe the switch, mainly in the United States, but occasionally elsewhere.

Multicoupler A tuned device that couples 2 or more (usually 16 in cellular) channels into or out of one feeder or antenna.

Multipath The interference patterns created by the addition of signals from more than one path. Virtually all mobile systems, including cellular, operate in a multipath environment; as distinct from point-to-point systems, which are line-of-sight, and usually there is only one original path.

NAM Number Assignment Module. A PROM 32×8 bits long that contains subscriber, system, and options details about a cellular telephone.

NAMPS Narrow-band AMPS. A Motorola system to split the AMPS 30-kHz channels into three 10-kHz channels, in a manner similar to the two-to-one split of the 25-kHz JTACS band in Japanese systems.

Nanosecond 1/1,000,000,000 part of a second. The prefix *nano* is used before a number of units like volts, meters, amperes, etc., to signify this same fractional part.

NiCad or NICAD A nickel–cadmium battery. A rechargeable battery of the type commonly found in cellular handhelds and other mobile handheld radios.

NMT Nordic Mobile Telephone. Can be NMT450, NMT470, NMT900, where the numbers indicate the frequency in MHz.

OEM Original Equipment Manufacturer. A third-party manufacturer that specializes in producing hardware for other companies, and generally has no identifiable products of its own. Increasingly the way most equipment is being made today.

Omni-Directional Antenna An antenna radiating energy equally in all directions (horizontally) around it.

OTSAP Over-the Air Service Provisioning. This is the activation of a mobile including all validation procedures over-the-air.

PA Power Amplifier. The power output amplifier. In cellular circles this usually means the final RF stage. Also often refers to the output stage of the RF part of a transmitter.

Packet Switched A method of taking a data stream and breaking it up into packets, which are sent individually to the receiving end where they are re-formed into the original message.

PACS   Personal Access Communications System. A US cordless telephone system similar to PHS and DECT that is based on 32-kbits/s ADPCM channels.

PCM   Pulsed Code Modulation. A digital transmission that uses a number of channels over the same bearer in different timeslots.

PCN   Personal Communications Network. Cellular networks operating at 1800–2200 MHz. These are virtually identical to their 800- and 900-MHz counterparts, except that the mobile power levels are usually lower (about one-half).

PCS   Personal Cellular System. The same as PCN. The preferred term in the United States.

Phase-Locked Loop   A circuit that can lock onto a desired frequency from a reference using a phase comparator.

PHP   Personal Handy Phone. A Japanese cordless phone technology.

Picocell   A very small cell, typically of about 100 meters coverage, smaller than a microcell (200–500 meters coverage).

PM   Phase Modulation. An analog modulation form related to FM in which the phase of the carrier is varied with the amplitude of the modulating signal.

PMID   Passive Intermodulation Distortion. Intermodulation distortion caused by passive components such as connectors and filters.

PMR   Public Mobile Radio (two-way radio).

Point-to-Point   A link between two and only two stations.

Point-to-Multipoint   A link from a point (or station) to many other stations. For example, a WLL system may be based on a transmitter that uses an omnidirectional antenna (the point) to communicate to hundreds of subscribers (the multipoints). Also subscriber's wireless interconnection to the PSTN.

Polarization   The orientation of the electric field of the antenna. Polarizations are vertical for signals leaving a vertical antenna (e.g., a handheld with the antenna held vertically). The polarization of the signal changes as it propagates due to reflection and refractions. Polarizations are commonly vertical, horizontal, cross-polarized (usually meaning two polarizations, each 45 degrees from the vertical and horizontal), or circular (meaning that the polarization spirals).

Portable   A handheld cellular telephone.

POT   Plain Old Telephone.

ppm   Parts per million.

Preselector Filter   The first filter in the RF path, the purpose of which is to pass the required RX bandwidth and reject other signals (in particular the transceiver's own TX).

Preferred Mode   In multimode mobiles, this is the default mode. For example, a CDMA/AMPS systems may first attempt to establish a CDMA connection before trying AMPS.

Primary Battery   A battery that is not meant to be recharged, as for example, the common carbon batteries used for radios and toys.

PSTN   Public Switched Telephone Network. The telephone system, including the whole network.

PTT   Push To Talk. A radio switch (usually part of the microphone) that must be pushed before the user can transmit. Usual in two-way radio (PMR).

Radome   A covering for an antenna that is transparent to RF.

Rayleigh Fading   Another term for multipath fading. Rayleigh was the first to do a serious study of interference patterns (using light, but the principle is the same).

Refraction   Propagation other than in a straight line due to bending of the path by some material medium (for example, air or water). Refraction occurs when a change in density of the medium exists.

Repair   Restoring a unit to working order by replacing or reconfiguring some internal component, usually at the board or component level. This usually involves bench work with test equipment.

Repeater   A device that can be either active or passive that receives an incoming signal and relays it on.

Reseller   A person or organization that sells cellular services to the public on networks owned by third parties.

RF   Radio Frequencies. Varies from 10 kHz to 300,000 MHz.

RHN   Repeater hybrid network. Mostly refers to CDMA networks that can be as much as 60 percent plus repeaters.

Roaming   Using a cellular phone through a system other than the usual "home" switch.

RSA   Rural Service Area (U.S.). Small cellular service areas.

RX   Receiver.

SA   Selective Access. The term for the scrambling on the GPS systems to limit civilian accuracy to about 100 meters. This is no longer activated.

SDR   Software Dependent Radio. Essentially, DSP based radio systems.

Secondary Battery   A battery that produces electrical energy by a chemical process that is reversible. Broadly a rechargeable battery.

Sector Antenna   A directional antenna that produces coverage of one or more sectors of the total base-station coverage.

Signal-To-Noise Ratio (SNR)   The power ratio between the received signal source and the noise source, usually expressed in dB of the wanted signal to the noise.

SIM CARD   A GSM subscriber's identity module. The card that holds this ID information as well as other subscriber data.

SINAD   Similar to signal-to-noise, but it adds the distortion products (Signal-to-Noise And Distortion) to the noise power.

Sleep Mode   A technique used in digital systems to conserve power, particularly for handhelds. Typically the mobile will search for incoming calls on startup, and then put itself in sleep mode (turn off all non-essential hardware) for a known period until the next broadcast of incoming call data. An accurate clock in the mobile is required for this.

Smart Antennas   Antennas that process the signals they receive in phase or amplitude (or both), in a manner which changes the effective antenna pattern.

SMS   Smart Measurement Systems. These systems are hybrid of top end PC controllers and test equipment. Using AI and/or fuzzy logic, these test sets are designed to do more than just measure. In production and development, they will monitor, report, and alert the users to trends and problems.

Soft Handoff   A make-before-break handoff that ensures signal path continuity through the handoff process.

Spectral Regrowth   A rather quaint name for the intermodulation distortion products produced by adjacent carriers in a linear amplifier that appear in-band.

Spread Spectrum   A wide-band radio service that can be CDMA, frequency-hopping, or chirp, where the carrier frequency is swept with each data burst.

Standby Time   The number of hours that a freshly charged battery can run a mobile on recieve only (no calls).

SS7, or Signaling System No. 7   The current standard for interswitch common-channel signaling.

Switch   In cellular radio, the connecting switch between the telephone network and the radio base station. Also called the MSC, Mobile Exchange, MTSC, MTSO, MTX, and other names.

TACS   Total Access Cellular System.

Talktime   The total time that a fully charged battery will permit the mobile to stay connected in the in-use mode.

TDMA   Time Division Multiple Access. A digital (usually radio) system that allows a number of users to use the same system by being dynamically assigned a particular timeslot on request. Often used to describe rural radio telephone systems that use this mode. In the United States, only for cellular people, it usually means DAMPS.

TDD   Time Division Duplex. A duplex system where the same frequency can be used for transmit and receive, by separating the TX and RX into different timeslots, as is done in DECT.

TIA   Telecommunications Industry Association. A US public standards-making group.

Tower   A self-supporting structure intended to hold an antenna(s), as distinct from guyed structures.

Traffic   Calls in progress. Measured in Erlangs; one call for one hour equals one Erlang.

Transceiver   A transmitter and receiver in one unit, such as a mobile telephone, or walkie-talkie.

Transportable   A cellular mobile telephone, complete with battery pack, that enables portable operation. This unit can also be used in a vehicle, running from the vehicle battery. It is usually significantly larger and has higher output power than a true portable. These were common in the 1980s and early 1990s but are now virtually extinct.

Tri-Mode Mobile   Refers to mobiles that can operate in three modes. Usually CDMA/DAMPS/AMPS, GSM800/900/1900 or E-TDMA/DAMPS/AMPS.

TSP   Telecom Service Providers.

TX   Transmitter.

UHF   Ultra-High Frequency. The radio frequency band from 300 MHz to 3000 MHz.

Umbrella Cell   A cell that covers a large area and may have one or more small cells within it. Most successfully done with NMT systems.

UMTS   Universal Mobile Telecommunications System. The European standard for 3G wireless services.

UPT   Universal Personal Telecommunications. Number portability, allowing the user to have one number for all services.

USDC   United States Digital Cellular. Another name for the US TDMA system referred to in this book as DAMPS.

UTC   Universal Coordinated Time.

Valve-Regulated Lead Acid (VRLA)   This is the formal name for sealed lead acid batteries. The batteries are pressurized and designed to promote oxygen recombination at the negative electrode to minimize water loss.

VLSI   Very Large-Scale Integrated Circuit. These were the main component parts of cellular systems in the early 1990s. A VLSI can perform a range of functions, but usually of a common type, such as the whole RF section, a logic controller, or a signaling interpreter.

VSELP Vector Sum Excited Linear Predictive. A CODEC as used in IS-136.

VPN Virtual Private Network. The ability to add things like Centrex and other private network facilities to a public network.

WAP Wireless Application Port.

Wavelength The distance from a point on a radio wave to the same point on the next wave. Cellular wavelengths (800/900 MHz) are about 0.3 meter long. For any frequency the wavelength = 300 meters/freq. (MHz).

W-CDMA Wide-band CDMA. Refers to wide-band relative to cdmaOne, usually meaning a 5-MHz or more bandwidth.

Wireline Conventional PSTN operation where most subscribers are connected by cable.

WDM Wave Division Multiplex. Mostly used for putting more than one carrier on an optical fiber.

X25 An international standard for packet data communications on the PSTN that can use analog lines.

INDEX